Organic Chemistry of Explosives

Organic Chemistry of Explosives

Dr. Jai Prakash Agrawal
CChem FRSC (UK)
Former Director of Materials
Defence R&D Organisation
DRDO Bhawan, New Delhi, India
email: jpa@vsnl.com

Dr. Robert Dale Hodgson
Consultant Organic Chemist,
Syntech Chemical Consultancy,
Morecambe, Lancashire, UK
Website: http://www.syntechconsultancy.co.uk
email: rdhodgson2001@yahoo.com

BICENTENNIAL 1807 WILEY 2007 BICENTENNIAL

John Wiley & Sons, Ltd

Telephone (+44) 1243 779777

Email (for orders and customer service enquiries): cs-books@wiley.co.uk
Visit our Home Page on www.wiley.com

Reprinted with corrections June 2007

Other Wiley Editorial Offices

John Wiley & Sons Inc., 111 River Street, Hoboken, NJ 07030, USA

Jossey-Bass, 989 Market Street, San Francisco, CA 94103-1741, USA

Wiley-VCH Verlag GmbH, Boschstr. 12, D-69469 Weinheim, Germany

John Wiley & Sons Australia Ltd, 42 McDougall Street, Milton, Queensland 4064, Australia

John Wiley & Sons (Asia) Pte Ltd, 2 Clementi Loop #02-01, Jin Xing Distripark, Singapore 129809

John Wiley & Sons Canada Ltd, 6045 Freemont Blvd, Mississauga, Ontario, L5R 4J3, Canada

Wiley also publishes its books in a variety of electronic formats. Some content that appears in print may not be available in electronic books.

Library of Congress Cataloging-in-Publication Data

Agrawal, J. P.
Organic chemistry of explosives / J. P. Agrawal and R. D. Hodgson.
p. cm.
Includes bibliographical references and index.
ISBN-13: 978-0-470-02967-1 (cloth : alk. paper)
ISBN-10: 0-470-02967-6 (cloth : alk. paper)
1. Explosives. I. Hodgson, R. D. II. Title.
TP270.A36 2006
662′.201547—dc22

2006022827

British Library Cataloguing in Publication Data

A catalogue record for this book is available from the British Library

ISBN-13 978-0-470-02967-1 (HB)
ISBN-10 0-470-02967-6 (HB)

Caution! Read This

The information given in this book is believed to be accurate and includes nothing not already in the public domain. John Wiley & Sons and the authors of this book accept no responsibility and cannot be held liable for any damage to property, injury, illness or death of a person or persons resulting from the misuse of this information. Energetic materials are extremely dangerous and should only be prepared by persons skilled in this area and licensed to do so, even then, do not use information directly from this book, please consult the primary research papers. Finally, a word of warning to those individuals that misuse science for malicious purposes or those foolish enough to attempt illegal and dangerous experiments. Making explosives or procuring chemicals for the synthesis of explosives without license is both dangerous and a very serious criminal offence. We must emphasize that John Wiley & Sons and the authors cannot be held responsible for the actions of irresponsible individuals in any shape or form.

Contents

Foreword

In the past a significant amount of research worldwide was directed at the synthesis of new energetic compounds as potential explosives or propellant ingredients. This research involved the synthesis of many different classes of energetic compounds, including heterocycles, nitrohydrocarbons, nitrate esters, nitramines and caged compounds. The research in this area has been reviewed many times in the past but these reviews usually concentrated on one class of energetic compounds, e.g. nitroalkenes or nitroazoles, and except for possibly Urbanski's volumes on the *Chemistry and Technology of Explosives*, a comprehensive study of energetic compound synthesis has not been undertaken.

The *Organic Chemistry of Explosives* by J. P. Agrawal and R. D. Hodgson is a comprehensive study of the various methods to synthesize the different classes of energetic compounds along with methods to synthesize the various explosophoric groups that predominate the field. It is intended to read like a tutorial on energetic compound synthesis, providing a historical perspective of the various synthetic methods used for energetic compound synthesis, along with enough details and discussion to understand the nuances of energetic compound synthesis.

The *Organic Chemistry of Explosives* also provides a perspective on the possible applications of various energetic compounds, why they are interesting as explosives or propellant ingredients, and what advantages and disadvantages they might have compared to other energetic compounds. Finally, it provides insight into the many factors an energetic compound synthetic chemist must consider when designing new target compounds and presents the various criteria (performance, ease of synthesis, cost, sensitivity to external stimuli, and chemical and thermal stability) that define whether a given energetic compound will be useful to the energetic materials community.

Dr. Philip F. Pagoria
Energetic Materials Center
Lawrence Livermore National Laboratory
Livermore, CA 94550
USA

Preface

Explosives have attracted a lot of unwanted publicity over the years for their misuse in the taking of life and the destruction of property. Explosives are perceived by most as materials of fear and at no time is this more prevalent than in times of war. Although such concerns and views are not unfounded, there is a bigger picture. More explosives have been used in times of peace than in all of the wars and conflicts put together. How many of the great engineering achievements would have been possible if not for the intervention of explosives? Blasting and quarrying have allowed the construction of our transport links, supplied the rock and raw materials for our buildings, and enabled the extensive mining of minerals and other essential materials. Explosives are in fact no more than a tool and remain as some of the most fascinating products of chemistry.

Much of the information concerning the synthesis of organic explosives, and energetic materials in general, can be found in the form of papers and reviews in academic chemistry journals. The *Journal of Energetic Materials* (USA); *Propellants, Explosives, Pyrotechnics* (Germany); *Combustion, Explosion and Shockwaves* (Russia) and *Explosives Engineering* (UK) are specialized journals for reporting the recent advances in the synthesis and technology of energetic materials. The mainstream organic chemistry journals occasionally report on the synthesis of energetic materials if that work has a general significance to organic chemistry. *Chemical Abstracts* is an invaluable and up to date source of information on patents and publications relating to advances in energetic materials chemistry and technology. Further, there are some national/international societies/associations in this field and their main task is to organize annual conferences/seminars, which provide a forum to scientists, engineers, technologists and academicians working in this area to exchange information on the latest developments.

Tenny L. Davis first published his two volumes of *The Chemistry of Powder and Explosives* in 1941 and 1943, and these were subsequently merged into a single volume. This useful work gives an overview of energetic materials synthesis in the early years. During and after the Second World War much research was pooled into the science of energetic materials, and consequently, the number of reported organic compounds with explosive properties increased enormously together with our knowledge of this subject. Tadeusz Urbański, a Polish chemist at the Institute of Organic Chemistry and Technology, Technical University in Warsaw, published the four volume series of *Chemistry and Technology of Explosives* over the years of 1964, 1965, 1967 and 1984. This work is a wealth of knowledge for anything from the industrial and laboratory synthesis of explosives to their physical, chemical, thermal and explosive properties. Unfortunately, this text is now out of print and over 20 years old. As the number of reported energetic materials continues to grow at a rapid rate, and while a number of excellent reviews have been published to fill this knowledge gap, there is still no single text available which is completely devoted to the synthesis of energetic materials from the simplest mixed acid nitration to the synthesis of modern high performance explosives via dinitrogen pentoxide nitration methodology.

For a long time, a reference/text book has been needed which provides detailed information on the synthetic routes to a wide range of energetic materials. The objective of this book is to fill this gap in the literature. The *Organic Chemistry of Explosives* is a text of pure chemistry which condenses together all the synthetic methods and routes available for the synthesis of organic explosives into one volume. This book is a reference source for chemists working in the field of energetic materials and all those with an interest in the chemistry of nitramines, nitro compounds, nitrate esters and nitration in general. We assume the readers to be new to the chemistry of explosives and so discuss everything from the simplest mixed acid nitration of toluene to the complex synthesis of caged nitro compounds. In doing so, we believe students with a sound knowledge of the basics of organic chemistry will also find this book of value.

While writing this book our approach has been to focus on synthetic methods and use individual synthesis to supplement the discussion rather than bombarding the readers with a near endless list of syntheses. This strikes at the fundamental principles used for energetic materials synthesis and we believe this will be more helpful to the readers. This brings us to the most important class of reaction used for energetic materials synthesis: that of nitration, which is the most widely studied and well understood of any reaction class in organic chemistry. A considerable proportion of this book is devoted to nitration. The books/papers/reviews listed under Acknowledgements were invaluable in the writing of this manuscript and we would recommend the reading of these for further understanding and details of nitration chemistry.

The *Organic Chemistry of Explosives* is split into nine well-defined chapters, based on the observation that explosive properties are imparted into a compound by the presence of certain functional groups. Chapters 1, 3, 4 and 5 discuss the methods which can be used to introduce *C*-nitro, *O*-nitro, and *N*-nitro functionality into organic compounds; the advantages and disadvantages of each synthetic method or route is discussed, together with the scope and limitations, aided with numerous examples in the form of text, reaction diagrams and tables. Chapters 2, 6 and 7 discuss the synthesis of energetic compounds in the form of polynitropolycycloalkanes, caged and strained nitramines, and *N*-heterocycles respectively. Chapter 8 discusses the synthesis of explosives containing functionality less widely encountered, including: organic azides, peroxides, diazophenols, and energetic compounds derived from guanidine and its derivatives. In the end, Chapter 9 gives an account of nitration with dinitrogen pentoxide and its likely significance for the futuristic synthesis of energetic materials.

We have tried to be as thorough as possible to include all relevant information related to the synthesis of organic explosives and although no attempt has been made to discuss the synthesis of every organic explosive ever made, there are several hundred compounds discussed in the text, enough to give the reader a sound knowledge of the synthesis of explosives. It would be quite impossible to cover all the available literature on the synthesis of explosives in a single volume text and it is just possible that some synthetically important papers might have been overlooked and we apologize for this. The readers are requested to inform us about such omissions which would be greatly appreciated and included in the next edition of this book.

We hope that this book will contribute to provide organic chemists with a comprehensive knowledge of the synthetic routes to explosives and especially those that form the basis of worldwide chemical industries.

J. P. Agrawal
R. D. Hodgson

Abbreviations

Ac	Acetyl [CH_3CO]
ADN	Ammonium dinitramide
ADNBF	7-Amino-4,6-dinitrobenzofuroxan
n-Am	*normal*-Amyl [$CH_3(CH_2)_4$]
AMMO	3-Azidomethyl-3-methyloxetane
ANF	3-Amino-4-nitrofurazan
ANFO	Ammonium nitrate-fuel oil explosive
ANPy	2,6-Diamino 3,5-dinitropyridine
ANPyO	2,6-Diamino-3,5-dinitropyridine-1-oxide
ANPz	2,6-Diamino-3,5-dinitropyrazine
ANTA	3-Amino-5-nitro-1,2,4-triazole
ANTZ	4-Amino-5-nitro-1,2,3-triazole
aq	Aqueous
ARDEC	US Army Research, Development and Engineering Center
ATA	4-Amino-1,2,4-triazole
ATNI	Ammonium 2,4,5-trinitroimidazole
ATTz	6-Amino-1,2,4-triazolo[4,3-*b*][1,2,4,5]tetrazine
Aza-TACOT	Tetranitrodipyridotetraazapentalene
AZTC	1-(Azidomethyl)-3,5,7-trinitro-1,3,5,7-tetraazacyclooctane
B:	Base
BAEA	Bis(2-azidoethyl)apidate
BAMO	3,3-Bis(azidomethyl)oxetane
BDNPA	Bis(2,2-dinitropropyl)acetal
BDNPF	Bis(2,2-dinitropropyl)formal
Bicyclo-HMX	2,4,6,8-Tetranitro-2,4,6,8-tetraazabicyclo[3.3.0]octane
Bn	Benzyl [$PhCH_2$]
BOC	*tertiary*-Butoxycarbonyl [$(CH_3)_3COCO$]
BPABF	4,4′-Bis(picrylamino)-3,3′-bifurazan
BPAF	3,4-Bis(picrylamino)furazan
BSX	1,7-Diacetoxy-2,4,6-trinitro-2,4,6-triazaheptane
BT	1,2,4-Butanetriol
BTATNB	1,3-Bis(1,2,4-triazol-3-amino)-2,4,6-trinitrobenzene (SDATO)
BTDAONAB	*N*,*N*′-Bis(1,2,4-triazol-3-yl)-4,4′-diamino-2,2′,3,3′,5,5′,6,6′-octanitroazobenzene
BTTN	1,2,4-Butanetriol trinitrate
BTX	1-Picryl-5,7-dinitro-2*H*-benzotriazole

i-Bu	*iso*-Butyl [$(CH_3)_2CHCH_2$]
n-Bu	*normal*-Butyl [$CH_3(CH_2)_3$]
s-Bu	*secondary*-Butyl [$CH_3CH_2CH(CH_3)$]
t-Bu	*tertiary*-Butyl [$(CH_3)_3C$]
Bu-NENA	*N*-Butyl-2-nitroxyethylnitramine
CAN	Ceric ammonium nitrate
cat	Catalytic
CL-20	2,4,6,8,10,12-Hexanitro-2,4,6,8,10,12-hexaazaisowurtzitane (HNIW)
CMDB	Composite modified double-base (propellant)
conc	Concentrated
m-CPBA	*meta*-Chloroperoxybenzoic acid
d	Density
DAAF	4,4′-Diamino-3,3′-azoxyfurazan
DAAT	3,3′-Azo-bis(6-amino-1,2,4,5-tetrazine)
DAAzF	4,4′-Diamino-3,3′-azofurazan
DABCO	1,4-Diazabicyclo[2.2.2]octane
DABF	4,4′-Diamino-3,3′-bifurazan
DADE	1,1-Diamino-2,2-dinitroethylene (FOX-7)
DADN	1,5-Diacetyl-3,7-dinitro-1,3,5,7-tetraazacyclooctane
DADNBF	5,7-Diamino-4,6-dinitrobenzofuroxan
DAF	3,4-Diaminofurazan
DAHNS	3,3′-Diamino-2,2′,4,4′,6,6′-hexanitrostilbene
DANTNP	4,6-Bis(3-amino-5-nitro-1*H*-1,2,4-triazole-1-yl)-5-nitropyrimidine
DAPT	3,7-Diacetyl-1,3,5,7-tetraazabicyclo[3.3.1]nonane
DATB	1,3-Diamino-2,4,6-trinitrobenzene
DATH	1,7-Diazido-2,4,6-trinitro-2,4,6-triazaheptane
DATNT	3,5-Diamino-2,4,6-trinitrotoluene
DB	Double-base (propellant)
DBU	1,8-Diazabicyclo[5.4.0]undec-7-ene
DDF	4,4′-Dinitro-3,3′-diazenofuroxan (DNAF)
DDNP	2-Diazo-4,6-dinitrophenol (DINOL)
dec	Decomposition temperature
DEG	Diethylene glycol
DEGBAA	Diethylene glycol bis(azidoacetate) ester
DEGDN	Diethylene glycol dinitrate (DGDN)
DERA	Defence Evaluation and Research Agency
DFAP	1,1,3,5,5-Pentanitro-1,5-bis(difluoramino)-3-azapentane
DGDN	Diethylene glycol dinitrate (DEGDN)
DIAD	Diisopropyl azodicarboxylate
dil	Dilute
DINA	*N*-Nitrodiethanolamine dinitrate
DINGU	1,4-Dinitroglycouril
DINOL	2-Diazo-4,6-dinitrophenol (DDNP)
DIPAM	3,3′-Diamino-2,2′,4,4′,6,6′-hexanitrobiphenyl
DiTeU	Bis(2,2,2-trinitroethyl)urea
DMDO	Dimethyldioxirane
DME	1,2-Dimethyoxyethane

DMF	Dimethylformamide
DMSO	Dimethylsulfoxide
DNABF	4,4′-Dinitro-3,3′-azoxy-bis(furazan)
C-DNAT	5,5′-Dinitro-3,3′-azo-1,2,4-triazole
N-DNAT	1,1′-Dinitro-3,3′-azo-1,2,4-triazole
DNAzBF	4,4′-Dinitro-3,3′-azo-bis(furazan)
DNBF	4,6-Dinitrobenzofuroxan
DNBT	5,5′-Dinitro-4,4′-bis(1,2,3-triazole)
DNF	3,4-Dinitrofurazan
DNFX	3,4-Dinitrofuroxan
2,4-DNI	2,4-Dinitroimidazole
DNNC	1,3,5,5-Tetranitrohexahydropyrimidine
DNPP	3,6-Dinitropyrazolo[4,3-*c*]pyrazole
DNT	Dinitrotoluene
DNTZ	4,5-Dinitro-1,2,3-triazole
DNU	*N,N′*-Dinitrourea
DPO	2,5-Dipicryl-1,3,4-oxadiazole
DPT	1,5-Dinitroendomethylene-1,3,5,7-tetraazacyclooctane
EDNA	Ethylenedinitramine
EGBAA	Ethylene glycol bis(azidoacetate) ester
EGDN	Ethylene glycol dinitrate
EIDS	Extremely insensitive detonating substance
eq	Equivalent
Estane	Poly(urethane-ester-MDI) binder (Goodrich)
Et	Ethyl [CH_3CH_2]
EWG	Electron withdrawing group
Explosive D	Ammonium picrate
FEFO	Bis(2-fluoro-2,2-dinitroethyl)formal
FLSC	Flexible linear shaped charge
FOX-7	1,1-Diamino-2,2-dinitroethylene (DADE)
FOX-12	*N*-Guanylurea salt of dinitramide
g	Gas phase
GAP	Glycidyl azide polymer
GC	Gas chromatography
gem	Geminal
GLYN	Glycidyl nitrate
GTN	Glyceryl trinitrate (nitroglycerine)
HAB	Hexakis(azidomethyl)benzene
HBIW	2,4,6,8,10,12-Hexabenzyl-2,4,6,8,10,12-hexaazaisowurtzitane
Hexyl	2,2′,4,4′,6,6′-Hexanitrodiphenylamine
HHTDD	2,6-Dioxo-1,3,4,5,7,8-hexanitrodecahydro-1*H*,5*H*-diimidazo[4,5-*b*:4′,5′-*e*]pyrazine
HK-55	2,4,6-Trinitro-2,4,6,8-tetraazabicyclo[3.3.0]octane-3-one
HK-56	2,5,7-Trinitro-2,5,7,9-tetraazabicyclo[4.3.0]nonane-8-one
HMMO	3-Hydroxymethyl-3-methyloxetane
HMPA	Hexamethylphosphoramide
HMTD	Hexamethylenetriperoxidediamine

HMX	1,3,5,7-Tetranitro-1,3,5,7-tetraazacyclooctane
HNAB	2,2′,4,4′,6,6′-Hexanitroazobenzene
HNF	Hydrazinium nitroformate
HNFX	3,3,7,7-Tetrakis(difluoramino)octahydro-1,5-dinitro-1,5-diazocine
HNIW	2,4,6,8,10,12-Hexanitro-2,4,6,8,10,12-hexaazaisowurtzitane (CL-20)
HNS	2,2′,4,4′,6,6′-Hexanitrostilbene
HNTP	4 -(2′,3′,4′,5′-Tetranitropyrrole)-3,5-dinitro-1,2,4-triazole
Hr	Hour
HTPB	Hydroxy-terminated polybutadiene
ICI	Imperial Chemical Industries
IHE	Insensitive high explosive
IHNX	2,4-Bis(5-amino-3-nitro-1,2,4-triazolyl)pyrimidine
IR	Infrared spectroscopy
K-55	2,4,6,8-Tetranitro-2,4,6,8-tetraazabicyclo[3.3.0]octane-3-one
K-56	2,5,7,9-Tetranitro-2,5,7,9-tetraazabicyclo[4.3.0]nonane-8-one (TNABN)
K-6	1,3,5-Trinitro-2-oxo-1,3,5-triazacyclohexane (Keto-RDX)
Kel-F800	Copolymer of vinylidene fluoride and hexafluoropropylene or chlorotrifluoroethylene (3M Company)
Keto-RDX	1,3,5-Trinitro-2-oxo-1,3,5-triazacyclohexane (K-6)
LANL	Los Alamos National Laboratory
LAX-112	3,6-Diamino-1,2,4,5-tetrazine-2,5-dioxide
LDA	Lithium diisopropylamine
liq	Liquid phase
LLM-101	3,6-Dinitro-1,2,4,5-cyclohexanetetrol 1,4-dinitrate
LLM-105	2,6-Diamino-3,5-dinitropyrazine-1-oxide
LLM-116	4-Amino-3,5-dinitropyrazole
LLM-119	1,4-Diamino-3,6-dinitropyrazolo[4,3-*c*]pyrazole
LLNL	Lawrence Livermore National Laboratory
LOVA	Low vulnerability ammunition
m-	Meta
MDI	4,4′-Methylenebis(phenyl isocyanate)
Me	Methyl [CH_3]
MEK	Methylethylketone
Methyl Tris-X	2,4,6-Tris(2-methyl-2-nitroxyethylnitramino)-1,3,5-triazine
mins	Minutes
MNT	Mononitrotoluene
m.p.	Melting point
MPa	Mega Pascal
Ms	Mesyl or methanesulfonate [CH_3SO_2]
MTN	Metriol trinitrate or 1,1,1-tris(hydroxymethyl)ethane trinitrate
NAWC	Naval Air Warfare Center
NBS	*N*-Bromosuccinimide
NC	Nitrocellulose
NENA	Nitroxyethylnitramines
NENO	*N,N′*-Dinitro-*N,N′*-bis(2-hydroxyethyl)oxamide dinitrate
NG	Nitroglycerine
NHTPB	Nitrated hydroxy-terminated polybutadiene

NIMMO	3-Nitratomethyl-3-methyloxetane
NMHP	5-(Nitratomethyl)-1,3,5-trinitrohexahydropyrimidine
NMP	*N*-Methylpyrrolidinone
NMR	Nuclear magnetic resonance
NONA	2,2′,2″,4,4′,4″,6,6′,6″-Nonanitro-*m*-terphenyl
NOTO	5-[4-Nitro-(1,2,5)oxadiazolyl]-5*H*-[1,2,3]triazolo[4,5-*c*][1,2,5]oxadiazole
Ns	Nosyl or 4-nitrobenzenesulphonyl [4-$NO_2C_6H_4SO_2$]
NSWC	Naval Surface Warfare Center
NTO	3-Nitro-1,2,4-triazol-5-one
Nu	Nucleophile
o-	Ortho
ONC	Octanitrocubane
o/*p*-	Ortho/para ratio
p-	Para
PADNT	4-Picrylamino-2,6-dinitrotoluene
PADP	2,6-Bis(picrylazo)-3,5-dinitropyridine
PANT	4-Picrylamino-5-nitro-1,2,3-triazole
PAT	5-Picrylamino-1,2,3,4-tetrazole
PATO	3-Picrylamino-1,2,4-triazole
PBX	Plastic bonded explosive
PCC	Pyridinium chlorochromate
PCX	3,5-Dinitro-3,5-diazapiperidinium nitrate
PDADN	2,2-Bis(azidomethyl)-1,3-propanediol dinitrate
PDDN	1,2-Propanediol dinitrate
Pentryl	1-(2-Nitroxyethylnitramino)-2,4,6-trinitrobenzene
PETKAA	Pentaerythritol tetrakis(azidoacetate) ester
PETN	Pentaerythritol tetranitrate
Ph	Phenyl [C_6H_5]
Picramide	2,4,6-Trinitroaniline
Picryl	2,4,6-Trinitrophenyl [2,4,6-$(NO_2)_3C_6H_2$]
PL-1	2,4,6-Tris(3′,5′-diamino-2′,4′,6′-trinitrophenylamino)-1,3,5-triazine
PNP	Polynitrophenylene
Poly[AMMO]	Poly[3-azidomethyl-3-methyloxetane]
Poly[BAMO]	Poly[3,3-bis(azidomethyl)oxetane]
Poly-CDN	Nitrated cyclodextrin polymers
Poly[GYLN]	Poly[glycidyl nitrate]
Poly[NIMMO]	Poly[3-nitratomethyl-3-methyloxetane]
PPA	Polyphosphoric acid
i-Pr	*iso*-Propyl [$(CH_3)_2CH$]
n-Pr	*normal*-Propyl [$CH_3CH_2CH_2$]
PRAN	2-(5-Amino-3-nitro-1,2,4-triazolyl)-3,5-dinitropyridine
Pyr	Pyridine
PYX	2,6-Bis(picrylamino)-3,5-dinitropyridine
R	Alkyl group (unless otherwise stated)
RDX	1,3,5-Trinitro-1,3,5-triazacyclohexane
R-salt	1,3,5-Trinitroso-1,3,5-triazacyclohexane
SAT	5,5′-Styphnylamino-1,2,3,4-tetrazole

SDATO	1,3-Bis(1,2,4-triazol-3-amino)-2,4,6-trinitrobenzene (BTATNB)
SNPE	Societe Nationale des Poudres et Explosifs
Styphnic acid	2,4,6-Trinitroresorcinol
TACOT	Tetranitrodibenzotetraazapentalene
TADBIW	2,6,8,12-Tetraacetyl-4,10-dibenzyl-2,4,6,8,10,12-hexaazaisowurtzitane
TAIW	2,6,8,12-Tetraacetyl-2,4,6,8,10,12-hexaazaisowurtzitane
TAT	1,3,5,7-Tetraacetyl-1,3,5,7-tetraazacyclooctane
TATB	1,3,5-Triamino-2,4,6-trinitrobenzene
TATP	Triacetone triperoxide
TAX	1-Acetyl-3,5-dinitro-1,3,5-triazacyclohexane
TBDMS	*tertiary*-Butyldimethylsilyl [Me_3CSiMe_2]
TBHP	*tertiary*-Butylhydroperoxide
TBS	*tertiary*-Butyldimethylsilyl [Me_3CSiMe_2]
TBTDO	1,2,3,4-Tetrazino[5,6-*f*]benzo-1,2,3,4-tetrazine 1,3,7,9-tetra-*N*-oxide
TEG	Triethylene glycol
TEGDN	Triethylene glycol dinitrate
Tert	Tertiary
Tetryl	*N*,2,4,6-Tetranitro-*N*-methylaniline
TEX	4,10-Dinitro-4,10-diaza-2,6,8,12-tetraoxaisowurtzitane
Tf	triflyl or trifluoromethanesulfonyl [CF_3SO_2]
TFA	Trifluoroacetic acid
TFAA	Trifluoroacetic anhydride
THF	Tetrahydrofuran
THP	Tetrahydropyran
TIPS	Triisopropylsilyl [$(Me_2CH)_3Si$]
TMHI	1,1,1-Trimethylhydrazinium iodide
TMNTA	1,1,1-Tris(hydroxymethyl)nitromethane tris(azidoacetate) ester
TMS	Trimethylsilyl [$(CH_3)_3Si$]
TNABN	2,5,7,9-Tetranitro-2,5,7,9-tetraazabicyclo[4.3.0]nonane-8-one (K-56)
TNAD	*trans*-1,4,5,8-Tetranitro-1,4,5,8-tetrazadecalin
TNAZ	1,3,3-Trinitroazetidine
TNB	2,4,6-Trinitrobenzene
TNBT	2,2′,4,4′-Tetranitro-bis(1,3,4-triazole)
TNC	1,3,5,7-Tetranitrocubane
TNENG	*N*-Nitro-*N*′-(2,2,2-trinitroethyl)guanidine
TNFX	3,3-Bis(difluoramino)octahydro-1,5,7,7-tetranitro-1,5-diazocine
TNGU	1,3,4,6-Tetranitroglycouril
TNHP	1,3,5-Trinitrohexahydropyrimidine
TNI	2,4,5-Trinitroimidazole
TNPDU	2,4,6,8-Tetranitro-2,4,6,8-tetraazabicyclo[3.3.1]nonane-3,7-dione or tetranitropropanediurea
TNT	2,4,6-Trinitrotoluene
TNX	2,4,6-Trinitroxylene
TPM	2,4,6-Tris(picrylamino)-1,3,5-triazine
TRAT	1,3,5-Triacetyl-1,3,5-triazacyclohexane
Tris-X	2,4,6-Tris(2-nitroxyethylnitramino)-1,3,5-triazine
Triton-B	Benzyltrimethylammonium hydroxide

Ts	Tosyl or 4-toluenesulphonyl [4-$MeC_6H_4SO_2$]
UDMH	Unsymmetrical dimethylhydrazine (Me_2NNH_2)
vic	Vicinal
Viton	Copolymer of vinylidene fluoride and perfluoropropylene (DuPont)
VNS	Vicarious nucleophilic substitution
VOD	Velocity of detonation

Acknowledgements

While writing this book we have consulted innumerable research papers and books listed under references. It is not possible to thank all writers/researchers/scientists individually, but we are grateful to all those who have contributed to the cause of explosives and high energy materials in any way.

The research and writing of this book would have been considerably more difficult if not for the texts and reviews of some researchers/scientists. We found the following books/papers invaluable and would like to express our sincere thank to their authors, contributors and editors:

1. G. S. Lee, A. R. Mitchell, P. F. Pagoria and R. D. Schmidt, "A Review of Energetic Materials Synthesis," *Thermochim. Acta.*, 2002, **384**, 187–204.

2. N. Ono, *The Nitro Group in Organic Synthesis, Organic Nitro Chemistry Series*, Wiley-VCH, Weinheim (2001).

3. I. J. Dagley and R. J. Spear, "Synthesis of Organic Energetic Compounds," in *Organic Energetic Compounds.*, Ed. P. L. Marinkas, Nova Science Publishers Inc., New York, Chapter 2, 47–163 (1996).

4. *Nitration: Recent Laboratory and Industrial Developments, ACS Symposium Series 623*, Eds. L. F. Albright, R. V. C. Carr and R. J. Schmitt, American Chemical Society, Washington, DC (1996).

5. *Kirk-Othmer Encyclopedia of Chemical Technology,* 4th Edn, Vol. 10, Ed. M. Grayson, Wiley-Interscience, New York, 1–125 (1993).

6. *Chemistry of Energetic Materials*, Eds. D. R. Squire and G. A. Olah, Academic Press, San Diego, CA (1991).

7. *Nitro Compounds: Recent Advances in Synthesis and Chemistry, Organic Nitro Chemistry Series*, Eds. H. Feuer and A. T. Neilsen, VCH Publishers, New York (1990).

8. G. A. Olah, R. Malhotra and S. C. Narang, *Nitration: Methods and Mechanisms*, Wiley-VCH, Weinheim (1989).

9. T. Urbański, *Chemistry and Technology of Explosives,* Vol. 1 (1964), Vol. 2 (1965), Vol. 3 (1967), Vol. 4 (1984), Pergamon Press, Oxford.

10. R. J. Spear and W. S. Wilson, "Recent Approaches to the Synthesis of High Explosive and Energetic Materials," *J. Energ. Mater.*, 1984, **2**, 61–149.

11. R. G. Coombes, "Nitro and Nitroso Compounds," in *Comprehensive Organic Chemistry: The Synthesis and Reactions of Organic Compounds,* Vol. 2, Ed. I. O. Sutherland, Pergamon Press, Oxford, Chapter 7, 305–381 (1979).

12. *Industrial and Laboratory Nitrations, ACS Symposium Series 22*, Eds. L. F. Albright and C. Hanson, American Chemical Society, Washington, DC (1976).

13. *The Chemistry of the Nitro and Nitroso Groups, Part 2, Organic Nitro Chemistry Series*, Ed. H. Feuer, Wiley-Interscience, New York (1970).

14. *The Chemistry of the Nitro and Nitroso Groups, Part 1, Organic Nitro Chemistry Series*, Ed. H. Feuer, Wiley-Interscience, New York (1969).

15. F. G. Borgardt, P. Noble. Jr and W. L. Reed, "Chemistry of the Aliphatic Polynitro Compounds and their Derivatives," *Chem. Rev.*, 1964, **64**, 19–57.

16. P. A. S. Smith, "Esters and Amides of Nitrogen Oxy-acids,"*Open Chain Nitrogen Compounds,* Vol. 2., Benjamin, New York, Chapter 15, 455–513 (1966).

17. Nitro Paraffins (Ed. H. Feuer), *Tetrahedron*, 1963, **19**, Suppl. 1.

18. N. Kornblum, "The Synthesis of Aliphatic and Alicyclic Nitro Compounds," *Org. React.*, 1962, **12**, 101–156.

19. A. V. Topchiev, *Nitration of Hydrocarbons and Other Organic Compounds*, Translated from Russian by C. Matthews, Pergamon Press, London (1959).

20. T. L. Davis, *Chemistry of Powder and Explosives,* Coll. Vol., Angriff Press, Hollywood, CA (reprinted 1992, first printed 1943).

Our special thanks go to the Directors, Officers and Staff of the British Library Document Centre and the Libraries of the Universities of Lancaster, Leeds and York, UK and the Defence Scientific Information and Documentation Centre (DESIDOC), Delhi and the High Energy Materials Research Laboratory (HEMRL), Pune, India without whose support this book would not have seen the light of the day.

The authors are also grateful to the following copyright owners for their permission to reproduce tables and text from their publications: American Chemical Society (*The Journal of Organic Chemistry*; *Industrial and Laboratory Nitrations, ACS Symposium Series 22*; and *Nitration: Recent Laboratory and Industrial Developments, ACS Symposium Series 623*) and Infomedia India Ltd (*Chemical World*).

Dr. Agrawal would like to thank his wife Sushma and Dr. Hodgson his mother Sheila and father Dale for their patience, understanding and valuable support during the writing of this book.

Finally our thanks are due to Mr. Paul Deards, Commissioning Editor (Physical Sciences), John Wiley & Sons, Ltd, The Atrium, Southern Gate, Chichester, West Sussex, PO19 8SQ, UK for his support and valuable suggestions.

Background

In the simplest terms, an explosive is defined as a substance, which on initiation by friction, impact, shock, spark, flame, heating, or any simple application of an energy pulse, undergoes a rapid chemical reaction evolving a large amount of heat and so exerting a high pressure on its surroundings. The vast majority of explosives release gaseous products on explosion but this is not an essential requirement as in the case of some metal acetylides.

Explosives, propellants and pyrotechnics belong to a broad group of compounds and compositions known as energetic materials. Many organic explosives consist of a carbon core incorporating covalently bonded oxidiser groups such as nitro, nitramine, nitrate ester etc. These groups, containing bonds like N-N and N-O, have two or more atoms covalently bonded with non-bonding electrons present in p-orbitals. This creates electrostatic repulsion between the atoms, and consequently, many explosives have a positive heat of formation. On explosion an internal redox reaction occurs where these bonds break and form gaseous products, like N_2 and CO_2, where the non-bonding electrons are tied up in stable π-bonds.

All explosives can be classified as either low or high explosives. Low explosives, also known as propellants, while still containing the oxygen needed for their combustion, are at most combustible materials which undergo deflagration by a mechanism of surface burning. Low explosives can still explode under confinement but this is a consequence of the increase in pressure caused by the release of gaseous products. Some low explosives can also detonate under confinement if initiated by the shock of another explosive. Low explosives include substances like gunpowder, smokeless powder and gun propellants. High explosives, on the other hand, need no confinement for explosion, for their chemical reactions are far more rapid and undergo the physical phenomenon of detonation. In these materials the chemical reaction follows a high-pressure shock wave which propagates the reaction as it moves through the explosive substance. High explosives typically detonate at a rate between 5500–9500 m/s and this velocity of detonation (VOD) is used to compare the performance of different explosives. High explosives include compounds like TNT, NG, RDX and HMX.

Another way to classify an explosive is how sensitive it is to mechanical or thermal stimuli. The sensitivity of explosives to stimuli is a broad spectrum, but those explosives that readily explode from light to modest mechanical stimuli are designated as primary explosives, while those explosives which need the shock of an explosion or a high energy impulse are known as secondary explosives or simply high explosives. Primary explosives usually explode on the application of heat, whereas some secondary explosives will simply burn in small enough quantities. A number of sensitivity tests have been designed to determine the sensitivity of a given explosive to thermal and mechanical stimuli. Some materials are very near the crossover between a primary and secondary explosive. Primary explosives, also known as initiators, are classified as to their effectiveness in causing the detonation of another explosive. Some primary explosives are poor initiators while others are powerful initiators and have found use in detonators. Brisance is another term used in the science of explosives and refers to the "shattering" power of an explosive. This has a direct relation to the detonation pressure or detonation velocity.

Advances in technology mean that energetic materials are required for even more demanding applications. Research in explosive technologies is now heavily focused on the design and synthesis of explosives for specialized applications. Explosives of high thermal stability and those with a low sensitivity to impact or friction are particularly desirable. Agrawal (*Prog. Energy Combust. Sci.*, 1998, **24**, 1) has proposed a different way of classifying explosives while reviewing the recent developments in this field. This is based on a single most important property of an explosive/high energy material, and accordingly, classifies the explosives reported so far in the literature in the following manner.

1. "Thermally Stable" or "heat-resistant" explosives
2. High-performance explosives i.e. high density and high velocity of detonation (VOD) explosives
3. Melt-castable explosives
4. Insensitive high explosives (IHEs)
5. Energetic plasticizers and binders for explosive and propellant compositions
6. Energetic materials synthesized by using dinitrogen pentoxide (N_2O_5)

Oxygen balance (OB) is defined as the ratio of the oxygen content of a compound to the total oxygen required for the complete oxidation of all carbon, hydrogen and other oxidisable elements to CO_2, H_2O, etc and is used to classify energetic materials as either oxygen deficient or oxygen rich. Most energetic materials are oxygen deficient.

All the terms discussed so far try to classify an explosive from its physical or explosive properties. The classification of explosives from a chemical viewpoint is of course more relevant to this book. Explosives can be classified according to the functionality they contain, and in particularly, the functional groups that impart explosive properties to a compound. Plets (*Zh. Obshch. Khim.*, 1953, **5**, 173) divided explosives into the following eight classes depending on the groups they contained; each group is known as an "explosophore".

1. $-NO_2$ and $-ONO_2$ in both inorganic and organic substances
2. $-N{=}N-$ and $-N{=}N{=}N-$ in inorganic and organic azides and diazo compounds.
3. $-NX_2$, where X = halogen
4. $-N{=}C$ in fulminates
5. $-OClO_2$ and $-OClO_3$ in inorganic and organic chlorates and perchlorates respectively.
6. $-O-O-$ and $-O-O-O-$ in inorganic and organic peroxides and ozonides respectively.
7. $-C{\equiv}C-$ in acetylene and metal acetylides.
8. M–C metal bonded with carbon in some organometallic compounds.

Most organic explosives contain nitrate ester, nitramine, or aliphatic or aromatic *C*-nitro functionality. Explosives containing azide, peroxide, azo functionality etc are a minor class and amount to less than 4-5% of the total number of known explosives. Since this book is focused on discussing the various ways in which each of these functional groups can be incorporated into a compound, the organization of Chapters and Sections in this book is straightforward and is as follows:

1. Aliphatic *C*-nitro groups
2. Nitrate ester groups
3. Aromatic *C*-nitro groups
4. Nitramine, nitramide and nitrimine groups
5. Nitrogen heterocycles
6. Other groups, including: azide, peroxide, diazophenols, and nitrogen-rich compounds derived from guanidine derivatives.

1

Synthetic Routes to Aliphatic *C*-Nitro Functionality

1.1 INTRODUCTION

The nitro group, whether attached to aromatic or aliphatic carbon, is probably the most widely studied of the functional groups and this is in part attributed to its use as an 'explosophore' in many energetic materials.[1] The chemistry of the nitro group has been extensively reviewed in several excellent works including in a functional group series.[2–8]

A comprehensive discussion of the synthetic methods used to introduce the nitro group into aliphatic compounds, and its diverse chemistry, would require more space than available in this book. While every effort has been made to achieve this, some of these methods are given only brief discussion because they have not as yet found use for the synthesis of energetic materials, or their use is limited in this respect. The nature of energetic materials means that methods used to introduce polynitro functionality are of prime importance and so these are discussed in detail. Therefore, this work complements the last major review on this subject.[9]

The chemical properties of the nitro group have important implications for the synthesis of more complex and useful polynitroaliphatic compounds and so these issues are discussed in relation to energetic materials synthesis.

RCH_2NO_2	$R^1R^2CHNO_2$	$R^1R^2R^3CNO_2$
Primary nitroalkane	Secondary nitroalkane	Tertiary nitroalkane
$RCH(NO_2)_2$	$R^1R^2C(NO_2)_2$	$R^1R^2CHC(NO_2)_3$
Terminal *gem*-dinitroalkane	Internal *gem*-dinitroalkane	Trinitromethyl

Figure 1.1

Aliphatic nitroalkanes can be categorized into six basic groups: primary, secondary and tertiary nitroalkanes, terminal and internal *gem*-dinitroalkanes, and trinitromethyl compounds. Primary and secondary nitroalkanes, and terminal *gem*-dinitroalkanes, have acidic protons and find particular use in condensation reactions for the synthesis of more complex and

Organic Chemistry of Explosives J. P. Agrawal and R. D. Hodgson

functionalized compounds, of which some find application as energetic plasticizers and polymer precursors. Tertiary nitroalkanes and compounds containing internal *gem*-dinitroaliphatic functionality exhibit high thermal and chemical stability and are frequently present in the energetic polynitropolycycloalkanes discussed in Chapter 2. The chemical stability of these various groups is discussed in Section 1.13.

1.2 ALIPHATIC *C*-NITRO COMPOUNDS AS EXPLOSIVES

Nitromethane is not usually regarded as an explosive, but its oxygen balance suggests otherwise, and under certain conditions and with a strong initiator this compound can propagate its own detonation. Nitromethane has been used in combination with ammonium nitrate for blasting. Although this explosive is more powerful than conventional ammonium nitrate-fuel oil (ANFO) it is considerably more expensive. Other simple aliphatic nitroalkanes have less favorable oxygen balances and will not propagate their own detonation.

Polynitroaliphatic compounds have not found widespread use as either commercial or military explosives. This is perhaps surprising considering the high chemical and thermal stability of compounds containing internal *gem*-dinitroaliphatic functionality. In fact, many polynitroaliphatic compounds are powerful explosives, for example, the explosive power of 2,2-dinitropropane exceeds that of aromatic *C*-nitro explosives like TNT. Tetranitromethane, although not explosive on its own, contains a large amount of available oxygen and forms powerful explosive mixtures with aromatic hydrocarbons like toluene. The problem appears to be one of cost and availability of raw materials. Most commercial and military explosives in widespread use today contain nitrate ester, nitramine or aromatic *C*-nitro functionality because these groups are readily introduced into compounds with cheap and readily available reagents like mixed acid (sulfuric and nitric acids mixture). However, sometimes other factors can outweigh the cost of synthesis if a compound finds specialized use. Over the past few decades there has been a demand for more powerful explosives of high thermal and chemical stability. Such criteria are met in the form of polynitrocycloalkanes, which are a class of energetic materials discussed in Chapter 2. These compounds have attracted increased interest in the aliphatic *C*-nitro functionality which may result in the improvement of or discovery of new methods for its incorporation into compounds.

Improved methods for the synthesis of building blocks like 2-fluoro-2,2-dinitroethanol and 2,2-dinitropropanol have resulted in some polynitroaliphatic compounds finding specialized application. Bis(2-fluoro-2,2-dinitroethyl)formal (FEFO) and a 1:1 eutectic mixture of bis(2,2-dinitropropyl)formal (BDNPF) and bis(2,2-dinitropropyl)acetal (BDNPA) have both found use as plasticizers in energetic explosive and propellant formulations.

1.3 DIRECT NITRATION OF ALKANES

Nitroalkanes can be formed from the direct nitration of aliphatic and alicyclic hydrocarbons with either nitric acid[10] or nitrogen dioxide[11] in the vapour phase at elevated temperature. These reactions have achieved industrial importance but are of no value for the synthesis of nitroalkanes on a laboratory scale, although experiments have been conducted on a small scale in sealed tubes.[12–14]

The vapour phase nitration of hydrocarbons proceeds via a radical mechanism[3,15] and so it is found that tertiary carbon centres are nitrated most readily, followed by secondary and primary centres which are only nitrated with difficulty. With increased temperature these reactions become less selective; at temperatures of 410–430 °C hydrocarbons often yield a complex mixture of products. At these temperatures alkyl chain fission occurs and nitroalkanes of shorter chain length are obtained along with oxidation products. An example is given by Levy and Rose[16] who nitrated propane with nitrogen dioxide at 360 °C under 10 atmospheres of pressure and obtained a 75–80 % yield of a mixture containing: 20–25 % nitromethane, 5–10 % nitroethane, 45–55 % 2-nitropropane, 20 % 1-nitropropane and 1 % 2,2-dinitropropane.

The nitration of moderate to high molecular weight alkane substrates results in very complex product mixtures. Consequently, these reactions are only of industrial importance if the mixture of nitroalkane products is separable by distillation. Polynitroalkanes can be observed from the nitration of moderate to high molecular weight alkane substrates with nitrogen dioxide. The nitration of aliphatic hydrocarbons has been the subject of several reviews.[15,17]

Both nitric acid and nitrogen dioxide, in the liquid and vapour phase, have been used for the nitration of the alkyl side chains of various alkyl-substituted aromatics without affecting the aromatic nucleus.[13,18] Thus, treatment of ethylbenzene with nitric acid of 12.5 % concentration in a sealed tube at 105–108 °C is reported to generate a 44 % yield of phenylnitroethane.[13] The nitration of toluene with nitrogen dioxide at a temperature between 20–95 °C yields a mixture of phenylnitromethane and phenyldinitromethane with the proportion of the latter increasing with reaction temperature.[18]

The nitration of aliphatic hydrocarbons with dinitrogen pentoxide[19] and nitronium salts[20] has been described. Topchiev[21] gives an extensive discussion of works related to hydrocarbon nitration conducted prior to 1956.

1.4 ADDITION OF NITRIC ACID, NITROGEN OXIDES AND RELATED COMPOUNDS TO UNSATURATED BONDS

1.4.1 Nitric acid and its mixtures

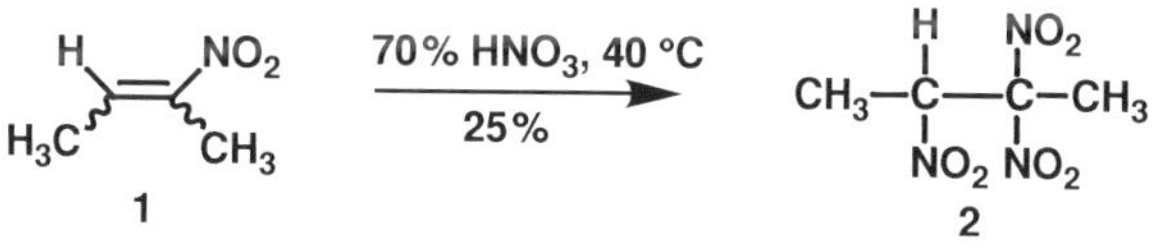

H, NO_2, H_3C, CH_3 1 — 70% HNO_3, 40 °C, 25% → CH_3–C(H)(NO_2)–C(NO_2)(NO_2)–CH_3 2

Figure 1.2

Alkenes can react with nitric acid, either neat or in a chlorinated solvent, to give a mixture of compounds, including: *vic*-dinitroalkane, β-nitro-nitrate ester, *vic*-dinitrate ester, β-nitroalcohol, and nitroalkene products.[21–26] Cyclohexene reacts with 70 % nitric acid to yield a mixture of 1,2-dinitrocyclohexane and 2-nitrocyclohexanol nitrate.[23] Frankel and Klager[24] investigated the reactions of several alkenes with 70 % nitric acid, but only in the case of 2-nitro-2-butene (1) was a product identified, namely, 2,2,3-trinitrobutane (2).

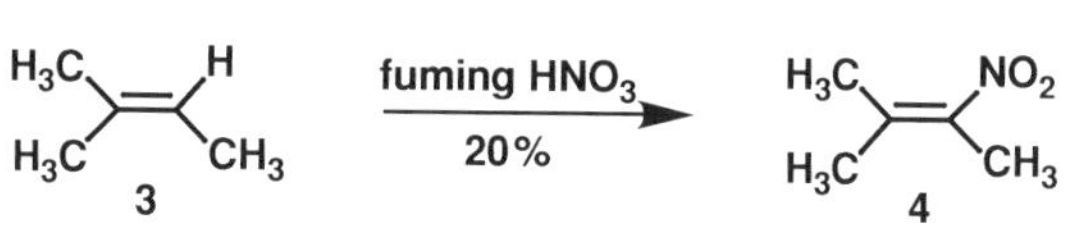

H_3C, H, H_3C, CH_3 3 — fuming HNO_3, 20% → H_3C, NO_2, H_3C, CH_3 4

Figure 1.3

The reaction of fuming nitric acid with 2-methyl-2-butene (3) is reported to yield 2-methyl-3-nitro-2-butene (4).[26] The reaction of alkenes with fuming nitric acid, either neat or in chlorinated solvents, is an important route to unsaturated nitrosteroids, which assumedly arise from the dehydration of β-nitroalcohols or the elimination of nitric acid from β-nitro-nitrate esters.[25,27] Temperature control in these reactions is important if an excess of oxidation by-products is to be avoided.

Mixed acid has been reported to react with some alkenes to give β-nitro-nitrate esters amongst other products.[26]

Solutions of acetyl nitrate, prepared from fuming nitric acid and acetic anhydride, can react with alkenes to yield a mixture of nitro and nitrate ester products, but the β-nitroacetate is usually the major product.[28–30] Treatment of cyclohexene with this reagent is reported to yield a mixture of 2-nitrocyclohexanol nitrate, 2-nitrocyclohexanol acetate, 2-nitrocyclohexene and 3-nitrocyclohexene.[29,30] β-Nitroacetates readily undergo elimination to the α-nitroalkenes on heating with potassium bicarbonate.[5] β-Nitroacetates are also reduced to the nitroalkane on treatment with sodium borohydride in DMSO.[31]

Solutions of acetyl nitrate have also been used for the synthesis of α-nitroketones from enol esters and ethers.[30,32]

The reaction of alkynes with nitric acid or mixed acid is generally not synthetically useful. An exception is the reaction of acetylene with mixed acid or fuming nitric acid which leads to the formation of tetranitromethane.[33a] A modification to this reaction uses a mixture of anhydrous nitric acid and mercuric nitrate to form trinitromethane (nitroform) from acetylene.[34] Nitroform is produced industrially via this method in a continuous process in 74 % yield.[34] The reaction of ethylene with 95–100 % nitric acid is also reported to yield nitroform (and 2-nitroethanol).[33b,c] The nitration of ketene with fuming nitric acid is reported to yield tetranitromethane.[35] Tetranitromethane is conveniently synthesized in the laboratory by leaving a mixture of fuming nitric acid and acetic anhydride to stand at room temperature for several days.[33d]

1.4.2 Nitrogen dioxide

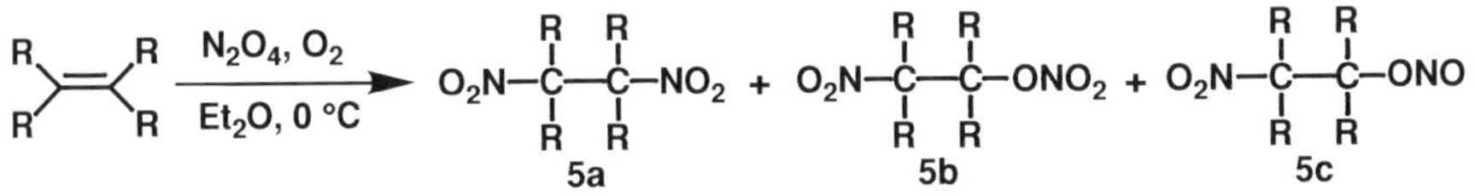

$$R_2C{=}CR_2 \xrightarrow[\text{Et}_2\text{O, 0 °C}]{N_2O_4,\ O_2} O_2N{-}CR_2{-}CR_2{-}NO_2\ (\mathbf{5a}) + O_2N{-}CR_2{-}CR_2{-}ONO_2\ (\mathbf{5b}) + O_2N{-}CR_2{-}CR_2{-}ONO\ (\mathbf{5c})$$

Figure 1.4

The addition of nitrogen oxides and other sources of NO_2 across the double bonds of alkenes is an important route to nitro compounds. Alkenes react with dinitrogen tetroxide in the presence of oxygen to form a mixture of *vic*-dinitro (5a), β-nitro-nitrate ester (5b) and β-nitro-nitrite ester (5c) compounds; the nitrite ester being oxidized to the nitrate ester in the presence of excess dinitrogen tetroxide.[36] A stream of oxygen gas is normally bubbled through the reaction mixture to expel nitrous oxide formed during the reaction and so prevent more complex mixtures being formed. These reactions can be synthetically useful for the synthesis of *vic*-dinitroalkanes because nitrate and nitrous ester by-products are chemically unstable and are readily hydrolyzed to the corresponding β-nitroalcohol on treatment with methanol.

1,2-Dinitroethane and 1,2-dinitrocyclohexane can be formed in this way from the corresponding alkenes in 42 % and 37 % yield respectively.[36a]
The addition of dinitrogen tetroxide across the double bonds of electron deficient fluorinated alkenes is a particularly useful route to *vic*-dinitro compounds where yields are frequently high;[8,37] tetrafluoroethylene gives a 53 % yield of 1,2-dinitro-1,1,2,2-tetrafluoroethane.[38]

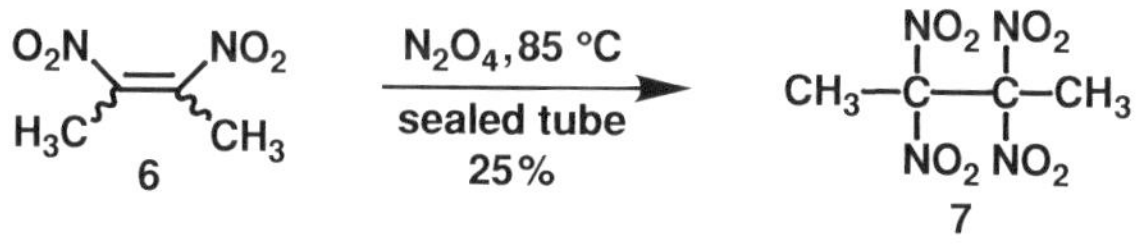

O_2N NO_2 H_3C CH_3 6 — N_2O_4, 85 °C, sealed tube, 25% → $CH_3-C(NO_2)_2-C(NO_2)_2-CH_3$ 7

Figure 1.5

The reaction of α-nitroalkenes with nitrogen dioxide or its dimer, dinitrogen tetroxide, has been used to synthesize polynitroalkanes. Thus, the reaction of dinitrogen tetroxide with 2,3-dinitro-2-butene (6) and 3,4-dinitro-3-hexene is reported to yield 2,2,3,3-tetranitrobutane (7, 25 %) and 3,3,4,4-tetranitrohexane (32 %) respectively.[39]

Additions of dinitrogen tetroxide across C–C double bonds are selective. The β-nitro-nitrates formed from terminal alkenes have the nitro group situated on the carbon bearing the most hydrogen and this is irrespective of neighbouring group polarity.[36] Altering reaction conditions and stoichiometry enables the preferential formation of β-nitro-nitrates over *vic*-dinitroalkanes, which, although inherently unstable, provide a synthetically useful route to α-nitroalkenes via base-catalyzed elimination.[40] β-Nitro-nitrates are reduced to the nitroalkane on treatment with sodium borohydride in ethanol.[41] β-Nitro-nitrates also undergo facile hydrolysis to the β-nitroalcohol, and conversion of the latter to the methanesulfonate[42] or acetate,[5] followed by reaction with triethylamine or potassium bicarbonate respectively, yields the α-nitroalkene. The reaction of alkenes with dinitrogen tetroxide in the presence of iodine yields β-nitroalkyl iodides, which on treatment with sodium acetate also yield α-nitroalkenes. 1,4-dinitro-2-butene has been prepared in this way from butadiene.[5] The synthesis of α-nitroalkene has been recently reviewed by Ono.[2]

The reaction of alkenes with nitrogen oxides and other nitrating agents have been extensively discussed by Olah,[3] Topchiev,[21] and in numerous reviews.[43]

The reaction of alkynes with dinitrogen tetroxide is less synthetically useful as a route to nitro compounds. The reaction of 3-hexyne with dinitrogen tetroxide yields a mixture of *cis*- and *trans*-3,4-dinitro-3-hexene (4.5 % and 13 % respectively), 4,4-dinitro-3-hexanone (8 %), 3,4-hexanedione (16 %) and propanoic acid (6 %).[44] 2-Butyne forms a mixture containing both *cis*- and *trans*-2,3-dinitro-2-butene (7 % and 34 % respectively).[44]

1.4.3 Dinitrogen pentoxide

Alkenes react with dinitrogen pentoxide in chlorinated solvents to give a mixture of β-nitro-nitrate, *vic*-dinitro, *vic*-dinitrate ester and nitroalkene compounds.[45a,b] At temperatures between −30 °C and −10 °C the β-nitro-nitrate is often the main product. The β-nitro-nitrates are inherently unstable and readily form the corresponding nitroalkenes.[40] Propylene reacts with dinitrogen pentoxide in methylene chloride between −10 °C and 0 °C to form a mixture of 1-nitro-2-propanol nitrate (27 %) and isomeric nitropropenes (12 %). The same reaction with cyclohexene is more complicated.[45a]

At temperatures between 0 °C and 25 °C the *vic*-dinitrate ester is often observed in the product mixture and can be the major product in some cases.[45c–e] The synthesis of *vic*-dinitrate esters via this route is discussed in Section 3.6.2. Fischer[46] has given a comprehensive review of work relating to the mechanism of dinitrogen pentoxide addition to alkenes.

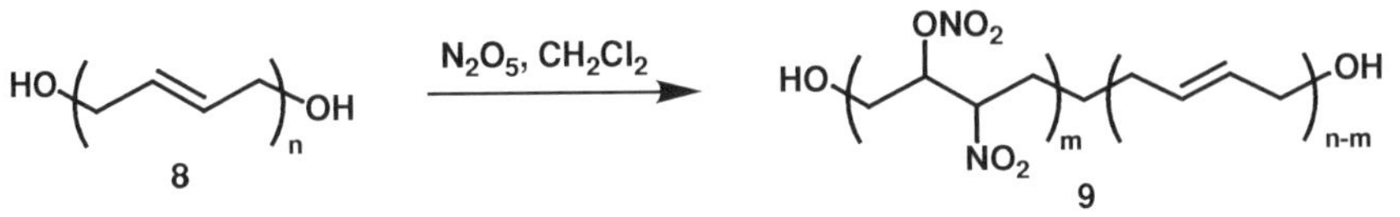

Figure 1.6

Hydroxy-terminated polybutadiene (8) (HTPB) has been treated with dinitrogen pentoxide in methylene chloride. The product (9) is an energetic oligomer but is unlikely to find application because of the inherent instability of β-nitronitrates.[47] Initial peroxyacid epoxidation of some of the double bonds of HTPB followed by reaction with dinitrogen pentoxide yields a product containing *vic*-dinitrate ester groups and this product (NHTPB) is of much more interest as an energetic binder (see Section 3.10).[47]

1.4.4 Nitrous oxide and dinitrogen trioxide

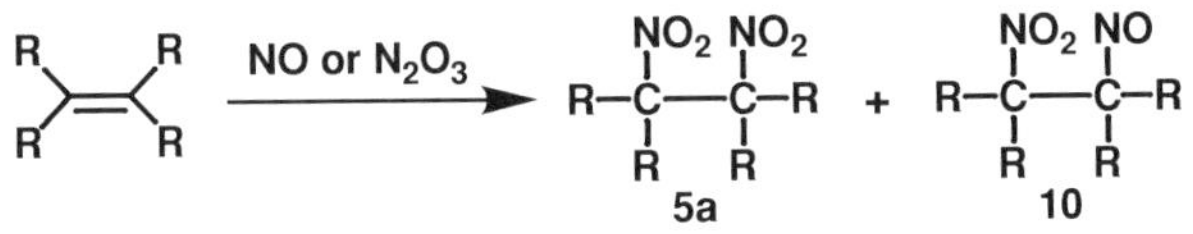

Figure 1.7

The addition of nitrous oxide (NO) or dinitrogen trioxide (N_2O_3) across the double bond of an alkene usually generates a mixture of dinitro (5a) and nitro-nitroso (10) alkanes.[48,49] The reaction of tetrafluoroethylene with dinitrogen trioxide is reported to give 1,2-dinitrotetrafluoroethane and 1-nitro-2-nitrosotetrafluoroethane in 8 % and 42 % yield respectively;[48] the same reaction with nitrous oxide leading to increased yields of 15 % and 68 % respectively.[49] When an excess of nitrous oxide or dinitrogen trioxide is used in these reactions the *vic*-dinitroalkane is usually the main product.[49]

1.4.5 Other nitrating agents

Alkenes react with nitryl chloride to give β-nitroalkyl chlorides, β-chloroalkyl nitrites and *vic*-dichloroalkane products.[50] Nitryl chloride reacts with enol esters to give α-nitroketones.[32b]

A process known as alkene nitrofluorination has been extensively used for the synthesis of β-nitroalkyl fluorides. Reagents used generate the nitronium cation in the presence of fluoride anion, and include: HF/HNO_3,[51] $HF/HNO_3/FSO_3H$,[52] NO_2F,[53] SO_2/NO_2BF_4[54] and HF/pyridine/NO_2BF_4.[55]

A mixture of silver nitrite and iodine reacts with alkenes to give β-nitroalkyl iodides,[56] and therefore, provides a convenient route to α-nitroalkenes.[5] Treatment of alkenes with ammonium nitrate and trifluoroacetic anhydride in the presence of ammonium bromide, followed by treatment of the resulting β-nitroalkyl bromide with triethylamine, is also a general route to α-nitroalkenes.[57]

The reaction of alkenes with nitronium salts proceeds through a nitrocarbocation. The product(s) obtained depends on both the nature of the starting alkene and the conditions used.[3,58]

α-Nitroketones have been synthesized from the reactions of silyl enol ethers with nitronium tetrafluoroborate[59] and tetranitromethane in alkaline media.[60] The reaction of enol acetates with trifluoroacetyl nitrate, generated *in situ* from ammonium nitrate and trifluoroacetic anhydride, also yields α-nitroketones.[61]

1.5 HALIDE DISPLACEMENT

1.5.1 Victor Meyer reaction

One of the most important reactions for the laboratory synthesis of primary aliphatic nitro compounds was discovered by V. Meyer and O. Stüber[62] in 1872 and involves treating alkyl halides with a suspension of silver nitrite in anhydrous diethyl ether. Benzene, hexane and petroleum ether have also been used as solvents for these reactions which are usually conducted between 0 °C and room temperature in the absence of light.

Primary alkyl iodides and bromides are excellent substrates for the Victor Meyer reaction, providing a route to both substituted and unsubstituted nitroalkanes (Table 1.1).[63,65,70,71] The formation of the corresponding nitrite ester is a side-reaction and so the nitroalkane is usually isolated by distillation when possible. The reaction of primary alkyl chlorides with silver nitrite is too slow to be synthetically useful. Secondary alkyl halides and substrates with branching on

$$R{-}X \quad + \quad AgNO_2 \xrightarrow[X\,=\,I\text{ or }Br]{Et_2O} R{-}NO_2 \quad + \quad AgX$$

Table 1.1
Synthesis of nitroalkanes and their derivatives from the reaction of alkyl halides with silver nitrite under the Victor Meyer conditions

Alkyl halide	Yield (%) of nitroalkane	Yield (%) of nitrite ester	Ref.
$CH_3CH_2CH_2Br$	67	19	63
$(CH_3)_2CHBr$	19–26	24–34	64
$CH_3CH_2CH_2CH_2Br$	73	13	65
$CH_3CH_2CHBrCH_3$	19–24	27–37	64
$HOCH_2CH_2I$	62	---	66
$H_2C{=}CHCH_2Br$	55	---	67
ICH_2CO_2Et	77	---	68
p-$O_2NC_6H_4CH_2Br$	75	5	69
p-$MeOC_6H_4CH_2Br$	26	55	69

$$XCH_2(CH_2)_nCH_2X \xrightarrow[\text{or } AgNO_2,\ Et_2O]{NaNO_2,\ DMF,\ \text{phloroglucinol},\ (NH_2)_2CO} O_2NCH_2(CH_2)_nCH_2NO_2$$

X = I or Br

Table 1.2
Synthesis of dinitroalkanes from the reaction of dihaloalkanes with silver nitrite and sodium nitrite (ref. 75–79)

Homologue	Yield (%) of dinitroalkane		
n =	diiodide/$AgNO_2$	dibromide/$AgNO_2$	dibromide/$NaNO_2$
1	37[75]	20[78]	6[79]
2	46[75]	37[78]	33[79]
3	45[75]	53[78]	29[79]
4	48[75]	---	42[79]
5	60[76]	---	---
6	---	---	---
7	---	---	---
8	50[77]	100[78]	---

the carbon chain give much lower yields of nitro compound.[64] Reactions with such substrates are much slower and nitrite ester formation is much more of a problem. In fact, the nitrite ester and the corresponding alkene can be the main products of the reaction of a secondary alkyl halide with silver nitrite. Consequently, the Victor Meyer reaction is not considered useful for the synthesis of secondary nitroalkanes. Nitrate ester by-products[72,73] can arise from either disproportionation of silver nitrite,[72] a process accelerated by heat and light, or from dehydrohalogenation[64] of the alkyl halide substrate.

The Victor Meyer reaction is remarkably versatile and tolerant of many functional groups, providing a route to arylnitromethanes[69] from benzyl iodides and bromides, α-nitroesters[68] from α-iodoesters, and nitro-substituted epoxides like 1-nitro-2,3-epoxypropane, 1-nitro-2,3-epoxybutane, and 3-nitro-1,2-epoxybutane from the corresponding iodoepoxides[74] (Table 1.1). The Victor Meyer reaction has also been used for the synthesis of numerous α,ω-dinitroalkanes from the corresponding α,ω-diiodoalkanes[75–77] and α,ω-dibromoalkanes[78] where yields of 37–50 % are reported (Table 1.2). 1,5-Dinitropentane (62 %), 1,8-dinitrooctane (64 %) and 1,10-dinitrodecane have recently been synthesized from the corresponding α,ω-diiodoalkanes using water as the reaction solvent.[80] Kornblum[81] has compiled a comprehensive list of reactions illustrating the versatility of the Victor Meyer reaction.

With regard to the mechanism of the Victor Meyer reaction, it is thought that both S_N2 and S_N1 transition states play a role in product ratios.[69,82] Both transition states are known to be highly dependent on the nature of the substrate, the solvent and the reaction conditions used.[83] Which mechanism or combination of mechanisms is operating is important from a practical point of view; the more S_N1 character in these reactions the more nitrite ester formed via attack on an intermediate carbocation. Unsurprisingly, the reaction of silver nitrite with tertiary alkyl halides is not a viable route to tertiary nitroalkanes.[64] The formation of the silver halogen bond is a strong driving force for the Victor Meyer reaction, and consequently, sulfonate esters, and other substrates containing equally good leaving groups, fail to react with silver nitrite.[69]

1.5.2 Modified Victor Meyer Reaction

The synthesis of aliphatic nitro compounds from the reaction of alkyl halides with alkali metal nitrites was discovered by Kornblum and co-workers[84,85] and is known as the modified Victor Meyer reaction or the Kornblum modification. The choice of solvent in these reactions is crucial when sodium nitrite is used as the nitrite source. Both alkyl halide and nitrite anion must be in solution to react, and the higher the concentration of nitrite anion, the faster the reaction. For this reason, both DMF and DMSO are widely used as solvents, with both able to dissolve appreciable amounts of sodium nitrite. Although sodium nitrite is more soluble in DMSO than DMF the former can react with some halide substrates.[86] Urea is occasionally added to DMF solutions of sodium nitrite to increase the solubility of this salt and hence increase reaction rates. Other alkali metal nitrites can be used in these reactions, like lithium nitrite,[87] which is more soluble in DMF than sodium nitrite but is also less widely available.

$$RR'CHNO_2 + RR'CHONO_2 \xrightarrow{NO_2^-} RR'CHOH + RR'C(NO)NO_2 \qquad \text{(Eq. 1.1)}$$

Figure 1.8

Reaction time is extremely important in avoiding the side reaction illustrated in Eq. (1.1), where the nitroalkane product reacts with nitrite anion and any nitrite ester, formed as by-product, to give a pseudonitrole.[88] The reaction of sodium nitrite with alkyl halides is much faster than this competing nitrosation side reaction, even so, prompt work-up on reaction completion is essential for obtaining good yields.

As a rule, the reaction of primary alkyl iodides with sodium nitrite in DMF takes about 2.5 hours for completion but this is raised to about 6 hours for primary alkyl bromides.[81] The addition of urea cuts these reaction times by approximately half.[80] The reaction of alkyl chlorides with alkali metal nitrites is too slow to be synthetically useful. Secondary alkyl bromides and alicyclic iodides require the addition of a nitrite ester scavenger such as phloroglucinol.[84] Under these conditions secondary nitroalkanes and their derivatives are formed in good yield,[84,85] an advantage over the use of silver nitrite which is generally limited to primary alkyl halides. Nitroalkane products containing electron-withdrawing groups that significantly increase the acidity of the nitro group, i.e. α-nitroesters, are particularly susceptible to nitrosation and also require the addition of a nitrite ester scavenger.[89,90] The reaction of tertiary alkyl halides with sodium nitrite is not a feasible synthetic route to nitro compounds.[84]

The modified Victor Meyer reaction has been used successfully for the synthesis of many substituted and unsubstituted, primary and secondary, nitro compounds from the corresponding alkyl iodides, bromides, sulfonate esters etc. (Table 1.3).[81,84] Arylnitromethanes are synthesized in good yield from benzylic halides if reactions are conducted at subambient temperatures.[84,92] α-Nitroesters[89,90] and β-nitroketones[91] are obtained from the corresponding α-haloesters and β-haloketones. Some substrates containing a strong electron-withdrawing group on the same carbon as the halogen i.e. α-haloketones are destroyed under the modified Victor Meyer conditions and so the use of silver nitrite is preferred.

Silver nitrite gives significantly higher yields of nitro compounds from primary alkyl halides, and consequently, the synthesis of α,ω-dinitroalkanes from the reaction of α,ω-dihaloalkanes with sodium nitrite[79] is inferior to the same reaction with silver nitrite[75–78] (Table 1.2). However, the use of a solvent system composed of DMSO and MEK is reported to considerably improve the yields of α,ω-dinitroalkane when using sodium nitrite.[93]

Table 1.3
Synthesis of nitroalkanes and their derivatives from the reaction of alkyl halides with sodium nitrite under the modified Victor Meyer conditions

Substrate	Solvent	Yield (%) of nitroalkane	Ref.
$CH_3CH_2CH_2CH_2SO_2CH_3$	DMF	46	84
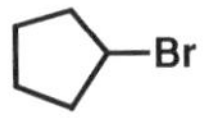 Cyclopentyl-Br	DMF	57	84
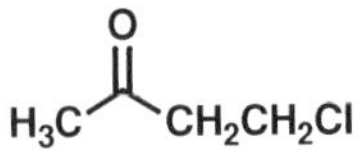 $H_3C(C{=}O)CH_2CH_2Cl$	DMF	47	91
$CH_3CHBrCO_2Et$	DMSO	66	89
$PhCH_2Br$	DMF	55	84
$(CH_3)_2C(CN)Br$	DMF	52	87
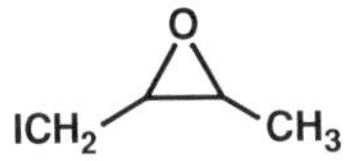 ICH_2-epoxide-CH_3	DMF	70	74

The reaction of α-halocarboxylic acids with sodium nitrite has been used to synthesize nitromethane, nitroethane and nitropropane, although the reaction fails for higher nitroalkanes.[94]

A number of other reactions have been reported which use nitrite anion as a nucleophile, including: (1) reaction of alkyl halides with potassium nitrite in the presence of 18-crown-6,[95] (2) reaction of alkyl halides with nitrite anion bound to amberlite resins,[96] (3) synthesis of 2-nitroethanol from the acid-catalyzed ring opening of ethylene oxide with sodium nitrite,[97] and (4) reaction of primary alkyl chlorides with sodium nitrite in the presence of sodium iodide.[98]

1.5.3 Ter Meer reaction

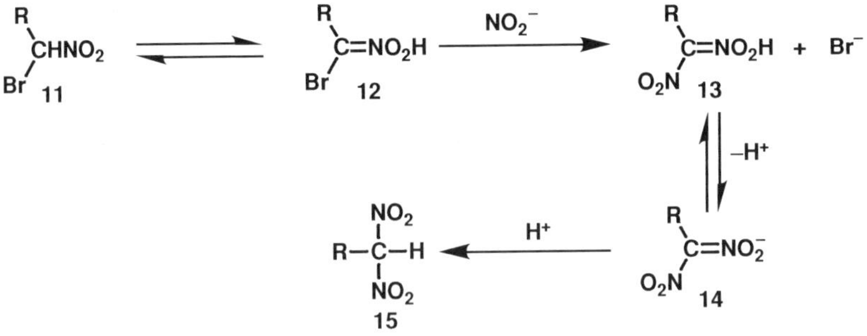

Figure 1.9

The Ter Meer reaction[99] provides a convenient route to terminal *gem*-dinitroaliphatic (15) compounds via the displacement of halogen from terminal α-halonitroalkanes (11) with nitrite anion in alkaline solution. These reactions are believed[100] to go via the tautomeric α-halonitronic acid (12) rather than the α-halonitroalkane (11). The synthesis of internal *gem*-dinitroaliphatic compounds from the corresponding internal α-halonitroalkanes is not possible via this route. The α-halonitroalkane substrates for the Ter Meer reaction are readily obtained from the bromination[62] or chlorination[101] of primary nitroalkanes in alkaline solution, or from the reaction of halogenating agents with oximes followed by oxidation of the resulting α-halonitrosoalkane.[102,103]

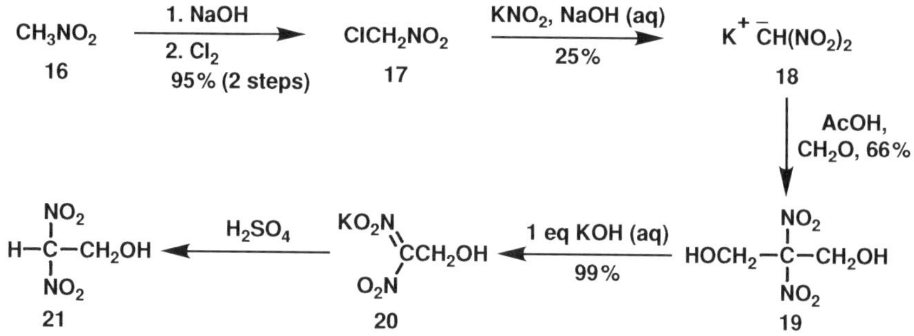

$$\underset{\mathbf{16}}{CH_3NO_2} \xrightarrow[\text{2. } Cl_2,\ 95\%\ (2\ steps)]{\text{1. NaOH}} \underset{\mathbf{17}}{ClCH_2NO_2} \xrightarrow[25\%]{KNO_2,\ NaOH\ (aq)} \underset{\mathbf{18}}{K^+\ {}^-CH(NO_2)_2} \xrightarrow{AcOH,\ CH_2O,\ 66\%} \underset{\mathbf{19}}{HOCH_2-C(NO_2)_2-CH_2OH}$$

$$\underset{\mathbf{19}}{HOCH_2-C(NO_2)_2-CH_2OH} \xrightarrow[99\%]{1\ eq\ KOH\ (aq)} \underset{\mathbf{20}}{(KO_2N)(O_2N)C{=}CH_2OH} \xrightarrow{H_2SO_4} \underset{\mathbf{21}}{H-C(NO_2)_2-CH_2OH}$$

Figure 1.10

The Ter Meer reaction has been used for the synthesis of the potassium salt of dinitromethane (18) by treating chloronitromethane (17) with potassium nitrite in aqueous sodium hydroxide solution.[104] The potassium salt of dinitromethane is a dangerous shock sensitive explosive and should not be isolated; treatment of (18) with formaldehyde in acidic solution forms the more manageable bis-methylol derivative, 2,2-dinitro-1,3-propanediol (19).[104] Partial hydrolysis of (19) with potassium hydroxide, followed by acidification, yields 2,2-dinitroethanol (21).[105] 2,2-Dinitroethanol has also been synthesized from the alkaline bromination of 2-nitroethanol followed by nitrite displacement and subsequent acidification of the resulting potassium 2,2-dinitroethanol.[106] 2,2-Dinitro-1,3-propanediol and 2,2-dinitroethanol are useful derivatives of dinitromethane and find extensive use in condensation and esterification reactions for the synthesis of energetic oligomers and plasticizers.

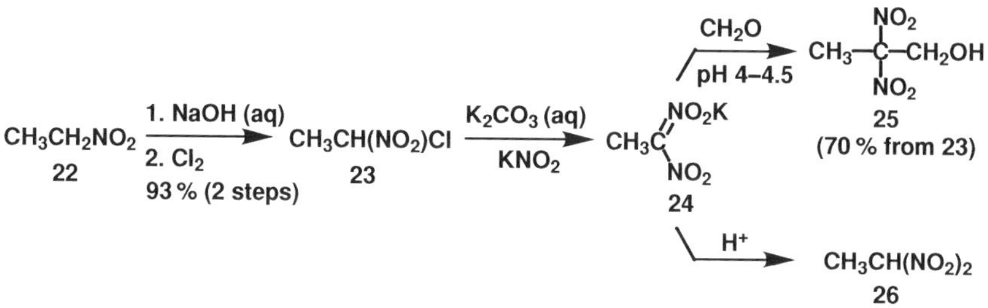

$$\underset{\mathbf{22}}{CH_3CH_2NO_2} \xrightarrow[\text{2. } Cl_2,\ 93\%\ (2\ steps)]{\text{1. NaOH (aq)}} \underset{\mathbf{23}}{CH_3CH(NO_2)Cl} \xrightarrow[KNO_2]{K_2CO_3\ (aq)} \underset{\mathbf{24}}{CH_3C(NO_2)(=NO_2K)}$$

$$\mathbf{24} \xrightarrow[pH\ 4\text{–}4.5]{CH_2O} \underset{\mathbf{25}\ (70\ \%\ from\ 23)}{CH_3-C(NO_2)_2-CH_2OH} \qquad \mathbf{24} \xrightarrow{H^+} \underset{\mathbf{26}}{CH_3CH(NO_2)_2}$$

Figure 1.11

Both 1,1-dinitroethane (26) and 1,1-dinitropropane, and their methylol derivatives, 2,2-dinitropropanol (25) and 2,2-dinitro-1-butanol respectively, have been synthesized via the Ter Meer reaction.[99,107] The formal and acetal of 2,2-dinitropropanol in the form of a 1:1 eutectic is an energetic plasticizer and so the synthesis of 2,2-dinitropropanol has been the subject of much investigation.[107] On a pilot plant scale 2,2-dinitropropanol (25) is synthesized in 60 % overall yield from nitroethane (22). Thus, treatment of 1-chloronitroethane (23) with potassium nitrite in aqueous potassium carbonate forms the potassium salt of 1,1-dinitroethane (24), which on acidification, followed by *in situ* reaction with formaldehyde, yields 2,2-dinitropropanol (25). Note, while dinitromethane is unstable at room temperature other terminal *gem*-dinitroalkanes, like 1,1-dinitroethane (26) and 1,1-dinitropropane, are perfectly stable.

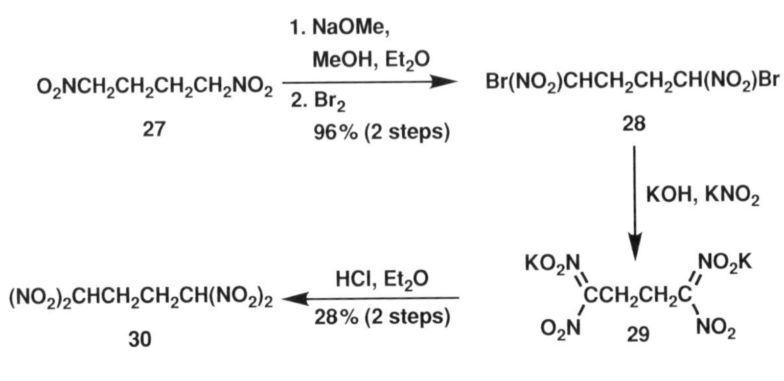

Figure 1.12

The Ter Meer reaction has been used to synthesize $\alpha,\alpha,\omega,\omega$-tetranitroalkanes from the corresponding α,ω-dihalo-α,ω-dinitroalkanes.[108] Thus, treatment of 1,4-dibromo-1,4-dinitrobutane (28) under the Ter Meer conditions yields the dinitronate salt of 1,1,4,4-tetranitrobutane (29); acidification of the latter yields 1,1,4,4-tetranitrobutane (30).[108]

The Ter Meer reaction has not been widely exploited for the synthesis of *gem*-dinitroaliphatic compounds. This is partly because the Kaplan–Shechter oxidative nitration (Section 1.7) is more convenient, but also because of some more serious limitations. The first is the inability to synthesize internal *gem*-dinitroaliphatic compounds; functionality which shows high chemical stability and is found in many cyclic and caged energetic materials. Secondly, the *gem*-nitronitronate salts formed in the Ter Meer reactions often need to be isolated to improve the yield and purity of the product. Dry *gem*-nitronitronate salts are hazardous to handle and those from nitroalkanes like 1,1,4,4-tetranitrobutane are primary explosives which can explode even when wet. Even so, it is common to use conditions that lead to the precipitation of *gem*-nitronitronate salts from solution, a process that both drives the reaction to completion and also provides isolation and purification of the product salt by simple filtration. Purification of *gem*-nitronitronate salts by filtration from the reaction liquors, followed by washing with methanol or ethanol to remove occluded impurities, has been used, although these salts should never be allowed to completely dry.

The choice of base used in the Ter Meer reaction is important for two reasons. First, studies have found that strong bases, such as alkali metal hydroxides, inhibit the reaction and promote side-reactions, whereas the weaker alkali metal carbonates generally give higher yields.[107] Secondly, if the *gem*-nitronitronate salt needs to be purified by filtration it should be sparingly soluble in the reaction solvent and both the reaction solvent and the counterion of the *gem*-nitronitronate salt affect this solubility.[107] Use of the potassium salt is advantageous for aqueous systems where the *gem*-nitronitronate salts are usually only sparingly soluble, whereas the sodium salt can be used for nonaqueous reactions.[107]

It must be emphasized that *gem*-nitronitronate salts should never be stored on safety grounds. These salts readily react with formaldehyde to give the methylol derivatives which are more stable and less hazardous to handle. The latter are often used directly in condensation reactions where treatment with aqueous base forms the *gem*-nitronitronate salt *in situ*.

1.5.4 Displacements using nitronate salts as nucleophiles

Nitronate salts can react with alkyl halides to yield polynitroaliphatic compounds with varying degrees of success. The main by-products of these reactions arise for competitive *O*-alkylation. Alkyl nitrates are formed as by-products when the nitroform anion is used in these reactions.[109]

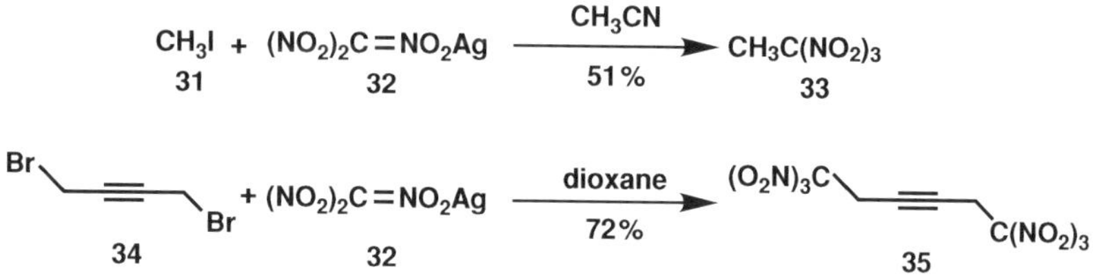

$$\underset{31}{CH_3I} + \underset{32}{(NO_2)_2C{=}NO_2Ag} \xrightarrow[51\,\%]{CH_3CN} \underset{33}{CH_3C(NO_2)_3}$$

$$\underset{34}{BrCH_2C{\equiv}CCH_2Br} + \underset{32}{(NO_2)_2C{=}NO_2Ag} \xrightarrow[72\%]{dioxane} \underset{35}{(O_2N)_3CCH_2C{\equiv}CCH_2C(NO_2)_3}$$

Figure 1.13

Methyl iodide (31) reacts with silver nitroform (32) in acetonitrile to give a 51 % yield of 1,1,1-trinitroethane (33).[109] The potassium salt of nitroform in acetone has been used for the same reaction.[110] Yields between 28 % and 65 % have been reported for the reaction of silver nitroform in acetonitrile with higher molecular weight alkyl iodides.[109] The choice of solvent is important in some reactions, for example, silver nitroform reacts with 1,4-dibromo-2-butyne (34) in solvents like dioxane[111] and acetone[111] to give 1,1,1,6,6,6-hexanitro-3-hexyne (35) in approximately 72 % yield, whereas the same reaction in acetonitrile[109] is reported to give a mixture of compounds.

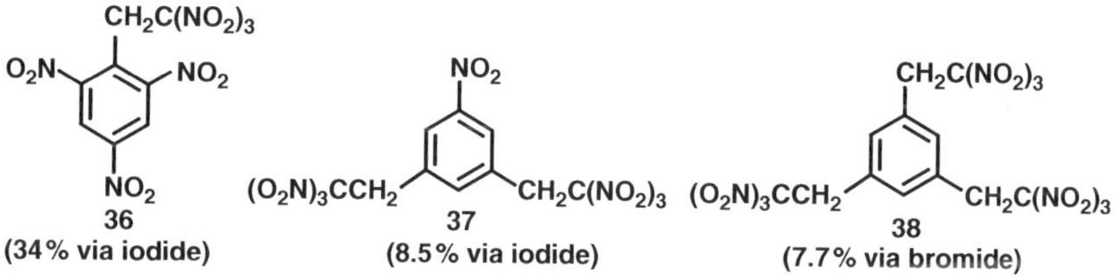

$CH_2C(NO_2)_3$, O_2N, NO_2, NO_2
36
(34 % via iodide)

NO_2, $(O_2N)_3CCH_2$, $CH_2C(NO_2)_3$
37
(8.5 % via iodide)

$CH_2C(NO_2)_3$, $(O_2N)_3CCH_2$, $CH_2C(NO_2)_3$
38
(7.7 % via bromide)

Figure 1.14

Poor to modest yields of trinitromethyl compounds are reported for the reaction of silver nitroform with substituted benzyl iodide and bromide substrates. Compounds like (36), (37), and (38) have been synthesized via this route; these compounds have much more favourable oxygen balances than TNT and are probably powerful explosives.[112] The authors noted that considerable amounts of unstable red oils accompanied these products. The latter are attributed to *O*-alkylation, a side-reaction favoured by an S_N1 transition state and typical of reactions involving benzylic substrates and silver salts. Further research showed that while silver nitroform favours *O*-alkylation, the sodium, potassium and lithium salts favour *C*-alkylation.[113] The synthesis and chemistry of 1,1,1-trinitromethyl compounds has been extensively reviewed.[9,114,115] The alkylation of nitronate salts has been the subject of an excellent review by Nielsen.[116]

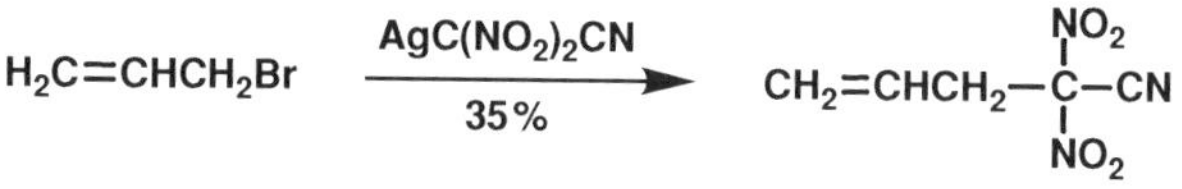

$$H_2C{=}CHCH_2Br \xrightarrow[35\%]{AgC(NO_2)_2CN} CH_2{=}CHCH_2{-}C(NO_2)_2{-}CN$$

Figure 1.15

The reactions of alkyl halides with the silver salt of dinitroacetonitrile have been shown to be of limited use for the synthesis of polynitroaliphatic compounds.[109,117] These reactions give a mixture of *C*-, *N*- and *O*-alkylation products with product distribution highly dependent on the nature of the substrate.[109,117]

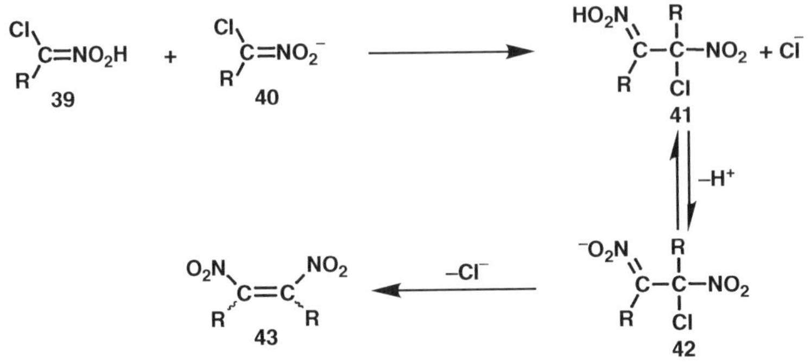

Figure 1.16

The addition of α-chloronitroalkanes to solutions of alkali metal hydroxide has been used for the synthesis of some 1,2-dinitroethylene derivatives (43).[118,119] These reactions involve attack of the nitronate salt (40) on the *aci*-form (39) of the unreacted *gem*-chloronitroalkane followed by formal loss of hydrogen chloride. 2,3-Dinitro-2-butene and 3,4-dinitro-3-hexene (45) are formed in this way from 1-chloro-1-nitroethane and 1-chloro-1-nitropropane (44) respectively.[118]

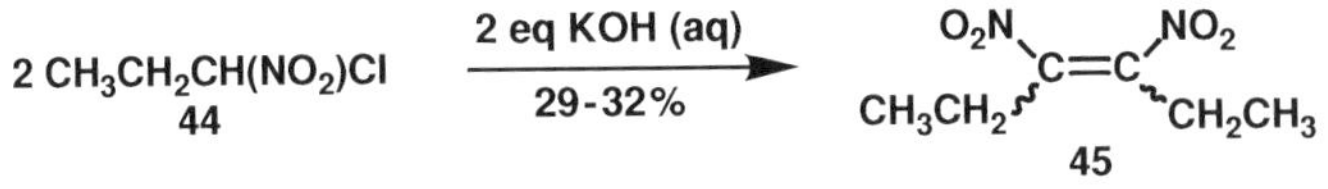

Figure 1.17

In a related reaction to that described above, nitronate salts have been reacted with α-chloronitroalkanes as a route to polynitroaliphatic compounds.[120] 2,3-Dimethyl-2,3-dinitrobutane (48) is formed from the reaction of the nitronate salt of 2-nitropropane (46) with 2-chloro-2-nitropropane (47).[120] A modification to the original process uses nitronate salts in the presence of iodine to form an α-iodonitroalkane *in situ*.[121]

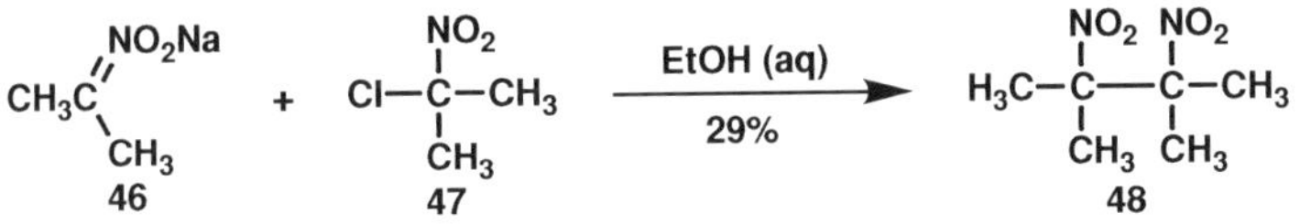

Figure 1.18

1.6 OXIDATION AND NITRATION OF C–N BONDS

1.6.1 Oxidation and nitration of oximes

Methods used for the oxidation or nitration of C–N bonds have found wide use for the synthesis of novel polynitropolycycloalkanes as can be seen in Chapter 2. The conversion of readily available ketone groups to nitro and *gem*-dinitroaliphatic functionality is usually via the oxime and this is an important route used in energetic materials synthesis.

1.6.1.1 Scholl reaction[122]

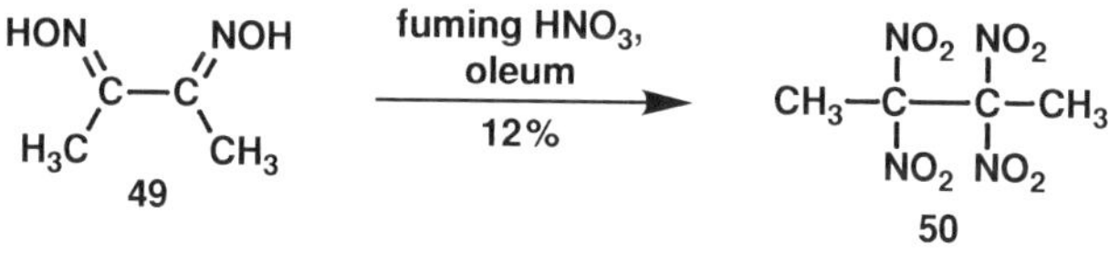

HON NOH H_3C CH_3 49 — fuming HNO_3, oleum, 12% → CH_3–C(NO_2)$_2$–C(NO_2)$_2$–CH_3 50

Figure 1.19

One route to *gem*-dinitroalkanes involves the tandem nitration–oxidation of oximes.[122] The nitration of dimethylglyoxime (49) with anhydrous mixed acid is reported to give 2,2,3,3-tetranitrobutane (50) in 12 % yield.[39] However, ammonium nitrate and nitric acid in methylene chloride is a more commonly used reagent, but the product, usually a pseudonitrole, needs treatment with hydrogen peroxide to yield the *gem*-dinitro compound. The nitric acid used in these reactions often contains 12–24 % dissolved dinitrogen tetroxide ('red fuming nitric acid') and so the reaction has similarities with the Ponzio reaction (Section 1.6.1.2). Some energetic materials recently synthesized with this reagent are illustrated in Table 1.4 (also see Chapter 2).

Table 1.4
Synthesis of energetic polynitropolycycloalkanes via the Scholl reaction

Oxime	Conditions/reagents	Product	Ref.
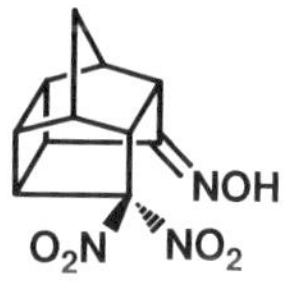 NOH, O_2N, NO_2	1. 98% red HNO_3, NH_4NO_3, CH_2Cl_2, reflux 2. 30% H_2O_2 (aq) reflux 31% (2 steps)	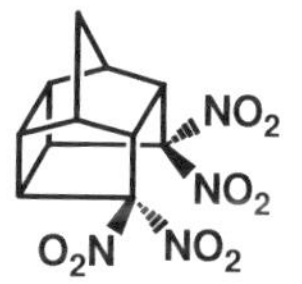 NO_2, NO_2, O_2N, NO_2	123
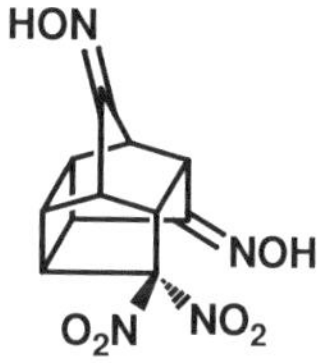 HON, NOH, O_2N, NO_2	1. 98% red HNO_3, urea, NH_4NO_3, CH_2Cl_2, reflux 2. 30% H_2O_2 (aq) reflux 19% (2 steps)	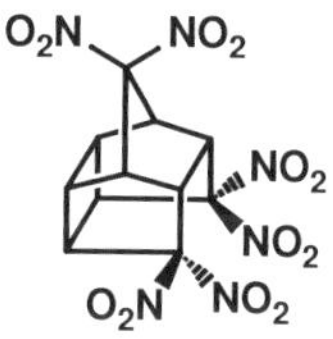 O_2N, NO_2, NO_2, NO_2, O_2N, NO_2	124
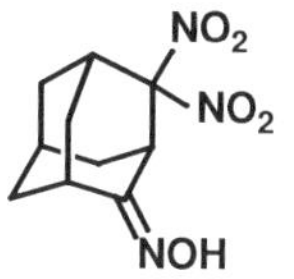 NO_2, NO_2, NOH	98% HNO_3, urea, CH_2Cl_2, NH_4NO_3 reflux, 29%	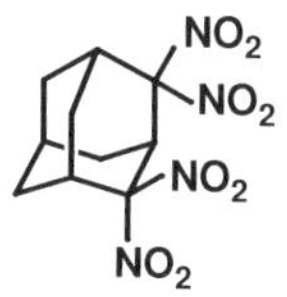 NO_2, NO_2, NO_2, NO_2	125
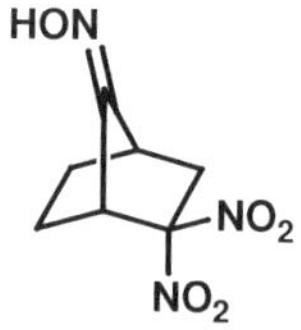 HON, NO_2, NO_2	1. 98% red HNO_3, CH_2Cl_2, urea, NH_4NO_3, reflux 2. 30% H_2O_2 (aq) reflux 52% (2 steps)	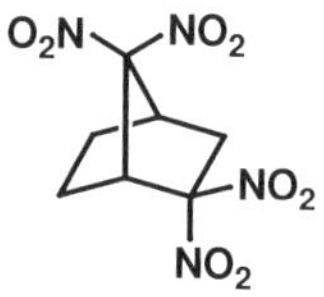 O_2N, NO_2, NO_2, NO_2	126

Yields are frequently moderate for Scholl reactions. Aldoximes are not usually compatible with these harsh reaction conditions and are very sensitive to factors such as temperature and reaction time. Consequently, oxidation to the corresponding carboxylic acid is a major side-reaction. However, both the ketoxime (51) and the aldoxime (53) are reported to give good yields of the corresponding *gem*-dinitro compounds, (52) and (54) respectively, on treatment with absolute nitric acid in methylene chloride followed by hydrogen peroxide.[127]

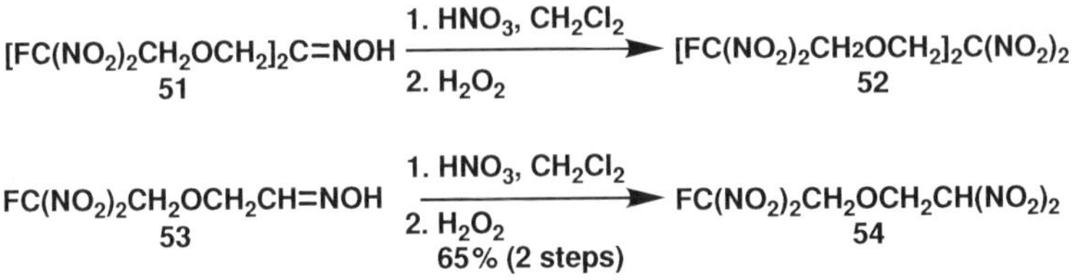

$[FC(NO_2)_2CH_2OCH_2]_2C{=}NOH$ (51) → (1. HNO_3, CH_2Cl_2; 2. H_2O_2) → $[FC(NO_2)_2CH_2OCH_2]_2C(NO_2)_2$ (52)

$FC(NO_2)_2CH_2OCH_2CH{=}NOH$ (53) → (1. HNO_3, CH_2Cl_2; 2. H_2O_2; 65% (2 steps)) → $FC(NO_2)_2CH_2OCH_2CH(NO_2)_2$ (54)

Figure 1.20

The nitration of oximinocyanoacetic acid esters with mixed acid at room temperature yields dinitrocyanoacetic acid esters which are precursors to dinitroacetonitrile and its salts.[128] α-Oximinonitriles can be nitrated to α,α-dinitronitriles with anhydrous nitric acid containing ammonium nitrate, and then subjected to ammonolysis with aqueous ammonia to give the ammonium salt of the corresponding *gem*-dinitroalkane.[129]

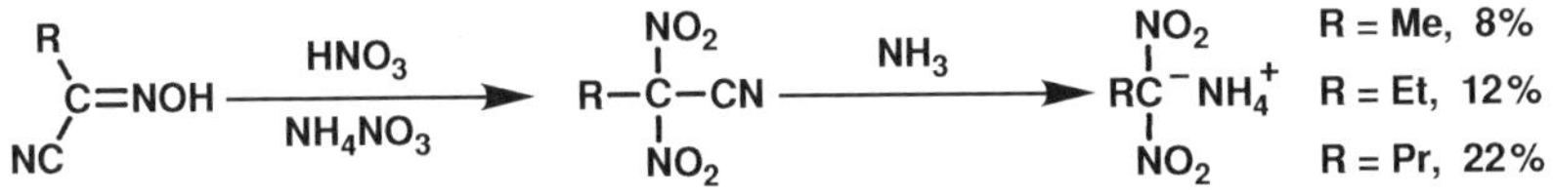

$R(NC)C{=}NOH$ → (HNO_3, NH_4NO_3) → $R{-}C(NO_2)_2{-}CN$ → (NH_3) → $RC^-(NO_2)_2\ NH_4^+$

R = Me, 8%
R = Et, 12%
R = Pr, 22%

Figure 1.21

A series of α,α-dinitroesters have been synthesized from the reaction of α-oximinoesters with cold nitric acid containing ammonium nitrate, followed by oxidation of the resulting pseudonitroles with oxygen.[130]

1.6.1.2 Ponzio reaction

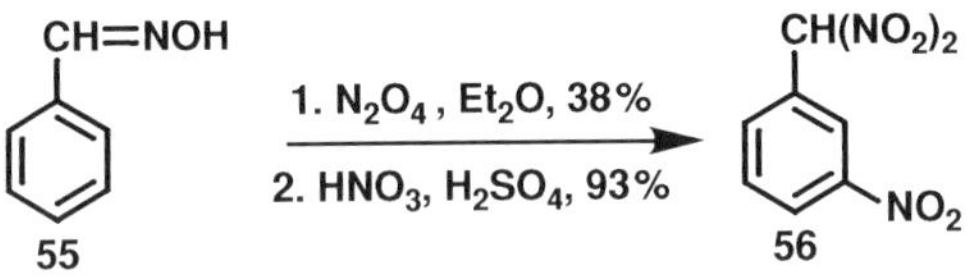

$C_6H_5CH{=}NOH$ (55) → (1. N_2O_4, Et_2O, 38%; 2. HNO_3, H_2SO_4, 93%) → $3\text{-}O_2NC_6H_4CH(NO_2)_2$ (56)

Figure 1.22

The Ponzio reaction[131] provides a useful route to *gem*-dinitro compounds and involves treating oximes with a solution of nitrogen dioxide or its dimer in diethyl ether or a chlorinated solvent. The Ponzio reaction works best for aromatic oximes where the synthesis of many substituted aryldinitromethanes have been reported.[132] Compound (56), an isomer of TNT, is formed from the reaction of dinitrogen tetroxide with the oxime of benzaldehyde (55) followed by mono-nitration of the aromatic ring with mixed acid.[133] Yields are usually much lower for aliphatic aldoximes and ketoximes.[134b,135] The parent carbonyl compound of the oxime is usually the major by-product in these reactions.

Dinitrogen pentoxide has also been used to synthesize aryldinitromethanes from arylaldoximes in yields between 20 % and 60 %.[134] Aromatic ring nitration can also occur during these reactions.[134] Dinitrogen pentoxide in chloroform converts α-chlorooximes to the corresponding α-chloro-α,α-dinitro compounds.[135]

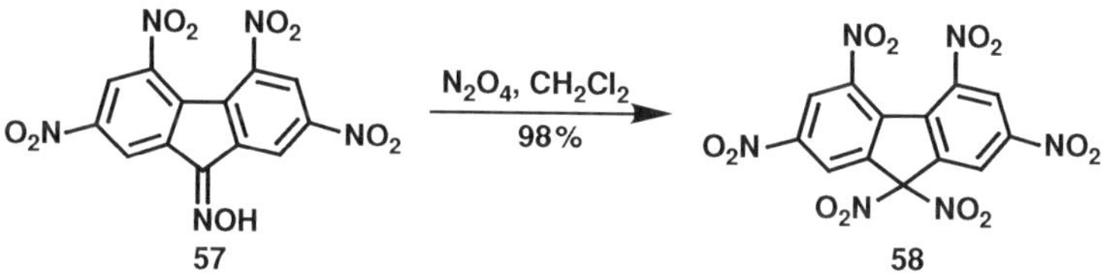

N_2O_4, CH_2Cl_2 98%

57 58

Figure 1.23

Millar and co-workers[136] conducted an extensive reinvestigation into the Ponzio reaction using both dinitrogen pentoxide and dinitrogen tetroxide in different solvents. The mechanism of the Ponzio reaction was studied together with correlations between substrate structure and product yield/purity. The thermally stable explosive 2,4,5,7,9,9-hexanitrofluorene (58) was synthesized from fluoren-9-one in 81 % yield via initial ring nitration with mixed acid, followed by oxime formation and treatment of the resulting product (57) with dinitrogen tetroxide in methylene chloride.[136,137] Millar and co-workers[136] also synthesized 2-(dinitromethyl)-4-nitrophenol, a potential substitute for picric acid, via the reaction of 2-hydroxybenzaldehyde oxime with dinitrogen tetroxide in acetonitrile (21 %). 1,1-Dinitro-1-(4-nitrophenyl)ethane was also synthesized from 4-nitroacetophenone oxime.

1.6.1.3 Peroxyacid oxidation

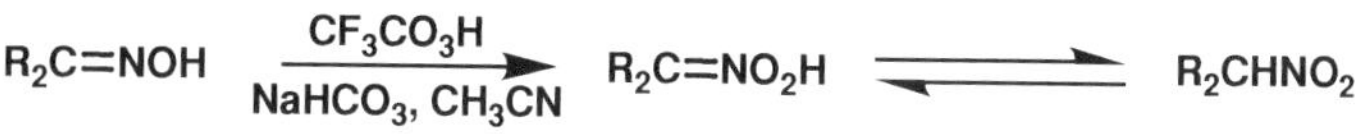

$$R_2C{=}NOH \xrightarrow[NaHCO_3,\ CH_3CN]{CF_3CO_3H} R_2C{=}NO_2H \rightleftharpoons R_2CHNO_2$$

Figure 1.24

The direct oxidation of an oxime to a nitro group can be achieved with peroxytrifluoroacetic acid, a reagent usually formed *in situ* from the reaction of 90 % hydrogen peroxide with either trifluoroacetic acid or trifluoroacetic anhydride.[138,139] Reactions are commonly conducted in acetonitrile at gentle reflux in the presence of sodium bicarbonate for aliphatic oximes, or alternatively, in the presence of a sodium hydrogen phosphate buffer for aromatic and alicyclic oximes. Urea is frequently added to these reactions to scavenge any oxides of nitrogen formed. Emmons and Pagano[138] used this route to synthesize a number of primary and secondary nitro compounds from the corresponding aldoximes and ketoximes respectively in yields of 40–80 %. This method fails for some sterically hindered oximes.

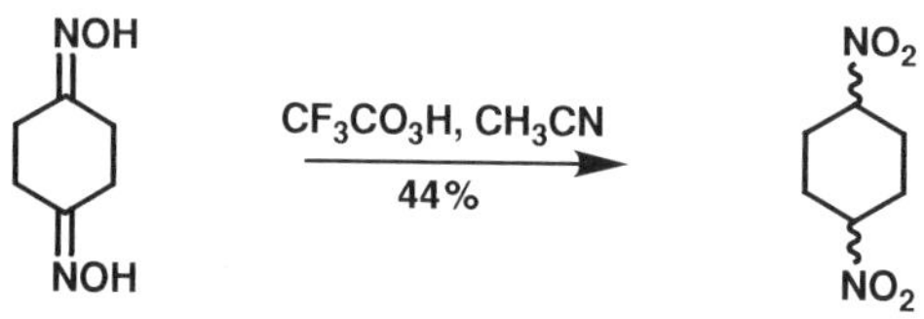

CF_3CO_3H, CH_3CN 44%

Figure 1.25

Table 1.5
Synthesis of energetic polynitropolycycloalkanes via the oxidation of oximes with peroxytrifluoroacetic acid

Oxime	Conditions/reagents	Product	Ref.
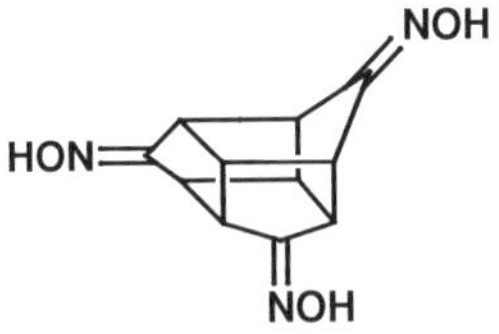	$(CF_3CO)_2O$, 90% H_2O_2, CH_3CN, $NaHCO_3$, urea 70–75 °C, 35%	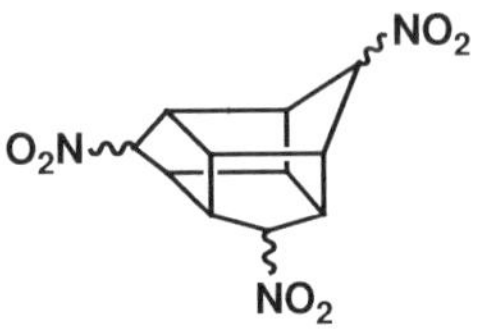	123
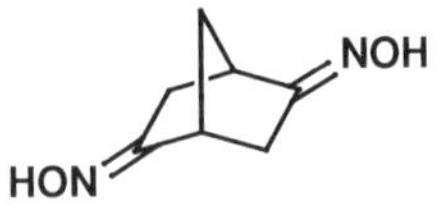	$(CF_3CO)_2O$, 90% H_2O_2, CH_3CN, Na_2HPO_4, 75 °C, 65%		141
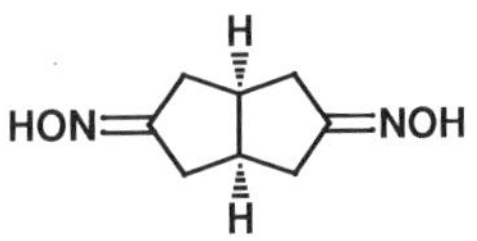	$(CF_3CO)_2O$, 90% H_2O_2, CH_3CN, $NaHCO_3$, 75 °C, 60%	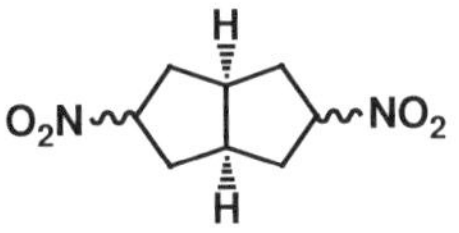	141

Nielsen[140] used the same chemistry to synthesize both 1,3- and 1,4-dinitrocyclohexanes from the corresponding dioximes. The peroxytrifluoroacetic acid reagent has been used in reaction routes to a number of highly energetic polynitropolycycloalkanes as illustrated in Table 1.5 (see also Chapter 2).

Some recent advances have been reported in oxime oxidation, including the *in situ* generation of peroxytrifluoroacetic acid from the reaction of urea hydrogen peroxide complex with TFAA in acetonitrile at 0 °C.[142] This method gives good yields of nitroalkanes from aldoximes but fails with ketoximes.

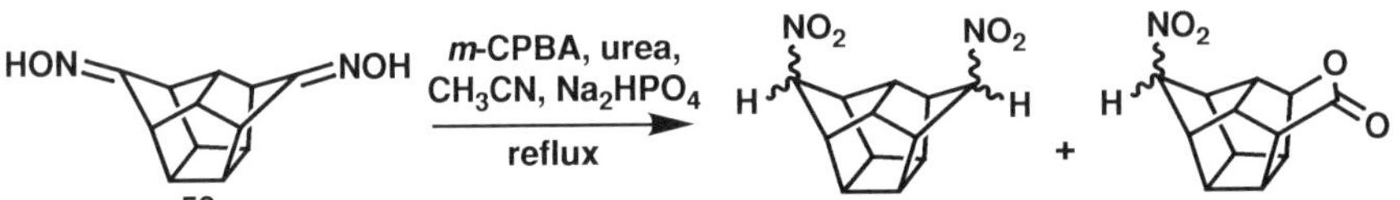

m-CPBA, urea, CH_3CN, Na_2HPO_4 reflux

59 → 60 (23%) (mixture of isomers) + 61 (15%)

Figure 1.26

Peroxyacids are powerful oxidants and so side reactions are to be expected. Ketones are known to undergo Baeyer–Villiger[143] oxidation to the carboxylic acid ester on treatment with peroxyacids and so these by-products can be observed if oxime hydrolysis occurs during the oxidation. Paquette and co-workers[144] observed such a by-product (61) during the oxidation of the dioxime (59) to the dinitro compound (60) with *m*-CPBA in hot acetonitrile.

Peroxyacetic acid generated *in situ* from sodium perborate and glacial acetic acid has been used for oxime to nitro group conversion.[145] Peroxyimidic acid generated from acetonitrile and hydrogen peroxide has found similar use.[146] An Mo(IV) peroxy complex has been reported for the oxidation of both ketoximes and aldoximes.[147]

1.6.1.4 Halogenation–oxidation–reduction route

$$R_2C{=}NOH \xrightarrow[Br_2,\ NaHCO_3]{NBS\ or} R_2C(Br)NO\ (62) \xrightarrow[or\ HNO_3,\ H_2O_2]{CF_3CO_3H\ or\ O_2} R_2C(Br)NO_2\ (63) \xrightarrow[or\ KOH\ (aq)]{NaBH_4,\ MeOH} R_2C(H)NO_2$$

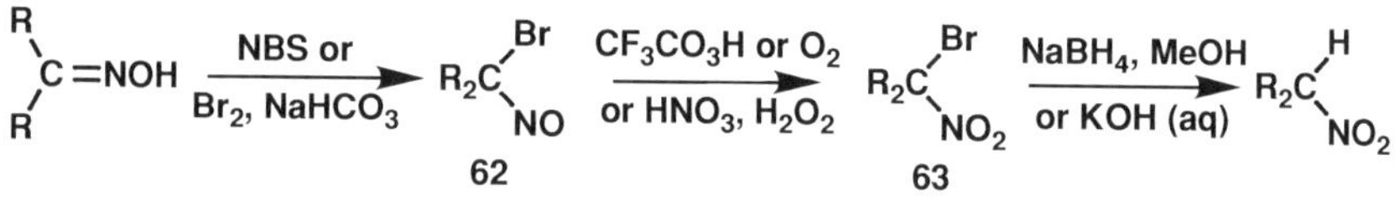

Figure 1.27

A much studied and still widely used reaction for the conversion of oximes to nitroalkanes involves treating the former with a halogen or source of halogen, followed by oxidation of the resulting α-bromonitrosoalkane (62) to an α-bromonitroalkane (63) which is then reduced to a secondary nitroalkane.[102,103,148] Although a lengthy route for this functional group conversion, the reaction proves extremely reliable and finds particular use for sterically hindered substrates.

Initial halogenation of the oxime can use chlorine,[101] hypobromite,[103] bromine,[62a,d] NBS,[103,148] or N-bromoacetamide.[148] Oxidation of the α-halonitrosoalkane can be achieved with nitric acid,[103] nitric acid–hydrogen peroxide,[148] atmospheric oxygen,[102] ozone,[149] or a peroxyacid.[140] Reduction of the α-halonitroalkane is achieved with sodium borohydride[103,148] or by catalytic hydrogenation,[149] although potassium hydroxide in ethanol[102] has been used for the conversion.

The original procedure for the bromination–oxidation–reduction route used bromine in aqueous potassium hydroxide, followed by oxidation with nitric acid–hydrogen peroxide and reduction with alkaline ethanol.[102] This procedure was improved by using NBS in aqueous sodium bicarbonate for the initial oxime bromination, followed by oxidation with nitric acid and final reduction of the α-bromonitroalkane with sodium borohydride in methanol. It is possible to convert oximes to nitroalkanes via this procedure without isolating or purifying any of the intermediates. This procedure is reported to give yields of between 10 and 55 % for a range of oxime to nitroalkane conversions.[103,148]

The bromination–oxidation–reduction route has been used in the syntheses of many energetic polynitropolycycloalkanes. Some of these reactions are illustrated in Table 1.6 (see also Chapter 2). A common strategy in these reactions is to use the oxime functionality to incorporate the nitro group, followed by oxidative nitration to *gem*-dinitro functionality via the Kaplan–Shechter reaction. This has been used in the case of 2,5-dinitronorbornane to synthesize 2,2,5,5-tetranitronorbornane.[126]

Research has focused on improving the efficiency of the halogenation–oxidation–reduction route by using reagents that perform the halogenation–oxidation in one step. Hypochlorous acid-hypochlorite[152,153] and hypobromous acid-hypobromite[154] systems have also been explored for the direct conversion of oximes to α-bromonitroalkanes and α-chloronitroalkanes respectively. Some N-haloheterocycles[155] have been reported to affect direct oxime to α-halonitroalkane conversion, and on some occasions, the use of NBS[126,152] or the free halogens[156] also leads to α-halonitroalkanes. A mixture of oxone and sodium chloride as a suspension in chloroform is reported as a one-pot method for the direct conversion of oximes to α-chloronitroalkanes.[157]

1.6.2 Oxidation of amines

The direct oxidation of an amino group to a nitro group is a desirable route to nitro compounds. The oxidation of tertiary amines with potassium permanganate has been known for some

Table 1.6
Synthesis of energetic polynitropolycycloalkanes via oxime halogenation

Oxime	Conditions/reagents	Product	Ref.
	1. Cl_2, CH_2Cl_2 2. NaOCl, Bu_4NHSO_4 3. Zn, NH_2OH.HCl, THF (aq) 20% (3 steps)		150
	1. Br_2, DMF, $NaHCO_3$, 2. TFAA, 85% H_2O_2, 24% (2 steps) 3. $NaBH_4$, EtOH, 70%		151
	1. NBS, $NaHCO_3$, dioxane (aq), 24% 2. $NaBH_4$, THF (aq), 66%		152
	1. NBS, $NaHCO_3$, dioxane (aq), 36% 2. $NaBH_4$, EtOH 3. AcOH (aq) 98% (2 steps)		126

time.[158] These reactions are often conducted in water, or in acetone–water mixtures for higher molecular weight amines.[158] Magnesium sulfate is frequently used as an additive in these reactions to control solution pH. 1,3,5,7-Tetranitroadamantane (71) has been obtained via the permanganate oxidation of the hydrochloride salt of 1,3,5,7-tetraaminoadamantane (70) (Table 1.7). The 45 % yield for this reaction reflects a relative yield of 82 % for the oxidation of each of the four amino groups.[159]

While primary aliphatic amines are converted to nitro compounds on reaction with ozone these reactions are accompanied by numerous by-products.[160] Such side-reactions are largely suppressed by first dissolving the amine onto silica gel followed by passing a stream of 3 % ozone in oxygen through the solid at –78 °C under anhydrous conditions, where yields of between 60 and 70 % are reported.[161] This route has been used to synthesize the energetic cyclopropane (65) from the diamine (64) (Table 1.7).[162]

Amines containing both primary and secondary alkyl groups are oxidized to nitroalkanes with peroxyacetic acid.[163] *m*-Chloroperoxybenzoic acid (*m*-CPBA) in chlorinated solvents at elevated temperature has found similar use.[164] The latter reagent in 1,2-dichloroethane has been used for the synthesis of both 1,3-dinitrocyclobutane (67)[150] (38 %) and 1,4-dinitrocubane (69)[165] from the corresponding diamines, (66) and (68), respectively (Table 1.7); further elaboration of the former allowing the synthesis of 1,1,3,3-tetranitrocyclobutane.[150] Peroxyacetic acid has been used for the oxidation of the tetraamine (70) but the crude product contained some nitroso functionality and needed treating with ozone for full conversion to (71) (Table 1.7).[166] Peroxytrifluoroacetic acid is not effective for the oxidation of aliphatic amines to nitro compounds.[163] The trifluoroacetic acid formed during these reactions protonates the amine

Table 1.7
Synthesis of energetic polynitropolycycloalkanes via amine oxidation

Amine	Conditions/reagents	Product	Ref.
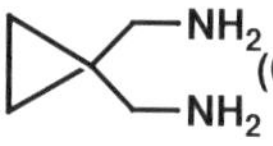 NH_2, NH_2 (64)	O_3 (excess), silica gel, -78 °C 20 - 28%	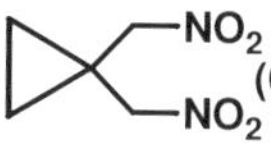 NO_2, NO_2 (65)	162
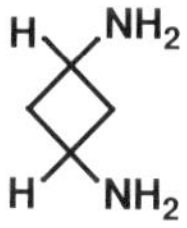 H, NH_2, H, NH_2 (66)	$ClCH_2CH_2Cl$, *m*-CPBA, reflux 38%	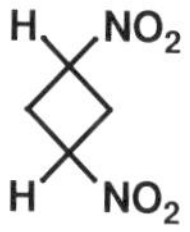 H, NO_2, H, NO_2 (67)	150
 H_2N, NH_2 (68)	$ClCH_2CH_2Cl$, *m*-CPBA, reflux 40%	 O_2N, NO_2 (69)	165
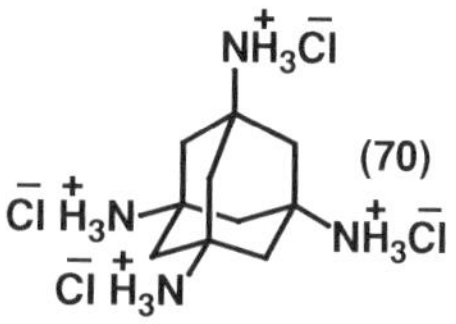 $\overset{+}{N}H_3\overset{-}{Cl}$, $\overset{-}{Cl}\,\overset{+}{H_3}N$, $\overset{+}{N}H_3\overset{-}{Cl}$, $\overset{-}{Cl}\,\overset{+}{H_3}N$ (70)	1. NaOH (aq) 2. $MgSO_4$, $KMnO_4$ 45% (2 steps)	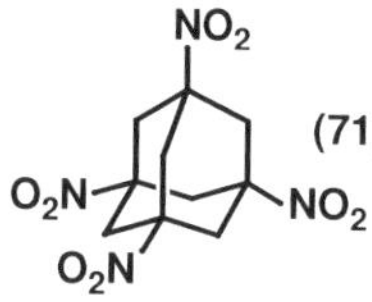 NO_2, O_2N, NO_2, O_2N (71)	159
	DMDO, $(CH_3)_2CO$, 91%		170

and prevents oxidation, whereas use of a buffer in these reactions results in acylation of the amine.[163] However, peroxytrifluoroacetic acid is an extremely useful reagent for the synthesis of polynitroarylenes from the oxidation of nitroanilines (see Section 4.7.1.5).[167]

Primary, secondary and tertiary aliphatic amines are efficiently converted to nitro compounds in 80–90% yield with dimethyldioxirane,[168] a reagent prepared[169] from the reaction of oxone ($2KHSO_5$-$KHSO_4$-K_2SO_4) with buffered acetone. Dimethyldioxirane (DMDO) has been used for the synthesis of 1,3,5,7-tetranitroadamantane (71) from the corresponding tetraamine as the tetrahydrochloride salt (70) and is an improvement over the initial synthesis[159] using permanganate anion (Table 1.7).[170] Oxone is able to directly convert some aromatic amines into nitro compounds.[171]

A recent method reported for the oxidation of primary aliphatic amines to nitro compounds uses *tert*-butylhydroperoxide (TBHP) and catalytic zirconium tetra-*tert*-butoxide in presence of molecular sieves.[172]

1.6.3 Nitration of nitronate salts

Nitronate salts and the tautomeric *aci*-form of nitroalkanes, known as nitronic acids, are converted to *gem*-dinitro compounds on treatment with dinitrogen tetroxide.[173–175] Novikov and co-workers[173] synthesized phenyldinitromethane by treating phenylnitromethane with dinitrogen tetroxide in ether and later [174] reported the synthesis of some substituted phenyltrinitromethanes from the direct nitration of the nitronate salts of phenylnitromethanes.

Phenyltrinitromethanes are similarly obtained from the nitration of *gem*-nitronitronate salts with a solution of dinitrogen tetroxide in ether. 1,1,1-Trinitroethane (73) can be formed in this way from the potassium salt of 1,1-dinitroethane (24).[176] Nitrolic acids, the products formed

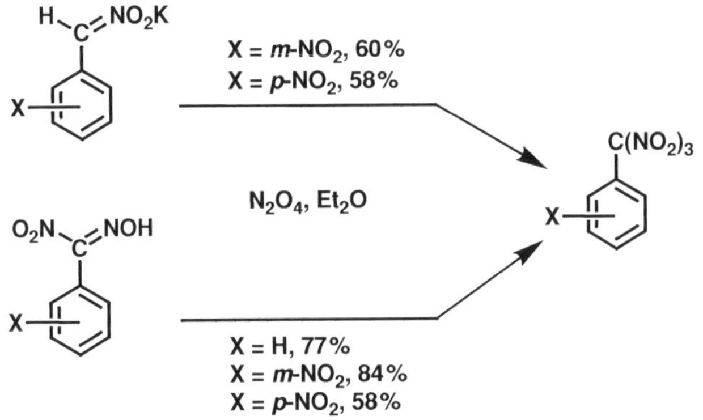

X = m-NO$_2$, 60%
X = p-NO$_2$, 58%

N$_2$O$_4$, Et$_2$O

X = H, 77%
X = m-NO$_2$, 84%
X = p-NO$_2$, 58%

Figure 1.28

by treating primary nitronic acids or their salts (nitronates) with nitrous acid, also undergo nitration to 1,1,1-trinitromethyl compounds.[173,174,177]

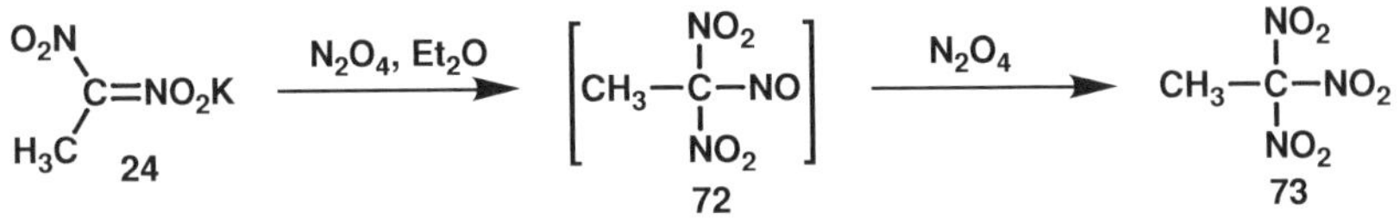

Figure 1.29

Mixed acid or anhydrous nitric acid has been used for the nitration of *gem*-nitronitronate salts to the corresponding 1,1,1-trinitromethyl compounds.[178] A convenient route to hexanitroethane (75) involves treating the dipotassium salt of 1,1,2,2-tetranitroethane (74) with mixed acid; the nitration proceeding via electrophilic addition of the nitronium cation to the bis-nitronitronate ion.[178]

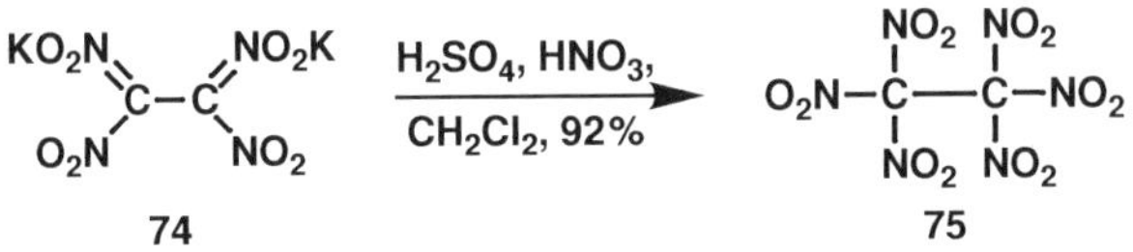

Figure 1.30

Tetranitromethane is an electrophilic source of nitronium cation and has been used in alkaline solution for the nitration of 1,1-dinitropropane to 1,1,1-trinitropropane (32 %), and for nitration of 1-phenyl-3-nitropropane to 1-phenyl-3,3-dinitropropane (32 %) and its further nitration to 1-phenyl-3,3,3-trinitropropane (33 %).[179]

Olsen and co-workers[180] reported the nitration of secondary nitroalkanes to *gem*-dinitro compounds with nitronium tetrafluoroborate in acetonitrile at −40 °C. Yields are lower compared to the Kaplan–Shechter reaction and significant amounts of pseudonitroles are formed, but this is possibly due to impure reagent.

Nitryl fluoride has been used for the nitration of terminal *gem*-dinitro compounds to the corresponding 1,1,1-trinitromethyl compounds.[181]

1.6.4 Oxidation of pseudonitroles

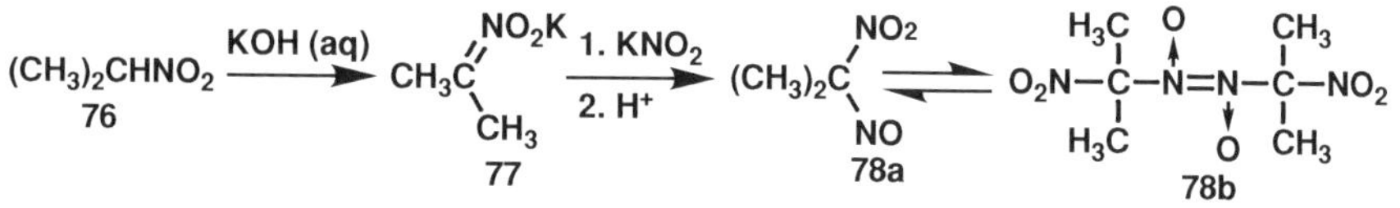

Figure 1.31

The tautomeric nitronic acids of secondary nitroalkanes or their nitronate salts react with nitrous acid or alkali metal nitrites to yield pseudonitroles.[182–184] These pseudonitroles are often isolated as their colourless dimers (78b) but are deep blue in monomeric form (78a). Primary nitroalkanes also form pseudonitroles (80b) but these rapidly isomerise to the nitrolic acid (80a).[182,183] Reactions are commonly conducted by slowly acidifying a mixture containing the nitronate salt and the metal nitrite, during which, the nitronic acid reacts with the nitrite anion. These reactions, first discovered by Meyer,[182] have been used to prepare 2-nitroso-2-nitropropane (78a) and acetonitrolic acid (80a) from 2-nitropropane (76) and nitroethane (22) respectively.[182]

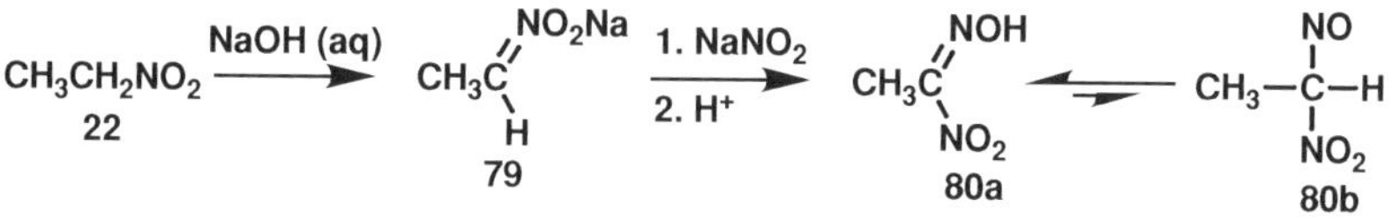

Figure 1.32

Pseudonitrole or nitrolic acid formation can be a side-reaction during the acidification of nitronate salts, particularly if the acid addition is slow. This process has been studied, optimized, and patented as a route to these compounds.[184]

Nitrolic acids undergo oxidation–nitration on treatment with dinitrogen tetroxide to the corresponding trinitromethyl compounds. Dinitroalkanes are obtained on oxidation of pseudonitroles and nitrolic acids with reagents such as chromium trioxide in acetic acid, dichromate, hydrogen peroxide, nitric acid, oxygen, and peroxyacids. Dinitrogen pentoxide in chlorinated solvents has been used for the oxidation of pseudonitroles to internal *gem*-dinitroalkanes.[185] The oxidation of acetonitrolic acid and 2-nitroso-2-nitropropane has been used to synthesize 1,1-dinitroethane[186] and 2,2-dinitropropane[187,188] respectively.

1.6.5 Oxidation of isocyanates

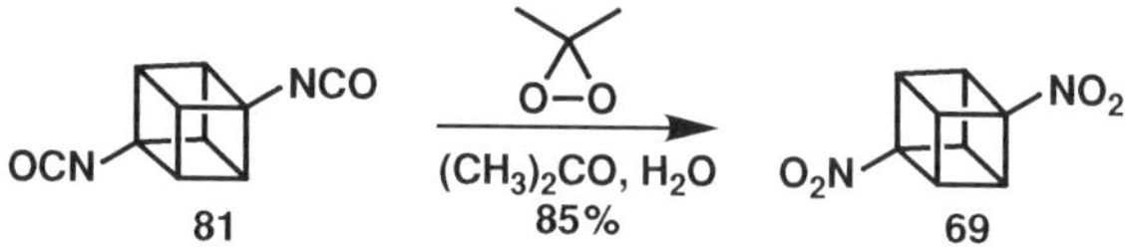

Figure 1.33

Dimethyldioxirane is a powerful oxidant prepared by reacting acetone with $KHSO_5$;[168] the latter is commercially available as a triple salt under the trade name of *oxone*. 1,4-Dinitrocubane

(69) is formed in 85 % yield via the oxidation of the corresponding bis-isocyanate (81) with dimethyldioxirane in wet acetone.[189] Similar chemistry has been used to prepare 1,3,5,7-tetranitrocubane and 1,3-dinitrocubane from the corresponding isocyanates (see Section 2.4).[190] Although dimethyldioxirane is a very efficient reagent for the oxidation of aliphatic isocyanates to nitro compounds, it has been relatively unexplored. These reactions are only limited by the lack of available methods for isocyanate formation and their commercial availability.

1.6.6 Oxidation of nitrosoalkanes

Nitroso compounds are formed during the addition of nitrous oxide,[48,49] dinitrogen trioxide,[48,49] and nitrosyl halides[50] to alkenes, and in some cases, from incomplete oxidation of amines[166] with peroxyacids like peroxyacetic acid. Quenching of carbanions with nitrosyl halides is also a route to nitroso compounds.[190] A full discussion on this subject is beyond the scope of this work and so the readers are directed to the work of Boyer.[191]

Oxidation of aliphatic nitroso functionality is usually facile but is not widely used in energetic materials synthesis. The following reagents have been used in these conversions: oxygen,[192] hydrogen peroxide,[193] nitrous oxide,[194] dinitrogen tetroxide,[195] chromium trioxide,[196] alkaline permanganate,[197] alkaline hypochlorite,[198] ozone,[166] ammonium persulfate,[198] peroxyacids[199] etc.[190]

1.7 KAPLAN–SHECHTER REACTION

$R_2C{=}NO_2^- Ag^+$ —AgNO₂→ [complex **82**]$^-$ Ag^+ → $R_2C(NO_2)_2$ + 2 Ag^0

Figure 1.34 Proposed mechanism for oxidative nitration (ref. 105)

Oxidative nitration, a process discovered by Kaplan and Shechter,[105] is probably the most efficient and useful method available for the synthesis of *gem*-dinitroaliphatic compounds from the corresponding nitroalkanes. The process, which is an electron-transfer substitution at saturated carbon, involves treatment of the nitronate salts of primary or secondary nitroalkanes with silver nitrate and an inorganic nitrite in neutral or alkali media. The reaction is believed[105,200] to proceed through the addition complex (82) which collapses and leads to oxidative addition of nitrite anion to the nitronate and reduction of silver from Ag^{+1} to Ag^0. Reactions proceed rapidly in homogeneous solution between 0 and 30 °C.

$CH_3CH_2NO_2$ (**22**) → 1. NaOH (aq); 2. $NaNO_2$, 2 $AgNO_3$; 3. AcOH, CH_2O; 80% (3 steps) → $CH_3C(NO_2)_2CH_2OH$ (**25**)

Figure 1.35

A range of primary and secondary nitroalkanes and their derivatives have been converted to the corresponding *gem*-dinitroalkanes via oxidative nitration, including: the conversion of nitroethane, 1-nitropropane, 2-nitropropane and 2-nitro-1,3-propanediol to 1,1-dinitroethane (78 %), 1,1-dinitropropane (86 %), 2,2-dinitropropane (93 %) and 2,2-dinitro-1,3-propanediol (77 %) respectively.[105] The silver nitrate used in these reactions can be recovered quantitatively on a laboratory scale and this has led to a study where oxidative nitration has been considered for the large-scale production of 2,2-dinitropropanol (25) from the nitroethane (22).[107]

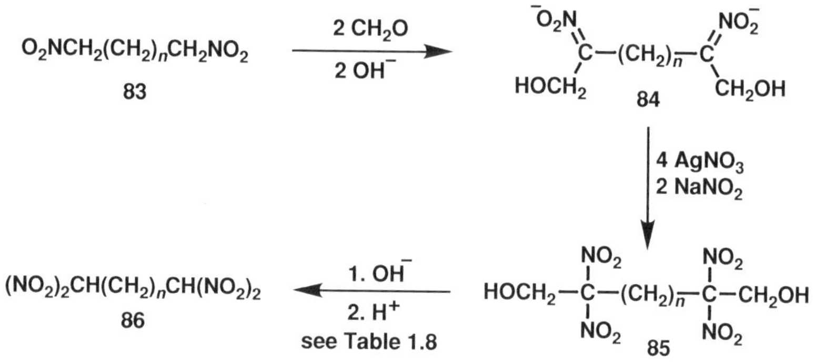

$O_2NCH_2(CH_2)_nCH_2NO_2$ (83) $\xrightarrow[2\ OH^-]{2\ CH_2O}$ $(^-O_2N{=})(HOCH_2)C{-}(CH_2)_n{-}C({=}NO_2^-)(CH_2OH)$ (84) $\xrightarrow[2\ NaNO_2]{4\ AgNO_3}$ $HOCH_2{-}C(NO_2)_2{-}(CH_2)_n{-}C(NO_2)_2{-}CH_2OH$ (85) $\xrightarrow[2.\ H^+]{1.\ OH^-}$ (see Table 1.8) $(NO_2)_2CH(CH_2)_nCH(NO_2)_2$ (86)

Figure 1.36

Feuer and co-workers[108] used oxidative nitration for the synthesis of $\alpha,\alpha,\omega,\omega$-tetranitroalkanes from the corresponding α,ω-dinitroalkanes (Table 1.8). However, this fails for α,ω-dinitroalkanes in which the nitro groups are not separated by at least three methylene units. Accordingly, oxidative nitration fails for both 1,3-dinitropropane ($n = 1$) and 1,4-dinitrobutane ($n = 2$) and gives a low yield of 1,1,5,5-tetranitropentane ($n = 3$) from 1,5-dinitropentane (Table 1.8). Feuer and co-workers[108] later discovered that the bis-methylol derivatives (84) of α,ω-dinitroalkanes (83) give good yields of product (85) when the separation between nitro groups is two methylene units or greater; the methylol groups can be removed *in situ* via base-catalyzed demethylolation, with loss of formaldehyde, to yield the $\alpha,\alpha,\omega,\omega$-tetranitroalkane (86) (Table 1.8). These reactions still fail for the bis-methylol derivative of 1,3-dinitropropane ($n = 1$) but give a 49 % yield of 1,1,4,4-tetranitrobutane ($n = 2$) after demethylolation of the corresponding bis-methylol derivative, namely, 2,2,5,5-tetranitro-1,6-hexanediol.

Oxidative nitration has a number of advantages over pre-existing routes to *gem*-dinitroalkanes, including:

- Reactions are successful for hindered compounds; 3,3-dinitro-2-butanol is obtained from the oxidative nitration of 3-nitro-2-butanol.[105]
- Oxidative nitration avoids the isolation of *gem*-nitronitronate salts, which are often unstable explosives with a high sensitivity to impact and friction.
- Oxidative nitration has been modified to an electrolytic process.[200]
- Good yields of internal *gem*-dinitroalkanes are attainable, whereas the Ter Meer reaction fails for the synthesis of this class of compounds.
- Oxidative nitration is a one step process from nitroalkane to *gem*- dinitroalkane, whereas the Ter Meer reaction requires two steps (initial halogenation followed by halide displacement with nitrite anion).

Table 1.8
Synthesis of tetranitroalkanes (86) via the oxidative nitration of dinitroalkanes (83) and their bis-methylol derivatives (84)

	$(NO_2)_2CH(CH_2)_nCH(NO_2)_2$	
Homologue	Yield (%) of tetranitroalkane (86)	
$n =$	via the dinitroalkane (83)	via the bis-methylol (84)
1	0	0
2	0	49
3	10	25
4	84	70
5	89	---

Source: Reprinted with permission from C. E. Colwell, H. Feuer, G. Leston and A. T. Nielsen, *J. Org. Chem.*, 1962, **27**, 3598; Copyright 1962 American Chemical Society.

Oxidative nitration is not effective for the synthesis of *gem*-dinitroaliphatic compounds containing an electron-withdrawing group α to the carbon bearing the nitro groups. Oxidative nitration is not successful for the conversion of terminal *gem*-dinitro compounds into 1,1,1-trinitromethyl derivatives.

A major drawback of the Kaplan–Shechter reaction is the use of expensive silver nitrate as one of the reagents, which prevents scale up to an industrial capacity. Urbański and co-workers[201] modified the process by showing that the silver nitrate component can be replaced with an inorganic one-electron transfer agent like ferricyanide anion. In a standard procedure the nitroalkane or the corresponding nitronate salt is treated in alkaline media with potassium

Table 1.9
Synthesis of energetic polynitropolycycloalkanes via oxidative nitration

Substrate	Conditions/reagents	Product	Ref.
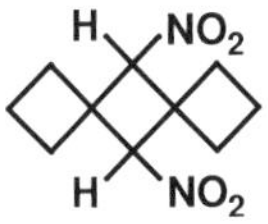	NaOH, $NaNO_2$, $K_3Fe(CN)_6$, $Na_2S_2O_8$, dioxane (aq), 64%	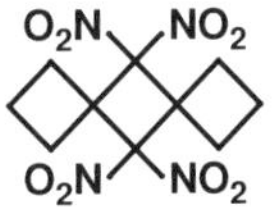	150
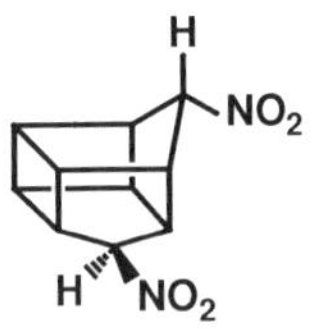	NaOH, $NaNO_2$, $K_3Fe(CN)_6$, MeOH (aq), 65%	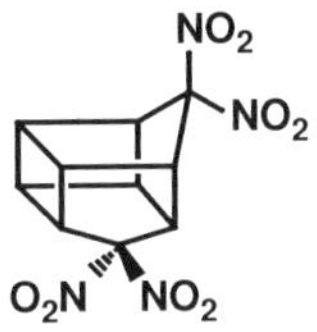	151
	NaOH, $NaNO_2$, $K_3Fe(CN)_6$, CH_2Cl_2, H_2O, MeOH 83–91%	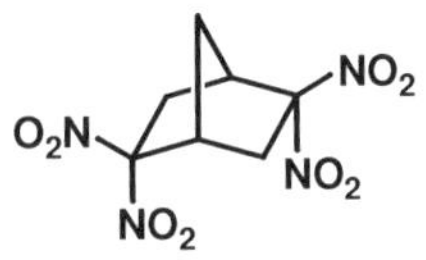	141, 126
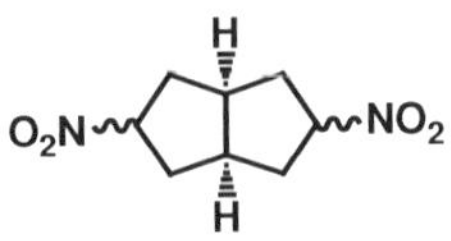	NaOH, $NaNO_2$, $K_3Fe(CN)_6$, CH_2Cl_2, H_2O, 76%	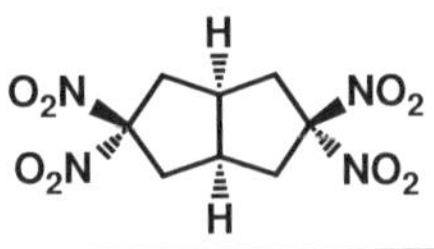	141

ferricyanide oxidant in the presence of sodium nitrite. The process was later optimized and its scope expanded by Kornblum and co-workers.[202] In a further modification to this procedure Grakauskas and co-workers[203] showed that a catalytic amount of ferricyanide can be used in conjunction with a stoichiometric amount of persulfate anion as co-oxidant. These modifications often lead to improved yields compared to the original method. The inexpensive reagents used in these reactions, and the high yields frequently obtained, makes this method a very valuable route to *gem*-dinitroaliphatic compounds. The importance of oxidative nitration as a route to *gem*-dinitroaliphatic compounds is reflected in its widespread use for the synthesis of numerous energetic compounds, like those illustrated in Table 1.9. Oxidative nitration has been used in the reported[204] synthesis of a powerful melt-castable explosive, TNAZ (89).

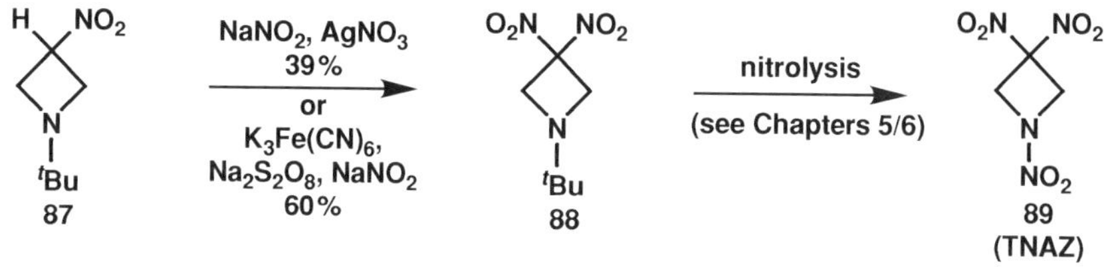

H, NO_2; N; ^tBu; 87

$NaNO_2$, $AgNO_3$ 39% or $K_3Fe(CN)_6$, $Na_2S_2O_8$, $NaNO_2$ 60%

O_2N, NO_2; N; ^tBu; 88

nitrolysis (see Chapters 5/6)

O_2N, NO_2; N; NO_2; 89 (TNAZ)

Figure 1.37

1.8 NITRATION OF COMPOUNDS CONTAINING ACIDIC HYDROGEN

1.8.1 Alkaline nitration

The alkaline nitration of compounds containing acidic hydrogen is a valuable route to aliphatic nitro compounds. In these reactions a base is used to remove an acidic proton from the substrate, which is then treated with a source of $-NO_2$. Substrates used in these reactions usually have an electron-withdrawing or resonance-stabilizing group positioned α to a proton(s), and these include: aliphatic and alicyclic ketones, nitriles, carboxylic acid esters, sulfonate esters, *N,N*-dialkylamides etc. Proton removal from these substrates may be reversible or irreversible depending on the pK_a of the base compared to the acidity of the substrates proton(s). Bases used for these reactions include: alkali metal alkoxides, sodium hydride, alkali metal amides, lithium bases etc. The nitrating agents range from alkyl nitrate esters to nitrogen oxides.

1.8.1.1 Alkaline nitration with nitrate esters

Alkaline nitration with alkoxide bases and nitrate esters was first explored by Endres and Wislicenus[205] who synthesized phenylnitromethane by treating ethyl phenylacetate with potassium ethoxide in ethanol, followed by addition of ethyl nitrate and hydrolysis–decarboxylation of the resulting α-nitroester with aqueous acid. Phenylnitromethane is synthesized in a similar way via alkaline nitration of benzyl cyanide, followed by treatment of the resulting α-nitronitrile with aqueous base.[206] Wieland and co-workers[207] used alkali metal alkoxides and nitrate esters for the nitration of cyclic ketones but the yields and purity of product are often poor.[208]

Feuer and co-workers[209] conducted extensive studies into alkaline nitration with nitrate esters, exploring the effect of base, time, stoichiometry, concentration, solvent, and temperature on yields and purity. Reactions are generally successful when the substrate α-proton acidity is in the 18–25 pK_a range. Alkoxide bases derived from simple primary and secondary aliphatic alcohols are generally not considered compatible in reactions using alkyl nitrates. Optimum conditions for many of these reactions use potassium *tert*-butoxide and amyl nitrate in THF at −30 °C, although in many cases potassium amide in liquid ammonia at −33 °C works equally well.

Feuer used both reagents to nitrate a number of cyclic ketones (90) to the dipotassium salts of the corresponding α,α'-dinitrocycloketones (91).[210] Although such salts can be isolated it is important to note that α,α'-dinitroketones are intrinsically unstable – direct acidification of these salts causes decomposition with the evolution of nitrogen oxides. In fact, nitro compounds derived from the nitration of active methylene groups are often unstable. This is a direct consequence of having an electron-withdrawing or resonance-stabilizing group α to a potential leaving group; both groups being able to stabilize any anion formed on decomposition.

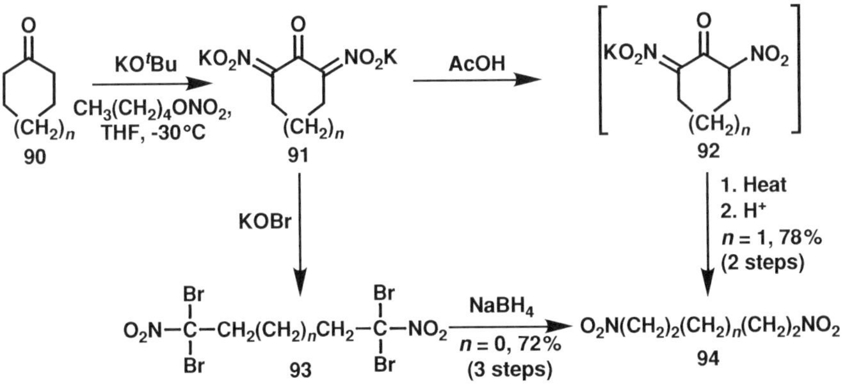

Figure 1.38

The salts of α,α'-dinitrocycloketones (91) are readily converted into α,ω-dinitroalkanes. Klager[211] developed a method whereby the alkali metal salt of the α,α'-dinitrocycloketone (91) is treated with alkaline hypobromite to form the corresponding tetrabromide (93) via a process of ring opening and formal loss of carbon monoxide; selective reduction of the latter with sodium borohydride yields the corresponding α,ω-dinitroalkane (94). Feuer and co-workers[212,213] described an improved method whereby the dipotassium salts of α,α'-dinitrocycloketones (91) are partially acidified with acetic acid to give the corresponding mono-potassium salts (92) which undergo spontaneous hydrolytic ring opening and yield α,ω-dinitroalkanes (94) on further acidification. 1,4-Dinitrobutane (72 %), 1,5-dinitropentane (78 %) and 1,6-dinitrohexane (75 %) have been synthesized from cyclopentane, cyclohexane, and cycloheptane respectively, via these routes.[211–213]

Treatment of acyclic ketones with one equivalent of potassium amide base in liquid ammonia, followed by acidification, yields the α-nitroketone and products resulting from fragmentation.[214] The same reactions with cyclic ketones, again, only using one equivalent of *tert*-butoxide[214] or potassium amide[215] base, generates α-nitrocycloketones and ω-nitrocarboxylic esters. α-Nitrocycloketones derived from the nitration of α,α'-dialkylcycloketones cannot form a stable anion and so cleavage to the ω-nitrocarboxylic ester predominants.[214]

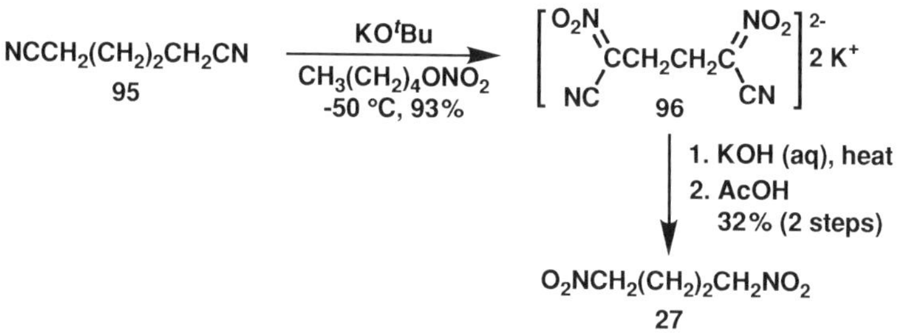

$NCCH_2(CH_2)_2CH_2CN$ **95** → (KO^tBu; $CH_3(CH_2)_4ONO_2$, -50 °C, 93%) → $[O_2N(NC)C{=}CH_2CH_2C{=}(NO_2)CN]^{2-}$ $2K^+$ **96** → (1. KOH (aq), heat; 2. AcOH; 32% (2 steps)) → $O_2NCH_2(CH_2)_2CH_2NO_2$ **27**

Figure 1.39

Feuer and co-workers[216] extended their studies to the alkaline nitration of α,ω-dinitriles. Nitration with potassium *tert*-butoxide and amyl nitrate in THF at $-30\,^{\circ}C$ yields the corresponding dipotassium salt of the α,ω-dinitro-α,ω-dinitrile. The nitronate salts from these reactions are isolated via methanol-induced precipitation from the aqueous reaction liquors, a process which also separates the product from impurities. These salts undergo hydrolysis on treatment with aqueous potassium hydroxide, and subsequent acidification yields the corresponding α,ω-dinitroalkane. This route has been used to synthesize 1,4-dinitrobutane (27) from apidonitrile (95) in 30 % overall yield.

Feuer and co-workers[217] also nitrated ring-substituted toluenes to the corresponding arylnitromethanes with potassium amide in liquid ammonia. Sulfonate esters[218] and *N,N*-dialkylamides[219] undergo similar nitration; the latter isolated as their α-bromo derivatives. Alkaline nitration of ethyl and *tert*-butyl carboxylic esters with potassium amide in liquid ammonia yields both the α-nitroester and the corresponding nitroalkane from decarboxylation.[220]

Treating the dianion of a carboxylic acid with an alkyl nitrate leads to an α-nitrocarboxylic acid which readily undergoes decarboxylation to the corresponding nitroalkane.[221] This method is particularly useful for the synthesis of arylnitromethanes containing electron-donating groups.[221]

1.8.1.2 Alkaline nitration with cyanohydrin nitrates

Acetone cyanohydrin nitrate, a reagent prepared from the nitration of acetone cyanohydrin with acetic anhydride-nitric acid,[222] has been used for the alkaline nitration of alkyl-substituted malonate esters.[222] In these reactions sodium hydride is used to form the carbanions of the malonate esters, which on reaction with acetone cyanohydrin nitrate form the corresponding nitromalonates. The use of a 100 % excess of sodium hydride in these reactions causes the nitromalonates to decompose by decarboxylation to the corresponding α-nitroesters. Alkyl-substituted acetoacetic acid esters behave in a similar way and have been used to synthesize α-nitroesters.[223] Yields of α-nitroesters from both methods average 50–55 %.

In a related reaction arylacetonitriles are nitrated to the corresponding α-nitronitriles, which on alkaline hydrolysis and subsequent acidification, yield arylnitromethanes.[223] This method has been used to convert phenylacetonitrile to phenylnitromethane in 70 % overall yield without the need for intermediate purification.

Acetone cyanohydrin nitrate is, however, of limited use for the nitration of many carbanions and is rapidly destroyed in the presence of alkoxides.[224] Its use for the nitration of weakly basic carbanions such as malonate arises from the perturbation of a side-reaction observed with alkyl nitrates, namely, the carbanion attacking the carbon atom of alkyl nitrate instead of nitrogen, and thus giving alkylation instead of nitration. Alkaline nitrations with acetone cyanohydrin nitrate compliment reactions using alkoxide base and nitrate esters; the latter are generally not applicable to substrates with proton acidities below pK_a 16.

1.8.1.3 Alkaline nitration with tetranitromethane

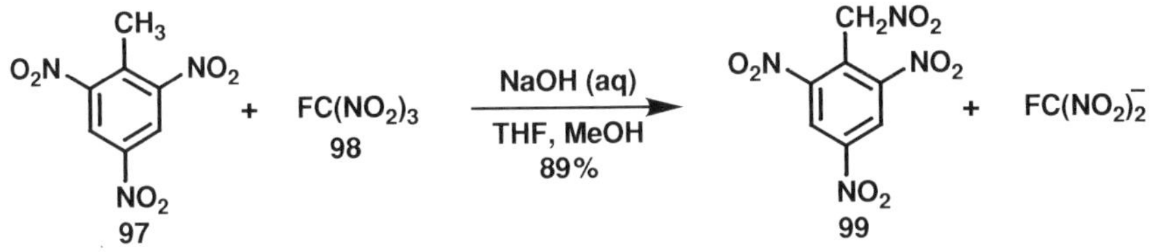

Figure 1.40

The carbon–nitrogen bonds of tetranitromethane are very electron deficient and prone to attack by nucleophiles. Consequently, tetranitromethane and its derivatives behave like nitrating agents in alkaline solution.[179] Fluorotrinitromethane (98) behaves in a similar way and has been used for the nitration of the acidic methyl group of 2,4,6-trinitrotoluene (97) (TNT) to give α,2,4,6-tetranitrotoluene (99) in 89 % yield.[225]

1.8.1.4 Alkaline nitration with nitrogen oxides and related compounds

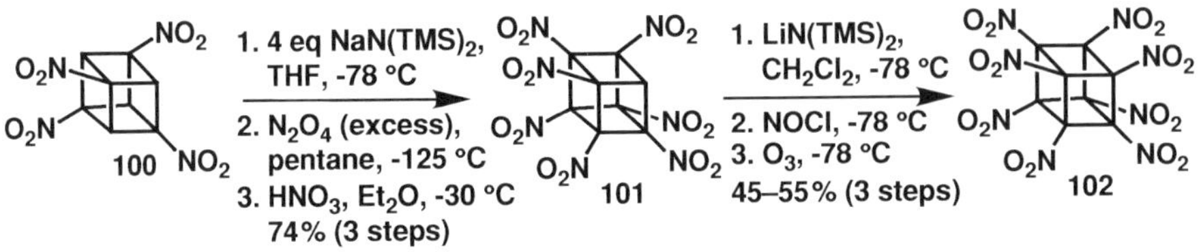

Figure 1.41

Eaton and co-workers[226] synthesis of octanitrocubane (102) (ONC) is a milestone in high-energy materials synthesis. The explosive performance of octanitrocubane is predicted to be very high with a detonation velocity of 9900 m/s.[227] This pioneering synthesis involves, as a key step, the alkaline nitration of 1,3,5,7-tetranitrocubane (100). In this reaction hexamethyldisilazide anion is used to deprotonate positions β to three nitro groups and the resulting tetra-anion is quenched/nitrated with excess dinitrogen tetroxide in pentane to give a 74 % yield of a mixture containing 95 % heptanitrocubane (101) and 5 % hexanitrocubane. This reaction involves nitration of the sodium anion of 1,3,5,7-tetranitrocubane at the melting interface (~105 °C) between frozen THF and dinitrogen tetroxide in pentane. This is a new nitration technique known as 'interfacial nitration' and probably proceeds via a radical process.[226] Heptanitrocubane (101) is converted to octanitrocubane (102) by further reaction with hexamethyldisilazide anion, followed by quenching with nitrosyl chloride and oxidation with ozone.[226] This example of β-deprotonation-nitration is only feasible because of the combined inductive effect of three nitro groups and the increased sp^2 character of these positions as a result of the strained cubane core. Even then, the low acidity of the β-proton is reflected in the need for a fairly strong base (hexamethyldisilazide in THF $\sim$ pK_a of 25.8[228]).

1.8.2 Acidic nitration

Electrophilic nitrations of aliphatic nitriles,[117] carboxylic acids,[229] carboxylic esters,[230] and β-diketones[231] have been reported. The nitration of 2-alkyl-substituted indane-1,3-diones with nitric acid, followed by alkaline hydrolysis, is a standard laboratory route to primary nitroalkanes.[231]

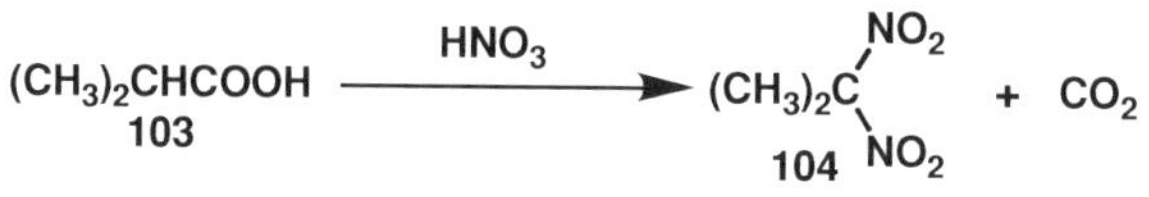

$$\underset{\mathbf{103}}{(CH_3)_2CHCOOH} \xrightarrow{HNO_3} \underset{\mathbf{104}}{(CH_3)_2C(NO_2)_2} + CO_2$$

Figure 1.42

Treatment of some carboxylic acids with nitric acid is a route to *gem*-dinitroalkanes, as in the case of *iso*-butyric acid (103), which undergoes nitration-decarboxylation on treatment with nitric acid to give 2,2-dinitropropane (104).[232] Yields are often poor for this type of reaction.

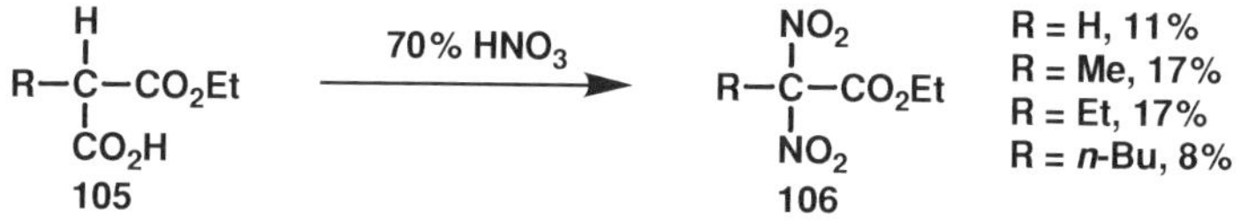

$$\underset{\mathbf{105}}{R{-}CH(CO_2H){-}CO_2Et} \xrightarrow{70\%\ HNO_3} \underset{\mathbf{106}}{R{-}C(NO_2)_2{-}CO_2Et}$$

R = H, 11%
R = Me, 17%
R = Et, 17%
R = *n*-Bu, 8%

Figure 1.43

The half esters of malonic acid (105) yield α,α-dinitroesters (106) on nitration-decarboxylation with nitric acid, although yields are often poor.[229] Treatment of these α,α-dinitroesters with hydrazine hydrate or alkali metal hydroxides yields the corresponding *gem*-dinitroalkanes.[229]

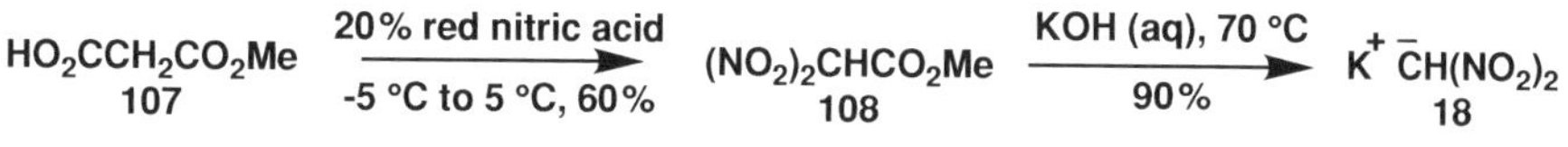

$$\underset{\mathbf{107}}{HO_2CCH_2CO_2Me} \xrightarrow[-5\ ^\circ C\ \text{to}\ 5\ ^\circ C,\ 60\%]{20\%\ \text{red nitric acid}} \underset{\mathbf{108}}{(NO_2)_2CHCO_2Me} \xrightarrow[90\%]{KOH\ (aq),\ 70\ ^\circ C} \underset{\mathbf{18}}{K^+\ {}^-CH(NO_2)_2}$$

Figure 1.44

A potential industrial route to potassium dinitromethane (18) involves treatment of methyl malonate (107) with red fuming nitric acid to give methyl α,α-dinitroacetate (108), followed by hydrolysis-decarboxylation with aqueous potassium hydroxide.[233] Dinitromethane is a precursor to 2,2-dinitroethanol and 2,2-dinitro-1,3-propanediol, both of which are useful in addition and esterification reactions for the production of energetic oligomers and plasticizers.

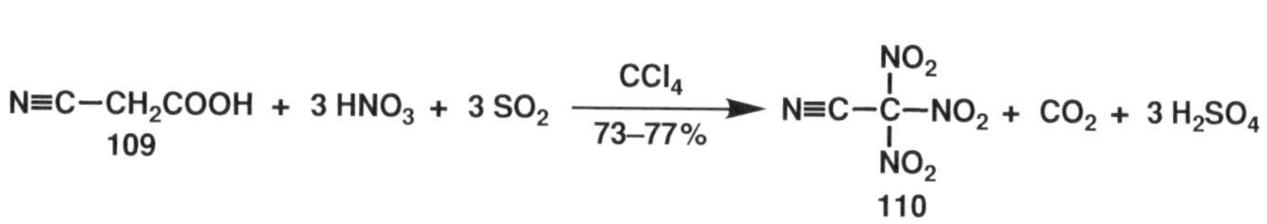

$$\underset{\mathbf{109}}{N{\equiv}C{-}CH_2COOH} + 3\,HNO_3 + 3\,SO_2 \xrightarrow[73\text{–}77\%]{CCl_4} \underset{\mathbf{110}}{N{\equiv}C{-}C(NO_2)_3} + CO_2 + 3\,H_2SO_4$$

Figure 1.45

Trinitroacetonitrile (110), a precursor to dinitroacetonitrile and its derivatives, can be synthesized from the nitration of cyanoacetic acid (109) with a solution of sulfur dioxide and absolute nitric acid in carbon tetrachloride.[117] This method is particularly attractive because the trinitroacetonitrile can be kept as a solution in carbon tetrachloride without isolation; trinitroacetonitrile is hazardous to handle and its vapours are both toxic and lachrymatory.

1.9 OXIDATIVE DIMERIZATION

$$2\ R^1R^2C{=}NO_2^- + S_2O_8^{2-} \xrightarrow{H^+} O_2N{-}C(R^1)(R^2){-}C(R^1)(R^2){-}NO_2 + 2\ SO_4^- \quad \text{(Eq. 1.2)}$$

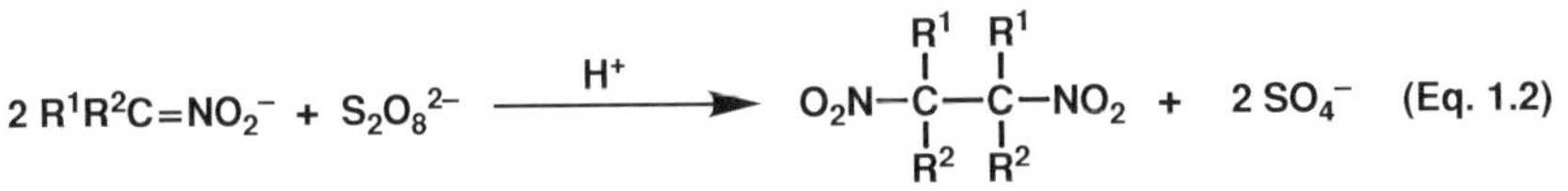

111

Figure 1.46

Kaplan and Shechter[234] found that certain oxidants react with the nitronate salts of secondary nitroalkanes to yield *vic*-dinitroalkanes (111) in a reaction referred to as oxidative dimerization. These reactions are believed to involve transfer of an electron from the secondary alkyl nitronate to the oxidant with the production of a nitroalkyl radical. The radical can then dimerize to the corresponding *vic*-dinitroalkane (111) (Equation 1.2) or lose nitric oxide to form a ketone via the Nef reaction (Equation 1.3). Unfortunately, formation of the ketone is a major side-reaction during oxidative dimerization and is often the major product.

$$R^1R^2C{=}NO_2^- + 2\ S_2O_8^{2-} + 2\ H_2O \xrightarrow{H^+} R^1R^2C{=}O + 4\ H^+ + NO_3^- + 4\ SO_4^{2-} \quad \text{(Eq. 1.3)}$$

Figure 1.47

Studies into oxidative dimerization have shown that only the persulfate anion is of synthetic value in these reactions. Reaction pH is also crucial; with reactions proceeding fastest when a pH of 7.2–9.4 is maintained.[234] The reaction medium becomes more acidic as the oxidation progresses and needs either buffering or the slow addition of alkali throughout the reaction. If the reaction medium is allowed to become acidic then the starting nitroalkane is regenerated and the Nef reaction predominates.

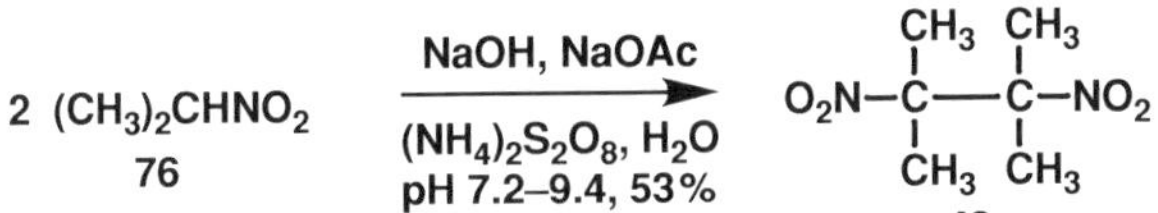

$$2\ (CH_3)_2CHNO_2\ (\mathbf{76}) \xrightarrow[(NH_4)_2S_2O_8,\ H_2O,\ pH\ 7.2\text{–}9.4,\ 53\%]{NaOH,\ NaOAc} O_2N{-}C(CH_3)_2{-}C(CH_3)_2{-}NO_2\ (\mathbf{48})$$

Figure 1.48

Oxidative dimerization gives reasonable yields of *vic*-dinitroalkanes for some substrates; 2,3-dimethyl-2,3-dinitrobutane (48, 53 %) and 3,4-dimethyl-3,4-dinitrohexane (37 %) are obtained from 2-nitropropane (76) and 2-nitrobutane respectively.[234] However, oxidative dimerization fails to convert 1,1-dinitroethane and trinitromethane into 2,2,3,3-tetranitrobutane and hexanitroethane respectively. Additionally, oxidative dimerisation is not a feasible route for the synthesis of *vic*-dinitroalkanes from primary nitroalkanes. Although oxidative dimerization is limited in scope, and yields are often poor, the starting materials are usually inexpensive.

1.10 ADDITION AND CONDENSATION REACTIONS

Primary and secondary nitroalkanes, and substrates containing terminal *gem*-dinitroaliphatic functionality, have one or more acidic α-protons, a consequence of inductive and resonance effects imposed by the nitro group. As a result, such compounds can behave like carbanions and participate in a number of addition and condensation reactions which are typical of substrates like ketones, aldehydes, and β-ketoesters. Such reactions are extremely useful for the synthesis of functionalized polynitroaliphatic compounds which find potential use as explosives, energetic oligomers and plasticizers.

Addition and condensation reactions employing nitroform or its methylol derivative, 2,2,2-trinitroethanol, have been used to synthesize a huge number of compounds containing the trinitromethyl group. Such compounds often have a very favourable oxygen balance which contributes to explosive performance. However, compounds containing the trinitromethyl group often exhibit an unacceptably high sensitivity to shock and impact. Kamlet[235] proposed that this high sensitivity was due to the restricted rotation of the C–NO_2 groups. This theory proposes that if rotation is restricted, energy normally dissipated through a combination of bond rotation and bonded atom vibrational modes is more likely to result in bond breaking. In any energetic material the sensitivity to impact is related to the weakest bond(s) in that molecule, and so, in many energetic compounds, the trinitromethyl is the trigger for decomposition. It is not surprising that research in this area rapidly slumped after these findings. The chemistry and synthesis of trinitromethyl containing compounds has been the subject of several excellent reviews.[9,114,115]

2-Fluoro-2,2-dinitroethanol, the methylol derivative of fluorodinitromethane, has been used extensively for the synthesis of fluorodinitromethyl compounds. The fluorine atom is similar in size to that of a proton and so rotation in the fluorodinitromethyl group is much less hindered compared to the trinitromethyl group. Consequently, these compounds are far less sensitive to impact than trinitromethyl compounds but only slightly less energetic. The explosive performance of fluorodinitromethyl compounds has been reviewed.[236]

2-Fluoro-2,2-dinitroethanol is synthesized on a large scale[237] from nitroform in a reaction involving fluorination[238] followed by *in situ* reduction with alkaline hydrogen peroxide[239] in the presence of formaldehyde. The intermediate in this reaction, fluorotrinitromethane, is also synthesized by treating tetranitromethane with potassium fluoride in DMF.[240] Fluorodinitromethane has been synthesized by treating 2,2-dinitro-1,3-propanediol with sodium hydroxide followed by fluorination,[238] or from the direct fluorination of potassium dinitromethane.[233] The salts of fluorodinitromethane are dangerous, and so, its *in situ* formation in addition reactions is greatly preferred. Fluorodinitromethyl compounds have also been synthesized from the fluorination of *gem*-nitronitronate salts.[239a] The synthesis of energetic fluorodinitromethyl compounds has been reviewed.[37,241,242]

1.10.1 1,2-Addition reactions

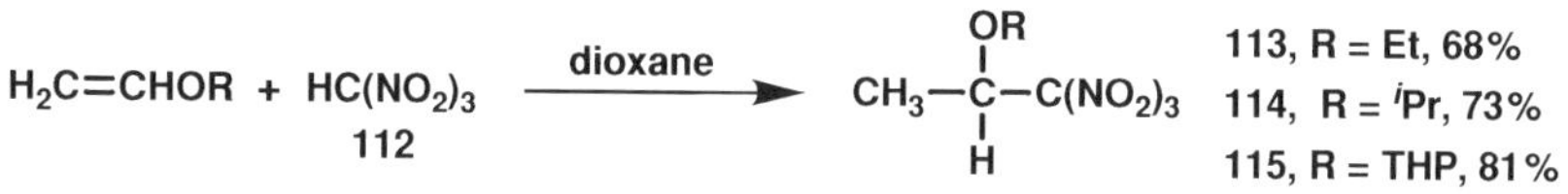

$$H_2C{=}CHOR + \underset{\mathbf{112}}{HC(NO_2)_3} \xrightarrow{\text{dioxane}} CH_3{-}\overset{OR}{\underset{H}{C}}{-}C(NO_2)_3$$

113, R = Et, 68%
114, R = ^{i}Pr, 73%
115, R = THP, 81%

Figure 1.49

The 1,2-addition of nitroform across the double bonds of vinyl ethers has been used to synthesize a large number of α-trinitromethyl ethers; (113–115) have been synthesized from the reaction of nitroform (112) with the corresponding vinyl ethers.[243] The synthesis of α-trinitromethyl ethers from the reaction of an acetal with nitroform involves 1,2-addition via an intermediate oxonium cation; (117) is synthesized in 85 % yield from the reaction of nitroform with the acetal (116).[244] α-Trinitromethyl ethers have been synthesized in a one-pot process from nitroform in the presence of an aldehyde and alcohol, in which case, the acetal is formed *in situ*.[245]

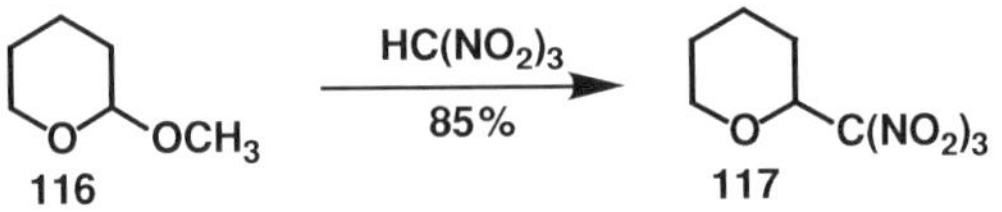

OCH_3 (116) $\xrightarrow[85\%]{HC(NO_2)_3}$ $C(NO_2)_3$ (117)

Figure 1.50

Shackelford and co-workers[246] studied the 1,2-addition of 2,2-dinitropropanol, 2,2,2-trinitroethanol, and 2-fluoro-2,2-dinitroethanol across the double bonds of vinyl ethers. These reactions are Lewis acid catalyzed because of the weak nucleophilic character of alcohols which contain two or three electron-withdrawing groups on the carbon β to the hydroxy functionality. Base catalysis is precluded since alkaline conditions lead to deformylation with the formation of formaldehyde and the nitronate salt.

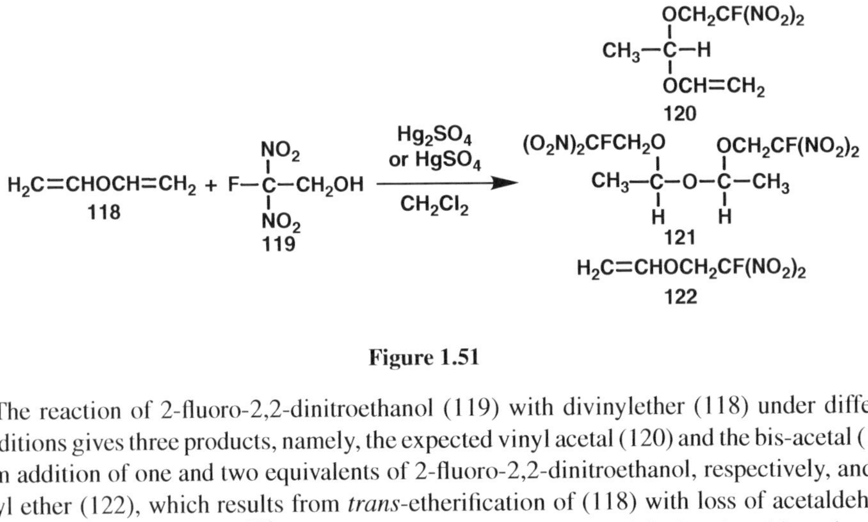

$H_2C{=}CHOCH{=}CH_2$ (118) + $F{-}C(NO_2)_2{-}CH_2OH$ (119) $\xrightarrow[CH_2Cl_2]{Hg_2SO_4 \text{ or } HgSO_4}$ $CH_3{-}CH(OCH_2CF(NO_2)_2)(OCH{=}CH_2)$ (120) + $(O_2N)_2CFCH_2O{-}CH(CH_3){-}O{-}CH(CH_3){-}OCH_2CF(NO_2)_2$ (121) + $H_2C{=}CHOCH_2CF(NO_2)_2$ (122)

Figure 1.51

The reaction of 2-fluoro-2,2-dinitroethanol (119) with divinylether (118) under different conditions gives three products, namely, the expected vinyl acetal (120) and the bis-acetal (121) from addition of one and two equivalents of 2-fluoro-2,2-dinitroethanol, respectively, and the vinyl ether (122), which results from *trans*-etherification of (118) with loss of acetaldehyde. Shackelford and co-workers[246] found that by altering the nature of the Lewis acid catalyst and the reaction stoichiometry they were able to alter the distribution ratio of these products.

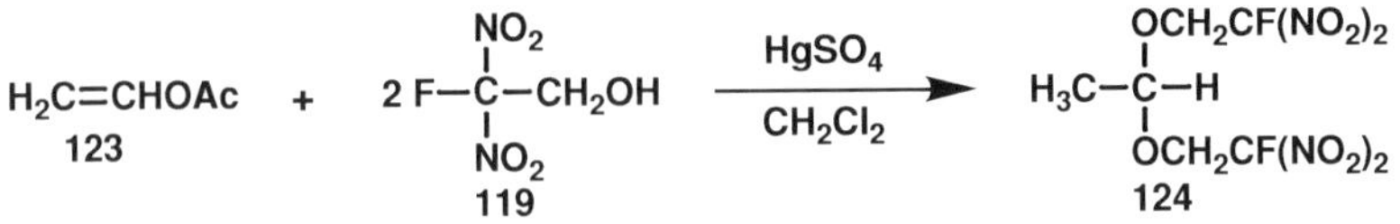

$H_2C{=}CHOAc$ (123) + 2 $F{-}C(NO_2)_2{-}CH_2OH$ (119) $\xrightarrow[CH_2Cl_2]{HgSO_4}$ $H_3C{-}CH(OCH_2CF(NO_2)_2)_2$ (124)

Figure 1.52

The *trans*-etherification of vinyl ethers is not uncommon under Lewis acid catalysis. 1,1-Bis(2-fluoro-2,2-dinitroethyl)ethyl acetal (124) is obtained on reaction of two equivalents of 2-fluoro-2,2-dinitroethanol (119) with vinyl acetate (123) in the presence of mercuric sulfate.[247]

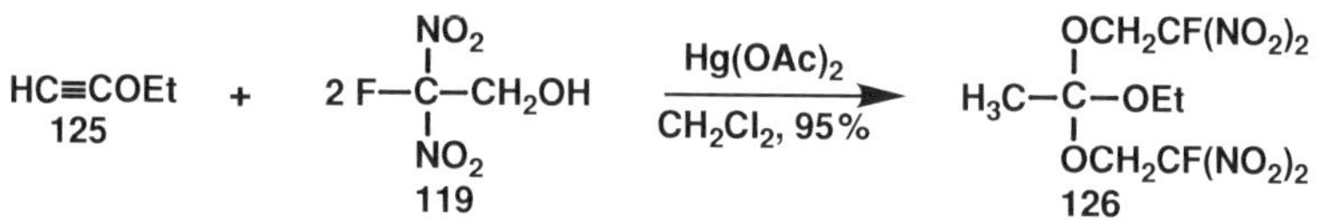

Figure 1.53

The acetylenic bond of propargyl ethers can react with polynitroaliphatic alcohols, as in the case of ethoxyacetylene (125), which reacts with two equivalents of 2-fluoro-2,2-dinitroethanol (119) to give the orthoester (126).[248]

1.10.2 1,4-Addition reactions

1.10.2.1 Michael reaction

Conjugate 1,4-addition of a nucleophile to the double bond of an electron deficient α,β-unsaturated compound is known as the Michael reaction and ranks as one of the most important carbon–carbon bond forming reactions in synthetic organic chemistry. The species undergoing addition to the double bond is known as a Michael donor and includes substrates which are capable of forming carbanions i.e. those with acidic protons. Nitroalkanes are good Michael donors and literature examples of their use in the synthesis of functionalized polynitroaliphatic compounds are extensive.[249–262] These include nitronate anions generated from nitroform, fluorodinitromethane, primary nitroalkanes, secondary nitroalkanes and compounds containing terminal *gem*-dinitroaliphatic functionality. The electron deficient alkene in these reactions is known as the Michael acceptor and includes α,β-unsaturated ketones, aldehydes, carboxylic acids, esters, amides, cyanides etc. Nitroalkenes are excellent Michael acceptors and important from the view of energetic materials synthesis and so these reactions are discussed separately in Section 1.10.2.2.

Michael reactions are base catalyzed and reversible, and so it is common to use either the nitronate salt of the nitroalkane substrate or the nitroalkane in the presence of a catalytic amount of alkali metal hydroxide, alkoxide or amine base.

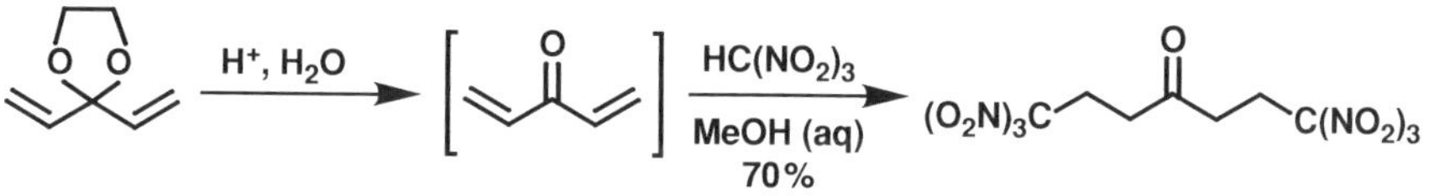

Figure 1.54

The product of a Michael addition depends on the number of acidic protons present in the nitroalkane substrate. Nitroform, which has one acidic proton, can only react with one equivalent of Michael acceptor.[252–258] Nitroform is a strong acid and sufficiently dissociated in solution so that it can be used in addition reactions without a base catalyst. The reaction of nitroform with unsaturated ketones has been investigated by Gilligan and Graff[256] and used to synthesize a number of trinitromethyl-based explosives.

Frankel[258] reported the reaction of nitroform with acrylic acid and its esters. Methyl 4,4,4-trinitrobutyrate (127), the product obtained from nitroform and methyl acrylate, has been used

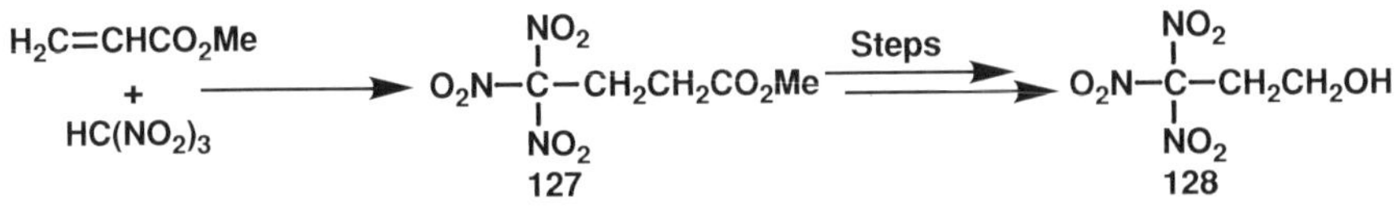

$H_2C{=}CHCO_2Me + HC(NO_2)_3 \longrightarrow O_2N{-}C(NO_2)_2{-}CH_2CH_2CO_2Me$ (127) $\xrightarrow{\text{Steps}} O_2N{-}C(NO_2)_2{-}CH_2CH_2OH$ (128)

Figure 1.55

for the synthesis of 3,3,3-trinitro-1-propanol (128) through a several step synthesis.[263] 4,4,4-Trinitrobutyric acid has been used for the same purpose.[264] Ross and co-workers[265] reported a number of reactions of nitroform with unsaturated aldehydes.

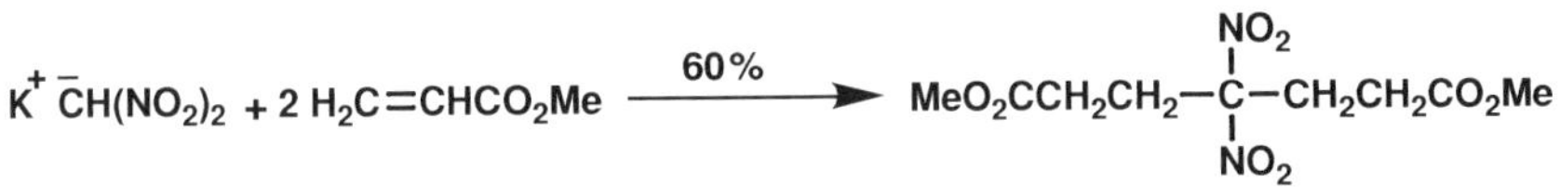

$K^+\,{}^-CH(NO_2)_2 + 2\,H_2C{=}CHCO_2Me \xrightarrow{60\%} MeO_2CCH_2CH_2{-}C(NO_2)_2{-}CH_2CH_2CO_2Me$

Figure 1.56

$MeO_2C(CH_2)_2{-}CH(NO_2){-}(CH_2)_2{-}CH(NO_2){-}(CH_2)_2CO_2Me$ (129)

$MeO_2C(CH_2)_2{-}C(NO_2)[(CH_2)_2CO_2Me]{-}(CH_2)_2{-}C(NO_2)[(CH_2)_2CO_2Me]{-}(CH_2)_2CO_2Me$ (130)

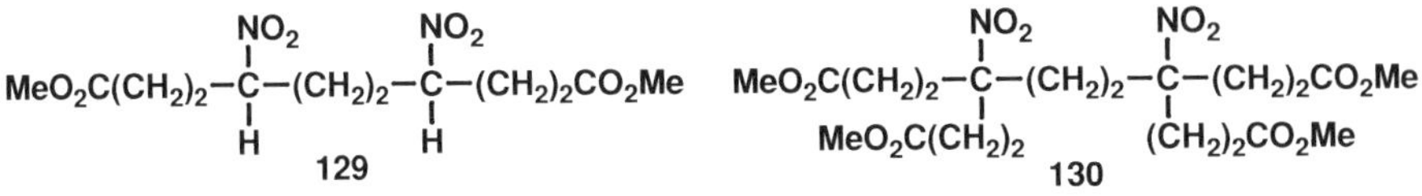

Figure 1.57

Dinitromethane has two acidic protons and reacts with Michael acceptors to form bis-adducts.[251,254,255] Secondary nitroalkanes can only react with one equivalent of Michael acceptor. In the absence of steric effects primary nitroalkanes usually react with two equivalents of Michael acceptor to form bis-adducts. Depending on the reaction stoichiometry, 1,4-dinitrobutane can be reacted with methyl acrylate to form either the bis-adduct (129) or the tetra-adduct (130) in good yield.[249]

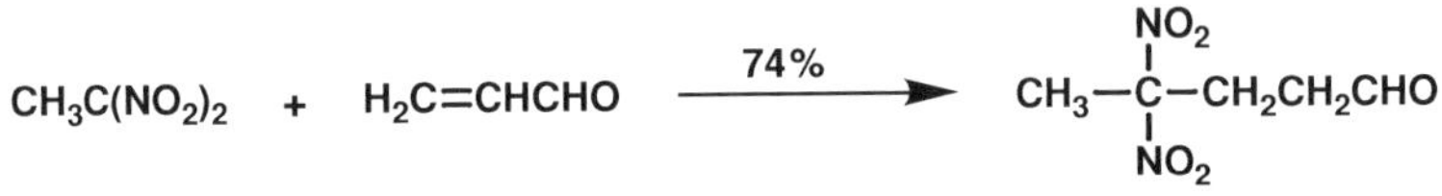

$CH_3C(NO_2)_2 + H_2C{=}CHCHO \xrightarrow{74\%} CH_3{-}C(NO_2)_2{-}CH_2CH_2CHO$

Figure 1.58

Aliphatic compounds containing terminal *gem*-dinitro functionality form adducts with Michael acceptors.[117,253–255,259,260] Of particular interest is the reaction of $\alpha,\alpha,\omega,\omega$-tetranitroalkanes with Michael acceptors.[250] Most $\alpha,\alpha,\omega,\omega$-tetranitroalkanes will react with two equivalents of Michael acceptor to form bis-adducts, like in the case of 1,1,4,4-tetranitrobutane, which reacts with two equivalents of methyl vinyl ketone, methyl acrylate, acrylonitrile etc.[250] The influence of steric effects becomes apparent with $\alpha,\alpha,\gamma,\gamma$-tetranitroalkanes, like 1,1,3,3-tetranitropropane, which can form either mono-adducts or bis-adducts depending on the Michael acceptor used; 1,1,3,3-tetranitropropane will only react with one equivalent of methyl acrylate and the sole product of this reaction is methyl 4,4,6,6-tetranitrohexanoate.[250]

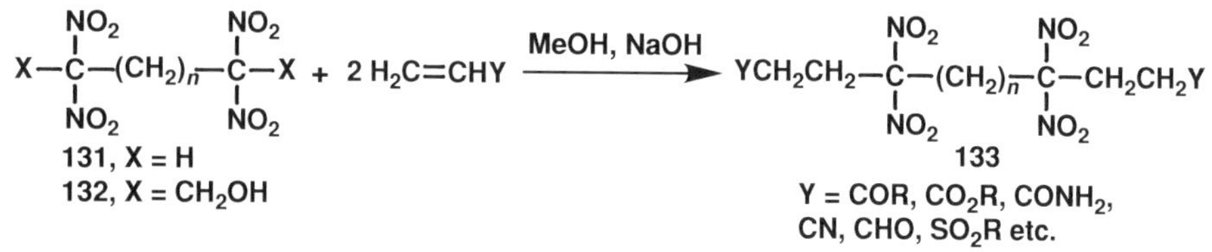

$$X{-}C(NO_2)_2{-}(CH_2)_n{-}C(NO_2)_2{-}X + 2\,H_2C{=}CHY \xrightarrow{\text{MeOH, NaOH}} YCH_2CH_2{-}C(NO_2)_2{-}(CH_2)_n{-}C(NO_2)_2{-}CH_2CH_2Y$$

131, X = H
132, X = CH_2OH

133
Y = COR, CO_2R, $CONH_2$, CN, CHO, SO_2R etc.

Figure 1.59

Many of the Michael reactions involving the addition of polynitroaliphatic compounds to Michael acceptors use the corresponding methylol derivatives, which are deformylated in the presence of base to give the nitronate anion and formaldehyde. Such procedures are primarily for safety reasons because the nitronate salts of many polynitroaliphatic compounds are shock sensitive explosives. Feuer and co-workers[250] investigated the Michael reactions of numerous α,β-unsaturated compounds with $\alpha,\alpha,\omega,\omega$-tetranitroalkanes (131) and their bis-methylol (132) derivatives. Interest in the Michael adducts derived from $\alpha,\alpha,\omega,\omega$-tetranitroalkanes (133) partly stems from their potential use for the synthesis of energetic oligomers;[250] simple functional group conversion of the terminal appendages giving rise to alcohol, carboxylic acid, isocyanate functionality etc.

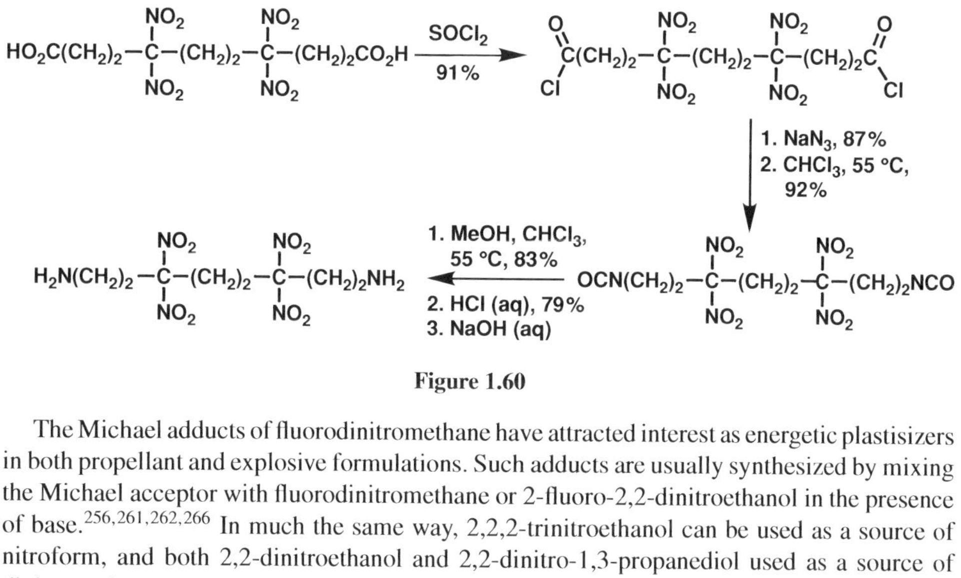

$$HO_2C(CH_2)_2{-}C(NO_2)_2{-}(CH_2)_2{-}C(NO_2)_2{-}(CH_2)_2CO_2H \xrightarrow[91\%]{SOCl_2} ClC(O)(CH_2)_2{-}C(NO_2)_2{-}(CH_2)_2{-}C(NO_2)_2{-}(CH_2)_2C(O)Cl$$

1. NaN_3, 87%
2. $CHCl_3$, 55 °C, 92%

$$OCN(CH_2)_2{-}C(NO_2)_2{-}(CH_2)_2{-}C(NO_2)_2{-}(CH_2)_2NCO \rightarrow H_2N(CH_2)_2{-}C(NO_2)_2{-}(CH_2)_2{-}C(NO_2)_2{-}(CH_2)_2NH_2$$

1. MeOH, $CHCl_3$, 55 °C, 83%
2. HCl (aq), 79%
3. NaOH (aq)

Figure 1.60

The Michael adducts of fluorodinitromethane have attracted interest as energetic plastisizers in both propellant and explosive formulations. Such adducts are usually synthesized by mixing the Michael acceptor with fluorodinitromethane or 2-fluoro-2,2-dinitroethanol in the presence of base.[256,261,262,266] In much the same way, 2,2,2-trinitroethanol can be used as a source of nitroform, and both 2,2-dinitroethanol and 2,2-dinitro-1,3-propanediol used as a source of dinitromethane.

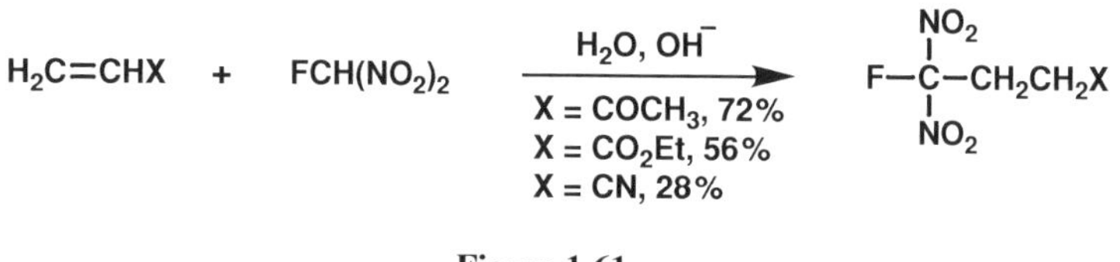

$$H_2C{=}CHX + FCH(NO_2)_2 \xrightarrow{H_2O,\ OH^-} F{-}C(NO_2)_2{-}CH_2CH_2X$$

X = $COCH_3$, 72%
X = CO_2Et, 56%
X = CN, 28%

Figure 1.61

The observant may ask, 'Why does the alkoxide anion of the nitroalcohol not add to the Michael acceptor?' The addition of alkoxide anions to Michael acceptors is well known, but alcohols with two or more strong electron-withdrawing groups on the β-carbon atom to the alcohol functionality make the corresponding alkoxide a very weak nucleophile. However, there are cases of the 2-fluoro-2,2-dinitroethoxide anion (134) undergoing Michael 1,4-addition with some very reactive Michael acceptors.[261]

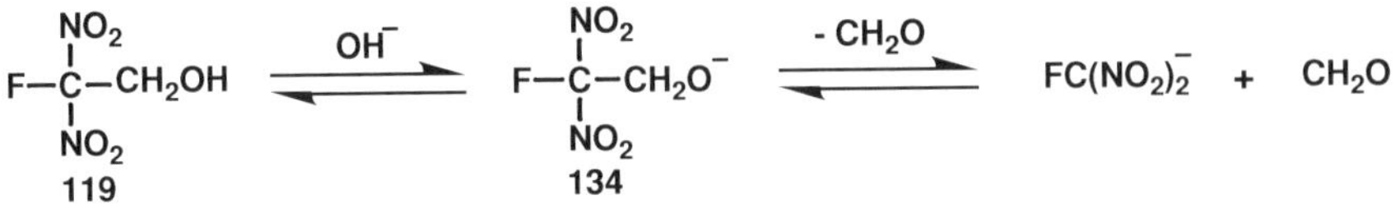

$$\underset{\textbf{119}}{FC(NO_2)_2CH_2OH} \underset{}{\overset{OH^-}{\rightleftharpoons}} \underset{\textbf{134}}{FC(NO_2)_2CH_2O^-} \overset{-\,CH_2O}{\rightleftharpoons} FC(NO_2)_2^- + CH_2O$$

Figure 1.62

1.10.2.2 Additions to nitroalkenes

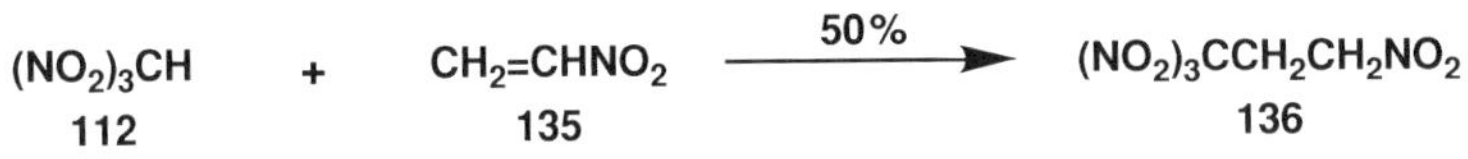

$$\underset{\textbf{112}}{(NO_2)_3CH} + \underset{\textbf{135}}{CH_2{=}CHNO_2} \xrightarrow{50\%} \underset{\textbf{136}}{(NO_2)_3CCH_2CH_2NO_2}$$

Figure 1.63

The conjugate 1,4-addition of nitronate anions and other nucleophiles to α-nitroalkenes constitutes an important method for the synthesis of polynitroaliphatic compounds.[250,253,258,260,267–269] Nitroform (112) reacts with nitroethene (135) and 2-nitropropene to yield 1,1,1,3-tetranitropropane[258] (136) and 1,1,1,3-tetranitrobutane[267] respectively. A number of examples of additions of 1,1-dinitroethane, 1,1-dinitropropane and 1,1-dinitrobutane to nitroalkenes have been reported.[260,267,269c] Feuer and co-workers[250] reported the synthesis of 1,3,3,6,6,8-hexanitrooctane (137) from the reaction of 1,1,4,4-tetranitrobutane (30) with nitroethene under basic conditions.

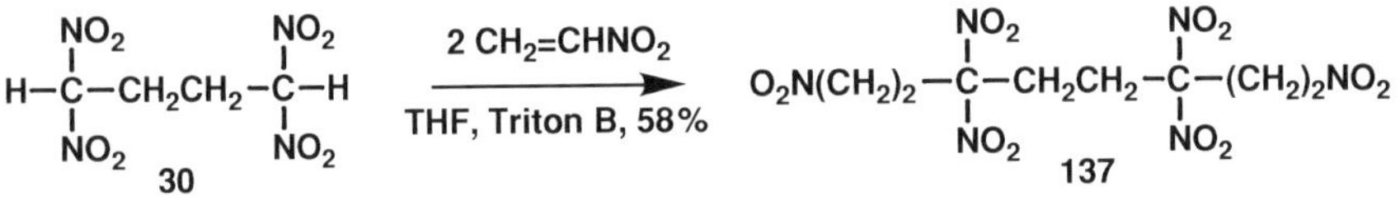

$$\underset{\textbf{30}}{H{-}C(NO_2)_2{-}CH_2CH_2{-}C(NO_2)_2{-}H} \xrightarrow[\text{THF, Triton B, 58\%}]{2\ CH_2{=}CHNO_2} \underset{\textbf{137}}{O_2N(CH_2)_2{-}C(NO_2)_2{-}CH_2CH_2{-}C(NO_2)_2{-}(CH_2)_2NO_2}$$

Figure 1.64

Synthetic routes to α-nitroalkenes have been discussed in previous sections. General routes include: (1) treating β-nitroacetates with alkali metal acetates, carbonates or bicarbonates,[5,268,270] (2) elimination of water from β-nitroalcohols via heating with phthalic anhydride[271] or in the presence of a base,[40,272,273] and (3) degradation of the Mannich products derived from a primary nitroalkane, formaldehyde, and a secondary amine.[274]

Direct Michael addition of secondary nitroalkanes to α-nitroalkenes gives acceptable yields of addition product, however, in general, primary nitroalkanes give much poorer yields of product. Low to moderate yields in both cases are mainly due to the tendency of α-nitroalkenes to polymerize before and during a reaction. More common is the *in situ* generation of the α-nitroalkenes in these addition reactions. Feuer and Miller[268] discovered that 2-nitroalkyl acetates (β-nitroacetates) react with base, in the form of sodium acetate or the sodium salt of a nitroalkane, to generate the corresponding α-nitroalkene. Michael adducts are formed if

Table 1.10 Michael 1,4-addition of nitroalkanes and nitramines with 2-nitroacetates (nitroalkene precursors)

Pseudo acid	Product	Yield (%)
	Michael additions with 2-nitrobutyl acetate	
1-nitropropane	3,5-dinitroheptane	15, 19[a], 13[b]
2-nitropropane	2-methyl-2,4-dinitrohexane	55
1,1-dinitroethane	2,2,4-trinitrohexane	65
ethylene dinitramine	5,8-diaza-3,5,8,10-tetranitrododecane	81
	Michael additions with 3-nitro-2-butyl acetate	
2-nitropropane	2,3-dimethyl-2,4-dinitropentane	33[a,b], 57[b,c]
ethylene dinitramine	4,7-diaza-3,8-dimethyl-2,4,7,9-tetranitrodecane	21
	Michael additions with 1,6-diacetoxy-2,5-dinitrohexane	
2-nitropropane	2,9-dimethyl-2,4,7,9-tetranitrodecane	98
1,1-dinitroethane	3,3,5,8,10,10-hexanitrododecane	94
1-nitraminobutane	5,12-diaza-5,7,10,12-tetranitrohexadecane	88

[a] Solvent was anhydrous THF. [b] 100% excess of the salt of the psuedo acid used (no sodium acetate). [c] Solvent was *tert*-butanol. [d] Equivalent amounts of reactants and sodium acetate used unless otherwise stated. [e] Reactions conducted in aqueous methanol unless otherwise stated.

Source: Reprinted with permission from H. Feuer and R. Miller, *J. Org. Chem.*, 1961, **26**, 1348; Copyright 1961 American Chemical Society.

a nitroalkane or another nucleophile is present during this *in situ* α-nitroalkene formation. 2-Nitrobutyl acetate, 3-nitro-2-butyl acetate and 1,6-diacetoxy-2,5-dinitrohexane are precursors to 2-nitro-1-butene, 2-nitro-2-butene and 2,5-dinitro-1,5-hexadiene respectively. In the presence of sodium acetate these nitroalkene precursors have been used to synthesize a variety of polynitroaliphatic compounds in good to high yield (Table 1.10).[268]

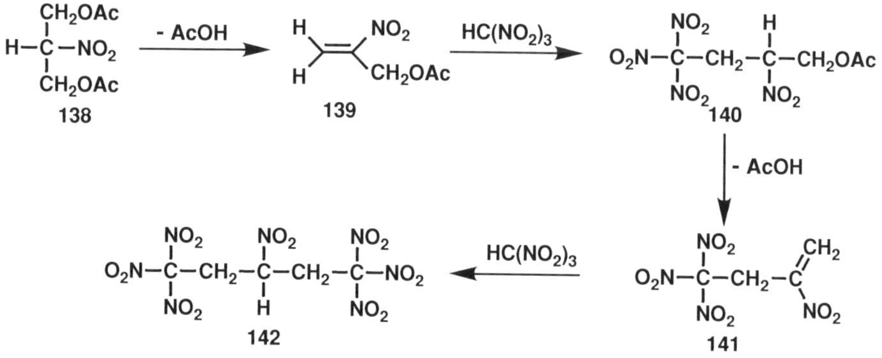

Figure 1.65

Frankel[258] synthesized 2-nitro-3-acetoxy-1-propene (139) by heating 1,3-diacetoxy-2-nitropropane (138) with sodium acetate under reduced pressure. The reaction of 2-nitro-3-acetoxy-1-propene with 1,1-dinitroethane yields 2,2,4,6,6-pentanitroheptane. The same reaction with nitroform provides 1,1,1,3,5,5,5-heptanitropentane (142), a powerful explosive (VOD ~ 9230 m/s) with an excellent oxygen balance.[258,274,275] The synthesis of 1,1,1,3,5,5,5-heptanitropentane from 2-nitro-1,3-propanediol[276] and 2-nitro-1-propen-3-ol[277] has also been reported and involves a similar mechanism.

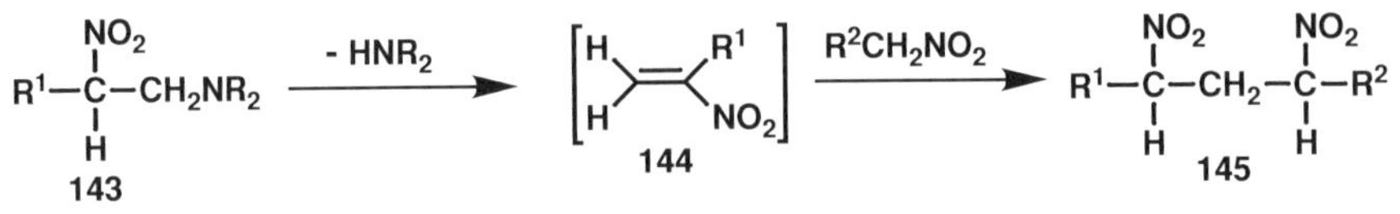

Figure 1.66

Some 1,3-dinitroalkanes (145) have been synthesized from the reaction of nitroalkanes with α-nitroalkenes (144) generated *in situ* from the decomposition of Mannich bases (143) derived from primary nitroalkanes.[274] Reported yields for these reactions are low and the formation of by-products limits the feasibility of the method.

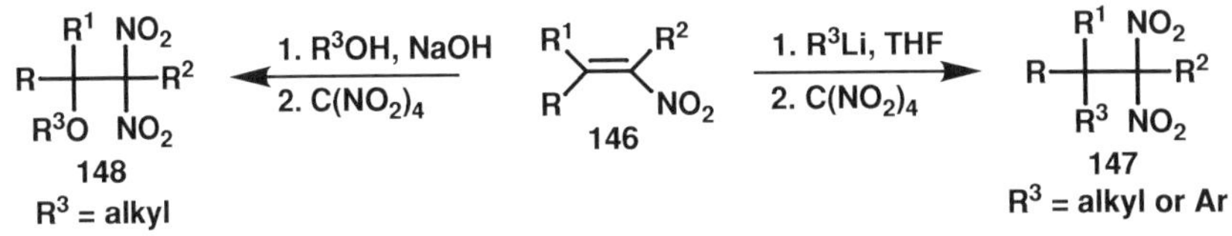

Figure 1.67

Nielsen and Bedford[279] synthesized *gem*-dinitroalkanes (147) from the Michael addition of organolithium reagents to α-nitroalkenes (146) followed by quenching of the resulting nitronate anion with tetranitromethane. The same reaction using alkoxides as bases provides β-alkoxy-*gem*-dinitroalkanes (148).[279]

1.10.2.3 Dinitroethylation

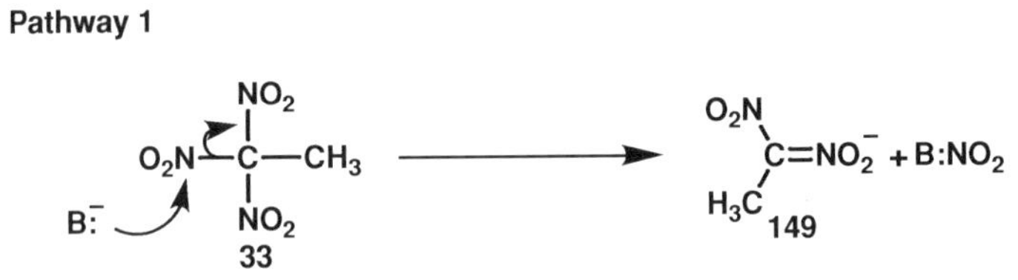

Figure 1.68

Shechter and Zeldin[280] discovered that 1,1,1-trinitroethane (33) can undergo two reactions on treatment with base. First, and typical of the chemistry of 1,1,1-trinitromethyl compounds, the base can attack one of the electron deficient nitro groups of 1,1,1-trinitroethane (33) and form the 1,1-dinitroethane anion (149) (Pathway 1). More unusual is the second reaction where the base abstracts a hydrogen from 1,1,1-trinitroethane (33), followed by formal elimination of nitrous acid and the formation of the reactive intermediate, 1,1-dinitroethene (150), which can react further with any nucleophiles present in a Michael 1,4-addition reaction (Pathway 2).

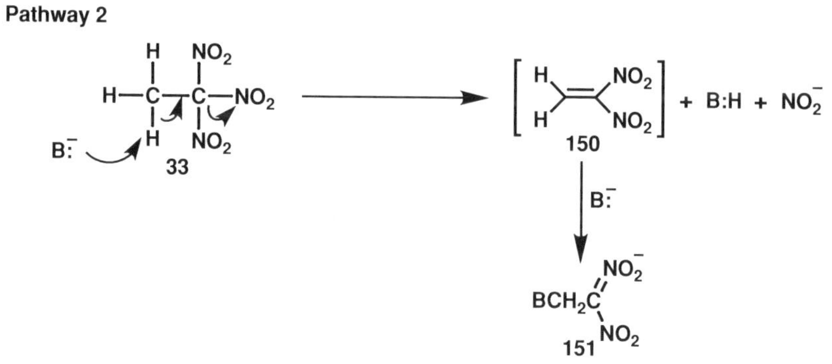

Pathway 2

$$B^{-} + H_3C{-}C(NO_2)_3\ (\mathbf{33}) \longrightarrow [H_2C{=}C(NO_2)_2]\ (\mathbf{150}) + B{:}H + NO_2^{-}$$

$$[H_2C{=}C(NO_2)_2]\ (\mathbf{150}) \xrightarrow{B^{-}} BCH_2C(NO_2)=NO_2^{-}\ (\mathbf{151})$$

Figure 1.69

Reactions are very dependent on the nature of the base and the reaction conditions used, for example, reaction of 1,1,1-trinitroethane with aqueous potassium hydroxide, or hydroxylamine in methanolic potassium methoxide, gives high yields of potassium 1,1-dinitroethane.[280] However, reaction of 1,1,1-trinitroethane with potassium ethoxide, potassium methoxide and ethanolic potassium cyanide is reported to give 2,2-dinitroethylether, methyl-2,2-dinitroethylether and 3,3-dinitropropionitrile respectively, all in approximately 80 % yield via the 1,4-addition of ethoxide, methoxide and cyanide anion to 1,1-dinitroethene respectively.[280]

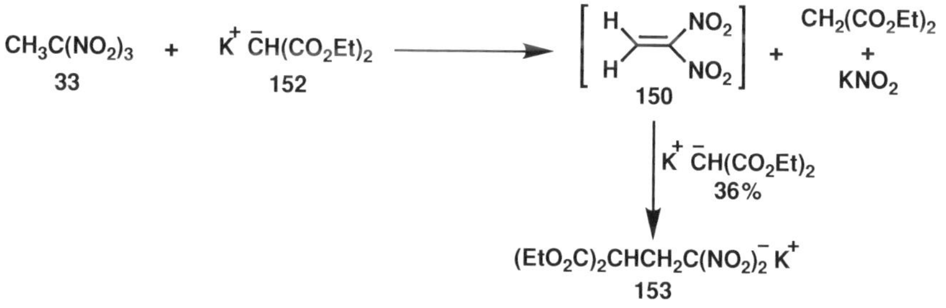

$$CH_3C(NO_2)_3\ (\mathbf{33}) + K^{+}\ {}^{-}CH(CO_2Et)_2\ (\mathbf{152}) \longrightarrow [H_2C{=}C(NO_2)_2]\ (\mathbf{150}) + CH_2(CO_2Et)_2 + KNO_2$$

$$[H_2C{=}C(NO_2)_2]\ (\mathbf{150}) \xrightarrow[36\%]{K^{+}\ {}^{-}CH(CO_2Et)_2} (EtO_2C)_2CHCH_2C(NO_2)_2^{-}K^{+}\ (\mathbf{153})$$

Figure 1.70

Other bases found to react with 1,1,1-trinitroethane via formation of 1,1-dinitroethene include: trimethylamine, guanidine and diethylmalonate anion (152), the latter forming (153) in 36 % yield. Shechter and Zeldin[280] found no correlation as to why some bases react with 1,1,1-trinitroethane so differently to others but noted that simple alkoxides, aliphatic amines, guanidine, cyanide and malonate anions reacted via the 1,1-dinitroethene pathway.

Frankel[281] discovered a similar reaction to the base-induced formation of 1,1-dinitroethene from 1,1,1-trinitroethane; treatment of 2-bromo-2,2-dinitroethyl acetate (154) with potassium

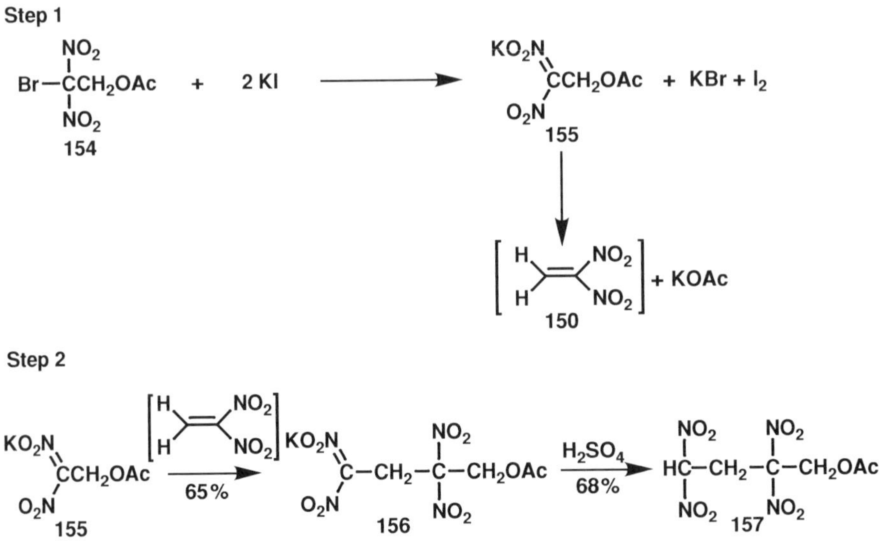

Figure 1.71

iodide gave the potassium salt of 2,2,4,4-tetranitrobutyl acetate (156) in 65 % yield, which on acidification with mineral acid, yielded 2,2,4,4-tetranitrobutyl acetate (157). Frankel[281] explained the unusual result by also postulating the formation of 1,1-dinitroethene, a highly reactive intermediate capable of undergoing Michael type 1,4-addition with any unreacted nitronate anion (155) present in solution. The generality of this reaction, known as 'dinitroethylation', for the synthesis of *gem*-dinitroaliphatic compounds is further illustrated by the formation of the potassium salt of 1,1,3,3-tetranitrobutane when 2-bromo-2,2-dinitroethyl acetate is treated with potassium iodide and the sodium salt of 1,1-dinitroethane.[281]

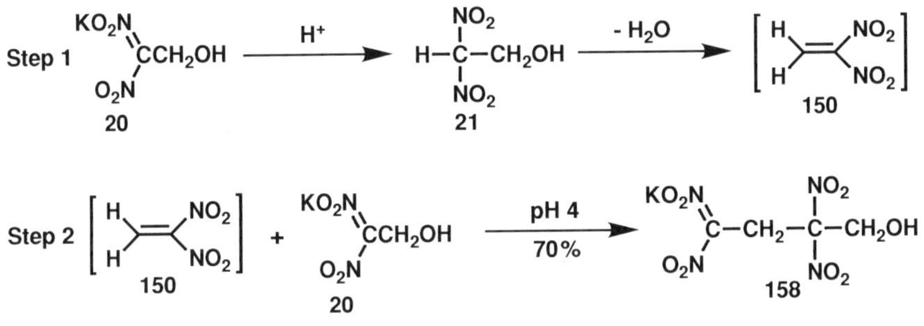

Figure 1.72

The formation of 1,1-dinitroethene (150) as an intermediate also accounts for the formation of potassium 2,2,4,4-tetranitrobutanol (158) when a solution of potassium 2,2-dinitroethanol (20) is partially acidified.[259] Klager and co-workers[259] postulated that 1,1-dinitroethene is formed via elimination of water from 2,2-dinitroethanol and this rapidly undergoes Michael 1,4-addition with any potassium 2,2-dinitroethanol still present in the reaction mixture. Demethylolation of (158) with potassium hydroxide yields the dipotassium salt of 1,1,3,3-tetranitropropane.

1.10.3 Mannich reaction

$$R_2NH + CH_2O \rightleftharpoons R_2NCH_2OH \rightleftharpoons R_2\overset{+}{N}{=}CH_2 + H_2O \quad \text{(Eq. 1.4)}$$

$$R_2\overset{+}{N}{=}CH_2 + \overset{-}{C}(NO_2)_3 \rightleftharpoons R_2NCH_2C(NO_2)_3 \quad \text{(Eq. 1.5)}$$

Figure 1.73

The Mannich reaction is an excellent route to polynitroaliphatic amines and their derivatives. β-Nitroalkylamines are formed from the reaction of an amine and aldehyde in the presence of a nitroalkane (Equations 1.4 and 1.5).[282–296] A large number of these reactions use nitroform,[285–287] fluorodinitromethane,[37,256,261,262] or their methylol derivatives, 2,2,2-trinitroethanol[284,291] and 2-fluoro-2,2-dinitroethanol,[37,256,261,262] to synthesize the corresponding trinitromethyl and fluorodinitromethyl derivatives respectively.

Primary and secondary nitroalkanes, dinitromethane,[282] and terminal *gem*-dinitroaliphatic compounds like 1,1-dinitroethane,[284,288] all contain acidic protons and have been used to generate Mannich products. Formaldehyde is commonly used in these reactions although the use of other aliphatic aldehydes has been reported.[282] The nitroalkane component is frequently generated *in situ* from its methylol derivative, a reaction which also generates formaldehyde. Ammonia,[282,289–291] aliphatic amines,[282–289] hydrazine,[288,289] and even urea[291] have been used as the amine component of Mannich reactions.

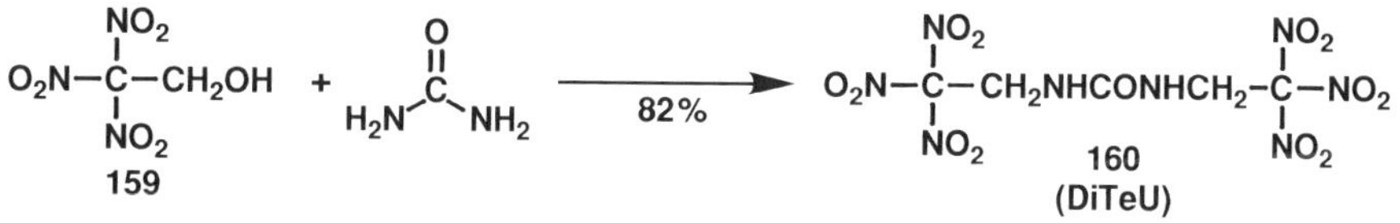

$$(O_2N)_3C{-}CH_2OH\ (\mathbf{159}) + H_2NC(O)NH_2 \xrightarrow{82\%} (O_2N)_3C{-}CH_2NHCONHCH_2{-}C(NO_2)_3\ (\mathbf{160},\ \text{DiTeU})$$

Figure 1.74

$$HC(NO_2)_3\ (\mathbf{112}) \xrightarrow[\text{or } (CH_2)_6N_4,\ 60\text{–}70\%]{NH_3,\ CH_2O,\ 90\%} HN[CH_2C(NO_2)_3]_2\ (\mathbf{161})$$

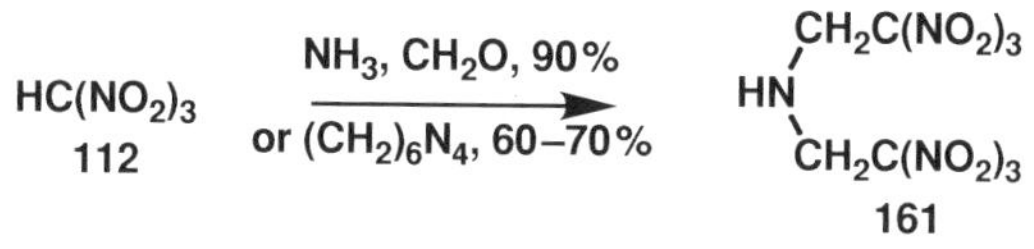

Figure 1.75

The powerful explosive, bis(2,2,2-trinitroethyl)urea (160) (DiTeU), is synthesized from the reaction of 2,2,2-trinitroethanol (159) with urea, or from the direct reaction of nitroform with formaldehyde and urea.[291] Bis(2,2,2-trinitroethyl)amine (161) has been synthesized from the reaction of 2,2,2-trinitroethanol with ammonia and also from the reaction of nitroform (112) with formaldehyde and ammonia, or hexamine.[291,296]

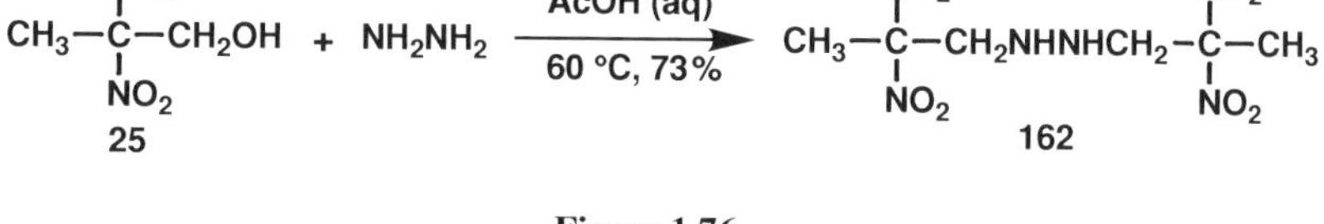

$$CH_3{-}C(NO_2)_2{-}CH_2OH\ (\mathbf{25}) + NH_2NH_2 \xrightarrow[60\ ^\circ C,\ 73\%]{AcOH\ (aq)} CH_3{-}C(NO_2)_2{-}CH_2NHNHCH_2{-}C(NO_2)_2{-}CH_3\ (\mathbf{162})$$

Figure 1.76

Frankel and Klager[289] have reported using the Mannich reaction for the condensation of 2,2-dinitroalkanols with ammonia and hydrazine. This method was used to synthesize 2,2,6,6-tetranitro-4-azaheptane (100 %) and bis(2,2-dinitropropyl)hydrazine (162) (73 %) from the reaction of 2,2-dinitropropanol (25) with ammonia and hydrazine hydrate respectively. This work was later extended to using polynitroaliphatic amines and diamines.[284]

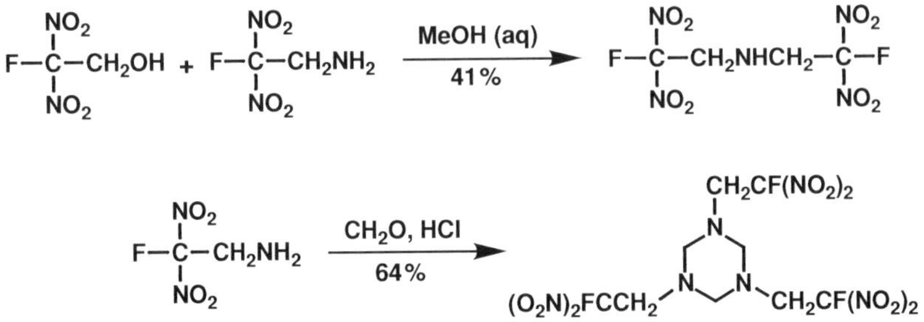

$$FC(NO_2)_2CH_2OH + FC(NO_2)_2CH_2NH_2 \xrightarrow[41\%]{MeOH\,(aq)} FC(NO_2)_2CH_2NHCH_2C(NO_2)_2F$$

$$FC(NO_2)_2CH_2NH_2 \xrightarrow[64\%]{CH_2O,\ HCl} \text{1,3,5-tris}[CH_2CF(NO_2)_2]\text{-hexahydro-1,3,5-triazine}$$

Figure 1.77

Mannich bases derived from polynitroalkanes are usually unstable because of the facile reverse reaction leading to stabilized nitronate anions. The nitration of Mannich bases to nitramines enhances their stability by reducing the electron density on the amine nitrogen through delocalization with the nitro group. The nitration of Mannich bases has been exploited for the synthesis of numerous explosives, some containing both C–NO_2 and N–NO_2 functionality.[293,295,297] Three such compounds, (163), (164) and (165), are illustrated below and others are discussed in Section 6.10.

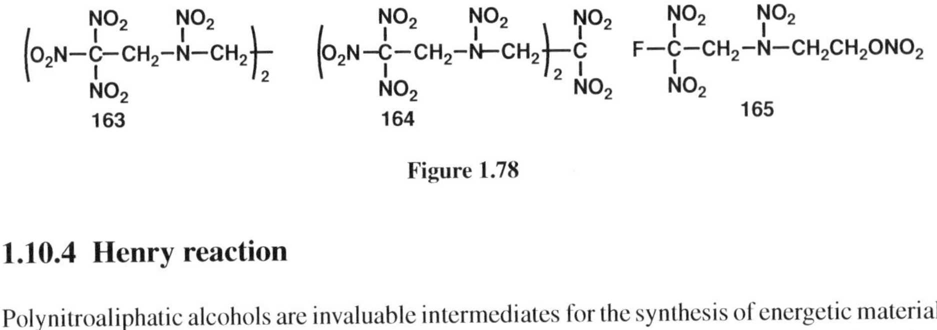

$$\left(O_2N{-}C(NO_2)_2{-}CH_2{-}N(NO_2){-}CH_2\right)_2 \quad \mathbf{163}$$

$$\left(O_2N{-}C(NO_2)_2{-}CH_2{-}N(NO_2){-}CH_2\right)_2C(NO_2)_2 \quad \mathbf{164}$$

$$F{-}C(NO_2)_2{-}CH_2{-}N(NO_2){-}CH_2CH_2ONO_2 \quad \mathbf{165}$$

Figure 1.78

1.10.4 Henry reaction

Polynitroaliphatic alcohols are invaluable intermediates for the synthesis of energetic materials (see Section 1.11). The most important route to β-nitroalcohols is via the Henry reaction where a mixture of the aldehyde and nitroalkane is treated with a catalytic amount of base, or the nitronate salt of the nitroalkane is used directly, in which case, on reaction completion, the reaction mixture is acidified with a weak acid. Reactions are reversible and in the presence of base the salt of the nitroalkane and the free aldehyde are reformed. This reverse reaction is known as demethylolation if formaldehyde is formed.

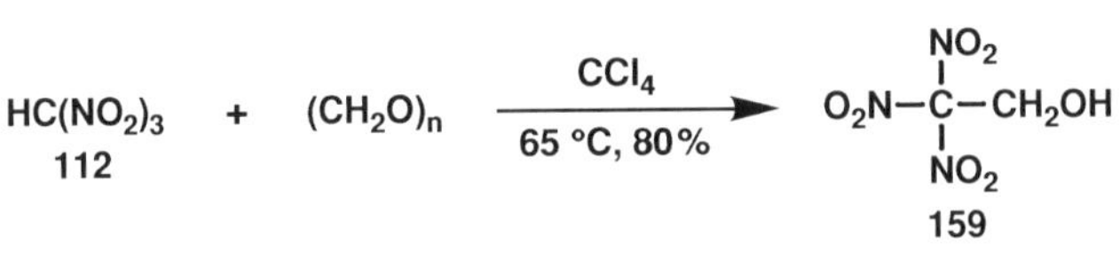

$$\underset{\mathbf{112}}{HC(NO_2)_3} + (CH_2O)_n \xrightarrow[65\,^\circ C,\ 80\%]{CCl_4} \underset{\mathbf{159}}{O_2N{-}C(NO_2)_2{-}CH_2OH}$$

Figure 1.79

Formaldehyde is the most important aldehyde used in Henry reactions in relation to energetic materials synthesis. Nitroform (112) reacts with formaldehyde in the form of trioxane or formalin to yield 2,2,2-trinitroethanol (159).[298,299] The Henry reaction of nitroform with aldehydes other than formaldehyde gives products which are not isolable.

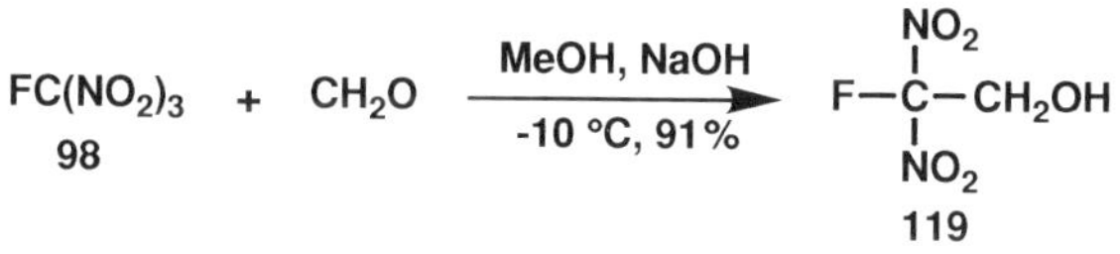

$$\underset{98}{FC(NO_2)_3} + CH_2O \xrightarrow[-10\,°C,\ 91\%]{MeOH,\ NaOH} \underset{119}{F\text{—}C(NO_2)_2\text{—}CH_2OH}$$

Figure 1.80

2-Fluoro-2,2-dinitroethanol (119) is rarely synthesized from fluorodinitromethane but is reportedly synthesized at Rockwell International, Rocketdyne Division in large quantities from the reaction of fluorotrinitromethane (98) with an alkaline solution of formalin.[300] The same reaction with alkaline hydrogen peroxide and formalin is also used for 2-fluoro-2,2-dinitroethanol production.[239] Unlike nitroform, fluorodinitromethane forms stable products with aldehydes other than formaldehyde.[301–305] Products obtained from the reaction of fluorodinitromethane with dialdehydes contain two hydroxy groups and may find use for the synthesis of energetic polymers.[303–305]

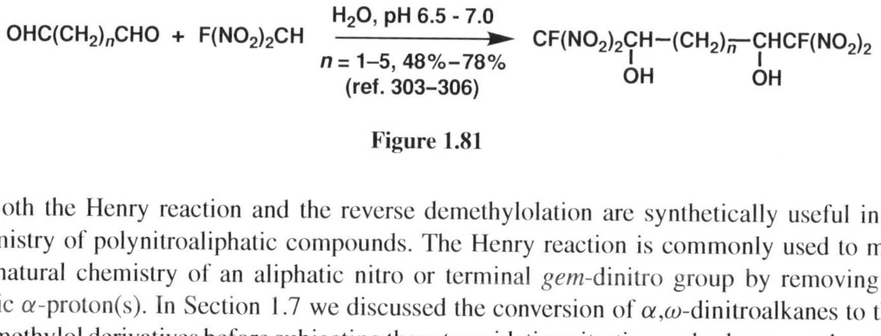

$$OHC(CH_2)_nCHO + F(NO_2)_2CH \xrightarrow[n = 1\text{–}5,\ 48\%\text{–}78\%\ (\text{ref. }303\text{–}306)]{H_2O,\ pH\ 6.5 - 7.0} CF(NO_2)_2CH(OH)\text{–}(CH_2)_n\text{–}CH(OH)CF(NO_2)_2$$

Figure 1.81

Both the Henry reaction and the reverse demethylolation are synthetically useful in the chemistry of polynitroaliphatic compounds. The Henry reaction is commonly used to mask the natural chemistry of an aliphatic nitro or terminal *gem*-dinitro group by removing the acidic α-proton(s). In Section 1.7 we discussed the conversion of α,ω-dinitroalkanes to their bis-methylol derivatives before subjecting them to oxidative nitration and subsequent demethylolation with base, a procedure which results in the formation of $\alpha,\alpha,\omega,\omega$-tetranitroalkanes.[108]

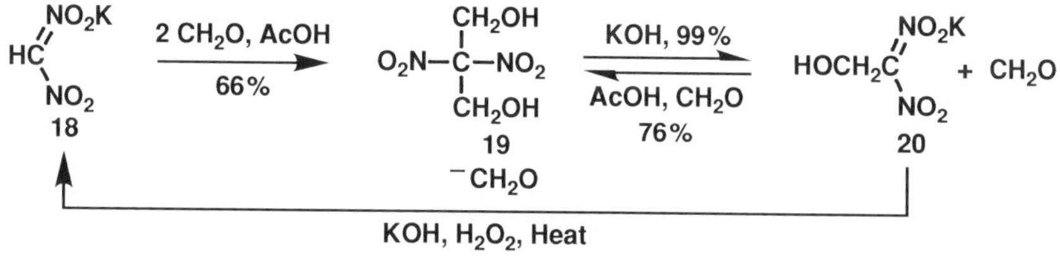

$$\underset{18}{HC(NO_2)=NO_2K} \xrightarrow[66\%]{2\ CH_2O,\ AcOH} \underset{19}{O_2N\text{—}C(CH_2OH)_2\text{—}NO_2} \underset{AcOH,\ CH_2O,\ 76\%}{\overset{KOH,\ 99\%,\ -CH_2O}{\rightleftharpoons}} \underset{20}{HOCH_2C(NO_2)=NO_2K} + CH_2O$$

$$20 \xrightarrow{KOH,\ H_2O_2,\ Heat} 18$$

Figure 1.82

Many of the nitronate salts of polynitroaliphatic compounds, particularly salts of *gem*-nitronitronates, exhibit properties similar to known primary explosives. Consequently, the storage of such salts is highly dangerous. Treatment of these nitronate salts with formaldehyde yields the corresponding methylol derivative via the Henry condensation. These methylol derivatives are much safer to store than the parent nitronate salts and are readily converted back to the nitronate on treatment with base. Thus, treatment of the potassium salt of dinitromethane (18) with an excess of formaldehyde in the presence of acetic acid yields the bis-methylol derivative, 2,2-dinitro-1,3-propanediol (19).[104,106] Treatment of 2,2-dinitro-1,3-propanediol (19) with one equivalent of potassium hydroxide leads to demethylolation and yields the potassium salt of 2,2-dinitroethanol (20).[104] Further demethylolation of potassium 2,2-dinitroethanol (20) to potassium dinitromethane (18) is only affected with an excess of base or with hot alkaline hydrogen peroxide solution.[104]

The use of polynitroaliphatic alcohols as sources of the corresponding nitronate anions is common in addition reactions. However, polynitroaliphatic alcohols are useful in their own right. The hydroxy functionalities of 2,2-dinitroethanol, 2,2-dinitro-1,3-propanediol and 2,2-dinitropropanol[107,306] are very versatile and have been extensively used for the synthesis of energetic plasticizers, polymers, explosives and oxidizers in propellants (Section 1.11). Diols obtained from the reactions of $\alpha,\alpha,\omega,\omega$-tetranitroalkanes with formaldehyde are particularly useful for the synthesis of energetic polymers based on ester and carbamate linkages.[108,250,259,281] α, ω-Dinitroalkanes can react with either two or four equivalents of formaldehyde to form diols or tetrols respectively; good yields can be obtained in both cases by varying reaction conditions and the base used.[307]

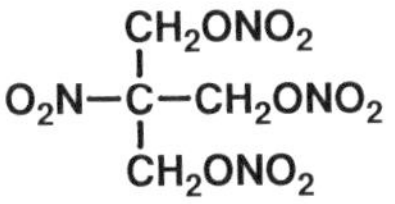

$O_2N-C(CH_2ONO_2)_3$ **166** $\qquad$ $O_2N-C(CH_3)(CH_2ONO_2)_2$ **167**

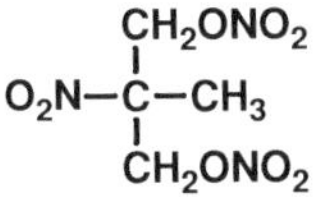

Figure 1.83

Henry reactions have been extensively exploited for the synthesis of nitrate ester explosives. The condensation of nitroalkanes with aldehydes, followed by esterification of the hydroxy groups with nitric acid, leads to a number of nitrate ester explosives (see Chapter 3). The two examples given above (166 and 167) are synthesized from the *O*-nitration of the polyols obtained from the condensation of formaldehyde with nitromethane[308] and nitroethane[309] respectively.

1.11 DERIVATIVES OF POLYNITROALIPHATIC ALCOHOLS

Polynitroaliphatic alcohols containing nitro groups on the carbon β to the hydroxy functionality are less basic than their alkyl counterparts. This decreased basicity of the hydroxy group makes reactions such as esterification, acetal formation and alkylation much slower than usual, and in some cases, these reactions may not proceed without catalysts. To add to the problem, normal base catalysts cannot be used in conjunction with 2,2-dinitroalkanols and 1,1,1-trinitro-2-alkanols because of their facile dissociation in alkaline solution.

The weak nucleophilic nature of polynitroaliphatic alcohols means that reactions often need to be catalyzed by Brønsted acids or Lewis acids. The following methods are commonly used for the esterification of polynitroaliphatic alcohols: (1) heating a solution of the alcohol and acid in the presence of sulfuric acid with Dean–Stark removal of water;[310] (2) using the acid chloride or anhydride in the presence of aluminium chloride;[311,312] (3) reacting the acid and alcohol in the presence of TFAA or PPA as condensing agents;[310,313,314] (4) *trans*-esterification of the methyl ester of the acid component with the alcohol in the presence of sulfuric acid or oleum;[310] (5) using the alcohol and acid chloride in the presence of pyridine, trialkylamines or potassium carbonate;[315,316] (6) reacting the acid chloride and the alcohol without solvent;[298,317,318] and (7) direct esterification of the alcohol with phosphorous trichloride, chlorosulfonic acid or anhydrous nitric acid.[311,314]

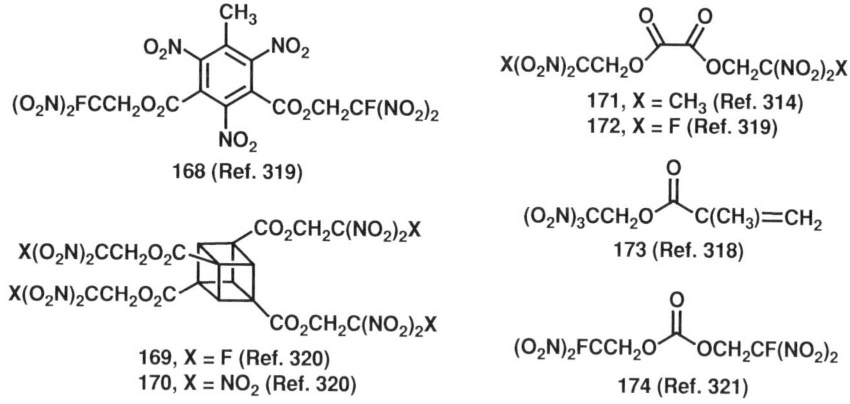

CH_3 O_2N NO_2 $(O_2N)_2FCCH_2O_2C$ $CO_2CH_2CF(NO_2)_2$ NO_2
168 (Ref. 319)

$X(O_2N)_2CCH_2O$ O O $OCH_2C(NO_2)_2X$
171, X = CH_3 (Ref. 314)
172, X = F (Ref. 319)

$X(O_2N)_2CCH_2O_2C$ $X(O_2N)_2CCH_2O_2C$ $CO_2CH_2C(NO_2)_2X$ $CO_2CH_2C(NO_2)_2X$
169, X = F (Ref. 320)
170, X = NO_2 (Ref. 320)

$(O_2N)_3CCH_2O$ O $C(CH_3){=}CH_2$
173 (Ref. 318)

$(O_2N)_2FCCH_2O$ O $OCH_2CF(NO_2)_2$
174 (Ref. 321)

Figure 1.84

A huge number of ester and carbonate derivatives of polynitroaliphatic alcohol have been synthesized; driven by the search for new explosives and energetic plasticizers and oxidizers for propellant and explosive formulations. Most of these are derived from 2-fluoro-2,2-dinitroethanol[319–321] and 2,2,2-trinitroethanol[298,310,311,318] and have excellent oxygen balances. Some examples are illustrated above (168–174) but more comprehensive lists can be found in numerous reviews.[37,114,241,242] Direct esterification of polynitroaliphatic alcohols with nitric acid, mixed acid, or acetic anhydride–nitric acid has been used as a route to mixed polynitroaliphatic–nitrate ester explosives.[311]

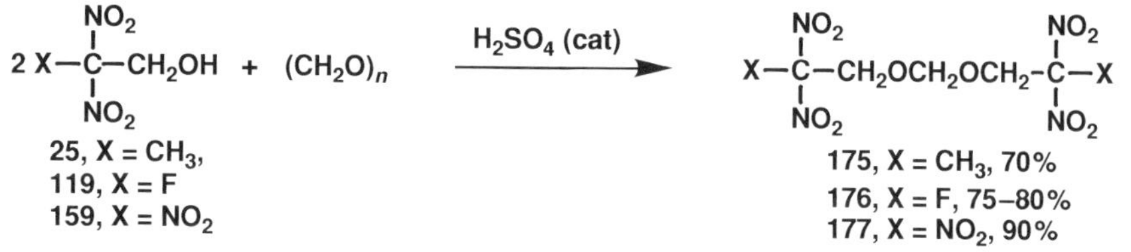

$$2\,X{-}C(NO_2)_2{-}CH_2OH + (CH_2O)_n \xrightarrow{H_2SO_4\ (cat)} X{-}C(NO_2)_2{-}CH_2OCH_2OCH_2{-}C(NO_2)_2{-}X$$

25, X = CH_3,
119, X = F
159, X = NO_2

175, X = CH_3, 70%
176, X = F, 75–80%
177, X = NO_2, 90%

Figure 1.85

$$HOCH_2{-}C(NO_2)_2{-}CH_2OH + (CH_2O)n \xrightarrow[69\%]{AcOH,\ H_2SO_4\ (cat)} \mathbf{178}$$

19

O_2N NO_2 O O
178

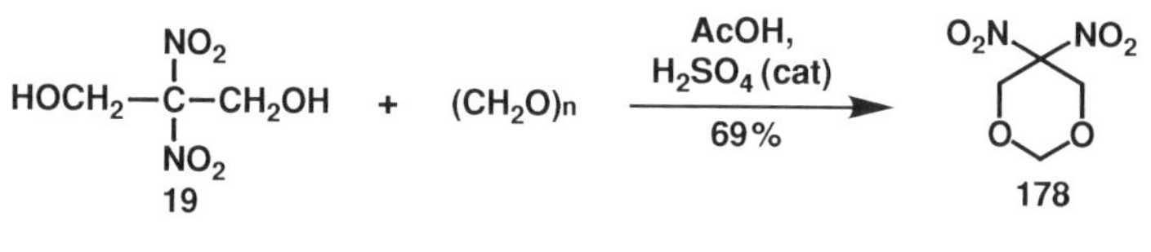

Figure 1.86

Formals and acetals prepared from the reaction of polynitroaliphatic alcohols with formaldehyde and acetaldehyde have found use as explosive plastisizers for nitrocellulose and in plastic bonded explosives (PBXs). Formals of polynitroaliphatic alcohols are commonly prepared via reaction with trioxane or paraformaldehyde in the presence of sulfuric acid as a condensing agent. Bis(2,2-dinitropropyl)formal (175) is prepared from the reaction of trioxane with 2,2-dinitropropanol (25).[322,323] The reaction of 2,2,2-trinitroethanol (159) and 2,2-dinitro-1,3-propanediol (19) with formaldehyde in the presence of sulfuric acid yields bis(2,2,2-trinitroethyl)formal (177)[322] and the 1,3-dioxane (178)[322,324] respectively. Bis(2-fluoro-2,2-dinitroethyl)formal (176) (FEFO), an energetic plastisizer used in some high energy PBXs, is prepared in 75–80 % yield from the condensation of 2-fluoro-2,2-dinitroethanol (119) with formaldehyde in the presence of concentrated sulfuric acid.[237,312,325–327]

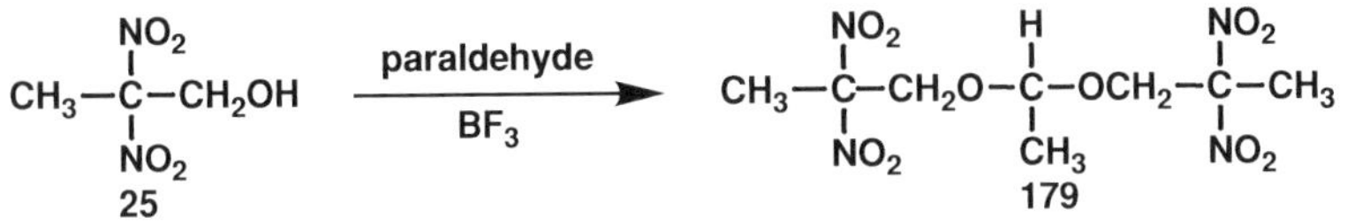

$$CH_3-C(NO_2)_2-CH_2OH \ (\mathbf{25}) \xrightarrow[BF_3]{\text{paraldehyde}} CH_3-C(NO_2)_2-CH_2O-CH(CH_3)-OCH_2-C(NO_2)_2-CH_3 \ (\mathbf{179})$$

Figure 1.87

Sulfuric acid cannot be used for the synthesis of acetals and so bis(2,2-dinitropropyl)acetal (179) is prepared from the reaction of paraldehyde with 2,2-dinitropropanol (25) in the presence of boron trifluoride.[322,323] A 50:50 eutectic mixture of bis(2,2-dinitropropyl)formal (175) and bis(2,2-dinitropropyl)acetal (179) has found use as an energetic liquid plastisizer for nitrocellulose.

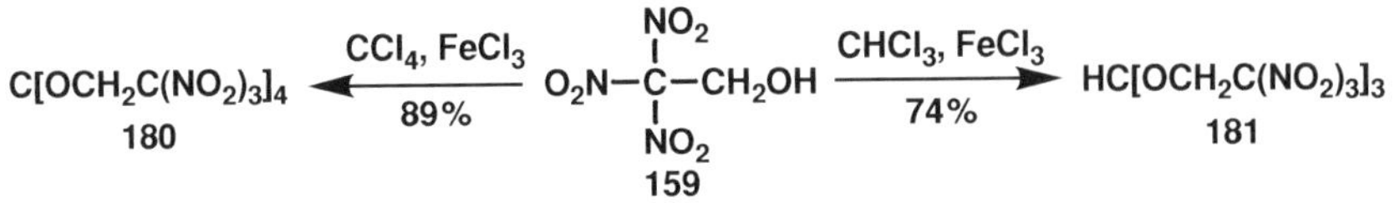

$$C[OCH_2C(NO_2)_3]_4 \ (\mathbf{180}) \xleftarrow[89\%]{CCl_4,\ FeCl_3} O_2N-C(NO_2)_2-CH_2OH \ (\mathbf{159}) \xrightarrow[74\%]{CHCl_3,\ FeCl_3} HC[OCH_2C(NO_2)_3]_3 \ (\mathbf{181})$$

Figure 1.88

Orthoesters of polynitroaliphatic alcohols have been synthesized in the presence of metal chloride Lewis acid catalysts. Tetrakis(2,2,2-trinitroethyl)orthocarbonate (180) and tris(2,2,2-trinitroethyl)orthoformate (181) are obtained from the reaction of 2,2,2-trinitroethanol (159) with carbon tetrachloride and chloroform, respectively, in the presence of anhydrous ferric chloride.[328,329] Analogous reactions with 2-fluoro-2,2-dinitroethanol have been reported.[329]

The weak nucleophilic nature of polynitroaliphatic alcohols is also reflected in their slow reactions with isocyanates to yield carbamates. These reactions often need the presence of Lewis acids like ferric acetylacetonate or boron trifluoride etherate.[330] The reaction of bifunctional isocyanates with polynitroaliphatic diols has been used to synthesize energetic polymers.[330]

Alkylations and other reactions using polynitroaliphatic alcohols as nucleophiles usually require Lewis acid catalysts. A comprehensive review of the chemistry of fluoronitro compounds, including 2-fluoro-2,2-dinitroethanol, has been given by Adolph and Koppes.[37] Many more examples of the use of polynitroaliphatic alcohols as nucleophiles and in other reactions can be found in numerous reviews.[37,114,241,242]

1.12 MISCELLANEOUS

1.12.1 1,1-Diamino-2,2-dinitroethylenes

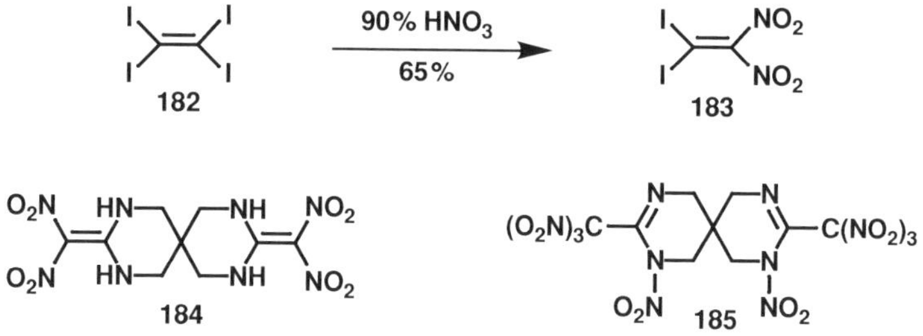

Figure 1.89

The nitration of 1,1,2,2-tetraiodoethylene (182) with 90 % nitric acid provides 1,1-diiododinitroethylene (183) in good yield.[331] Baum and co-workers[331] studied the chemistry of 1,1-diiododinitroethylene and found that displacement of the two vinylic iodide groups is effected by treatment with simple aliphatic amines, diamines and anilines to give 1,1-diamino-2,2-dinitroethylenes in excellent yield. Cyclic products are obtained from the reaction of 1,1-diiododinitroethylene with 1,2-diaminoethane (ethylenediamine), 1,3-diaminopropane and 1,4-diaminobutane. The spirocycle (184) is synthesized from the reaction of 1,1-diiododinitroethylene (183) with 2,2-bis(methylamino)-1,3-diaminopropane.

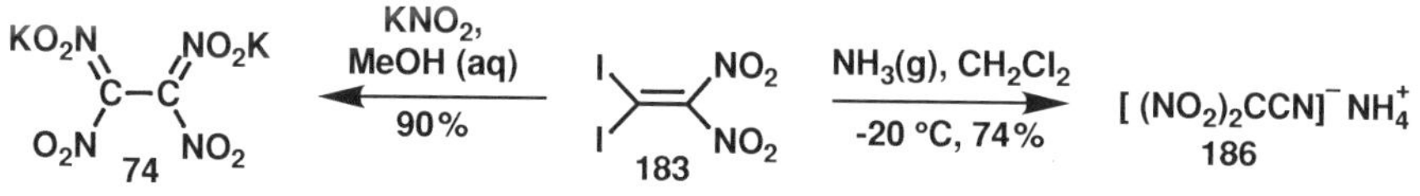

Figure 1.90

The reaction of 1,1-diiododinitroethylene (183) with potassium nitrite in aqueous methanol, and with anhydrous ammonia in methylene chloride, provides a route to the dipotassium salt of 1,1,2,2-tetranitroethane (74) (90 %) and the ammonium salt of dinitroacetonitrile (186) (74 %) respectively.[331]

Further work by Baum and co-workers[332] showed that the nitration of 1,1-diamino-2,2-dinitroethylenes with trifluoroacetic anhydride and nitric acid in methylene chloride yields 1,1,1-trinitromethyl derivatives via addition of nitronium ion to the double bond of the enamine; such treatment also resulting in the *N*-nitration of the products. In this way, trinitromethyl derivatives like (185) and (188) are obtained. Further treatment of these trinitromethyl derivatives with aqueous potassium iodide results in reductive denitration and the formation

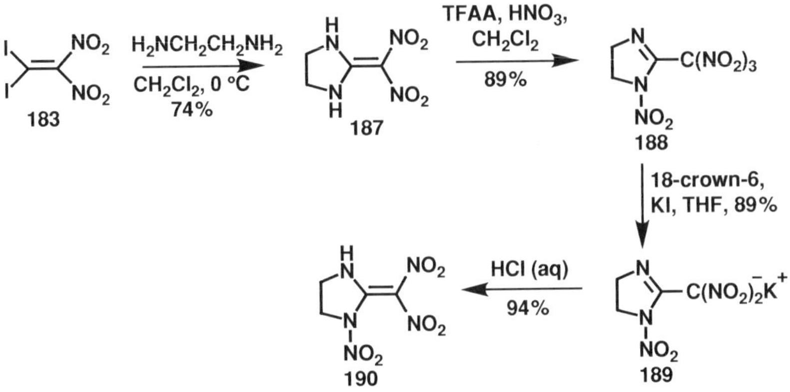

Figure 1.91

of nitronitronate salts (189), which on acidification, yield the *N*-nitro derivatives of the initial 1,1-diamino-2,2-dinitroethylenes (190).

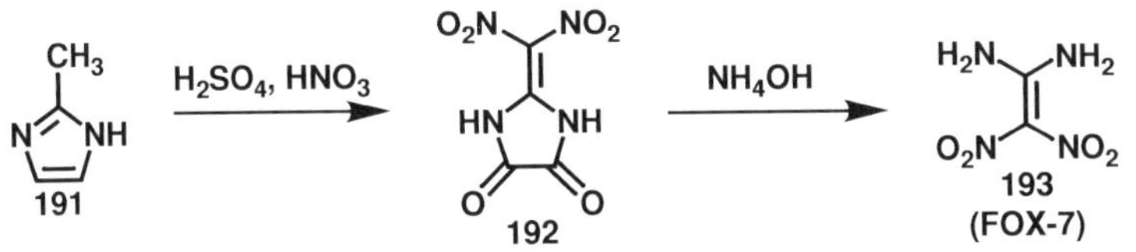

Figure 1.92

1,1-Diamino-2,2-dinitroethylenes have recently attracted interest as energetic explosives. 1,1-Diamino-2,2-dinitroethylene (193) (DADE or FOX-7) is a high energy material with a lower impact sensitivity than HMX and well suited for use in high explosive formulations.[333] FOX-7 (193) was first synthesized by Latypov and co-workers[334] who nitrated 2-methyl-4-nitroimidazole with mixed acid to give a mixture of parabanic acid and 2-(dinitromethylene)-4,5-imidazolidinedione (192), followed by reaction of the latter with aqueous ammonia to yield FOX-7 (193). 2-(Dinitromethylene)-4,5-imidazolidinedione (192) has also been prepared from the low temperature nitration of 2-methylimidazole (191).[335] The condensation of acetamidine hydrochloride with diethyl oxalate in methanol yields a mixture of 2-methylene-4,5-imidazolidinedione and 2-methoxy-2-methyl-4,5-imidazolidinedione; nitration of the latter and subsequent treatment with aqueous ammonia at pH 8–9 also provides FOX-7 in 50 % overall yield.[334] FOX-7 has been synthesized on a pilot plant scale in Sweden[336] by the nitration of 2-methyl-4,6-dihydroxypyrimidine.[337] The conversion of 2-(dinitromethylene)-4,5-imidazolidinedione to FOX-7 has been studied by Cai and co-workers[338] who achieved hydrolysis with water instead of ammonia and also found that hydrolysis with a carboxylic acid produces a larger crystal size which is more suitable for direct use in explosive formulations.

1.12.2 Other routes to aliphatic nitro compounds

A number of routes to aliphatic nitro compounds are not used for energetic materials synthesis but are included here for completeness.

Bachmann and Biermann[339] reported the synthesis of nitroalkanes from the thermolysis of acyl nitrates. The thermolysis of nitrite and nitrate esters over an asbestos catalyst is also reported to yield nitroalkanes.[340]

Olah and co-workers[341] reported the synthesis of nitroalkanes and nitroalkenes from the nitrodesilylation of alkylsilanes and allylsilanes, respectively, with nitronium salts.

Nitroacetylenes are generally unstable and very explosive and so they have been little studied. Schmitt and co-workers[342] used the nitrodesilylation of trialkylsilylacetylenes with both nitronium salts and nitryl fluoride to obtain nitroacetylenes. Dinitrogen pentoxide has also been used for the nitrodesilylation of trialkylsilylacetylenes.[343] Nitrodestannylation of allylsilanes has also been reported.[344]

Diels–Alder reactions using highly reactive polynitroalkenes have been reported. These include cycloaddition reactions with 1,1-dinitroethene,[106,345] 1,1,2,2-tetranitroethylene[346] and various fluoro-1,2-dinitroethylenes.[347]

1.12.3 Selective reductions

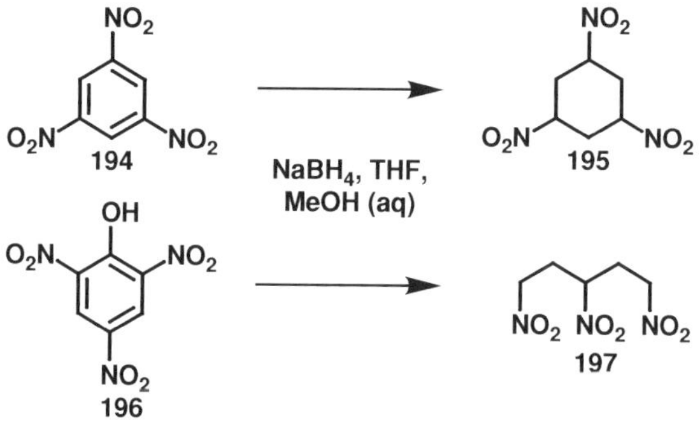

Figure 1.93

Reagents like lithium aluminium hydride and hydrogen over palladium readily reduce the aliphatic nitro group to the corresponding amino group. Sodium borohydride will reduce many functional groups but leaves both aromatic and aliphatic nitro groups intact. Sodium borohydride has been used for the selective reduction of polynitroaliphatic aldehydes,[348] ketones,[348] esters[349] and acid chlorides[310] to the corresponding polynitroaliphatic alcohols. Sodium borohydride has been used for the reduction of the aromatic rings of 1,3,5-trinitrobenzene (194) and picric acid (196) to yield 1,3,5-trinitrocyclohexane[350] (195) and 1,3,5-trinitropentane[351] (197) respectively.

1.13 CHEMICAL STABILITY OF POLYNITROALIPHATIC COMPOUNDS

The stability of polynitroaliphatic compounds to acids, bases and nucleophiles is often linked to the presence of an acidic α-proton(s) which may allow various resonance structures to lead to rearrangement or decomposition. Additionally, the presence of two or more nitro groups on the same carbon atom greatly increases the susceptibility of the carbon–nitrogen bonds to nucleophilic attack.

Table 1.11 Reactions of nitroalkanes and polynitroalkanes with mineral acid

Nitroalkane	Hydrolysis product
RCH_2NO_2	RCOOH
$R^1R^2CHNO_2$	No reaction
$R^1R^2C{=}NO_2H$	$R^1R^2C{=}O$
$R^1R^2C(NO_2)_2$	No reaction
$RCH(NO_2)_2$	RCOOH
$R^1R^2CHC(NO_2)_3$, C-H not activated	No reaction
$R^1R^2CHC(NO_2)_3$, C-H activated	$R^1R^2C{=}O$

Source: Reprinted with permission from J. C. Dacons, M. J. Kamlet and L. A. Kaplan, *J. Org. Chem.*, 1961, **26**, 4371; Copyright 1961 American Chemical Society.

1.13.1 Reactions with mineral acids

Polynitroalkanes are generally stable to mineral acids, although hydrolysis can occur with prolonged heating, depending on the arrangement of the nitro groups within the compound. On treatment with mineral acid primary aliphatic nitro groups are hydrolyzed to the corresponding carboxylic acid via an intermediate hydroxamic acid, whereas secondary nitro groups are unaffected.[352] The presence of an acidic proton in relation to compound stability is illustrated by the inertness of the internal *gem*-dinitroaliphatic group towards hot mineral acids, whereas a terminal *gem*-dinitroaliphatic group is converted to the corresponding carboxylic acid.[352] The same resistance to mineral acid hydrolysis is shown by tertiary nitroalkanes. Trinitromethyl groups are also stable to acid hydrolysis with the exception that an electron-withdrawing group or resonance-stabilizing group is not present on the β-carbon atom. In such cases an acidic proton β to the trinitromethyl group allows acid hydrolysis to lead to the formation of a carbonyl group with the degradation of one carbon atom from the compounds skeleton. A summary of the reactions of nitroalkanes and polynitroalkanes with mineral acid is given in Table 1.11.[352]

Acidification of the nitronate salts of polynitroalkanes can be complicated by the fact that some polynitroaliphatic compounds are unstable, as in the case of dinitromethane and 1,1,2,2-tetranitroethane, where both decompose readily at ambient temperature. The nitronate salts of both primary and secondary aliphatic nitro groups are decomposed to carbonyl compounds on acidification with mineral acid, a synthetic process known as the Nef reaction.[353] *gem*-Nitronitronate salts form the *gem*-dinitroaliphatic compound on acidification with mineral acid.

1.13.2 Reactions with base and nucleophiles

Both primary and secondary aliphatic nitro groups form nitronate salts on reaction with base. Terminal *gem*-dinitroaliphatic groups form the corresponding nitronitronate salts. Internal *gem*-dinitroaliphatic groups lack an acidic proton and cannot form nitronate salts. The nitro groups in compounds containing trinitromethyl groups are especially electron deficient and susceptible to attack by both bases and nucleophiles. The reaction of trinitromethyl compounds with base generates the corresponding *gem*-nitronitronate salt.

The effect of having multiple, powerful electron-withdrawing groups on the same carbon is seen in the extreme case of tetranitromethane with its readiness to lose a nitro group on

Table 1.12 Reactions of nitroalkanes and polynitroalkanes with alkali base

Nitroalkane	Product
$C(NO_2)_4$	$M^+\bar{C}(NO_2)_3$
$HC(NO_2)_3$	$M^+\bar{C}(NO_2)_3$
RCH_2NO_2	Nitronate salt
$R^1R^2CHNO_2$	Nitronate salt
$RCH(NO_2)_2$	Nitronate salt
$R^1R^2C(NO_2)_2$	No reaction
$R^1R^2CHC(NO_2)_3$	$R^1R^2CHC(NO_2)_2^- M^+$

treatment with both base and nucleophile. In fact, treatment of tetranitromethane with alkali hydroxides or aqueous ammonia provides a convenient route to nitroform salts.[354] This electron deficiency of the nitro groups in tetranitromethane finds use in the alkaline nitration of compounds with active methylene groups.[179] The point is further illustrated by the slow decomposition of tetranitromethane in the presence of water, giving nitroform and nitric acid as products. Trinitromethyl groups will also slowly decompose in the presence of hot water. A summary of the reactions of nitroalkanes and polynitroalkanes with alkali base is given in Table 1.12.

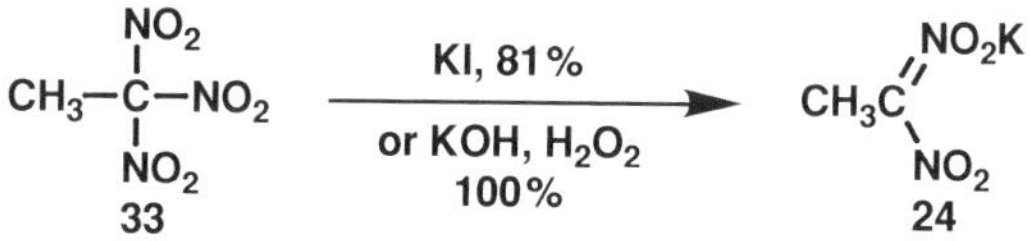

$CH_3C(NO_2)_3$ (33) → (KI, 81 % or KOH, H_2O_2 100 %) → $CH_3C(=NO_2K)NO_2$ (24)

Figure 1.94

As discussed above, the nitro groups of tetranitromethane and trinitromethyl compounds are susceptible to nucleophilic attack. Both potassium iodide[355] and alkaline hydrogen peroxide[356] affect the reductive denitration of trinitromethyl groups to *gem*-nitronitronates; 1,1,1-trinitroethane (33) is quantitatively reduced to potassium 1,1-dinitroethane (24) on treatment with alkaline hydrogen peroxide.[356] Nucleophiles such as potassium fluoride in DMF can displace nitrite anion from tetranitromethane.[357,358] Various nucleophiles, including azide,[359] chloride,[358] fluoride[359] and ethoxide[359] have been used to displace one of the nitro groups from fluorotrinitromethane.

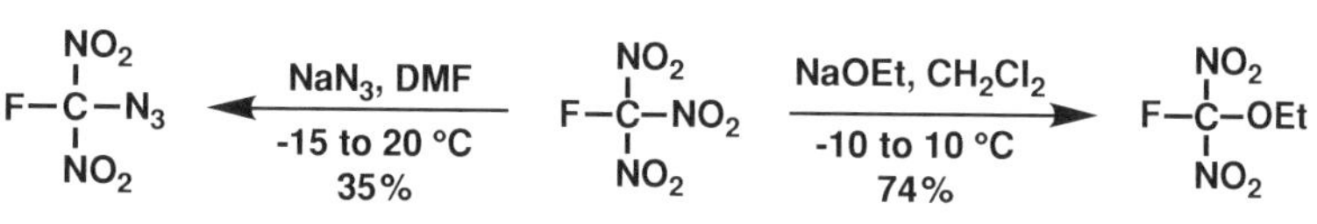

$FC(NO_2)_2N_3$ ← (NaN_3, DMF, -15 to 20 °C, 35%) ← $FC(NO_2)_3$ → (NaOEt, CH_2Cl_2, -10 to 10 °C, 74%) → $FC(NO_2)_2OEt$

Figure 1.95

The carbon–halogen bonds of 1-halo-1,1-dinitroaliphatic compounds are particularly electron deficient and susceptible to nucleophilic attack. This kind of reaction is synthetically useful in the chemistry of terminal *gem*-dinitroaliphatic compounds. Some *gem*-nitronitronate salts of polynitroalkanes are exceptionally sensitive to mechanical stimuli and explode readily. However, these salts are readily converted into the more predictable 1-halo-1,1-dinitro derivatives, which in turn, are reconverted to the original *gem*-nitronitronate salts on treatment with aqueous potassium iodide.

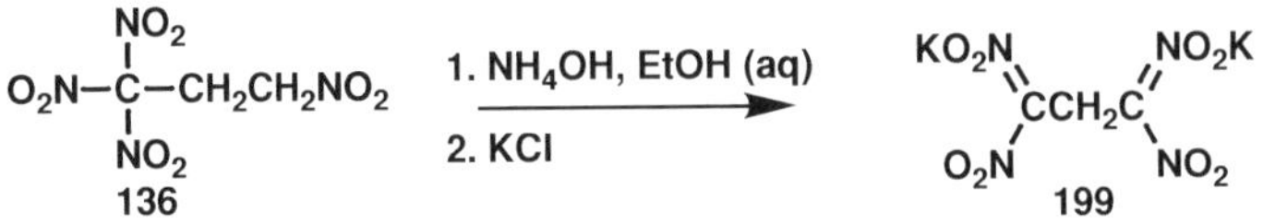

Figure 1.96

In Section 1.10.2.3 we observed that a base can react with 1,1,1-trinitromethyl compounds to either remove an acidic proton or act as a nucleophile to displace a nitro group. Trinitromethyl compounds can also undergo rearrangement reactions is the presence of a base or nucleophile. 1,1,1,3-Tetranitropropane (136) undergoes an internal redistribution of nitro groups on treatment with aqueous alkali or ammonium hydroxide to give 1,1,3,3-tetranitropropane, which is isolated as its sparingly soluble di-potassium salt (199) on adding an aqueous solution of potassium chloride to the reaction mixture.[253,360] This type of rearrangement occurs with other trinitromethyl derivatives of structure (198) and in this way, the potassium salt of 1,1,3,3-tetranitrobutane (201) is obtained from 1,1,1,3-tetranitrobutane (200).[361]

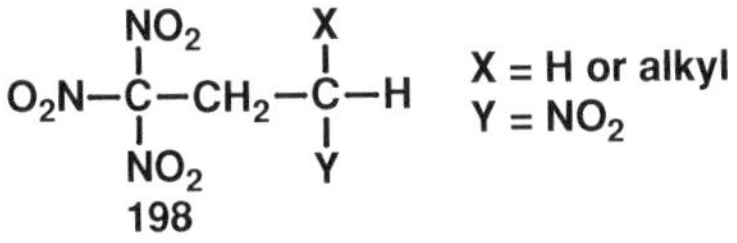

Figure 1.97

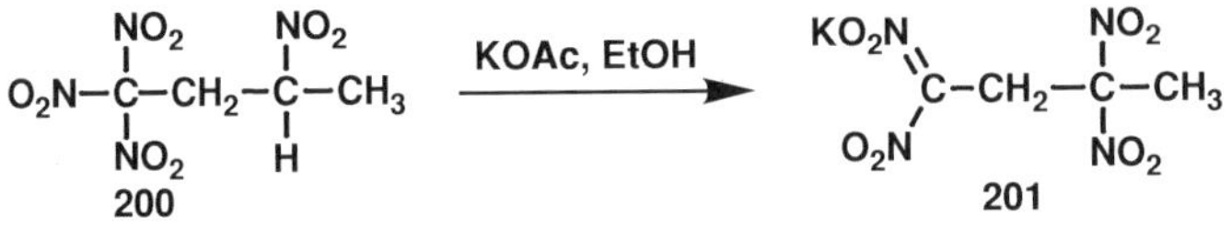

Figure 1.98

2,2,2-Trinitrochloroethane (202) and 2,2,2-trinitroethyl acetate (203) also undergo nitro group rearrangement in the presence of potassium nitrite to give the di-potassium salt of 1,1,2,2-tetranitroethane (74) in both cases.[362]

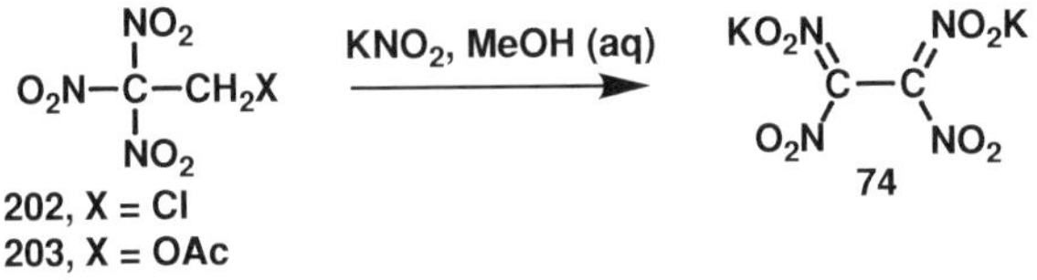

Figure 1.99

REFERENCES

1. (a) T. Urbański, *Chemistry and Technology of Explosives*, Vol. 1, Pergamon Press, Oxford (1964); (b) T. Urbański, *Chemistry and Technology of Explosives*, Vol. 4, Pergamon Press, Oxford (1984); (c) *Kirk-Othmer Encyclopedia of Chemical Technology*, 4th Edn, Vol. 10, Ed. M. Grayson, Wiley-Interscience, New York, 1–125 (1993).
2. N. Ono, *The Nitro Group in Organic Synthesis., Organic Nitro Chemistry Series.*, Wiley-VCH, Weinheim, Chapter 2, 3–29 (2001).
3. G. A. Olah, R. Malhotra and S. C. Narang, *Nitration: Methods and Mechanisms*, Wiley-VCH, Weinheim., Chapter 4, 219–311 (1989).
4. R. G. Coombes, in *Comprehensive Organic Chemistry: The Synthesis and Reactions of Organic Compounds*, Ed. I. O. Sutherland, Pergamon Press, Oxford, 325–356 (1979).
5. H. G. Padeken, O. von Schickh and A. Segnitz, in *Houben-Weyl, Methoden der Organischen Chemie, Band 10/1*, Ed. E. Muller, Georg Thieme Verlag, Stuttgart (1971).
6. H. O. Larson, in *The Chemistry of the Nitro and Nitroso Groups, Part 1, Organic Nitro Chemistry Series*, Ed. H. Feuer., Wiley-Interscience, Weinheim, Chapter 6, 301–348 (1969).
7. N. Kornblum, *Organic Reactions*, 1962, **12**, 101.
8. H. Feuer in *The Chemistry of Amino, Nitroso and Nitro Compounds and Their Derivatives*, Ed. S. Patai. Chemistry of Functional Groups Series, Supplement F, Chapter 19, John Wiley & Sons, Ltd, Chichester (1982).
9. F. G. Borgardt, P. Noble. Jr and W. L. Reed, *Chem. Rev.*, 1964, **64**, 19.
10. (a) H. B. Hass, E. B. Hodge and B. M. Vanderbilt, *Ind. Eng. Chem.*, 1936, **28**, 339; (b) H. B. Hass and J. A. Patterson, *Ind. Eng. Chem.*, 1938, **30**, 67; (c) H. B. Hass, H. J. Hibshmann and E. H. Pierson, *Ind. Eng. Chem.*, 1940, **32**, 427; (d) L. G. Alexander and H. B. Hass, *Ind. Eng. Chem.*, 1949, **41**, 2266; (e) H. B. Hass and H. Shechter, *Ind. Eng. Chem.*, 1947, **39**, 817; (f) H. B. Hass and H. Shechter, *J. Am. Chem. Soc.*, 1953, **75**, 1382.
11. (a) A. I. Titov, *Zh. Obshch. Khim.*, 1937, **7**, 1695; 1940, **10**, 1878; 1941, **11**, 1125; 1946, **16**, 1896; 1946, **16**, 1902; 1947, **17**, 385; 1949, **19**, 517; 1950, **20**, 521; 1952, **22**, 1329; 1954, **24**, 78; (b) A. I. Titov, *Usp. Khim.*, 1952, **21**, 881; 1958, **27**, 508; (c) M. Sloń and T. Urbański, *Roczniki. Chem.*, 1936, **16**, 466; 1937, **17**, 161; (d) M. Sloń and T. Urbański, *Compt. Rend (C).*, 1936, **203**, 620; 1937, **204**, 870; (e) H. Baldock, N. Levy and C. W. Scaife, *J. Chem. Soc.*, 1949, 2627; (f) G. Geisler, *Angew. Chem.*, 1955, **67**, 270.
12. C. Grundmann and H. Haldenwanger, *Angew. Chem.*, 1950, **62**, 556.
13. M. I. Konovalov, *Zh. Russk. Khim. Obshch.*, 1899, **31**, 255.
14. M. Kolinsky, S. Švastal and O. Wichterle, *Chem. Listy.*, 1954, **48**, 87.
15. A. I. Titov, *Tetrahedron*, 1963, **19**, 557.
16. N. Levy and I. D. Rose, *Chem. Soc. Quart. Rev.*, 1947, **1**, 358.
17. (a) F. Asinger, *Paraffins: Chemistry and Technology*, Pergamon Press, Oxford, 365–482 (1968); (b) A. P. Ballod and V. Ya. Shtern, *Russ. Chem. Rev.*, 1976, **45**, 721.
18. (a) N. G. Laptev and A. I. Titov, *Zh. Obshch. Khim.*, 1948, **18**, 741; 1949, **19**, 267; (b) A. I. Titov, *Zh. Obshch. Khim.*, 1948, **18**, 465, 473.
19. N. V. Schchitov and A. I. Titov, *Dokl. Akad. Nauk SSSR.*, 1951, **81**, 1085.
20. H. C. H. Lin and G. A. Olah, *J. Am. Chem. Soc.*, 1971, **93**, 1259.
21. A. V. Topchiev, *Nitration of Hydrocarbons and Other Organic Compounds*, Translated from Russian by C. Matthews, Pergamon Press, London (1959).
22. (a) F. Rahn and H. Wieland, *Chem. Ber.*, 1921, **54**, 1770; (b) A. Kekule, *Chem. Ber.*, 1869, **2**, 329; (c) N. L. Drake and E. P. Kohler, *J. Am. Chem. Soc.*, 1923, **45**, 1281; (d) R. Ansshütz and A. Gilbert, *Chem. Ber.*, 1921, **54**, 1854; 1924, **57**, 1697.
23. R. W. Long, *US Pat.* 2 551 027 (1951); *Chem. Abstr.*, 1951, **45**, 7293a.
24. M. B. Frankel and K. Klager, *J. Org. Chem.*, 1958, **23**, 494.

25. L. F. Fieser and M. Fieser, *Steroids*, Reinholds Publishing Corporation, New York., 43–44 (1959).
26. G. H. Carlson and A. Michael, *J. Am. Chem. Soc.*, 1935, **57**, 1268.
27. (a) A. Bowers, H. J. Ringold and M. B. Sánchez, *J. Am. Chem. Soc.*, 1959, **81**, 3702; (b) A. Bowers, L. C. Ibáñez and H. J. Ringold, *J. Am. Chem. Soc.*, 1959, **81**, 3707; (c) C. E. Anagnostopoulos and L. F. Fieser, *J. Am. Chem. Soc.*, 1954, **76**, 532; (d) J. R. Bull, E. R. H. Jones and G. D. Meakins, *J. Chem. Soc.*, 1965, 2601.
28. F. G. Bordwell and E. W. Garbisch Jr, *J. Org. Chem.*, 1962, **27**, 2322; 1963, **28**, 1765.
29. F. G. Bordwell and E. W. Garbisch Jr, *J. Am. Chem. Soc.*, 1960, **82**, 3588.
30. A. A. Griswold and P. S. Starcher, *J. Org. Chem.*, 1966, **31**, 357.
31. G. B. Bachman and R. J. Maleski, *J. Org. Chem.*, 1972, **37**, 2810.
32. (a) R. H. Fischer and H. M. Weitz, *Synthesis*, 1980, **4**, 261; (b) G. B. Bachman and T. Hokama, *J. Org. Chem.*, 1960, **25**, 178.
33. (a) K. F. Hager, *Ind. Eng. Chem.*, 1949, **41**, 2168; (b) P. V. McKee and K. J. Orton, *J. Chem. Soc.*, 1920, 783; (c) P. V. McKee, *J. Chem. Soc.*, 1927, 962; (d) P. Liang, in *Organic Syntheses*, Coll. Vol. III. Ed. E. C. Horning, John Wiley & Sons Inc., New York., 803 (1955).
34. A. Wetterholm, *Tetrahedron*, 1963, **19**, Suppl. 1, 155.
35. G. Darzen and M. Levy, *Compt. Rend. (C)*, 1949, **229**, 1081.
36. (a) N. Levy and C. W. Scaife, *J. Chem. Soc.*, 1946, 1093, 1100; (b) N. Levy, C. W. Scaife and A. E. Wilder-Smith, *J. Chem. Soc.*, 1946, 1096; 1948, 52; (c) H. Baldock, N. Levy and C. W. Scaife, *J. Chem. Soc.*, 1949, 2627; (d) E. Gudriniece, O. Nieland and G. Vanags, *Zh. Obshch. Khim.*, 1954, **24**, 1863; *Chem. Abstr.*, 1955, **49**, 13128; (e) W. K. Seifert, *J. Org. Chem.*, 1963, **28**, 125; (f) A. L. Daulton and R. B. Kaplan, *J. Am. Chem. Soc.*, 1952, **74**, 3052; (g) F. Conrad and H. Shechter, *J. Am. Chem. Soc.*, 1953, **75**, 5610.
37. H. G. Adolph and W. M. Koppes, in *Nitro Compounds: Recent Advances in Synthesis and Chemistry, Organic Nitro Chemistry Series*, Eds. H. Feuer and A. T. Neilsen, Wiley-VCH, Weinheim, Chapter 4, 367–605 (1990).
38. (a) P. L. Barrick, D. D. Coffman, W. E. Hanford, M. S. Raasch and G. W. Rigby, *J. Org. Chem.*, 1949, **14**, 747; (b) R. N. Haszeldine, *J. Chem. Soc.*, 1953, 2075.
39. D. E. Bisgrove, L. B. Clapp and C. E. Grabiel, *J. Am. Chem. Soc.*, 1955, **77**, 1293.
40. W. K. Seifert, *J. Org. Chem.*, 1963, **28**, 125.
41. J. M. Larkin and K. L. Kreuz, *J. Org. Chem.*, 1971, **36**, 2574.
42. J. Melton and J. E. McMurry, *J. Org. Chem.*, 1975, **40**, 2138.
43. (a) A. V. Stepanov and V. V. Veselovsky, *Russ. Chem. Rev.*, 2003, **72**, 327; (b) J. P. Adams and D. S. Box, *J. Chem. Soc. Perkin Trans. 1.*, 1999, 749; (c) J. P. Adams and J. R. Paterson, *J. Chem. Soc. Perkin Trans. 1.*, 2000, 3695; (d) J. H. Boyer, in *The Chemistry of the Nitro and Nitroso Groups, Part 1, Organic Nitro Chemistry Series*, Ed. H. Feuer, Wiley-Interscience, Weinheim, 229 (1969); (e) H. O. Larson, in *The Chemistry of the Nitro and Nitroso Groups, Part 1, Organic Nitro Chemistry Series*, Ed. H. Feuer., Wiley-Interscience, Weinheim, Chapter 6, 301–348 (1969).
44. W. D. Emmons and J. P. Freeman, *J. Am. Chem. Soc.*, 1957, **79**, 1712.
45. (a) W. D. Emmons and T. E. Stevens, *J. Am. Chem. Soc.*, 1957, **79**, 6008; (b) Ya. N. Dem'yanov, *Ct. Rd. Acad. Sci. USSR.*, 1930A, 447; (c) H. Akimoto, H. Bandow and M. Okuda, *J. Phys. Chem.*, 1980, **84**, 3604; (d) H. Akimoto, H. Bandow, M. Hoshine, G. Inove, T. Ogata, M. Okuda, F. Sakamaki and T. Tezuka, *Kokuritsu Kogai Kerkyusho Kenkyo Hokoku*, 1979, **9**, 29; (e) J. H. Canfield and G. H. Rohrback, *US Pat.* 3 729 501 (1973).
46. J. W. Fischer, in *Nitro Compounds: Recent Advances in Synthesis and Chemistry, Organic Nitro Chemistry Series*, Eds. H. Feuer and A. T. Neilsen, Wiley-VCH, Weinheim 315–325 (1990).
47. M. E. Colclough and N. C. Paul, in *Nitration: Recent Laboratory and Industrial Developments, ACS Symposium Series 623*, Eds. L. F. Albright, R. V. C. Carr and R. J. Schmitt, American Chemical Society, Washington, DC, Chapter 10, 97–103 (1996).
48. S. S. Dubov, V. A. Ginsburg, S. P. Marakov, N. F. Prirezentseva and N. P. Rodionova, *J. Gen. Chem. USSR*, 1960, **30**, 2388.

49. J. M. Birchall, A. J. Bloom, R. N. Haszeldine and C. J. Willis, *J. Chem. Soc.*, 1962, 3021.
50. (a) H. Petri, *Z. Anorg. Allg. Chem.*, 1948, **257**, 180; (b) C. C. Price and C. A. Sears, *J. Am. Chem. Soc.*, 1953, **75**, 3275.
51. B. Baasner, H. Hagemann and E. Klauke, *Ger. Pat.* 3 305 202 A1 (1984).
52. L. S. German, I. L. Knunyants and I. N. Rozhkov, *Izv. Akad. Nauk SSSR, Ser. Khim.*, 1963, 1946; Engl. Transl., 1794.
53. R. A. Bekker, B. L. Dyatkin and I. L. Knunyants, *Dokl. Akad. Nauk SSSR.*, 1966, **168**, 1319; Engl. Transl., 622.
54. M. M. Guseinov, I. G. Murakulov, V. A. Smit and A. G. Talybov, *Izv. Akad. Nauk SSSR, Ser. Khim.*, 1982, 654; Engl. Transl., 581.
55. I. Kerekes, M. Nojima, G. A. Olah, J. A. Olah, Y. D. Vanker and J. T. Welch, *J. Org. Chem.*, 1979, **44**, 3872.
56. A. Hassner, J. E. Kropp and G. J. Kent, *J. Org. Chem.*, 1969, **34**, 2628.
57. G. Kumaravel, *Ph.D. Thesis*, Indian Institute of Technology, Kanpur (1988).
58. B. V. Gidaspov, E. L. Golod, Yu. V. Guk and M. A. Ilyushin, *Russ. Chem. Rev.*, 1983, **52**, 284.
59. (a) M. M. Krayushkin, S. S. Novikov, V. V. Sevost'yanov, I. Sh. Shvarts and V. N. Yarovenko, *Bull. Acad. Sci. USSR., Div. Chem. Sci.*, 1976, **25**, 1589; (b) P. Dampawan and W. W. Zajac Jr, *J. Org. Chem.*, 1982, **47**, 1176.
60. J. K. Kochi and R. Rathore, *J. Org. Chem.*, 1996, **61**, 627.
61. P. Dampawan and W. W. Zajac Jr, *Synthesis*, 1983, 545.
62. (a) V. Meyer and O. Stüber, *Chem. Ber.*, 1872, **5**, 203, 399, 514; (b) V. Meyer and A. Rilliet, *Chem. Ber.*, 1872, **5**, 1029; (c) C. Chojnacki and V. Meyer, *Chem. Ber.*, 1872, **5**, 1034; (d) V. Meyer, *Liebigs Ann. Chem.*, 1874, **171**, 1.
63. H. Adkins and R. B. Reynolds, *J. Am. Chem. Soc.*, 1929, **51**, 279.
64. S. A. Herbert Jr, N. Kornblum, R. A. Smiley, B. Taub, H. E. Ungnade and A. M. White, *J. Am. Chem. Soc.*, 1955, **77**, 5528.
65. N. Kornblum, B. Taub and H. E. Ungnade, *J. Am. Chem. Soc.*, 1954, **76**, 3209.
66. P. J. Hartman and W. E. Noland, *J. Am. Chem. Soc.*, 1954, **76**, 3227.
67. N. Kornblum, *Org. React.*, 1962, **12**, 101.
68. M. E. Chalmers, R. Daniels and N. Kornblum, *J. Am. Chem. Soc.*, 1955, **77**, 6654.
69. R. K. Blackwood, D. C. Iffland, N. Kornblum and R. A. Smiley, *J. Am. Chem. Soc.*, 1955, **77**, 6269.
70. N. L. Drake and C. W. Plummer, *J. Am. Chem. Soc.*, 1954, **76**, 2720.
71. N. Kornblum and H. E. Ungnade, in *Organic Syntheses*, Coll. Vol. IV, Ed. N. Rabjohn, John Wiley & Sons Inc, New York, 724 (1963).
72. D. C. Iffland, N. Kornblum, N. N. Lichtin and J. T. Patton, *J. Am. Chem. Soc.*, 1947, **69**, 307.
73. N. Kornblum, J. B. Nordmann and J. T. Patton, *J. Am. Chem. Soc.*, 1948, **70**, 746.
74. L. M. Andreeva, O. M. Lerner, V. V. Perekalin and I. F. Sokovishina, *Zh. Org. Khim.*, 1965, **1**, 636; *Chem. Abstr.*, 1965, **63**, 5573.
75. H. Feuer and G. Leston, *Org. Synth.*, 1954, **34**, 37.
76. Danziger and Von Braun, *Chem. Ber.*, 1913, **46**, 103.
77. Sobecki and Von Braun, *Chem. Ber.*, 1911, **44**, 2531.
78. A. A. Fainzilberg, S. S. Novikov and L. V. Okhlobystina, *Izv. Akad. Nauk SSSR, Otd. Khim. Nauk.*, 1962, 517; *Chem. Abstr.*, 1962, **57**, 14920h.
79. J. K. Stille and E. D. Vessel, *J. Org. Chem.*, 1960, **25**, 478.
80. R. Ballini, L. Barboni and G. Giarlo, *J. Org. Chem.*, 2004, **69**, 6907.
81. N. Kornblum, *Organic Reactions*, 1962, **12**, 101.
82. B. M. Graybill, G. S. Hammond, M. F. Hawthorne and J. H. Waters, *J. Am. Chem. Soc.*, 1960, **82**, 704.
83. P. Sykes, *A Guidebook to Mechanism in Organic Chemistry*, 6th Edn, Longman, Harlow, Essex, 77–100 (1986).

84. R. K. Blackwood, G. E. Graham, N. Kornblum, H. O. Larson, D. D. Mooberry and E. P. Oliveto, *Chem. Ind. (London)*, 1955, 443; *J. Am. Chem. Soc.*, 1956, **78**, 1497.
85. N. Kornblum and J. W. Powers, *J. Org. Chem.*, 1957, **22**, 455.
86. G. J. Anderson, W. J. Jones, N. Kornblum, H. O. Larson, O. Levand, J. W. Powers and W. M. Weaver, *J. Am. Chem. Soc.*, 1957, **79**, 6562.
87. W. M. Weaver, *Ph.D. Thesis*, Purdue University (1958).
88. R. K. Blackwood, N. Kornblum and D. D. Mooberry, *J. Am. Chem. Soc.*, 1956, **78**, 1501.
89. R. K. Blackwood, N. Kornblum and D. D. Mooberry, *J. Am. Chem. Soc.*, 1957, **79**, 2507.
90. R. K. Blackwood and N. Kornblum, in *Org. Synth.*, Coll. Vol. IV. Ed. N. Rabjohn, John Wiley & Sons, Inc., New York, 454 (1963).
91. R. Fusco and S. Rossi, *Chem. Ind. (London)*, 1957, 1650.
92. N. Kornblum and W. M. Weaver, *J. Am. Chem. Soc.*, 1958, **80**, 4333.
93. K. Fukui, H. Kitano, H. Takayama and S. Yoneda, *Kogyo. Kagaku. Zasshi.*, 1961, **64**, 1153; *Chem. Abstr.*, 1962, **57**, 3265i.
94. (a) H. Kolbe, *J. Prakt. Chem.*, 1872, **5**, 427; (b) H. Reinhenkel and W. Treibs, *Chem. Ber.*, 1954, **87**, 341.
95. M. E. Childs and W. P. Weber, *J. Org. Chem.*, 1976, **41**, 3486.
96. S. Colonna and G. Gelbard, *Synthesis*, 1977, 113.
97. H. N. Lee and R. W. Van House, *US Pat.* 3 426 084 (1969); *Chem. Abstr.*, 1969, **70**, 67594k.
98. A. J. Blake, E. C. Boyd, R. O. Gould and R. M. Paton, *J. Chem. Soc. Perkin Trans. 1*, 1994, 2841.
99. E. Ter Meer, *Liebigs Ann. Chem.*, 1876, **181**, 1.
100. M. F. Hawthorne, *J. Am. Chem. Soc.*, 1956, **78**, 4980.
101. D. R. Levering, *J. Org. Chem.*, 1962, **27**, 2930.
102. M. O. Foster, *J. Chem. Soc.*, 1899, 1141; 1900, 254.
103. G. X. Criner and D. C. Iffland, *J. Am. Chem. Soc.*, 1953, **75**, 4047.
104. G. B. Bachman, H. Feuer and J. P. Kispersky, *J. Am. Chem. Soc.*, 1951, **73**, 1360.
105. R. B. Kaplan and H. Shechter, *J. Am. Chem. Soc.*, 1961, **83**, 3535.
106. E. E. Hamel, M. H. Gold and K. Klager, *J. Org. Chem.*, 1957, **22**, 1665.
107. J. S. Dehn, E. E. Hamel, J. A. Love, J. J. Scigliano and A. H. Swift, *I and EC Product, Research and Development*, 1962, **1**, 108.
108. C. E. Colwell, H. Feuer, G. Leston and A. T. Nielsen, *J. Org. Chem.*, 1962, **27**, 3598.
109. W. D. Emmons, B. M. Graybill, G. S. Hammond, M. F. Hawthorne, C. O. Parker and J. H. Waters, *Tetrahedron*, 1963, **19**, Suppl. 1, 177.
110. J. R. Autera, T. C. Castorina, S. Helf and F. S. Holahan, *J. Am. Chem. Soc.*, 1962, **84**, 756.
111. P. O. Tawney, *US Pat.* 3 040 105 (1962); *Chem. Abstr.*, 1962, **57**, 13609f.
112. W. S. Reich, G. G. Rose and W. Wilson, *J. Chem. Soc.*, 1947, 1234.
113. V. I. Ereshko, A. A. Fainzilberg and S. A. Shevelev, *Izv. Akad. Nauk SSSR, Ser. Khim* (Engl. Transl.), 1976, 2535.
114. L. A. Kaplan, in *The Chemistry of the Nitro and Nitroso Groups, Part 2, Organic Nitro Chemistry Series*, Ed. H. Feuer, Wiley-Interscience, New York., Chapter 5, 289–328 (1970).
115. (a) Nitro Paraffins (Ed. H. Feuer), *Tetrahedron*, 1963, **19**, Suppl. 1; (b) Nitro Compounds (Ed. T. Urbański), *Tetrahedron*, 1964, **20**, Suppl. 1.
116. A. T. Nielsen, in *The Chemistry of the Nitro and Nitroso Groups, Part 1, Organic Nitro Chemistry Series*, Ed. H. Feuer., Wiley-Interscience, New York, Chapter 7, 349–486 (1969).
117. W. D. Emmons, K. S. McCallum, A. S. Pagano, C. O. Parker and H. A. Rolewicz, *Tetrahedron*, 1962, **17**, 89.
118. D. E. Bisgrove, J. F. Brown Jr and L. B. Clapp, in *Organic Syntheses*, Coll. Vol. IV, Ed. N. Rabjohn., John Wiley & Sons, Inc., New York, 372 (1963).
119. (a) C. D. Nenitzescu, *Chem. Ber.*, 1929, **62**, 2669; (b) E. M. Nygaard and T. T. Noland, *US Pat.* 2 396 282 (1946); *Chem. Abstr.*, 1946, **40**, 3126.
120. H. B. Hass and L. W. Seigle, *J. Org. Chem.*, 1940, **5**, 100.

121. (a) D. A. Isacescu and C. D. Nenitzescu, *Chem. Ber.*, 1930, **63**, 2484; (b) C. D. Nenitzescu, *Chem. Ber.*, 1929, **62**, 2669; (c) C. Dale and R. L. Shriner, *J. Am. Chem. Soc.*, 1936, **58**, 1502; (d) G. B. Brown and R. L. Shriner, *J. Org. Chem.*, 1937, **2**, 376.
122. R. Scholl, *Chem. Ber.*, 1888, **21**, 506.
123. G. S. Annapurna, G. V. Madhava Sharma, A. P. Marchand and P. R. Pednekar, *J. Org. Chem.*, 1987, **52**, 4784.
124. B. E. Arney Jr, P. R. Dave and A. P. Marchand, *J. Org. Chem.*, 1988, **53**, 443.
125. H. L. Ammon, C. S. Choi, P. R. Dave and M. Ferraro, *J. Org. Chem.*, 1990, **55**, 4459.
126. R. P. Kashyap, A. P. Marchand, R. Sharma, W. H. Watson and U. R. Zope, *J. Org. Chem.*, 1993, **58**, 759.
127. V. Grakauskas, *J. Org. Chem.*, 1973, **38**, 2999.
128. C. O. Parker, *Tetrahedron*, 1962, **17**, 109.
129. L. W. Kissinger and H. E. Ungnade, *J. Org. Chem.*, 1960, **25**, 1471.
130. L. W. Kissinger and H. E. Ungnade, *J. Org. Chem.*, 1959, **24**, 666.
131. H. Krauch and W. Kunz, in *Organic Named Reactions*, John Wiley and Sons, Inc., New York, 363 (1964).
132. (a) E. Bamberger and R. Seligman, *Chem. Ber.*, 1902, **35**, 3884; (b) L. I. Khmelnitskii, O. V. Lebedev and S. S. Novikov, *Izv. Akad. Nauk SSSR, Otd. Khim. Nauk.*, 1961, 477; *Chem. Abstr.*, 1961, **55**, 23389; (c) L. I. Khmelnitskii, O. V. Lebedev and S. S. Novikov, *Zh. Obshch. Khim.*, 1958, **28**, 2303; *Chem. Abstr.*, 1959, **53**, 3111; (d) L. I. Khmelnitskii, O. V. Lebedev and S. S. Novikov, *Izv. Akad. Nauk SSSR, Ser. Khim.*, 1960, 1783; *Chem. Abstr.*, 1961, **55**, 19833; (e) J. L. Riebsomer, *Chem. Rev.*, 1945, **36**, 157; (f) G. Lust and M. M. Frojmovic, *Can. J. Chem.*, 1968, **46**, 3719.
133. (a) L. Crombie and B. S. Roughley, *Tetrahedron*, 1986, **42**, 3147; (b) W. V. E. Doering and L. F. Fieser, *J. Am. Chem. Soc.*, 1946, **68**, 2252; (c) A. Massa and M. Milone, *Gazz. Chim. Ital.*, 1940, **70**, 196.
134. (a) L. I. Bagal, G. I. Kolesetskaya and I. V. Tselinskii, *Zh. Org. Khim.*, 1970, **6**, 334; (b) O. A. Luk'yanov and G. V. Pokhvisneva, *Izv. Acad. Nauk SSSR, Ser. Khim.*, 1991, 2148; (c) O. A. Luk'yanov and T. I. Zhiguleva, *Izv. Akad. Nauk SSSR, Ser. Khim.*, 1982, 1423.
135. O. A. Luk'yanov and T. S. Zhiguleva, *Izv. Akad. Nauk SSSR, Ser. Khim.*, 1982, 1423.
136. R. G. Coombes, P. J. Honey and R. W. Millar, in *Nitration: Recent Laboratory and Industrial Developments, ACS Symposium Series 623*, Eds. L. F. Albright, R. V. C. Carr and R. J. Schmitt, American Chemical Society, Washington, DC, Chapter 13, 134–150 (1996).
137. P. J. Honey, *M. Phil. Thesis*, Hatfield Polytechnic, UK (1991).
138. W. D. Emmons and A. S. Pagano, *J. Am. Chem. Soc.*, 1955, **77**, 4557.
139. M. F. Hawthorne, *J. Am. Chem. Soc.*, 1957, **79**, 2510.
140. A. T. Nielsen, *J. Org. Chem.*, 1962, **27**, 1993.
141. R. Gilardi, G. A. Olah, P. Ramaiah and G. K. Surya Prakash, *J. Org. Chem.*, 1993, **58**, 763.
142. R. Ballini, E. Marcantoni and M. Petrini, *Tetrahedron Lett.*, 1992, **33**, 4835.
143. (a) C. H. Hassall, *Org. React.*, 1957, **9**, 73; (b) G. R. Krow, *Org. React.*, 1993, **43**, 251.
144. P. Engel, K. Nakamura and L. A. Paquette, *Chem. Ber.*, 1986, **119**, 3782.
145. C. S. Lee, G. A. Olah, G. K. Surya Prakash and P. Ramaiah, *Synlett.*, 1992, 337.
146. B. Aebischer and A. Vasella, *Helv. Chim. Acta.*, 1983, **66**, 789.
147. F. P. Ballistreri, E. Barbuzzi, G. A. Tomaselli and R. M. Toscano, *Synlett.*, 1996, 1093.
148. D. C. Iffland and T.-F. Yen, *J. Am. Chem. Soc.*, 1954, **76**, 4083.
149. M. W. Barnes and J. M. Patterson, *J. Org. Chem.*, 1976, **41**, 733.
150. T. G. Archibald, K. Baum, M. C. Cohen and L. C. Garver, *J. Org. Chem.*, 1989, **54**, 2869.
151. S. Chander Suri and A. P. Marchand, *J. Org. Chem.*, 1984, **49**, 2041.
152. T. G. Archibald and K. Baum, *J. Org. Chem.*, 1988, **53**, 4645.
153. E. G. Corey and H. Estreicher, *Tetrahedron Lett.*, 1980, **21**, 1117.
154. H. H. Raman and S. Ranganathan, *Tetrahedron*, 1974, **30**, 63.
155. T. R. Walters, J. M. Woods and W. W. Zajac Jr, *J. Org. Chem*, 1991, **56**, 316.

156. L. A. Paquette, C. -C. Shen and L. M. Waykole, *J. Org. Chem.*, 1988, **53**, 4969.
157. P. Ceccherell, M. Curini, F. Epifano, M. C. Marcotullio and O. Rosati, *Tetrahedron Lett.*, 1998, **39**, 4385.
158. (a) R. J. Clutter, N. Kornblum and W. J. Jones, *J. Am. Chem. Soc.*, 1956, **78**, 4003; (b) E. Smulders and H. Stetter, *Chem. Ber.*, 1971, **104**, 917.
159. E. E. Gilbert and G. P. Sollott, *J. Org. Chem.*, 1980, **45**, 5405.
160. P. S. Bailey and J. E. Keller, *J. Org. Chem.*, 1968, **33**, 2680.
161. E. Keinan and Y. Mazur, *J. Org. Chem.*, 1977, **42**, 844.
162. P. J. Carroll, P. A. Kondracki and P. A. Wade, *J. Am. Chem. Soc.*, 1991, **113**, 8807.
163. W. D. Emmons, *J. Am. Chem. Soc.*, 1957, **79**, 5528.
164. (a) P. Hofer, L. Milewich and C. H. Robinson, *J. Org. Chem.*, 1966, **31**, 524; (b) W. T. Borden and K. E. Gilbert, *J. Org. Chem.*, 1979, **44**, 659.
165. J. Alster, P. E. Eaton, E. E. Gilbert, J. J. Pluth, G. D. Price, B. K. Ravi Shanker and O. Sandus, *J. Org. Chem.*, 1984, **49**, 185.
166. T. R. Walters, J. M. Woods and W. W. Zajac Jr, *J. Org. Chem.*, 1989, **54**, 2468.
167. (a) W. D. Emmons, *J. Am. Chem. Soc.*, 1954, **76**, 3470; (b) L. I. Khmelnitskii, S. S. Novikov and T. S. Novikova, *Izv. Akad. Nauk SSSR, Ser. Khim.*, 1962, 516.
168. R. Jeyaraman, L. Mohan and R. W. Murray, *Tetrahedron Lett.*, 1986, **27**, 2335.
169. R. Jeyaraman and R. W. Murray, *J. Org. Chem.*, 1985, **50**, 2847.
170. L. Mohan, R. W. Murray and S. N. Rajadhyaksha, *J. Org. Chem.*, 1989, **54**, 5783.
171. K. R. Beck Jr, A. E. Moormann and D. L. Zabrowksi, *Tetrahedron Lett.*, 1988, **29**, 4501.
172. K. Krohn and J. Kupke, *Eur. J. Org. Chem.*, 1998, 679.
173. L. I. Khmelnitskii, O. V. Lebedev and S. S. Novikov, *Zh. Obshch. Khim.*, 1958, **28**, 2303; *Chem. Abstr.*, 1959, **53**, 3111.
174. L. I. Khmelnitskii, O. V. Lebedev and S. S. Novikov, *Izv. Akad. Nauk SSSR, Ser. Khim.*, 1960, 1783, 2019; *Chem. Abstr.*, 1961, **55**, 19833.
175. Yu. P. Egorov, L. I. Khmelnitskii, O. V. Lebedev and S. S. Novikov, *Zh. Obshch. Khim.*, 1958, **28**, 2305; *Chem. Abstr.*, 1959, **53**, 4112.
176. (a) A. Hantzsch, *Chem. Ber.*, 1899, **32**, 637; (b) K. Auwers, *Chem. Ber.*, 1929, **62**, 2296.
177. (a) V. V. Smirnov and A. I. Titov, *Dokl. Akad. Nauk SSSR.*, 1952, **83**, 243; *Chem. Abstr.*, 1953, **47**, 4298; (b) L. I. Khmelnitskii, O. V. Lebedev and S. S. Novikov, *Izv. Akad. Nauk SSSR, Otd. Khim. Nauk.*, 1961, 477; *Chem. Abstr.*, 1961, **55**, 23389.
178. (a) F. G. Borgardt, J. A. Gallaghan, C. J. Hoffman, P. Noble Jr and W. L. Reed, *A.I.A.A.J.*, 1963, **1**, 395; (b) F. G. Borgardt, P. Noble Jr and A. K. Seeler, *J. Org. Chem.*, 1966, **31**, 2806.
179. C. W. Plummer, *US Pat.* 2 991 315 (1956); *Chem. Abstr.*, 1962, **56**, 2330e.
180. D. W. Fish, E. E. Hamel and R. E. Olsen, in *Advanced Propellant Chemistry, Advances in Chemistry Series No. 54*, Ed. R. F. Gould, American Chemical Society, Washington, DC, Chapter 6, 48–54 (1966).
181. L. T. Eremenko, B. S. Fedorov and R. G. Gafurov, *Izv. Akad. Nauk SSSR, Ser. Khim* (Engl. Transl.), 1971, **20**, 1501; 1974, **23**, 879.
182. (a) V. Meyer, *Chem. Ber.*, 1872, **5**, 203; 1873, **6**, 1492; 1874, **7**, 425; (b) V. Meyer, *Liebigs Ann. Chem.*, 1875, **175**, 88; (c) J. Locher and V. Meyer, *Chem. Ber.*, 1874, **7**, 670, 1510.
183. (a) L. Semper and H. Weiland, *Chem. Ber.*, 1906, **39**, 2522; (b) R. Libers and W. E. Noland, *Tetrahedron*, 1963, **19**, Suppl. 1, 23; (c) M. O. Foster, *J. Chem. Soc.*, 1900, 251.
184. (a) E. M. Nygaard, *US Pat.* 2 401 267 (1946); *Chem. Abstr.*, 1946, **40**, 6092; (b) T. T. Noland and E. M. Nygaard, *US Pat.* 2 401 269 (1946); *Chem. Abstr.*, 1946, **40**, 6093; (c) J. H. McCracken, T. T. Noland and E. M. Nygaard, *US Pat.* 2 370 185 (1945); *Chem. Abstr.*, 1945, **39**, 3551.
185. O. A. Luy'yanov, S. S. Nivikov and V. A. Tartakovskii, *USSR Pat.* 320 479 (1971).
186. E. Sakellarious and H. Wieland, *Chem. Ber.*, 1919, **52**, 904.
187. R. B. Bishop and W. I. Denton, *Ind. Eng. Chem.*, 1948, **40**, 381.
188. J. Zublin, *Chem. Ber.*, 1877, **10**, 2083.

189. P. E. Eaton and G. E. Wicks, T. T. Noland and E. M. Nygaard, *J. Org. Chem.*, 1988, **53**, 5353.
190. P. E. Eaton, R. Gilardi, J. Hain, N. Kanomata, J. Li, K. A. Lukin and E. Punzalan, *J. Am. Chem. Soc.*, 1997, **119**, 9591.
191. J. H. Boyer, in *The Chemistry of the Nitro and Nitroso Groups, Part 1, Organic Nitro Chemistry Series*, Ed. H. Feuer, Wiley-Interscience, New York, Chapter 5, 215–299 (1969).
192. J. Jander and R. N. Haszeldine, *Naturwissenschaften.*, 1953, **40**, 579.
193. (a) G. B. Fazekas and G. A. Takacs, *J. Photochem.*, 1983, **21**, 9; (b) J. Banus, *J. Chem. Soc.*, 1953, 3755.
194. J. M. Birchall, A. J. Bloom, R. N. Haszeldine and C. J. Willis, *J. Chem. Soc.*, 1962, 3021.
195. R. A. Bekker, B. L. Dyatkin and I. L. Knunyants, *Dokl. Akad. Nauk USSR*, 1966, **168**, 1319; Engl. Transl., 622.
196. G. H. Crawford, J. R. Lacher, J. D. Park and A. P. Stefani, *J. Org. Chem.*, 1961, **26**, 3316.
197. V. A. Ginsburg, L. L. Martynova and M. N. Vasil'eva, *Zh. Org. Khim.*, 1971, **7**, 2074; Engl. Transl., 2154.
198. L. Poizat and A. Seyerwetz, *Compt. Rend (C).*, 1909, **148**, 1110.
199. (a) J. O. Edwards, K. M. Ibne-Rasa and C. G. Lauro, *J. Am. Chem. Soc.*, 1963, **85**, 1165; (b) J. H. Boyer and S. E. Ellzey. Jr, *J. Org. Chem.*, 1959, **24**, 2038.
200. D. R. Levering and C. M. Wright, *Tetrahedron*, 1963, **19**, Suppl. 1, 3.
201. Z. Matacz, H. Piotrowska and T. Urbański, *Pol. J. Chem.*, 1979, **53**, 187.
202. W. J. Kelly, N. Kornblum and H. K. Singh, *J. Org. Chem.*, 1983, **48**, 332.
203. K. Baum, L. C. Garver and V. Grakauskas, *J. Org. Chem.*, 1985, **50**, 1699.
204. (a) T. G. Archibald, K. Baum, C. George and R. Gilardi, *J. Org. Chem.*, 1990, **55**, 2920; (b) T. G. Archibald, S. G. Bott, A. P. Marchand and D. Rajagopal, *J. Org. Chem.*, 1995, **60**, 4943; (c) K. Hayashi, T. Kumagai and Y. Nagao, *Heterocycles.*, 2000, **53**, 447; (d) T. G. Archibald, M. D. Coburn and M. A. Hiskey, *Waste Management.*, 1997, **17**, 143.
205. A. Endres and W. Wislicenus, *Chem. Ber.*, 1902, **35**, 1755.
206. F. H. Babers and A. P. Black, in *Organic Syntheses, Coll. Vol. II*, Ed. A. H. Blatt, John Wiley & Sons, Inc., New York, 512 (1943).
207. J. J. Chavin, P. Garbsch and H. Wieland, *Liebigs Ann. Chem.*, 1928, **461**, 295.
208. H. Feuer, C. Savides and J. W. Shepherd, *J. Am. Chem. Soc.*, 1956, **78**, 4364.
209. H. Feuer, in *Industrial and Laboratory Nitrations, ACS Symposium Series 22*, Ed. L. F. Albright and C. Hanson, American Chemical Society, Washington, DC, 160–175 (1976).
210. H. Feuer, C. Savides and J. W. Shepherd, *J. Am. Chem. Soc.*, 1956, **78**, 4364.
211. K. Klager, *J. Org. Chem.*, 1955, **20**, 646.
212. R. S. Anderson and H. Feuer, *J. Am. Chem. Soc.*, 1961, **83**, 2960.
213. R. S. Anderson, H. Feuer and A. M. Hall, *J. Org. Chem.*, 1971, **36**, 140.
214. H. Feuer and P. M. Pivawer, *J. Org. Chem.*, 1966, **31**, 3152.
215. H. Feuer, S. Golden, A. M. Hall and R. L. Reitz, *J. Org. Chem.*, 1968, **33**, 3622.
216. H. Feuer and C. Savides, *J. Am. Chem. Soc.*, 1959, **81**, 5826.
217. (a) H. Feuer and J. P. Lawrence, *J. Org. Chem.*, 1972, **37**, 3662; (b) H. Feuer and H. Friedman, *J. Org. Chem.*, 1975, **40**, 187.
218. M. Auerbach and H. Feuer, *J. Org. Chem.*, 1970, **35**, 2551.
219. H. Feuer and E. F. Vincent. Jr, *J. Org. Chem.*, 1964, **29**, 939.
220. H. Feuer and R. P. Monter, *J. Org. Chem.*, 1969, **34**, 991.
221. (a) P. E. Pfeffer and L. S. Silbert, *Tetrahedron Lett.*, 1970, **11**, 699; (b) V. M. Baghdanov and F. M. Hauser, *J. Org. Chem.*, 1988, **53**, 2872.
222. W. D. Emmons and J. P. Freeman, *J. Am. Chem. Soc.*, 1955, **77**, 4387.
223. W. D. Emmons and J. P. Freeman, *J. Am. Chem. Soc.*, 1955, **77**, 4391.
224. W. D. Emmons and J. P. Freeman, *J. Am. Chem. Soc.*, 1955, **77**, 4673.
225. I. Angres, L. A. Kaplan and M. E. Sitzman, *J. Org. Chem.*, 1977, **42**, 563.
226. (a) P. E. Eaton, R. Gilardi, J. Hain, N. Kanomata, J. Li, K. A. Lukin and E. Punzalan, *J. Am. Chem. Soc.*, 1997, **119**, 9591; (b) P. E. Eaton, R. Gilardi and M. -X. Zhang, *Angew. Chem. Int. Ed.*, 2000,

39, 401; (c) P. E. Eaton, N. Gelber, R. Gilardi, S. Iyer, R. Surapaneni and M. -X. Zhang, *Propell. Explos. Pyrotech.*, 2002, **27**, 1.
227. R. D. Gilardi and J. Karle, in *Chemistry of Energetic Materials*, Eds. D. R. Squire and G. A. Olah, Academic Press, San Diego, CA, 2 (1991).
228. R. R. Fraser, T. S. Mansour and S. Savard, *J. Org. Chem.*, 1985, **50**, 3232.
229. L. W. Kissinger and H. E. Ungnade, *J. Org. Chem.*, 1958, **23**, 1340.
230. V. M. Belikov, L. V. Ershova, V. N. Goditidze and S. S. Novikov, *Izv. Akad. Nauk USSR, Otd. Khim. Nauk.*, 1959, 943; *Chem. Abstr.*, 1960, **54**, 259g.
231. L. Zalvkaev and E. Vanag, *J. Gen. Chem. USSR*, 1956, **26**, 657.
232. J. Bredt, *Chem. Ber.*, 1881, **14**, 1780; 1882, **15**, 2318.
233. V. Grakauskas and A. M. Guest, *J. Org. Chem.*, 1978, **43**, 3485.
234. R. B. Kaplan and H. Shechter, *J. Am. Chem. Soc.*, 1953, **75**, 3980.
235. M. J. Kamlet, *NAVORD Rep. 6206*, US Naval Ordnance Lab, Whiteoak, Maryland (1959).
236. H. G. Adolph and M. J. Kamlet, *Proc.* 7th *International Symposium on Detonation*, US Naval Academy, Annapolis, Maryland, Vol. 1, 60 (1981).
237. K. Klager and R. R. Rindone, 'Development of an efficient Process to Manufacture Bis(2-fluoro-2,2-dinitroethyl)formal (FEFO)', 18th *International Annual Conference of ICT*, Karlsruhe, Germany, 28/1–28/14 (1987).
238. V. Grakauskas and K. Baum, *J. Org. Chem.*, 1968, **33**, 3080.
239. (a) A. L. Fridman, F. A. Gabitov, E. A. Ponomareva, Yu. N. Senichev, *Zh. Obshch. Khim.*, 1968, **38**, 1902; Engl. Tranl., 1850; (b) M. J. Kamlet and H. G. Adolph, *J. Org. Chem.*, 1968, **33**, 3073; (c) H. G. Adolph, *US Pat.* 3 446 857 (1969).
240. M. A. Besprozvannyi, A. A. Fainzilberg, G. Kh. Khismamutdinov, M. Sh. Lvova, V. O. Slovetskii and O. G. Usyshkin, *Izv. Akad. Nauk USSR, Ser. Khim.* (Engl. Transl.), 1971, **20**, 2397.
241. R. J. Spear and W. S. Wilson, 'Recent Approaches to the Synthesis of High Explosive and Energetic Materials', *J. Energ. Mater.*, 1984, **2**, 61–149.
242. I. J. Dagley and R. J. Spear, 'Synthesis of Organic Energetic Compounds', in *Organic Energetic Compounds*, Ed. P. L. Marinkas., Nova Science Publishers Inc., New York, Chapter 2, 47–163 (1996).
243. H. L. Cates Jr and H. Shechter, *J. Org. Chem.*, 1961, **26**, 51.
244. A. L. Krieger, C. S. Rondestvedt and M. Stiles, *Tetrahedron*, 1963, **19**, Suppl. 1, 197.
245. P. O. Tawney, *US Pat.* 3 050 565 (1962); *Chem. Abstr.*, 1963, **58**, 1350a.
246. R. E. Cochoy, R. R. McGuire and S. A. Shackelford, *J. Org. Chem.*, 1992, **57**, 2950.
247. K. Baum and V. Grakauskas, 'Research in Fluoro-Nitro Compounds', *Report 2099, Nov 7, 1961, Contract Nonr-2655(00)*, Available from Defence Technical Information Center, Cameron Station, Alexandria, VA 22304-6145.
248. R. E. Cochoy, R. R. McGuire and S. A. Shackelford, *J. Org. Chem.*, 1990, **55**, 1401.
249. H. Feuer and R. Harmetz, *J. Org. Chem.*, 1961, **26**, 1061.
250. H. Feuer, G. Leston, R. Miller and A. T. Nielsen, *J. Org. Chem.*, 1963, **28**, 339.
251. R. D. Geckler, M. H. Gold and L. Herzog, *J. Am. Chem. Soc.*, 1951, **73**, 749.
252. H. Feuer and U. E. Lynch-Hart, *J. Org. Chem.*, 1961, **26**, 391.
253. J. C. Dacons, J. C. Hoffsommer and M. J. Kamlet, *J. Org. Chem.*, 1961, **26**, 4881.
254. I. S. Korsakova, S. S. Novikov and M. A. Yatskovskaya, *Dokl. Akad. Nauk USSR*, 1958, **118**, 954.
255. N. N. Bulatova, I. S. Korsakova and S. S. Novikov, *Zh. Obshch. Khim.*, 1959, **29**, 3659.
256. W. H. Gilligan and M. Graff, *J. Org. Chem.*, 1968, **33**, 1247.
257. H. Feuer, S. M. Pier and E. H. White, *J. Org. Chem.*, 1961, **26**, 1639.
258. M. B. Frankel, *Tetrahedron*, 1963, **19**, Suppl. 1, 213.
259. E. Hamel, J. P. Kispersky and K. Klager, *J. Org. Chem.*, 1961, **26**, 4368.
260. H. Shechter and L. Zeldin, *J. Am. Chem. Soc.*, 1951, **73**, 1276.
261. K. Baum and V. Grakauskas, *J. Org. Chem.*, 1969, **34**, 3927.
262. K. Baum and A. M. Guest, *Synthesis*, 1979, 311.
263. H. G. Adolph, *US Pat.* 453 675 (1983); *Chem. Abstr.*, 1983, **99**, 194427g.

264. H. G. Adolph, W. M. Koppes and M. E. Sitzmann, *J. Chem. Eng. Data.*, 1986, **31**, 119.
265. D. L. Ross, C. L. Coon, M. E. Hill and R. L. Simon, *J. Chem. Eng. Data.*, 1968, **13**, 437.
266. D. L. Ross, C. L. Coon and M. E. Hill, *US Pat.* 3 759 998 (1973).
267. (a) K. K. Babievskii, I. S. Korsakova and S. S. Novikov, *Izv. Akad. Nauk USSR, Otd. Khim. Nauk.*, 1959, 1480; 1959, 1847; (b) I. S. Ivanova, Y. V. Konnova and S. S. Novikov, *Izv. Akad. Nauk USSR, Otd. Khim. Nauk.*, 1962, 2078.
268. H. Feuer and R. Miller, *J. Org. Chem.*, 1961, **26**, 1348.
269. (a) E. D. Bergmann, D. Ginsburg and R. Pappo, *Organic Reactions*, 1959, **10**, 179; (b) C. T. Bahner and H. T. Kite, *J. Am. Chem. Soc.*, 1949, **71**, 3597; (c) K. Klager, *J. Org. Chem.*, 1955, **20**, 650; (d) J. Hine and L. A. Kaplan, *J. Am. Chem. Soc.*, 1960, **82**, 2915.
270. G. Rutz and E. Schmidt, *Chem. Ber.*, 1928, **61**, 2142.
271. G. D. Buckley and C. W. Scaife, *J. Chem. Soc.*, 1947, 1471.
272. H. B. Fraser and G. A. R. Kon, *J. Chem. Soc.*, 1934, 604.
273. J. F. Bourland and H. B. Hass, *US Pat.* 2 343 256 (1944).
274. (a) W. E. Hamlin and H. R. Synder, *J. Am. Chem. Soc.*, 1950, **72**, 5082; (b) M. T. Atwood and G. B. Bachman, *J. Am. Chem. Soc.*, 1956, **78**, 484.
275. K. Klager and R. M. Smith, *Propell. Explos. Pyrotech.*, 1983, **8**, 25.
276. L. O. Atovmyan, L. T. Eremenko, M. A. Fadeev, N. I. Golovina, G. V. Oreshko and N. G. Zhitomirskaya, *Izv. Akad. Nauk USSR, Ser. Khim.*, 1984, 549; *Chem. Abstr.*, 1984, **101**, 191084s.
277. L. T. Eremenko and G. V. Oreshko, *Izv. Akad. Nauk USSR, Ser. Khim.*, 1989, 1107; *Chem. Abstr.*, 1990, **112**, 35245s.
278. H. Feuer and S. Markofsky, *J. Org. Chem.*, 1964, **29**, 929.
279. C. D. Bedford and A. T. Nielsen, *J. Org. Chem.*, 1978, **43**, 2460.
280. H. Shechter and L. Zeldin, *J. Am. Chem. Soc.*, 1957, **79**, 4708.
281. M. B. Frankel, *J. Org. Chem.*, 1958, **23**, 813.
282. A. A. Fainzilberg, V. I. Gulevskaya, S. S. Novikov and S. N. Shvedova, *Izv. Akad. Nauk USSR, Otd. Khim. Nauk.*, 1960, 2056.
283. G. B. Bachman, H. Feuer and W. May, *J. Am. Chem. Soc.*, 1954, **76**, 5124.
284. M. B. Frankel and K. Klager, *J. Chem. Eng. Data.*, 1962, **7**, 412.
285. H. Feuer and U. E. Lynch-Hart, *J. Org. Chem.*, 1961, **26**, 587.
286. H. Feuer and U. E. Lynch-Hart, *J. Org. Chem.*, 1961, **26**, 391.
287. H. Feuer and W. A. Swarts, *J. Org. Chem.*, 1962, **27**, 1455.
288. E. E. Hamel, *Tetrahedron*, 1963, **19**, Suppl. 1, 85.
289. M. B. Frankel and K. Klager, *J. Am. Chem. Soc.*, 1957, **79**, 2953.
290. K. Klager, *J. Org. Chem.*, 1958, **23**, 1519.
291. (a) R. Schenck and A. Wetterholm, *US Pat.* 2 731 460 (1956); *Chem. Abstr.*, 1956, **50**, 7125g; (b) A. Wetterholm, *Tetrahedron*, 1963, **19**, Suppl. 1, 331.
292. K. Shino, *Tokyo Kogyo Shikensho Hokoku*, 1970, **65**, 46; *Chem. Abstr.*, 1971, **74**, 140812u.
293. X. Chen, 'Synthesis and Properties of Some *N,N'*-Bis(2,2-dinitropropyl)alkylene dinitramines', in *Proc. International Symposium on Pyrotechnics and Explosives*, China Academic Publishers, Beijing, China, 177 (1987).
294. W. H. Gilligan, *J. Org. Chem.*, 1971, **36**, 2138.
295. L. T. Eremenko, D. A. Nesterenko and N. S. Satsibullina, *Izv. Akad. Nauk USSR, Ser. Khim* (Engl. Transl.), 1970, 1261.
296. W. J. Murray and C. W. Sauer, *US Pat.* 3 006 957 (1957); *Chem. Abstr.*, 1962, **56**, 2330c.
297. M. B. Frankel and E. F. Witucki, *US Pat.* 4 701 557 (1987); *Chem. Abstr.*, 1988, **108**, 97345a.
298. N. S. Marans and R. P. Zelinski, *J. Am. Chem. Soc.*, 1950, **72**, 5329.
299. H. Feuer and T. Kucera, *J. Org. Chem.*, 1960, **25**, 2069.
300. As reported by H. G. Adolph and W. M. Koppes, in *Nitro Compounds: Recent Advances in Synthesis and Chemistry., Organic Nitro Chemistry Series*, Ed. H. Feuer and A. T. Neilsen, Wiley-VCH, Weinheim, 440 (1990).

301. L. T. Eremenko and F. Ya. Natsibullin, *Izv. Akad. Nauk USSR, Ser. Khim.*, 1969, 1331; Engl. Transl., 1227.
302. L. T. Eremenko and G. V. Oreshko, *Izv. Akad. Nauk USSR, Ser. Khim.*, 1969, 1765; Engl. Transl., 1634.
303. H. G. Adolph, *J. Org. Chem.*, 1970, **35**, 3188.
304. H. G. Adolph and M. J. Kamlet, *US Pat.* 3 553 273 (1971).
305. H. G. Adolph, W. M. Koppes and M. E. Sitzmann, *J. Chem. Eng. Data.*, 1986, **31**, 119.
306. L. W. Kissinger and H. E. Ungnade, *Tetrahedron*, 1963, **19**, Suppl. 1, 121.
307. C. E. Colwell, H. Feuer and A. T. Nielsen, *Tetrahedron*, 1963, **19**, Suppl. 1, 57.
308. (a) L. Henry, *Compt. Rend (C)*, 1895, **121**, 210; (b) J. A. Wyler, *US Pat.* 2 231 403 (1941); *Chem. Abstr.*, 1941, **35**, 3265.
309. H. B. Hass and B. M. Vanderbilt, *Ind. Eng. Chem.*, 1940, **32**, 34.
310. M. H. Gold and K. Klager, *Tetrahedron*, 1963, **19**, Suppl. 1, 77.
311. M. B. Frankel, *J. Org. Chem.*, 1962, **27**, 331.
312. H. G. Adolph and M. J. Kamlet, *J. Org. Chem.*, 1969, **34**, 45.
313. R. E. Cochoy and R. R. McGuire, *J. Org. Chem.*, 1972, **37**, 3041.
314. L. W. Kissinger, W. E. McQuistion and M. Schwartz, *J. Org. Chem.*, 1961, **26**, 5203.
315. V. Grakauskas, *J. Org. Chem.*, 1970, **35**, 3030.
316. T. M. Benziger, L. W. Kissinger, R. K. Rohwer and H. E. Ungnade, *J. Org. Chem.*, 1963, **28**, 2491.
317. A. V. Fokin, V. A. Komarov, L. D. Kuznetsova, A. I. Rapkin, Yu. N. Studnev, *Izv. Akad. Nauk USSR, Ser. Khim.*, 1976, 489; Engl. Transl., 472.
318. H. Feuer, H. B. Hass and R. D. Lowrey, *J. Org. Chem.*, 1960, **25**, 2070.
319. M. B. Frankel, N. N. Ogimachi, G. L. Rowley and E. F. Witucki, *J. Chem. Eng. Data*, 1971, **16**, 373.
320. J. C. Bottaro, P. E. Penwell and R. J. Schmitt, *'Synthesis of Cubane Based Energetic Materials, Final Report, December 1989'*, SRI International, Menlo Park, CA [AD-A217 147/8/XAB].
321. H. M. Peters and R. L. Simon Jr, *US Pat Appl.* 641 320 (1975); *Chem. Abstr.*, 1977, **87**, 67846u.
322. M. E. Hill and K. G. Shipp, *J. Org. Chem.*, 1966, **31**, 853.
323. L. W. Kissinger and H. E. Ungnade, *J. Org. Chem.*, 1966, **31**, 369.
324. T. M. Benziger, L. W. Kissinger and R. K. Rohwer, *Tetrahedron*, 1963, **19**, Suppl. 1, 317.
325. L. T. Eremenko, Y. F. Natsibullin and G. V. Oreshko, *Izv. Akad. Nauk USSR, Ser. Khim.*, 1970, 2556; Engl. Trans., 2400.
326. M. E. Hill and K. G. Shipp, *US Pat.* 3 526 667 (1970).
327. K. Baum, V. Grakauskas, F. E. Martin and N. W. Thomas, *'Research in Nitropolymers and their Application to solid Propellants', Report 1877, Oct 13, 1960, Contract Nonr-2655(00)*, Available from Defence Technical Information Center, Cameron Station, Alexandria, VA 22304-6145.
328. F. I. Dubovitskii, G. B. Manelis and G. Nazin, *Dokl. Akad. Nauk USSR.*, 1968, **177**, 1128, 1387; *Izv. Akad. Nauk USSR, Ser. Khim.*, 1968, **383**, 2628, 2629 and 2631.
329. M. E. Hill, M. J. Kamlet and K. G. Shipp, *US Pat.* 3 388 147 (1968).
330. H. L. Herman, in *Encyclopaedia of Explosives and Related Items, Vol. 8*, Ed. S. M. Kaye, ARRADCOM, Dover, New Jersey, N144 (1978).
331. T. G. Archibald, K. Baum, S. S. Bigelow, J. L. Flippen-Anderson, C. George, R. Gilardi and N. V. Nguyen, *J. Org. Chem.*, 1992, **57**, 235.
332. K. Baum, J. L. Flippen-Anderson, C. George and R. Gilardi and N. V. Nguyen, *J. Org. Chem.*, 1992, **57**, 3026.
333. U. Bemm, J. Bergman, A. Langlet, H. Östmark, U. Wellmar and N. Wingborg, 11th *International Detonation Symposium*, Snowmass, CO, 1998, 807–812.
334. U. Bemm, J. Bergman, A. Langlet, N. V. Latypov and U. Wellmar, *Tetrahedron*, 1998, **54**, 11525.
335. H. Q. Cai, B. B. Cheng, J. S. Li, Y. J. Shu and W. F. Yu, *Acta Chim. Sinica.*, 2004, **62**, 295.
336. U. Bemm, H. Bergman, P. Goede, M. Hihkio, E. Holmgren, M. Johansson, L. Karlsson, A. Langlet, N. V. Latypov, H. Östmark, A. Pettersson, M. -L. Pettersson, H. Stenmark, N. Wingborg and C. Vorde, 32nd *ICT International Annual Conference on Energetic Materials*, Karlsruhe, Germany, 2001, 26/1–21.

337. A. Astratev, D. Dashko, A. Marshin, A. Stepanov and N. Urazgildeev, *Russ. J. Org. Chem.*, 2001, **37**, 729.
338. H. Q. Cai, B. B. Cheng, H. H. Huang, J. S. Li and Y. J. Shu, *J. Org. Chem.*, 2004, **69**, 4369.
339. G. B. Bachman and T. F. Biermann, *J. Org. Chem.*, 1970, **35**, 4229.
340. G. B. Bachman and N. W. Connon, *J. Org. Chem.*, 1969, **34**, 4121.
341. G. A. Olah and C. Rochin, *J. Org. Chem.*, 1987, **52**, 701.
342. (a) C. D. Bedford, J. C. Bottaro, R. Malhotra and R. J. Schmitt, *J. Org. Chem.*, 1987, **52**, 2294; (b) C. D. Bedford and R. J. Schmitt, *Synthesis*, 1986, 132.
343. S. C. Motte, H. G. Viehe and J. Volker, *Chimia*, 1975, **29**, 5.
344. (a) E. J. Corey and H. Estreicher, *Tetrahedron Lett.*, 1980, **21**, 1113; (b) V. Jäger, J. C. Motte and H. G. Viehe, *Chimica*, 1975, **29**, 516; (c) V. V. Korol'kov, A. N. Nesmeyanov and T. L. Tolstoya, *Dokl. Akad. Nauk USSR*, 1978, **241**, 1103.
345. H. Feuer, C. B. Lawyer and R. Miller, *J. Org. Chem.*, 1961, **26**, 1357.
346. K. Baum and T. S. Griffin, *J. Org. Chem.*, 1980, **45**, 2880.
347. (a) L. T. Eremenko and G. V. Oreshko, *Izv. Akad. Nauk USSR, Ser. Khim.*, 1987, 1429; (b) L. T. Eremenko, G. V. Lagodzinskaya and G. V. Oreshko, *Izv. Akad. Nauk USSR, Ser. Khim.*, 1989, 709.
348. D. E. Ley, H. Shechter and L. Zeldin, *J. Am. Chem. Soc.*, 1952, **74**, 3664.
349. H. Feuer and T. J. Kucera, *J. Org. Chem.*, 1960, **25**, 2069.
350. R. Schmitz and T. Severin, *Chem. Ber.*, 1962, **95**, 1417.
351. M. Adam and T. Severin, *Chem. Ber.*, 1963, **96**, 448.
352. J. C. Dacons, M. J. Kamlet and L. A. Kaplan, *J. Org. Chem.*, 1961, **26**, 4371.
353. W. E. Noland, *Chem. Rev.*, 1955, **55**, 137.
354. A. Hantzsch and A. Rinckenberger, *Chem. Ber.*, 1899, **32**, 628.
355. D. J. Glover and M. J. Kamlet, *J. Org. Chem.*, 1961, **26**, 4734.
356. D. J. Glover, *Tetrahedron*, 1963, **19**, Suppl. 1, 219.
357. A. L. Fridman, F. M. Mukhametshin and V. D. Surkov, *Zh. Org. Khim.*, 1971, **7**, 2003.
358. A. A. Fainzilberg, G. Kh. Khisamutdinov, M. Sh. Lvova and V. I. Slovetskii, *Izv. Akad. Nauk USSR, Ser. Khim.*, 1971, 1073.
359. H. G. Adolph and M. J. Kamlet, *J. Org. Chem.*, 1968, **33**, 3073.
360. K. K. Babievskii, A. A. Fainzilberg, I. S. Korsakova, S. S. Novikov and S. A. Shevelev, *Dokl. Akad. Nauk USSR*, 1959, **124**, 589; *Chem. Abstr.*, 1959, **53**, 11206.
361. K. K. Babievskii, A. A. Fainzilberg, I. S. Korsakova, S. S. Novikov and S. A. Shevelev, *Dokl. Akad. Nauk USSR*, 1960, **132**, 846; *Chem. Abstr.*, 1960, **54**, 20841.
362. F. G. Borgardt, P. Noble Jr and A. K. Seeler, *J. Org. Chem.*, 1966, **31**, 2806.

2

Energetic Compounds 1: Polynitropolycycloalkanes

This chapter is an extension of Chapter 1 and discusses the more recent research into energetic compounds which contain strained or caged alicyclic skeletons in conjunction with *C*-nitro functionality. This chapter complements Chapter 1 by providing case studies which show how the same methods and principles that introduce *C*-nitro functionality into simple aliphatic compounds can be used as part of complex synthetic routes towards caged polynitrocycloalkanes. The chemistry used for the synthesis of caged structures can be complex but the introduction of *C*-nitro functionality follows the same principles as discussed in Chapter 1. It is suggested that chemists who are not familiar with this field of chemistry consult Chapter 1 before reading this chapter.

2.1 CAGED STRUCTURES AS ENERGETIC MATERIALS

Many explosives in frequent use today derive their energy solely from the heat released on the combustion of the carbon skeleton. Considerable research efforts have focused on synthesizing explosives containing strained or caged structures which derive their energy from both the heat of combustion of the carbon skeleton and the relief of molecular strain. Energetic materials with caged structures also benefit from a decrease in molecular motion which often leads to higher crystal density and a corresponding increase in explosive performance. Of equal importance is the thermal and chemical stability of these new explosives; ideally, an increase in molecular energy should not compromise stability.

The sensitivity of an explosive to impact and friction is a key factor in deciding whether it finds practical use. Most developed countries have an ongoing program to gradually replace current explosives and propellants with insensitive materials, a process which will greatly reduce the risk of accidental detonation. At present these programs are still in their early phases.

The weakest bonds in an explosive will often determine its sensitivity to impact and such bonds are usually present in the 'explosophoric' groups. Steric and electronic factors also play an important role. Unsurprisingly, factors which increase explosive performance usually have a detrimental effect on stability and sensitivity, and so a compromise must be made. As the database of energetic materials and their properties is ever increasing this task becomes

Organic Chemistry of Explosives J. P. Agrawal and R. D. Hodgson

more of a science than a balancing act. Correlations between structure and properties such as thermal, chemical and impact sensitivity have advanced significantly as a result of such databases. Target compounds are increasingly chosen for synthesis as a result of theoretical performance calculations and computer-aided design. Such calculations allow the prediction of density, detonation velocity and pressure, and heat of formation.

Preliminary evaluations of polynitropolycyclic compounds reveal that this class of energetic materials is relatively powerful and shock insensitive, and so, well suited for use in future explosive and propellant formulations.

Energetic materials with strained or caged structures are often much more difficult to synthesize compared to their open chain counterparts. This presents a further challenge to researchers of new energetic materials – while new compounds can be synthesized on a laboratory scale, and their properties and performance tested, the complexity of the synthetic routes may render their use as explosives nonfeasible. This particularly applies to polynitropolycyclic hydrocarbons because the direct nitration of these hydrocarbons is not a feasible route of introducing nitro groups without considerable decomposition.

2.2 CYCLOPROPANES AND SPIROCYCLOPROPANES

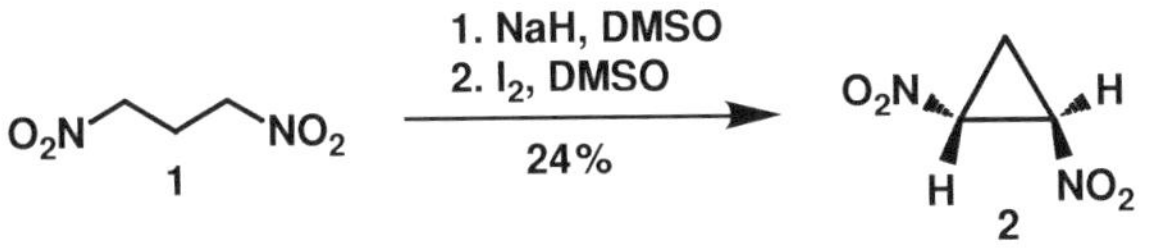

1. NaH, DMSO
2. I_2, DMSO
24%

Figure 2.1

$CH_3CH_2NO_2$ (3) → LiOCH$_3$, MeOH → $[CH_3CH{=}\overset{-}{N}O_2\overset{+}{Li}]$ (4) → 1. (2-nitropropene), 2. AcOH, 71% → 5 → 1. NaH, DMSO; 2. I_2, DMSO; 36% (2 steps) → 6

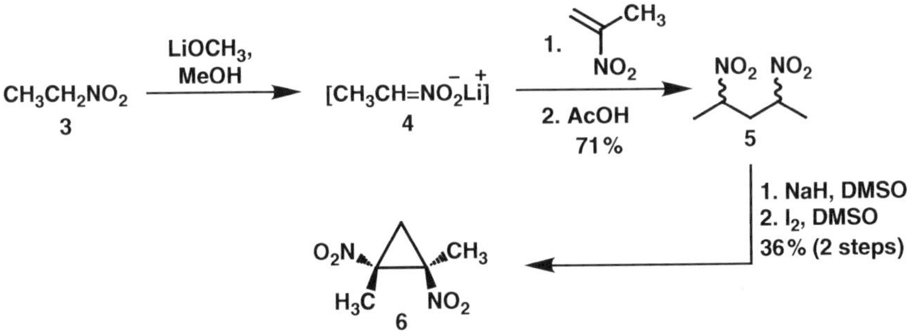

Figure 2.2

The heat of formation of cyclopropane is approximately 276 KJ/mol with a corresponding bond strain energy of 230 KJ/mol. Consequently, polynitro derivatives of cyclopropane and spirocyclopropane constitute a class of low molecular weight energetic materials. Wade and co-workers[1] synthesized some of these compounds using an oxidative cyclization of the corresponding open chain 1,3-dinitronate dianions with iodine in dimethyl sulfoxide. In this way, *trans*-1,2-dinitrocyclopropane (2), *trans*-1,2-dimethyl-1,2-dinitrocyclopropane (6), and *trans*-1,2-diethyl-1,2-dinitrocyclopropane were prepared from 1,3-dinitropropane (1), 2,4-dinitropentane (5), and 3,5-dinitroheptane respectively. Of the compounds investigated, only the *trans*-isomers were isolated, possibly as a result of thermodynamic control where the two electronegative nitro groups are positioned *anti*-periplanar to one another. *Trans*-1,2-dinitrocyclopropane (2) was found to have a density of 1.59 g/cm^3, although its chemical and thermal stability was not investigated.

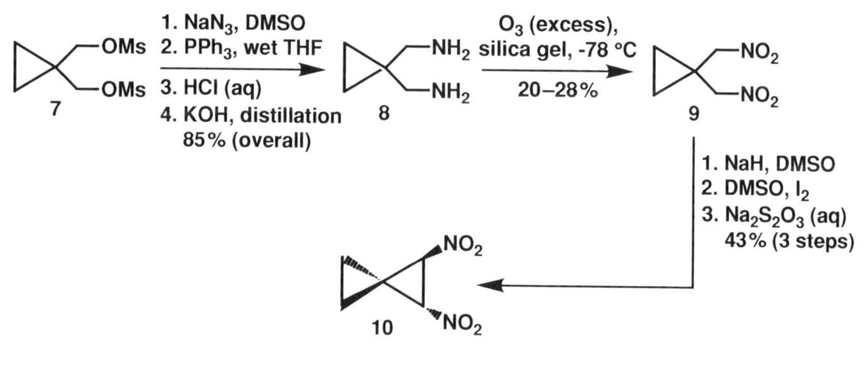

Figure 2.3

Wade and co-workers[2] used a similar strategy for the synthesis of *trans*-1,2-dinitrospiropentane (10), which is prepared in 43 % yield by treating the dianion of 1,1-*bis*(nitromethyl)-cyclopropane (9) with iodine in DMSO; the latter prepared by treating the corresponding diamine (8) with excess ozone while absorbed onto the surface of silica gel.

H
NO_2
11
LDA, THF, -80 °C
NO_2
NO_2
12

Figure 2.4

The fused dicyclopropane (12) has been synthesized by treating nitrocyclopropane (11) with lithium diisopropylamine in THF at low temperature.[3]

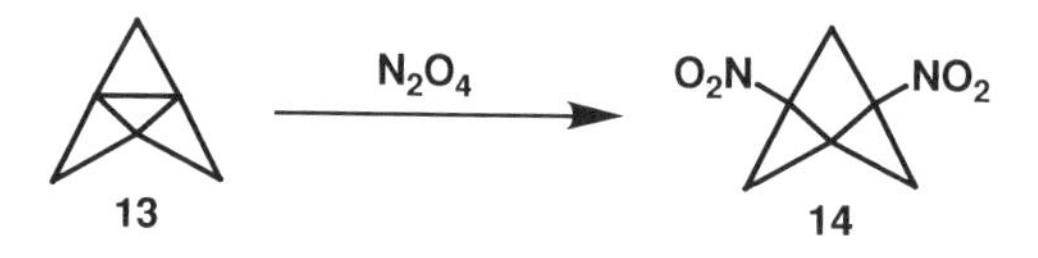

Figure 2.5

1,3-Dinitrobicyclo[1.1.1]pentane (14) is the major product from treating [1.1.1]propellane (13) with dinitrogen tetroxide.[4]

2.3 CYCLOBUTANES AND THEIR DERIVATIVES

Archibald and co-workers[5] have explored the synthesis of polynitrocyclobutanes and their derivatives. The synthesis of these compounds via the nucleophilic substitution of cyclobutyl halides with nitrite anion was ruled out at an early stage because displacement in this system is too slow for practical use. This is a consequence of the molecular strain in the cyclobutane ring, which causes carbon atoms to deviate from sp^3 hybridization towards sp^2 character.

Archibald and co-workers[5] found that aminocyclobutanes could be oxidized to the corresponding nitrocyclobutanes in moderate yield when using *m*-chloroperoxybenzoic acid (*m*-CPBA) as oxidant. Using this strategy, 1,3-dinitrocyclobutane (16) was prepared from 1,3-diaminocyclobutane (15) in 38 % yield. Interestingly, 1,3-dinitrocyclobutane (16) is obtained as a mixture of isomers from the crude reaction mixture but this completely epimerizes to the *cis*-isomer on purification by flash chromatography on silica gel.

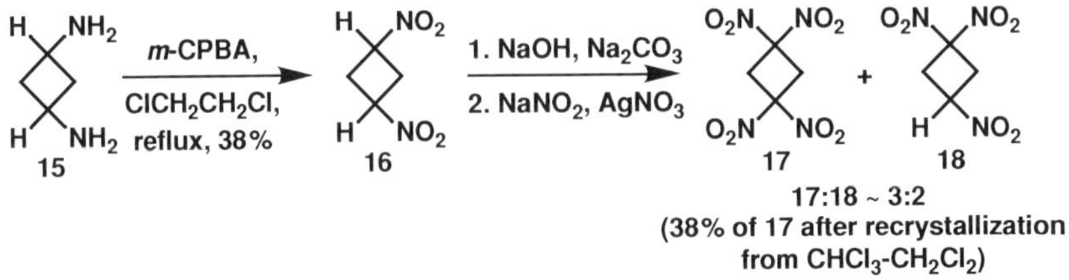

Figure 2.6

The conversion of 1,3-dinitrocyclobutane (16) to 1,1,3,3-tetranitrocyclobutane (17) is complicated by the instability of the disodium salt of 1,3-dinitrocyclobutane in strongly basic solution or at temperatures above 5 °C. This prevents oxidative nitration with sodium nitrite and potassium ferricyanide in the presence of aqueous sodium hydroxide. However, oxidative nitration of the disodium salt of 1,3-dinitrocyclobutane with a mixture of silver nitrate and sodium nitrite generated a 3:2 mixture of 1,1,3,3-tetranitrocyclobutane (17) and 1,1,3-trinitrocyclobutane (18), from which the former was isolated in 38 % yield after fractional recrystallization from chloroform–methylene chloride. Attempts to convert 1,1,3-trinitrocyclobutane (18) to 1,1,3,3-tetranitrocyclobutane (17) fail under the conditions of oxidative nitration. This is possibly due to the instability of the anion of 1,1,3-trinitrocyclobutane in aqueous solution, which may also account for the relatively low yield for the conversion of (16) to (17). 1,1,3,3-Tetranitrocyclobutane (17) is found to have a crystal density of 1.83 g/cm^3, and although the compound rapidly degrades in alkaline solution, it is thermally stable up to its melting point of 165 °C.

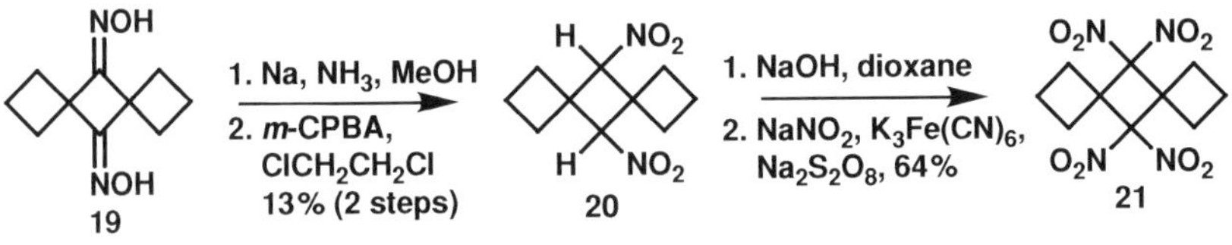

Figure 2.7

Archibald and co-workers[5] used a similar strategy of amine oxidation, followed by oxidative nitration, for the conversion of 5,10-diaminodispiro[3.1.3.1]decane to 5,5,10,10-tetranitrodispiro[3.1.3.1]decane (21). 5,10-Diaminodispiro[3.1.3.1]decane was prepared from the reduction of the corresponding oxime (19) with sodium in liquid ammonia–methanol. 5,10-Dinitrodispiro[3.1.3.1]decane (20) undergoes oxidative nitration to give 5,5,10,10-tetranitrodispiro[3.1.3.1]decane (21) in 64 % yield.

2,5,8,10-Tetranitrodispiro[3.1.3.1]decane (24) was obtained as a mixture of isomers by treating the oxime (22) with chlorine in methylene chloride, followed by oxidation with hypochlorite and reductive dehalogenation of the resulting *gem*-chloronitro intermediate (23) with zinc in alkaline solution.[5] No reference is made to the attempted oxidative nitration of (24) to 2,2,5,5,8,8,10,10-octanitrodispiro[3.1.3.1]decane.

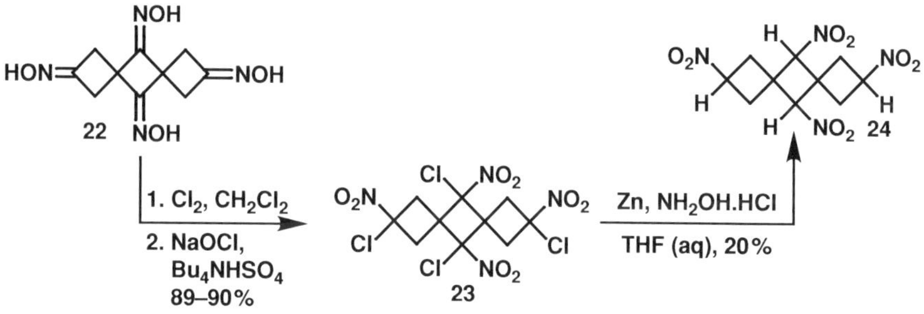

Figure 2.8

2.4 CUBANES

Of the various caged structures investigated for the synthesis of energetic compounds some have more internal strain than others. The cubane skeleton is highly energetic ($\Delta H_f \sim 620$ KJ/mol) and shows a high degree of molecular strain. Consequently, the nitro derivatives of cubane exhibit much higher performance than those of say, adamantane, which show little to no molecular strain. While this lack of molecular strain is reflected in higher thermal stability, the nitro derivatives of cubane are generally thermally stable. Additionally, the higher decomposition temperatures of some nitro derivatives of cubane, as compared to cubane itself, could infer that the electron-withdrawing nitro groups stabilize the cubane system and enhance thermal stability.[6]

Interest in polynitrocubanes surfaced when studies[7] predicted these compounds to have high crystal densities coupled with explosive performances significantly greater than standard *C*-nitro explosives like TNT. These studies were correct – highly nitrated cubanes are now known to constitute a class of high-energy shock insensitive explosives.

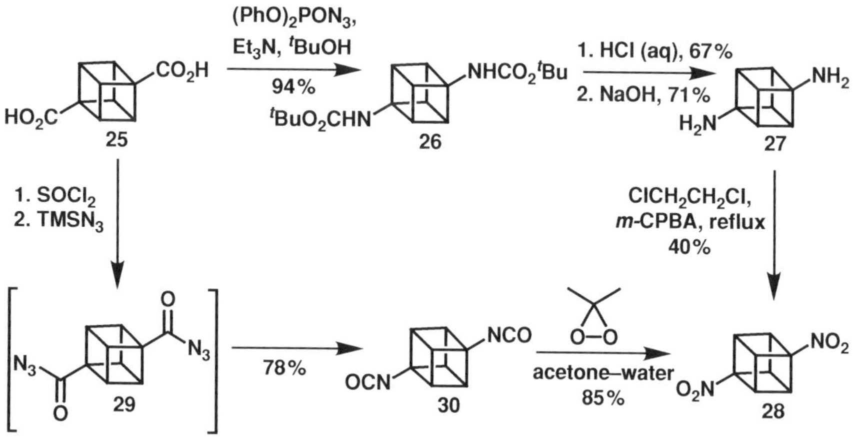

Figure 2.9

1,4-Dinitrocubane (28) has been synthesized by Eaton and co-workers[6,8] via two routes both starting from cubane-1,4-dicarboxylic acid (25). The first of these routes uses diphenylphosphoryl azide in the presence of a base and *tert*-butyl alcohol to effect direct conversion of the carboxylic acid (25) to the *tert*-butylcarbamate (26). Hydrolysis of (26) with mineral acid, followed by direct oxidation of the diamine (27) with *m*-CPBA, yields 1,4-dinitrocubane (28).[6] Initial attempts to convert cubane-1,4-dicarboxylic acid (25) to 1,4-diaminocubane (27) via a Curtius rearrangement of the corresponding diacylazide (29) were abandoned due to the extremely explosive nature of the latter.[6] However, subsequent experiments showed that treatment of the acid chloride of cubane-1,4-dicarboxylic acid with trimethylsilyl azide allows the formation of the diisocyanate (30) without prior isolation of the dangerous diacylazide (29) from solution.[8] Oxidation of the diisocyanate (30) to 1,4-dinitrocubane (28) was achieved with dimethyldioxirane in wet acetone.[8] Dimethyldioxirane is also reported to oxidize both the diamine (27) and its hydrochloride salt to 1,4-dinitrocubane (28) in excellent yield.[8]

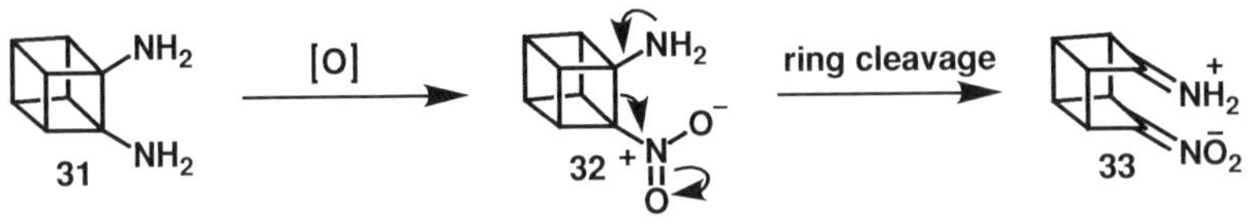

Figure 2.10

1,3-Dinitrocubane is prepared by either of the methods mentioned above.[9] The synthesis of 1,2-dinitrocubane from the oxidation of 1,2-diaminocubane (31) is complicated by cleavage of the cubane ring – when 1,2-diaminocubane (31) is oxidized, 1-amino-2-nitrocubane (32) is formed as a halfway intermediate, and the consequence of having an electron-donating group vicinal to a carbon bearing a electron-withdrawing group is cleavage of the cubane ring.[10] This is obviously a problem which must be considered when more highly nitrated derivatives of cubane are synthesized.

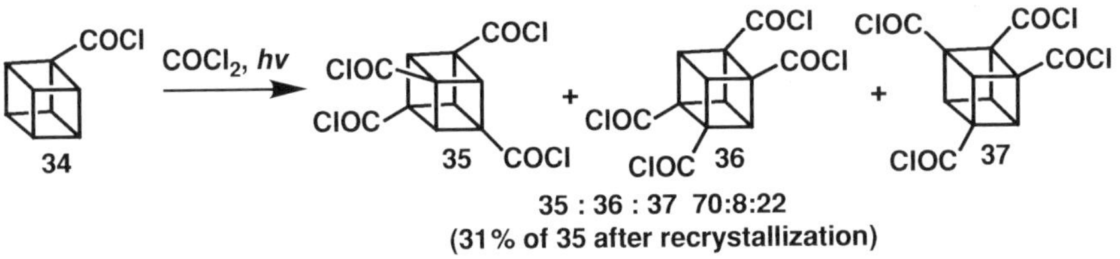

Figure 2.11

Eaton and co-workers also reported the synthesis of 1,3,5-trinitrocubane[11] and 1,3,5,7-tetranitrocubane (39)[10,11]. The required tri- and tetra-substituted cubane precursors were initially prepared via stepwise substitution of the cubane core using amide functionality to permit *ortho*-lithiation of adjacent positions.[11] The synthesis of precursors like cubane-1,3,5,7-tetracarboxylic acid was long and inefficient by this method and required the synthesis of toxic organomercury intermediates. Bashir-Hashemi[12,13] reported an ingenious route to cubane-1,3,5,7-tetracarboxylic acid chloride (35) involving photochemical chlorocarbonylation of cubane carboxylic acid chloride (34) with a mercury lamp and excess oxalyl chloride. Under optimum conditions this reaction is reported to give a 70:8:22 isomeric mixture of 35:36:37

respectively. Fractional recrystallization of this isomeric mixture from diethyl ether affords the 1,3,5,7-isomer (35) in 31 % yield.

Eaton and co-workers[10,11] used cubane-1,3,5,7-tetracarboxylic acid chloride (35) as a precursor to 1,3,5,7-tetranitrocubane (39) (TNC). Treatment of cubane-1,3,5,7-tetracarboxylic acid chloride (35) with trimethylsilyl azide gave the dangerously explosive tetraacylazide, which generates the tetraisocyanate (38) on heating in chloroform at reflux. Isolation of the tetraacylazide (38) from solution is a dangerous operation and limits the scale of this reaction. Acylazide formation is generally much safer if conducted in dilute solution, although reactions are then too slow to be useful. Eaton[10] found that acylazide isolation was not necessary if azide formation was preformed with trimethylsilyl azide in the presence of a catalytic amount of 2,6-di-*tert*-butylpyridine hydrochloride. Under these conditions Curtius rearrangement to the tetraisocyanate (38) can be conducted *in situ*. Oxidation of the tetraisocyanate (38) with dimethyldioxirane in wet acetone proceeds smoothly to give 1,3,5,7-tetranitrocubane (39) in 47 % yield overall yield from the acid chloride (35).[10,11] 1,3,5,7-Tetranitrocubane has a density of 1.814 g/cm^3, is highly energetic, and doesn't melt until 270 °C.

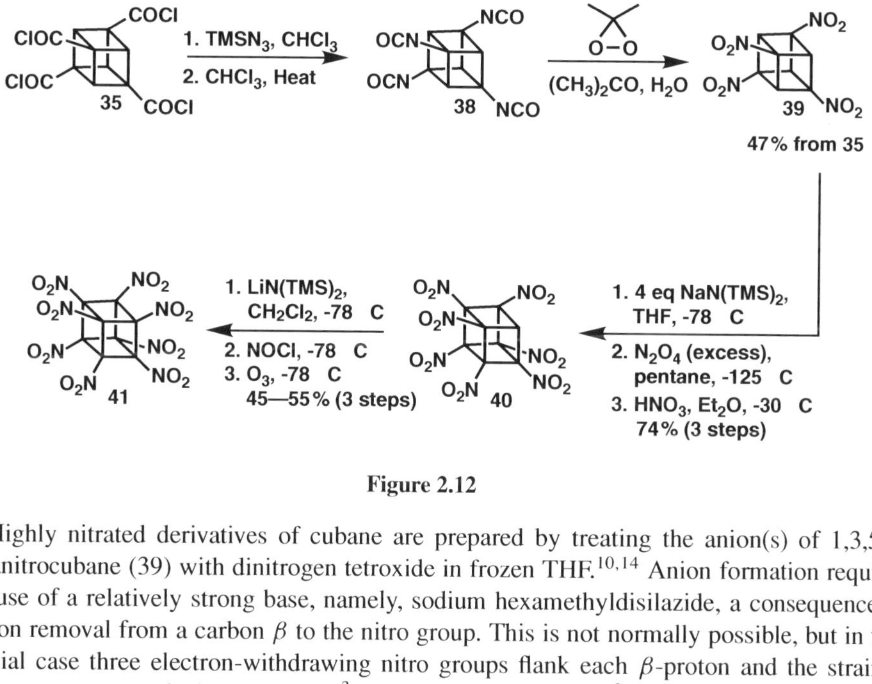

Figure 2.12

Highly nitrated derivatives of cubane are prepared by treating the anion(s) of 1,3,5,7-tetranitrocubane (39) with dinitrogen tetroxide in frozen THF.[10,14] Anion formation requires the use of a relatively strong base, namely, sodium hexamethyldisilazide, a consequence of proton removal from a carbon β to the nitro group. This is not normally possible, but in this special case three electron-withdrawing nitro groups flank each β-proton and the strained cubane core causes deviation from sp^3 hybridization towards sp^2 character. Treating 1,3,5,7-tetranitrocubane (39) with four equivalents of base, followed by reaction with frozen N_2O_4 in tetrahydrofuran–pentane, gives heptanitrocubane (40) in 74 % yield. Further nitration to octanitrocubane (41) (ONC) required the anion of (40) to be treated with nitrosyl chloride followed by oxidation with ozone.

Octanitrocubane (ONC) has a density of 1.979 g/cm^3, a calculated[15] heat of formation of 594 kJ/mol, and a decomposition temperature above 200 °C. The explosive performance of octanitrocubane (41) from theoretical calculations is predicted to be extremely high. The most recent theoretical estimate[14b,16] of VOD is 9900 m/s, making this compound one of the most powerful explosives synthesized to date. Surprisingly, the density of heptanitrocubane

(2.028 g/cm^3) is higher than octanitrocubane, and so this compound, which is easier to synthesize, may prove a more powerful explosive than the latter from the point of view of application.

2.5 HOMOCUBANES

Marchand and co-workers conducted extensive studies into the synthesis of polynitropolycyclic caged compounds, including those containing: pentacyclo[5.4.0.0^{2,6}.0^{3,10}.0^{5,9}]undecane,[17–19] pentacyclo[6.3.0.0^{2,6}.0^{3,10}.0^{5,9}]undecane[17] (D_3-trishomocubane), and pentacyclo[5.3.0.0^{2,6}.0^{3,10}.0^{4,8}]decane[19–22] (1,3-bishomocubane) skeletons. The D_3-trishomocubane and bishomocubane skeletons are less strained than cubane and so their polynitro derivatives are less energetic but generally show higher thermal stability.

A widely used strategy for the synthesis of polynitropolycyclic caged compounds is to first synthesize the appropriate caged polycycle to contain either di- or tri-ketone functionality, which can subsequently be transformed into nitro or *gem*-dinitro functionality via the oxime. The methods available for oxime to nitro group conversion are usually robust. Direct oxidation of an oxime to a nitro group can be achieved with varying success using peroxyacids (Section 1.6.1.3). Another common method uses bromination-oxidation of the oxime followed by reductive debromination with sodium borohydride (Section 1.6.1.4). An oxime can usually undergo direct conversion to a *gem*-dinitro group via oxidation-nitration with red fuming nitric acid followed by *in situ* oxidation with hydrogen peroxide (Section 1.6.1.1). Oxidative nitration is usually the method of choice for the conversion of an existing nitro group to a *gem*-dinitro group (Section 1.7).

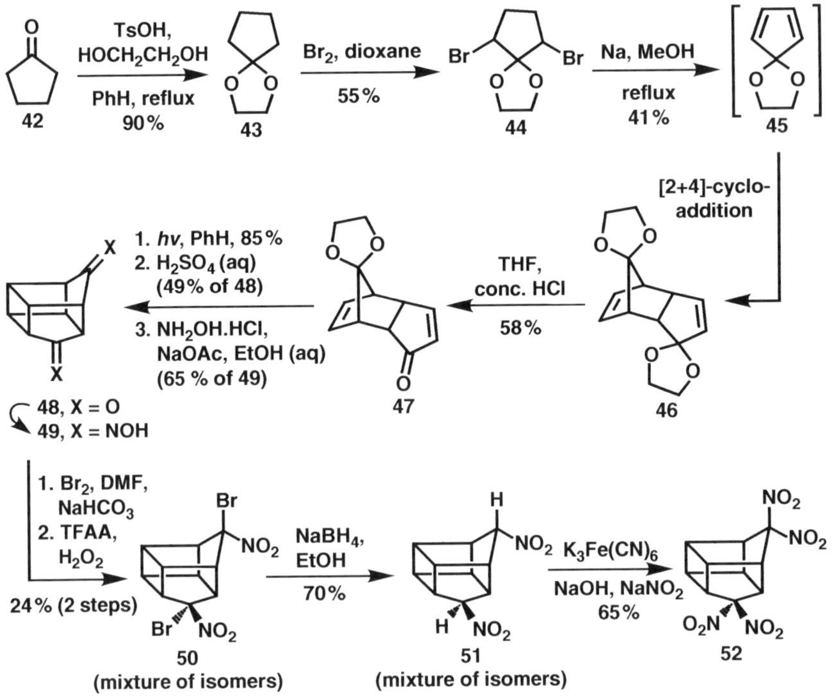

Figure 2.13

Marchand and co-workers[20] synthesis of 5,5,9,9-tetranitropentacyclo[5.3.0.0^{2,6}.0^{3,10}.0^{4,8}] decane (52) required the dioxime of pentacyclo[5.3.0.0^{2,6}.0^{3,10}.0^{4,8}]decane-5,9-dione (49) for the incorporation of the four nitro groups. Synthesis of the diketone precursor (48) was achieved in only five steps from cyclopentanone. Thus, acetal protection of cyclopentanone with ethylene glycol, followed by α-bromination, and dehydrobromination with sodium in methanol, yielded the reactive intermediate (45), which underwent a spontaneous Diels–Alder cycloaddition to give (46). Selective acetal deprotection of (46) was followed by a photo-initiated intramolecular cyclization and final acetal deprotection with aqueous mineral acid to give the diketone (48). Derivatization of the diketone (48) to the corresponding dioxime (49) was followed by conversion of the oxime groups to *gem*-dinitro functionality using standard literature procedures.

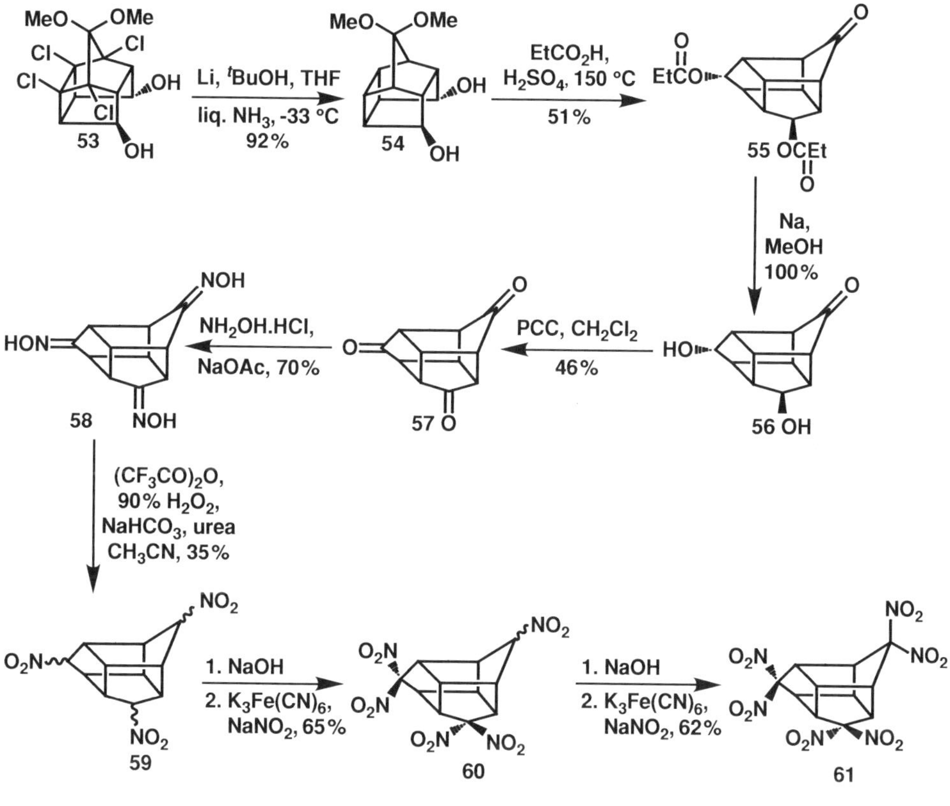

Figure 2.14

Triketone (57), a key intermediate in the synthesis of 4,4,7,7,11,11-hexanitropentacyclo [6.3.0.0^{2,6}.0^{3,10}.0^{5,9}]undecane (61) (D_3-hexanitrotrishomocubane), has been synthesized independently by both Marchand and co-workers,[17] and Fessner and Prinzach[23]. Marchand and co-workers[17] prepared the trioxime (58) from the corresponding triketone (57). Oxidation of (58) with peroxytrifluoroacetic acid in acetonitrile provides a direct route to the trinitro derivative (59) in 35 % yield, this yield reflecting an efficiency of 70 % for the oxidation of each oxime group. Subsequent oxidative nitration of (59) with sodium nitrite and potassium ferricyanide in aqueous sodium hydroxide yields the target D_3-hexanitrotrishomocubane (61).

Tests on D_3-hexanitrotrishomocubane show it to be both less sensitive and significantly more powerful than TNT whilst exhibiting high thermal stability.

Polynitro derivatives of pentacyclo[5.4.0.0^{2,6}.0^{3,10}.0^{5,9}]undecane[17–19] have attracted interest as potential high-energy explosives. Molecular strain in this caged system could arise from both the constrained norbornyl moiety and the cyclobutane ring. Additional strain would be expected from nonbonding interactions if the 8-*endo* and 11-*endo* positions were substituted with *gem*-dinitro groups.

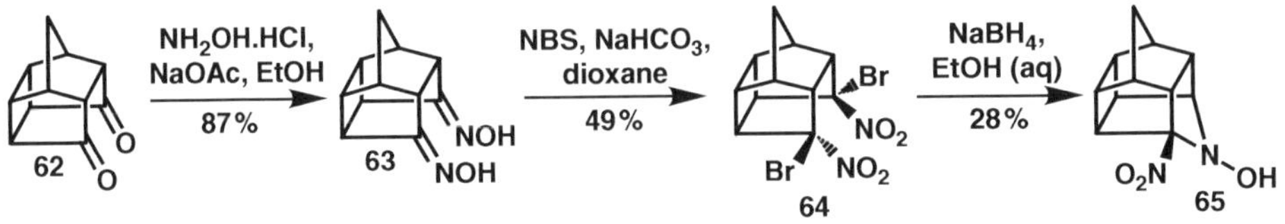

Figure 2.15

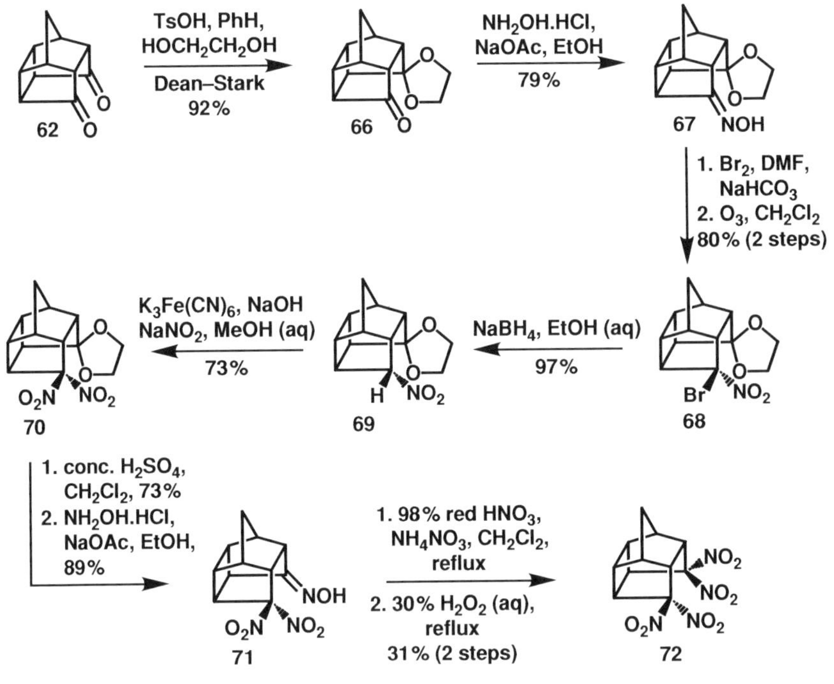

Figure 2.16

Marchand and co-workers have provided synthetic routes to both 8,8,11,11-tetranitro- (72)[17] and 4,4,8,8,11,11-hexanitro- (80)[18] pentacyclo[5.4.0.0^{2,6}.0^{3,10}.0^{5,9}]undecanes. Initial attempts to synthesize target (72) from the dioxime (63) failed when it was found that sodium borohydride reduction of the *gem*-bromonitro intermediate (64) gave the aza-heterocycle (65) as the major product. Consequently, an indirect route was explored where one of the two ketone groups of (62) is protected as an acetal (66) while the other ketone group is converted to a *gem*-dinitro group via the oxime (67). Deprotection of the acetal (70), followed by elaboration of the ketone functionality to a *gem*-dinitro group, yields 8,8,11,11-tetranitropentacyclo-[5.4.0.0^{2,6}.0^{3,10}.0^{5,9}]undecane (72).

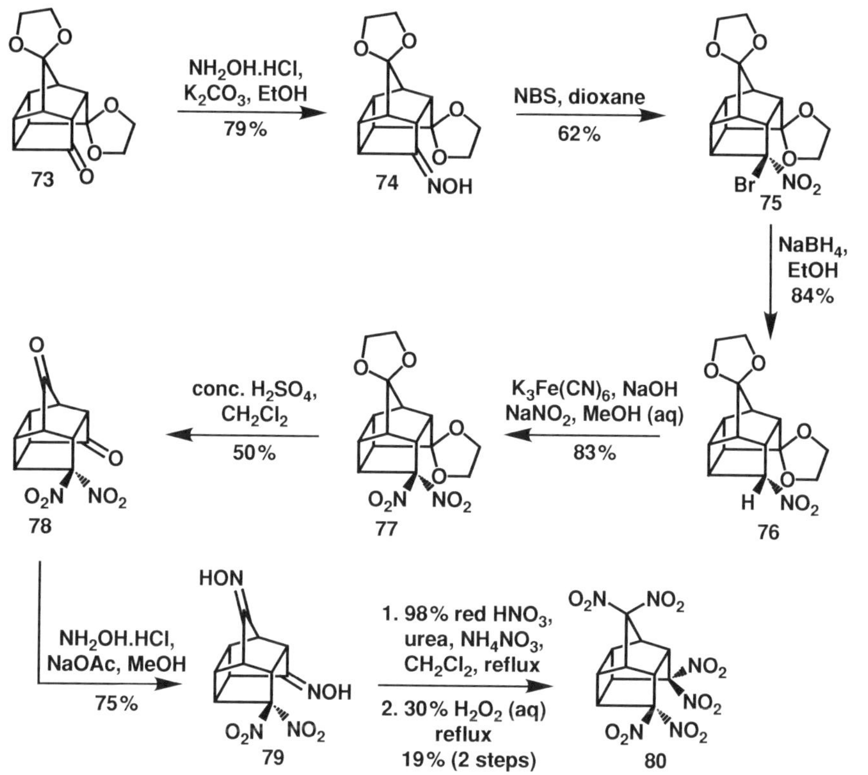

Figure 2.17

Marchand and co-workers[18] anticipated that aza-heterocycle formation would be a problem for the synthesis of 4,4,8,8,11,11-hexanitropentacyclo[5.4.0.0^{2,6}.0^{3,10}.0^{5,9}]undecane (80) and so they used a similar strategy to that in figure 2.16 whereby one of the ketone groups of (73), in either the 5- or 11-position, is protected as an acetal. A similar elaboration of the ketone groups afforded the hexanitro derivative (80).

Chinese chemists[24] have reported the synthesis of pentacyclo[4.3.0.0^{3,8},0^{4,7}]nonane-2,4-bis(trinitroethyl ester) (88). This compound may find potential use as an energetic plastisizer in futuristic explosive and propellant formulations. The synthesis of (88) uses widely available hydroquinone (81) as a starting material. Thus, bromination of (81), followed by oxidation, Diels–Alder cycloaddition with cyclopentadiene, and photochemical [2 + 2] cycloaddition, yields the dione (85) as a mixture of diastereoisomers, (85a) and (85b). Favorskii rearrangement of this mixture yields the dicarboxylic acid as a mixture of isomers, (86a) and (86b), which on further reaction with thionyl chloride, followed by treating the resulting acid chlorides with 2,2,2-trinitroethanol, gives the energetic plastisizer (88) as a mixture of isomers, (88a) and (88b). Improvements in the synthesis of nitroform, and hence 2,2,2-trinitroethanol, makes the future application of this product attractive.

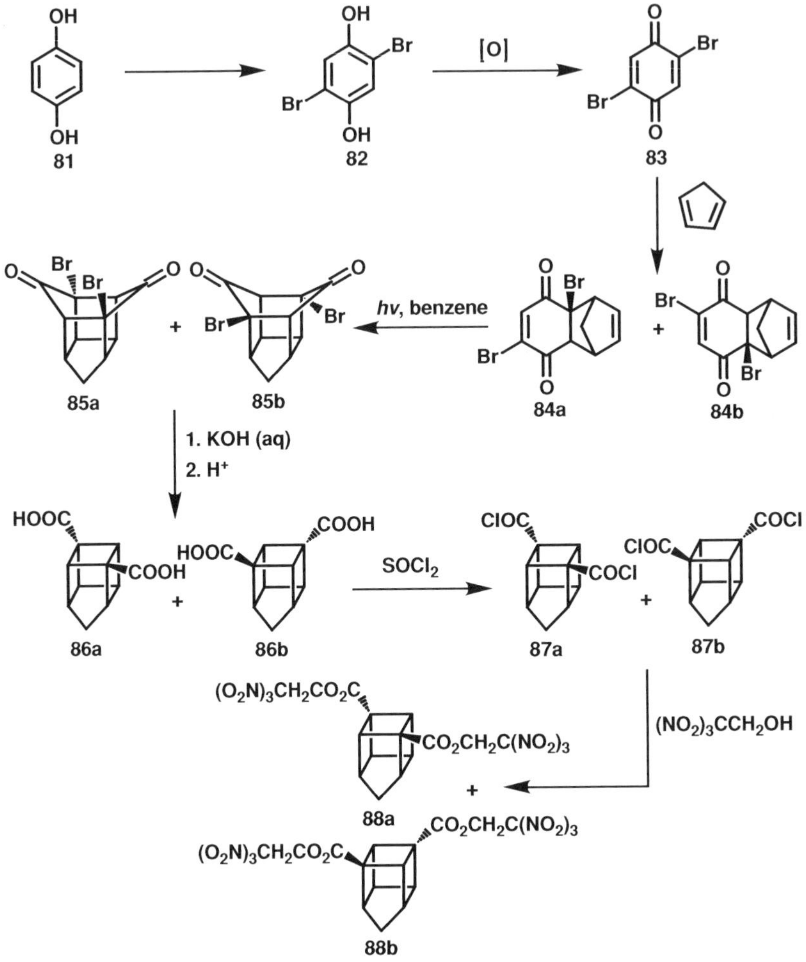

Figure 2.18

2.6 PRISMANES

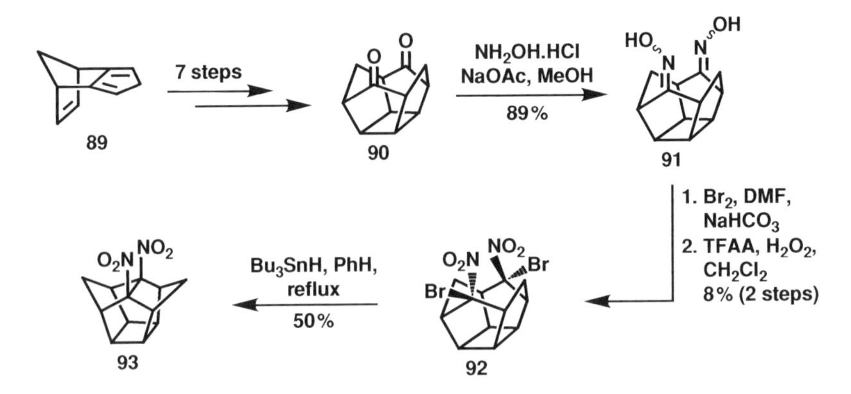

Figure 2.19

The search for energetic compounds with high crystal densities has focused attention on the polynitro derivatives of 1,3-bishomopentaprismane. However, the synthesis and incorporation of this compact core into explosives is complex and synthetically challenging. In general, complex caged structures require multiple, nontrivial synthetic steps, often with one or more of these using a photoinitiated cyclization or cycloaddition step. The 1,3-bishomopentaprismane core is no exception. The 2,6-dinitro isomer (93) requires the synthesis of [4]-peristylane-2,6-dione (90), which is prepared in no fewer than seven steps from the cycloadduct (89).[25] Incorporation of the two nitro groups into the [4]-peristylane core uses standard robust chemistry, the *gem*-bromonitro intermediate (92) being formed from the bromination-oxidation of the dioxime (91). The 1,3-bishomopentaprismane core is formed from the reaction of (92) with tributyltin hydride in a radical induced C–C bond formation.[26]

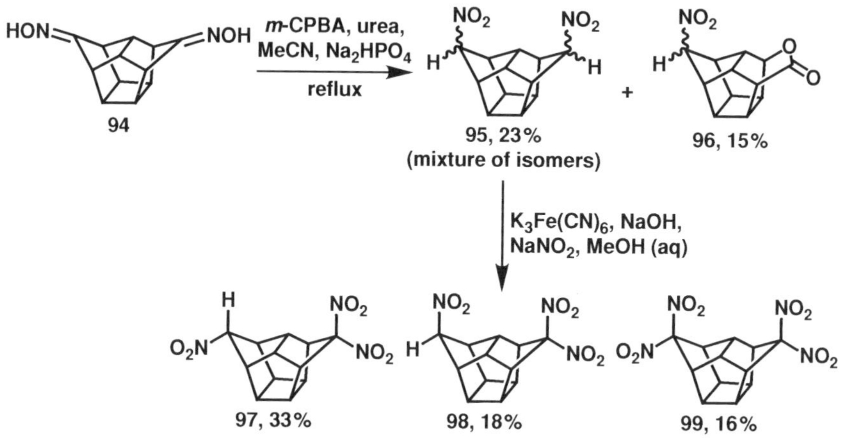

Figure 2.20

Paquette and co-workers[27] synthesized the 5,11-dinitro isomer of 1,3-bishomopentaprismane (95) by treating the dioxime (94) with a buffered solution of *m*-CPBA in refluxing acetonitrile. A significant amount of lactone by-product (96) is formed during this step and may account for the low isolated yield of (95). Oxidative nitration of (95) with sodium nitrite and potassium ferricyanide in alkaline solution yields a mixture of isomeric trinitro derivatives, (97) and (98), in addition to the expected 5,5,11,11-tetranitro derivative (99), albeit in low yield. Incomplete reactant to product conversion in this reaction may result from the low solubility of either (97) or (98) in the reaction medium, and hence, incomplete formation of the intermediate nitronate anions.

2.7 ADAMANTANES

The highly rigid skeleton of adamantane results in much higher crystal densities compared to its open chain counterparts, and hence, higher performance for its nitro derivatives. The adamantane core shows little to almost no strain and so some of its polynitro derivatives show exceptionally high thermal stability. Research has focused on placing nitro groups on the tertiary carbon bridgehead positions of adamantane, and forming *gem*-dinitro derivatives at the methylene carbon positions. Both tertiary nitro and internal *gem*-dinitro functionalities show high chemical stability.

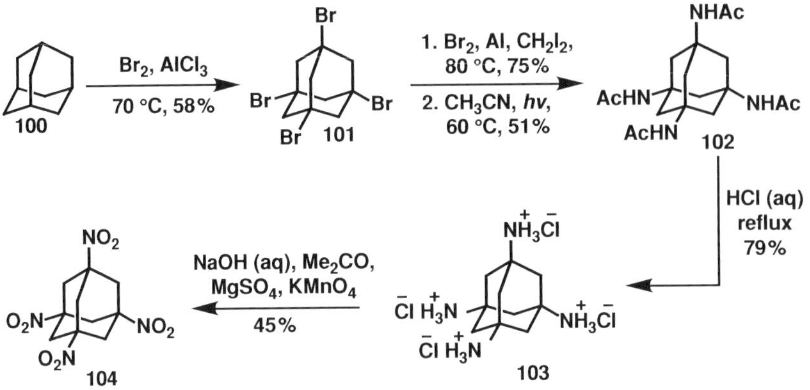

Figure 2.21

Sollett and Gilbert[28] working for ARDEC (US Army Research, Development and Engineering Center) first reported the synthesis of 1,3,5,7-tetranitroadamantane (104) in 1980. Their synthesis starts directly from adamantane (100), which on halogenation with bromine and aluminium chloride yields 1,3,5,7-tetrabromoadamantane (101). Previous attempts for a direct $Br \rightarrow NH_2$ conversion have been reported[29] without success. A more indirect route for this conversion involves halogen exchange of the tetrabromide (101) to the tetraiodide, followed by a photo-induced Ritter reaction to give the tetraacetamide (102), which on acid catalyzed hydrolysis yields the tetrahydrochloride salt of 1,3,5,7-tetraaminoadamantane (103).[28] The synthesis is complete by treating (103) with an aqueous acetone solution of potassium permanganate which generates 1,3,5,7-tetranitroadamantane (104) in 45 % yield; this reflecting a relative yield of 82 % for the oxidation of each of the four amino groups. Surprisingly, while the permanganate oxidation of tertiary amines has been known for some time, this method had never been used to synthesize polynitro compounds until this example was reported. 1,3,5,7-Tetranitroadamantane shows high thermal stability (m.p. 361 °C) and a similar high chemical stability would be expected.

The synthesis of 1,3,5,7-tetranitroadamantane (104) from 1,3,5,7-tetraaminoadamantane (103) has been improved upon by the use of dimethyldioxirane[30] (91 %), and also, by using a mixture of sodium percarbonate and N,N,N',N'-tetraacetylethylenediamine in a biphasic solvent system, followed by treating the crude product with ozone (91 %)[31]; the latter involving the *in situ* generation of peroxyacetic acid.

Archibald and Baum[32] reported the synthesis of 2,2,6,6-tetranitroadamantane (109). Their synthesis starts from the dioxime of 2,6-adamantanedione (105), which on reaction with a buffered solution of NBS yields the bromonitro intermediate (106), surprisingly, without the need of an oxidant for nitroso to nitro group conversion. Reaction of (106) with sodium borohydride yields 2,6-dinitroadamantane (107). Attempted oxidative nitration of the anion of (107) with sodium nitrite and silver nitrate was unsuccessful, even though the same reaction with 2-nitroadamantane gave 2,2-dinitroadamantane in 89 % yield. However, the alkaline nitration of (107) with tetranitromethane in ethanolic potassium hydroxide gave 2,2,6,6-tetranitroadamantane (109) in 68 % yield along with a 20 % yield of 6,6-dinitro-2-adamantanone (108); the latter probably arising from a Nef reaction. 2,2,6,6-Tetranitroadamantane, like its 1,3,5,7-isomer, is reported to exhibit high thermal stability.

Archibald and Baum[32] tried to apply the same strategy used to prepare 2,2,6,6-tetranitroadamantane (109) to synthesize 2,2,4,4-tetranitroadamantane (117) from the dioxime of

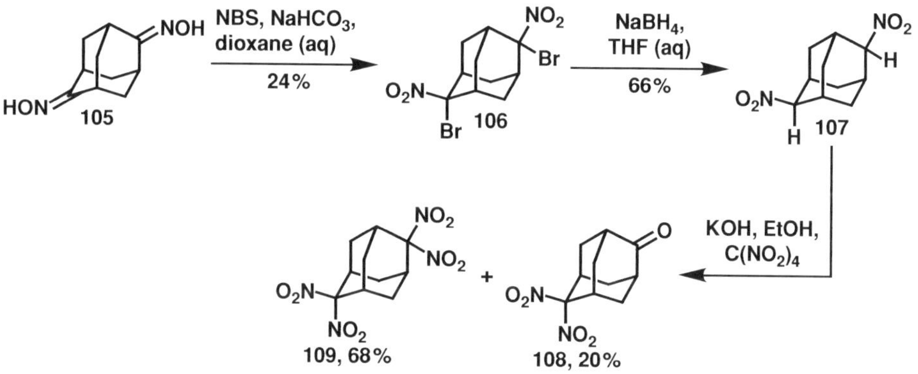

Figure 2.22

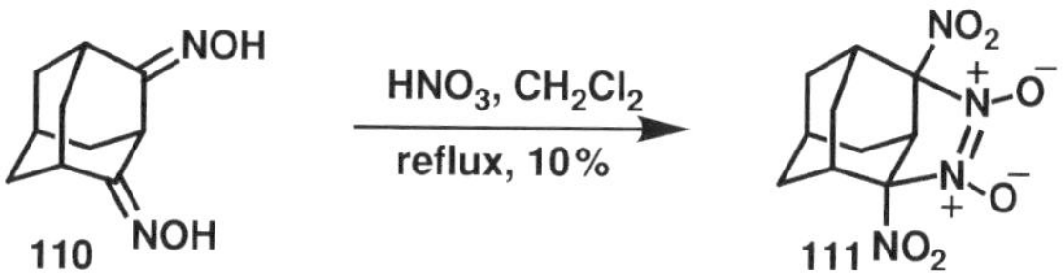

Figure 2.23

2,4-adamantanedione (110). However, this synthesis was unsuccessful, possibly due to severe steric crowding at the 2- and 4-positions. Archibald and Baum[32] explored the feasibility of using pure nitric acid in methylene chloride for the direct oxidation-nitration of oximes. This strategy failed when applied to the dioxime of 2,6-adamantanedione (105) and led to oxime hydrolysis. The same strategy applied to the dioxime of 2,4-adamantanedione (110) gave a 10 % yield of 2,4-dinitro-2,4-dinitrosoadamantane as its internal dimer (111). Attempted oxidation of (111) to 2,2,4,4-tetranitroadamantane (117) was unsuccessful. These results have implications for the synthesis of highly nitrated adamantanes.

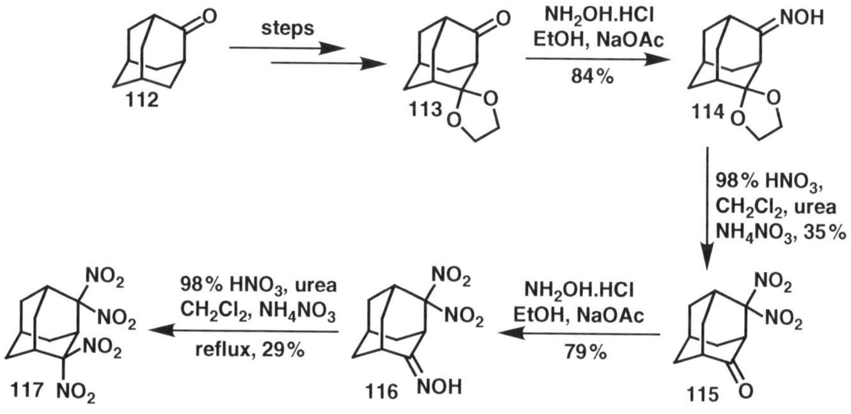

Figure 2.24

Dave and co-workers[33] have reported a successful synthesis of 2,2,4,4-tetranitroadamantane (117) which uses the mono-protected diketone (113) as a key intermediate. In this synthesis (113) is converted to the oxime (114) and then treated with ammonium nitrate and nitric acid in methylene chloride to yield the *gem*-dinitro derivative (115). This nitration-oxidation step also removes the acetal-protecting group to leave the second ketone group free. Formation of the oxime (116) from ketone (115), followed by a similar nitration-oxidation with nitric acid and ammonium nitrate, yields 2,2,4,4-tetranitroadamantane (117). In this synthesis the protection strategy enables each carbonyl group to be treated separately and thus prevents the problem of internal nitroso dimer formation.

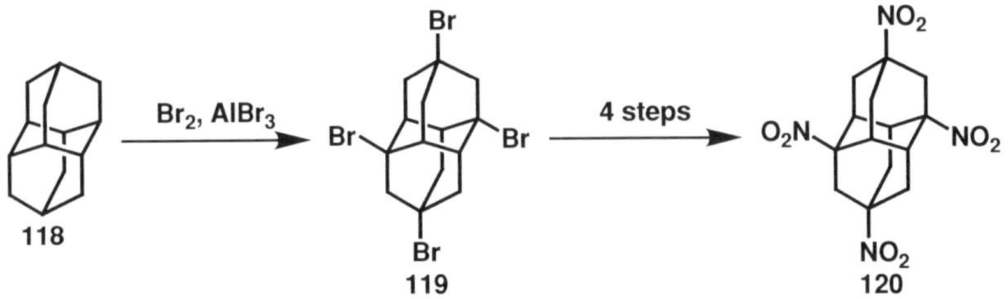

Figure 2.25

Sollett and Gilbert[34] reported the synthesis of 1,4,6,9-tetranitrodiamantane (120) using an identical strategy to that used for the synthesis of 1,3,5,7-tetranitroadamantane. It is worth noting that chemists at ARDEC have made significant contributions in the field of polynitropolycyclic hydrocarbons. An interesting theoretical study using thermodynamic calculations was used to predict the properties of various caged polynitroalkanes and to determine the number of nitro groups needed in these compounds for optimum performance.[35]

2.8 POLYNITROBICYCLOALKANES

2.8.1 Norbornanes

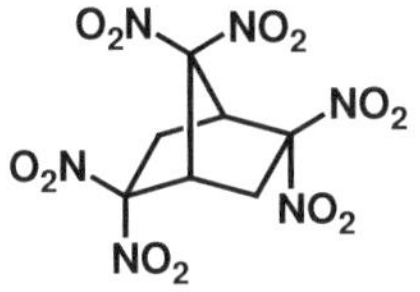

Figure 2.26

Polynitro derivatives of norbornane have been explored as a class of energetic materials. Of particular interest in this area are derivatives like 2,2,5,5,7,7-hexanitronorbornane (121), which has an excellent carbon to nitro group ratio. At present, only the 2,2,5,5-tetranitro (127) and 2,2,7,7-tetranitro (136) isomers of norbornane have been synthesized, with the hexanitro isomers remaining elusive.

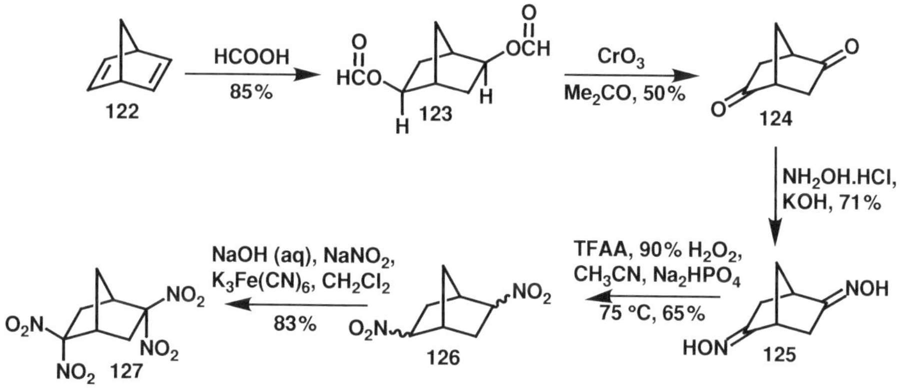

Figure 2.27 'Olah and co-workers' synthesis of 2,2,5,5-tetranitronorbornane[36]

Olah and co-workers[36] reported the synthesis of 2,2,5,5-tetranitronorbornane (127) from 2,5-norbornadiene (122). In this synthesis formylation of (122) with formic acid yields the diformate ester (123), which on treatment with chromium trioxide in acetone yields 2,5-norbornadione (124). Formation of the dioxime (125) from 2,5-norbornadione (124) is followed by direct oxidation to 2,5-dinitronorbornane (126) with peroxytrifluoroacetic acid generated *in situ* from the reaction of 90 % hydrogen peroxide with TFAA. Oxidative nitration of 2,5-dinitronorbornane (126) with sodium nitrite and potassium ferricyanide in alkaline solution generates 2,2,5,5-tetranitronorbornane (127) in excellent yield.

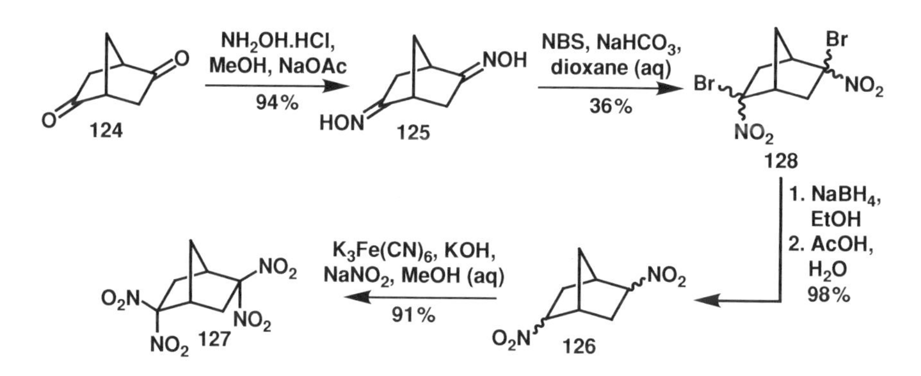

Figure 2.28 'Marchand and co-workers' synthesis of 2,2,5,5-tetranitronorbornane[37]

Marchand and co-workers[37] reported the synthesis of 2,2,5,5-tetranitronorbornane (127) at the same time as Olah[36] and used the same dioxime (125) as a key intermediate. Marchand and co-workers synthesized 2,5-dinitronorbornane (126) via bromination-oxidation of the dioxime (125) followed by reductive debromination of the *gem*-bromonitro derivative (128). Oxidative nitration was used to convert 2,5-dinitronorbornane (126) to 2,2,5,5-tetranitronorbornane (127).

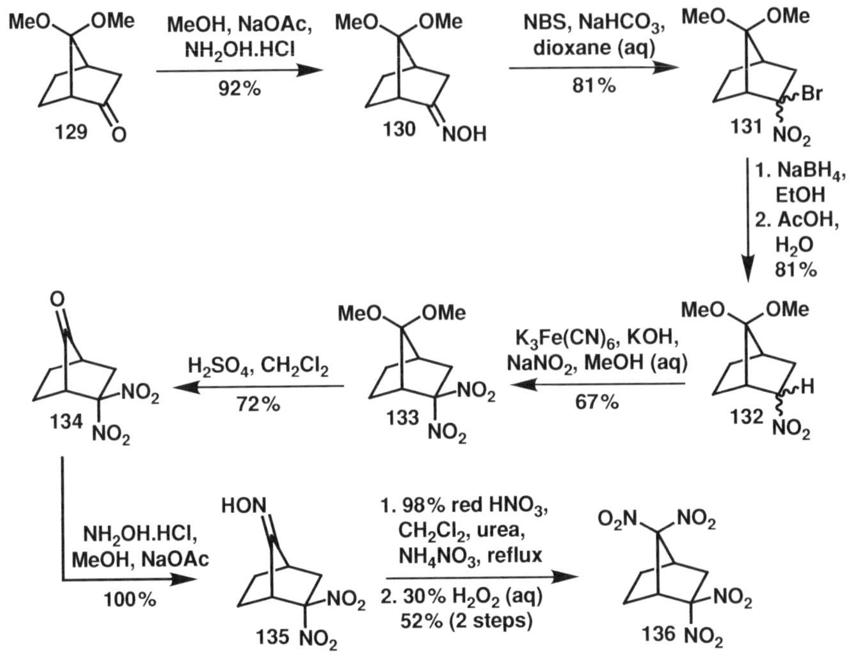

Figure 2.29

Marchand and co-workers[37] also reported the synthesis of 2,2,7,7-tetranitronorbornane (136). This synthesis is much longer and more indirect than the synthesis of the 2,2,5,5-isomer because strained 1,3-diones like 2,7-norbornadione are susceptible to ring cleavage under both acidic and basic conditions, a process known as Haller–Bauer cleavage. Marchand and co-workers strategy to the 2,2,7,7-isomer (136) uses a derivative of 2,7-norbornadione where one of the ketone groups is protected. The methyl acetal (129) was used for this purpose, which on derivatization to the corresponding oxime (130), followed by bromination-oxidation, reductive debromination and oxidative nitration, yields the *gem*-dinitro derivative (133). Ring cleavage is no longer a problem at this point in the synthesis and so (133) is hydrolyzed to the ketone (134) under acid catalysis. Subsequent oxime formation, followed by an oxidation-nitration step, yields 2,2,7,7-tetranitronorbornane (136).

2.8.2 Bicyclo[3.3.0]octane

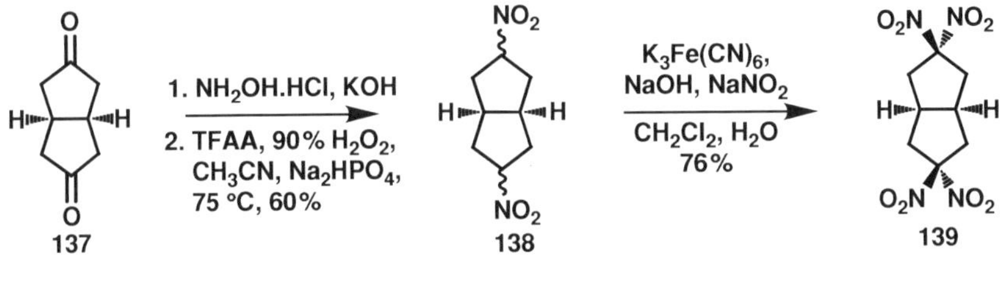

Figure 2.30

Olah and co-workers[36] reported the synthesis of 3,3,7,7-tetranitro-*cis*-bicyclo[3.3.0]octane (139) from the diketone (137). In this synthesis treatment of the dioxime of (137) with a buffered solution of peroxytrifluoroacetic acid gives 3,7-dinitro-*cis*-bicyclo[3.3.0]octane (138), which on oxidative nitration yields (139) in 76 % yield.

2.8.3 Bicyclo[3.3.1]nonane

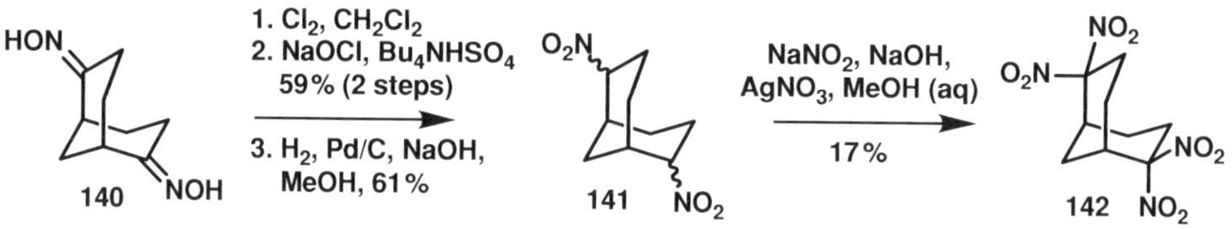

Figure 2.31

Baum and Archibald[32] reported the synthesis of 2,2,6,6-tetranitrobicyclo[3.3.1]nonane (142). This synthesis starts from the dioxime (140), which on halogenation with chlorine, followed by oxidation with hypochlorite and reductive dehalogenation with hydrogen in the presence of palladium on carbon, yields 2,6-dinitrobicyclo[3.3.1]nonane (141). Oxidative nitration of (141) with sodium nitrite and silver nitrate under alkaline conditions yields 2,2,6,6-tetranitrobicyclo[3.3.1]nonane (142). The greater molecular freedom in 2,2,6,6-tetranitrobicyclo[3.3.1] nonane (crystal density-1.45 g/cm^3) compared to the isomeric 2,2,6,6-tetranitroadamantane (crystal density-1.75 g/cm^3) is reflected is their considerably different crystal densities.

REFERENCES

1. P. J. Carroll, W. P. Dailey and P. A. Wade, *J. Am. Chem. Soc.*, 1987, **109**, 5452.
2. P. J. Carroll, P. A. Kondracki and P. A. Wade, *J. Am. Chem. Soc.*, 1991, **113**, 8807.
3. J. D. Dunitz, Y. Kai, H. O. Kalinowski, P. Knochel, S. Kwaitkowski, J. F. M. Oth and D. Seebach, *Helv. Chim. Acta.*, 1982, **65**, 137.
4. S. T. Waddell and K. B. Wiberg, *J. Am. Chem. Soc.*, 1990, **112**, 2194.
5. T. G. Archibald, K. Baum, M. C. Cohen and L. C. Garver, *J. Org. Chem.*, 1989, **54**, 2869.
6. J. Alster, P. E. Eaton, E. E. Gilbert, J. J. Pluth, G. D. Price, B. K. Ravi Shankar and O. Sandus, *J. Org. Chem.*, 1984, **49**, 185.
7. (a) H. H. Cady, Estimation of the Density of Organic Explosives from their Structural Formulas, *Report #LA-7760-MS*, Los Alamos National Laboratory, New Mexico (1979); (b) M. E. Grice, J. S. Murray and P. Politzer, in *Decomposition, Combustion and Detonation Chemistry of Energetic Materials*, Eds. T. B. Brill, T. P. Russell, W. C. Tao and R. B. Wardle, Materials Research Society, Pittsburgh, PA, 1996, **418**, 55.
8. P. E. Eaton and G. E. Wicks, *J. Org. Chem.*, 1988, **53**, 5353.
9. Work conducted by G. W. Griffin, as reported by A. P. Marchand, *Tetrahedron*, 1988, **44**, 2377.
10. P. E. Eaton, R. Gilardi, J. Hain, N. Kanomata, J. Li, K. A. Lukin and E. Punzalan, *J. Am. Chem. Soc.*, 1997, **119**, 9591.
11. P. E. Eaton, R. Gilardi and Y. Xiong, *J. Am. Chem. Soc.*, 1993, **115**, 10195.
12. A. Bashir-Hashemi, *Angew. Chem. Int. Ed.*, 1993, **32**, 612.
13. H. Ammon, A. Bashir-Hashemi, N. Gelber and J. Li, *J. Org. Chem.*, 1995, **60**, 698.

14. (a) P. E. Eaton, R. Gilardi and M. -X. Zhang, *Angew. Chem. Int. Ed.*, 2000, **39**, 401; (b) P. E. Eaton, N. Gelber, R. Gilardi, S. Iyer, S. Rao and M. -X. Zhang, *Propell. Explos. Pyrotech.*, 2002, **27**, 1.
15. A. M. Astakhov, A. Yu. Babushkin and R. S. Stepanov, *Combustion, Explosion and Shock Waves*, 1998, **34**, 85.
16. R. D. Gilardi and J. Karle, in *Chemistry of Energetic Materials*, Ed. D. R. Squire and G. A. Olah, Academic Press, San Diego, 2 (1991).
17. G. S. Annapurna, G. V. Madhava Sharma, A. P. Marchand and P. R. Pednekar, *J. Org. Chem.*, 1987, **52**, 4784.
18. B. E. Arney Jr, P. R. Dave and A. P. Marchand, *J. Org. Chem.*, 1988, **53**, 443.
19. B. E. Arney Jr, P. R. Dave, J. L. Flippen-Anderson, C. George, R. Gilardi, A. P. Marchand and D. Rajapaksa, *J. Org. Chem.*, 1989, **54**, 1769.
20. S. Chander Suri and A. P. Marchand, *J. Org. Chem.*, 1984, **49**, 2041.
21. A. P. Marchand and D. Sivakumar Reddy, *J. Org. Chem.*, 1984, **49**, 4078.
22. (a) H. L. Ammon, G. S. Annapurna, J. L. Flippen-Anderson, C. George, R. Gilardi, A. P. Marchand and V. Vidyasagar, *J. Org. Chem.*, 1987, **52**, 4781; (b) S. Chander Suri and A. P. Marchand, *J. Org. Chem.*, 1984, **49**, 2041.
23. W.-D. Fessner and H. Prinzbach, *Tetrahedron*, 1986, **42**, 1797.
24. X. Guan, Z. Su and Y. Yu, Synthesis of the Pentacyclo[4.3.0.$0^{2,5}$.$0^{3,8}$.$0^{4,7}$]nonane-2,4-bis (trinitroethyl ester), in *Proc. 17th International Pyrotechnics Seminar (Combined with 2rd Beijing International Symposium on Pyrotechnics and Explosives)*, Beijing Institute Technical Press, Beijing, China, 224 (1991).
25. (a) A. R. Browne, C. W. Doecke, L. A. Paquette and R. V. Williams, *J. Am. Chem. Soc.*, 1983, **105**, 4113; (b) A. R. Browne, C. W. Doecke, J. W. Fischer and L. A. Paquette, *J. Am. Chem. Soc.*, 1985, **107**, 686.
26. P. Engel, J. W. Fischer and L. A. Paquette, *J. Org. Chem.*, 1985, **50**, 2524.
27. P. Engel, K. Nakamura and L. A. Paquette, *Chem. Ber.*, 1986, **119**, 3782.
28. E. E. Gilbert and G. P. Sollott, *J. Org. Chem.*, 1980, **45**, 5405.
29. M. Krause and H. Stetter, *Liebig Ann. Chem.*, 1968, **60**, 717.
30. L. Mohan, R. W. Murray and S. N. Rajadhyaksha, *J. Org. Chem.*, 1989, **54**, 5783.
31. T. R. Walters, J. M. Woods and W. W. Zajac Jr, *J. Org. Chem.*, 1989, **54**, 2468.
32. T. G. Archibald and K. Baum, *J. Org. Chem.*, 1988, **53**, 4645.
33. H. L. Ammon, C. S. Choi, P. R. Dave and M. Ferraro, *J. Org. Chem.*, 1990, **55**, 4459.
34. E. E. Gilbert and G. P. Sollott, *US Pat.*, 4 535 193 (1985).
35. J. Alster, E. E. Gilbert, O. Sandus, N. Slagg and G. P. Sollott, *J. Energ. Mater.*, 1986, **4**, 5.
36. R. Gilardi, G. A. Olah, P. Ramaiah and G. K. Surya Prakash, *J. Org. Chem.*, 1993, **58**, 763.
37. R. P. Kashyap, A. P. Marchand, R. Sharma, W. H. Watson and U. R. Zope, *J. Org. Chem.*, 1993, **58**, 759.

3

Synthetic Routes to Nitrate Esters

This chapter discusses the various synthetic methods available for the synthesis and incorporation of nitrate ester functionality into organic compounds. Nitrate esters are organic esters of nitric acid and as such, their synthesis from the parent alcohol or polyol is fairly straightforward. However, the nitrate esters of many polyols are highly sensitive to mechanical stimuli and so their synthesis, on both a laboratory and industrial scale, has to be treated as significantly more dangerous than the synthesis of some other explosives. Many of the modern laboratory methods for the preparation of nitrate esters focus on reducing this danger.

Low molecular weight nitrate esters, even those derived from polyols like glycerol, are volatile and toxic. This toxicity originates from their biological activity as vasodilators and has seen their extensive use in the treatment of angina pectoris. The use of nitrate esters in organic synthesis is not extensive or particularly notable. However, the nitrate ester group is an important 'explosophore' and this energetic group is found in numerous commercial and military explosives, some of which are synthesized on an enormous scale.

3.1 NITRATE ESTERS AS EXPLOSIVES

Nitrate esters[1] are an important class of explosives for both commercial and military use. Perhaps the most well known is glyceryl trinitrate (GTN) (1) which is commonly known as nitroglycerine (NG). Nitroglycerine is produced commercially from the nitration of glycerol with mixed acid. Pure nitroglycerine is a pale yellow oily liquid which freezes at temperatures below 13 °C. In the pure state nitroglycerine shows an unacceptably high sensitivity to impact and mechanical stimuli. The findings of Alfred Nobel, that nitroglycerine can be absorbed onto porous materials, solved the problems associated with a liquid explosive of high sensitivity. Nitroglycerine has been mixed with many absorbents and additives to give gelatinized explosives like the many forms of dynamite, gelignite, blasting gelatin, and propellants like cordite and ballistite. Nitroglycerine is one of the few explosives with a positive oxygen balance and so it is often mixed with carbonaceous material or oxygen deficient explosives like nitrocellulose. Despite nitroglycerine having many unfavourable properties synonymous with nitrate esters, it is a powerful high explosive of high brisance (VOD $\sim$ 7750 m/s at $d = 1.59$ g/cm^3). Nitroglycerine finds extensive use in gelatinized commercial explosives and as a component of double-base (DB) gun and rocket propellants for military applications. Recently, the use

Organic Chemistry of Explosives J. P. Agrawal and R. D. Hodgson

of nitroglycerine has been reported for composite modified double-base (CMDB) propellants intended for use in some missiles.

O_2NO ... ONO_2, ONO_2
1
glyceryl trinitrate, GTN
(nitroglycerine, NG)

O_2NO ... ONO_2
2
ethylene glycol dinitrate
(EGDN)

CH_2ONO_2
$O_2NOCH_2-C-CH_2ONO_2$
CH_2ONO_2
3
pentaerythritol tetranitrate
(PETN)

Figure 3.1

Ethylene glycol dinitrate (EGDN) (2) is very similar to nitroglycerine in physical appearance, being a viscous pale yellow to colourless oil. It is a powerful explosive (VOD ~ 7800–8000 m/s at $d = 1.49$ g/cm^3) with a perfect oxygen balance for the full combustion of its aliphatic skeleton. It is considered more stable than nitroglycerine and less sensitive to impact but more volatile. Ethylene glycol dinitrate is mainly used in mixtures with nitroglycerine for the manufacture of nonfreezing dynamites and as an energetic plasticizer.

Pentaerythritol tetranitrate (PETN) (3) is a powerful explosive which exhibits considerable brisance on detonation (VOD ~ 8310 m/s at $d = 1.77$ g/cm^3). It is the most stable and least reactive of the common nitrate ester explosives. The relatively high sensitivity of PETN to friction and impact means that it is usually desensitized with phlegmatizers like wax and the product is used in detonation cord, boosters and as a base charge in detonators. Pentaerythritol tetranitrate can be mixed with synthetic polymers to form plastic bonded explosives (PBXs) like detasheet and Semtex-1A. A cast mixture of PETN and TNT in equal proportions is known as pentolite and has seen wide use as a military explosive and in booster charges. The physical, chemical and explosive properties of PETN commend its use as a high explosive.

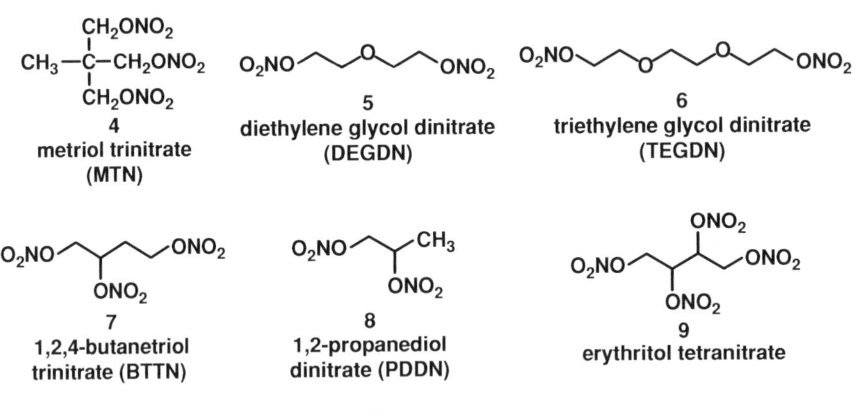

Figure 3.2

The nitrate esters of metriol (1,1,1-trimethylolethane), diethylene glycol (DEG), triethylene glycol (TEG) and 1,2,4-butanetriol (BT) i.e. metriol trinitrate (MTN, VOD ~ 7050 m/s) (4), diethylene glycol dinitrate (DEGDN/DGDN, VOD ~ 6750 m/s) (5), triethylene glycol dinitrate (TEGDN) (6) and 1,2,4-butanetriol trinitrate (BTTN) (7) respectively, were also synthesized by the reaction of the parent alcohol with mixed acid around the time of discovery of nitroglycerine.

The main aim of synthesis and characterization of these nitrate esters was to explore their use as additives to nitroglycerine to lower its freezing point and also, at the same time, to reduce the impact and friction sensitivities of nitroglycerine resulting in its safer handling. Many liquid nitrate esters are excellent energetic plasticizers. Triethylene glycol dinitrate (TEGDN) (6) and 1,2,4-butanetriol trinitrate (BTTN) (7) are currently being investigated in the UK, the USA and France as alternatives to nitroglycerine in propellant and explosive formulations.[2] 1,2-Propanediol dinitrate (PDDN) (8) has found practical use as a high-energy monopropellant for marine applications i.e. torpedoes.[1,3]

Erythritol tetranitrate (9), the product from the *O*-nitration of erythritol, is an explosive equal in power to nitroglycerine and is a solid. Unfortunately, erythritol is not a commercially available chemical, the main source being from some species of seaweed. Consequently, erythritol tetranitrate has never found use as a practical explosive.

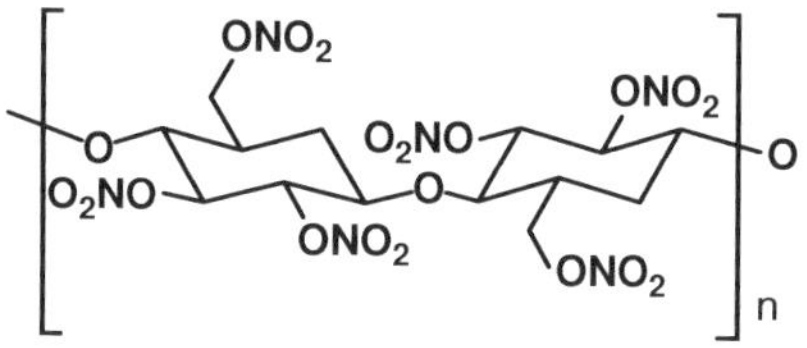

10
nitrocellulose (NC)

Figure 3.3

Nitrocellulose (10), the product from nitrating cellulose in its various forms, is of vast importance in the explosives industry as the main component of many modern gun and rocket propellants, and as a binder for explosive compositions. Each monosaccharide unit of cellulose contains three hydroxy groups available for *O*-nitration. The trinitrate corresponds to the fully nitrated product containing 14.14 % nitrogen, but this degree of nitration is difficult to attain. There are many types of nitrocellulose, all with different nitrogen content, which in turn, affects physical properties such as solubility in organic solvents and affinity for water. Two main grades of nitrocellulose are used in the explosives industry. Nitrocellulose of 12.2 % nitrogen is commonly used in gelatinized explosives like the gelatine dynamites, whereas the product containing 13.45 % nitrogen (guncotton) finds wide use in double-base propellants. The extensive use of nitrocellulose in explosive technologies arises from the wide availability of the raw materials, namely, mixed acid and cellulose, and the fact that nitrocellulose dissolves in some solvents and liquid explosives to form gels – it is essentially the first energetic binder to have been synthesized.

Nitrate esters are amongst the most powerful explosives known. However, their use for some applications is limited. Like most esters they are readily hydrolyzed in the presence of acid or base. Nitrate esters will gradually release nitric acid on prolonged exposure to water or moisture, leading to an auto-catalyzed decomposition, and thus preventing long-term storage. Nitrate esters are generally much more sensitive to mechanical stimuli and shock compared to their *C*-nitro and *N*-nitro counterparts.

Urbański[1] has given a comprehensive account of the use of nitrate esters as commercial and military explosives and their detailed industrial synthesis in Volumes 2 and 4 of *Chemistry and Technology of Explosives*.

3.2 NITRATION OF THE PARENT ALCOHOL

3.2.1 *O*-Nitration with nitric acid and its mixtures

The direct action of nitric acid and its mixtures on the parent alcohol is by far the most important method for the production of nitrate esters on both an industrial and laboratory scale.[4] While such reactions are essentially esterifications they are commonly referred to as *O*-nitrations because the reaction mechanism, involving substitution of hydrogen for a nitro group, is not dissimilar to other nitrations and frequently involves the same nitrating species.

$$H_2SO_4 + HONO_2 \rightleftharpoons H_2\overset{+}{O}NO_2 + HSO_4^- \quad \text{(Eq. 3.1)}$$

$$H_2SO_4 + H_2\overset{+}{O}NO_2 \rightleftharpoons NO_2^+ + HSO_4^- + H_3O^+ \quad \text{(Eq. 3.2)}$$

$$2\,H_2SO_4 + HONO_2 \rightleftharpoons NO_2^+ + 2\,HSO_4^- + H_3O^+ \quad \text{(Eq. 3.3)}$$

Figure 3.4

Mixed acid generated from sulphuric and nitric acids still remains the most important reagent for the industrial production of nitrate esters and explosives in general. The mixed acid contains many nitrating species including the powerful nitronium electrophile (NO_2^+) (Equations 3.1, 3.2 and 3.3). For a long time it was thought that the sulfuric acid in mixed acid acted solely as a dehydrating agent to mop up water formed during the nitration. It is now known that sulfuric acid, a stronger acid than nitric acid, protonates the latter to form the nitracidium cation ($H_2ONO_2^+$) which can lose water to form the nitronium ion; nitric acid acts as a base in this respect. Water is formed during *O*-nitration with the effect of diluting the acid mixture and hence shifting the equilibrium towards the starting material. In this respect the sulfuric acid does serve as a dehydrating agent. It is believed that the heat of solution liberated from the dilution of the mixed acid is responsible for the exothermic nature of *O*-nitration with mixed acid.

Mixed acid has been used to synthesize many commercially important nitrate ester explosives from the parent polyols including: nitroglycerine (NG), ethylene glycol dinitrate (EGDN), diethylene glycol dinitrate (DEGDN), triethylene glycol dinitrate (TEGDN), metriol trinitrate (MTN), 1,2,4-butanetriol trinitrate (BTTN), 1,2-propanediol dinitrate (PDDN) etc.[1] Glycerol is a by-product of soap manufacture and is also synthesized from propylene procured from the petrochemical industry. Ethylene glycol, which is also used to synthesize diethylene glycol and triethylene glycol, is synthesized from ethylene gas, also from the petrochemical industry. Higher polyols like erythritol and mannitol and other naturally occurring sugars are far less commercially available in quantity and this has played a part in why their nitrate esters have not found wide use as explosives. Cellulose is widely available in nature, and after suitable processing, is *O*-nitrated with mixed acid for the manufacture of nitrocellulose. An enormous amount of research has been focused on the relationship between mixed acid composition, reaction time and temperature, and the nitrogen content of the nitrocellulose produced.[5]

Mixed acid is by no means the perfect nitrating agent. This reagent is highly acidic, oxidizing and unselective. Attempts at the selective nitration of glycerol with mixed acid inevitably lead to a mixture of products, and substrates like glycidol cannot be nitrated with mixed acid due to acid-catalyzed ring opening.[6] Nitrations with mixed acid and nitric acid are exothermic and on a large scale there is always the problem of thermal runaway and potential explosion. Consequently, on an industrial scale, the mixed acid nitration of polyols requires strict control, including: remote handling, elaborate reactors and blast-proof buildings. Product separation is a frequent problem with the mixed acid nitration of polyols. The mixed acid residue and the aqueous washings often contain considerable amounts of dissolved nitrate ester, presenting both a safety and a waste problem; ethylene glycol dinitrate is soluble in water to the extent of 0.5 g per 100 ml.[7] If the acid is to be recycled for other nitrations, additional plant is needed to remove or destroy nitrate ester residue. Additionally, products need washing free from occluded acid and may need additional purification steps for complete stabilization.

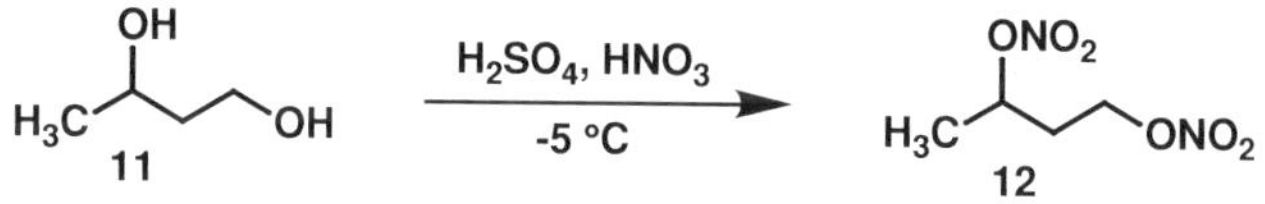

Figure 3.5

However, given all these faults, the low cost and efficiency of mixed acid as an *O*-nitrating agent is unparalleled on an industrial scale. Substrates containing both primary and secondary hydroxy groups are usually nitrated efficiently with mixed acid; work-up and purification is facile and the nitrate ester product is generally obtained in high yield. If nitration temperature is controlled, by-product formation from oxidation is not usually a problem. Some substrates are, however, much more susceptible to oxidation than others and this is the case for substrates containing methylene groups next to a hydroxy group.[8] Accordingly, substrates like 1-propanol, 1,3-propanediol, and 1,3-butanediol (11) should be nitrated at temperatures below 0 °C. 2-Nitro-1-propen-3-ol has been *O*-nitrated with mixed acid at low temperature.[9] Mixed acid or nitric acid cannot be used for the *O*-nitration of tertiary alcohols because the products, tertiary nitrate esters, are unstable under the reaction conditions.

The presence of nitrous acid during *O*-nitration is very undesirable, rendering the nitrate ester product and the nitration reaction unstable through the formation of nitrite esters. The addition of a trace amount of urea usually prevents nitrite ester formation. Marken and co-workers[10] used mixed acid with methylene chloride as co-solvent for the nitration of polyols. A redox probe was used to monitor the concentration of nitrite anion present in solution and hence regulate the substrate addition so as to prevent the formation of significant amounts of unstable nitrite esters. The procedure has been used to synthesize quantities of nitrate ester up to 450 g, both in high yield and of analytical quality, without the need for potentially hazardous purification procedures. 1,3-Propanediol dinitrate (89 %), diethylene glycol dinitrate (69 %), nitroglycerine (90 %), erythritol tetranitrate (89 %), and mannitol hexanitrate (23 %) have been prepared from the parent polyols via this method. Methylene chloride used in these reactions allows better temperature control and the nitrate ester products can be isolated in solution which is safer.

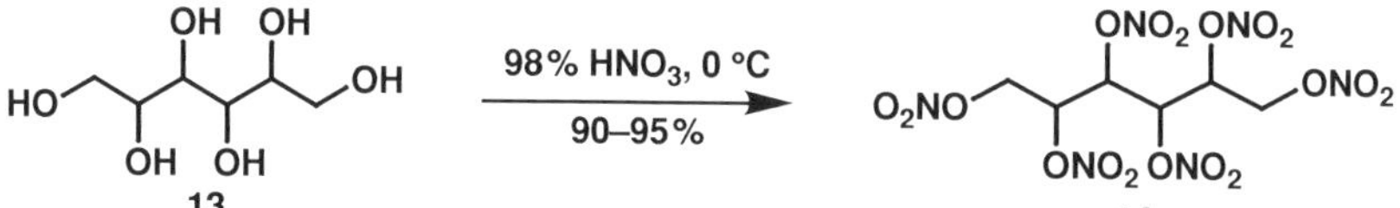

Figure 3.6

While nitric acid is a good solvent for many organic substrates, sulfuric acid is not, and its presence often reduces the solubility of polyols in the mixed acid, leading to heterogeneous suspensions. This is the case for many solid polyols, particularly sugars which form heterogeneous 'pastey masses' on addition to mixed acid. This has meant, for some substrates, nitric acid is the nitrating agent of choice. Solid polyols like pentaerythritol, erythritol and mannitol (13) are commonly *O*-nitrated with fuming or anhydrous nitric acid. In a typical procedure, dry air is bubbled through anhydrous nitric acid to remove any oxides of nitrogen present, followed by the addition of a trace of urea to remove any nitrous acid present. The so-called 'white nitric acid' is then cooled to 0 °C and the substrate, either as a liquid or finely powdered solid, added slowly to an excess of the acid and stirred for a short time. The solution is poured into a large excess of water and after extraction or filtration the nitrate ester is usually procured in excellent yield. In these cases the use of excess nitric acid is essential to ensure complete *O*-nitration of the substrate. Pentaerythritol tetranitrate, erythritol tetranitrate and mannitol hexanitrate (14) can be prepared in ~ 95 % yield using this simple method.[11] Many nitrated sugars have been prepared using either fuming nitric acid[12] or mixed acid[13] at 0–10 °C.

Commercial 70 % nitric acid can be used for the *O*-nitration of low molecular weight alcohols like ethanol and 2-propanol. The nitrate ester products are isolated from the cautious distillation of a mixture of the alcohol and excess 70 % nitric acid.[14] The presence of urea in these reactions is very important for the destruction of nitrous acid and its omission can lead to very violent fume-off. However, this method is not recommended on safety grounds. Using temperatures above ambient for the *O*-nitration of alcohols, with either nitric acid or mixed acid, is dangerous and greatly increases the risk of explosion.

A mixture of fuming nitric acid in acetic anhydride is a common nitrating agent for the synthesis of nitrate esters from alcohols and polyols.[15] The reaction of fuming nitric acid with acetic anhydride can be violent, and as such, the components are usually mixed between 0 and 5 °C. It is generally accepted that these mixtures contain acetyl nitrate. Acetyl nitrate is a weak nitrating agent but in the presence of a strong acid like nitric acid this can ionize to nitronium and acetate ions. Acetyl nitrate reagent has been extensively used for the nitration of sugars.[16] The nitration of unsaturated alcohols with nitric acid in acetic anhydride has been studied.[17] In most cases yields are low (10–30 %) because acetyl nitrate can add to the double bond to form β-nitroacetates (Section 1.3.1). The reaction of the corresponding unsaturated halide derivatives with silver nitrate in acetonitrile is a more useful method for the synthesis of unsaturated nitrate esters (Section 1.4.1).[18]

More obscure reagents used for *O*-nitration include mixtures of phosphoric and nitric acids, and anhydrous nitric acid in which phosphorous pentoxide has been dissolved.[19] The latter mixture contains dissolved dinitrogen pentoxide and is a powerful nitrating agent. A mixture of anhydrous nitric acid containing catalytic amounts of boron trifluoride has been reported to lead to the rapid *O*-nitration of alcohols.[20]

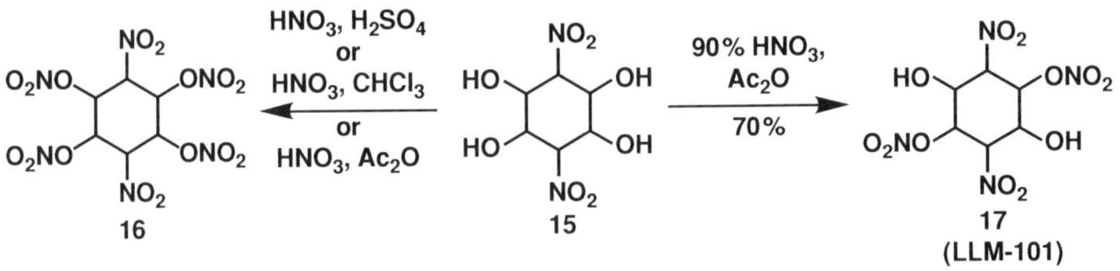

Figure 3.7

An interesting example of the effectiveness of different reagents for *O*-nitration can be seen during the synthesis of *neo*-inositol-based nitrate ester explosives. 1,4-Dideoxy-1,4-dinitro-*neo*-inositol (15), a compound readily prepared from the condensation of nitromethane and glyoxal in the presence of base,[21] undergoes conversion to the tetranitrate ester (16) on treatment with reagents like mixed acid, nitric acid in chloroform, and 98 % nitric acid in acetic anhydride.[22] The tetranitrate ester (16) is a shock sensitive compound of little use as a practical explosive. Treatment of 1,4-dideoxy-1,4-dinitro-*neo*-inositol (15) with a less powerful nitrating agent, namely, a solution of 90 % nitric acid in acetic anhydride, yields the dinitrate ester (17) (LLM-101), a compound with significantly better properties.[22]

3.2.2 *O*-Nitration with dinitrogen tetroxide

$$R{-}OH + N_2O_4 \xrightarrow[\text{or } CH_2Cl_2,\ 0\ ^\circ C \text{ to } 5\ ^\circ C]{\text{Neat, } -170\ ^\circ C \text{ to } 20\ ^\circ C} R{-}ONO + HNO_3 \quad \text{(Eq. 3.4)}$$

Figure 3.8

Dinitrogen tetroxide reacts with simple alcohols in the gas and liquid phase to yield the corresponding nitrite ester as the major product together with trace amounts of oxidation products (Equation 3.4).[23–26] This is the case for neat reactions and those conducted in methylene chloride between subambient and ambient temperatures.

$$R{-}O^-Na^+ + N_2O_4 \xrightarrow[-75\ ^\circ C]{ROH,\ CH_2Cl_2} R{-}ONO_2 + NaNO_2 \quad \text{(Eq. 3.5)}$$

Figure 3.9

In the presence of strong base, i.e. the alkali metal salt of the alcohol, and at low temperatures, dinitrogen tetroxide behaves as a nitrating agent and the corresponding nitrate ester can be obtained in high yield (Equation 3.5).[25,26] Reactions need to be conducted at temperatures of −75 °C or lower and a solution of the nitrogen oxide in methylene chloride is used. The alkali metal salts of higher alcohols (above C_4) are not completely homogeneous in methylene chloride at these low temperatures and so reactions do not proceed satisfactorily.[26]

3.2.3 *O*-Nitration with dinitrogen pentoxide

Dinitrogen pentoxide is a universal nitrating agent with many advantages over conventional nitrations using mixed acid (see Chapter 9). Amongst the advantages are: (1) faster reactions, (2) easier temperature control, (3) easier product isolation, (4) higher purity products, and (5) absence of large amounts of acid waste for disposal. *O*-Nitrations with dinitrogen pentoxide (Equation 3.6) are noted to be clean with an absence of oxidation by-products.[26–28]

$$R{-}OH + N_2O_5 \longrightarrow R{-}ONO_2 + HNO_3 \quad \text{(Eq. 3.6)}$$

Figure 3.10

New and improved routes for the industrial synthesis of dinitrogen pentoxide mean that its use is increasing and for many applications it may replace the use of mixed acid in the near future.[27,28] Dinitrogen pentoxide can be prepared by: (1) ozonolysis of dinitrogen tetroxide,[29] (2) electrolysis of nitric acid–dinitrogen tetroxide solutions,[30] and (3) dehydration of nitric acid.[31,32]

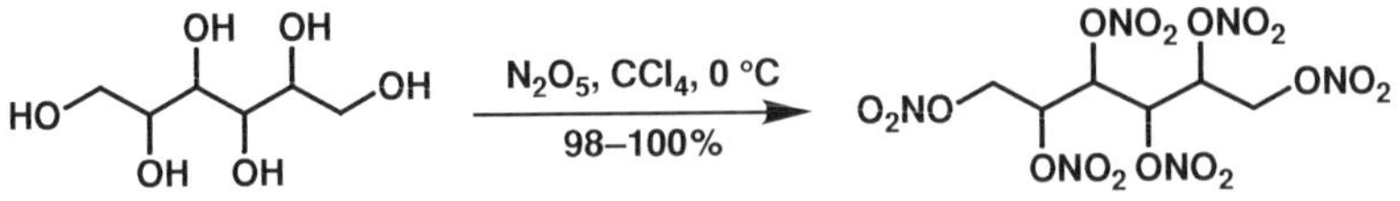

N_2O_5, CCl_4, 0 °C
98–100%
13 → 14

Figure 3.11

Passing a mixed stream of ozone and dinitrogen tetroxide through a solution of the alcohol or suspension of the polyol in methylene chloride or chloroform usually gives the nitrate ester in excellent yield.[26] The substrate may only show sparing solubility in the chlorinated solvent but reactions are still fast and efficient. This is the case with mannitol (13), which can be suspended in carbon tetrachloride and nitrated to mannitol hexanitrate (14) in near quantitative yield.[33]

The nitration of carbohydrates with dinitrogen pentoxide is usually conducted between 10 and 15 °C and sometimes with added sodium fluoride, an acid-binding agent.[32] Acid removal is not usually a problem with most substrates, but can be a big problem with higher carbohydrates such as cellulose and starch which stubbornly retain occluded acid. The use of sodium fluoride circumvents such problems. The effectiveness of dinitrogen pentoxide as a nitrating agent is illustrated by the facile conversion of cellulose to nitrocellulose of 13.9–14.0 % nitrogen content.[34]

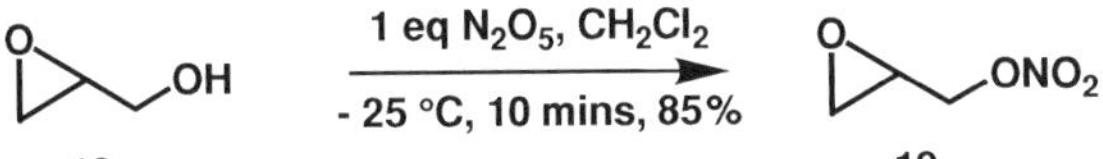

1 eq N_2O_5, CH_2Cl_2
- 25 °C, 10 mins, 85%
18 → 19

Figure 3.12

O-Nitration with dinitrogen pentoxide is exceptionally fast, a point illustrated during the synthesis of glycidyl nitrate (19) from glycidol (18); although nitric acid is formed during the reaction, initial *O*-nitration is much faster than the acid-catalyzed epoxide ring opening.[35] This method can be used for the industrial synthesis of glycidyl nitrate,[36] which is used as a precursor to energetic polymers/binders[36,37] (Section 3.10).

Dinitrogen pentoxide has been used for the synthesis of other energetic nitrate esters. Lustig and co-workers[38] synthesized *gem*-dinitrates by treating acetaldehyde and trifluoroacetaldehyde with dinitrogen pentoxide under super-cooled conditions. Peroxynitrates[39] and peroxyacid nitrates[40] have also been prepared by treating the corresponding hydroperoxides and peroxyacids with dinitrogen pentoxide at subambient temperatures. The poor stability of these products prevents their use as practical energetic materials.

3.2.4 *O*-Nitration with nitronium salts

$$ROH + NO_2BF_4 \longrightarrow RONO_2 + HBF_4 \quad \text{(Eq. 3.7)}$$

Figure 3.13

Nitronium tetrafluoroborate is a very efficient reagent for the *O*-nitration of alcohols and glycols containing both primary and secondary hydroxy functionality (Equation 3.7).[41–43] Yields of nitrate ester are high and frequently quantitative. *O*-Nitration with nitronium salts produces a strong acid and so an acid-binding agent should be present for acid sensitive substrates.

Commercially available nitronium tetrafluoroborate contains between 5 and 15 % nitrosonium tetrafluoroborate[44] and is difficult to purify due to its poor solubility in many organic solvents. The synthesis of high purity nitronium tetrafluoroborate has been described.[45–47] Nitronium hexafluorophosphate is less widely available but can be purified by recrystallization from nitromethane.[48]

Allylic and homoallylic alcohols are particularly susceptible to oxidation and dehydration. Hiskey and Oxley[49] successfully nitrated both allyl alcohol and 1-buten-4-ol with nitronium tetrafluoroborate in diethyl ether at −71 °C.

3.2.5 Transfer nitration

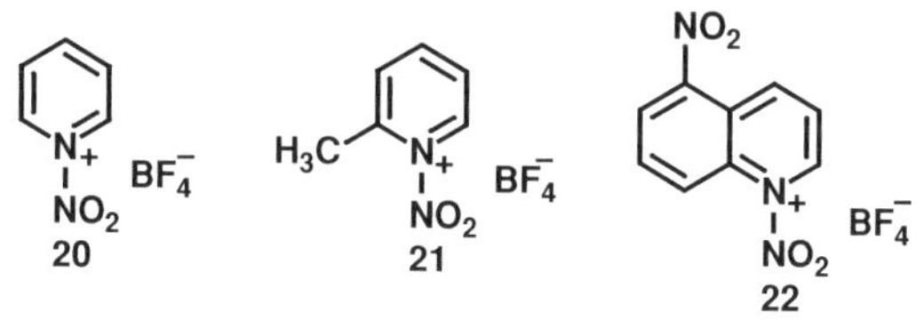

Figure 3.14

Olah and co-workers[50–53] conducted a comprehensive study into the use of *N*-nitropyridinium salts for nitration. Such salts are easily prepared from the slow addition of the appropriate heterocyclic base to an equimolar suspension of nitronium tetrafluoroborate in acetonitrile. Olah studied the effect on nitration of changing both the structure of the heterocyclic base and the counter ion. Three of these salts (20, 21, 22) illustrated above have been synthesized and used for the *O*-nitration of alcohols with success.[50] Transfer nitrations with *N*-nitropyridinium salts are particularly useful for the preparation of nitrate esters from acid-sensitive alcohols and polyols because conditions are essentially neutral.

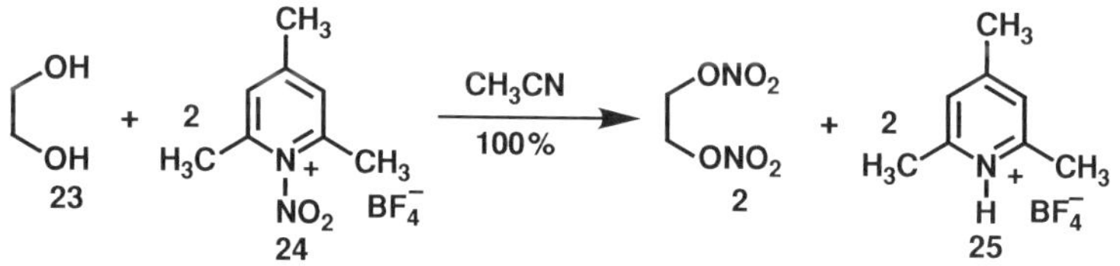

Figure 3.15

N-Nitrocollidinium tetrafluoroborate (24), generated from 2,4,6-collidine (2,4,6-trimethylpyridine) and nitronium tetrafluoroborate, has been used with great success for the transfer nitration of alcohols.[53] Olah and co-workers[53] used this reagent for the *O*-nitration of primary, secondary and tertiary alcohols. This work extended to the synthesis of polynitrate esters; nitroglycerine, ethylene glycol dinitrate (2) and 1,4-butanediol dinitrate have been formed from glycerol, ethylene glycol (23) and 1,4-butanediol respectively. For safety reasons none of these products were isolated from solution. *N*-Nitrocollidinium tetrafluoroborate is a less reactive *O*-nitrating agent than nitronium tetrafluoroborate but offers better control of conditions and yields are usual higher for difficult substrates – *N*-nitrocollidinium tetrafluoroborate[53] reacts with 1-adamantanol to give an 82 % yield of the corresponding nitrate ester, whereas the same reaction with nitronium tetrafluoroborate[54] yields only 2 % nitrate ester.

3.2.6 Other *O*-nitrating agents

3.2.6.1 Boron trifluoride hydrate–potassium nitrate

Olah and co-workers[55] reported using a nitrating agent composed of boron trifluoride hydrate and potassium nitrate for the *O*-nitration of alcohols. The method provides a route to nitrate esters of high purity and essentially free from nitrite esters. This reagent is also nonoxidizing. Olah found that a range of primary and secondary alcohols could be converted to their nitrate esters with this reagent: 2-methyl-1-propyl nitrate (62 %), *n*-pentyl nitrate (75 %), cyclopentyl nitrate (60 %), *n*-butyl nitrate (48 %) and 1-methylheptyl nitrate (79 %) are a few of the nitrate esters prepared from the parent alcohols. The reagent is not of general application for tertiary alcohol *O*-nitration; although 1-adamantanol nitrate is obtained in 86 % from the corresponding alcohol, *tert*-butyl alcohol undergoes dehydration and polymerization. The acidity of the reagent varies considerably depending on the degree of hydration of the boron trifluoride. It has been used for the *C*-nitration of some deactivated aromatic substrates.[56]

3.2.6.2 Lithium nitrate–trifluoroacetic anhydride

A mixture of anhydrous lithium nitrate and trifluoroacetic anhydride in acetonitrile in the presence of sodium carbonate has been used to convert alcohols to nitrate esters for a range of peptide, carbohydrate and steroid substrates.[57] Yields are good to high but products need purification to remove trifluoroacetate ester impurities, which can be significant in the absence of the carbonate. A similar system used for the nitration of electron-rich aromatic heterocycles employs trifluoroacetic anhydride with ammonium or potassium nitrate.[58]

3.2.6.3 Thionyl nitrate and thionyl chloride nitrate

Thionyl chloride nitrate and thionyl nitrate are prepared *in situ* from the reaction of a solution of thionyl chloride in THF with one and two equivalents of silver nitrate respectively, during which time silver chloride precipitates from solution.[59]

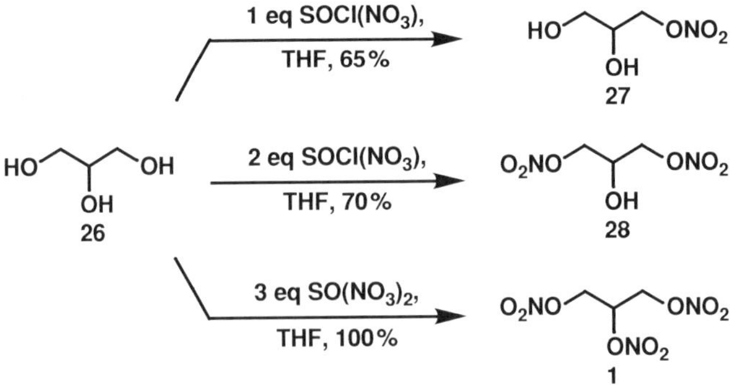

HO–CH2–CH(OH)–CH2–OH (26)
→ 1 eq $SOCl(NO_3)$, THF, 65% → HO–CH2–CH(OH)–CH2–ONO_2 (27)
→ 2 eq $SOCl(NO_3)$, THF, 70% → O_2NO–CH2–CH(OH)–CH2–ONO_2 (28)
→ 3 eq $SO(NO_3)_2$, THF, 100% → O_2NO–CH2–CH(ONO_2)–CH2–ONO_2 (1)

Figure 3.16

Both reagents are efficient for the *O*-nitration of primary and secondary hydroxy groups.[59] Thionyl chloride nitrate can be used for the selective nitration of primary hydroxy groups in the presence of secondary hydroxy groups. This is illustrated in the case of glycerol (26), which on treatment with one and two equivalents of thionyl chloride nitrate in THF, forms glyceryl-1-nitrate (27) and glyceryl-1,3-dinitrate (28) respectively. Thionyl nitrate is a less selective reagent. Treatment of glycerol (26) with three equivalents of thionyl nitrate in THF yields nitroglycerine (1) in quantitative yield. Both reagents are suitable for the *O*-nitration of acid sensitive substrates but have found little use, probably because more convenient *O*-nitrating agents are available for such substrates.

3.3 NUCLEOPHILIC DISPLACEMENT WITH NITRATE ANION

3.3.1 Metathesis between alkyl halides and silver nitrate

The reaction of alkyl halides with silver nitrate constitutes an extremely useful method for the synthesis of high purity nitrate esters on a laboratory scale.[60–68] The driving force for these reactions is the formation of the insoluble silver halide. Reactions have been conducted under homogenous[61,63] and heterogeneous conditions.[60,68] For the latter a solution of the alkyl halide in an inert solvent like benzene or ether is stirred with finely powdered silver nitrate. However, this method has been outdated and reactions are now commonly conducted under homogeneous conditions using acetonitrile as solvent.

The scope of these reactions is illustrated by the variety of functionalized nitrate esters which can be prepared (Table 3.1). In general, primary and secondary alkyl iodides and bromides

$$\underset{X = Cl, Br, I}{RX} \xrightarrow[CH_3CN]{AgNO_3} RONO_2 + AgX$$

Table 3.1
Synthesis of nitrate esters from the reaction of alkyl halides with silver nitrate

Entry	Substrate	Product	Yield (%)	Ref.
1	CH_3CH_2I	$CH_3CH_2ONO_2$	72	60
2	$HOCH_2CH_2Br$	$HOCH_2CH_2ONO_2$	70	61
3	$N{\equiv}CCH_2I$	$N{\equiv}CCH_2ONO_2$	32	61
4	epibromohydrin (oxirane–CH_2Br)	glycidyl nitrate (oxirane–CH_2ONO_2)	72	62
5	2,4,6-trinitrobenzyl bromide (CH_2Br; O_2N, NO_2, NO_2)	2,4,6-trinitrobenzyl nitrate (CH_2ONO_2; O_2N, NO_2, NO_2)	64	63
6	$ClCH_2OCH_2Cl$	$O_2NOCH_2OCH_2ONO_2$	82	64
7	$F{-}C(NO_2)_2{-}CH_2OCH_2Cl$	$F{-}C(NO_2)_2{-}CH_2OCH_2ONO_2$	---	65

give excellent yields of the corresponding nitrate ester. Allylic, benzylic and tertiary chlorides also give nitrate esters in good yield. The nucleophilic substitution of alkyl halides with silver nitrate is a very attractive method for the synthesis of some acid-sensitive nitrate esters, like glycidyl nitrate, which is prepared in 72 % yield from the reaction of epibromohydrin with silver nitrate (Table 3.1, Entry 4).[62] The direct nitration of glycidol with nitric acid or mixed acid is not possible due to acid-catalyzed ring opening of the epoxide functionality. Russian chemists[64,65] have reported the synthesis of energetic hemiformal nitrates from the metathesis of the corresponding chloromethyl ethers with silver nitrate (Table 3.1, Entries 6 and 7).

$$R{-}OH \xrightarrow[Et_2O,\ CH_3CN]{Ph_3P,\ I_2,\ Imidazole} [R{-}I] \xrightarrow{AgNO_3} R{-}ONO_2 \qquad (Eq.\ 3.8)$$

Figure 3.17

Tojo and co-workers[69] reported a ‘one-pot’ synthesis of alkyl nitrates from alcohols via the alkyl iodide; the alcohol is treated with a mixture of iodine, triphenylphosphine and imidazole in diethyl ether–acetonitrile, and the resulting alkyl iodide is reacted *in situ* with silver nitrate (Equation 3.8). Reported yields for primary alcohols are good to excellent but yields are lower for secondary alcohols.

3.3.2 Decomposition of nitratocarbonates

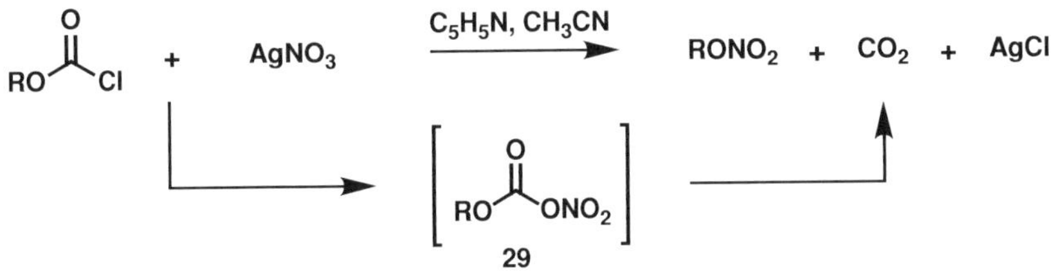

$$RO{-}C(=O){-}Cl + AgNO_3 \xrightarrow{C_5H_5N,\ CH_3CN} RONO_2 + CO_2 + AgCl$$

(via intermediate $[RO{-}C(=O){-}ONO_2]$ **29**)

Figure 3.18

An extremely mild method for the synthesis of nitrate esters from easily oxidized or acid-sensitive alcohols involves the decomposition of a nitratocarbonate (29).[70,71] The nitratocarbonate is prepared *in situ* from metathesis between a chloroformate (reaction between phosgene and an alcohol) and silver nitrate in acetonitrile in the presence of pyridine at room temperature. Under these conditions the nitratocarbonate readily decomposes to yield the corresponding nitrate ester and carbon dioxide. Few examples of these reactions are available in the literature[70,71] and they are limited to a laboratory scale.

3.3.3 Displacement of sulfonate esters with nitrate anion

Voelter and co-workers[72] synthesized the nitrate esters of some carbohydrates by treating the corresponding trifluoromethanesulfonate (triflate) esters with tetrabutylammonium nitrate. Yields were generally excellent. The same authors describe a one-pot process for the synthesis of nitrate esters from carbohydrates. In these reactions the intermediate triflate ester is formed *in situ* and reacted directly with tetrabutylammonium nitrate without prior isolation.

3.3.4 Displacements with mercury (I) nitrate

$$RBr + HgNO_3 \xrightarrow{DME} RONO_2 + HgBr \quad \text{(Eq. 3.9)}$$

Figure 3.19

McKillop and Ford[73] synthesized a range of primary and secondary alkyl nitrates in excellent yields by treating alkyl bromides with mercury (I) nitrate in 1,2-dimethoxyethane at reflux (Equation 3.9). This method has been used to synthesize substituted nitrate esters from both α-bromocarboxylic acid and α-bromoketone substrates. Unlike metathesis with silver salts, which are widely known to promote S_N1 reactions, this method is not useful for the synthesis of nitrate esters from tertiary alkyl halides.

3.4 NITRATE ESTERS FROM THE RING-OPENING OF STRAINED OXYGEN HETEROCYCLES

The strained rings of epoxides and oxetanes are susceptible to nucleophilic attack. In this section we discuss the reactions of these oxygen heterocycles with nitrogen oxides and other reagents as a route to nitrate esters.

3.4.1 Ring-opening nitration of epoxides

3.4.1.1 Reaction of epoxides with dinitrogen tetroxide

H H O 30 $\xrightarrow[CH_2Cl_2]{N_2O_4}$ ONO ONO$_2$ 31

$\xrightarrow{H_2O}$ HO ONO$_2$ 32

$\xrightarrow[N_2O_5 \text{ or } O_3]{[O]}$ O$_2$NO ONO$_2$ 2

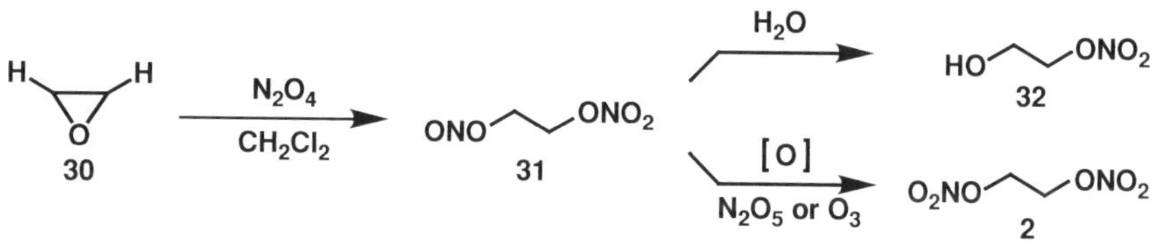

Figure 3.20

The reaction of epoxides with dinitrogen tetroxide in chlorinated solvents has been studied with the conclusion that under these conditions dinitrogen tetroxide has only weak nitrating ability.[74] These reactions yield unstable nitrate-nitrite esters which on aqueous work-up undergo hydrolysis to yield the corresponding β-hydroxy-nitrate ester; 2-nitratoethanol (32) can be formed in this way from ethylene oxide (30).[75] Nitrate-nitrite esters like (31) undergo efficient oxidation to the corresponding *vic*-dinitrate esters on treatment with reagents like ozone and dinitrogen pentoxide.[26,75] However, the same transformation of epoxide to *vic*-dinitrate ester can be achieved in one step with dinitrogen pentoxide and in much higher yield (Section 3.4.1.2).

3.4.1.2 Vicinal dinitrate esters from the reaction of epoxides with dinitrogen pentoxide

Millar and co-workers[75] found that treating epoxides with dinitrogen pentoxide in inert chlorinated solvents between 0 and 20 °C leads to nitrative ring opening and the formation of *vic*-dinitrate esters. These reactions are very chemically efficient, resulting in complete utilization of nitrogen by incorporating the whole molecule of N_2O_5 into the dinitrate ester product. No nitric acid is formed and if conditions are kept anhydrous the process constitutes a nonacidic nitrating method. Reactions are much less exothermic and easier to control compared to conventional nitrations with mixed acid.

In many cases almost quantitative yields are reported for the formation of *vic*-dinitrate esters from the reaction of simple alkyl and dialkyl epoxides with dinitrogen pentoxide (Table 3.2). Some of the products formed include: ethylene glycol dinitrate (2) (96 %), 1,2-propanediol dinitrate (8) (96 %), 2,3-butanediol dinitrate (94 %) and 1,2-butanediol dinitrate (96 %). Reaction times are of the order of 5–15 minutes.

Some functionalized epoxides react much slower with dinitrogen pentoxide, especially substrates containing oxygen functionality near to the epoxide group. This effect is seen during the reaction of glycidol with dinitrogen pentoxide, which undergoes rapid *O*-nitration to give glycidyl nitrate, followed by a much slower reaction to open the epoxide ring. Slow epoxide ring-opening allows polymer formation to compete with *O*-nitration. Consequently, Millar and

Table 3.2
Synthesis of vicinal dinitrate esters from the ring opening nitration of epoxides with dinitrogen pentoxide (Ref. 75)

Entry	Epoxide	Mole ratio N_2O_5 : epoxide	Nitrate ester	Yield (%)
1	(30)	1.1 : 1.0	(2)	96
2	(33)	1.1 : 1.0	(8)	96
3	(34)	1.1 : 1.0	(35)	85
4	(18)	3.0 : 1.0[a]	(1)	73
5	(36)	3.0 : 1.0[a]	(9)	55
6	(37)	2.15 : 1.0	(38)	31

[a] 1 mole equivalent of aluminium chloride added.

co-workers[75] found it necessary to add aluminium chloride to the reaction mixtures of some substrates. Glycidol (18) and butadiene diepoxide (36) react with dinitrogen pentoxide in the presence of 1.0 mole equivalent of aluminium chloride to give nitroglycerine (1) and erythritol tetranitrate (9) respectively (Table 3.2, Entries 4 and 5). The synthesis of erythritol tetranitrate via this method is attractive because the scarcity of erythritol prevents its direct nitration with mixed acid on an industrial scale.

Research efforts are ongoing into the use of dinitrogen pentoxide for the industrial synthesis of nitrated hydroxy-terminated polybutadiene (NHTPB) from epoxidated HTPB (Section 3.10).[76] The reaction of aziridines with dinitrogen pentoxide is an important route to 1,2-nitramine-nitrates and these reactions are discussed in Section 5.8.1.[77]

3.4.1.3 *Reaction of epoxides with nitrate anion under acidic conditions*

30 → HNO_3 (aq), NH_4NO_3 or HNO_3, $CHCl_3$ 58% → 32 ($HOCH_2CH_2ONO_2$)

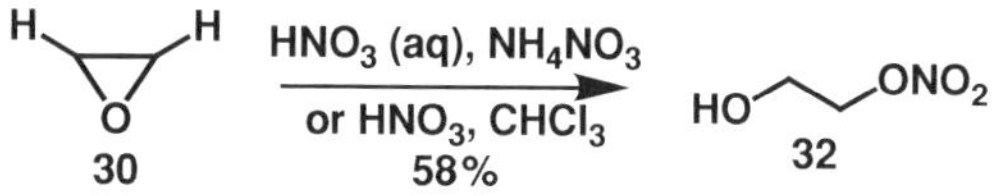

Figure 3.21

Nichols and co-workers[78,79] studied the formation of β-hydroxy nitrate esters via the acid-catalyzed ring-opening of epoxides with nitrate anion. Reactions were conducted in aqueous media using solutions of ammonium nitrate is nitric acid and under nonaqueous conditions using solutions of pure nitric acid in chloroform. 2-Nitratoethanol (32) is formed in ~58 % yield from ethylene oxide (30) using either of these methods.[78]

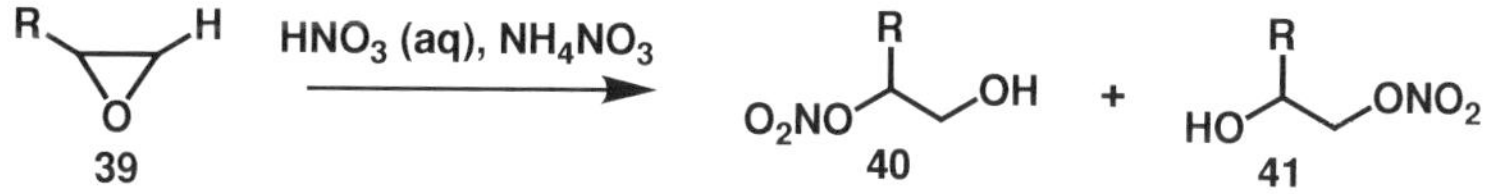

39 → HNO_3 (aq), NH_4NO_3 → 40 + 41

Figure 3.22

Unsymmetrical epoxides (39) can form two isomers, (40) and (41), on reaction with nitrate anion and so raise the issue of regioselectivity. Under acidic conditions terminal epoxides are found to predominantly yield the primary nitrate ester (41); although this is not clear cut and propylene oxide is reported to yield an ill defined mixture of isomers.[78] A comprehensive study on the regioselectivity of epoxide opening with nitrate anion under acidic conditions was conducted on glycidol.[79]

3.4.1.4 *Single electron transfer nitration of epoxides with ceric ammonium nitrate*

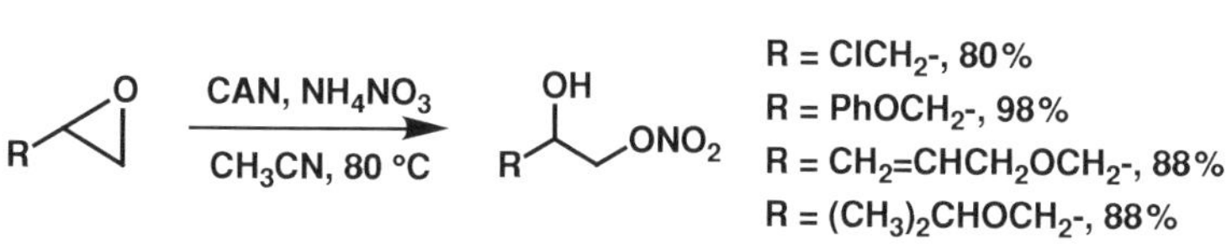

CAN, NH_4NO_3, CH_3CN, 80 °C

R = $ClCH_2$-, 80 %
R = $PhOCH_2$-, 98 %
R = CH_2=$CHCH_2OCH_2$-, 88 %
R = $(CH_3)_2CHOCH_2$-, 88 %

Figure 3.23

Ceric ammonium nitrate (CAN) has been used for the synthesis of β-hydroxy nitrate esters from epoxides.[80–82] These reactions are performed in the presence of nitrate sources like ammonium nitrate or n-tetrabutylammonium nitrate, or with CAN alone, in acetonitrile between room temperature and reflux.[80,81] Yields of β-hydroxy nitrate esters are high to excellent for a range of epoxide substrates. Reactions are generally very stereoselective for unsymmetrical epoxides carry either activating or deactivating groups α to the epoxy group; epichlorohydrin and 3-phenoxy-1-epoxypropane both react to give the primary nitrate esters.[80] Poor selectivity is observed for unsymmetrical epoxides where electronic and steric effects are counterbalanced i.e. for 1-epoxypropane.[80] Lower yields are realized for more hindered epoxides and tertiary centres can lead to undesirable side-reactions.[81]

3.4.1.5 Other reactions

A suspension of thallium (III) nitrate in hexane reacts with epoxides to give the corresponding β-hydroxy nitrate esters in good yield.[83] The same reagent in acetonitrile has been used to synthesize α-nitratoketones[84] from substituted acetophenones, 1,2-dinitrate esters from alkenes,[85] and 1,3-dinitrates from ring-opening nitration of cyclopropanes.[85]

3.4.2 1,3-Dinitrate esters from the ring-opening nitration of oxetanes with dinitrogen pentoxide

N_2O_4, CH_2Cl_2 → $O_2NO{-}C(R^1)(H){-}C(R^2)(R^3){-}C(H)(H){-}ONO_2$

R^1, R^2, R^3 = H, 88%
R^1 = H, R^2, R^3 = Me, 73%
R^1 = H, R^2, R^3 = -$(CH_2)_5$-, 86%
R^1 = Me, R^2, R^3 = Et, 17%

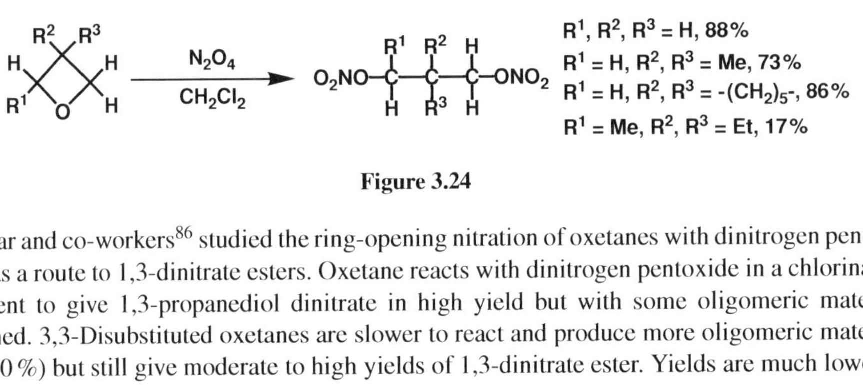

Figure 3.24

Millar and co-workers[86] studied the ring-opening nitration of oxetanes with dinitrogen pentoxide as a route to 1,3-dinitrate esters. Oxetane reacts with dinitrogen pentoxide in a chlorinated solvent to give 1,3-propanediol dinitrate in high yield but with some oligomeric material formed. 3,3-Disubstituted oxetanes are slower to react and produce more oligomeric material (5–30 %) but still give moderate to high yields of 1,3-dinitrate ester. Yields are much lower if the oxetane substrate contains 2-alkyl substituents.

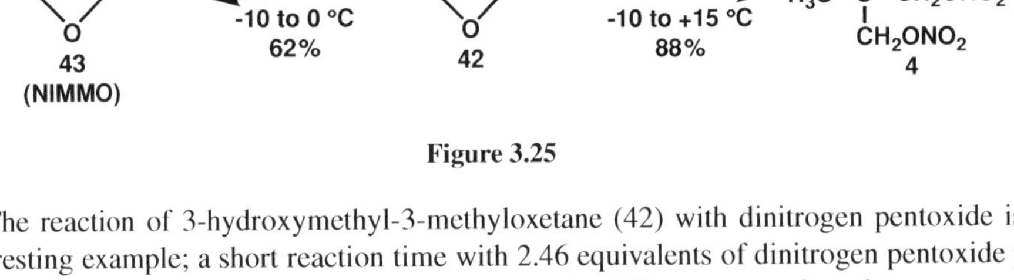

43 (NIMMO) ← (2.46 eq N_2O_5, $CHCl_3$, 0.5 hr, -10 to 0 °C, 62%) ← 42 → (4.0 eq N_2O_5, $CHCl_3$, 28 hr, -10 to +15 °C, 88%) → $H_3C{-}C(CH_2ONO_2)_3$ 4

Figure 3.25

The reaction of 3-hydroxymethyl-3-methyloxetane (42) with dinitrogen pentoxide is an interesting example; a short reaction time with 2.46 equivalents of dinitrogen pentoxide producing 3-(nitratomethyl)-3-methyloxetane (NIMMO) (43), an energetic polymer precursor, as the sole product (62 %), whereas longer reaction times with 4.0 equivalents of dinitrogen pentoxide yields the high explosive, metriol trinitrate (4) as the sole product (88 %).

The use of oxetanes for the synthesis of polynitrate esters is generally of less value than the use of epoxides, which are readily available from the epoxidation of alkenes.[87] The analogous reaction of azetidines with dinitrogen pentoxide is a route to 1,3-nitramine-nitrates and these reactions are discussed in Section 5.8.2.[88]

3.4.3 Other oxygen heterocycles

Saturated cyclic ethers with ring sizes greater than four exhibit much less internal strain and so the driving force for ring cleavage is much lower; tetrahydropyran and oxepane react slowly with dinitrogen pentoxide in chlorinated solvent to yield complex mixtures resulting from oxidation and oligomerization.[89]

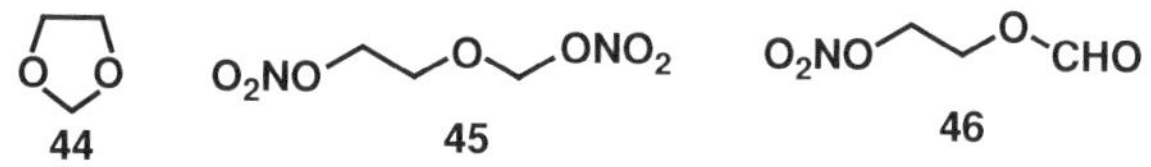
O O
44
O2NO
O
ONO2
45
O2NO
O
CHO
46

Figure 3.26

Cyclic formals react with dinitrogen pentoxide in chlorinated solvent to yield unstable but interesting ring-opened products, including hemiformal nitrates; 1,3-dioxolane (44) reacts to yield a mixture of hemiformal nitrate (45) and formal ether (46) products.[89] Similar products are formed from acyclic formals and dinitrogen pentoxide.[89]

3.5 NITRODESILYLATION

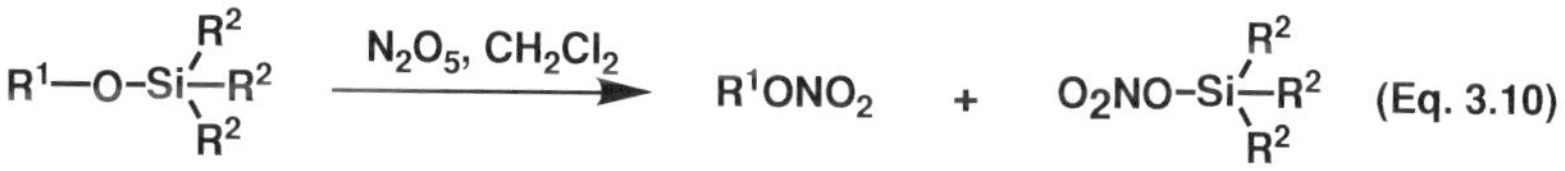
R1—O-Si—R2
N2O5, CH2Cl2
R1ONO2
+
O2NO-Si—R2
(Eq. 3.10)

Figure 3.27

Millar and co-workers[90] treated a number of silyl ethers with a solution of dinitrogen pentoxide in methylene chloride at subambient temperature and obtained the corresponding nitrate esters through silicon–oxygen heteroatom cleavage (Equation 3.10). These reactions, known as nitrodesilyations, are an important route to nitrate esters and very amenable to the synthesis of polynitrate-based high energy materials for explosives and propellants. The same reactions with silylamines yield nitramines (Section 5.7).

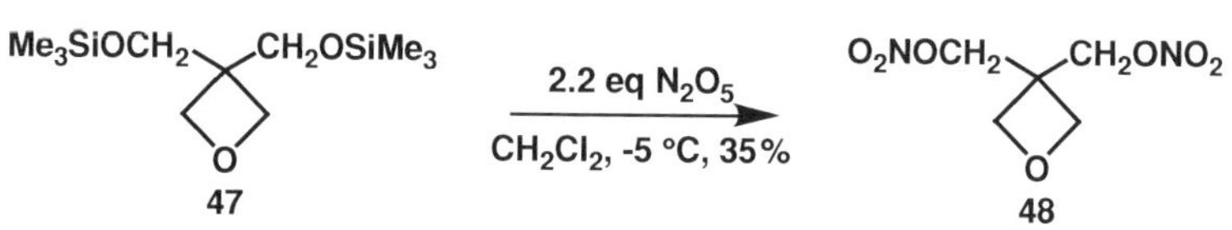
Me3SiOCH2
CH2OSiMe3
O
47
2.2 eq N2O5
CH2Cl2, -5 °C, 35%
O2NOCH2
CH2ONO2
O
48

Figure 3.28

Trimethylsilyl ethers and other simple *n*-alkylsilyl ethers give high yields of nitrate ester products and reactions are clean. Nitrate esters prepared in this way include: 2-ethylhexanol nitrate (92 %), 2-octanol nitrate (88 %), 1-decanol nitrate (87 %) and 1,6-hexanediol dinitrate (83 %). Yields are notably lower for compounds with silyl groups containing secondary (TIPS) and tertiary (TBDMS) alkyl groups. This in no way detracts value from the method; trimethylsilyl chloride, the reagent most likely to be used to synthesize these substrates on an industrial scale, is the cheapest and most readily available of the chlorosilanes. Under the conditions of nitrodesilylation, the nitrate ester is obtained as a solution in methylene chloride and so this method must be regarded as safer than conventional *O*-nitration with mixed acid. This reaction is suitable for the synthesis of products containing acid-sensitive functionality such as the oxetane (48), an energetic polymer precursor, provided that anhydrous conditions are maintained; the low yield in this reaction is due to the neopentyl-type steric hindrance at the reacting centres and not due to ring cleavage. The reaction of dinitrogen pentoxide with cyclic silyl ethers is more complex and depends on the nature of the substrate and the reaction stoichiometry.

3.6 ADDITIONS TO ALKENES

3.6.1 Nitric acid and its mixtures

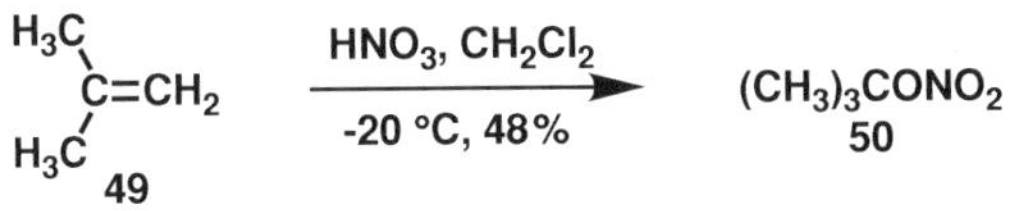

$$(H_3C)_2C{=}CH_2 \ \textbf{49} \xrightarrow[-20\ ^\circ C,\ 48\%]{HNO_3,\ CH_2Cl_2} (CH_3)_3CONO_2 \ \textbf{50}$$

Figure 3.29

Solutions of nitric acid in chlorinated solvents can add to some alkenes to give nitrate esters. Some tertiary nitrate esters can be prepared in this way; isobutylene (49) reacts with fuming nitric acid of 98.6 % concentration in methylene chloride to give *tert*-butyl nitrate (50).[91] However, the products obtained depend on both the substrate and the reaction conditions; β-nitro-nitrate esters, *vic*-dinitrate esters, β-nitroalcohols and nitroalkenes have been reported as products with other alkenes.[92] Oxidation products like carboxylic acids are also common, especially at elevated temperatures and in the presence of oxygen.[93] The reaction of alkenes with fuming nitric acid is an important route to unsaturated nitrosteroids, which assumedly arise from the dehydration of β-nitroalcohols or the elimination of nitric acid from β-nitro-nitrate esters.[94]

Mixed acid has been reported to react with some alkenes to give β-nitro-nitrate esters amongst other products.[91]

Solutions of acetyl nitrate at subambient temperature can react with alkenes to yield a mixture of nitro and nitrate ester products. Cyclohexene forms a mixture of 2-nitrocyclohexanol nitrate, 2-nitrocyclohexanol acetate, 2-nitrocyclohexene and 3-nitrocyclohexene.[95] This illustrates one of the problems of allylic and homoallylic alcohol *O*-nitration with this reagent.

3.6.2 Nitrogen oxides

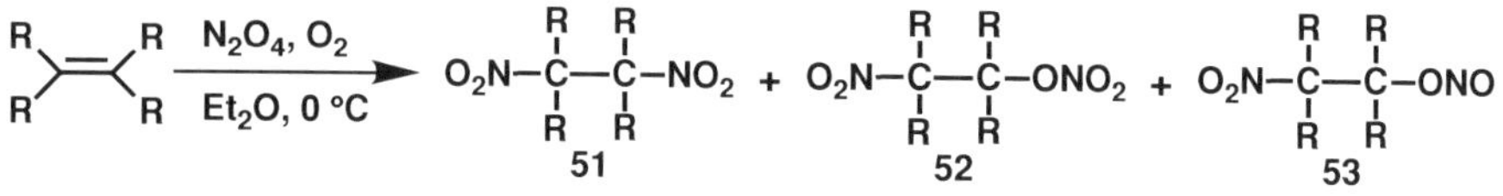

$$R_2C{=}CR_2 \xrightarrow[\text{Et}_2\text{O, 0 °C}]{N_2O_4,\ O_2} O_2N{-}CR_2{-}CR_2{-}NO_2\ (\mathbf{51}) + O_2N{-}CR_2{-}CR_2{-}ONO_2\ (\mathbf{52}) + O_2N{-}CR_2{-}CR_2{-}ONO\ (\mathbf{53})$$

Figure 3.30

Alkenes react with dinitrogen tetroxide in ether at 0 °C in the presence of oxygen to form a mixture of *vic*-dinitroalkane (51), β-nitro-nitrate (52) and β-nitro-nitrite (53) compounds, the nitrite being oxidized to the nitrate ester in the presence of excess dinitrogen tetroxide.[96] A stream of oxygen gas is normally bubbled through the reaction mixture to expel nitrous oxide formed during the reaction. Nitrous oxide can add to alkenes to form nitroso compounds and so its presence can lead to even more complex mixtures being formed. β-Nitroso-nitrates have been observed in some cases.[97] Detailed discussions of these reactions have been given by Olah[4d], Topchiev[23], and in numerous reviews[98] and mechanistic studies.[99] Additions of dinitrogen tetroxide across C–C double bonds are selective – the β-nitro-nitrate formed has the nitrate ester group situated on the carbon bearing the least hydrogens i.e. the most substituted carbon.[96] In general, difficulties in separation of such mixtures limit their synthetic value in relation to nitrate ester synthesis. However, such reactions have found preparative value for the synthesis of some dinitroalkanes[100] and nitroalkenes[101] (Section 1.4).

Alkenes react with dinitrogen pentoxide in chlorinated solvents at 0–25 °C to give a mixture of *vic*-dinitrate ester, *vic*-dinitroalkane and β-nitro-nitrate ester compounds.[102] Propylene reacts with dinitrogen pentoxide in the gas phase to give 1,2-propanediol dinitrate as one of the products.[103] Dinitrate ester products are also observed during the reactions of ethylene and *cis*-2-butene with dinitrogen pentoxide.[104] The same reactions at lower temperatures (−10 to −30 °C) produce the expected product of N_2O_5 addition, the β-nitro-nitrate ester, as the main product, along with some nitroalkene from elimination of nitric acid from the latter.[105] The reaction of hydroxy-terminated polybutadiene (HTPB) with dinitrogen pentoxide in methylene chloride yields an energetic oligomer containing β-nitro-nitrate ester functionality.[76] The inherent instability of β-nitro-nitrate esters limits their use for practical application as energetic materials.

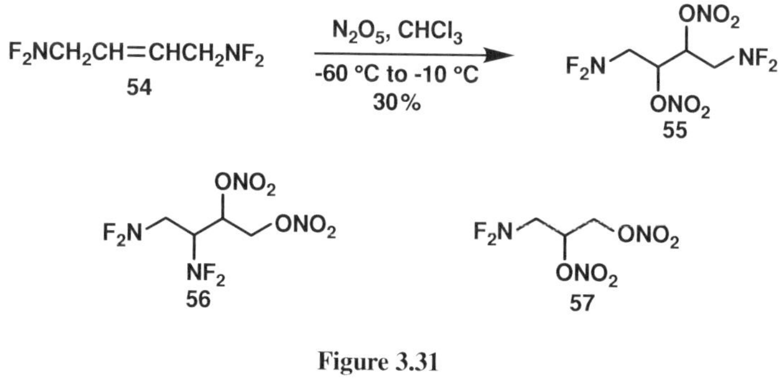

$$F_2NCH_2CH{=}CHCH_2NF_2\ (\mathbf{54}) \xrightarrow[\text{-60 °C to -10 °C, 30\%}]{N_2O_5,\ CHCl_3} \mathbf{55}$$

55: F_2N, ONO_2, ONO_2, NF_2; 56: F_2N, ONO_2, ONO_2, NF_2; 57: F_2N, ONO_2, ONO_2

Figure 3.31

Canfield and Rohrback[106] reported on the reaction of some electron-deficient difluoroaminoalkenes with dinitrogen pentoxide in chloroform at subambient temperatures. Contrary to previous work that the β-nitro-nitrate ester should be the main product, the corresponding *vic*-dinitrate esters were isolated from these reactions; 1,4-bis(*N*,*N*-difluoroamino)-2,3-butanediol dinitrate (55), 3,4-bis(*N*,*N*-difluoroamino)-1,2-butanediol dinitrate (56) and 3-(*N*,*N*-difluoroamino)-1,2-propanediol dinitrate (57) were isolated in 30 %, 27 %, and 55 % yields respectively, from the parent alkenes.

3.6.3 Metal salts

A suspension of thallium (III) nitrate in pentane at room temperature can react with alkenes to give *vic*-dinitrate esters.[85] Cyclohexene reacts with this reagent to give 1,2-cyclohexanediol dinitrate (85 %) (as a mixture of isomers) and 15 % cyclopentanecarboxaldehyde (hydride shift in the dethallation step). Some alkenes react extremely slowly with this reagent e.g. isomeric 5-decenes.

3.6.4 Halonitroxylation

A number of reagents have been reported to add across the unsaturated bonds of alkenes to give β-haloalkyl nitrates, a process known as halonitroxylation. The reaction of alkenes with a mixture of iodine and dinitrogen tetroxide in chloroform can give high yields of the corresponding β-iodoalkyl nitrate ester.[107] The same reaction with bromine yields β-bromoalkyl nitrate esters, although the *vic*-dibromoalkane is a major by-product.[107] Iodonium nitrate, a reagent prepared from the reaction of silver nitrate with iodine monochloride in chloroform/pyridine, can add to the double bond of an alkene to give any of three products, including: β-iodoalkyl nitrate esters, β-iodoalkyl pyridinium nitrates, or alkenyl pyridinium nitrates depending on the substrate.[108] Bromonium nitrate, a reagent prepared from the reaction of bromine with silver nitrate in chloroform/pyridine, reacts with alkenes to form β-bromoalkyl nitrate esters or β-bromoalkyl pyridinium nitrates depending on the substrate and conditions.[109] Other reagents used for the synthesis of β-haloalkyl nitrate esters include: mercury (II) nitrate and iodine,[110] mercury (II) nitrate and bromine,[110,111] and alkali metal nitrates and bromine water[112]. The reaction of an alkene with mercury (II) nitrate and a halogen is a process known as nitratomercuriation and also yields β-haloalkyl nitrate esters.[110,111]

3.7 DEAMINATION

Barton and Narang[113] have prepared nitrate esters by treating primary and secondary alkylamines with dinitrogen tetroxide in the presence of an amidine base like DBU. Wudl and Lee[114] conducted deamination reactions without any amidine base and reported much lower yields of nitrate ester product. The use of an amidine base is not necessary if the amine substrate is first *N*-protected as its TMS derivative and then treated with dinitrogen tetroxide.[115] The reaction of primary amines and nitramines with a solution of dinitrogen pentoxide in carbon tetrachloride at −30 °C is reported to lead to nitrate esters through deamination.[116]

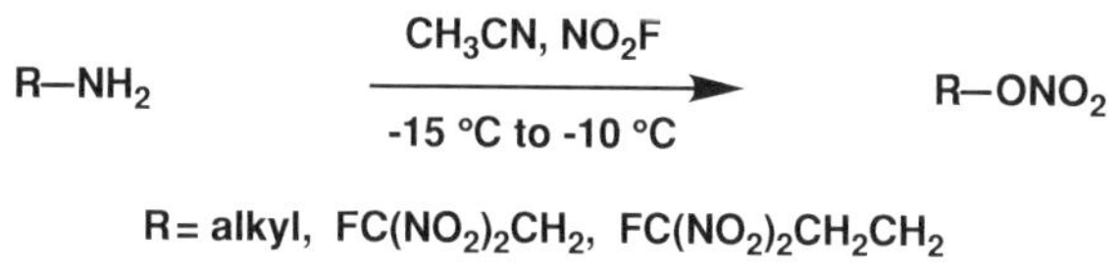

$$R\text{—}NH_2 \xrightarrow[-15\,^\circ C \text{ to } -10\,^\circ C]{CH_3CN,\ NO_2F} R\text{—}ONO_2$$

$R = $ alkyl, $FC(NO_2)_2CH_2$, $FC(NO_2)_2CH_2CH_2$

Figure 3.32

Eremenko and co-workers[117] used nitryl fluoride for the deamination of amines at subambient temperatures in acetonitrile. The same reaction occurs with primary nitramines and their alkali metal salts; bis-nitramines react to give the corresponding bis-nitrate esters.[117,118] The reaction of primary aliphatic amines and nitramines with nitronium salts also leads to deamination and the formation of alkyl nitrates.[117,119]

Katritsky and co-workers[120] reacted primary amines with the triflate salt of 5,6,8,9-tetrahydro-7-phenyl-dibenzo[*c*, *h*]acridine to generate the corresponding *N*-substituted acridinium triflate salts, which on refluxing in dioxane in the presence of benzyltrimethylammonium nitrate, yield the corresponding nitrate esters.

3.8 MISCELLANEOUS METHODS

- Cyclopropanes are reported to react with a suspension of thallium (III) nitrate in pentane at room temperature to give 1,3-dinitrate esters.[85] This reaction is interesting but its scope cannot be accessed because only one example appears in the literature.
- Cyclopropane is reported to react with dinitrogen pentoxide in methylene chloride at subambient temperature to yield 3-nitro-1-propanol nitrate.[121]
- Treatment of primary nitramines with absolute nitric acid yields the corresponding nitrate ester and nitrous oxide.[122]
- Alkyl hydroperoxides react with nitrous acid to give the corresponding nitrate ester.[123]
- Acyl nitrates react with low molecular weight aliphatic alcohols to give the alkyl nitrate ester and the corresponding carboxylic acid.[124]
- 2-Nitratoalkyl perchlorates have been synthesized from the cleavage of epoxides with nitronium perchlorate. These materials are extremely dangerous with little value as practical explosives.[125]
- Nitrite esters can be oxidised to nitrate esters, usually in excellent yields, on treatment with ozone or dinitrogen pentoxide.[26,75]
- Dinitrogen pentoxide reacts with alkanes in carbon tetrachloride at 0 °C via a radical mechanism to give nitration products which can include nitrate esters.[126–129] Reactions of alkanes with dinitrogen pentoxide in nitric acid are complex and of little synthetic value.[130] 1-Adamantyl nitrate is one of the products obtained from the photochemical irradiation of a solution of adamantane and dinitrogen pentoxide in methylene chloride.[131]
- Propyleneimine is reported to react with a solution of absolute nitric acid in methylene chloride to yield 1,2-propanediol dinitrate, presumably via the intermediate primary nitramine.[132]

3.9 SYNTHETIC ROUTES TO SOME POLYOLS AND THEIR NITRATE ESTER DERIVATIVES

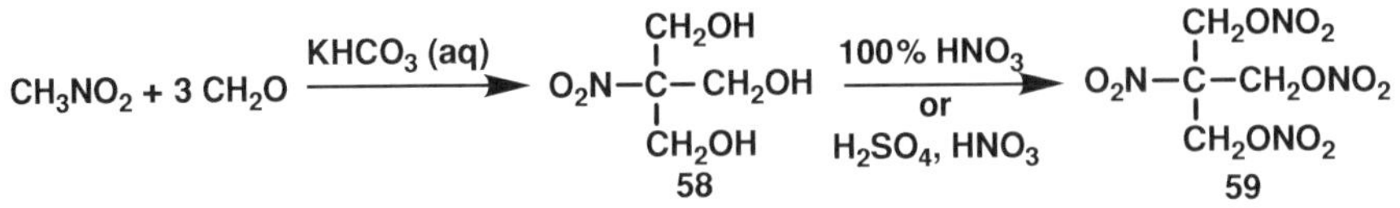

$$CH_3NO_2 + 3\,CH_2O \xrightarrow{KHCO_3\,(aq)} O_2N{-}C(CH_2OH)_3\ (\mathbf{58}) \xrightarrow[H_2SO_4,\,HNO_3]{100\%\,HNO_3\ \text{or}} O_2N{-}C(CH_2ONO_2)_3\ (\mathbf{59})$$

Figure 3.33

$$CH_3CH_2NO_2 + 2\,CH_2O \xrightarrow{KHCO_3\,(aq)} CH_3{-}C(NO_2)(CH_2OH)_2\ (\mathbf{60}) \xrightarrow[H_2SO_4,\,HNO_3]{100\%\,HNO_3\ \text{or}} CH_3{-}C(NO_2)(CH_2ONO_2)_2\ (\mathbf{61})$$

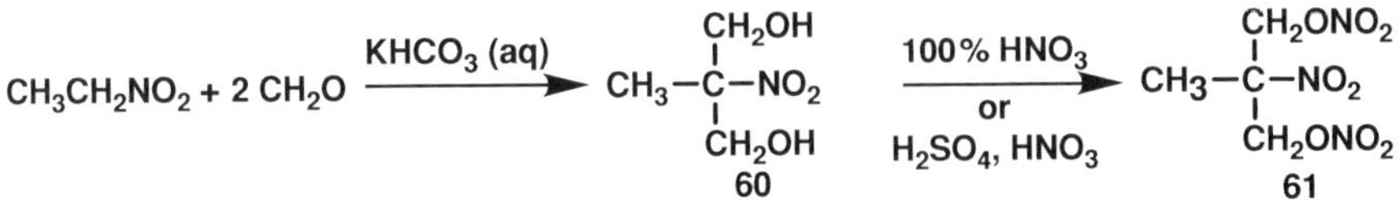

Figure 3.34

The reaction between formaldehyde and compounds containing acidic protons is probably the most important route to polyols. Some of these polyols have been *O*-nitrated and used as practical explosives. The condensation of nitromethane and nitroethane with excess formaldehyde in the presence of potassium hydrogen carbonate yields tris(hydroxymethyl)nitromethane (58)[133] and 1,1-bis(hydroxymethyl)nitroethane (60)[134] respectively. The nitration of (58) and (60) with either absolute nitric acid or mixed acid gives the secondary high explosives, (59)[135] and (61)[136] respectively.

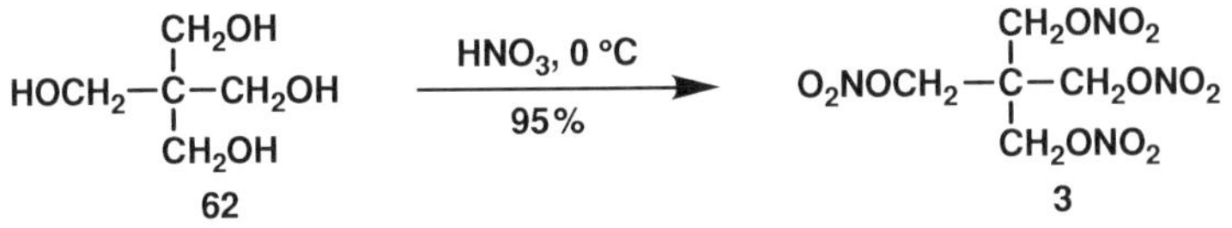

$$HOCH_2{-}C(CH_2OH)_3\ (\mathbf{62}) \xrightarrow[95\%]{HNO_3,\ 0\,^\circ C} O_2NOCH_2{-}C(CH_2ONO_2)_3\ (\mathbf{3})$$

Figure 3.35

The condensation of acetaldehyde with excess formaldehyde in the presence of aqueous calcium hydroxide yields pentaerythritol (62);[137] esterification of the latter with absolute nitric acid yields the powerful explosive, pentaerythritol tetranitrate (PETN) (3).[11]

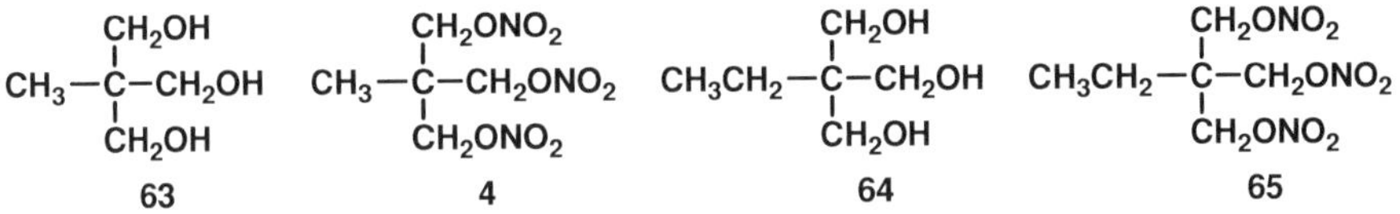

$$CH_3{-}C(CH_2OH)_3\ (\mathbf{63}) \quad CH_3{-}C(CH_2ONO_2)_3\ (\mathbf{4}) \quad CH_3CH_2{-}C(CH_2OH)_3\ (\mathbf{64}) \quad CH_3CH_2{-}C(CH_2ONO_2)_3\ (\mathbf{65})$$

Figure 3.36

1,1,1-Tris(hydroxymethyl)ethane (metriol) (63) and 1,1,1-tris(hydroxymethyl)propane (64) are commercially available and yield the trinitrate esters (4)[138] (metriol trinitrate) and (65)[139] on nitration with mixed acid. Metriol trinitrate (MTN) is a practical explosive and is attracting renewed interest as a promising alternative to nitroglycerine in propellant and explosive formulations. Metriol trinitrate, like nitroglycerine, is an excellent plasticizer for nitrocellulose.

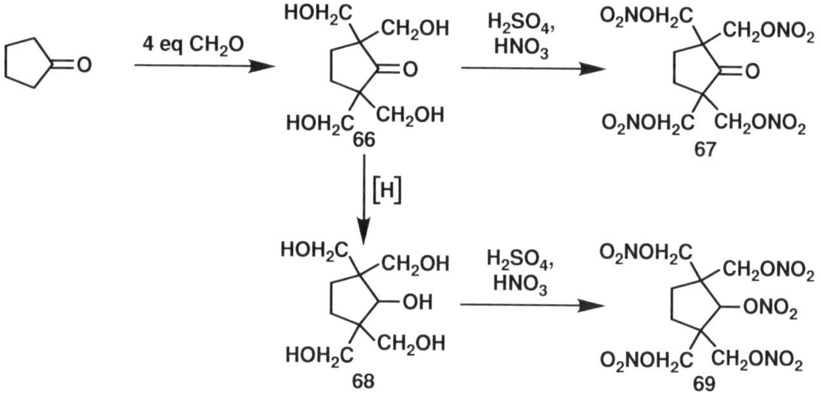

Figure 3.37

Friederich and Flick[140] synthesized the tetranitrate ester (67) by condensing cyclopentanone with four equivalents of formaldehyde followed by nitration of the condensation product (66) with mixed acid. Reduction of the ketone group of (66), followed by nitration, yields the pentanitrate ester (69). A similar route was used to synthesize analogous nitrate esters from cyclohexanone.[140]

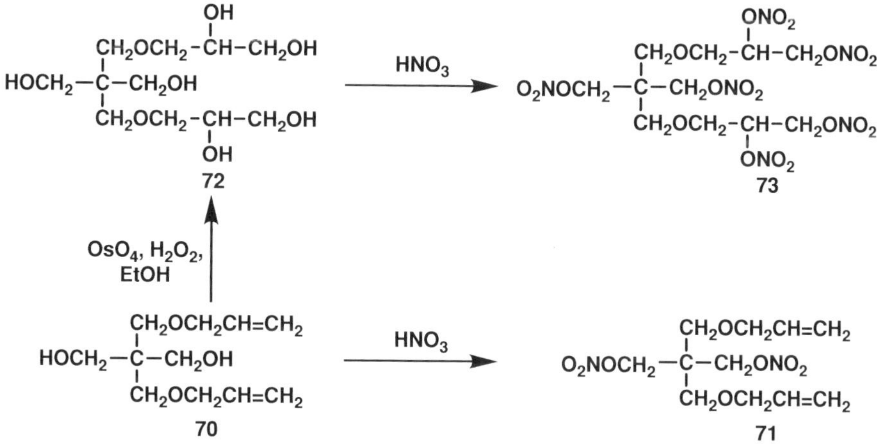

Figure 3.38

The dihydroxylation of alkenes is a useful strategy for the synthesis of polyols and these can be nitrated to the corresponding nitrate esters. Evans and Gallaghan[141] synthesized both the mono- (74) and di- (70) allyl ethers of pentaerythritol and used these for the synthesis of some novel nitrate ester explosives.

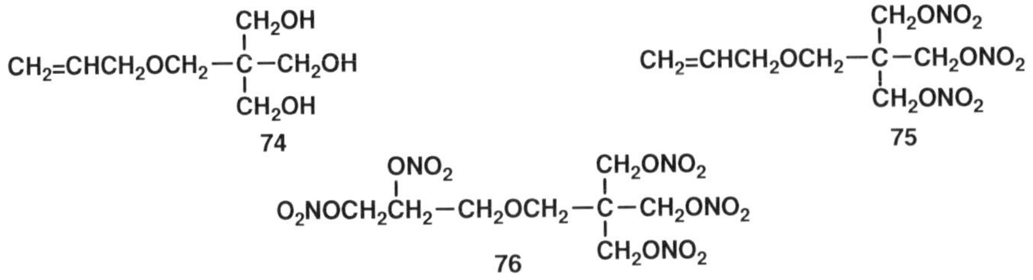

Figure 3.39

Dihydroxylation of the allyl groups of (70) with hydrogen peroxide and catalytic osmium tetroxide, followed by *O*-nitration of the product (72), yields the hexanitrate ester (73). Similar treatment of the mono-allyl ether (74) affords the pentanitrate ester (76). Evans and Gallaghan[141] also *O*-nitrated the hydroxy groups of (70) and (74) to yield the dinitrate and trinitrate esters, (71) and (75), respectively. The dinitrate ester (71) may find use as a monomer for the synthesis of energetic binders.

Many polynitrate esters are powerful explosives but have unfavourable physical properties. Consequently, a considerable amount of work has been directed towards changing the properties of pre-existing explosives. Such work serves to improve the properties of pre-existing explosives by increasing thermal stability, reducing sensitivity to shock, or lowering the melting point so that melt casting of charges becomes possible.

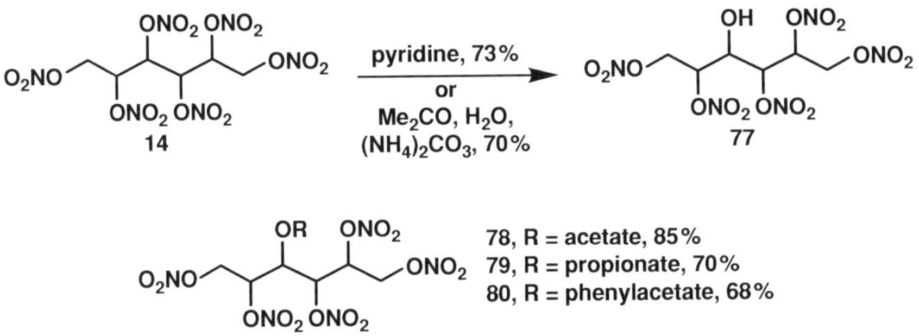

Figure 3.40

Interest has focused on derivatives of mannitol hexanitrate (14) as potential explosives because although this nitrate ester is a powerful explosive it has some property characteristics of a primary explosive. Treatment of mannitol hexanitrate (14) with pyridine[142,143] or ammonium carbonate[143] in aqueous acetone leads to a very selective denitration with the formation of mannitol-1,2,3,5,6-pentanitrate (77). Marans and co-workers[143] synthesized the acetate (78), the propionate (79), and the phenylacetate (80) derivatives of mannitol-1,2,3,5,6-pentanitrate and all have significantly lower melting points than mannitol hexanitrate. The incorporation of such groups can also help to increase the solubility of an explosive in the melt of another explosive.

Pentaerythritol tetranitrate (PETN) is a powerful high explosive with importance for both commercial and military applications. It is therefore unsurprising that work has been focused

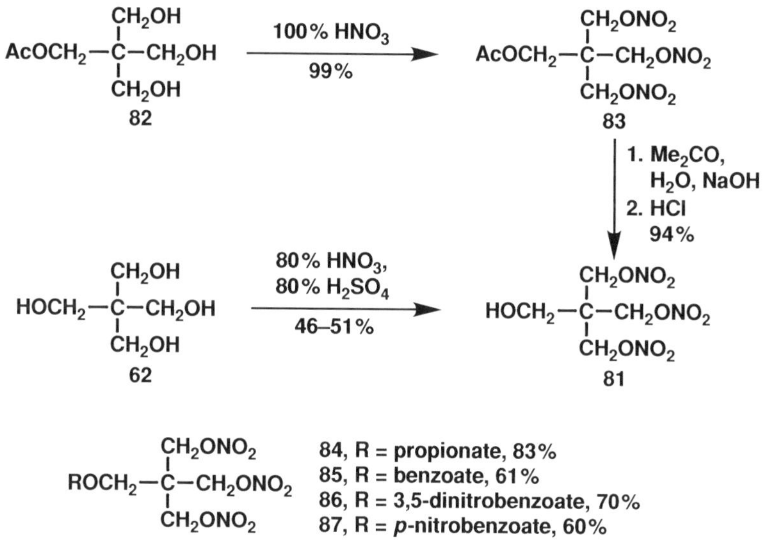

$AcOCH_2C(CH_2OH)_3$ (82) $\xrightarrow[99\%]{100\%\ HNO_3}$ $AcOCH_2C(CH_2ONO_2)_3$ (83)

(83) $\xrightarrow{1.\ Me_2CO,\ H_2O,\ NaOH;\ 2.\ HCl;\ 94\%}$ $HOCH_2C(CH_2ONO_2)_3$ (81)

$HOCH_2C(CH_2OH)_3$ (62) $\xrightarrow[46\text{–}51\%]{80\%\ HNO_3,\ 80\%\ H_2SO_4}$ $HOCH_2C(CH_2ONO_2)_3$ (81)

$ROCH_2C(CH_2ONO_2)_3$

84, R = propionate, 83%
85, R = benzoate, 61%
86, R = 3,5-dinitrobenzoate, 70%
87, R = *p*-nitrobenzoate, 60%

Figure 3.41

on the synthesis of nitrate ester analogues of pentaerythritol. Pentaerythritol trinitrate (81) is a key starting material for the synthesis of many of these analogues. Pentaerythritol trinitrate (81) has been synthesized via the controlled nitration of pentaerythritol (62) with mixed acid. The yields for this nitration average 48 % after the removal of PETN.[144] Pentaerythritol trinitrate (81) is also obtained from the nitration of pentaerythritol mono-acetate ester (82) followed by selective hydrolysis of the acetate group.[145]

Marans and co-workers[145] used pentaerythritol trinitrate (81) to synthesize a number of aryl and alkyl esters, including the formate, propionate (84), oxalate, succinate, benzoate (85), 3,5-dinitrobenzoate (86), and *ortho*-, *meta*-, and *para*- (87) nitrobenzoate esters. The *para*-nitrobenzoate ester (87) has also been prepared from the nitration of pentaerythritol mono-benzoate with mixed acid.[146]

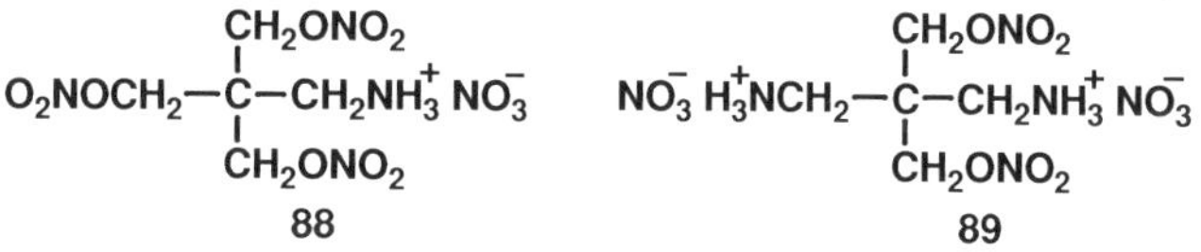

$O_2NOCH_2C(CH_2ONO_2)_2CH_2NH_3^+\ NO_3^-$ (88)

$NO_3^-\ H_3N^+CH_2C(CH_2ONO_2)_2CH_2NH_3^+\ NO_3^-$ (89)

Figure 3.42

Hiskey and co-workers[147] prepared a series of *O*-nitro amino nitrates by reacting the parent aminoalcohols with absolute nitric acid followed by precipitation of the salt with an organic solvent. Two of these compounds, (88) and (89), show similar thermal stability to PETN but are less sensitive to impact. Such compounds show a useful solubility in water and have potential for use as sensitizers in explosive slurries.

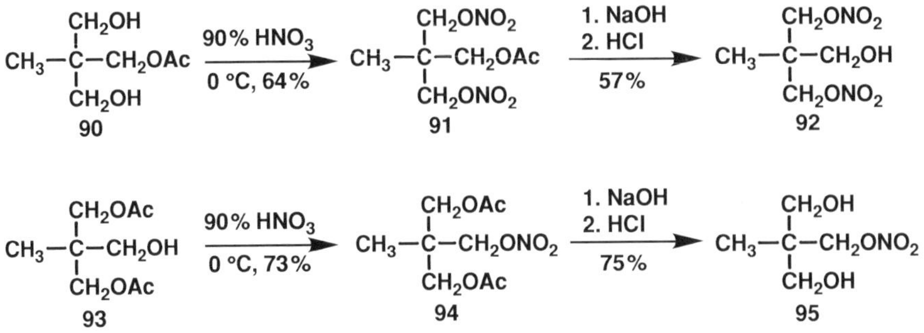

$CH_3-C(CH_2OH)_2-CH_2OAc$ (90) —[90% HNO_3, 0 °C, 64%]→ $CH_3-C(CH_2ONO_2)_2-CH_2OAc$ (91) —[1. NaOH, 2. HCl, 57%]→ $CH_3-C(CH_2ONO_2)_2-CH_2OH$ (92)

$CH_3-C(CH_2OAc)_2-CH_2OH$ (93) —[90% HNO_3, 0 °C, 73%]→ $CH_3-C(CH_2OAc)_2-CH_2ONO_2$ (94) —[1. NaOH, 2. HCl, 75%]→ $CH_3-C(CH_2OH)_2-CH_2ONO_2$ (95)

Figure 3.43

Marans and Preckel[148] synthesized both the mononitrate (95) and the dinitrate (92) esters of metriol by using a similar strategy to that used for pentaerythritol trinitrate. Thus, nitration of both the mono- (90) and the di- (93) acetate esters of metriol, followed by selective hydrolysis of the acetate groups, yields (92) and (95) respectively; the latter could be useful as a monomer for the synthesis of energetic polyurethane polymers.

3.10 ENERGETIC NITRATE ESTERS

Nitrate esters are a class of powerful explosives and this is mainly attributed to their better oxygen balance compared to aromatic nitro compounds. However, far fewer examples of energetic nitrate esters are available compared to energetic nitramines, *C*-nitro compounds and *N*-heterocycles. This is undoubtedly due to the presence of the $-O-NO_2$ bond, which is weaker than the $-N-NO_2$ and $-C-NO_2$ bonds, resulting in higher sensitivity to mechanical and thermal stimuli. Most modern research is heavily focused on synthesizing insensitive energetic materials and the nitrate ester group is not always conducive to this.

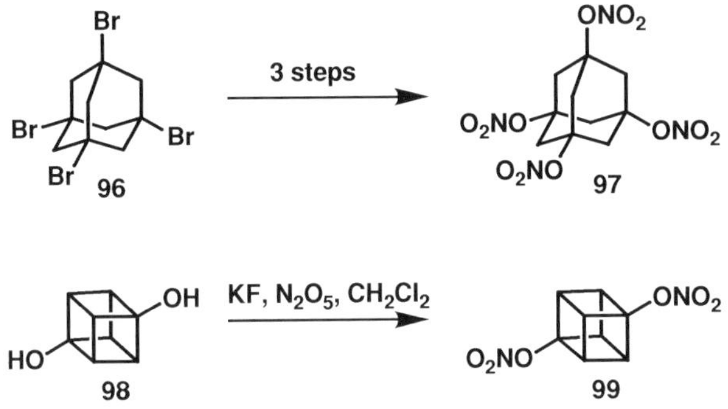

Br, Br, Br, Br (96) —[3 steps]→ ONO_2, O_2NO, ONO_2, O_2NO (97)

HO, OH (98) —[KF, N_2O_5, CH_2Cl_2]→ O_2NO, ONO_2 (99)

Figure 3.44

Some examples of nitrate ester incorporation into caged molecules have been reported: 1,3,5,7-tetranitroxyadamantane (97)[149] has been synthesized in three steps from 1,3,5,7-tetrabromoadamantane (96) and 1,4-dinitroxycubane (99)[150] has been synthesized from the nitration of the corresponding diol (98).

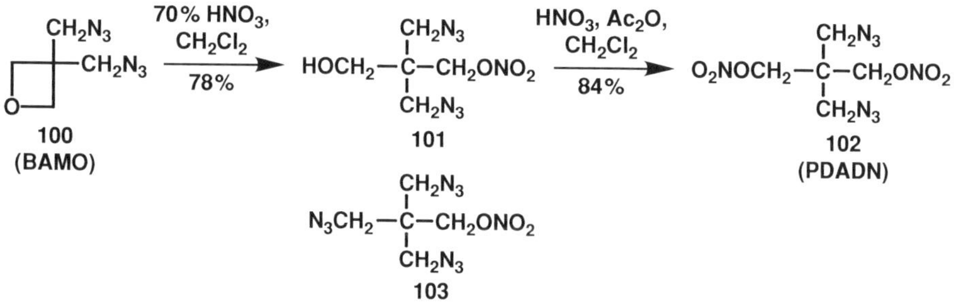

Figure 3.45

The incorporation of nitrate ester functionality into molecules containing other 'explosophores' is a common strategy for increasing oxygen balance and improving explosive performance. Pentaerythritol diazido dinitrate (102) (PDADN) has been synthesized by the ring cleavage of 3,3-bis(azidomethyl)oxetane (100) [BAMO, an energetic monomer to Poly(BAMO) energetic binder] with 70 % nitric acid, followed by nitration of the product (101) with fuming nitric acid in acetic anhydride.[151] The thermal properties of PDADN have been investigated.[152] Pentaerythritol triazide mononitrate (103) has also been synthesized by *O*-nitration of the corresponding alcohol with acetyl nitrate.[151]

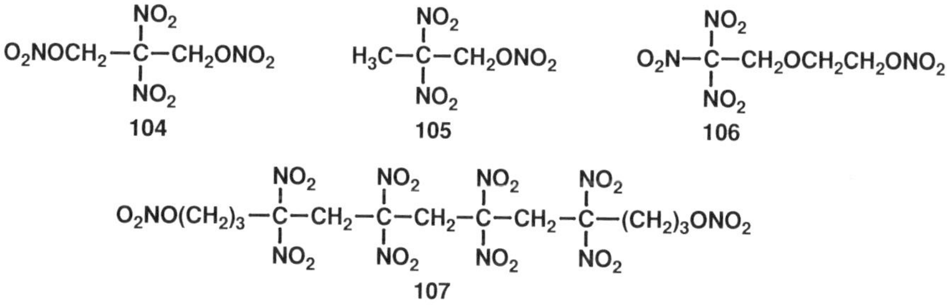

Figure 3.46

The Henry condensation of nitroform and terminal dinitromethyl compounds with formaldehyde and other aldehydes, followed by nitration of the resulting alcohol functionality, has been used to synthesize numerous explosives. The nitrate esters (104)[153], (105)[153], (106)[154] and (107)[155] have been synthesized from the action of absolute nitric acid on the parent alcohols. In a similar manner, NMHP (109) is synthesized from the condensation of TNHP (108) with formaldehyde, followed by *O*-nitration with absolute nitric acid.[156]

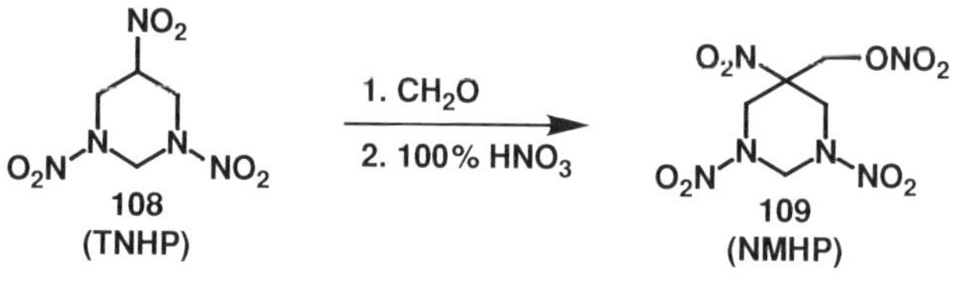

Figure 3.47

A number of secondary high explosives containing both nitramine and nitrate ester functionality have been reported. Aliphatic examples include: *N*-nitrodiethanolamine dinitrate (DINA) (110), prepared from the nitration of diethanolamine with nitric acid–acetic anhydride in the presence of zinc chloride,[157,158] and *N*,*N*′-dinitro-*N*,*N*′-bis(2-hydroxyethyl)oxamide dinitrate (NENO) (111), prepared from the mixed acid nitration of *N*,*N*′-bis(2-hydroxyethyl) oxamide[157,159].

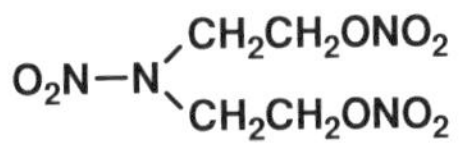

110 (DINA) 111 (NENO) 112 (Tris-X)

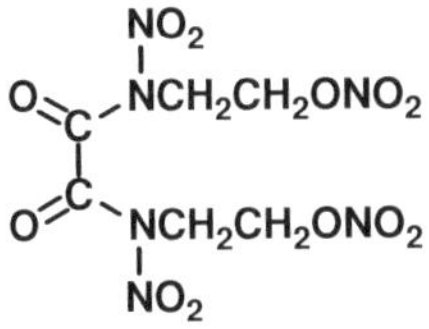

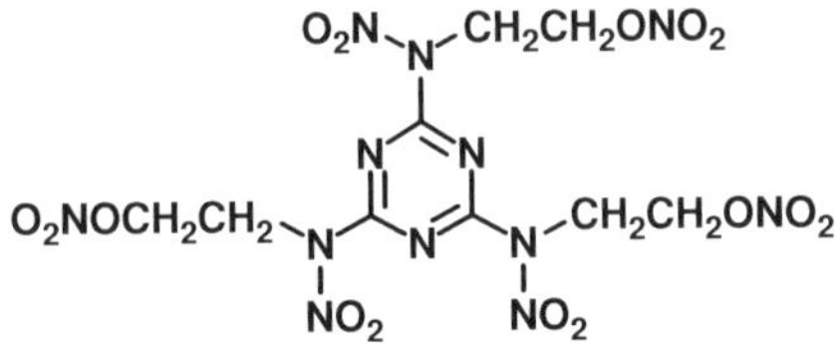

Figure 3.48

A number of nitramine-nitrate explosives have been prepared by Millar and co-workers from the action of dinitrogen pentoxide on aziridines and azetidines (Section 5.8).[77,88] Millar and co-workers used their aziridine ring-opening nitration methodology (Section 5.8.1) to synthesize the high performance melt-castable nitramine-nitrate explosive known as Tris-X (112).[160]

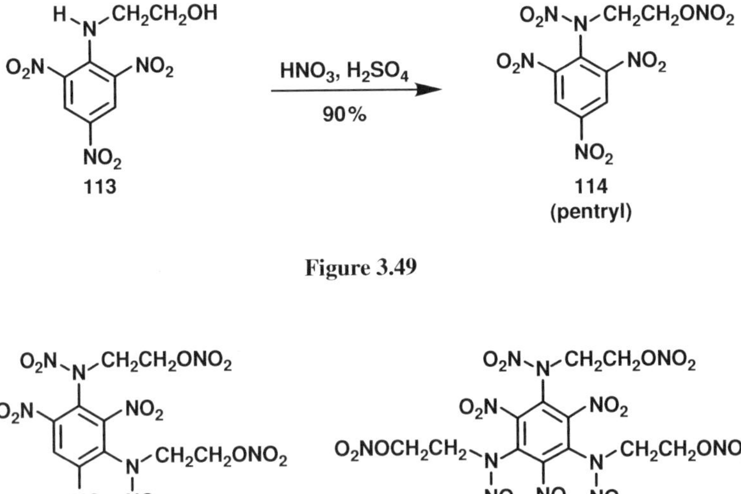

113 → (HNO_3, H_2SO_4, 90%) → 114 (pentryl)

Figure 3.49

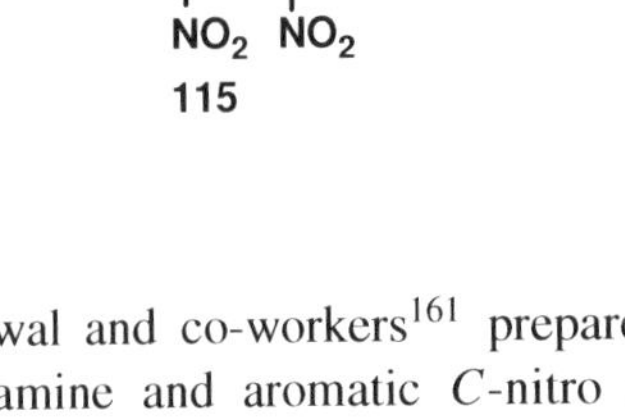

115 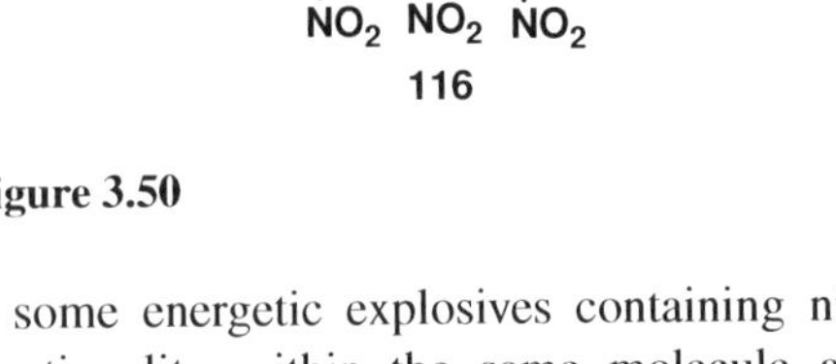 116

Figure 3.50

Agrawal and co-workers[161] prepared some energetic explosives containing nitrate ester, nitramine and aromatic *C*-nitro functionality within the same molecule and studied their thermal and explosive properties; 1-(2-nitroxyethylnitramino)-2,4,6-trinitrobenzene

(114) (pentryl), 1,3-bis(2-nitroxyethylnitramino)-2,4,6-trinitrobenzene (115) and 1,3,5-tris (2-nitroxyethylnitramino)-2,4,6-trinitrobenzene (116) have been prepared by condensing picryl chloride, styphnyl chloride and 1,3,5-trichloro-2,4,6-trinitrobenzene respectively with ethanolamine, followed by nitration with mixed acid. All three explosives were found to exhibit respectable performance [calculated VOD – 8100 m/s (114), 8500 m/s (115) and 8650 m/s (116)].

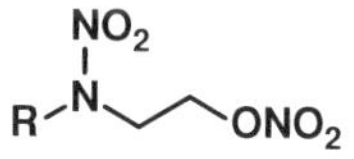

Figure 3.51

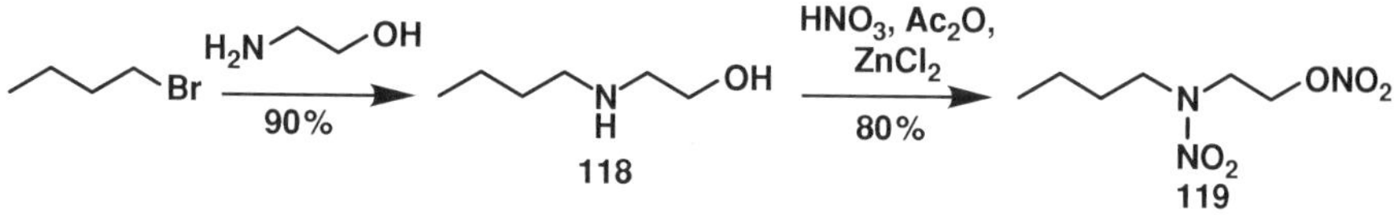

Figure 3.52

Nitroxyethylnitramines of general structure (117) are known as NENAs and are conveniently prepared from the nitrative cleavage of *N*-alkylaziridines with dinitrogen pentoxide in chlorinated solvents or from the nitration of the parent aminoalcohol. These compounds find use as energetic plasticizers in explosives and propellants; Bu-NENA (R = *n*-Bu) is a component of some LOVA (low vulnerability ammunition) propellants.[119,162] Bu-NENA (119) is synthesized in high yield from the chloride-catalyzed nitration of *N*-butylethanolamine (118).[163]

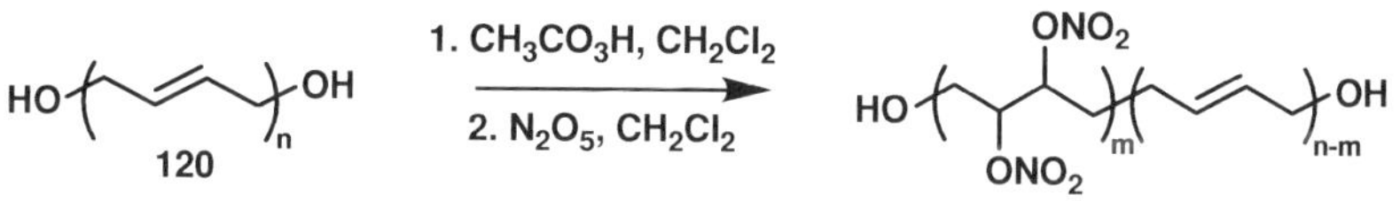

Figure 3.53

A number of energetic polymers containing nitrate ester functionality have received attention for use as binders in high performance propellant and explosive formulations. Nitrated hydroxy-terminated polybutadiene (121) (NHTPB) is synthesized from the epoxidation of the double bonds of HTPB (120) oligomers with peroxyacetic acid, followed by ring-opening nitration of the resulting epoxide functionality with dinitrogen pentoxide in methylene chloride.[76] Nitrated HTPB with 10 % of the double bonds converted to dinitrate ester functionality is a usable pre-polymer with a viscosity significantly low enough to permit easy processing and high solids loading. Such pre-polymers can be cured with aliphatic or aromatic diisocyanates to give energetic binders. The number of double bonds converted to dinitrate ester functionality depends on the epoxide content of the intermediate polymer. The epoxidation step must be conducted with care to avoid oxidative side-reactions and chain cleavage.

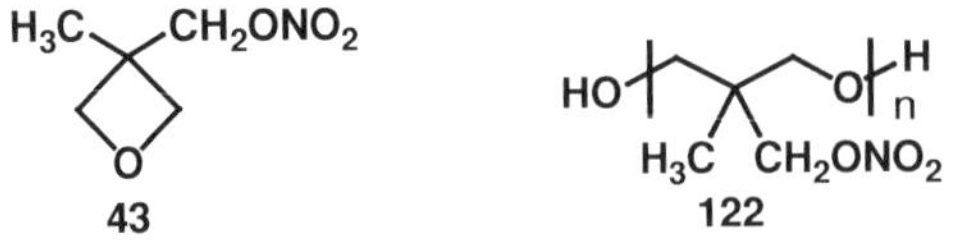

H_3C CH_2ONO_2 (oxetane)
43
(NIMMO)

HO–[…O]$_n$–H, H_3C CH_2ONO_2
122
Poly[NIMMO]

Figure 3.54

3-Nitratomethyl-3-methyloxetane (NIMMO) (43) is synthesized from the selective nitration of the hydroxy group of 3-hydroxymethyl-3-methyloxetane with dinitrogen pentoxide in an inert solvent.[36a] The cationic polymerization of NIMMO using an initiator system of boron trifluoride and a diol yields the energetic polymer poly[NIMMO] (122).[37] Poly[NIMMO] generated by this process is a viscous liquid with a very low sensitivity to impact and well suited for use as an energetic binder for rocket propellants and plastic bonded explosives (PBXs). The synthesis and scale-up of NIMMO and its polymerization to poly[NIMMO] was pioneered by chemists at DERA (British Defense Evaluation and Research Agency). Chemists at DERA have also developed a plastic bonded explosive called CPX-413 which is based on poly[NIMMO]/HMX/NTO/plasticizer and is ranked as an extremely insensitive detonating substance (EIDS).

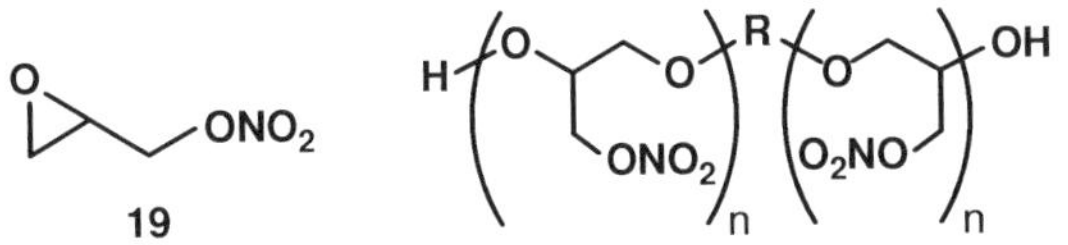

O, ONO_2
19
(GLYN)

H, O, R, O, OH, ONO_2, O_2NO, n, n
123
Poly[GLYN]

Figure 3.55

Glycidyl nitrate (GLYN) (19) is synthesized in high yield and purity from the selective nitration of glycidol with dinitrogen pentoxide in an inert solvent.[35,36] The cationic polymerization of glycidyl nitrate is more difficult than the polymerization of NIMMO and requires a strong mineral acid like tetrafluoroboric acid.[37] The product, poly[GLYN] (123), is a low molecular weight hydroxy-terminated pre-polymer which reacts with diisocyanates to give energetic polyurethane polymers.[164] Willer and co-workers[165] have reported on an improved process for producing poly[GLYN] which is well suited for use as an oligomer in solid high-energy compositions. Willer and co-workers[166] have also described the use of low molecular weight poly[GLYN] for use as an energetic plasticizer. In this context, poly[GLYN] has a number of advantages over traditional nitrate ester plasticizers, including: low volatility, low T_g ($\sim -40\,^\circ C$), excellent miscibility with the binder and decreased plasticizer mobility. On the basis of performance and the ease with which poly[GLYN] is prepared via dinitrogen pentoxide technology, it seems likely that it will prove to be a world leader in the field of energetic polymers.

Recent interest has turned to nitrated cyclodextrin polymers (poly-CDN) for potential use in insensitive and minimum smoke producing propellants.[167] The synthesis, purification and characterization of the following polymers was studied in detail:[168]

(1) r-Cyclodextrin polymer cross-linked with 1-chloro-2,3-epoxypropane.

(2) r-Cyclodextrin polymer cross-linked with 4,4′-methylene-bis(phenyl isocyanate).

(3) A linear polymer with pendant r-cyclodextrins.

(4) A linear tube consisting of α-cyclodextrins cross-linked with 1-chloro-2,3-epoxypropane.

The polymers 1, 3 and 4 were *O*-nitrated with nitric acid to give products containing nitrogen contents of 11.6 %, 19.9 % and 9.55 % respectively. Polymer 2, on the other hand, was nitrated with dinitrogen pentoxide in liquid carbon dioxide. Evaluations of these energetic polymers indicate that polymer 1 is a possible candidate for use in insensitive munitions.

REFERENCES

1. (a) T. Urbański, *Chemistry and Technology of Explosives, Vol. 2*, Pergamon Press, Oxford (1965); (b) T. Urbański, *Chemistry and Technology of Explosives, Vol. 4*, Pergamon Press, Oxford, Chapter 10 (1984); (c) V. Linder, in *Kirk-Othmer Encyclopedia of Chemical Technology, 3rd Edn, Vol. 9*, Ed. M. Grayson, Wiley-Interscience, New York (1980).
2. (a) E. H. Zeigler, *European Pat.* 66 999 (1982); (b) G. Doriath, *J. Propulsion Power*, 1995, **4**, 870; (c) M. L. Chan, *US Pat.* 5 316 600 (1994).
3. *Encyclopedia of Explosives and Related Items, Vol. 8*, Ed. S. M. Kaye, ARRADCOM, Dover, New Jersey (1978).
4. (a) T. Urbański, *Chemistry and Technology of Explosives, Vol. 2*, Pergamon Press, Oxford, 20 (1965); (b) R. Boschan, R. W. Van Dolah and R. T. Merrow, *Chem. Rev.*, 1955, **55**, 485; (c) P. A. S. Smith, *Open Chain Nitrogen Compounds, Vol. 2*, Benjamin, New York, 483–490 (1966); (d) G. A. Olah, R. Malhotra and S. C. Narang, *Nitration: Methods and Mechanisms*, VCH Publishers, New York, 269–275 (1989); (e) T. L. Davis, *Chemistry of Powder and Explosives, Coll. Vol.*, Angriff Press, Hollywood, CA, Chapter 5, 191–286 (1943, reprinted 1991); (f) *Vogel's Textbook of Practical Organic Chemistry*, Ed. A. Vogel, Longman Group Ltd, London, 523 (1970).
5. (a) T. L. Davis, *Chemistry of Powder and Explosives, Coll. Vol.*, Angriff Press, Hollywood, CA, 256–269 (1943, reprinted 1991); (b) G. Lunge, *J. Am. Chem. Soc*, 1901, **23**, 527; (c) J. Barsha, in *High Polymer Series, Vol. 5*, Ed. E. Ott, Interscience, Chapter 8, Section B, 622–666 (1943).
6. T. L. Davis, *Chemistry of Powder and Explosives, Coll. Vol*, Angriff Press, Hollywood, CA, 215 (1943, reprinted 1991).
7. V. Linder, in *Kirk-Othmer Encyclopedia of Chemical Technology, 3rd Edn, Vol. 9*, Ed. M. Grayson, Wiley-Interscience, New York, 573 (1980).
8. (a) H. Toivonen, *Ann. Acad. Sci. Fennicae, Ser. AII*, 1956, No. 72; *Chem. Abstr.*, 1958, **52**, 2806; (b) E. A. Parker and R. L. Shriner, *J. Am. Chem. Soc.*, 1933, **55**, 766.
9. L. T. Eremenko and G. V. Oreshko, *Ivz. Akad. Nauk SSSR, Ser. Khim.*, 1989, 1107; *Chem. Abstr.*, 1990, **112**, 35245s.
10. M. W. Barnes, C. E. Kristofferson, A. Manzara, C. D. Marken and M. M. Roland, *Synthesis*, 1977, 484.
11. T. L. Davis, *Chemistry of Powder and Explosives, Coll. Vol.*, Angriff Press, Hollywood, CA, 235–238 and 278–281 (1943, reprinted 1991).
12. (a) D. O'Meara and D. M. Shepherd, *J. Chem. Soc*, 1955, 4232; (b) J. Dewar and G. Fort, *J. Chem. Soc*, 1944, 492 and 496; (c) J. Dewar, G. Fort and N. McArthur, *J. Chem. Soc.*, 1944, 499.
13. J. Honeyman and J. W. W. Morgan, *Adv. Carbohydr. Chem.*, 1957, **12**, 117.
14. T. L. Davis, *Chemistry of Powder and Explosives, Coll. Vol*, Angriff Press, Hollywood, CA, 194–195 (1943, reprinted 1991).
15. (a) R. Boschan, R. W. Van Dolah and R. T. Merrow, *Chem. Rev*, 1955, **55**, 485; (b) H. Laurent, G. Snatzke and R. Wiechert, *Tetrahedron*, 1969, **25**, 761; (c) H. J. Cook, S. M. David and F. Kaufman, *J. Am. Chem. Soc.*, 1952, **74**, 4997; (d) F. E. Behr and R. D. Campbell, *J. Org. Chem.*, 1973, **38**,

1183; (e) J. H. Johnson Jr and K. V. Rao, *Tetrahedron Lett.*, 1998, **39**, 4611; (f) N. Hussain, D. O. Morgan, J. A. Murphy and C. R. White, *Tetrahedron Lett.*, 1994, **35**, 5069.
16. (a) S. Erhardt and Stoss, *Bioorg. Med. Chem. Lett.*, 1991, **1(11)**, 629; (b) J. Honeyman and J. W. W. Morgan, *Chem. Ind (London)*, 1953, 1035; (c) J. Honeyman and T. C. Stening, *J. Chem. Soc.*, 1958, 537.
17. A. Chaney, G. H. McFadden and M. L. Wolfrom, *J. Org. Chem.*, 1960, **25**, 1079.
18. (a) W. D. Emmons, A. F. Ferris, I. G. Marks and K. W. McLean, *J. Am. Chem. Soc.*, 1953, **75**, 4078; (b) L. Held and D. N. Kevill, *J. Org. Chem.*, 1973, **38**, 4445.
19. W. deC. Crater, *Ind. Eng. Chem.*, 1948, **40**, 1627.
20. W. F. Anzilotti, G. F. Hennion and R. J. Thomas, *Ind. Eng. Chem.*, 1940, **32**, 408.
21. A. Dinwoddie and G. Fort, *Brit. Pat.* 1 107 907 (1966).
22. C. L. Coon, E. S. Jessop, A. R. Mitchell, F. Pagoria and R. D. Schmidt, in *Nitration: Recent Laboratory and Industrial Developments, ACS Symposium Series 623*, Eds. L. F. Albright, R. V. C. Carr and R. J. Schmitt, American Chemical Society, Washington, DC, Chapter 14, 151–164 (1996).
23. A. V. Topchiev, *Nitration of Hydrocarbons and Other Organic Compounds*, Translated from Russian by C. Matthews, Pergamon Press, London (1959).
24. P. Gray and A. D. Yoffe, *J. Chem. Soc.*, 1951, 1412.
25. W. R. Feldman and E. H. White, *J. Am. Chem. Soc.*, 1957, **79**, 5832.
26. G. B. Bachman and N. W. Connon, *J. Org. Chem.*, 1969, **34**, 4121.
27. J. W. Fischer, in *Nitro Compounds: Recent Advances in Synthesis and Chemistry, Organic Nitro Chemistry Series*, Eds. H. Feuer and A. T. Neilsen, Wiley-VCH, Weinheim Chapter 3, 267–365 (1990).
28. *Nitration: Recent Laboratory and Industrial Developments, ACS Symposium Series 623*, Eds. L. F. Albright, R. V. C. Carr and R. J. Schmitt, American Chemical Society, Washington, DC (1996).
29. (a) A. D. Harris, H. B. Jonassen and J. C. Trebellas, *Inorg. Synth*, 1967, **9**, 83; (b) R. W. Millar, N. C. Paul and D. H. Richards, *UK Pat. Appls.* 2 181 124 (1987) and 2 181 139 (1987); (c) T. E. Devendorf and J. R. Stacy, in *Nitration: Recent Laboratory and Industrial Developments, ACS Symposium Series 623*, Eds. L. F. Albright, R. V. C. Carr and R. J. Schmitt, American Chemical Society, Washington, DC, Chapter 8, 68–77 (1996).
30. (a) M. J. Rodgers and P. F. Swinton, in *Nitration: Recent Laboratory and Industrial Developments, ACS Symposium Series 623*, Eds. L. F. Albright, R. V. C. Carr and R. J. Schmitt, American Chemical Society, Washington, DC, Chapter 7, 58–67 (1996); (b) J. E. Harrar and R. K. Pearson, *J. Electrochem. Soc.*, 1983, **130**, 108.
31. E. Pokorny and F. Russ, *Monatsh. Chem.*, 1913, **34**, 1913.
32. G. V. Caesar and M. Goldfrank, *J. Am. Chem. Soc.*, 1946, **68**, 372.
33. W. E. Elias and L. D. Hayward, *Tappi J.*, 1958, **41**, 246.
34. (a) B. Vollmert, *Makromol. Chem.*, 1951, **6**, 78; (b) L. Brissard, J. Chedin and R. Dalmon, *Compt. Rend*, 1935, **201**, 664.
35. (a) A. Arber, G. Bagg, E. Colclough, H. J. Desai, R. W. Millar, N. C. Paul, D. Salter and M. J. Stewart, 'Novel Energetic Monomers and Polymers Prepared using Dinitrogen Pentoxide Chemistry', *Proc. 21st Annual Conference of ICT on Technology of Polymer Compounds and Energetic Materials*, Karlsruhe, Germany, 3–6 July, 1990; (b) M. E. Colclough, P. Golding, P. J. Honey, R. W. Millar, N. C. Paul, A. J. Sanderson and M. J. Stewart, *Phil. Trans. R. Soc. London*, 1992, **A339**, 305.
36. (a) N. C. Paul, in *Nitration: Recent Laboratory and Industrial Developments, ACS Symposium Series 623*, Eds. L. F. Albright, R. V. C. Carr and R. J. Schmitt, American Chemical Society, Washington, DC, Chapter 15, 165–173 (1996); (b) H. J. Desai, W. Leeming, D. H. Paterson and N. C. Paul, 'Scale-up of Polyglycidyl Nitrate Manufacture; Process Development and Assembly', *Proc. Joint International Symposium on Energetic Materials Technology*, New Orleans, LA, 4–7 October, 1992, American Defence Preparedness Association, Arlington, VA.
37. (a) A. J. Amass, A. V. Cunliffe, H. J. Desai, J. Hamid, J. Honey and M. J. Stewart, *Polymer*, 1996, **37**(15), 3461; (b) Y. -G. Cheun, J. R. Cho and J. S. Kim, 'An Improved Synthetic Method of

Poly(NIMMO) and PGN Prepolymers', in *Proc. International Symposium on Energetic Materials Technology*, 1995, 61–67, American Defence Preparedness Association, Arlington, VA.
38. F. H. Jarke, A. J. Kacmarek, M. Lustig, I. J. Solomon and J. Shamir, *J. Org. Chem.*, 1975, **40**, 1851.
39. (a) E. F. J. Duynstee, J. G. H. M. Housmans, J. Vleugels and W. Voskuil, *Tetrahedron Lett.*, 1973, **14**, 2275; (b) D. D. DesMarteau and F. A. Hohorst, *Inorg. Chem.*, 1974, **13**, 715.
40. R. Louw, G. J. Sluis and H. W. Vermeeren, *J. Am. Chem. Soc.*, 1975, **97**, 4396.
41. (a) G. A. Olah, R. Malhotra and S. C. Narang, *Nitration: Methods and Mechanisms*, Wiley-VCH, Wienheim, 271–273 (1989); (b) S. Kuhn, L. Noszko, G. A. Olah and M. Szelke, *Chem. Ber.*, 1956, **89**, 2374.
42. G. A. Olah, in *Chemistry of Energetic Materials*, Eds. G. A. Olah and D. R. Squire, Academic Press, Chapter 7, 139–204 (1991).
43. B. V. Gidaspov, E. L. Golod, Yu. V. Guk and M. A. Ilyushin, *Russ. Chem. Rev.*, 1983, **52(3)**, 284.
44. *Aldrich Catalogue of Fine Chemicals and Laboratory Equipment, 2003–2004*, Aldrich Chemical Company, Dorset, UK, 1354 (2003).
45. (a) S. J. Kuhn and G. A. Olah, *Chem. Ind (London)*, 1958, 98; (b) S. J. Kuhn, A. J. Mlinko and G. A. Olah, *J. Chem. Soc.*, 1956, 4257.
46. S. J. Kuhn and G. A. Olah, *J. Am. Chem. Soc.*, 1961, **83**, 4564.
47. S. J. Kuhn, *Can. J. Chem.*, 1962, **40**, 1660.
48. J. H. Ridd and T. Yoshida, in *Industrial and Laboratory Nitrations, ACS Symposium Series 22*, Eds. L. F. Albright and C. Hanson, American Chemical Society, Washington, DC, Chapter 6, 103–113 (1976).
49. M. A. Hiskey and J. C. Oxley, *J. Energ. Mater*, 1989, **7**, 199.
50. G. A. Olah, J. A. Olah and N. A. Overchuk, *J. Org. Chem.*, 1965, **30**, 3373.
51. C. A. Cupas, S. C. Narang, G. A. Olah, J. A. Olah and R. L. Pearson, *J. Am. Chem. Soc.*, 1980, **102**, 3507.
52. C. A. Cupas and R. L. Pearson, *J. Am. Chem. Soc.*, 1968, **90**, 4742.
53. C. A. Cupas, S. C. Narang, G. A. Olah and R. L. Pearson, *Synthesis*, 1978, 452.
54. G. A. Olah, in *Chemistry of Energetic Materials*, Eds. G. A. Olah and D. R. Squire, Academic Press, 191 (1991).
55. X. -Y. Li, G. A. Olah, G. K. Surya Prakash and Q. Wang, *Synthesis*, 1993, 207.
56. I. Bucsi, X. -Y. Li, G. A. Olah and Q. Wang, *Synthesis*, 1992, 1085.
57. L. Andersen, A. Gavrila and T. Skrydstrup, *Tetrahedron Lett.*, 2005, **46**, 6205.
58. J. V. Crivello, *J. Org. Chem.*, 1981, **46**, 3056.
59. G. H. Hakimalahi, A. Khalafi-Nehzad, H. Sharghi and H. Zarrinmayeh, *Helv. Chim. Acta*, 1984, **67**, 906.
60. J. W. Baker and D. M. Easty, *J. Chem. Soc.*, 1952, 1193.
61. W. D. Emmons, A. F. Ferris, I. G. Marks and K. W. McLean, *J. Am. Chem. Soc.*, 1953, **75**, 4078.
62. (a) J. U. Nef, *Liebigs Ann. Chem.*, 1904, **335**, 238; (b) W. Will, *Chem. Ber.*, 1908, **41**, 1117.
63. W. von E. Doering and L. F. Fieser, *J. Am. Chem. Soc.*, 1946, **68**, 2252.
64. L. T. Eremenko, V. N. Grebennikov, A. M. Korolev and G. M. Nazin, *Ivz. Akad. Nauk SSSR, Ser. Khim*, 1971, 627; *Chem. Abstr.*, 1971, **75**, 48216a.
65. L. T. Eremenko, G. V. Oreshko and L. B. Romanova, *Ivz. Akad. Nauk SSSR, Ser. Khim*, 1973, 2140; *Chem. Abstr.*, 1974, **80**, 14497f.
66. N. Kornblum, J. B. Nordmann and J. T. Patton, *J. Am. Chem. Soc.*, 1948, **70**, 746.
67. Fr. Fichter and A. Petrovich, *Helv. Chim. Acta*, 1941, **24**, 253.
68. (a) D. C. Iffland, N. Kornblum, N. N. Lichtin, and J. T. Patton, *J. Am. Chem. Soc.*, 1947, **69**, 307; (b) E. Grand, *Bull. Soc. Chim. Fr.*, 1950, 120.
69. L. Castedo, C. F. Marcos, M. Monteagudo and G. Tojo, *Synth. Commun.*, 1992, **22(5)**, 677.
70. G. A. Mortimer, *J. Org. Chem.*, 1962, **27**, 1876.
71. R. Boschan, *J. Am. Chem. Soc.*, 1959, **81**, 3341.
72. N. Afza, A. Latif, A. Malik and W. Voelter, *Liebigs Ann. Chem.*, 1929, **19**, 1985.

73. M. E. Ford and A. McKillop, *Tetrahedron*, 1974, **30**, 2467.
74. (a) J. Boileau and A. M. Pujo, *Compt. Rend.*, 1953, **237**, 1422; (b) J. Boileau and A. M. Pujo, *Mém. Poudres*, 1955, **37**, 35; (c) J. Boileau, C. Frejacques and A. M. Pujo, *Bull. Soc. Chim. Fr.*, 1955, 974.
75. (a) P. Golding, R. W. Millar, N. C. Paul and D. H. Richards, *Tetrahedron Lett.*, 1988, **29**, 2731; (b) P. Golding, R. W. Millar, N. C. Paul and D. H. Richards, *Tetrahedron*, 1993, **49**, 7037.
76. M. E. Colclough and N. C. Paul, in *Nitration: Recent Laboratory and Industrial Developments, ACS Symposium Series 623*, Eds. L. F. Albright, R. V. C. Carr and R. J. Schmitt, American Chemical Society, Washington, DC, Chapter 10, 97–103 (1996).
77. (a) P. Golding, R. W. Millar, N. C. Paul and D. H. Richards, *Tetrahedron Lett.*, 1988, **29**, 2735; (b) P. Golding, R. W. Millar, N. C. Paul and D. H. Richards, *Tetrahedron*, 1993, **49**, 7063.
78. J. D. Ingham, A. B. Magnusson and P. L. Nichols Jr, *J. Am. Chem. Soc.*, 1953, **75**, 4255.
79. J. D. Ingham and P. L. Nichols Jr, *J. Am. Chem. Soc.*, 1954, **76**, 4477.
80. N. Iranpoor and P. Salehi, *Tetrahedron*, 1995, **51(3)**, 909.
81. J. R. Hanson, M. Troussier, C. Uyanik and F. Viel, *J. Chem. Res. (S)*, 1998, 118.
82. R. Di Fabio, T. Rossi and R. J. Thomas, *Tetrahedron Lett.*, 1997, **38(2)**, 3587.
83. F. Lanciano and E. Mincione, *Tetrahedron Lett.*, 1980, **21**, 1149.
84. M. Edwards, R. P. Hug, A. Mckillop and D. W. Young, *J. Org. Chem.*, 1978, **43**, 3373.
85. (a) R. J. Bertsch and R. J. Ouellette, *J. Org. Chem.*, 1976, **41**, 2783; (b) R. J. Bertsch and R. J. Ouellette, *J. Org. Chem.*, 1974, **39**, 2755; (c) C. Brock, P. A. Crooks, W. J. Layton, P. Burn and S. L. Smith, *J. Org. Chem.*, 1985, **50**, 5372.
86. (a) P. Golding, R. W. Millar, N. C. Paul and D. H. Richards, *Tetrahedron Lett*, 1988, **29**, 2735; (b) P. Golding, R. W. Millar, N. C. Paul and D. H. Richards, *Tetrahedron*, 1993, **49**, 7051.
87. A. S. Rao, in *Comprehensive Organic Synthesis, Vol., 7*, Eds. I. Fleming, S. V. Ley and B. M. Trost, Pergamon Press, Section 3.1, 357–387 (1991).
88. (a) P. Golding, R. W. Millar, N. C. Paul and D. H. Richards, *Tetrahedron Lett*, 1988, **29**, 2735; (b) P. Golding, R. W. Millar, N. C. Paul and D. H. Richards, *Tetrahedron*, 1995, **51**, 5073.
89. R. W. Millar, N. C. Paul and D. H. Richards, *Tetrahedron Lett.*, 1989, **30**, 6431.
90. R. W. Millar and S. Philbin, *Tetrahedron*, 1997, **53**, 4371.
91. G. H. Carlson and A. Michael, *J. Am. Chem. Soc.*, 1935, **57**, 1268.
92. (a) E. Sakellarios and H. Wieland, *Chem. Ber.*, 1919, **52**, 898; 1920, **53**, 201; (b) F. Rahn and H. Wieland, *Chem. Ber.*, 1921, **54**, 1771; (c) A. Kekule, *Chem. Ber.*, 1869, **2**, 329; (d) N. L. Drake and E. P. Kohler, *J. Am. Chem. Soc.*, 1923, **45**, 1281; (e) R. Ansshütz and A. Gilbert, *Chem. Ber.*, 1921, **54**, 1854; 1924, **57**, 1697.
93. Y. L. Kniglyak and I. V. Martynov, *J. Gen. Chem. USSR*, 1965, **35**, 974.
94. (a) J. R. Bull, E. R. H. Jones and G. D. Meakins, *J. Chem. Soc.*, 1965, 2601; (b) L. F. Fieser and M. Fieser, *Steroids*, Reinhold, New York, 43, 44 and 545 (1959); (c) A. Bowers, M. B. Sánchez and H. J. Reingold, *J. Am. Chem. Soc.*, 1959, **81**, 3702; (d) C. E. Anagnostopoulos and L. F. Fieser, *J. Am. Chem. Soc.*, 1954, **76**, 532.
95. (a) A. A. Griswold and P. S. Starcher, *J. Org. Chem.*, 1966, **31**, 357; (b) F. G. Bordwell and E. W. Garbisch Jr, *J. Am. Chem. Soc.*, 1960, **82**, 3588.
96. (a) N. Levy and C. W. Scaife, *J. Chem. Soc.*, 1946, 1093 and 1100; (b) N. Levy, C. W. Scaife and A. E. Wilder-Smith, *J. Chem. Soc.*, 1946, 1096; 1948, 52; (c) H. Baldock, N. Levy and C. W. Scaife, *J. Chem. Soc.*, 1949, 2627; (d) E. Gudriniece, O. Nieland and G. Vanags, *Zh. Obshch. Khim*, 1954, **24**, 1863; *Chem. Abstr.*, 1955, **49**, 13128; (e) W. K. Seifert, *J. Org. Chem.*, 1963, **28**, 125; (f) F. Conrad and H. Shechter, *J. Am. Chem. Soc.*, 1953, **75**, 5610.
97. (a) A. A. Griswold and P. S. Starcher, *J. Org. Chem.*, 1966, **31**, 357; (b) N. V. Dormidontova, M. I. Faberov, V. A. Podgornova and B. F. Ustavshchikov, *Neftekhimiya*, 1965, **5**, 873; *Chem. Abstr.*, 1966, **64**, 7981.
98. (a) A. V. Stepanov and V. V. Veselovsky, *Russ. Chem. Rev.*, 2003, **72**(4), 327; (b) J. P. Adams and D. S. Box, *J. Chem. Soc. Perkin Trans. 1*, 1999, 749; (c) J. P. Adams and J. R. Paterson, *J. Chem.*

Soc. Perkin Trans. 1, 2000, 3695; (d) J. H. Boyer, in *The Chemistry of the Nitro and Nitroso Groups, Part 1, Organic Nitro Chemistry Series*, Ed. H. Feuer, Wiley-Interscience, New York, 229 (1969); (e) H. O. Larson, in *The Chemistry of the Nitro and Nitroso Groups, Part 1, Organic Nitro Chemistry Series*, Ed. H. Feuer, Wiley-Interscience, New York, 301–348 (1969).

99. (a) D. F. Church, J. W. Lightsey and W. A. Pryor, *J. Am. Chem. Soc.*, 1982, **104**, 6685; (b) O. H. Lerner and V. V. Perekalin, *Dokl. Akad. Nauk USSR*, 1959, **129**, 1303.
100. (a) N. Levy and C. W. Scaife, *J. Chem. Soc.*, 1946, 1093 and 1100; (b) N. Levy, C. W. Scaife and A. E. Wilder-Smith, *J. Chem. Soc.*, 1946, 1096; 1948, 52; (c) F. G. Borgardt, P. Noble Jr and W. L. Reed, *Chem. Rev.*, 1964, **64**, 19.
101. (a) W. D. Emmons and T. W. Stevens, *J. Am. Chem. Soc.*, 1958, **80**, 338; (b) H. G. Padeken, U. O. von Schickh and A. Segnitz, in *Houben-Weyl, Methoden der Organishen Chemie, Band 10/1*, Ed. E. Muller, Thieme, Stuttgart, 82 (1971).
102. Ya. N. Dem'yanov, *Ct. Rd. Acad. Sci. USSR*, **1930(A)**, 447.
103. H. Akimoto, H. Bandow and M. Okuda, *J. Phys. Chem.*, 1980, **84**, 3604.
104. H. Akimoto, H. Bandow, M. Hoshine, G. Inove, T. Ogata, M. Okuda and T. Tezuka, *Kokuritsu Kogai Kerkyusho Kenkyo Hokoku*, 1979, **9**, 29.
105. W. D. Emmons and T. E. Stevens, *J. Am. Chem. Soc.*, 1957, **79**, 6008.
106. J. H. Canfield and G. H. Rohrback, *US Pat.* 3 729 501 (1973).
107. G. B. Bachman and T. J. Logan, *J. Org. Chem.*, 1956, **21**, 1467.
108. (a) U. E. Diner and J. W. Lown, *J. Chem. Soc (D)*, 1970, 333; (b) A. V. Joshua and J. W. Lown, *J. Chem. Soc. Perkin Trans. 1*, 1973, 2680.
109. A. V. Joshua and J. W. Lown, *Can. J. Chem.*, 1977, **55**, 508.
110. J. Barluenga, J. M. Martinez-Gallo, C. Nájera and M. Yus, *J. Chem. Soc. Chem. Commun*, 1985, 1422.
111. A. J. Bloodworth and N. Cooper, *J. Chem. Soc. Chem. Commun*, 1986, 709.
112. A. W. Francis, *J. Am. Chem. Soc.*, 1925, **47**, 2347.
113. D. H. R. Barton and S. C. Narang, *J. Chem. Soc. Perkin Trans. 1*, 1977, 1114.
114. T. B. K. Lee and F. Wudl, *J. Am. Chem. Soc.*, 1971, **93**, 271.
115. T. B. K. Lee and F. Wudl, *J. Chem. Soc (D)*, 1970, 490.
116. W. D. Emmons, A. S. Pagano and T. E. Stevens, *J. Org. Chem.*, 1958, **23**, 311.
117. L. T. Eremenko, B. S. Fedorov and R. G. Gafurov, *Izv. Akad. Nauk USSR, Ser. Khim.*, 1977, 345.
118. L. T. Eremenko, B. S. Fedorov and R. G. Gafurov, *Izv. Akad. Nauk USSR, Ser. Khim*, 1971, 1501.
119. D. W. Fish, E. E. Hamel and R. E. Olsen, in *Advanced Propellant Chemistry, Advances in Chemistry Series No 54*, Ed. R. E. Gould, American Chemical Society, Washington, DC, Chapter 6, 48–54 (1966).
120. A. R. Katritsky and L. Marzorati, *J. Org. Chem.*, 1980, **45**, 2515.
121. P. Golding, P. J. Honey, R. W. Millar, N. C. Paul and D. H. Richards, *European. Pat. Appl.* 86 308 290 (1987); *Publication Serial* EP-0223440A; *US Pat. Appl.* 923 024 (1987).
122. A. P. Franchimont, *Rec. Trav. Chim. Pays-Bas*, 1910, **29**, 311.
123. A. Baeyer and V. Villiger, *Chem. Ber.*, 1901, **34**, 755.
124. F. E. Francis, *J. Chem. Soc.*, 1906, 1.
125. V. N. Kirin, A. S. Kozmin, N. M. Yureva and N. S. Zefirov, *Ivz. Akad. Nauk USSR, Ser. Khim*, 1983, 703.
126. N. V. Schchitov and A. I. Titov, *Dokl. Akad. Nauk USSR*, 1951, **81**, 1085.
127. J. C. D. Brand, *J. Am. Chem. Soc.*, 1955, **77**, 2703.
128. R. A. Ogg, *J. Chem. Phys.*, 1950, **18**, 572 and 770.
129. Von. J. Runge and W. Triebs, *J. Prakt. Chem.*, 1962, **15**, 146.
130. (a) M. L. Bender, J. Figueras and M. Kilpatrick, *J. Org. Chem.*, 1958, **23**, 410; (b) F. Asinger and K. Halcour, *Chem. Ber.*, 1961, **94**, 83.
131. S. Kojo, I. Tabushi and F. Yoshida, *Chem. Lett.*, 1974, 1431.

132. P. Golding, R. W. Millar, N. C. Paul and D. H. Richards, *Tetrahedron Lett*, 1991, **32**, 4985.
133. L. Henry, *Compt. Rend*, 1895, **121**, 210.
134. H. B. Bass and B. M. Vanderbilt, *Ind. Eng. Chem.*, 1940, **32**, 34.
135. Aubry, *Mém. Poudres*, 1932–33, **25**, 197.
136. (a) F. G. Bergheim, *US Pats.* 1 691 955 (1929) and 1 751 438 (1930); (b) J. A. Wyler, *US Pat.* 2 195 551 (1940); (c) L. Médard, *Mém. Poudres*, 1953, **35**, 59.
137. R. H. Barth, E. Berlow and J. E. Snow, *The Pentaerythritols*, Reinhold, New York, 2–24 (1958).
138. R. Colson, *Mém. Poudres*, 1948, **30**, 43.
139. C. P. Spaeth, *US Pat.* 1 883 044 (1933).
140. W. Friederich and K. Flick, *Ger. Pat.* 509 118 (1929).
141. R. Evans and J. A. Gallaghan, *J. Am. Chem. Soc.*, 1953, **75**, 1248.
142. L. D. Hayward, *J. Am. Chem. Soc.*, 1951, **73**, 1974.
143. D. E. Elrick, N. S. Marans and R. F. Preckel, *J. Am. Chem. Soc.*, 1954, **76**, 1373.
144. A. T. Camp, D. E. Elrick, N. S. Marans and R. F. Preckel, *J. Am. Chem. Soc.*, 1955, **77**, 751.
145. D. E. Elrick, N. S. Marans and R. F. Preckel, *J. Am. Chem. Soc.*, 1954, **76**, 1304.
146. A. G. Westfälisch. *Ger. Pats.* 638 432 (1936) and 638 433 (1936).
147. M. J. Hatch, M. A. Hiskey and J. C. Oxley, *Propell. Explos. Pyrotech.*, 1991, **16**, 40.
148. N. S. Marans and R. F. Preckel, *J. Am. Chem. Soc.*, 1954, **76**, 3223.
149. E. E. Gilbert, *US Pat.* 4 476 060 (1984).
150. J. C. Bottaro, E. Penwell and R. J. Schmitt, *'Synthesis of Cubane Based Energetic Materials, Final Report, December 1989'*, SRI International, Menlo Park, CA [AD-A217 147/8/XAB].
151. M. B. Frankel and E. R. Wilson, *J. Org. Chem.*, 1985, **50**, 3211.
152. N. Binge, S. Jirong, Y. Qingsen, H. Rongzu and G. Shaojun, *Thermochim. Acta*, 2000, **352–353**, 133.
153. T. M. Benziger, L. W. Kissinger, R. K. Rohwer and H. E. Ungnade, *J. Org. Chem.*, 1963, **28**, 2491.
154. H. G. Adolph and M. E. Sitzmann, *Statutory Invent. Regist.* US 644 (Cl. 558–483; CO7C77/00), 06 Jun 1989, Appl. 226, 335, 27 Jul 1988; *Chem. Abstr.*, 1989, **111**, 214120y.
155. M. B. Milton, *J. Org. Chem.*, 1962, **27**, 331.
156. H. H. Licht and H. Ritter, *Propell. Explos. Pyrotech*, 1985, **10**, 147.
157. T. Urbański, *Chemistry and Technology of Explosives, Vol. 3*, Pergamon Press, Oxford, 36–37 (1967).
158. W. J. Chute, K. G. Herring, L. E. Toombs and G. F. Wright, *Can. J. Res.*, 1948, **26B**, 89.
159. R. S. Stuart and G. F. Wright, *Can. J. Res.*, 1948, **26B**, 401.
160. (a) P. Golding, R. W. Millar, N. C. Paul and D. H. Richards, *Tetrahedron*, 1993, **49**, 7063; (b) P. Bunyan, P. Golding, R. W. Millar, N. C. Paul, D. H. Richards and J. A. Rowley, *Propell. Explos. Pyrotech.*, 1993, **18**, 55.
161. J. P. Agrawal, Mehilal, S. H. Sonawane and R. N. Surve, *J. Hazard. Mater.*, 2000, **77**, 11.
162. (a) L. A. Fang, S. Q. Hua, V. G. Ling and L. Xin, 'Preliminary Study on Bu-NENA Gun Propellants', *27th International Annual Conference of ICT*, Karlsruhe, Germany, June 25–28, 1996, 51; (b) N. F. Stanley and P. A. Silver, 'Bu-NENA Gun Propellants', *JANNAF Propulsion Meetings, Vol. 2*, 10 September 1990, 515; (c) R. A. Johnson and J. J. Mulley, 'Stability and Performance Characteristics of NENA Materials and Formulations', *Joint International Symposium on Energetic Materials Technology*, New Orleans, Louisiana, 5–7 October, 1992, 116.
163. M. M. Bhalerao, B. R. Gandhe, M. A. Kulkarni, K. C. Rao and A. K. Sikder, *Propell. Explos. Pyrotech.*, 2004, **29**, 93.
164. (a) M. E. Colclough, H. Desai, N. C. Paul and N. Shepherd, *J. Ballistics*, 1995, **12**, 169; (b) M. E. Colclough, H. Desai, P. Golding, R. W. Millar, N. C. Paul and M. J. Stewart, *Polym. Adv. Technol.*, 1994, **5**, 554; (c) H. Bull, W. B. H. Leeming, E. J. Marshall, N. C. Paul and M. J. Rodgers, 'An Investigation into Poly(GLYN) Cure Stability', *27th International Annual Conference of ICT*, Karlsruhe, Germany, June 25–28, 1996, 99.

165. R. S. Day, A. G. Stern and R. L. Willer, *US Pat.* 5 120 827 (1992).
166. R. S. Day, A. G. Stern and R. L. Willer, *US Pat.* 5 380 777 (1995).
167. (a) J. Consaga, *US Pat.* 728 918 (1992); *Chem. Abstr.*, 1992, **117**, 29937f; (b) J. P. Consaga and R. C. Gill, 'Synthesis and Use of Cyclodextrin Nitrate', *Proc. 29th International Annual Conference of ICT*, Karlsruhe, Germany, 1998, V5-1.
168. B. Kosowski, C. Meyersand, D. Robitelle, A. Ruebner and G. Statton, 'Cyclodextrin Polymer Nitrate', *Proc. 31st International Annual Conference of ICT*, Karlsruhe, Germany, 2000, V12-1.

4

Synthetic Routes to Aromatic *C*-Nitro Compounds

4.1 INTRODUCTION

The nitration of aromatic hydrocarbons is one of the most widely studied and well-documented reactions in organic chemistry. Aromatic nitro compounds are of huge industrial importance in the synthesis of pharmaceutical drugs, agrochemicals, polymers, solvents and perfumes, and for the synthesis of other industrially important chemicals containing amine and isocyanate functionality. However, early research into aromatic nitration was fuelled exclusively by their use as explosives and intermediates in the synthesis of dyestuffs. The former is the subject of this chapter.

While we believe our discussions of nitramine and nitrate ester synthesis to be comprehensive, it would be quite impossible to have a comprehensive discussion of aromatic nitration in this short chapter – published studies into aromatic nitration run into many tens of thousands. The purpose of this chapter is primarily to discuss the methods used for the synthesis of polynitroarylene explosives. Undoubtedly the most important and direct method for the synthesis of polynitroarylenes involves direct electrophilic nitration of the parent aromatic hydrocarbon. This work gives an overview of aromatic nitration but the discussion doesn't approach mechanistic studies in detail. Readers with more specialized interests in aromatic nitration are advised to consult several important works published in this area which give credit to this important reaction class.[1–12] The use of polynitroarylenes as explosives and their detailed industrial synthesis has been expertly covered by Urbański in Volumes 1 and 4 of *Chemistry and Technology of Explosives*.[13]

Nucleophilic aromatic substitution is a useful route to many functionalized polynitroarylenes whether the displacement is of a pre-existing nitro group or any other suitable leaving group on the aromatic ring. For this reason nucleophilic substitution is discussed in this chapter but only in the context of preparing polynitroarylene explosives (Section 4.8). The ease of nucleophilic displacement of nitro groups in polynitroarylenes has implications for the suitability of some compounds as explosives and so this is also discussed (Section 4.8.2).

Organic Chemistry of Explosives J. P. Agrawal and R. D. Hodgson

4.2 POLYNITROARYLENES AS EXPLOSIVES

Polynitroarylenes hold a central position in the field of explosives and for many years they have been an important class of explosives for military use. 2,4,6-Trinitrotoluene (TNT) (1) was first synthesized in 1863 and has found wide use as a secondary high explosive. It is by no means a high performance explosive (VOD $\sim$ 6940 m/s, $d = 1.64$ g/cm^3), but it has a combination of properties, such as relatively high chemical stability, moderate insensitivity to impact and friction, and a low enough melting point (80.8 °C) to permit melt casting of charges, which makes it suitable for mass use in munitions. Equally important, TNT is synthesized from readily available and cheap raw materials. The high oxygen deficiency of TNT (−74 %) has meant that melt-cast mixtures with ammonium nitrate, known as amatols, have found wide use in the past. The brisance of such mixtures is relatively low and so the powerful nitramine explosives RDX and HMX are currently the most widely used military explosives. TNT still finds use as a practical explosive but in mixtures where it is effectively an energetic binder for cast compositions. The cyclotols (RDX/TNT), Torpex (RDX/TNT/Al), pentolite (PETN/TNT) and PTX-1 (RDX/tetryl/TNT) are all secondary high explosive compositions based on TNT.

1,3,5-Trinitrobenzene (TNB) (2) is a more powerful explosive than TNT. However, the direct synthesis of TNB from benzene is not practical and the need for an indirect route for its synthesis makes its manufacture too expensive for use as a practical high explosive.

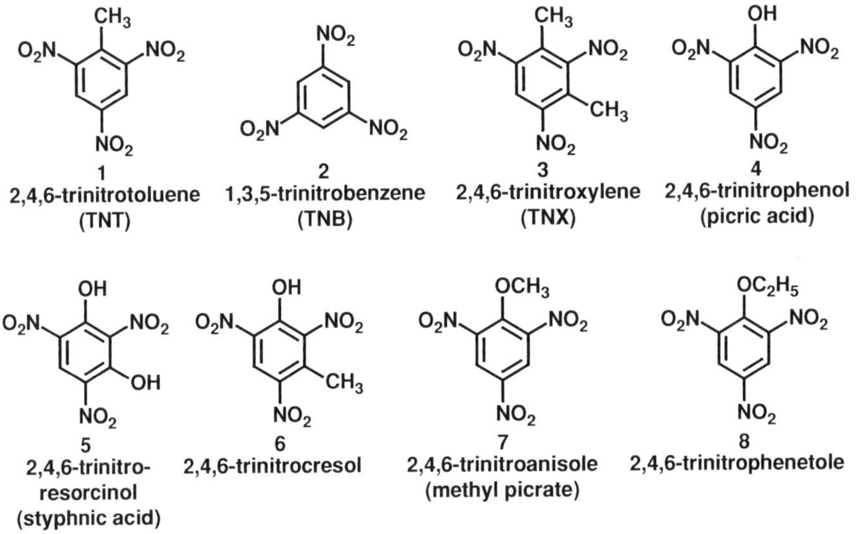

1 2,4,6-trinitrotoluene (TNT)
2 1,3,5-trinitrobenzene (TNB)
3 2,4,6-trinitroxylene (TNX)
4 2,4,6-trinitrophenol (picric acid)
5 2,4,6-trinitroresorcinol (styphnic acid)
6 2,4,6-trinitrocresol
7 2,4,6-trinitroanisole (methyl picrate)
8 2,4,6-trinitrophenetole

Figure 4.1

During the first half of the last century a large number of aromatic nitro compounds found limited use as secondary explosives. These included: 2,4,6-trinitroxylene (TNX) (3), 2,4,6-trinitrophenol (picric acid) (4), 2,4,6-trinitrocresol (6), 2,4,6-trinitroanisole (7), 2,4,6-trinitrophenetole (8), trinitronaphthalene (mixture of 1,3,5-, 1,4,5- and 1,3,8-isomers) (9), 2,2′,4,4′,6,6′-hexanitrodiphenylsulfide (10), 2,2′,4,4′,6,6′-hexanitrocarbanilide (11) and 2,2′,4,4′,6,6′-hexanitrodiphenylamine (hexyl) (12) amongst others. The use of such compounds as explosives was related to the availability of the starting aromatic compounds and their isolation from coal tar. The advent of the modern petrochemical industry means that availability of many aromatic chemicals is no longer a problem. However, the use of many polynitroarylenes as explosives is confined to history because of their moderate performance.

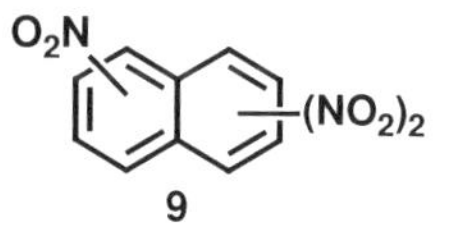

Figure 4.2

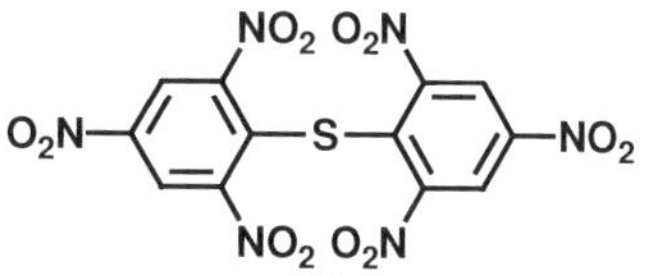

2,4,6-Trinitrophenol (4), commonly known as picric acid (VOD $\sim$ 7350 m/s, $d = 1.71$ g/cm^3), was once used as a military explosive although its highly acidic nature enables it to readily corrode metals. This kind of reaction has led to many fatal accidents, a consequence of some metal picrates being very sensitive primary explosives. The lead salt of picric acid is a dangerous explosive and should be avoided at all cost. In contrast, the ammonium (Explosive D, VOD $\sim$ 7050 m/s, $d = 1.60$ g/cm^3) and guanidine salts of picric acid are unusually insensitive to impact and have been used in armour piercing munitions.

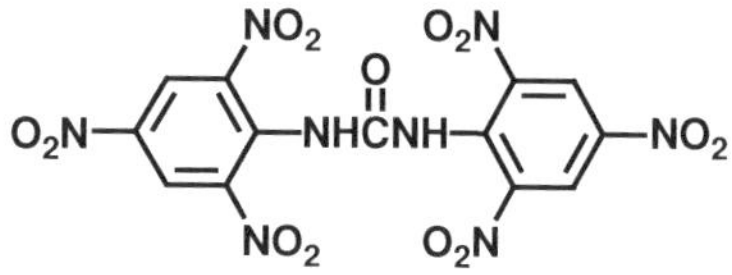

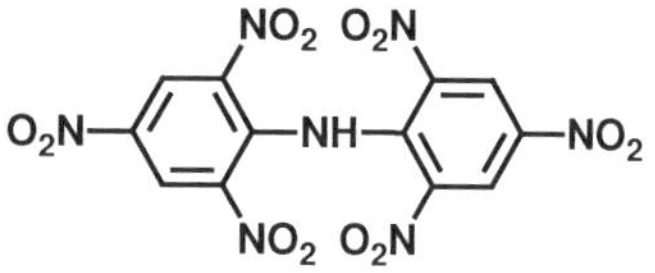

2,4,6-Trinitroresorcinol (5) (styphnic acid) has also seen limited use as an explosive because of its acidic properties and the relatively high cost of resorcinol. However, the lead salt of styphnic acid has found use as a primary explosive in detonators and primers.

The acidity of polynitrophenols is a direct consequence of the negative inductive effect of the nitro groups. This has implications for other polynitroarylenes as practical explosives. 2,4,6-Trinitroanisole (7), 2,4,6-trinitrophenetole (8), 2,4,6-trinitroaniline (picramide) and other 2,4,6-trinitrophenyl (picryl) compounds containing potential leaving groups undergo slow hydrolysis in the presence of water/moisture to generate picric acid or other polynitrophenols which can attack metal, and so, present a real risk when used in munitions destined for long-term storage.

This powerful inductive effect is seen in highly nitrated polynitroarylenes. Thus, benzene and its substituted derivatives containing four, five or six nitro groups on the same ring i.e. one or more nitro groups in a *meta* position, are all chemically unstable and readily lose a nitro group on hydrolysis or in similar nucleophilic reactions. Many highly nitrated polynitroarylenes are powerful explosives but are prevented from being practical explosives because of their poor chemical stability. These issues of chemical stability are discussed more fully in Sections 4.8.2 and 4.8.3.

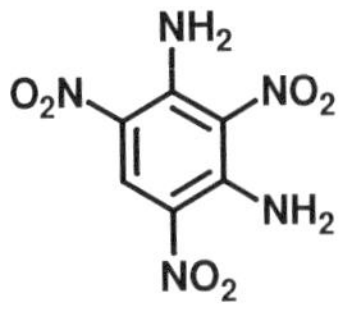

13
1,3-diamino-2,4,6-trinitrobenzene (DATB)

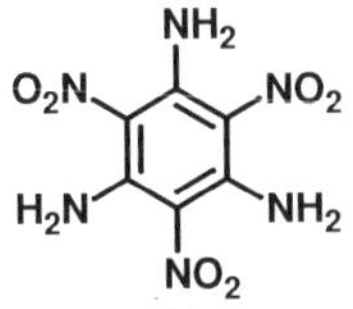

14
1,3,5-triamino-2,4,6-trinitrobenzene (TATB)

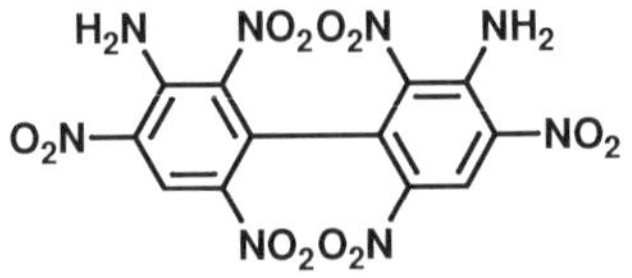

15
3,3'-diamino-2,2',4,4',6,6'-hexanitrobiphenyl (DIPAM)

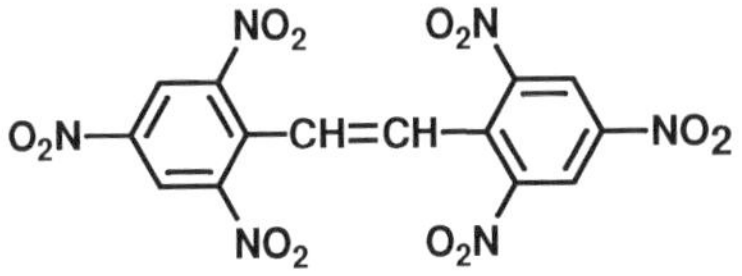

16
2,2',4,4',6,6'-hexanitrostilbene (HNS)

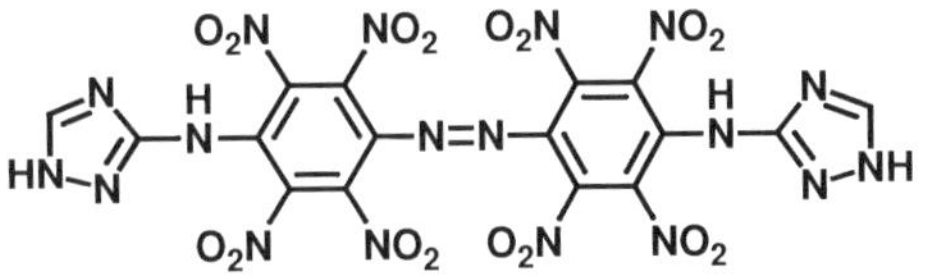

17
N,N'-bis(1,2,4-triazol-3-yl)-4,4'-diamino-2,2',3,3',5,5',6,6'-octanitroazobenzene (BTDAONAB)

Figure 4.3

Interest in polynitroarylenes has resumed over the past few decades as the demand for thermally stable explosives with a low sensitivity to impact has increased. This is mainly due to advances in military weapons technology but also for thermally demanding commercial applications i.e. oil well exploration, space programmes etc. Explosives like 1,3-diamino-2,4,6-trinitrobenzene (DATB) (13), 1,3,5-triamino-2,4,6-trinitrobenzene (TATB) (14), 3,3′-diamino-2,2′,4,4′,6,6′-hexanitrobiphenyl (DIPAM) (15), 2,2′,4,4′,6,6′-hexanitrostilbene (HNS, VOD $\sim$ 7120 m/s, $d = 1.70$ g/cm^3) (16) and *N,N′*-bis(1,2,4-triazol-3-yl)-4,4′-diamino-2,2′,3,3′,5,5′,6,6′-octanitroazobenzene (BTDAONAB) (17) fall into this class. TATB is the benchmark for thermal and impact insensitive explosives and finds wide use for military, space and nuclear applications.

Some liquid nitro compounds have found past use in explosive compositions. A mixture of 2,4-dinitroethylbenzene and 2,4,6-trinitroethylbenzene, known as K-10, currently finds use as an energetic plasticizer in some propellant formulations. K-10 plasticizer, also known as Rowanite 8001, is manufactured by Royal Ordnance in the UK and also finds use as a plasticizer in PBXs.

4.3 NITRATION

The direct nitration of aromatic substrates is usually the method of choice for the synthesis of aromatic nitro compounds on both industrial and laboratory scales. Other routes are usually only considered when the required product has an unusual substitution pattern or is so deactivated that nitration is exceptionally difficult. Many of these alternative methods are limited to a laboratory scale.

Olah[1] showed that nitrations can be split into the three categories of electrophilic, nucleophilic and free radical nitration. Free radical nitrations are extensively used for the industrial synthesis of low molecular weight nitroalkanes from aliphatic hydrocarbons. Nucleophilic nitration is the basis for a number of important methods for the synthesis of nitro and polynitro alkanes. Generally speaking only electrophilic nitration is of preparative importance for the synthesis of aromatic nitro compounds. In these reactions the nitration of aromatic substrates proceeds via electrophilic aromatic substitution and uses reagents that generate the electrophilic nitronium cation or other active electrophilic nitrating species. The aromatic substrate uses the electron density of the aromatic ring to attack the nitronium ion and, consequently, substituents on the aromatic ring which withdraw or release electron density have an enormous effect on substrate reactivity. The nitro group is usually introduced by substitution of hydrogen. Substitution of other atoms such as halogen is usually only useful in saturated aliphatic systems, although there are important exceptions. Detailed mechanistic studies on aromatic nitration can be found in several important works.[1,3,4,7–11,14,15]

When an aromatic substrate is nitrated there is also the issue of selectivity to be considered. Substituents which activate the aromatic ring towards electrophilic nitration direct substitution to the *o*- and *p*-positions, whereas deactivating substituents usually direct to *m*-positions. In short, this arises from partial charge localization at the *o*- and *p*-positions, so any increase or decrease in electron density is most significant at these positions. Mechanistic reasoning for these effects can be found in any standard organic chemistry textbook and so further discussion is not given. There are other selectivity issues which are more specific and strongly dependent on the nature of the nitrating agent used and the reaction conditions. Some of these effects are typical of certain classes of aromatic substrate and so these are discussed.

Nitrations giving a complex mixture of products are not useful in organic chemistry or for the synthesis of explosives, and so, another route to the required product should be considered which is more selective. Although it is acceptable for commercial explosives to contain a mixture of aromatic nitro compounds, military explosives are almost always single compounds with well-defined physical properties.

4.3.1 Nitration with mixed acid

Acid-catalyzed electrophilic nitration is the most common and important route to polynitroarylene explosives. These reactions usually employ nitric acid in the presence of a Brønsted or Lewis acid. While many nitrating agents have been reported[1] and these are discussed later, the most widely used reagent for aromatic nitration on both industrial and laboratory scales is a mixture of sulfuric and nitric acids known as 'mixed acid'. This mixture contains many nitrating species including the powerful nitronium electrophile (NO_2^+). For a long time it was thought that the sulfuric acid in mixed acid acted solely as a dehydrating agent to mop up water formed during the nitration. It is now known that sulfuric acid, a stronger acid than nitric acid, protonates the latter to form the nitracidium cation ($H_2ONO_2^+$) which can lose water to form the nitronium ion (Equations 4.1, 4.2 and 4.3); nitric acid acts as a base in this respect. In concentrated sulfuric acid, nitric acid exists almost entirely as nitronium ions.[15,16] Nitric acid can protonate itself, but even anhydrous nitric acid contains only 3–4 % of the nitrogen present as nitronium ion.[15,16] Consequently, the nitracidium cation is probably the active nitrating agent when concentrated nitric acid is used alone for the nitration of some of the more activated substrates.

$$H_2SO_4 + HONO_2 \rightleftharpoons H_2\overset{+}{O}NO_2 + HSO_4^- \quad \text{(Eq. 4.1)}$$

$$H_2SO_4 + H_2\overset{+}{O}NO_2 \rightleftharpoons NO_2^+ + HSO_4^- + H_3O^+ \quad \text{(Eq. 4.2)}$$

$$2\,H_2SO_4 + HONO_2 \rightleftharpoons NO_2^+ + 2\,HSO_4^- + H_3O^+ \quad \text{(Eq. 4.3)}$$

Figure 4.4

Nitrating agent, solvent, temperature, concentration and the ratio of substrate to nitrating agent must all be considered when an aromatic substrate is to be nitrated. Substituents which withdraw electron density from the aromatic ring make nitration more difficult, whereas those that release electron density through either inductive or resonance effects make nitration easier. Reaction rates between substrates can be several orders of magnitude different. Reactive substrates can be nitrated with mixed acid containing relatively large amounts of water, whereas less reactive substrates and substrates that require polynitration need treatment with more concentrated acids and usually at higher temperatures. Nitrations with mixed acid vary from the use of concentrated nitric acid or even dilute nitric acid in sulfuric acid to the use of oleum and fuming nitric acid at elevated temperatures. Commercial 'concentrated' nitric acid contains 70 % acid, whereas fuming nitric acid contains 90 %+ acid, but both are commonly used in nitrations. Pure nitric acid containing 98–99 % nitric acid is more expensive, less widely available and has a limited shelf-life. Anhydrous mixed acid is often prepared from fuming nitric acid and oleum. Oleum, or 'fuming sulfuric acid' as it is sometimes known, is sulfuric acid containing dissolved sulfur trioxide and the latter reacts with water formed during nitration to generate more sulfuric acid.

The rate of a nitration is fastest at the start of a reaction when a large excess of nitric acid is present. As the reaction progresses the water formed during nitration dilutes the mixed acid and slows the rate of reaction, and as such, it is common towards the end of a nitration, when most of the substrate has reacted, to heat the reaction to completion. This dilution of the acid with water is an important point and the amount of sulfuric acid used should be enough to take up all the water formed during the reaction; otherwise, nitration may be incomplete and result in an unfavourable mixture of product and starting material. Increased amounts of water in mixed acid rapidly reduce the concentration of nitronium ions.[17–19] When concentrated nitric acid is used for the nitration of some of the more reactive substrates a large excess of sulfuric is often used to compensate for the water present.

It is common practice during the mono- or di-nitration of a substrate, dependent on its reactivity, to use a small excess of nitric acid (1–5 %) in the mixed acid. The further nitration and polynitration of substrates often requires more vigorous conditions and a greater excess of nitric acid (5–100 %) is used in order to ensure complete nitration.

During most nitrations on both a laboratory and industrial scale the aromatic substrate is added slowly and in portions/aliquots to the mixed acid. This is desirable on safety grounds but has the problem that the first aliquots of substrate are in the presence of a large excess of nitric acid and this may lead to overnitration. In contrast, the last portions of substrate to be added may remain unreacted, although the presence of a small excess of nitric acid in the mixed acid often accounts for this problem. Adding the mixed acid to the substrate would avoid these problems but must be considered a more dangerous process and is rarely practical. A more suitable method involves the slow and simultaneous addition of both mixed acid and organic substrate to the reaction vessel. The advent of continuous flow reactors allows this on an industrial scale and increases safety. Studies show that optimum nitrating conditions are achieved when a constant ratio of water to sulfuric acid is maintained throughout a nitration.[20]

Nitrations with mixed acid are exothermic reactions. The polynitration of aromatic substrates frequently involves strong mixed acid and high temperatures. Heating organic substances with solutions containing strong oxidants like nitric acid always involve some danger and so it is customary that the temperature is kept as low as possible during the initial stages of nitration and raised later on when most of the substrate has reacted. More than often, substrate addition is regulated so that a stable temperature is maintained without the need for external heating or cooling. Temperature increase during nitration is often associated with an increase in undesirable by-products (Section 4.3.5). However, many deactivated substrates, particularly those already containing one or more nitro groups, react only slowly with mixed acid at ambient temperature.

While nitric acid is generally a good solvent for many organic compounds, the mixed acids commonly used in nitrations, especially those containing oleum, result in heterogeneous reactions, meaning that a high stir rate is important for uniform substrate nitration. When working with mole quantities of substrate a high stir rate also becomes an important safety point because heterogeneous nitrations containing large portions of unreacted substrate are susceptible to thermal runaway and the risk of explosion.

Experimentally, the use of mixed acid for the nitration of aromatic substrates is very convenient. Reactions are often quenched by the addition of water, where the product usually precipitates. Solids are simply filtered from the acid liquors and oils are either separated or extracted into organic solvents. However, on an industrial scale, these mixed acid nitrations create environmental problems from air and water pollution (Sections 4.3.5 and 4.8.2).

4.3.2 Substrate derived reactivity

4.3.2.1 Phenols and phenol ethers

Phenols and phenol ethers are nitrated with relative ease, and even though the deactivating effect of the nitro group means that progressive nitration becomes slower and more difficult, the introduction of three nitro groups into the aromatic ring can be achieved under fairly mild conditions. However, the direct nitration of substrates containing phenolic groups can be low yielding because of facile oxidation and the vigour of such reactions. Phenol itself is readily oxidized to oxalic acid on heating with concentrated nitric acid. The direct nitration of phenol with mixed acid provides low yields of picric acid along with much resinous matter, the acidic liquors also containing oxalic acid. With all phenolic substrates, the higher the temperature of the nitration, the more by-products formed.

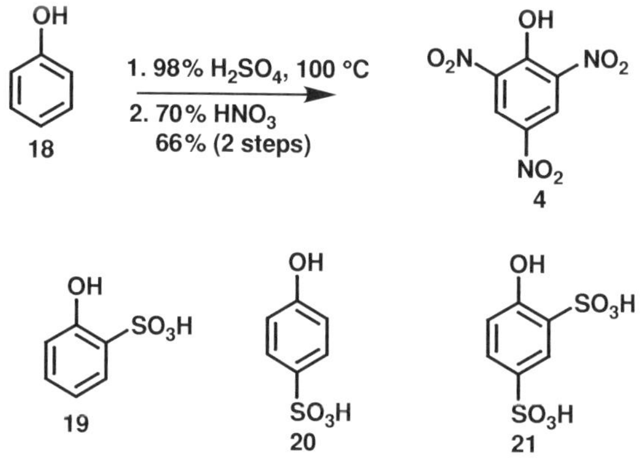

OH
18
1. 98% H_2SO_4, 100 °C
2. 70% HNO_3
66% (2 steps)
OH
O_2N
NO_2
NO_2
4

OH
SO_3H
19
OH
SO_3H
20
OH
SO_3H
SO_3H
21

Figure 4.5

A commonly used strategy for the higher nitration of phenolic substrates is to sulfonate the electron-rich aromatic ring before nitration. Sulfonic acid groups are electron withdrawing and moderate the nitration step as well as protect the substrate from oxidation. In these reactions a nitro group substitutes a sulfonic acid group and the now less reactive nitrophenol intermediates are nitrated much more smoothly. Such a procedure largely avoids side-reactions due to oxidation. On a laboratory scale, picric acid (4) is readily synthesized from phenol via this route. In a typical procedure, phenol (18) is heated on a steam bath with four equivalents of concentrated sulfuric acid to form a mixture containing both *o*- and *p*-phenolsulfonic acids (19 and 20) in addition to some 2,4-phenoldisulfonic acid (21); the latter predominates at higher temperatures. This mixture is then cooled and treated with concentrated nitric acid.[21,22] The same procedure using more dilute solutions of nitric acid (44–65 %) has been reported.[23] However, the direct nitration of 2,4-dinitrophenol, obtained from 2,4-dinitrochlorobenzene, provides a more practical route to picric acid on an industrial scale (Section 4.8.1.3).[23]

The reactivity of phenols to electrophilic nitration is illustrated further by the facile conversion of *m*-nitrophenol to 2,3,4,6-tetranitrophenol with anhydrous mixed acid.[24] The latter is a powerful explosive, but chemically unstable, like all polynitroarylenes containing a nitro group positioned *o*/*p*- to other nitro groups.

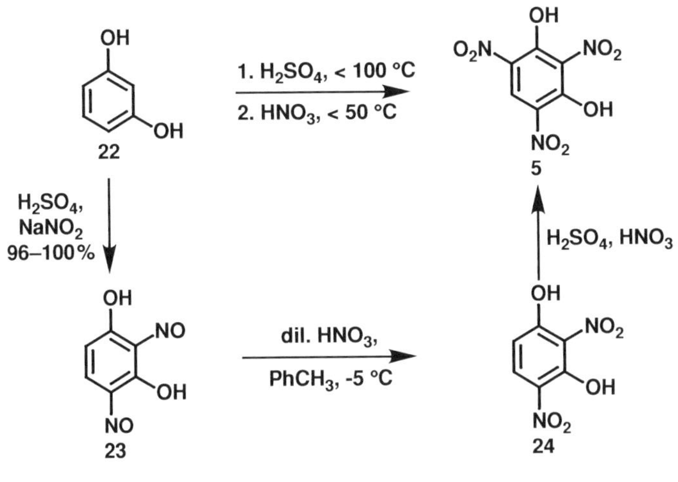

Figure 4.6

The sulfonation–nitration strategy also provides a route to styphnic acid (5) (2,4,6-trinitroresorcinol) from resorcinol (22) but the control of temperature in this reaction is very important.[25] The synthesis of styphnic acid (5) from the nitration of 2,4-dinitroresorcinol (24) with mixed acid or concentrated nitric acid is a higher yielding route. 2,4-Dinitroresorcinol (24) is conveniently prepared from the nitrosation[26a,b] of resorcinol (22) followed by oxidation[26c] of the resulting 2,4-dinitrosoresorcinol (23) with dilute nitric acid. 2,4-Dinitrosoresorcinol (23) also generates styphnic acid (5) on treatment with concentrated nitric acid.[27]

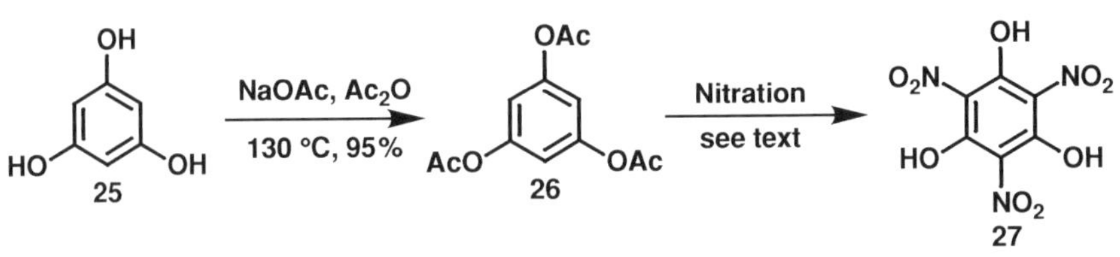

Figure 4.7

Phloroglucinol (25) is more susceptible to oxidation than both phenol and resorcinol. However, its direct nitration to 2,4,6-trinitrophloroglucinol (27) can be achieved by the slow addition of a nitrating agent composed of concentrated sulfuric acid and concentrated nitric acid to a solution of phloroglucinol dihydrate in concentrated sulfuric acid between 0 and 10 °C (72 %).[28–30] Acetylation of phloroglucinol (25) yields the triacetate (26) which moderates reactivity but still allows the synthesis of 2,4,6-trinitrophloroglucinol (27) on treatment with fuming nitric acid at 0 °C (53 %),[31] mixed acid at –10 °C (91 %),[30] or dinitrogen pentoxide in sulfuric acid (92 %).[30] 2,4,6-Trinitrophloroglucinol has also been obtained on treatment of phloroglucinol with dinitrogen pentoxide in sulfuric acid (74 %)[30] and via the oxidation of 1,3,5-trinitrosophloroglucinol with 65 % nitric acid (70 %).[30,32] 2,4,6-Trinitrophloroglucinol finds use in cap and percussion compositions and as a flash sensitizer in some detonators.[28]

m-Cresol is more susceptible to oxidation than phenol due to additional activation from the methyl group. This accounts for the lower yield of 2,4,6-trinitrocresol product on trinitration with mixed acid or concentrated nitric acid. For such substrates the use of sulfonation prior to nitration is essential. The di-nitration of 1-naphthol to 2,4-dinitro-1-naphthol also uses a sulfonation–nitration strategy because of the substrate's susceptibility to oxidation, prior heating of 1-napthol with concentrated sulfuric acid forming 1-napthol-2,4-disulphonic acid.

Phenol ethers, like the parent phenols, are reactive substrates. Phenol ethers like anisole and phenetole are readily nitrated to their picryl ethers, 2,4,6-trinitroanisole and 2,4,6-trinitrophenetole respectively, on treatment with mixed acid composed of concentrated nitric and sulfuric acids at 0 °C.[33] Such reactions are vigorous, prone to oxidative side-reactions, and pose a considerable safety risk. The direct nitration of 2,4-dinitrophenol ethers, obtained from the reaction of 2,4-dinitrochlorobenzene with alkoxides, provides a more practical route to picryl ethers on an industrial scale.[33]

4.3.2.2 Amines

Aromatic amines are exceptionally reactive to electrophilic nitration. Aniline is readily nitrated to 2,4,6-trinitroaniline (picramide) with mixed acid using acetic acid as solvent. When excess sulfuric acid is present, aniline is largely protonated, making nitration difficult and directing the incoming nitro group to the *m*-position to yield *m*-nitroaniline. The sensitivity of aniline to oxidative degradation and the formation of phenolic by-products in the presence of nitrous acid means that other indirect routes to picramide have been reported,[34] including treating picryl chloride with hydroxylamine,[35] and methylation of picric acid followed by treatment with ammonia.[30] Picramide is also prepared from the nitration of either *o*- or *p*-nitroacetanilides using a solution of potassium nitrate in concentrated sulfuric acid,[36] the acetamino behaving as a reactivity moderator and also protecting the amino group against oxidation. Picramide has no use as an explosive but may find use for the future industrial synthesis of TATB via vicarious nucleophilic amination (Section 4.8.4).[37]

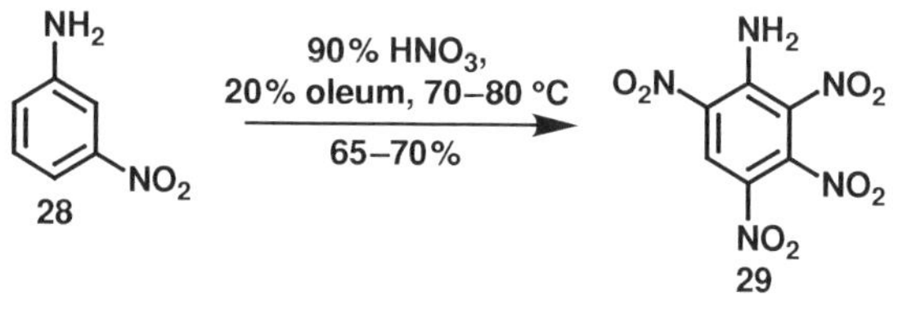

28 → 29: 90% HNO_3, 20% oleum, 70–80 °C, 65–70%

Figure 4.8

Nitration of *m*-nitroaniline (28) with fuming nitric acid and oleum yields 2,3,4,6-tetranitroaniline (29), a powerful but chemically unstable explosive.[38] 2,3,4,6-Tetranitroaniline readily reacts with a range of nucleophiles, including water to yield 3-amino-2,4,6-trinitrophenol.

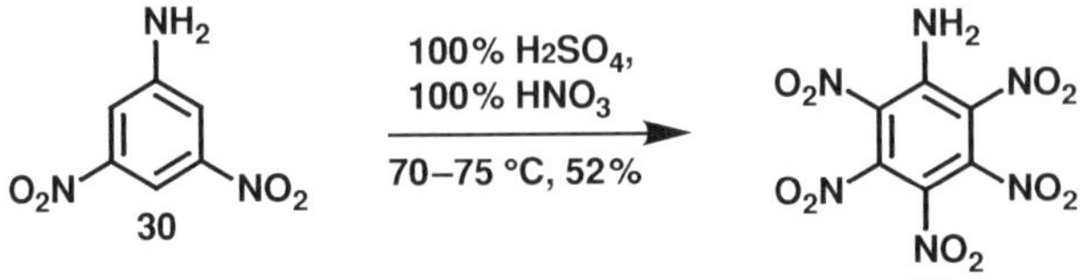

Figure 4.9

The remarkable ability of the amino group to promote electrophilic substitution is illustrated by the nitration of 3,5-dinitroaniline (30) to 2,3,4,5,6-pentanitroaniline (31) in 52 % yield when treated with anhydrous mixed acid.[39] The pre-existing nitro groups are advantageous by making the aromatic ring less prone to oxidation.

Nitrations of aromatic amines often involve the intermediate formation of *N*-nitramines, although these are rarely seen under the strongly acidic conditions of mixed acid nitration (Section 4.5). *N*,2,4,6-Tetranitro-*N*-methylaniline (tetryl) is an important secondary high explosive usually synthesized from the nitration of *N*,*N*-dimethylaniline or 2,4-dinitro-*N*-methylaniline.[40] The synthesis of tetryl is discussed in Section 5.14.

The high explosive known as hexyl (2,2′,4,4′,6,6′-hexanitrodiphenylamine) is synthesized in two steps from the nitration of 2,4-dinitrodiphenylamine. The first nitration uses 55 % nitric acid at elevated temperature for conversion to 2,2′,4,4′-tetranitrodiphenylamine. Introduction of two more nitro groups to yield hexyl requires a mixture of concentrated sulfuric and nitric acids, although the reaction can be conducted at room temperature (Section 4.8.1.3).[41]

4.3.2.3 Toluene

The methyl group of toluene makes nitration a relatively facile process. However, as more nitro groups are introduced, the aromatic ring becomes more electron deficient and deactivated towards electrophilic attack, and so requires more vigorous conditions for further nitration. The direct nitration of toluene to TNT with mixed acid is not industrially feasible. First, a very large excess of strong mixed acid or oleum would be required to compensate for the water formed during the mono- and di-nitration, and secondly, such a strong nitrating agent would lead to many by-products. Consequently, the nitration of toluene to 2,4,6-trinitrotoluene (TNT) is usually conducted in two or three steps.[13a] This is a common strategy for the polynitration of many aromatic substrates. Substituents which activate towards electrophilic substitution also make side-reactions like oxidation more facile. Mixed acid, particularly at high temperatures, is a strong oxidizing agent, and oleum, if present, can sulfonate activated substrates. Consequently, while a strong mixed acid may be needed for the introduction of the final nitro group into a polynitroarylene, this reagent is probably not suitable for initial nitration.

Industrial TNT production produces both atmospheric and water pollution. The spent acid from the three stages of mono-, di- and tri-nitration pose considerable disposal problems. On an industrial scale the mixed acid from previous di- and tri-nitrations is usually refortified with nitric acid and used for mono- and di-nitration respectively. Diluted sulfuric acid is often reconcentrated, but at considerable expense. Another source of pollution is from 'red water', the aqueous washing from sulphite treatment of crude TNT and resulting from the presence of unstable unsymmetrical isomers of TNT.

Hill and co-workers[42] studied the two stage nitration of toluene to TNT and showed that the production of *m*-isomers was mainly from the initial mono-nitration step. Hill and co-workers showed that the nitration of toluene to dinitrotoluene at temperatures above 40 °C always produces 3.5 % or more of *m*-isomers. In contrast, the nitration of pure *o*- or *p*-nitrotoluenes produces TNT with less than 0.1 % of *m*-isomers. Initial nitration of toluene to dinitrotoluene (mainly 2,4- and 2,6-isomers) is highly exothermic and an increase in temperature is known to increase the amount of *m*-isomers in the nitration product. Hill and co-workers showed that mixed acid containing up to 7 % water was as effective as oleum–nitric acid mixtures for nitrating toluene to dinitrotoluene, yields being >99 % after 1 hour of reaction time. Conducting these nitrations between 5 and −35 °C resulted in a reduction in *m*-isomer ratio from 2.06 % to 1.40 %. It was further shown that conducting the initial nitration of toluene at −10 °C, followed by further nitration under standard conditions, resulted in crude TNT containing only 1.8 % of *m*-isomers as compared to the usual 3.5 %. Hill and co-workers also nitrated toluene to dinitrotoluene with potassium nitrate in sulfuric acid at −10 °C (97 %) and with nitric acid in triflic acid at lower temperatures. The Olin Corporation[43] reports that this di-nitration can be achieved with nitric acid alone.

The nitration of dinitrotoluene to TNT requires the use of mixed acid fortified with oleum, and the use of elevated temperature, which produces its own pollutants and impurities, including tetranitromethane and nitrogen oxides. An excellent discussion of industrial TNT production has been given by Urbański.[13a]

4.3.2.4 Other alkylbenzenes

The presence of two methyl groups in *m*-xylene makes it a more reactive substrate than toluene, but as a consequence, this substrate is more susceptible to oxidation. Therefore, the nitration of *m*-xylene requires lower temperatures and the use of less concentrated mixed acid. In fact, the nitration of either 2,4- or 2,6-dinitro-*m*-xylenes to 2,4,6-trinitroxylene (TNX) can be achieved with mixed acid containing up to 10 % water.[44] TNX is a less powerful explosive than TNT and has a poor oxygen balance (−78.4 %).

4.3.2.5 Benzene

The synthesis of 1,3,5-trinitrobenzene (TNB) from the direct nitration of *m*-dinitrobenzene is very difficult. Desvergnes[45] reported a 71 % yield of TNB (2) on treatment of *m*-dinitrobenzene (32) with a large excess of mixed acid composed of anhydrous nitric acid and 60 % oleum at a reaction temperature of 110 °C for several days. Similar results are also reported from other sources.[46]

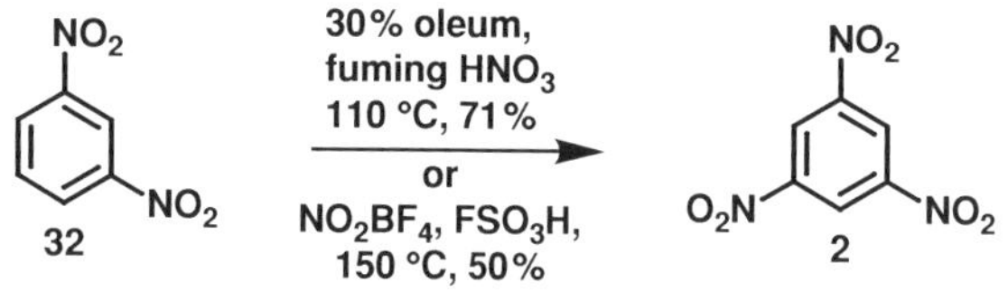

32 → 2: 30% oleum, fuming HNO_3, 110 °C, 71% or NO_2BF_4, FSO_3H, 150 °C, 50%

Figure 4.10

Olah and Lin[47] reported obtaining pure TNB (2) in 50 % yield from the nitration of *m*-dinitrobenzene (32) with nitronium tetrafluoroborate in fluorosulfuric acid at 150 °C; higher yields of TNB can be obtained by reducing reaction time but the product then contains some unreacted *m*-dinitrobenzene. A solution of dinitrogen pentoxide in sulfuric acid at 160 °C is also reported to effect the conversion of *m*-dinitrobenzene to TNB.[48] TNB is rarely prepared via these routes and is best obtained from TNT via an indirect route (Section 4.9).

4.3.2.6 Halobenzenes

Halogen substituents withdraw electron density from the aromatic nucleus but direct *o*/*p*-through resonance effects. The result is that halobenzenes undergo nitration with more difficulty relative to benzene. The nitration of chlorobenzene with strong mixed acid gives a mixture of 2,4- and 2,6-isomeric dinitrochlorobenzenes in which the former predominates.[49] The nitration of 2,4-dinitrochlorobenzene to 2,4,6-trinitrochlorobenzene (picryl chloride) requires an excess of fuming nitric acid in oleum at elevated temperature.[50,51] Both are useful for the synthesis of other polynitroarylene explosives but only 2,4-dinitrochlorobenzene finds industrial importance (Sections 4.8.1.2 and 4.8.1.3).

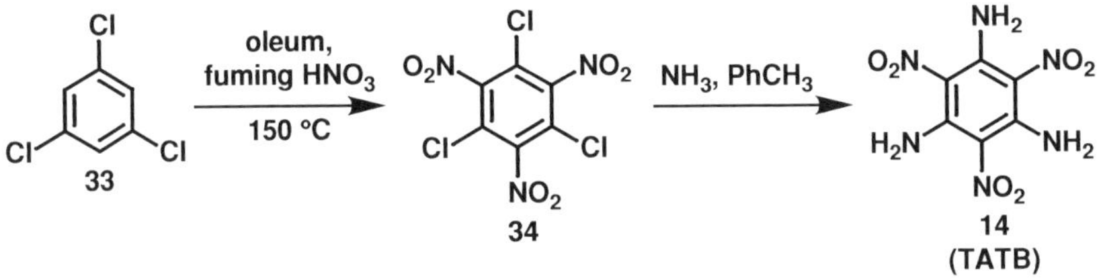

Figure 4.11

1,3-Dichloro-2,4,6-trinitrobenzene and 1,3,5-trichloro-2,4,6-trinitrobenzene (34) are produced on an industrial scale for the synthesis of the thermally stable explosives 1,3-diamino-2,4,6-trinitrobenzene (DATB) and 1,3,5-triamino-2,4,6-trinitrobenzene (TATB) (14) respectively. The nitration of both 1,3-dichlorobenzene[52] and 1,3,5-trichlorobenzene[53] (33) requires very harsh conditions, the latter requiring an excess of fuming nitric acid and oleum at 150 °C for 2.5 hours to yield a mixture containing the required 1,3,5-trichloro-2,4,6-trinitrobenzene (34) (89 %) and 11 % of trichloro- and tetrachloro-dinitrobenzene impurities. These impurities need separation before amminolysis to TATB (14) is conducted.[53] The nitration of 1,3,5-trichloro-2,4-dinitrobenzene to 1,3,5-trichloro-2,4,6-trinitrobenzene using potassium nitrate in sulfuric acid has been reported.[54] TATB has also been synthesized from the amminolysis[55] of 1,3,5-trifluoro-2,4,6-trinitrobenzene, but the synthesis of the latter, in two steps[56] from 1,3,5-trifluoro-2-nitrobenzene, also requires harsh nitrating conditions and the yield is low (~43 %).

4.3.2.7 Other substrates

The successive nitration of naphthalene allows the introduction of four nitro groups. Nitration of naphthalene with concentrated nitric acid mainly yields 1-nitronaphthalene and a small amount of 2-nitronaphthalene.[7] Nitration of pure 1-nitronaphthalene with mixed acid yields

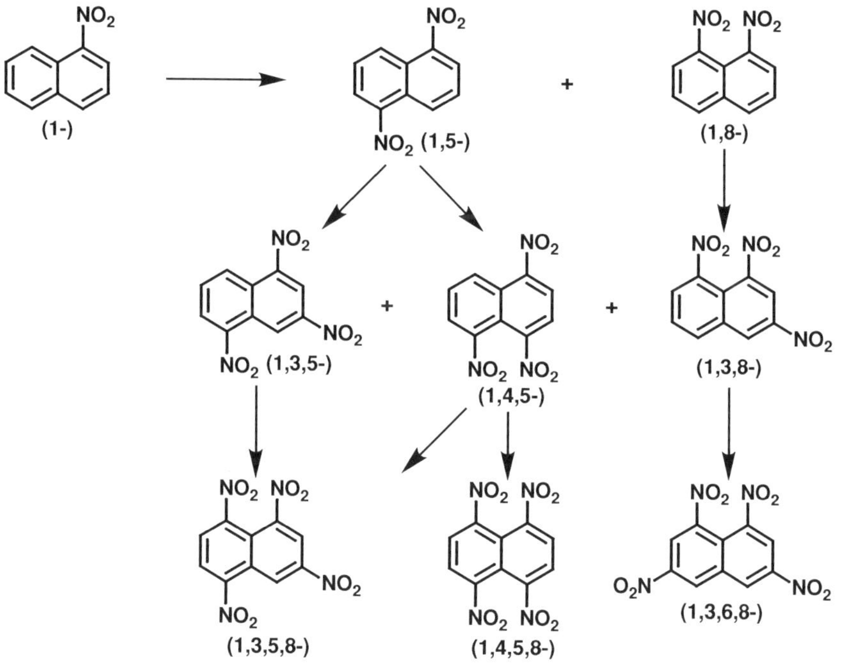

Figure 4.12 Nitration of 1-nitronaphthalene with mixed acid – Ref. 59

a mixture of 1,5- and 1,8-isomers in an approximate 35:65 ratio depending on the nitrating agent and conditions used.[57,58] Further nitration of 1,5-dinitronaphthalene yields both 1,3,5- and 1,4,5-trinitronaphthalenes, whereas 1,8-dinitronaphthalene predominantly yields 1,3,8-trinitronaphthalene.[58,59] Commercial trinitronaphthalene is thus largely a mixture of 1,3,5-, 1,4,5- and 1,3,8-isomers. In practice small amounts of other isomers are present and their ratio, relative to the main isomers, is heavily dependent on temperature and nitrating agent used.[58] Introduction of a further nitro group requires concentrated sulfuric acid and fuming nitric acid to yield a mixture of isomeric 1,3,5,8-, 1,3,6,8- and 1,4,5,8-tetranitronaphthalenes.[58,59] Direct nitration of 1-nitronaphthalene or dinitronaphthalenes to tetranitronaphthalenes requires an excess of strong mixed acid and increased temperature.[58–60] The most extensive study into naphthalene nitrations was conducted by Ward and co-workers who synthesized all ten possible isomers of dinitronaphthalene and studied their nitration to tri- and tetra-nitronaphthalenes.[57–59] These 'unnatural' isomers were largely synthesized by diazotization of the corresponding naphthylamines in the presence of excess nitrite anion and a copper salt (Section 4.6).[61–66]

Some substrates like benzoic acid, benzaldehyde etc. are so deactivated that direct nitration to their trinitro derivatives is not possible. Direct nitration of benzoic acid with excess fuming nitric and concentrated sulfuric acids at a temperature of 145 °C for several hours results in the formation of 3,5-dinitrobenzoic acid (54–58 %).[67] The use of oleum in such reactions can significantly reduce the rate of nitration due to carbonyl protonation (see Section 4.3.3). Consequently, indirect routes are used for the synthesis of polynitroarylenes like 2,4,6-trinitrobenzoic acid and 2,4,6-trinitrobenzaldehyde (Section 4.9).

Exact experimental conditions including the composition of nitrating agent used for the industrial synthesis of all common polynitroarylene explosives is given by Urbański.[13a]

Topchiev[7] has given an extensive discussion of nitrating conditions used, and the products obtained, for all manner of aromatic substrates.

Polynitroarylenes with unusual substitution patterns and those containing more than three nitro groups per benzene ring are usually not useful as practical explosives. Many of these compounds can be synthesized via indirect routes, including: arylamine and arylhydroxylamine oxidation, treating diazonium salts of arylamines with nitrite anion in the presence of copper salts, and Ullmann coupling of polynitrohalobenzenes. The latter is useful for the synthesis of high molecular weight polynitroarylenes (Section 4.10) and partially nitrated biphenyls.

4.3.3 Effect of nitrating agent and reaction conditions on product selectivity

The nature of the substituents presence in an aromatic substrate has a large effect on which positions are substituted on nitration. The isomeric ratio of products obtained on nitration is also dependent on the nitration conditions, nitrating agent, nitrating medium and its acidity, acid composition and concentration etc., but often to a lesser extent.

Substrates containing substituents with nonbonding electrons localized on heteroatoms i.e. phenol ethers, anilines, substituted anilines, acetanilides etc. often exhibit very large differences in product isomer ratio with change in nitration conditions. The nitration of such substrates with mixed acid[68–70] often yields higher than expected proportions of the *para* isomer, whereas solutions of nitric acid in acetic anhydride[69–71] are usually very selective for the formation of the *ortho* isomer. Such results are related to the nature of the nitrating medium, with high selectivity for *ortho* nitration observed in aprotic solvents and high *para* selectivity typical of nitration in protic solvents. A possible explanation[72] for these observations is that protic solvents, such as mixed acid, hinder *ortho* attack by strongly solvating the electron-rich heteroatom via hydrogen bond type interactions. On the other hand, substrates like phenol ethers and acetamides have the highest electron density located on the basic heteroatoms, and as such, on electronic grounds and in the absence of heteroatom solvation, substitution would be favoured at the *ortho* positions, and this is usually observed in aprotic solvents.[73] Other possible explanations and mechanisms have been reported.[8,70,71,73]

The localization of electron density on heteroatoms is seen during the nitration of anilines which are *N*-nitrated in aprotic media.[74] Anilines can give abnormal *o*/*p*-isomer ratios resulting from a process of *N*-nitration followed by rearrangement to the ring nitrated product, a process which often occurs *in situ* in the strongly acidic medium of mixed acid (Section 4.5).

The nitration of phenols can result in anomalous and large differences in product isomer ratios, showing a high dependence on both nitrating agent and reaction medium. Here the situation is complicated by the intervention of an alternative nitration mechanism – that of nitrous acid catalyzed nitration, which proceeds via *in situ* nitrosation–oxidation (see Section 4.4).

Changing the composition of a nitrating agent like mixed acid, resulting in a change of acidity, can have an effect on both regioselectivity and substrate reactivity.[75] The strongly acidic nature of sulfuric acid in mixed acid mixtures is observed to perturb the nitration of some substrates. This is most extreme in the case of some anilines, like *N*,*N*-dimethylaniline, where an excess amount of sulfuric acid results in amine protonation, deactivation of the aromatic ring, and a complete reversal of selectivity from *o*/*p*- to *m*-substitution.[76] Other substrates are similarly deactivated by excess sulfuric acid. A striking example is seen with benzoic acid which is nitrated approximately 20 times faster with mixed acid containing 95 % sulfuric acid as compared to 100 % sulfuric acid.[77] A major factor in this reduction of activity is the higher concentration of protonated carboxylic acid in the 100 % sulfuric acid. Studies have shown that nitration rates are at a maximum when conducted in 90 % sulfuric acid for a range of deactivated substrates.[77] The use of more concentrated sulfuric acid or oleum often retards the nitration of deactivated substrates by leading to an increase in the proportion of protonated substrate. The use of oleum in mixed acid also reduces substrate solubility, and hence, makes nitration more heterogeneous in nature, further reducing reaction rate. However, oleum is frequently needed in reactions where polynitration is required in order to mop up water formed during the reaction and keep the concentration of nitronium ions high. Other deactivated substrates containing carbonyl functionality i.e. carboxylic esters, aldehydes and ketones, are also affected by high sulfuric acid concentrations, which usually results in an increase in the proportion of *m*-isomer formed.

4.3.4 Other nitrating agents

Olah[1] has given an excellent review of the reagents used for acid-catalyzed electrophilic nitration, which supplements an earlier review.[78] These reagents are composed of either Brønsted acids or Lewis acids in the presence of a nitrating agent of general formula NO_2X which acts as a source of nitronium ions or other electrophilic nitrating species. Topchiev[7] gives an extensive discussion of the practical use of such reagents in nitrating aromatic substrates. This is a brief overview of an otherwise huge area of nitration chemistry.

4.3.4.1 Nitric acid

Nitric acid is a weak nitrating agent when used alone, although it is a strong enough acid to protonate itself and generate the nitracidium ion ($H_2ONO_2^+$). The latter is probably the active nitrating agent when concentrated or fuming nitric acids are used in nitrations and, consequently, only fairly activated substrates undergo nitration under these conditions. Draper and Ridd[79] nitrated a variety of electron-rich aromatic compounds with nitric acid of different strengths. Nitration of phenolic substrates with dilute nitric acid often involves a nitrosation–oxidation pathway and this is discussed further in Section 4.4. Other activated substrates like phenol ethers, anilines etc. are readily nitrated with dilute nitric acid.

Fuming nitric acid is generally a good solvent for most organic compounds but the addition of sulfuric acid often lowers substrate solubility. The use of fuming nitric acid either neat or as a solution in carbon tetrachloride, chloroform, nitromethane, acetic acid etc. is restricted to reactive substrates. Fuming nitric acid in acetic acid is a mild, not very active nitrating agent, but is useful because the oxidizing properties of nitric acid are largely suppressed in this medium.

Fuming nitric acid is a reasonable nitrating agent but its activity rapidly diminishes as water is formed during a nitration, and so it is commonly used in excess. Fuming nitric acid at elevated temperatures is reported to convert toluene to 2,4-dinitrotoluene.[43] The same reagent converts mesitylene to a mixture of 2,4-di- and 2,4,6-trinitromesitylenes.[80] Concentrated nitric acid heated to 100 °C is reported to give a mixture of isomeric 2,4- and 2,6-dinitroxylenes from *m*-xylene.[80]

4.3.4.2 Nitric acid in the presence of Lewis acids and Brønsted acids

Strong Brønsted acids promote the formation of nitronium ions when mixed with nitric acid. Perchloric,[81] hydrofluoric,[82] phosphoric,[83] polyphosphoric,[83] trifluoroacetic,[84] fluorosulfuric,[85,86] trifluoromethanesulfonic[86] (triflic) and methanesulfonic acids[87] have all been used in conjunction with nitric acid for aromatic nitration. Perchloric acid must be considered dangerous in the presence of organic compounds and finds little use. Trifluoroacetic acid can protonate nitric acid and form nitronium ions but has also found little use. Nitrations with phosphoric and polyphosphoric acids are frequently heterogeneous in nature and can give abnormal results. Nitric acid in the presence of solid supported acid catalysts[88] like Nafion-H, a perfluorinated resin containing sulfonic acid groups, has found some favour. Such resins are readily recycled and spent acid disposal is not an issue.

Nitric acid in the presence of 'super acids' like fluorosulfuric acid and triflic acid is particularly suitable for the polynitration of deactivated aromatic substrates. In such media nitric acid is completely ionized to nitronium ions, anhydrous nitric acid–triflic acid mixtures forming nitronium triflate and hydroxonium triflate.[86] In the presence of excess super acid these powerful nitrating agents may contain the protonitronium cation ($NO_2H_2^+$) as one of the active nitrating agents.[89] Reaction rates with such reagents can be extremely high, and so methylene chloride or other halogenated hydrocarbons can be used as co-solvents, which also aids substrate solubility.[86] Temperatures as low as $-110\,^{\circ}C$ are used for the mono-nitration of reactive substrates. Anhydrous nitric acid–triflic acid is more powerful than mixed acid and can be used in cases where substrate sulfonation or oxidation may be a problem. Fluorosulfuric acid can lead to side-reactions resulting from oxidation and sulfonation. Olah and co-workers[85] reported the use of this reagent at elevated temperature for the conversion of benzene to 1,3,5-trinitrobenzene. Nitric acid–superacid mixtures have been used for some difficult nitrolysis reactions (Section 5.6.1.1 and Section 6.5).[90]

Numerous Lewis acids promote the formation of nitronium ions when in the presence of nitric acid.[7] Nitric acid–boron trifluoride,[91] and the nitric acid–hydrogen fluoride–boron trifluoride reagents described by Olah[92] are practical nitrating agents; the latter provides a convenient preparation of nitronium tetrafluoroborate. Olah[1] reports that nitric acid–magic acid (FSO_3H-SbF_5) is extremely effective for the polynitration of aromatic substrates.

4.3.4.3 Nitric acid–mercuric nitrate

Nitric acid in the presence of catalytic amounts of mercury (II) nitrate reacts with some substrates, under certain conditions, to give substituted polynitrophenols. The first example, reported by Boeters and Wolffenstein[93] and known as 'oxynitration', involved treating benzene with 50–55 % nitric acid in the presence of mercury (II) nitrate. The product is a mixture of unreacted benzene, nitrobenzene, *m*-dinitrobenzene, 2,4-dinitrophenol and picric acid, from which the latter can be isolated by steam distillation of this crude mixture followed by recrystallization of the residue from hot water.

The mechanism of 'oxynitration' is generally accepted to involve the formation of phenyl mercuric nitrate which reacts with nitrogen oxides in the nitric acid to form a nitroso compound and then a diazonium salt; the latter forms a phenol under the aqueous conditions which is then further nitrated.[94] The use of more concentrated nitric acid favours a process of mercuration–nitration and suppresses the formation of phenols.[95,96]

Other examples of 'oxynitration' include the formation of: (1) 2,4,6-trinitro-*m*-cresol from toluene,[96,97] (2) 2,4-dinitronaphth-1-ol from naphthalene,[96,97] (3) 3-chloro-2,4,6-trinitrophenol from chlorobenzene[96,97] and (4) 3-hydroxy-2,4,6-trinitrobenzoic acid from benzoic acid.[93]

4.3.4.4 Nitric acid–acid anhydrides

A solution of nitric acid in acetic anhydride is a versatile reagent for nitrolysis reactions and for the *N*-nitration of amines (see Chapters 5 and 6). This reagent has also found considerable use for the *C*-nitration of aromatic substrates.[98]

$$HNO_3 + (CH_3CO)_2O \rightleftharpoons H_3C{-}C(=O){-}ONO_2 + CH_3CO_2H \quad \text{(Eq. 4.4)}$$

Figure 4.13

It is generally accepted that nitric acid–acetic anhydride mixtures contain acetyl nitrate (Equation 4.4). Acetyl nitrate is a weak nitrating agent but in the presence of a strong acid, like nitric acid, it can ionize to nitronium and acetate ions. Protonated acetyl nitrate could also be an active nitrating agent under these conditions.

The reaction of fuming nitric acid with acetic anhydride can be violent, and therefore the components are usually mixed between 0 to 5 °C. Acetic anhydride in these mixtures acts as a dehydrating agent, reacting with water formed during the nitration to generate acetic acid.

Solutions of nitric acid–acetic anhydride are mild, nonoxidizing, and can be used for the nitration of substrates prone to oxidation. High *o*/*p*-ratios are observed for the nitration of some activated substrates (Section 4.3.3).[69–71]

Nitric acid–trifluoroacetic anhydride mixtures are used extensively for nitrolysis[99] and *N*-nitration[100] reactions (see Chapters 5 and 6). The same is not true for aromatic nitrations. This reagent contains trifluoroacetyl nitrate, which can ionize to nitronium and trifluoroacetate ions in the presence of strong acid.

Nitric acid–triflic anhydride is a powerful nitrating agent and probably involves nitronium triflate as the active nitrating agent.[1]

4.3.4.5 Nitronium salts

Olah[1,101,102] has extensively reviewed the synthesis and use of nitronium salts in aromatic nitration and so only a brief discussion is given here. A review by Gidaspov[103] brought to light much important research conducted by Russian chemists in this area.

A convenient method for the preparation of nitronium tetrafluoroborate involves treating a mixture of absolute nitric acid[92] or an alkyl nitrate[104] with anhydrous hydrogen fluoride and an excess of boron trifluoride. Pure nitronium triflate can be synthesized by treating triflic anhydride[105] or triflic acid[106] with dinitrogen pentoxide in an inert solvent. Other methods for nitronium salt synthesis are discussed by Olah.[1]

Nitronium salts are efficient and powerful nitrating agents. Nitronium tetrafluoroborate is the most commonly used nitronium salt for nitration and is commercially available as a solid or as a solution in sulfolane in which the ions are highly solvated and exist as an ion pair. Nitronium tetrafluoroborate shows poor solubility in most organic solvents and so the more soluble nitronium hexafluorophosphate is sometimes preferred for nitrations. Olah and co-workers[1,101] have studied the nitration of a vast array of aromatics with nitronium salts. Solutions of nitronium salts in aprotic organic solvents are useful for the nitration of acid sensitive or readily oxidized substrates. Nitronium tetrafluoroborate has been used for the synthesis of picryl chloride (80 %) and picryl fluoride (40 %) from 2,4-dinitrochlorobenzene and 2,4-dinitrofluorobenzene respectively.[101]

Nitronium salts in sulfuric, triflic or fluorosulfuric acids are extremely reactive and well suited for the polynitration of deactivated substrates. Olah and Lin[47,107] studied the nitration of *m*-dinitrobenzene to 1,3,5-trinitrobenzene with a solution of nitronium tetrafluoroborate in fluorosulfuric acid at 150 °C. An optimum yield of 66 % was obtained after a reaction time of 3 hours. However, the crude reaction mixture was found to contain 17 % unreacted *m*-dinitrobenzene. After a reaction time of 3.8 hours the yield of 1,3,5-trinitrobenzene dropped to 50 % but the product was free from *m*-dinitrobenzene and was essentially pure.

4.3.4.6 *Nitrogen oxides*

Dinitrogen pentoxide has emerged as an immensely important nitrating agent. Its versatility is unique – in aprotic solvents it is nonacidic and mild, whereas in strong acids like sulfuric or nitric acids it is an extremely powerful nitrating agent. Dinitrogen pentoxide has the ability to effect the nitration of reactive, acid sensitive or easily oxidized aromatic substrates, and at the same time, moving to a different solvent allows the polynitration of deactivated substrates. Pollution created by mixed acid nitrations and the disposal problems posed by spent acid may mean that dinitrogen pentoxide surpasses mixed acid as the most important industrial nitrating agent. Nitrations with dinitrogen pentoxide are discussed in Chapter 9.

A solution of dinitrogen tetroxide in sulfuric acid is also a powerful nitrating agent. In this medium dinitrogen tetroxide is ionized to nitronium and nitrosonium ions.[7,108] Titov[109] reported using a solution of dinitrogen tetroxide in oleum for the nitration of nitrotoluene to dinitrotoluene and then to trinitrotoluene, the two separate steps proceeding in 98 % and 85 % yields respectively.

Dinitrogen tetroxide forms a stable complex with boron trifluoride but this is a weak nitrating agent.[110] Aromatic nitrations with other Lewis acids have been reported, including: $AlCl_3$, $FeCl_3$, $TiCl_4$, BCl_3, PF_3, BF_3, AsF_5 and SbF_5.[1,7,109,111]

Nitrogen dioxide in the presence of ozone has been used for aromatic nitrations.[112] Such conditions are useful for the nitration of reactive and acid sensitive substrates. Lewis acids have been used in ozone-mediated nitrations with nitrogen dioxide.[113]

Aromatic substrates containing Lewis basic substituents can undergo *ortho*-lithiation. Quenching these anions with dinitrogen tetroxide at low temperature is an example of nucleophilic aromatic nitration.[114] Similar examples have been reported with anions generated from Grignard reactions with arylhalides.[115]

4.3.4.7 *Metal nitrates in the presence of Lewis acids, Brønsted acids and acid anhydrides*

Solutions of sodium or potassium nitrates in concentrated sulfuric acid can be used as a substitute for anhydrous mixed acid.[116] 2,4,6-Trinitroaniline (picramide) has been synthesized via the addition of a solution of potassium nitrate in concentrated sulfuric acid to a solution of either *o*- or *p*-acetanilide in oleum.[36]

Topchiev[7] first reported on the use of metal nitrate–Lewis acid mixtures for aromatic nitration. Many of these nitrations are heterogeneous due to the poor solubility of metal nitrates in organic solvents.

Solutions of alkali metal or ammonium nitrates in trifluoroacetic anhydride are useful nitrating agents for a range of activated to moderately deactivated substrates and particularly useful for acid sensitive substrates.[117] Solutions of copper (II) nitrate and ferric (III) nitrate in acetic anhydride have been used for the nitration of some reactive aromatic substrates.[118] A large variety of aromatics substrates have been nitrated with tetramethylammonium nitrate and triflic anhydride in methylene chloride between −78 °C and room temperature.[119]

Metal nitrates supported on various acidic clays have been used as nitrating agents for some reactive aromatic substrates in attempts to improve product isomer ratios.[120]

4.3.4.8 Other nitrating agents

Alkyl nitrates in the presence of sulfuric acid[121] and Lewis acids,[7,122] like $SnCl_4$, $AlCl_3$, and BF_3, have been used as nitrating agents. Nitrations in the presence of Nafion-H acidic resin have also been reported.[123] Alkyl nitrates do not effect the nitration of aromatic substrates in the absence of an acid catalyst.

Nitryl halides effect the nitration of aromatic substrates in the presence of Lewis acids, although competing halogenation is a side-reaction in some cases.[124]

Tetranitromethane in alkaline solution has been used for the nitration of some electron-rich aromatic phenols and amines.[125]

Olah[126] has conducted extensive studies into transfer nitration with *N*-nitro heterocycles. Studies were conducted into the effect of the heterocycle and its counterion on reactivity towards aromatic substrates.

4.3.5 Side-reactions and by-products from nitration

Spent acid disposal and the emission of nitrogen oxides are general consequences of industrial nitrations with mixed acid. Side-reactions during nitration can also lead to undesirable pollutants and by-products.[127] Substrates containing strong electron-donating groups like phenols, phenol ethers and anilines, and some fused aromatics like naphthalene are all prone to oxidation during nitration. Aromatic substrates containing alkyl substituents are particularly susceptible to oxidation when nitrated. This is seen during the mixed acid nitration of toluene to nitrotoluene where the crude product contains small amounts of phenolic by-products.[128] Phenolic by-products are often observed during the nitration of aromatic amines as a result of diazotization with nitrogen oxides or nitrous acid present in the mixed acid.

Substituents like the nitro group protect against oxidation by withdrawing electron density from the aromatic ring. However, even compounds with relatively electron deficient aromatic rings will slowly oxidize on treatment with nitric acid or strong mixed acid at elevated temperatures. TNT and picric acid slowly form tetranitromethane on heating with concentrated nitric acid via rupture of the aromatic ring. In fact, tetranitromethane is formed during the nitration of dinitrotoluene to TNT and is present in the crude product. The harsh conditions needed for the nitration of chlorobenzene to picryl chloride leads to the formation of some chloropicrin, a highly toxic lachrymator.

1,3,5-Trinitrobenzene is present in crude TNT manufactured by mixed acid nitration and results from methyl group oxidation followed by decarboxylation.[129] In fact, a convenient method for the synthesis of 1,3,5-trinitrobenzene involves oxidation of 2,4,6-trinitrotoluene with a solution of sodium dichromate in sulfuric acid, followed by decarboxylation of the resulting 2,4,6-trinitrobenzoic acid in boiling water.[130] 1,3,5-Trinitrobenzene is prepared from 2,4,6-trinitro-*m*-xylene by a similar route.[131] 2,4,6-Trinitroanisole can be prepared from the nitration of methoxybenzoic acid (anisic acid).[132] Picric acid has been prepared in yields of 95 % by treating 4-hydroxybenzoic acid with 8 M nitric acid.[133] Such examples illustrate the vulnerability of ring functionality during polynitration.

As a general rule, the lower the temperature of a nitration, the fewer the oxidation by-products that are formed.

4.4 NITROSATION–OXIDATION

Phenolic groups make aromatic nitration a very facile process, so much so that the selective introduction of one or even two nitro groups into the aromatic nucleus can be a problem. The ease with which phenolic substrates undergo oxidative side-reactions can also pose a problem. Taking phenol as an example, the action of commercial 70 % nitric acid on this substrate readily introduces three nitro groups into the aromatic ring, although the process is accompanied by much oxidative degradation. The problem is more evident with polyhydric phenols like resorcinol, where the introduction of three nitro groups into the aromatic ring is very facile.

Nitrosation of phenolic substrates usually uses nitrous acid prepared *in situ* from a dilute mineral acid and an alkali metal nitrite.[134] In general, for every phenolic group present in a substrate an equal number of nitroso groups can be introduced into the aromatic ring; phenol, resorcinol and phloroglucinol react with nitrous acid to form 4-nitrosophenol, 2,4-dinitrosoresorcinol and 2,4,6-trinitrosophloroglucinol respectively.

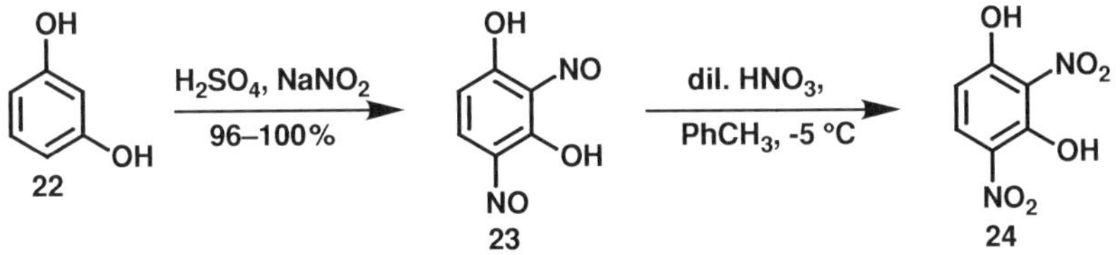

Figure 4.14

Resorcinol (22) is readily converted to 2,4-dinitrosoresorcinol (23) in quantitative yield by slowly adding an aqueous solution containing two mole equivalents of sodium nitrite to a solution of resorcinol that has been acidified with two mole equivalents of sulfuric acid.[26a,b] The 2,4-dinitrosresorcinol (23) from this process can be oxidized to 2,4-dinitroresorcinol (24) with a binary phase system of toluene and dilute nitric acid at −5 °C.[26c] Treatment of either 2,4-dinitrosoresorcinol or 2,4-dinitroresorcinol with concentrated nitric acid provides a convenient route to styphnic acid (2,4,6-trinitroresorcinol).[27]

Both 2,4-dinitrosoresorcinol (23) and 2,4-dinitroresorcinol (24) are important in the explosives industry. The lead salt of 2,4-dinitrosoresorcinol has a low ignition temperature and finds use in priming compositions and in electrical igniters. The lead salt of 2,4-dinitroresorcinol is a weak initiator but is found to exhibit high sensitivity to friction and stab action without being highly sensitive to impact, and as such, this compound has found use in primers.

2,4,6-Trinitrophloroglucinol has been synthesized via the nitrosation of phloroglucinol followed by nitric acid oxidation.[30,32] The direct nitration of phloroglucinol must be conducted at subambient temperature to avoid excessive oxidation, and even then, the yield rarely exceeds 70 %.[28–30]

Some phenolic substrates are readily nitrated with nitric acid of 5 % concentration or lower.[134] Such reactions are catalyzed by nitrous acid either already present in the nitric acid or from initial oxidation of the phenolic substrate. Reaction of the substrate with nitrous acid yields a nitrosophenol, which is then oxidized to a nitrophenol on reaction with nitric acid; the latter step regenerates nitrous acid and so the process is autocatalytic. The nitration of phenols via *in situ* nitrosation–oxidation with dilute nitric acid is often very selective for the *para*-isomer. The result is similar to that obtained in the phenol substrate is first treated with nitrous acid and the resulting nitrosophenol isolated and oxidized to the nitrophenol. Occasionally a catalytic amount of sodium nitrite is added to dilute nitric acid to increase *para*-selectivity.[135]

It is important to note that nitrous acid catalyzed nitrations are specific to certain groups of compounds. Nitrous acid can retard the nitration of many aromatic compounds by lowering the concentration of the nitronium cation. However, under the conditions described i.e. dilute nitric acid solutions, the concentration of active nitrating species (probably the nitracidium ion) is so low that nitrosation competes. The nitration of phenol ethers and other reactive compounds is also catalyzed by nitrous acid.

4.5 NITRAMINE REARRANGEMENT

The nitration of anilines with mixed acid often leads to the *N*-nitration of the amino group and the formation of nitramines. The nitramines formed in these reactions can undergo acid-catalyzed rearrangement to the ring nitrated aniline in a reaction known as the Bamberger rearrangement.[136] The Bamberger rearrangement probably proceeds via an electrocyclic rearrangement and is usually effected by suspending the substrate in neat sulfuric acid at ambient or subambient temperature. Reactions are usually very selective for the *o*-isomer. Significant amounts of the *p*-isomer can be formed at low solution acidities although such conditions generally result in much lower yields.[137]

Depending on the amine substrate and the nitrating conditions used, it is not uncommon for an intermediate nitramine to undergo direct rearrangement to the ring-nitrated product without prior isolation, in which case, the formation of the nitramine as an intermediate can only be postulated. Due to the high *o*-selectivity often observed with this type of reaction the *o*/*p*-ratio can be very different to that where the aromatic ring is directly nitrated.

Amine substrates whose rings are strongly deactivated with nitro groups are *N*-nitrated with relative ease; 2,4-dinitro-*N*-methylaniline undergoes *N*-nitration on treatment with 70 % nitric acid at room temperature.[138] It is known that the *N*-nitration of anilines is favoured by the presence of a large excess of nitric acid.

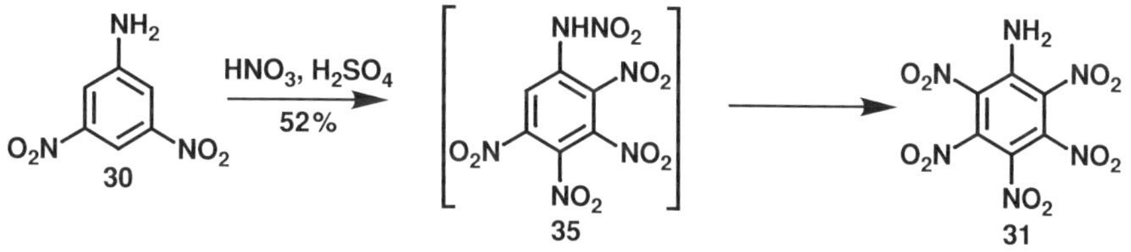

Figure 4.15

The nitration of 3,5-dinitroaniline (30) to 2,3,4,5,6-pentanitroaniline (31) is known to involve the formation of the intermediate nitramine *N*,2,3,4,5-pentanitroaniline (35).[39,139,140] It is also known that the metal salts of arylnitramines show more of a tendency to undergo Bamberger rearrangement than the free nitramines.[141] Nielsen and co-workers[139] modified the original synthesis[140] of 2,3,4,5,6-pentanitroaniline by adding a lead salt to facilitate the rearrangement of the intermediate nitramine to the ring nitrated product.

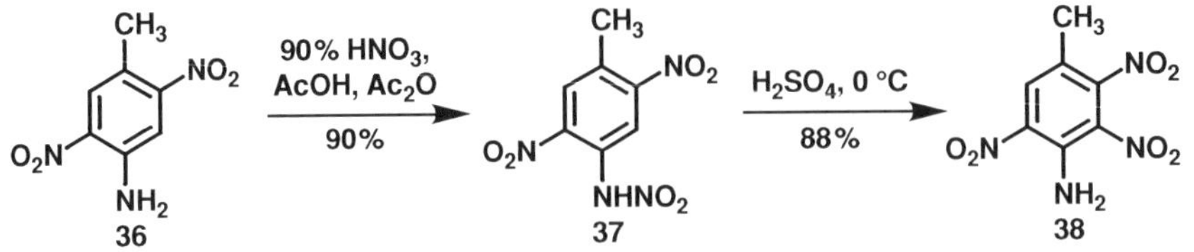

Figure 4.16

Chemists at the Naval Air Warfare Center (NAWC), Weapons Division, China Lake, have reported many examples of polynitroarylamine synthesis via Bamberger rearrangements of arylnitramines.[139,142–145] The nitration of 4-amino-2,5-dinitrotoluene (36) with a mixture of nitric acid and acetic anhydride in glacial acetic acid at room temperature yields the nitramine (37) which on treatment with neat sulfuric acid, provides 4-amino-2,3,5-trinitrotoluene (38) as the sole product.[145] Nitration of 3,4-dinitroaniline (39) with a solution of nitric acid in acetic anhydride yields *N*,3,4-trinitroaniline (40); acid-catalyzed rearrangement of the latter in neat sulfuric acid furnishes a 74 % yield of isomeric 2,3,4- (41) and 2,4,5- (42) trinitroanilines in a 4:6 ratio.[139] Accordingly, a mixture of products can be expected when an unsymmetrical arylnitramine has two unsubstituted *ortho* positions available.

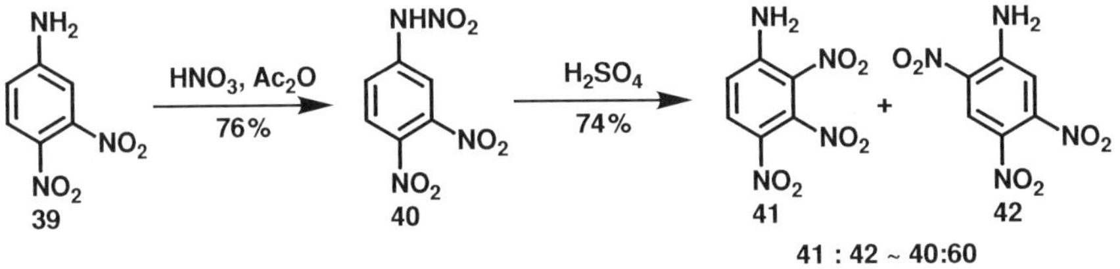

Figure 4.17

The rearrangement of arylnitramines is a synthetically useful reaction. However, on numerous occasions the formation of nitramines during the nitration of anilines can complicate a reaction. One such problem is the formation of zwitterionic diazo oxides, also known as diazophenols or diazonium phenolates. These can be formed from nitramines containing a pre-existing *o*-nitro group in a process resulting in formal loss of nitric acid. Many diazophenols are primary explosives and can contaminate products formed from the Bamberger rearrangement of nitramines; DINOL (2-diazo-4,6-dinitrophenol) (43) is a diazophenol and a primary explosive used in detonators.

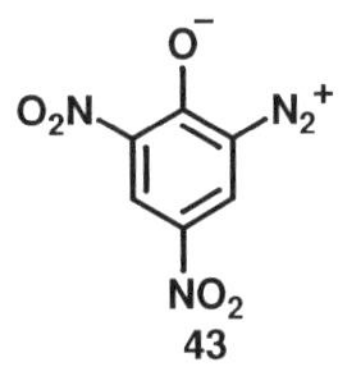

Figure 4.18

Reaction mixtures containing an arylnitramine with an *o*-nitro group require careful work-up because the formation of diazophenols is often a facile and sometimes spontaneous process. The attempted isolation of numerous nitramines from the nitration of aromatic amines has led to the isolation of the diazophenol as the sole product; the bis-nitramine (44) is found to decompose spontaneously at 25 °C in the solid state and in solution to give the diazophenol (45).[142]

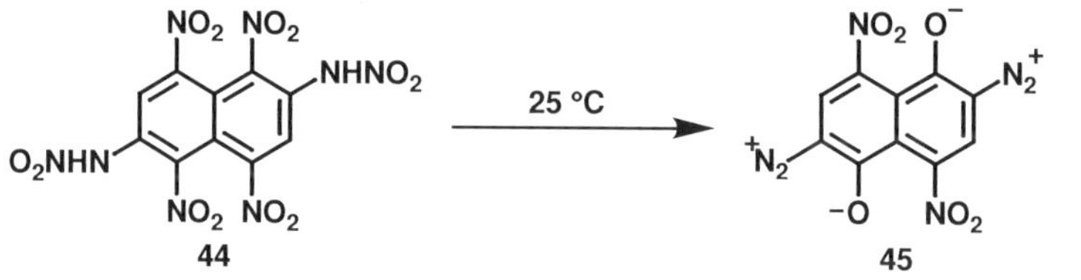

Figure 4.19

Diazophenol formation is most competitive when a nitramine substrate contains an electron-withdrawing nitro group *ortho* to the nitro group being displaced and hence *meta* to the nitramine functionality, assumedly because that site is then activated towards nucleophilic aromatic substitution.[144] Heating nitramines in inert chlorinated solvents also favours diazophenol formation but this is suppressed by using urea or sulfamic acid as additives.

The course of a reaction involving *N*-nitration of an aromatic amine is very dependent on both the substrate and the reaction conditions, i.e. nitrating agent, temperature, work-up etc. The nitration of 3-amino-2,5-dinitrotoluene with mixed acid in glacial acetic acid at 0 °C yields the corresponding diazophenol with no trace of the intermediate nitramine.[145] In contrast, nitration of the same substrate at ambient temperature with mixed acid composed of 90 % nitric acid and 96 % sulfuric acid yields the corresponding nitramine.[145]

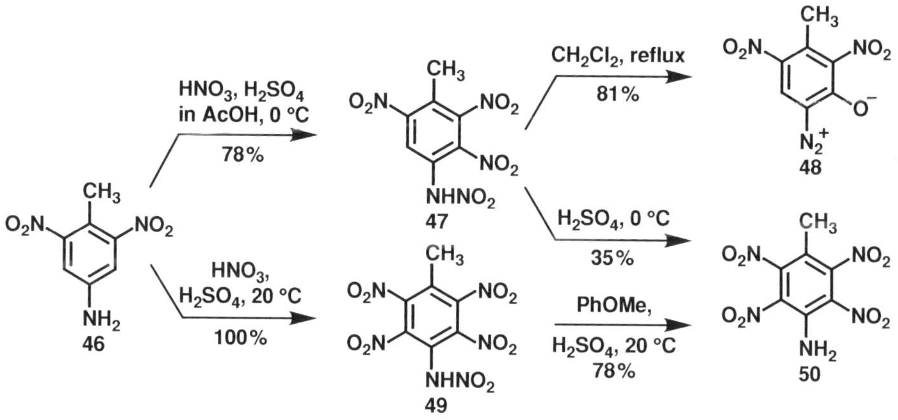

Figure 4.20

In some cases the Bamberger rearrangement of an arylnitramine can be low yielding and may not be the best route to a product. For example, 4-amino-2,3,5,6-tetranitrotoluene (50) is prepared in low yield via the acid-catalyzed rearrangement of the nitramine (47), the latter formed from the controlled nitration of 4-amino-2,6-dinitrotoluene (46) with mixed acid in glacial acid or with nitronium tetrafluoroborate in methylene chloride.[145] Alternatively, the same starting material, 4-amino-2,6-dinitrotoluene (46), can be fully nitrated to 4-amino-*N*,2,3,5,6-pentanitrotoluene (49), again with mixed acid, and the nitramino group subsequently cleaved with sulfuric acid in anisole.[145] This latter route provides the product, 4-amino-2,3,5,6-tetranitrotoluene (50), in a significantly higher overall yield.

4.6 REACTION OF DIAZONIUM SALTS WITH NITRITE ANION

Sandmeyer-type reactions are a useful route to polynitroarylenes with unusual substitution patterns. In these reactions an arylamine is treated with a source of nitrous acid to form an intermediate diazonium salt which is readily displaced on reaction with a suitable nucleophile. Many substituents can be incorporated into the aromatic ring via this method, including the nitro group.

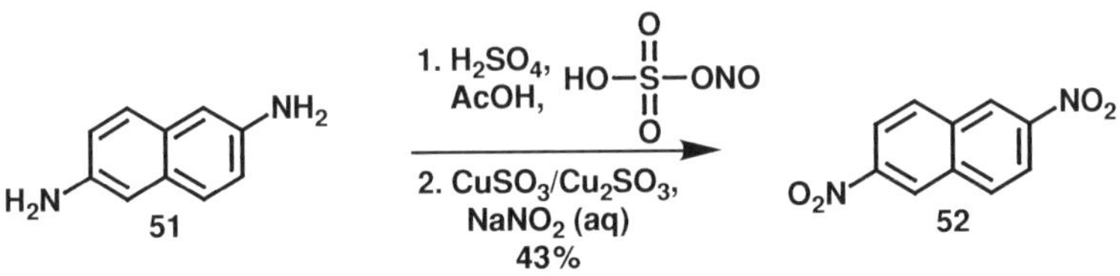

Figure 4.21

A number of different experimental procedures have been used to effect the conversion of a diazonium group to a nitro group. The most common of these treats the preformed diazonium salt with a cupric/cuprous sulfite catalyst in the presence of sodium nitrite; copper bronze or precipitated copper has also been used. These reactions only occur in neutral or basic solution. This can be achieved by first isolating the diazonium salt as the sulfate,[62] tetrafluoroborate,[146] or cobaltinitrite,[63] and washing the solid salt free from acid, or alternatively, adding calcium carbonate[64] or sodium hydrogen carbonate[65] to the reaction mixture to neutralize excess acid present. Chatt and Wynne[66] synthesized 2,6-dinitronaphthalene (52) by adding solid nitrosylsulfuric acid to a solution of 2,6-diaminonaphthalene (51) in sulfuric and acetic acids, followed by isolation of the intermediate diazonium salt as its solid sulfate salt and treatment with sodium nitrite in the presence of copper sulfite. The isolation of diazonium salts, especially those containing two or more nitro groups, must be considered hazardous and confined to a small scale. Experimentally, the most convenient method adds a solution of the diazonium sulfate to an aqueous solution of sodium hydrogen carbonate and excess sodium nitrite in the presence of a copper catalyst.[65] This method is only useful on a laboratory scale because of the large excess of reagents needed. It is not uncommon in these reactions to use up to 6000 mole % excess of sodium nitrite and up to 400 mole % excess of the copper salt.[72]

Figure 4.22

Nitration via diazotisation has been extensively used for the synthesis of isomeric dinitronaphthalenes.[61–66] Ward and co-workers[62] used nitration via diazotisation to prepare 3,3′,4,4′-tetranitrobiphenyl from 3,3′-dinitrobenzidine, and 3,4,5-trinitrotoluene from 3,5-dinitro-4-toluidine. Ward and Hardy[147] prepared 1,4,6-trinitronaphthalene from 4,7-dinitro-1-naphthylamine. Körner and Contardi used the nitrate salts of aryldiazonium compounds for the synthesis of polynitro derivatives of benzene[148] and toluene.[149–151] Accordingly,

1,2,4- and 1,2,3-trinitrobenzenes were formed from the isomeric 2,4- and 2,6-dinitroanilines respectively.[148] In a similar way, 3,4,5-,[149] 2,3,5-,[150] and 2,3,6-[151] trinitrotoluenes were prepared from the isomeric dinitrotoluidines. Holleman[152] used a similar route for the conversion of picramide (53) to 1,2,3,5-tetranitrobenzene (54) in 69 % yield. Such examples illustrate why nitration via diazotisation is a useful route to polynitroarylenes – products not available from the direct nitration of aromatic substrates can be synthesized.

4.7 OXIDATION OF ARYLAMINES, ARYLHYDROXYLAMINES AND OTHER DERIVATIVES

4.7.1 Oxidation of arylamines and their derivatives

The oxidation of an aromatic amino group to a nitro group is an attractive strategy for the synthesis of polynitroarylenes. Successive introduction of nitro groups into an aromatic ring reduces electron density and makes further electrophilic nitration more difficult. The amino group is a strongly activating group that will allow the polynitration of arylamines under mild nitrating conditions. The directing and activating properties of the amino group, coupled with its ability to be oxidized to a nitro group, has allowed the synthesis of many highly nitrated arylenes, and particularly those with unusual substitution patterns which cannot be prepared from direct nitration.

The reactivity of an arylamine to an electrophilic oxidant mainly depends on its nucleophilicity, which in turn, is loosely related to basicity. The basicity of an amino group is related to whether the other substituents on the aromatic ring are electron releasing or electron withdrawing. Electron-withdrawing groups, like the nitro group, reduce amine basicity and make oxidation more difficult. In general, the weaker the basicity of the amino group the more powerful the oxidant needed for amino to nitro group conversion. Of equal importance is a substrate's susceptibility to oxidative degradation. Many polynitroarylamines can be subjected to really quite harsh oxidizing conditions without noticeable degradation. Best results are usually achieved using a balancing act between the substrate and strength of oxidant used, so avoiding overoxidation but ensuring a high enough reactivity to effect substrate to product conversion in a reasonable time. Slow oxidations may allow side-reactions to predominate and lead to substrate degradation. If fused ring or aromatic methyl groups are present this is a potential problem.

Peroxyacids are the most widely used class of oxidant for aromatic amino to nitro group conversion and include: peroxydisulfuric, peroxymonosulfuric, peroxyacetic, peroxytrifluoroacetic and peroxymaleic acids. The oxidizing potential of the peroxyacid is, as a rule, proportional to the strength of the parent deoxy-acid. Dimethyldioxirane (DMDO) and ozone have also found use for amino to nitro group conversion.

4.7.1.1 Peroxydisulfuric acid

$$2\,H_2SO_4 + H_2O_2 \longrightarrow H_2S_2O_8 + 2\,H_2O \quad \text{(Eq. 4.5)}$$

$$H_2O + SO_3 \longrightarrow H_2SO_4 \quad \text{(Eq. 4.6)}$$

Figure 4.23

Peroxydisulfuric acid ($H_2S_2O_8$) is one of the most powerful peroxyacid oxidants available. Peroxydisulfuric acid is usually prepared *in situ* as a solution in sulfuric acid via the addition of 90–98 % hydrogen peroxide to an excess of oleum (Equations 4.5 and 4.6). This preparation now poses a problem because of the limited availability of high concentration hydrogen peroxide solutions to standard laboratories. Peroxydisulfuric acid is also generated *in situ* by passing ozone–oxygen mixtures through oleum (Equation 4.7).[39] Solutions of peroxydisulfuric acid are unstable and slowly decompose at room temperature. Water must be rigorously excluded because of the rapid formation of the much weaker oxidant, peroxymonosulfuric acid.

$$H_2SO_4 + SO_3 + O_3 \longrightarrow H_2S_2O_8 + O_2 \quad \text{(Eq. 4.7)}$$

Figure 4.24

The real potential of peroxydisulfuric acid for the synthesis of highly nitrated arylenes was brought to light by research conducted at the Naval Air Warfare Center, Weapons Division, China Lake.[39,139,143,145,153,154] Some of the highly nitrated polynitroarylenes synthesized during this study are illustrated in Table 4.1. Peroxydisulfuric acid prepared from both hydrogen peroxide and ozone has been used to synthesize tetranitrobenzenes[39,139] and pentanitrobenzene[139,153] from isomeric trinitroanilines and tetranitroanilines respectively (Table 4.1, Entries 2a and 2b). Tetranitrotoluenes[139,145] and pentanitrotoluene[143,39] have been synthesized from isomeric trinitrotoluidines and tetranitrotoluidines respectively (Table 4.1, Entries 3, 4a and 4b). Highly nitrated biphenyls, terphenyls and stilbenes have also been synthesized from the corresponding amino derivatives (Table 4.1, Entry 7).[139,154]

Reactions are usually conducted by adding concentrated hydrogen peroxide to a solution of the arylamine in oleum. Reactions are often slow at room temperature but can be increased by using sulfuric acid solutions containing a high concentration of hydrogen peroxide, and hence, peroxydisulfuric acid. This is usually achieved with >95 % hydrogen peroxide solution and oleum containing 30 % dissolved sulfur trioxide.

Some substrates show limited solubility in sulfuric acid solutions and this can affect the rate of oxidation. However, the main factor for slow amine oxidation is due to the high concentration of protonated amine under these highly acidic conditions. Under these conditions only weakly basic amines have a high enough concentration of unprotonated form to permit oxidation to occur. As a result, sulfuric acid solutions of peroxydisulfuric acid are only useful for the oxidation of very weakly basic amines. Peroxydisulfuric acid oxidizes trinitrotoluidines to tetranitrotoluenes (Table 4.1, Entry 3) but leaves the more basic dinitrotoluidines unaffected.[145] The opposite is true of peroxyacids like peroxytrifluoroacetic acid and so the reagents are very much complementary.

It is unsurprising that some substrates react with peroxydisulfuric acid faster when in 100 % sulfuric acid than in oleum. A striking example of this is illustrated by the oxidation of 2,6-dinitroaniline to 1,2,3-trinitrobenzene in 56 % yield when the sulfuric acid concentration is 96 % whereas in 20 % oleum this substrate is unaffected.[139]

The powerful explosive hexanitrobenzene (55) (experimental[155] VOD ~ 9500 m/s) is prepared from the oxidation of 2,3,4,5,6-pentanitroaniline (31) with peroxydisulfuric acid generated *in situ* from both oleum/hydrogen peroxide[139,153] and oleum/ozone[39] (Table 4.1, Entries 1a and 1b). Hexanitrobenzene is one of the few explosives containing no hydrogen atoms and belongs to a select group of organic explosives called 'nitrocarbons'. Decanitrobiphenyl (65) is another nitrocarbon explosive and is prepared from the oxidation of 4,4′-diamino-2,2′,3,3′,5,5′,6,6′-octanitrobiphenyl (64) with peroxydisulfuric acid (Table 4.1, Entry 7).[154]

Table 4.1 Oxidation of arylamines and their derivatives with peroxydisulfuric acid

Entry	Substrate	Conditions	Product	Yield (%)
1a	NH_2, O_2N, NO_2, O_2N, NO_2, NO_2 (31)	20% oleum, 98% H_2O_2, 25–30 °C, 24 hrs	NO_2, O_2N, NO_2, O_2N, NO_2, NO_2 (55)	58[139,153]
1b		30% oleum, O_3, 20–25 °C, 48 hrs		78[39]
2a	NH_2, O_2N, NO_2, NO_2 (53)	100% H_2SO_4, 98% H_2O_2, 25–30 °C, 24 hrs	NO_2, O_2N, NO_2, NO_2 (54)	93[139,153]
2b		10% oleum, O_3, 20–25 °C, 72 hrs		49[39]
3	CH_3, O_2N, NO_2, NH_2, NO_2 (56)	100% H_2SO_4, 98% H_2O_2	CH_3, O_2N, NO_2, NO_2, NO_2 (57)	82[139]
4a	CH_3, O_2N, NO_2, O_2N, NO_2, NH_2 (58)	20% oleum, 88% H_2O_2, 20–25 °C, 18 hrs	CH_3, O_2N, NO_2, O_2N, NO_2, NO_2 (59)	82[143]
4b		15% oleum, O_3, 20–25 °C, 48 hrs		57[39]
5	CH_3, O_2N, NO_2, Ac_2N, NAc_2, NO_2 (60)	30% oleum, 90% H_2O_2, 20–25 °C, 18 hrs	CH_3, O_2N, NO_2, O_2N, NO_2, NO_2 (61)	54[139]
6	NH_2, O_2N, NO_2, F, F, NO_2 (62)	30% oleum, 90% H_2O_2, 20–25 °C, 18 hrs	NO_2, O_2N, NO_2, F, F, NO_2 (63)	84[139]
7	O_2N, NO_2, NO_2, NO_2, H_2N, NH_2, O_2N, NO_2, NO_2, NO_2 (64)	30% oleum, 95% H_2O_2, 25 °C, 10–14 days	O_2N, NO_2, NO_2, NO_2, O_2N, NO_2, O_2N, NO_2, NO_2, NO_2 (65)	39[154]

Products from peroxydisulfuric acid oxidations are usually isolated in high yield and high purity with potential by-products such as azo, azoxy and nitroso compounds usually absent. Product isolation is usually facile; the product either precipitates from solution or can be extracted, with any unreacted amine remaining in the acid liquors. Peroxydisulfuric acid is, however, a very strong oxidant and some substrates are rapidly destroyed, which is the case for polynitrophenylenediamines.[139] The reactivity of such substrates can be moderated by prior protection as their *N*-acetyl or *N,N'*-diacetyl derivatives. Thus, pentanitrotoluene (61) can be prepared from 3,5-bis(diacetylamino)-2,4,6-trinitrotoluene (60) (Table 4.1, Entry 5).[139]

4.7.1.2 Peroxymonosulfuric acid (Caro's acid)

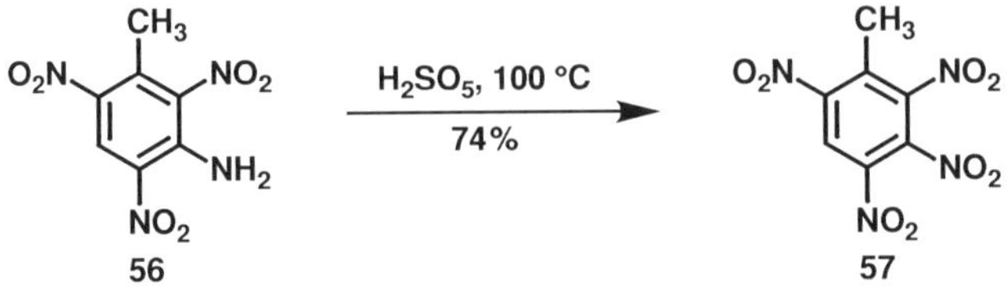

Figure 4.25

Peroxymonosulfuric acid (H_2SO_5) can be prepared by adding aqueous hydrogen peroxide, or ammonium or alkali metal persulfates to sulfuric acid solutions of various concentrations. The reagent is a weaker oxidant than peroxydisulfuric acid and often reacts with nitroanilines to give the corresponding nitroso compound.[156,157] Nitro compounds can be formed under more forcing conditions.[158–162] 3-Amino-2,4,6-trinitrotoluene (56) is oxidized to 2,3,4,6-tetranitrotoluene (57) in 74 % yield when heated to 100 °C with peroxymonosulfuric acid prepared from potassium persulfate and sulfuric acid.[152] However, 2,4,6-trinitroaniline (picramide) remains unchanged on treatment with peroxymonosulfuric acid.

4.7.1.3 Peroxytriflic acid

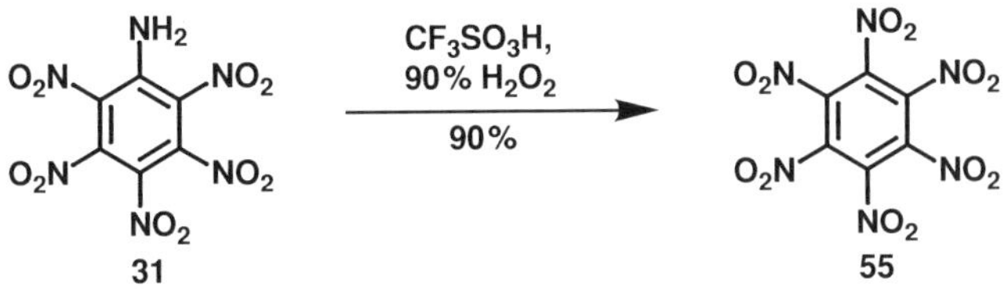

Figure 4.26

Peroxytriflic acid (CF_3SO_4H) is prepared *in situ* from the addition of 90–98 % hydrogen peroxide to an excess of triflic acid. Peroxytriflic acid was first reported by Nielsen and co-workers[139] and is possibly the most powerful peroxyacid known. The full potential of peroxytriflic acid has not been explored but it is reported to oxidize the weakly basic amine 2,3,4,5,6-pentanitroaniline (31) to hexanitrobenzene (55) in 90 % yield; peroxydisulfuric acid achieves the same conversion in only 58 % yield.[139]

4.7.1.4 Peroxyacetic acid

Peroxyacetic acid (CH_3CO_3H) is a weak oxidant. Strong solutions are potentially explosive and hazardous to store, although a 40 % solution in acetic acid is available commercially. Emmons[164] prepared anhydrous solutions of peroxyacetic acid from the reaction of 90 %+ hydrogen peroxide with acetic anhydride in chloroform containing a trace of sulfuric acid.

Table 4.2 Oxidation of arylamines with peroxycarboxylic acids

Entry	Substrate	Conditions	Product	Yield (%)
1	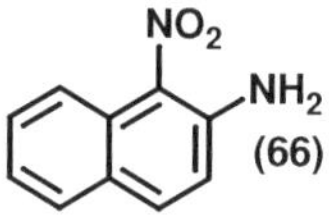 1-nitro-2-naphthylamine (NO_2, NH_2) (66)	CH_3CO_3H	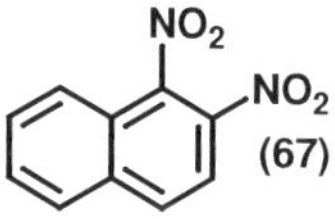 1,2-dinitronaphthalene (NO_2, NO_2) (67)	25[163]
2	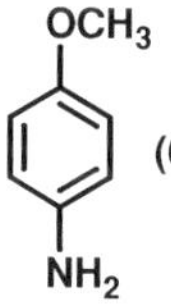 4-methoxyaniline (OCH_3, NH_2) (68)	$CHCl_3$, H_2SO_4 (cat), Ac_2O, 90% H_2O_2, reflux	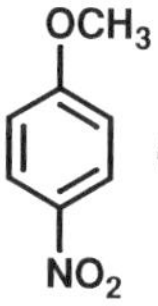 4-nitroanisole (OCH_3, NO_2) (69)	82[164]
3	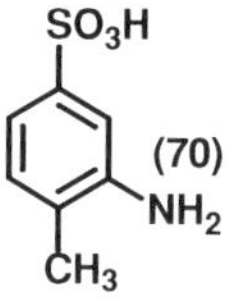 (SO_3H, NH_2, CH_3) (70)	1. 30% H_2O_2, CH_3CO_2H 70–75 °C 2. $NaHCO_3$	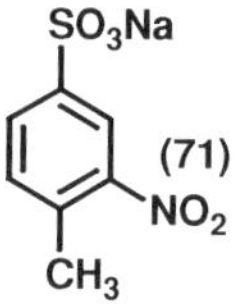 (SO_3Na, NO_2, CH_3) (71)	74[165]
4	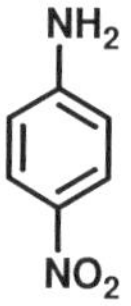 4-nitroaniline (NH_2, NO_2) (72)	maleic anhydride, 90% H_2O_2, CH_2Cl_2 (HO_2C-CH=CH-CO_3H)	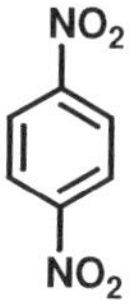 1,4-dinitrobenzene (NO_2, NO_2) (73)	87[166]
5a	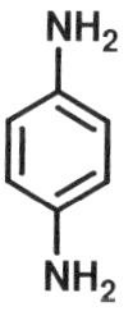 1,4-diaminobenzene (NH_2, NH_2) (74)	90% H_2O_2, $(CF_3CO)_2O$, CH_2Cl_2, reflux	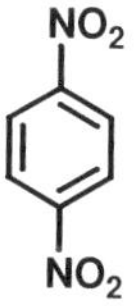 1,4-dinitrobenzene (NO_2, NO_2) (75)	86[167]
5b		90% H_2O_2, CF_3CO_2H, CH_2Cl_2, reflux		89[167]
6a	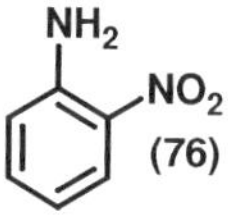 2-nitroaniline (NH_2, NO_2) (76)	90% H_2O_2, $(CF_3CO)_2O$, CH_2Cl_2, reflux	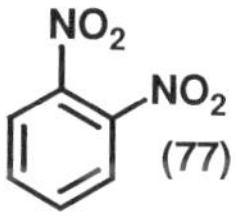 1,2-dinitrobenzene (NO_2, NO_2) (77)	92[167]
6b		90% H_2O_2, CF_3CO_2H, CH_2Cl_2, reflux		81[167]
7	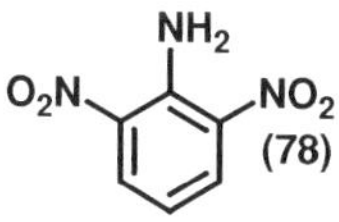 2,6-dinitroaniline (O_2N, NH_2, NO_2) (78)	CF_3CO_3H	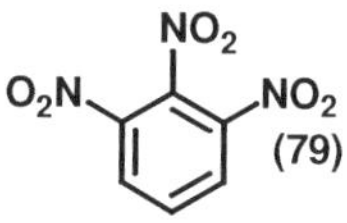 1,2,3-trinitrobenzene (O_2N, NO_2, NO_2) (79)	92[168]

These solutions readily oxidize nitroanilines to the corresponding dinitrobenzenes but are unable to oxidize amines of lower basicity like the dinitroanilines.[163,164] Peroxyacetic acid is well suited for the oxidation of arylamines containing strong electron-donating groups (Table 4.2, Entry 2). Stronger oxidants like peroxytrifluoroacetic acid rapidly degrade such substrates and give rise to various phenolic by-products. Peroxyacetic acid is also the reagent of choice for the oxidation of some condensed ring arylamines (Table 4.2, Entry 1).[163]

The reagent prepared from the reaction of 30 % hydrogen peroxide with glacial acetic acid also contains peroxyacetic acid but the main product of arylamine oxidation is usually the corresponding nitroso compound.[169,170] On heating with an excess of this reagent the nitro compound is usually obtained.[165,169]

Azo and azoxy compounds are potential by-products of peroxyacetic acid oxidations, particularly when the rate of oxidation is slow.[164,171] Competitive condensation reactions are usually avoided by using an excess of oxidant and also avoiding the presence of excess acid which retards amine oxidation. The latter is sometimes suppressed by using a sodium bicarbonate buffer.[171,172]

4.7.1.5 Peroxytrifluoroacetic acid

Trifluoroacetic acid (CF_3CO_3H) sits between peroxyacetic acid and peroxydisulfuric acid in oxidizing potential. Anhydrous solutions of peroxytrifluoroacetic acid in methylene chloride can be prepared by the addition of 90 %+ hydrogen peroxide to a solution of trifluoroacetic anhydride in methylene chloride containing a trace of sulfuric acid.[167] Solutions of peroxytrifluoroacetic acid prepared from less concentrated hydrogen peroxide solution or trifluoroacetic acid are less reactive to arylamines.

Emmons[167] found anhydrous solutions of peroxytrifluoroacetic acid in methylene chloride particularly clean and efficient for the oxidation of nitroanilines and phenylenediamines to the corresponding dinitrobenzenes, and for the conversion of dinitrotoluidines and dinitroanilines to the corresponding trinitrotoluenes and trinitrobenzenes respectively (Table 4.2, Entries 5a, 6a and 7). Blucher and co-workers,[173] and Novikov and co-workers,[168] have also reported the synthesis of polynitroarylenes via this method. Peroxytrifluoroacetic acid fails to convert weakly basic amines to their nitro compounds; trinitroanilines, like picramide, are not converted to tetranitrobenzenes.[167] Peroxytrifluoroacetic acid also has a high potential for overoxidation with substrates containing strong electron-donating groups or condensed rings.[167] However, the formation of azo and azoxy condensation products is generally not a problem with peroxytrifluoroacetic acid and many products are obtained in excellent yield and in a high state of purity.

4.7.1.6 Peroxymaleic acid

Peroxymaleic acid ($HO_2CHC{=}CHCO_3H$) is between peroxytrifluoroacetic acid and peroxyacetic acid in terms of oxidant strength. Peroxymaleic acid is prepared *in situ* from the reaction of 90 % hydrogen peroxide with maleic anhydride in methylene chloride.[166] Like peroxytrifluoroacetic acid, it degrades anilines containing strong electron-donating groups but it is more tolerant of arylamine substrates containing condensed rings. Peroxymaleic acid efficiently oxidizes nitroanilines to the corresponding dinitrobenzenes (Table 4.2, Entry 4).

4.7.1.7 Dimethyldioxirane (DMDO)

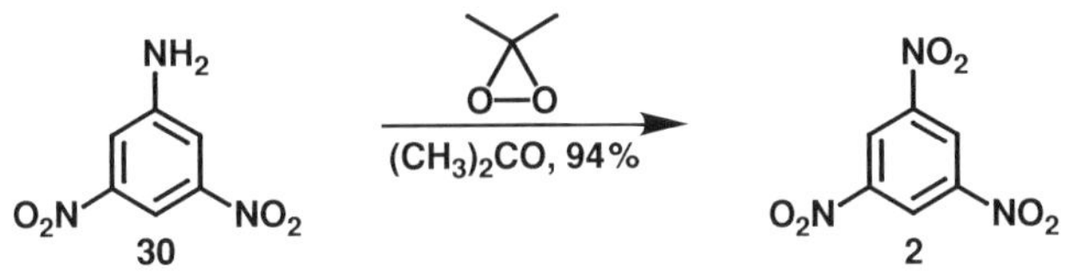

NH2
O2N NO2
30
O—O
(CH3)2CO, 94%
NO2
O2N NO2
2

Figure 4.27

Dimethyldioxirane is a powerful oxidant prepared as a solution in acetone from the distillation of a buffered solution of oxone ($2KHSO_5.KHSO_4.K_2SO_4$) in acetone.[174] Murray and co-workers[175,176] used dimethyldioxirane for the synthesis of numerous aromatic nitro compounds. This reagent efficiently converts isomeric nitroanilines and phenylenediamines to the corresponding dinitrobenzenes.[175] In the case of *p*-phenylenediamine the hydrochloride salt was used in the oxidation, demonstrating the high potential of dimethyldioxirane as an oxidant. Dimethyldioxirane has also been used to convert 3,5-dinitroaniline (30) to 1,3,5-trinitrobenzene (2) in excellent yield.[175] Despite the clear potential of dimethyldioxirane as an oxidant for amino to nitro group conversion it has not found wide use. Some moderately basic arylamines are converted to nitro compounds on direct treatment with oxone.[177]

4.7.1.8 Other oxidants

The oxidation of arylamines with ozone is generally not a feasible route to the corresponding nitro compound.[178] First absorbing the arylamine onto silica gel and then passing a stream of ozone over the surface has achieved some success.[179] *m*-Chloroperoxybenzoic acid (*m*-CPBA) in 1,2-dichloroethane at reflux has found limited use.[180] Other more obscure reagents have been described.[159,181,182]

Some reagents are milder and less powerful oxidants and have been used to oxidize arylamines to the corresponding nitroso compounds. These include 30 % hydrogen peroxide in acetic acid,[169,170] aqueous solutions of potassium permanganate,[183] and alkaline hypochlorite[184] amongst others. The hypochlorite oxidation of arylamines containing *o*-nitro substituents is reported to yield benzofuroxans.[185] For a discussion of the synthesis of aromatic nitroso compounds the readers are directed to a review by Boyer.[186]

Reagents effecting nitroso to nitro group conversion include: nitric acid,[187] acidic dichromate,[188] hydrogen peroxide,[189] permanganate,[190] chromium trioxide,[191] hypochlorite,[191] ozone,[192] and all other reagents previously discussed in this section for amino to nitro group conversion.[186]

4.7.2 Oxidation of arylhydroxylamines and their derivatives

The oxidation of arylhydroxylamines and their *O*-alkylated derivatives to the corresponding nitro compounds can be achieved with both ozone and nitric acid. This strategy is a useful alternative to the oxidation of arylamines.

Polynitroarylenes containing a variety of leaving groups are very susceptible to nucleophilic displacement (Section 4.8), and so, treatment of such substrates with hydroxylamine or an alkoxyamine usually generates the corresponding arylhydroxylamine or its *O*-alkyl derivative.

4.7.2.1 Ozone

Nielsen and co-workers[39] studied the oxidation of arylhydroxylamines and their *O*-methyl derivatives with ozone in inert solvents at subambient temperature. 1,2,3,5-Tetranitrobenzene (54) is formed in quantitative yield from the oxidation of both *N*-hydroxy-2,4,6-trinitroaniline

Table 4.3 Oxidation of arylhydroxylamines with ozone (ref. 39)

Entry	Substrate	Conditions	Product	Yield (%)
1	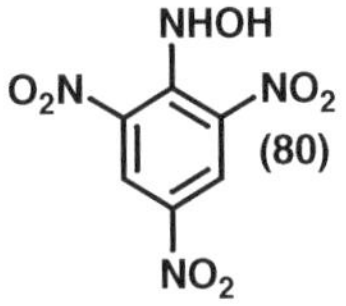 (80)	O_3, CH_2Cl_2, -10 °C, 15 mins	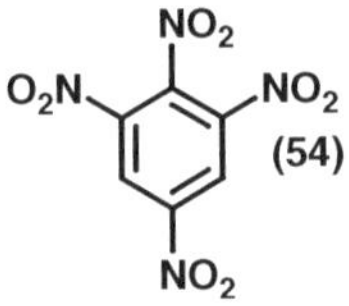 (54)	100
2	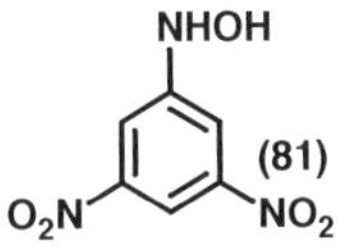 (81)	O_3, CH_2Cl_2, -35 °C, 25 mins	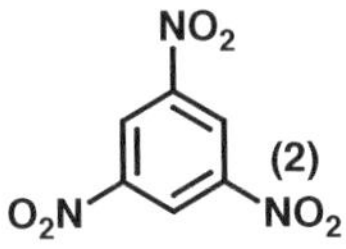 (2)	100
3	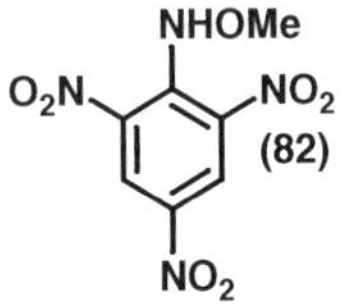 (82)	O_3, CH_2Cl_2, -10 °C, 75 mins	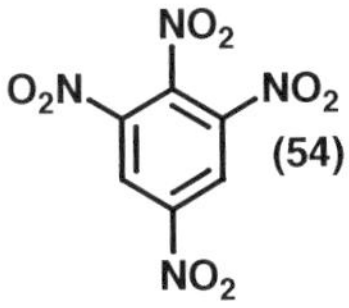 (54)	100
4	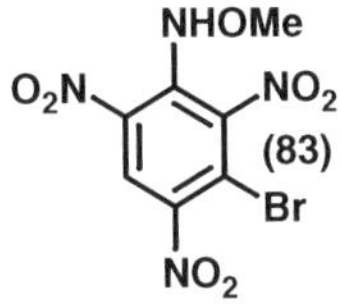 (83)	O_3, EtOAc, 25 °C, 120 mins	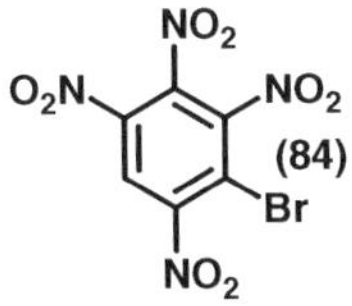 (84)	100

(80) and *N*-methoxy-2,4,6-trinitroaniline (82) with ozone (Table 4.3, Entries 1 and 3). Both ethyl acetate and carbon tetrachloride have been used as solvents for the oxidation of *N*-methoxy-3-bromo-2,4,6-trinitroaniline (83) to 1-bromo-2,3,4,6-tetranitrobenzene (84) (Table 4.3, Entry 4). Yields for these reactions are usually very high and reactions are extremely rapid even at the subambient temperatures used. The method fails to efficiently oxidize substrates containing two or more hydroxyamino or methoxyamino groups on the same aromatic ring.

4.7.2.2 Nitric acid

Both fuming (90 %+) and concentrated (70 %) nitric acids have been used for the oxidation of arylhydroxylamines to the corresponding nitro compounds. Reactions are conducted at elevated temperatures where the oxidizing potential of nitric acid is at its highest. Yields are generally poor to moderate.

Borsche[193] synthesized both 1,2,4- and 1,2,3-trinitrobenzenes by treating *N*-hydroxy-2,4-dinitroaniline and *N*-hydroxy-2,6-dinitroaniline, respectively, with fuming nitric acid. The method is very convenient for the synthesis of such substrates because the starting materials are readily obtainable from the reaction of hydroxylamine with the appropriate dinitrochlorobenzene isomer. Borsche[193,194] also synthesized 1,2,3,5-tetranitrobenzene (60 %)

from *N*-hydroxy-2,4,6-trinitroaniline. The synthesis of 1,2,4,5-trinitrobenzene from the nitric acid oxidation of 2,4-bis(hydroxyamino)-1,5-dinitrobenzene has also been reported.[195,196]

The harsh conditions employed for the nitric acid oxidation of arylhydroxylamines limits the scope of the reaction. Borsche and Feske[197] reported the synthesis of 2,3,4,6-tetranitrotoluene from the nitric acid oxidation of 3-(hydroxylamino)-2,4,6-trinitrotoluene, although a yield for this reaction was not given. Considering the susceptibility of the methyl group to oxidation the yield is expected to be low.

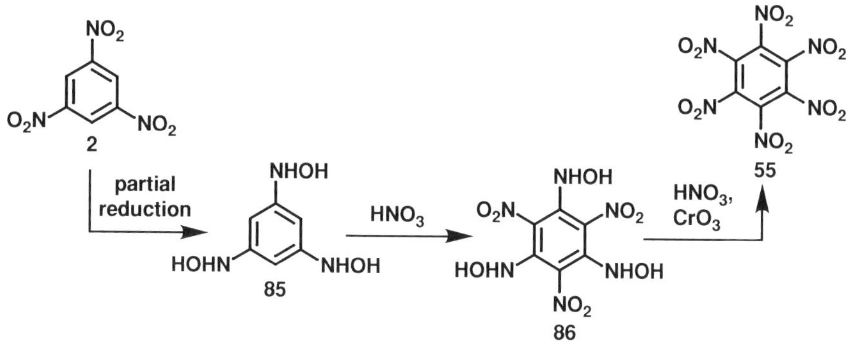

Figure 4.28

A synthesis of hexanitrobenzene (55) has been reported which involves hydroxyamino oxidation. Although no experimental details of this procedure are given the synthesis is reported to start from partial reduction of 1,3,5-trinitrobenzene (2) to 1,3,5-tris(hydroxyamino)benzene (85), followed by ring nitration and subsequent oxidation of the three hydroxyamino groups of (86) with nitric acid in the presence of chromium trioxide.[198] Nielsen and co-workers attempted to replicate this synthetic route but were unsuccessful.[153]

4.8 NUCLEOPHILIC AROMATIC SUBSTITUTION

The nitro group is one of the most potent substituents for withdrawing electron density from the aromatic nucleus. While such groups make electrophilic substitution more difficult they facilitate nucleophilic attack on the aromatic ring. Nitro groups introduce centres of low electron density at the *o*- and *p*-positions on the benzene ring, and consequently, halogens and other leaving groups in these positions are especially susceptible to nucleophilic attack. The effect of leaving group activation by *o*/*p*-nitro groups is seen in the case of 1,4-dichloro-2,6-dinitrobenzene, a substrate readily obtained from the nitration of *p*-dichlorobenzene.[199] In this substrate the 1-chloro group is flanked by two *o*-nitro groups and is readily displaced by a range of nucleophiles; the 4-chloro group is positioned *m*- to these nitro groups and is much less reactive towards nucleophilic attack.[200]

In general, the more nitro groups present on the aromatic ring the easier the leaving group displacement. Nucleophilic aromatic substitution is therefore a very important reaction in the chemistry of polynitroarylenes. While the use of such reactions has been extensive in the synthesis of explosives, the reaction also has important implications for the chemical stability of many polynitroarylenes (discussed in Section 4.8.2).

4.8.1 Displacement of halide

4.8.1.1 Substrate reactivity

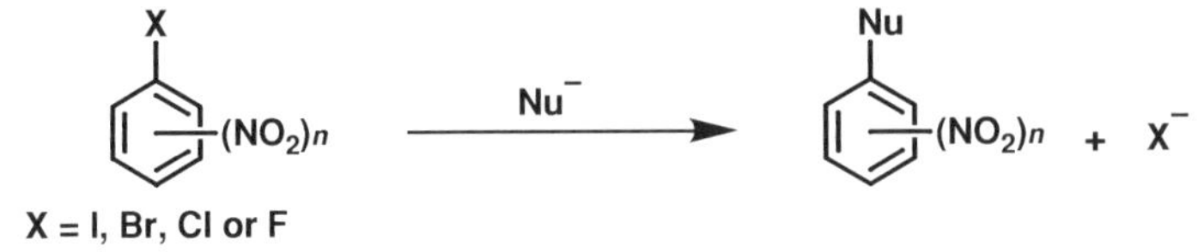

X ... $(NO_2)n$ —Nu^-→ Nu ... $(NO_2)n$ + X^-

X = I, Br, Cl or F

Figure 4.29

Many nucleophiles have been used for the displacement of halide anion from nitro-substituted halobenzenes, including: aliphatic and aromatic amines, ammonia, hydrazine, hydroxide, alkoxides, phenoxides, halides, azide, sulfides and many more. These reactions proceed via an addition–elimination mechanism through an intermediate Meisenheimer complex. The rate-determining step in these reactions is the formation of the Meisenheimer complex, and hence, any process which lowers the energy of this complex increases the reaction rate. Most organic textbooks state that the rate of nucleophilic aromatic substitution is directly related to the polarization ability of the halogen substituent, leading to a reactivity order of F > Cl > Br > I. This is usually the case when reactions are conducted in protic solvents with first row nucleophiles.[201] For example, it is found that 2,4-dinitrofluorobenzene is several thousand times more reactive towards simple oxygen nucleophiles compared to the corresponding chloro- and bromo-derivatives, with 2,4-dinitroiodobenzene showing the least reactivity.[202–204] However, such rules tend to cover over important and complex issues – there are many examples where the lighter halogens are the least reactive towards some nucleophiles.[205] Although the theory behind such a rule is very logical, in practice, there are other factors to consider which might completely alter the relative rates of such reactions. It is found that reaction rate is heavily dependent on the nature of the substrate and the nucleophile, in addition to the reaction temperature and solvent.[205] de Boer and Dirkx[205] have given a very comprehensive review of the activating effects of the nitro group in aromatic substitution.

4.8.1.2 The chemistry of picryl chloride

2,4,6-Trinitrochlorobenzene (picryl chloride) (87) can be prepared from the nitration of 2,4-dinitrochlorobenzene with nitronium tetrafluoroborate[101] or mixed acid composed of fuming nitric acid and oleum.[50,51] Picryl chloride is also synthesized from the reaction of phosphorous oxychloride with the pyridinium salt of picric acid.[206]

The presence of three nitro groups on the aromatic ring of picryl chloride makes the chloro group extremely reactive towards nucleophiles. Picryl chloride (87) is hydrolyzed to picric acid (4) in the presence of hot water or aqueous sodium hydroxide.[207] Aminolysis of picryl chloride in the presence of primary and secondary amines is complete in minutes at room temperature.[208] Picryl chloride is therefore a very useful starting material for the synthesis of a range of other picryl derivatives. The reaction of picryl chloride (87) with ammonia can be used to synthesize 2,4,6-trinitroaniline (53) (picramide).[209] Treatment of picryl chloride with alcohols under reflux forms picric acid and the alkyl chloride of the corresponding alcohol, whereas the same reaction in the presence of alkali metal hydroxides, or the alkoxide anion of

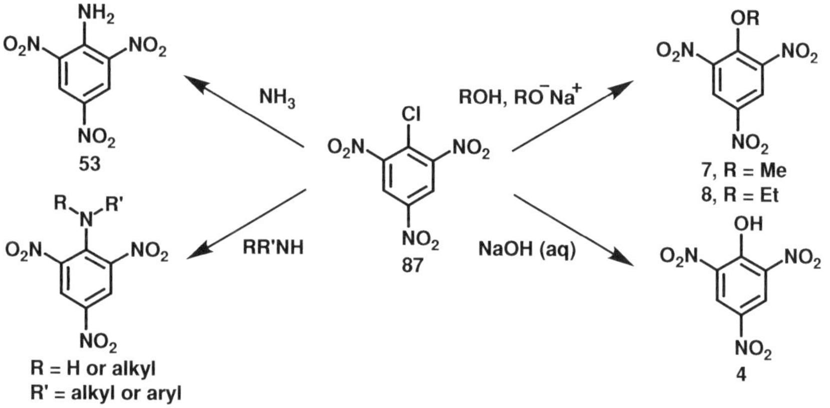

NH_2
O_2N NO_2
NO_2
53

NH_3

R R' N
O_2N NO_2
NO_2
R = H or alkyl
R' = alkyl or aryl

RR'NH

Cl
O_2N NO_2
NO_2
87

ROH, RO^-Na^+

NaOH (aq)

OR
O_2N NO_2
NO_2
7, R = Me
8, R = Et

OH
O_2N NO_2
NO_2
4

Figure 4.30

the alcohol, yields picryl ethers like 2,4,6-trinitroanisole (7) and 2,4,6-trinitrophenetole (8).[210] The reaction of picryl chloride with the potassium salt of methylnitramine yields the practical high explosive *N*,2,4,6-tetranitro-*N*-methylaniline (tetryl, Section 5.14).[211]

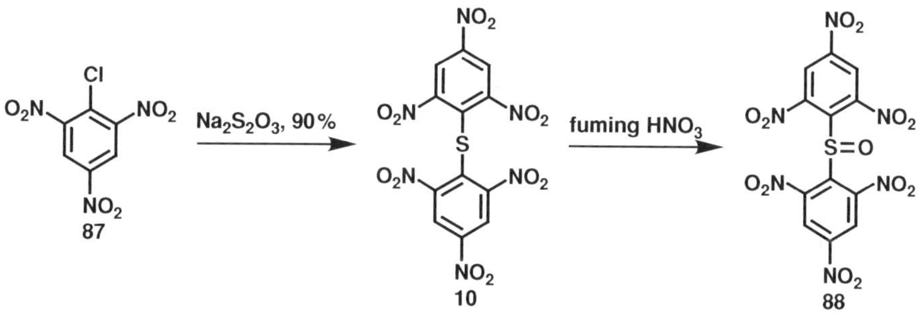

Cl
O_2N NO_2
NO_2
87

$Na_2S_2O_3$, 90%

NO_2
O_2N NO_2
S
O_2N NO_2
NO_2
10

fuming HNO_3

NO_2
O_2N NO_2
S=O
O_2N NO_2
NO_2
88

Figure 4.31

Many high explosives can be synthesized from the reaction of picryl chloride with various nucleophiles. 2,2′,4,4′,6,6′-Hexanitrodiphenylsulfide (10) can be prepared from the reaction of picryl chloride (87) with sodium thiosulfate in ethanol solution in the presence of magnesium carbonate.[212] Oxidation of (10) with fuming nitric acid forms 2,2′,4,4′,6,6′-hexanitrodiphenylsulfone (88).[212]

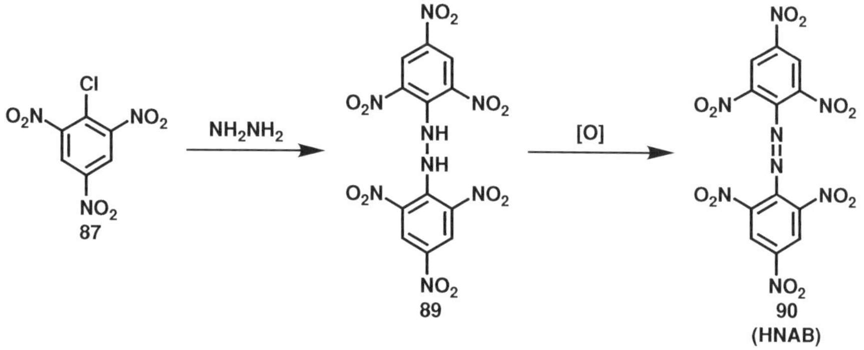

Cl
O_2N NO_2
NO_2
87

NH_2NH_2

NO_2
O_2N NO_2
NH
NH
O_2N NO_2
NO_2
89

[O]

NO_2
O_2N NO_2
N
N
O_2N NO_2
NO_2
90
(HNAB)

Figure 4.32

2,2′,4,4′,6,6′-Hexanitroazobenzene (HNAB) (90), an explosive exhibiting thermal stability[213] (m.p. 220 °C, VOD ~ 7250 m/s, $d = 1.77$ g/cm^3), is formed from the reaction of hydrazine with picryl chloride (87) followed by oxidation of the resulting hydrazide (89) with fuming nitric acid.[214] The commercial availability of picryl chloride limits such reactions. However, 2,4-dinitrochlorobenzene is widely available and can be used to synthesize many of the above products (Section 4.8.1.3).

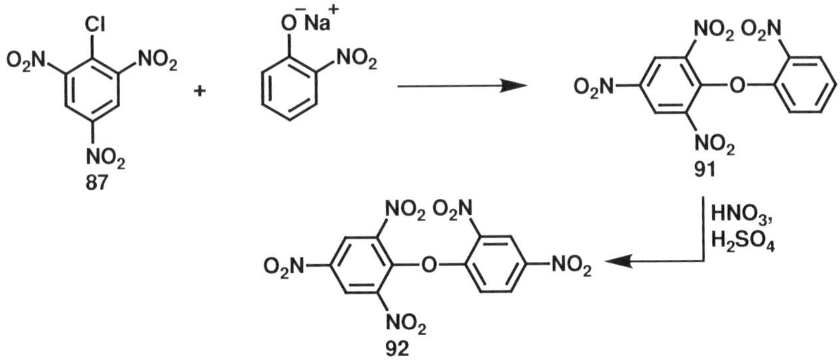

Figure 4.33

The harsh conditions needed to introduce five or more nitro groups into diphenyl ether lead to the destruction of the aromatic ring. Highly nitrated derivatives of diphenyl ether can be prepared by an indirect route; 2,2′,4,4′,6-pentanitrodiphenyl ether (92) is the product from the controlled nitration of (91), which is obtained from the reaction of picryl chloride (87) with sodium *o*-nitrophenolate.[215]

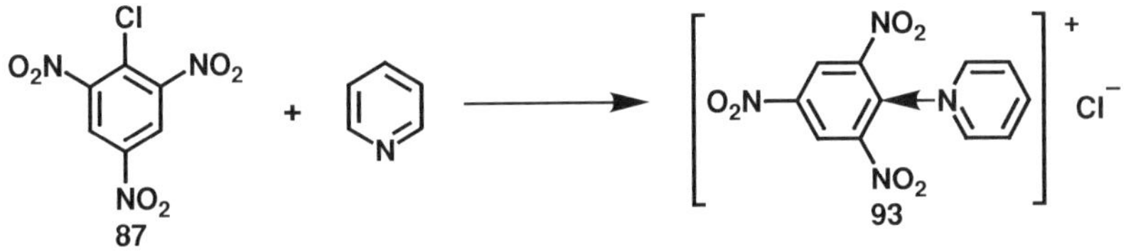

Figure 4.34

Picryl chloride (87) reacts with pyridine to form picrylpyridinium chloride (93), a useful intermediate which reacts with a range of nucleophiles[216]; 2,4,6-trinitrodiphenylamine and 2,4,6-trinitrodiphenylether can be formed from aniline and phenol respectively.[217] 2,4,6-Trinitrodiphenylamine is readily nitrated with mixed acid to the high explosive 2,2′,4,4′,6,6′-hexanitrodiphenylamine (hexyl).[218]

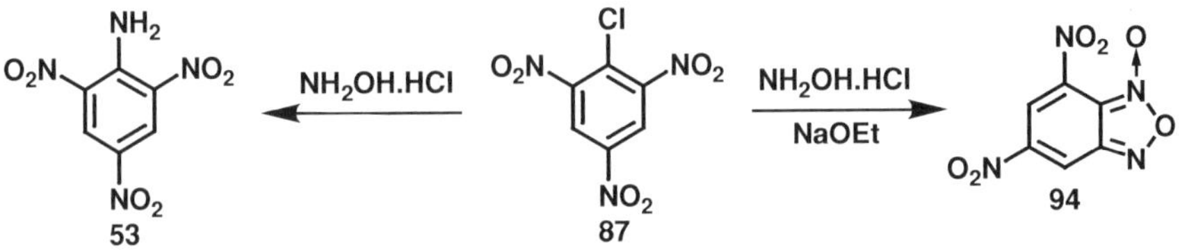

Figure 4.35

Picryl chloride (87) reacts with hydroxylamine hydrochloride to yield 2,4,6-trinitroaniline (53) (picramide) and not the expected *N*-hydroxy-2,4,6-trinitroaniline.[35] In contrast, the same reaction in the presence of sodium ethoxide is reported to yield 4,6-dinitrobenzofuroxan (94) via substitution of the halogen by hydroxylamine, followed by an internal redox reaction between the hydroxyamino group and one of the adjacent *o*-nitro groups.[219]

Picryl chloride has been used successfully in a number of copper-mediated Ullmann coupling reactions. 2,2′,4,4′,6,6′-Hexanitrobiphenyl has been synthesized by heating picryl chloride with copper powder.[220,221] The same reaction in the presence of a hydride source (hot aqueous alcohol) yields 1,3,5-trinitrobenzene (TNB).[221,222] The Ullmann reactions between picryl chloride and isomeric iodonitrobenzenes with copper bronze in DMF has been used to synthesize 2,2′,4,6-, 2,3′,4,6-, and 2,4,4′,6-tetranitrobiphenyls.[223]

4.8.1.3 The chemistry of 2,4-dinitrochlorobenzene

2,4-Dinitrochlorobenzene is an industrially important chemical synthesized from the nitration of chlorobenzene with mixed acid. The halogen atom of 2,4-dinitrochlorobenzene is activated by two *o*/*p*-nitro groups and is particularly reactive. Consequently, 2,4-dinitrochlorobenzene is used as a cheap and readily available starting material for the synthesis of many explosives.

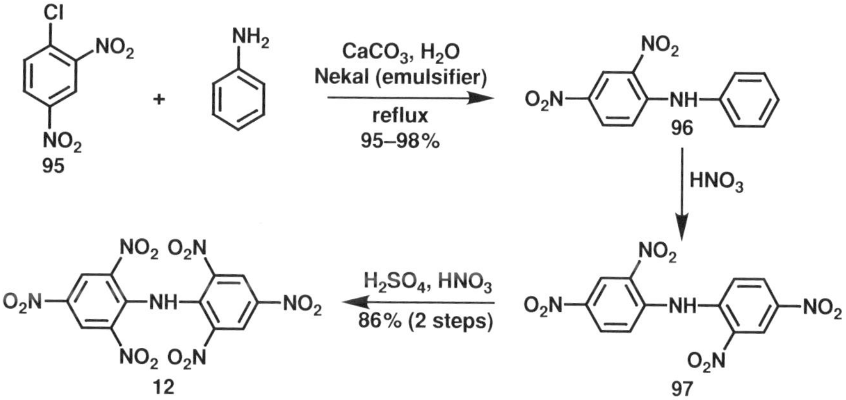

Cl
NO2
NO2
95
+
NH2
CaCO3, H2O
Nekal (emulsifier)
reflux
95–98%
NO2
O2N
NH
96
HNO3
NO2
O2N
NH
NO2
O2N
97
H2SO4, HNO3
86% (2 steps)
NO2 O2N
O2N
NH
NO2
NO2 O2N
12

Figure 4.36

The standard industrial and laboratory method for the synthesis of the high explosive known as hexyl (12) (2,2′,4,4′,6,6′-hexanitrodiphenylamine) involves treating 2,4-dinitrochlorobenzene (95) with aniline to produce 2,4-dinitrodiphenylamine (96), followed by a two-stage nitration.[41,224]

The direct nitration of phenol and other substrates containing electron-donating groups is often very vigorous and low yielding due to the formation of excessive by-products from oxidative degradation. A much safer and more convenient route involves nitrating a derivative of the substrate which already contained some nitro groups on the aromatic nucleus. The effect of these nitro groups would be to moderate the reaction and protect the substrate from oxidation. Picric acid (4) is formed in high yield from the mixed acid nitration of 2,4-dinitrophenol (98), which is synthesized from the reaction of 2,4-dinitrochlorobenzene (95) with aqueous sodium hydroxide.[225]

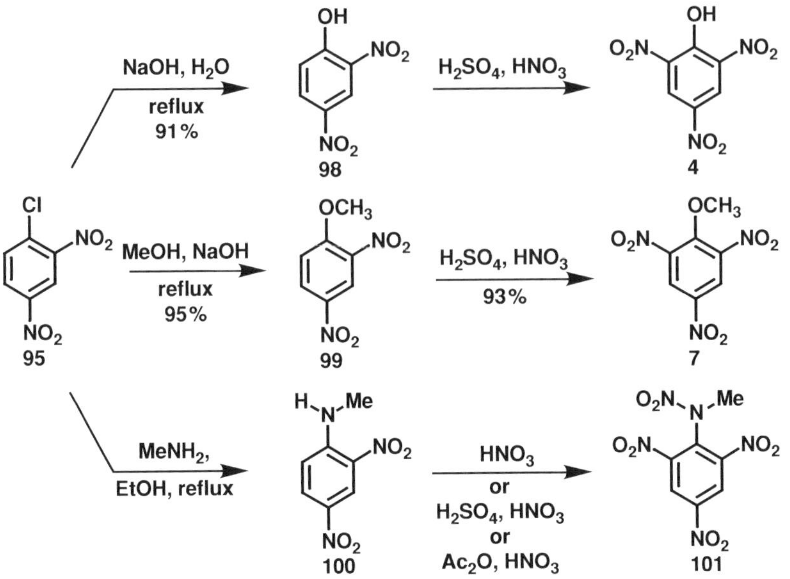

Figure 4.37

Other explosives, such as 2,4,6-trinitroanisole (7),[33b,226] 2,4,6-trinitroaniline (picramide),[227] and tetryl (101),[40,228] are conveniently prepared from the nitration of the corresponding 2,4-dinitro derivatives, which in turn, are prepared from the reaction of 2,4-dinitrochlorobenzene with the appropriate nucleophile.

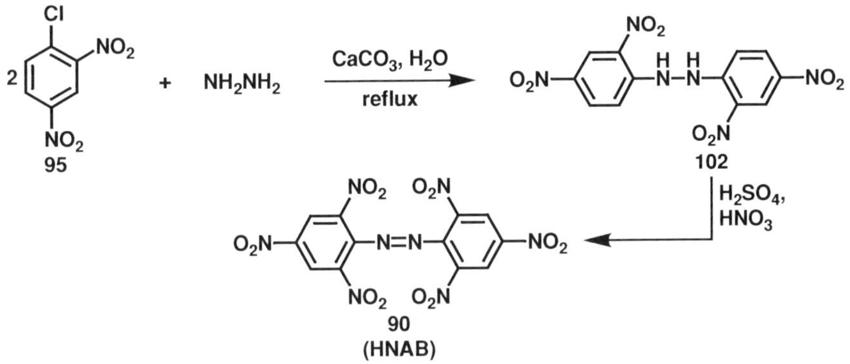

Figure 4.38

2,2′,4,4′,6,6′-Hexanitroazobenzene (HNAB) (90) can be synthesized from the nitration–oxidation of 2,2′,4,4′-tetranitrohydrazobenzene (102) with mixed acid, the latter synthesized from the reaction of aqueous hydrazine with 2,4-dinitrochlorobenzene (95) in the presence of an inorganic carbonate.[214]

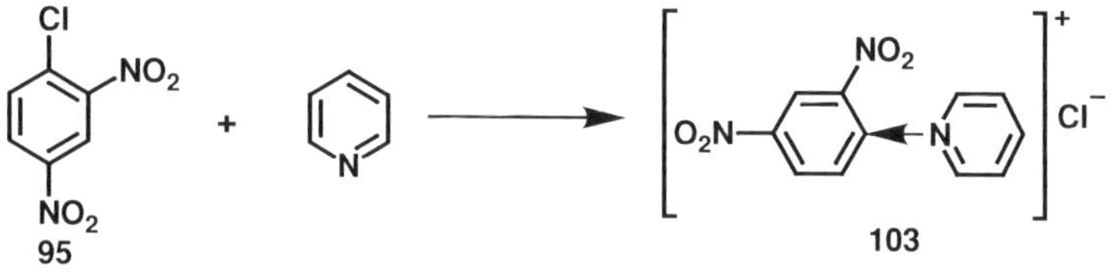

Figure 4.39

2,4-Dinitrochlorobenzene (95) reacts with pyridine to form 2,4-dinitrophenylpyridinium chloride (103),[229] a reactive intermediate which readily reacts with a variety of nucleophiles.[230] The reaction of (103) with hydrogen sulfide yields 2,2′,4,4′-tetranitrodiphenylsulfide (104),[231] which on nitration–oxidation with fuming nitric acid, yields 2,2′,4,4′,6,6′-hexanitrodiphenylsulfoxide (105).[212] The sulfide (104) is also formed from the reaction of two equivalents of 2,4-dinitrochlorobenzene (95) with sodium thiosulfate or sodium disulfide in aqueous ethanol.[212]

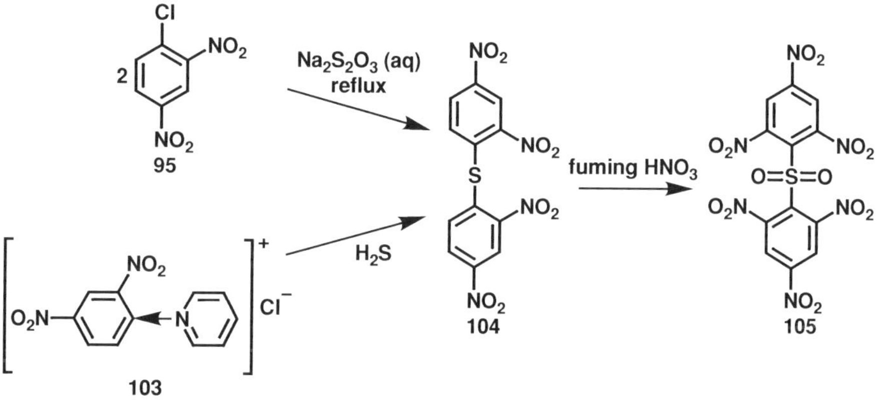

Figure 4.40

2,4-Dinitroiodobenzene is prepared by treating 2,4-dinitrochlorobenzene with sodium iodide in acetone.[232] 2,4-Dinitrohalobenzenes undergo Ullmann coupling on treatment with copper powder to yield 2,2′,4,4′-tetranitrobiphenyl.[233]

4.8.1.4 Thermally insensitive explosives from halide displacement with nitrogen nucleophiles

Analysis of the structure–properties relationships of many aromatic explosives has revealed that the presence of amino functionality adjacent to nitro groups has a marked effect on both thermal stability and sensitivity to impact.[234] The most striking example is seen during the sequential introduction of amino groups into 1,3,5-trinitrobenzene.[235] Accordingly, both 1,3-diamino-2,4,6-trinitrobenzene (DATB) and 1,3,5-triamino-2,4,6-trinitrobenzene (TATB) are significantly more thermally stable and less sensitive to impact than 2,4,6-trinitroaniline (picramide) and are designated as ‘heat resistant’ explosives. Such effects are attributed to intramolecular hydrogen bonding interactions between adjacent amino and nitro groups which also increases crystal density and improves explosive performance.[236] Much research has been focused into this area. Most of the methods for incorporating amino groups into potential explosives make use of the low electron density in polynitroarylenes and use amines and ammonia as nucleophiles to attack the corresponding halo derivatives.

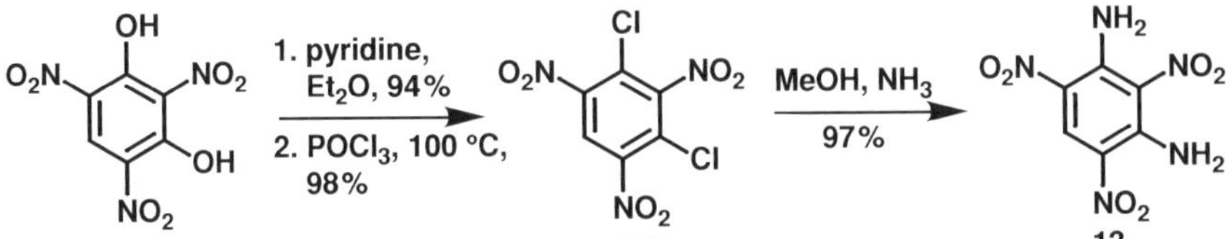

Figure 4.41

1,3-Diamino-2,4,6-trinitrobenzene (DATB) (13) is prepared in high yield by treating 1,3-dichloro-2,4,6-trinitrobenzene (106) with ammonia in methanol.[237] 1,3-Dichloro-2,4,6-trinitrobenzene (106) can be synthesized from the mixed acid nitration of 1,3-dichlorobenzene,[52] or more conveniently, by treating styphnic acid (5) with pyridine followed by refluxing the resulting dipyridinium styphnate with phosphorous oxychloride.[206b,237] DATB (m.p. 286 °C, VOD ~ 7500 m/s, $d = 1.84$ g/cm^3) was once widely used in the US for applications requiring high thermal stability and low sensitivity to impact. The use of DATB in plastic bonded compositions with Estane, Kel F and Viton A polymers has been patented.[238]

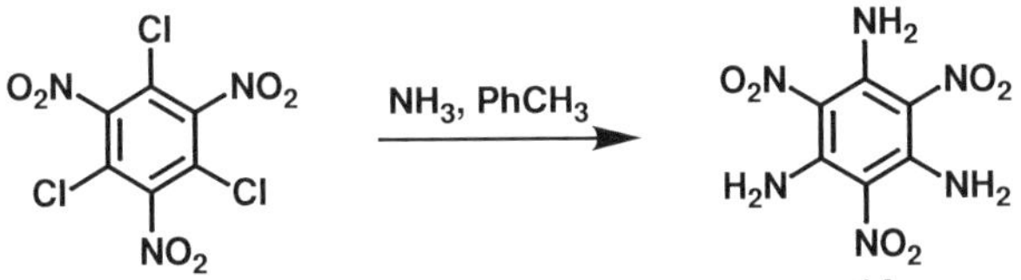

Figure 4.42

1,3,5-Triamino-2,4,6-trinitrobenzene (TATB) (14) is the most thermally stable and impact insensitive explosive in current use. TATB is synthesized on an industrial scale from the tri-nitration of 1,3,5-trichlorobenzene with strong mixed acid followed by nucleophilic displacement of the chloro groups of 1,3,5-trichloro-2,4,6-trinitrobenzene (34) with ammonia in toluene under autoclave.[53] The extra amino group in TATB (m.p. >350 °C with decomposition, VOD ~ 8000 m/s, $d = 1.94$ g/cm^3) compared to DATB decreases oxygen balance but increases density and results in higher overall performance. TATB finds extensive use in military applications requiring high thermal stability and a low sensitivity to impact and has replaced DATB in this respect. The importance of TATB as a thermally stable explosive has meant that its synthesis has been widely studied. The synthesis of TATB is discussed in more detail in Section 4.8.4.

More complex explosives incorporating amino groups have been prepared from the reaction of polynitroarylene halides with amine nucleophiles. Agrawal and co-workers[239] have synthesized PADNT (107) from the reaction of 4-amino-2,6-dinitrotoluene (46) with picryl chloride (87) in methanol; 4-amino-2,6-dinitrotoluene is synthesized from the reduction of

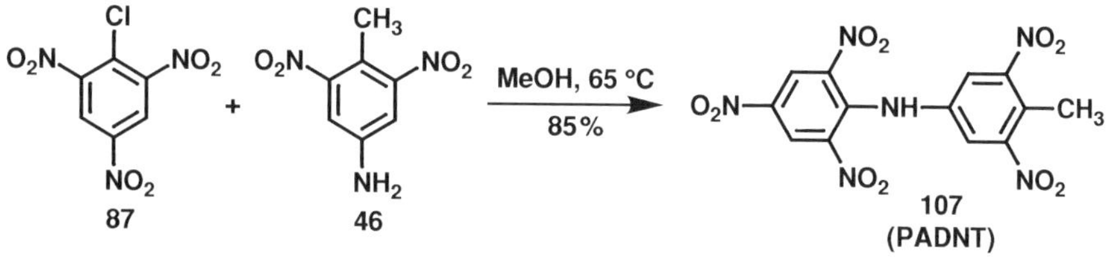

Figure 4.43

2,4,6-trinitrotoluene (α-TNT) with hydrogen sulfide in concentrated ammonium hydroxide.[240] PADNT shows considerable insensitivity to impact and friction and may find use as a new energetic ingredient for the development of safe and insensitive explosive and propellant formulations.

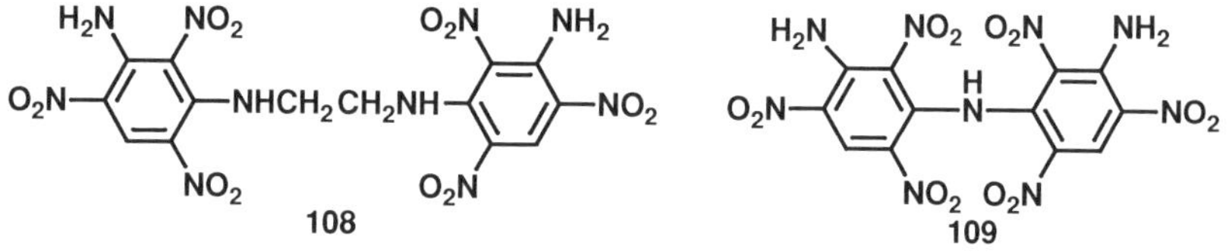

Figure 4.44

N,N′-Bis(3-aminopicryl)-1,2-ethanediamine (108) (m.p. 275 °C) is prepared from the reaction of ethylenediamine with two equivalents of 3-chloro-2,4,6-trinitroaniline.[241] The same chemists[241] reported 3,3′-diamino-2,2′,4,4′,6,6′-hexanitrodiphenylamine (109), a heat resistant explosive (m.p. 232–237 °C) prepared from the reaction of 1,3-dichloro-4,6-dinitrobenzene with 3-chloroaniline followed by mixed acid nitration and subsequent chloro group displacement with ammonia. The potassium salt of 3,3′-diamino-2,2′,4,4′,6,6′-hexanitrodiphenylamine shows very high thermal stability.[242]

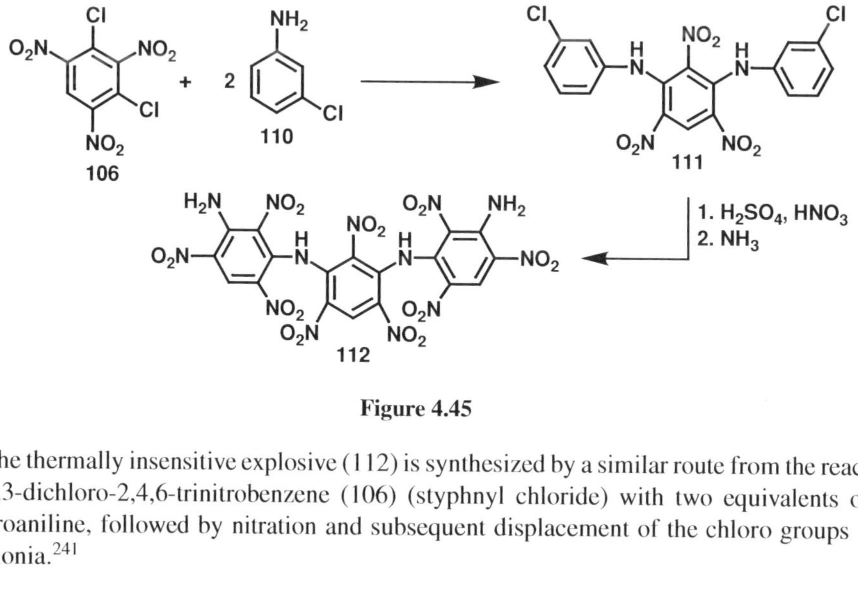

Figure 4.45

The thermally insensitive explosive (112) is synthesized by a similar route from the reaction of 1,3-dichloro-2,4,6-trinitrobenzene (106) (styphnyl chloride) with two equivalents of 3-chloroaniline, followed by nitration and subsequent displacement of the chloro groups with ammonia.[241]

2,4,6-Tris(picrylamino)-1,3,5-trinitrobenzene (113), the product from condensing three equivalents of aniline with 1,3,5-trichloro-2,4,6-trinitrobenzene followed by nitration with mixed acid, is a high-molecular weight explosive which shows good thermal stability (m.p. 234 °C) but is readily detonated.[243]

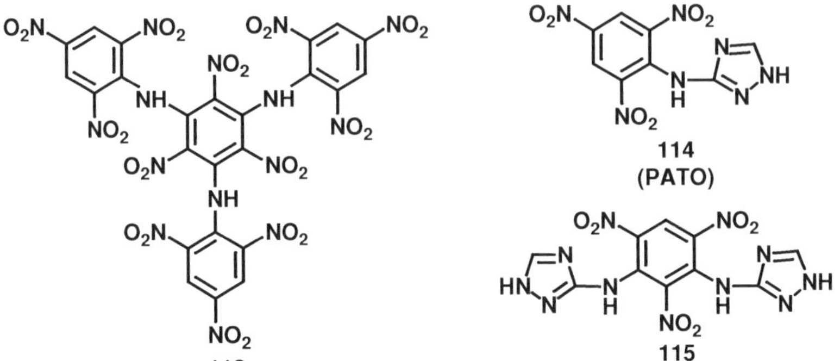

Figure 4.46

High thermal stability is observed in explosives where the amino functionality is part of a heterocyclic ring. A number of thermally stable explosives have been synthesized by condensing polynitroarylene halides with triazole rings. 3-Picrylamino-1,2,4-triazole (PATO) (114) is a thermally stable explosive (m.p. 310 °C, VOD ~ 7850 m/s, $d = 1.94$ g/cm^3) conveniently prepared from the reaction of picryl chloride with 3-amino-1,2,4-triazole.[244] Agrawal and co-workers[245] prepared a similar compound known as BTATNB (115) from the reaction of styphnyl chloride with two equivalents of 3-amino-1,2,4-triazole. Thermal and explosive performance data suggests that BTATNB is slightly more thermally stable than PATO and safer towards impact and friction. By following a similar strategy, Agrawal and co-workers reported a number of explosives possessing a wide spectrum of properties.

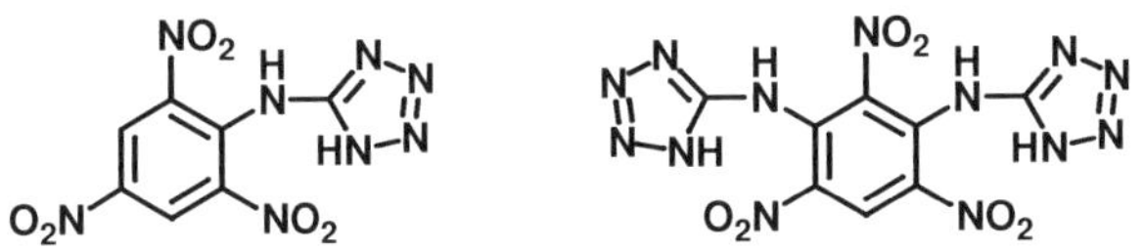

Figure 4.47

Similarly, 5-picrylamino-1,2,3,4-tetrazole (PAT) (116)[246,247] and 5,5′-styphnylamino-1,2,3,4-tetrazole (SAT) (117)[247] have been synthesized by condensing picryl chloride and styphnyl chloride with 5-amino-1,2,3,4-tetrazole in methanol respectively. A comparison of thermal and explosive properties of newly synthesized PAT (calculated VOD ~ 8126 m/s) and SAT (calculated VOD ~ 8602 m/s) reveals that PAT is more thermally stable than SAT but more sensitive to impact and friction.[247]

High molecular weight polynitroarylenes often show increased thermal stability. Thermal stability is further enhanced when amino functionality is positioned adjacent to nitro groups.

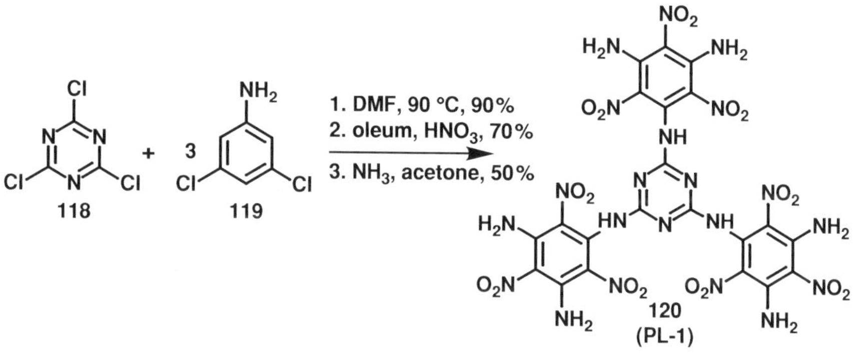

Figure 4.48

Agrawal and co-workers[248] synthesized 2,4,6-tris(3′,5′-diamino-2′,4′,6′-trinitrophenylamino)-1,3,5-triazine (PL-1) (120) by condensing three equivalents of 3,5-dichloroaniline with cyanuric chloride, followed by nitration with fuming nitric acid in oleum and final chloro group displacement with ammonia in acetone. PL-1 possesses a unique combination of moderate heat resistance, high density and insensitivity to impact and friction. A large number of heat resistant energetic materials based on heterocyclic rings are discussed in Chapter 7. The research conducted on heat resistant explosives has been reviewed by Dunstan,[249] Urbanski,[250] Lu,[241] and more recently by Agrawal.[251]

4.8.2 Nitro group displacement and the reactivity of polynitroarylenes

Nucleophiles can react with polynitroarylenes to displace one of the nitro groups. It is found that nitro groups positioned *o*- or *p*- to each other are chemically unstable and are readily displaced by a range of nucleophiles. Only nitro groups positioned *m*- to one another show relatively high chemical stability. Such observations result from the electron-withdrawing nitro group leading to a greater reduction in electron density at *o*- and *p*-positions relative to *m*-positions. The more nitro groups present on the aromatic ring the greater the tendency is for displacement with nucleophiles.

An unfortunate consequence of the high reactivity of some nitro groups is that many powerful explosives are too chemically reactive for both commercial and military use. This is the case with highly nitrated compounds containing four or more nitro groups per benzene ring. Some of these compounds are illustrated below; chemically unstable nitro groups i.e. those positioned *o*/*p*- to other nitro groups, are indicated by an asterisk.

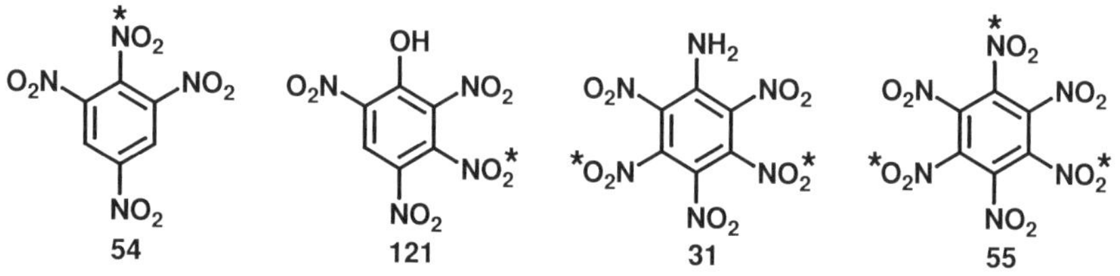

Figure 4.49

Highly nitrated derivatives of benzene readily react with water to form phenols. 1,2,3,5-Tetranitrobenzene (54) is readily converted to picric acid on reaction with hot water.[252] This type of reaction has practical concerns if such an explosive is used in a military context – picric acid forms dangerous picrates if allowed to come into contact with a metal surface i.e. the inside of a munition's shell. Other explosives like 2,3,4,6-tetranitrophenol (121) and 2,3,4,6-tetranitrotoluene react with water to form 2,4,6-trinitroresorcinol (styphnic acid) and 2,4,6-trinitro-*m*-cresol respectively. Some hindered nitro groups are displaced smoothly even at room temperature, and hexanitrobenzene (55), although a powerful explosive, is readily hydrolyzed to 2,4,6-trinitrophloroglucinol on reaction with water;[253] the latter is also formed from the reaction of water with 2,3,4,5,6-pentanitrophenol.[240]

The reactions of hexanitrobenzene[153] (55) and 2,3,4,5,6-pentanitroaniline[143] (31) with ammonia have been used to synthesize the thermally stable explosive 1,3,5-triamino-2,4,6-trinitrobenzene (TATB). Holmes and Flürschiem[140] have studied the reactions of 2,3,4,5,6-pentanitroaniline with nucleophiles.

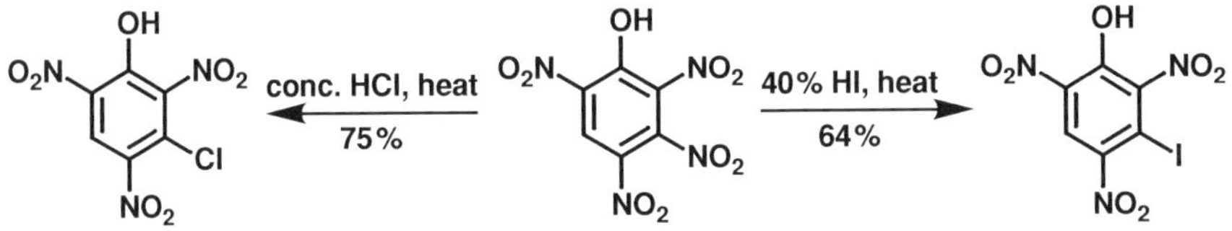

Figure 4.50

Orlova and co-workers[254] reacted 1,2,3,5-tetranitrobenzene with hydrochloric and hydrobromic acids to form picryl chloride and picryl bromide respectively. The same chemists treated 2,3,4,6-tetranitroaniline and 2,3,4,6-tetranitrophenol (121) with aqueous solutions of hydrogen halides to form 3-halo-2,4,6-trinitroanilines and 3-halo-2,4,6-trinitrophenols (122 and 123) respectively.

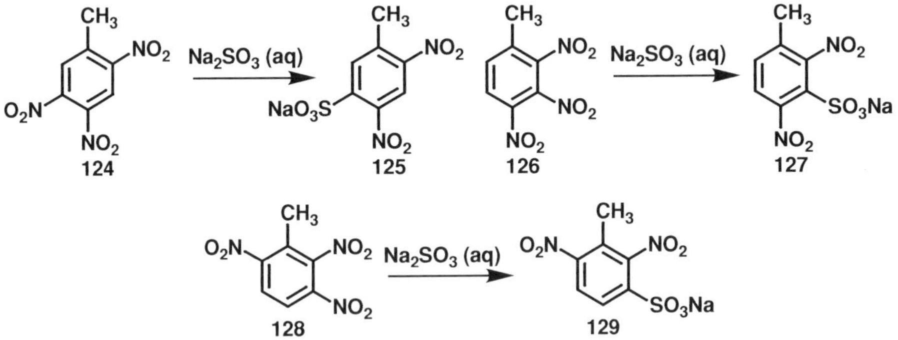

Figure 4.51

The reactivity of nitro groups positioned *o*/*p*- to other nitro groups has implications for the use of other polynitroarylenes as explosives. For example, of the numerous possible isomers of trinitrotoluene, only the symmetrical 2,4,6-isomer (α-TNT) is chemically stable enough for use as an explosive. Only in the case of the 2,4,6-isomer are the three nitro groups positioned *m*- to one another; all other isomers of trinitrotoluene contain either one or two nitro group in

o/p-positions to one another, making them considerably more reactive towards nucleophilic attack.[255]

The 2,4,5- (124), 2,3,4- (126), and 2,3,6- (128) isomers of trinitrotoluene are produced during the nitration of toluene and are present in the crude product.[255] Such isomers reduce the chemical stability of α-TNT and need to be removed from the product. The increased reactivity of these unsymmetrical isomers towards nucleophilic attack can be exploited in the purification of α-TNT. Thus, heating crude TNT with an aqueous solution of sodium sulfite will replace any reactive nitro groups present in the unsymmetrical isomers with a sodium sulfonate group (125, 127, 129). These sulfonates are water-soluble and are washed out of the product in the aqueous liquors.[256] It is important to remember that while polynitroarylenes like 2,4,6-trinitrotoluene show sufficient chemical stability for wide use as explosives they are still reactive towards nucleophilic attack, although much less so than their unsymmetrical isomers – nitro groups also reduce electron density at m-positions, but to a lesser extent than o/p-positions. The reaction of α-TNT with sodium sulfite is much slower than with its unsymmetrical isomers, but at elevated temperatures and with long reaction times the proportion of α-TNT lost as water-soluble sulfonate can be significant. The purification reaction also serves to remove other impurities present in crude TNT, namely, tetranitromethane, which also forms a water-soluble sulfonate on reaction with aqueous sodium sulfite.

In a similar manner, of the isomeric trinitrobenzenes, only the symmetrical 1,3,5-isomer shows sufficient chemical stability for use as an explosive. Even so, the aromatic ring of 1,3,5-trinitrobenzene is highly electron deficient and reaction with alkali metal carbonates or bicarbonates in aqueous boiling methanol yields 3,5-dinitroanisole.[257] Unsymmetrical isomers of trinitrobenzene are much more reactive than the 1,3,5-isomer, with only relatively mild conditions needed to effect the displacement of their nitro groups.[258]

It is found that some nucleophiles will displace an activated nitro group in preference to a halogen atom in a similarly activated position. Accordingly, 2,4,5-trinitrochlorobenzene reacts with ammonia to form 5-chloro-2,4-dinitroaniline.[259]

4.8.3 Displacement of other groups

The displacement of both halides and nitro groups from polynitroarylenes has been covered in Sections 4.8.1 and 4.8.2. The centres of low electron density induced by electron-withdrawing nitro groups also allow the displacement of many other groups, including hydrogen, alkoxy, aryloxy, sulfonate ester etc.

4.8.3.1 Displacement of hydrogen

A reaction of growing importance in energetic materials synthesis is that of amination by nucleophilic substitution of aromatic hydrogen. Such reactions usually involve treating aromatic nitro compounds with an aminating agent under basic conditions. The reaction of 1,3,5-trinitrobenzene (TNB) with hydroxylamine under basic conditions yields 2,4,6-trinitroaniline (picramide) and a small amount of 1,3-diamino-2,4,6-trinitrobenzene (DATB).[260] Mitchell and co-workers[261] found at elevated temperatures and in the presence of sodium methoxide in DMSO, picramide reacts with excess hydroxylamine to yield 1,3,5-triamino-2,4,6-trinitrobenzene (TATB). Such reactions involving hydrogen displacement come under the general classification of vicarious nucleophilic substitutions (VNS).[262]

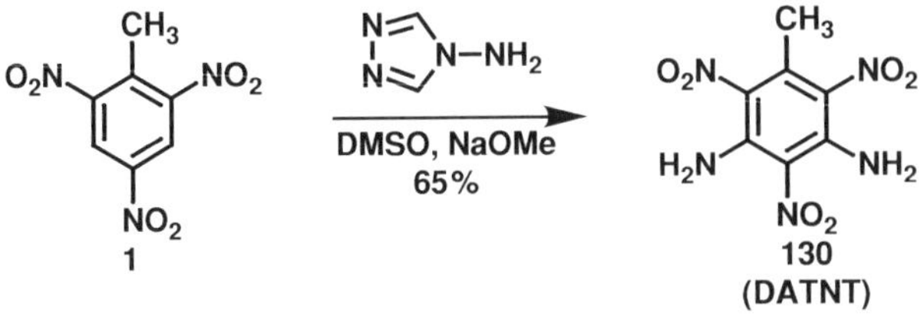

Figure 4.52

Other aminating agents have been studied in relation to efficiency and regioselectivity.[263] 4-Amino-1,2,4-triazole (ATA) and 1,1,1-trimethylhydrazinium iodide (TMHI) are efficient aminating agents; the latter is synthesized from the reaction of *N*,*N*-dimethylhydrazine with methyl iodide. Mitchell and co-workers[264] synthesized 3,5-diamino-2,4,6-trinitrotoluene (130) (DATNT) in 65 % yield by treating TNT (1) with 4-amino-1,2,4-triazole and sodium methoxide in DMSO. This synthesis is significantly more efficient than previous syntheses[265] of this compound. DATNT is considerably more thermally stable than TNT. Mitchell and co-workers[37] used the same methodology for the synthesis of TATB from TNB and picramide (see Section 4.8.4). Mitchell and co-workers[266] also reported the synthesis of DATB by treating TNB with two equivalents of TMHI. A series of 1-substituted 3,5-dinitrobenzenes were aminated by VNS in the same work.[266]

Hydrogen displacement from unsymmetrical substrates like 1,3-dinitrobenzene can produce a mixture of products. Reaction of 1,3-dinitrobenzene with hydroxylamine produces a mixture of 2,4-dinitroaniline and 2,6-dinitroaniline; 1,3-diamino-2,4-dinitrobenzene is formed if two equivalents of hydroxylamine are employed.[260]

The effect which amino functionality has on the thermal and impact sensitivity of polynitroarylenes (Section 4.8.1.4) makes amination by VNS a method with much future potential for energetic materials synthesis. Other carbon, nitrogen, oxygen and sulfur nucleophiles can displace aromatic hydrogen; examples with 1,3-dinitrobenzene[267] and 1,3,5-trinitrobenzene[268] are extensive.

4.8.3.2 Displacement of alkoxy and aryloxy groups

The displacement of alkoxy and aryloxy groups from polynitroarylenes is generally more difficult than the displacement of halide from similar substrates. This is a consequence of alkoxy groups being poorer leaving groups than halide anions. In these reactions the nucleophile should be a poorer leaving group or a better nucleophile than the one being displaced in order to drive the equilibrium towards the product.

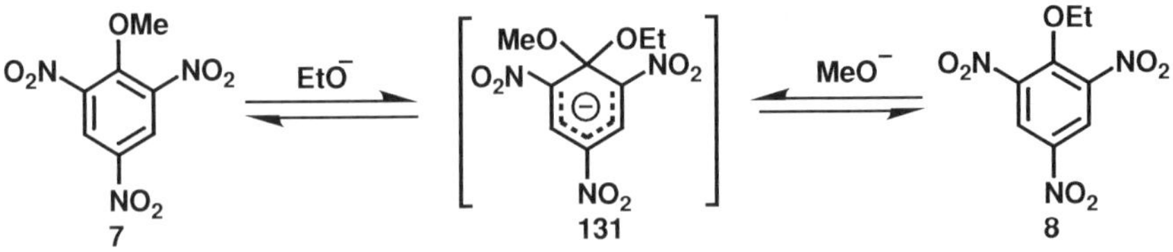

Figure 4.53

2,4,6-Trinitrophenetole (8) can be prepared by heating 2,4,6-trinitroanisole (7) in ethanol containing a catalytic amount of sodium ethoxide.[269] This reaction is so facile that methanol must not be used for the recrystallization of 2,4,6-trinitrophenetole and vice versa. The *trans*-etherification of 2,4,6-trinitroanisole with ethanol and 2,4,6-trinitrophenetole with methanol goes via the same Meisenheimer intermediate (131) which is stable enough to be isolated.[270] Both 2,4,6-trinitroanisole and 2,4,6-trinitrophenetole and similar picryl ethers slowly form picric acid in the presence of moisture, so inhibiting their use in munitions destined for long-term storage. The formation of 2,4,6-trinitroaniline from the reaction of 2,4,6-trinitroanisole or 2,4,6-trinitrophenetole with ammonia is too slow to be synthetically useful.

Aryloxy groups are much easier to displace compared to primary and secondary alkoxide anions and so, aryl ethers are generally more useful in displacement reactions. Amine nucleophiles react with unsymmetrical aryl ethers to form the amine of the heavier nitrated moiety.[271] Accordingly, 2,4,6-trinitrodiphenyl ether reacts with ammonia to expel phenoxide anion and form 2,4,6-trinitroaniline.

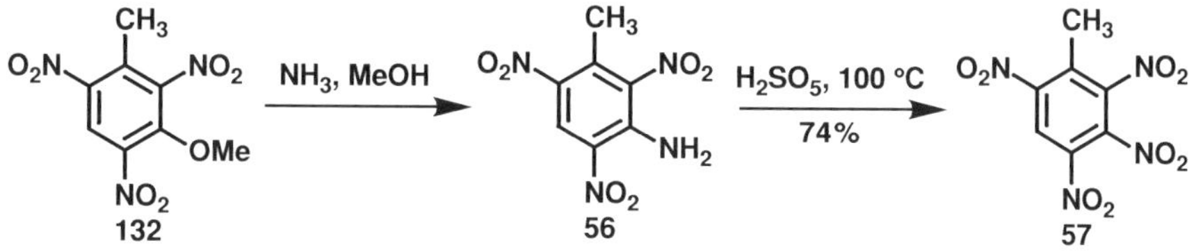

Figure 4.54

The displacement of alkoxy groups from polynitroarylenes has been used for the indirect synthesis of some highly nitrated polynitroarylenes. Holleman[272] synthesized 2,3,4,6-tetranitrotoluene (57) by treating 3-methoxy-2,4,6-trinitrotoluene (132) with ammonia in methanol, followed by oxidation of the resulting product (56) with peroxymonosulfuric acid.

4.8.3.3 Other displacements

Some synthetically useful reactions exploit the displacement of sulfonate esters from polynitroarylenes. The *p*-toluenesulfonate (tosyl) group is readily introduced into aromatic systems from the reaction of phenols with tosyl chloride. The tosyl group is an excellent leaving group and is readily displaced by a range of nucleophiles. When polynitrophenols are reacted with tosyl chloride in the presence of pyridine the intermediate tosylate is too reactive to be isolated, and instead, the tosylate reacts with the pyridinium chloride formed during the reaction to form an arylpyridinium salt.[273] Such salts are synthetically useful because the pyridinium group is readily displaced by various nucleophiles. The chemistry of picrylpyridinium chloride and 2,4-dinitrophenylpyridinium chloride has been discussed in Sections 4.8.1.2 and 4.8.1.3 respectively.

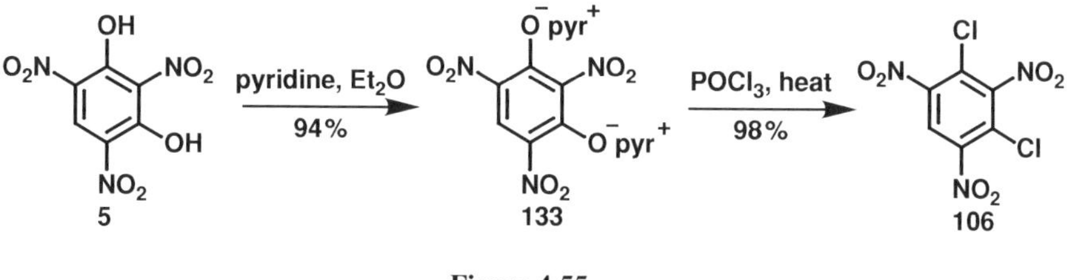

Figure 4.55

Warman and Siele[237] reported a high yielding route to 1,3-dichloro-2,4,6-trinitrobenzene (106) (styphnyl chloride) which involves treating styphnic acid (5) with two equivalents of pyridine followed by reacting the resulting pyridinium salt (133) with phosphorous oxychloride. 1,3-Dichloro-2,4,6-trinitrobenzene is an important precursor to the thermally stable explosive DATB (Section 4.8.1.4).

4.8.4 Synthesis of 1,3,5-triamino-2,4,6-trinitrobenzene (TATB)

2,4,6-Trinitro-1,3,5-triaminobenzene (14), chemically abbreviated to TATB, has emerged as an important modern explosive. TATB is very insensitive to impact and friction, exhibits high thermal stability, and is usable in the 260–290 °C range, but has no observable melting point (>350 °C with decomposition). Such properties are attributed to strong intermolecular and intramolecular hydrogen bonding between adjacent nitro and amino groups.[274] TATB finds extensive military use in low vulnerability munitions and in applications requiring high thermal stability i.e. warheads of high-speed missiles. Agencies like the American Department of Energy and Department of Defence have switched over from using conventional secondary high explosives in nuclear weapons to using TATB as a matter of safety.[275] The performance of TATB ($d = 1.94$ g/cm^3, VOD $\sim$ 8000 m/s) lags behind nitramine explosives like RDX ($d = 1.70$ g/cm^3, VOD $\sim$ 8440 m/s) and HMX ($d = 1.89$ g/cm^3, VOD $\sim$ 9110 m/s) but is considerably higher than other polynitroarylenes like TNT ($d = 1.55$ g/cm^3, VOD $\sim$ 6850 m/s), a consequence of increased crystal density due to the introduction of amino functionality. A large number of explosive compositions and plastic bonded explosives based on TATB have been reported for specialized applications and these have been recently reviewed by Agrawal.[276] At present the cost of TATB production is relatively high and prevents its use for commercial applications.

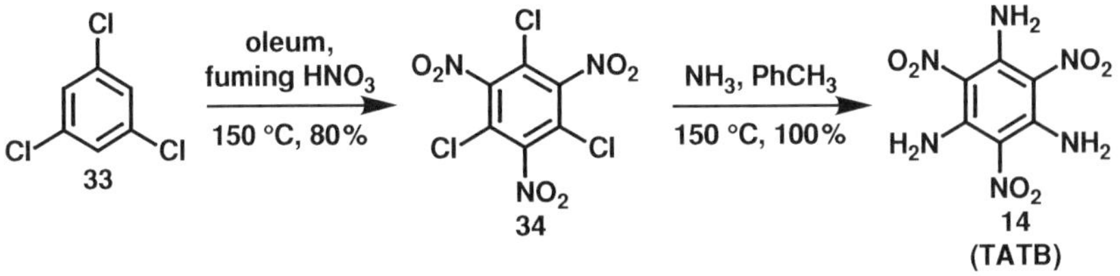

Figure 4.56

All reported syntheses of TATB to date involve the nitration of substrates containing leaving groups which are subsequently replaced by amino groups. The current industrial synthesis of TATB (14) involves the nitration of 1,3,5-trichlorobenzene (33) to 1,3,5-trichloro-2,4,6-trinitrobenzene (34) followed by reaction with ammonia in toluene under pressure.[53] Both nitration and amination steps require forced conditions with elevated temperatures.

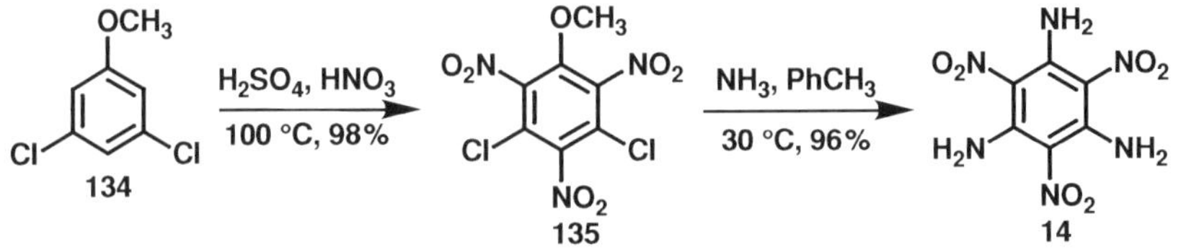

Figure 4.57

The tri-nitration of 3,5-dichloroanisole (134) followed by ammonolysis of the product (135) with ammonia is a much more facile route to TATB (14); although the cost and availability of (134) makes this route of academic interest only.[277]

The tri-nitration of 1,3-dimethoxybenzene with mixed acid, followed by amination, is a patented route to DATB.[278] A similar route to TATB employing 1,3,5-trimethoxybenzene and dinitrogen pentoxide only results in moderate yields.[30,279]

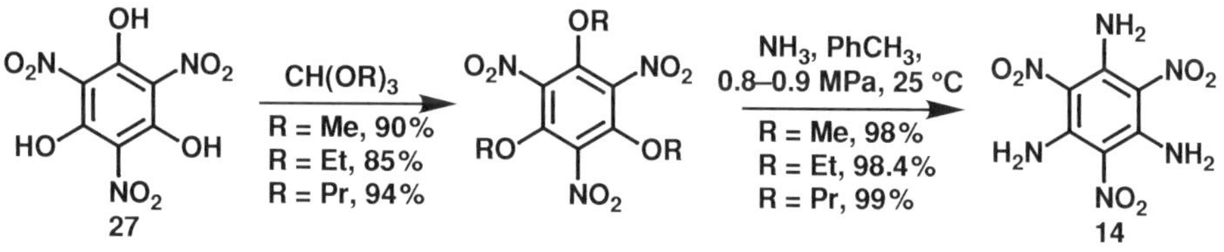

Figure 4.58

Bellamy and co-workers[30] conducted an extensive study into the synthesis of TATB from phloroglucinol. Phloroglucinol and its triacetate derivative were tri-nitrated to 2,4,6-trinitrophloroglucinol (27) with mixed acid or dinitrogen pentoxide in sulfuric acid. 2,4,6-Trinitrophloroglucinol (27) was alkylated with orthoformate esters to give the trimethoxy-, triethoxy- or tripropoxy-2,4,6-trinitrobenzenes; subsequent amination in toluene under pressure, or with liquid ammonia, gave TATB (14) in excellent yield in all three cases.

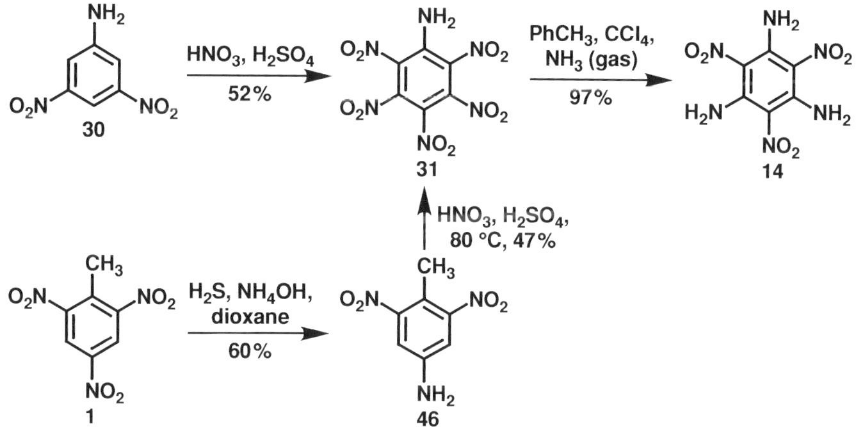

Figure 4.59

The synthesis of TATB (14) from the reaction of 2,3,4,5,6-pentanitroaniline (31) with ammonia has been reported.[140,143] In one route, 2,3,4,5,6-pentanitroaniline (31) is synthesized from the nitration of 3,5-dinitroaniline (30);[39,139,140] the latter is obtained from the selective reduction of TNB[280] or via a Schmidt reaction with 3,5-dinitrobenzoic acid.[281] Another route to 2,3,4,5,6-pentanitroaniline (31) involves the selective reduction of TNT (1) with hydrogen sulfide in ammonia followed by nitration of the resulting 4-amino-2,6-dinitrotoluene (46), during which the methyl group is lost by oxidation–decarboxylation.[143,282]

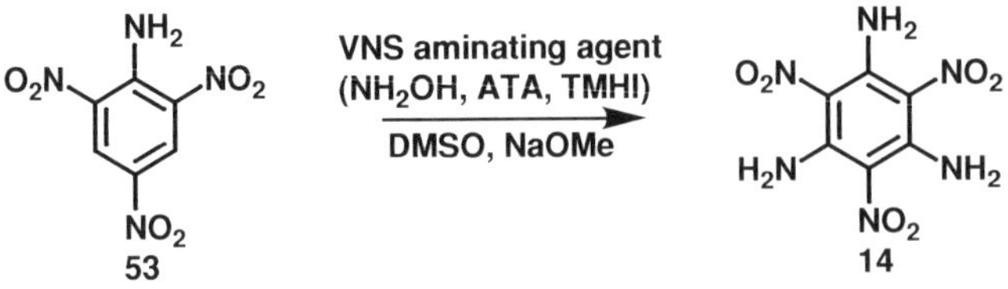

Figure 4.60

Recently, TATB (14) has been synthesized by treating TNB and picramide (53) with VNS aminating agents like 4-amino-1,2,4-triazole (ATA) and 1,1,1-trimethylhydrazinium iodide (TMHI) in the presence of sodium methoxide in DMSO.[37,283] Hydroxylamine has been used but requires elevated temperatures for the same reaction. The main advantage of these methods is the reduction in the production cost of TATB as a result of relatively inexpensive starting materials i.e. picramide (from Explosive D) and TMHI (from UDMH). Explosive D (ammonium picrate) and UDMH are readily available in large quantities in several advanced countries as a result of implementation of demilitarization programmes in these countries.[283]

4.9 THE CHEMISTRY OF 2,4,6-TRINITROTOLUENE (TNT)

2,4,6-Trinitrotoluene (TNT) is a readily available and useful precursor to many other polynitroarylenes. Many of the reactions utilizing TNT in this way make use of the acidity of the methyl group protons. This is itself a consequence of the strengthened hyperconjugation in TNT as a result of the large negative inductive effect generated by having three nitro groups on the aromatic nucleus.

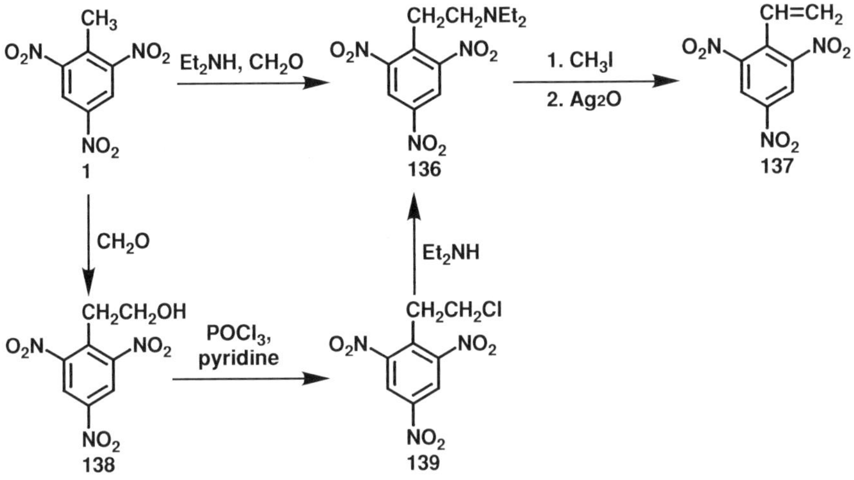

Figure 4.61

The acidity of the methyl group in TNT is illustrated by its ability to participate in Mannich-type reactions; in the same role as a ketone, the methyl group of TNT behaves as a nucleophile under basic conditions and can attack an iminium cation formed from the reaction of formaldehyde with an amine. The Mannich product from one such reaction, the tertiary amine (136) has been used for the synthesis of 2,4,6-trinitrostyrene (137).[284] In this reaction, treatment of the tertiary amine (136) with methyl iodide leads to a quaternary ammonium salt, which on treatment with silver (I) oxide leads to Hoffman elimination and the formation of (137). Urbañski and Bonecki[285] have reported the synthesis of 2,4,6-trinitrostyrene (137) using a very similar strategy; the base-catalyzed condensation of TNT (1) with formaldehyde[286] generates 2,4,6-trinitrophenethyl alcohol (138), which on reaction with phosphorous oxychloride, followed by diethylamine, yields the 2,4,6-trinitrostyrene precursor (136).

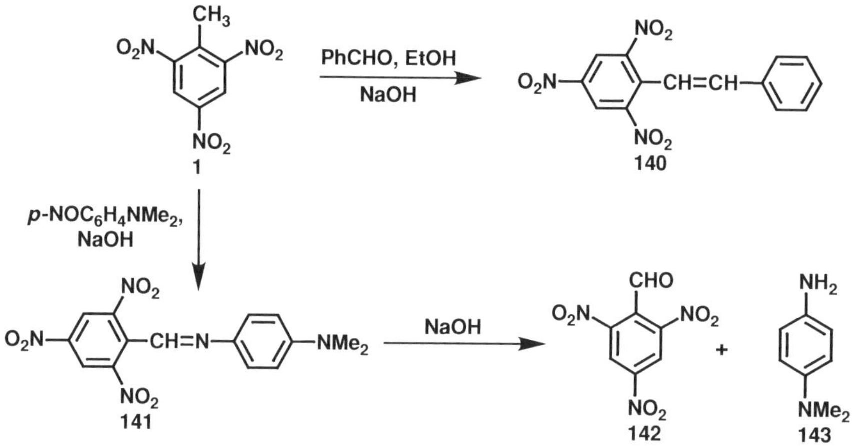

Figure 4.62

The methyl group of TNT will condense with aldehydes in a similar way to other substrates containing acidic protons. TNT (1) reacts exothermically with benzaldehyde in alkaline solution to form 2,4,6-trinitrostilbene (140).[287] 2,4,6-Trinitrobenzaldehyde (142) is synthesized from the base hydrolysis of the imine (141) formed from the condensation of TNT (1) with *p*-nitroso-*N*,*N*-dimethylaniline.[287]

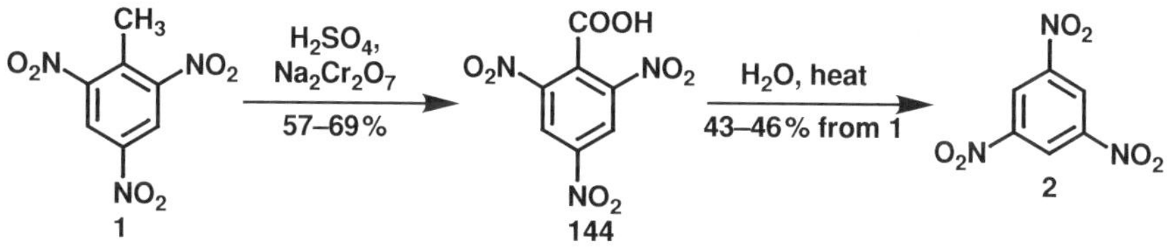

Figure 4.63

A synthetically important route to 1,3,5-trinitrobenzene on both laboratory and industrial scales utilizes TNT as a starting material. Thus, when TNT (1) is treated with an oxidizing mixture of sulfuric acid and sodium dichromate the product formed is 1,3,5-trinitrobenzoic acid (144), which is thermally unstable and loses carbon dioxide on heating as a suspension in boiling water to form 1,3,5-trinitrobenzene (2).[130]

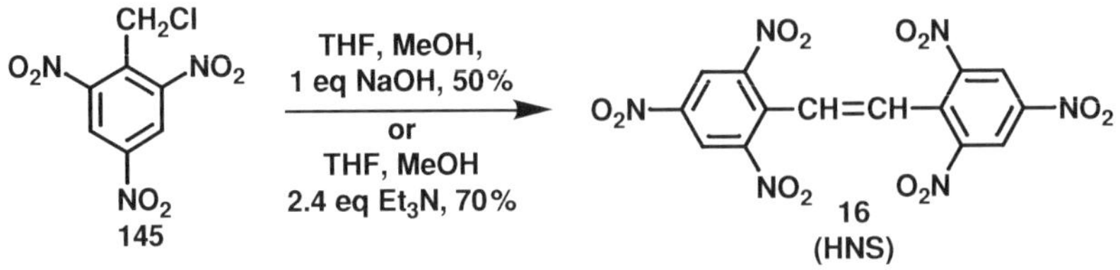

Figure 4.64

2,4,6-Trinitrobenzyl chloride is isolated in 85 % yield if a reaction mixture composed of TNT and 5 % aqueous sodium hypochlorite in methanol at 0 °C is quenched into dilute acid after 1 minute; longer reaction times at ambient temperature lead to the isolation of the heat resistant explosive 2,2′,4,4′,6,6′-hexanitrostilbene (HNS) in 42 % yield.[288] HNS (16) is also formed in 50 % yield from the reaction of 2,4,6-trinitrobenzyl chloride (145) with sodium hydroxide in THF–methanol.[288,289] Sollot[290] later found that the yield of 2,2′,4,4′,6,6′-hexanitrostilbene (16) could be increased to 70 % by treating the same substrate with 2.4 mole equivalents of triethylamine in the same solvent mixture.

The oxidative coupling of TNT with hypochlorite was shown to involve 2,2′,4,4′,6,6′-hexanitrodibenzyl as an intermediate, which can be isolated in 79 % yield by stopping the reaction at an early stage.[288,291] Treating TNT with a mixture of anhydrous copper (II) sulfate–pyridine complex in the presence of potassium hydroxide in ethanol–triglyme is also reported to yield 2,2′,4,4′,6,6′-hexanitrodibenzyl in 56 % yield.[292] Golding and Hayes[293] later investigated the oxidative coupling of TNT in the presence of base, oxygen and various transition metal catalysts.

TNT has been used as a starting material for the synthesis of 2,3,4,5,6-pentanitroaniline[143,282] (see Section 4.8.4), and hence, for the synthesis of hexanitrobenzene[39,139,153] via oxidation with peroxydisulfuric acid, and 1,3,5-triamino-2,4,6-trinitrobenzene (TATB) via nucleophilic displacement with ammonia.[140,143]

4.10 CONJUGATION AND THERMALLY INSENSITIVE EXPLOSIVES

Some high-molecular weight aromatic nitro compounds have been recognized for their high thermal stability. The effect on thermal stability of having amino functionality adjacent to nitro groups is discussed in Section 4.8.1.4. Conjugation between aromatic rings is also known to increase thermal stability in explosives.[251]

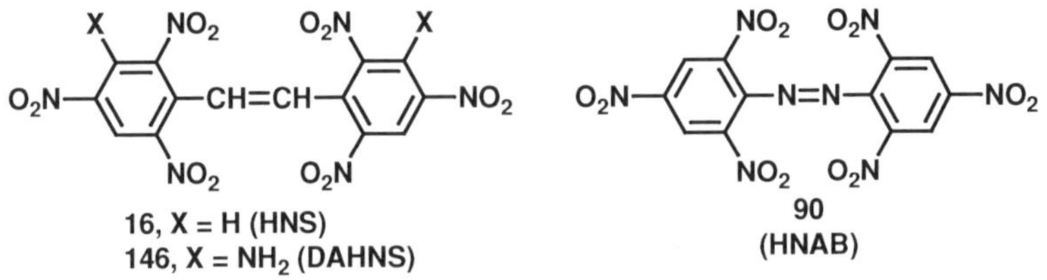

Figure 4.65

2,2′,4,4′,6,6′-Hexanitrostilbene (HNS, m.p. 316 °C, $d = 1.74$ g/cm^3, VOD ~ 7000 m/s) (16) shows higher than expected thermal stability due to conjugation between the aromatic rings. HNS finds wide application as a heat-resistant explosive and was used on the Apollo spaceship for stage separation.[294] The Atomic Weapons Research Establishment at Aldermaston in the UK has developed a plastic bonded explosive composed of HNS/Kel-F800 95/5 for specialized applications. HNS is synthesized industrially via the method of Shipp and Kaplan[288] where TNT undergoes oxidative coupling on treatment with sodium hypochlorite (see Section 4.9). Introduction of two amino groups into HNS in the 3- and 3′-positions gives 3,3′-diamino-2,2′,4,4′,6,6′-hexanitrostilbene (DAHNS) (146), an explosive with even higher thermal stability.[241]

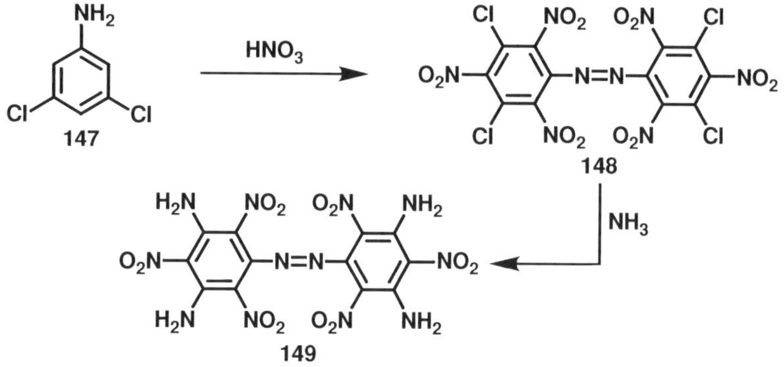

Figure 4.66

The conjugation in 2,2′,4,4′,6,6′-hexanitroazobenzene (HNAB) (90) is also reflected in its thermal stability (m.p. 220 °C). The synthesis of HNAB from picryl chloride and 2,4-dinitrochlorobenzene is discussed in Sections 4.8.1.2 and 4.8.1.3 respectively. 3,3′,5,5′-Tetraamino-2,2′,4,4′,6,6′-hexanitroazobenzene (149) has been synthesized by an unusual but efficient route which involves the nitration–oxidative coupling of 3,5-dichloroaniline (147) on treatment with nitric acid, followed by reaction of the resulting product, 3,3′,5,5′-tetrachloro-2,2′,4,4′,6,6′-hexanitroazobenzene (148), with ammonia.[295] Both the tetrachloro (148) and tetraamino (149) derivatives exhibit high thermal stability.

Agrawal and co-workers[296] have reported the synthesis of *N*,*N*′-bis(1,2,4-triazol-3-yl)-4,4′-diamino-2,2′,3,3′,5,5′,6,6′-octanitroazobenzene (17) (BTDAONAB) via nitration–oxidative coupling of 4-chloro-3,5-dinitroaniline (152) followed by nucleophilic displacement of the chloro groups with 3-amino-1,2,4-triazole. BTDAONAB has the unique distinction of being the most thermally stable explosive reported so far (DTA exotherm ~ 550 °C) as compared to well known thermally stable explosives such as TATB (~ 360 °C), TACOT (~ 410 °C), NONA (~ 440–450 °C), and PYX (~ 460 °C).

Amino derivatives of polynitrobiphenyls often exhibit high thermal stability. The thermally insensitive explosive (156) is synthesized in three synthetic steps from 3,5-dimethoxychlorobenzene (154) in a route employing nitration, Ullmann coupling and ammonolysis.[297] 3,3′-Diamino-2,2′,4,4′,6,6′-hexanitrobiphenyl (DIPAM) (157) is synthesized from 3-chloroanisole by a similar route.[298] DIPAM (m.p. 304 °C, $d = 1.82$ g/cm^3) is

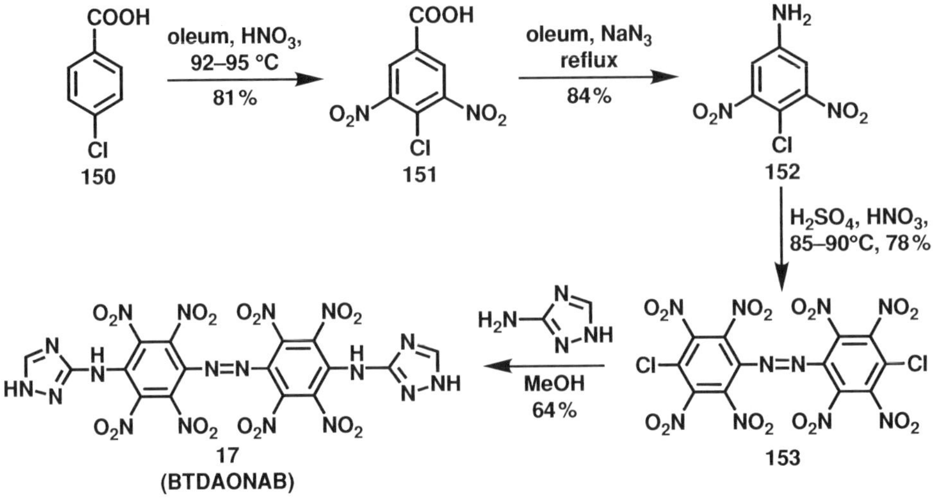

COOH
Cl
150
oleum, HNO_3,
92–95 °C
81 %
COOH
O_2N NO_2
Cl
151
oleum, NaN_3
reflux
84 %
NH_2
O_2N NO_2
Cl
152
H_2SO_4, HNO_3,
85–90°C, 78 %
O_2N NO_2 O_2N NO_2
Cl –N=N– Cl
O_2N NO_2 O_2N NO_2
153
H_2N N N N-NH
MeOH
64 %
O_2N NO_2 O_2N NO_2
N HN-N H N –N=N– H N N N-NH
O_2N NO_2 O_2N NO_2
17
(BTDAONAB)

Figure 4.67

Cl
MeO OMe
154
1. H_2SO_4, HNO_3
2. Cu powder,
xylene, reflux
NO_2
MeO OMe
O_2N NO_2
O_2N NO_2
MeO OMe
NO_2
155
NH_3, MeOH
NO_2
H_2N NH_2
O_2N NO_2
O_2N NO_2
H_2N NH_2
NO_2
156

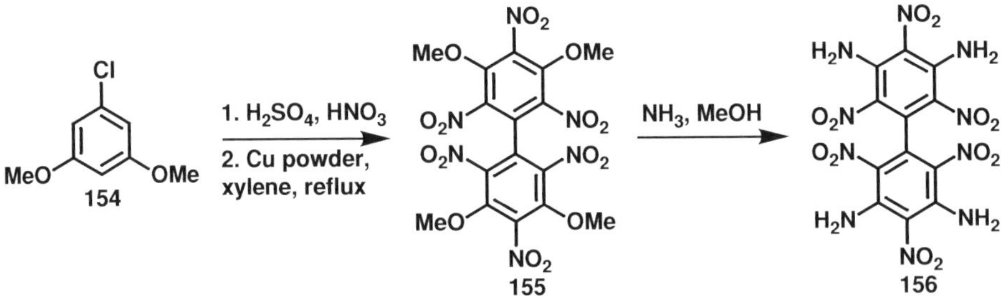

Figure 4.68

extremely insensitive to electrostatic discharge and has been used for seismic experiments on the moon.[299]

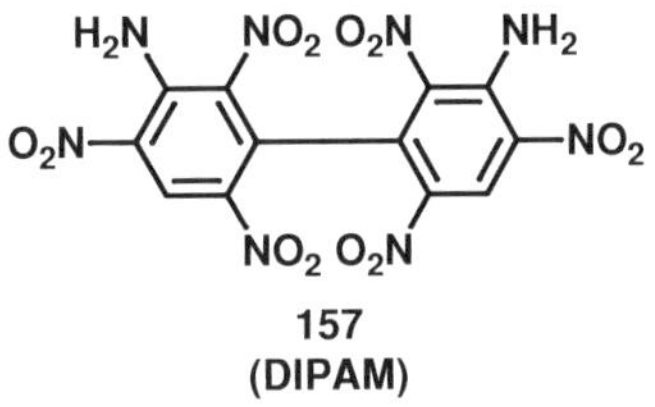

H_2N NO_2 O_2N NH_2
O_2N– –NO_2
NO_2 O_2N
157
(DIPAM)

Figure 4.69

High molecular weight often results in an increase in thermal stability, probably from the increase in melting point – decomposition is much more rapid in a melt than in the solid phase. 2,2′,2″,4,4′,4″,6,6′,6″-Nonanitro-*m*-terphenyl (NONA) (158) is synthesized from the Ullman coupling of styphnyl chloride (106) with two equivalents of picryl chloride (87).[300] NONA exhibits extremely high thermal stability (m.p. 440–450 °C with decomposition).

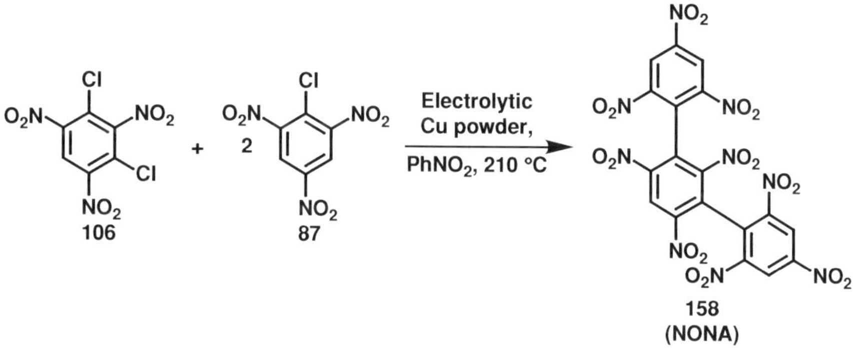

Figure 4.70

Polynitropolyphenylene (PNP) (159), which is a polymeric explosive, exhibits high thermal stability and possesses a low sensitivity to friction and impact. This polymeric mixture, synthesized[301] from the reaction of styphnyl chloride (106) with copper powder in nitrobenzene, has found use as a thermally stable binder in pyrotechnic compositions.[302,303]

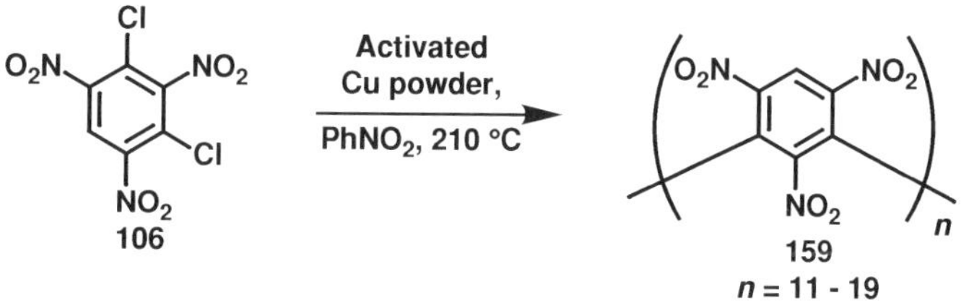

Figure 4.71

Similar routes have been used for the synthesis of other polynitrophenylenes. 1,3,5-Tris(2,4,6-trinitrophenyl)-2,4,6-trinitrobenzene (160) is synthesized from the reaction of 1,3,5-trichloro-2,4,6-trinitrobenzene with three equivalents of picryl chloride in the presence of activated copper powder in refluxing mesitylene.[304] 2,2″,4,4′,4″,6,6′,6″-Octanitro-*m*-terphenyl (161) has been synthesized from picryl chloride and 1,3-dichloro-4,6-dinitrobenzene.[305]

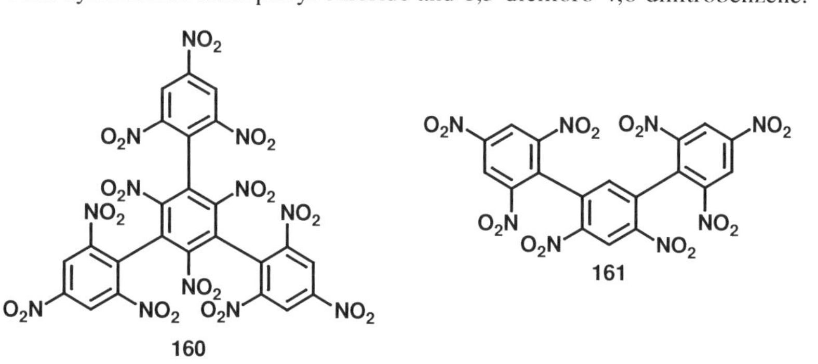

Figure 4.72

REFERENCES

1. G. A. Olah, R. Malhotra and S. C. Narang, *Nitration: Methods and Mechanisms*, Wiley-VCH, Weinheim (1989).
2. N. Ono, *The Nitro Group in Organic Synthesis, Organic Nitro Chemistry Series*, Wiley-VCH, Weinheim (2001).
3. *The Chemistry of the Nitro and Nitroso Groups, Part 1, Organic Nitro Chemistry Series*, Ed. H. Feuer, Wiley-Interscience, New York (1969).
4. *The Chemistry of the Nitro and Nitroso Groups, Part 2, Organic Nitro Chemistry Series*, Ed. H. Feuer, Wiley-Interscience, New York (1970).
5. *Nitro Compounds: Recent Advances in Synthesis and Chemistry, Organic Nitro Chemistry Series*, Eds. H. Feuer and A. T. Neilsen, Wiley-VCH, Weinheim (1990).
6. *Houben-Weyl, Methoden der Organischen Chemie, Band 10/1* (1971) and *E16D*/1 (1992), Ed. E. Muller, George Thieme Verlag, Stuttgart.
7. A. V. Topchiev, *Nitration of Hydrocarbons and Other Organic Compounds*, Translated from Russian by C. Matthews, Pergamon Press, London (1959).
8. P. De la Mare and J. H. Ridd, *Aromatic Substitution: Nitration and Halogenation*, Butterworth, London (1959).
9. K. Schofield, *Aromatic Nitration*, Cambridge University Press, Cambridge (1980).
10. J. G. Hoggett, R. B. Moodie, J. R. Penton and K. Schofield, *Nitration and Aromatic Reactivity*, Cambridge University Press, Cambridge (1971).
11. *Industrial and Laboratory Nitrations, ACS Symposium Series 22*, Eds. L. F. Albright and C. Hanson, American Chemical Society, Washington, DC (1976).
12. *Nitration: Recent Laboratory and Industrial Developments, ACS Symposium Series 623*, Eds. L. F. Albright, R. V. C. Carr and R. J. Schmitt, American Chemical Society, Washington, DC (1996).
13. (a) T. Urbański, *Chemistry and Technology of Explosives, Vol. 1*, Pergamon Press, Oxford (1964); (b) T. Urbański, *Chemistry and Technology of Explosives, Vol. 4*, Pergamon Press, Oxford (1984).
14. C. K. Ingold, *Structure and mechanism in Organic Chemistry, 2nd Edn*, Cornell University Press, Ithaca, New York (1969).
15. E. D. Hughes, C. K. Ingold and R. I. Reed, *J. Chem. Soc.*, 1950, 2400–2473.
16. C. K. Ingold and D. J. Millen, *J. Chem. Soc.*, 1950, 2612.
17. E. D. Hughes, C. K. Ingold and R. B. Pearson, *J. Chem. Soc.*, 1958, 4357.
18. J. Chedin and J. C. Pradier, *Hebd. Seanc. Acad. Sci (Paris)*, 1936, **203**, 722.
19. C. Hanson, M. W. T. Pratt and M. Sohrabi, in *Industrial and Laboratory Nitrations, ACS Symposium Series 22*, Ed. L. F. Albright and C. Hanson, American Chemical Society, Washington, DC, Chapter 15, 225–242 (1976).
20. J. J. Jacobs Jr, J. F. Levy and D. F. Othmer, *Ind. Eng. Chem.*, 1942, **34**, 286.
21. T. L. Davis, *Chemistry of Powder and Explosives, Coll. Vol.*, Angriff Press, Hollywood, CA, 161–162 (reprinted 1992, first printed 1943).
22. A. G. Vogel, *Vogel's Textbook of Practical Organic Chemistry, 3rd Edn*, Longmans, Experiment IV-110, 649 (1956).
23. T. Urbański, *Chemistry and Technology of Explosives, Vol. 1*, Pergamon Press, Oxford, 'Manufacture of Picric acid', Chapter XIV, 499–523 (1964).
24. (a) J. J. Blanksma, *Recl. Trav. Chim. Pays-Bas*, 1902, **21**, 256; (b) L. F. v Duin, *Recl. Trav. Chim. Pays-Bas*, 1920, **39**, 145.
25. T. Urbański, *Chemistry and Technology of Explosives, Vol. 1*, Pergamon Press, Oxford, 540–541 (1964).
26. (a) M. L. Nichols and W. R. Orndorff, *J. Am. Chem. Soc.*, 1923, **45**, 1536; (b) B. A. Bydal, *Org.*

Prep. Proc. Int., 1973, **5**, 271; (c) A. Gay-Lussac and J. Meniger, *Mém. Poudres*, 1958, **40**, 7; *Chem. Abstr.*, 1961, **55**, 3489.
27. (a) T. Kametani and K. Ogasawara, *Chem. Pharm. Bull*, 1967, **15**, 893; (b) T. Urbański, in *Reagents for Organic Synthesis, Vol. 3*, Eds. L. F. Fieser and M. Fieser, Wiley-Interscience, New York, 212 (1972).
28. J. P. Agrawal, Mehilal, A. K. Sikder, N. Sikder and D. V. Survase, *Ind. J. Eng. Mater. Sci.*, 2004, **11**, 59.
29. R. L. Atkins, A. A. Defusco and A. T. Nielsen, *Org. Prep. Proc. Int.*, 1982, **14**, 393; *US Pat.* 4 434 304 (1984).
30. A. J. Bellamy, P. Golding and S. J. Ward, *Propell. Explos. Pyrotech.*, 2002, **27**, 49.
31. (a) F. Moll and R. Nietzki, *Chem. Ber.*, 1893, **26**, 2185; (b) I. Erni, K. Hegetschweiler and W. Schneider, *Helv. Chim. Acta.*, 1990, **73**, 97.
32. (a) D. A. Salter and R. J. J. Simkins, *US Pat.* 3 933 926 (1976); (b) K. Freudenberg, H. Fikentscher and W. Wenner, *Liebigs Ann. Chem.*, 1925, **442**, 322.
33. (a) T. Urbański, *Chemistry and Technology of Explosives, Vol. 1*, Pergamon Press, Oxford, 547–548 (1964); (b) H. Martinsen, *Z. Phys. Chem.*, 1905, **50**, 425; 1907, **59**, 605.
34. (a) E. Y. Spencer and G. F. Wright, *Can. J. Res.*, 1946, **24B**, 204; (b) A. F. Hollemann, *Recl. Trav. Chim. Pays-Bays*, 1930, **49**, 112.
35. W. Borsche and D. Rantscheff, *Liebigs Ann. Chem.*, 1911, **379**, 161.
36. O. N. Witt and E. Witte, *Chem. Ber.*, 1908, **41**, 3090.
37. A. R. Mitchell, P. F. Pagoria and R. D. Schmidt, *US Pat.* 5 569 783 (1996); 5 633 406 (1997); 6 069 277 (2000).
38. (a) L. A. Kaplan, *US Pat.* 3 062 885 (1962); *Chem. Abstr.*, 1963, **58**, 5572b; (b) W. Coulson and A. Forster, *J. Chem. Soc.*, 1922, 1988; (c) C. F. V. v. Duin, *Recl. Trav. Chim. Pays-Bas*, 1917, **37**, 111; (d) B. Flürschiem and T. Simon, *J. Chem. Soc.*, 1910, 81; (e) B. Flürschiem and T. Simon, *Chem. Ind (London)*, 1921, 97.
39. R. L. Atkins, C. Bergens, A. T. Nielsen and W. S. Wilson, *J. Org. Chem.*, 1984, **49**, 503.
40. (a) T. Urbański, *Chemistry and Technology of Explosives, Vol. 3*, Pergamon Press, Oxford, 41–46 (1967); (b) H. H. Hodgson and J. Turner, *J. Chem. Soc.*, 1942, 584; (c) A. Semeńczuk and T. Urbański, *Bull. Acad. Polon. Sci., Cl. III*, 1957, **5**, 6491; 1958, **6**, 309.
41. (a) P. A. Dame and E. J. Hoffman, *J. Am. Chem. Soc.*, 1919, **41**, 1013; (b) *US Pat.* 1 326 947 (1920); *Chem. Abstr.*, 1920, **14**, 633.
42. W. G. Blucher, C. L. Coon, M. E. Hill, C. N. Marynowski, G. J. McDonald, H. M. Peters, D. L. Ross, R. L. Simon and W. Tolberg, in *Industrial and Laboratory Nitrations, ACS Symposium Series 22*, Eds. L. F. Albright and C. Hanson, American Chemical Society, Washington, DC, Chapter 17, 253–271 (1976).
43. B. T. Pennington and A. B. Quakenbush, (Olin Corporation) in *Nitration: Recent Laboratory and Industrial Developments, ACS Symposium Series 623*, Eds. L. F. Albright, R. V. C. Carr and R. J. Schmitt, American Chemical Society, Washington, DC, Chapter 19, 214–222 (1996).
44. (a) T. Urbański, *Chemistry and Technology of Explosives, Vol. 1*, Pergamon Press, Oxford, 398 (1964); (b) P. Aubertein and H. Lecorche, *Mém. Poudres*, 1948, **30**, 85; *Chem. Abstr.*, 1952, **46**, 919a.
45. L. Desvergnes, *Chim. Ind (Paris)*, 1931, **23**, 291.
46. (a) A. A. Pollitt and L. G. Radcliffe, *Chem. Ind (London)*, 1921, 45T; (b) P. Hepp, *Liebigs Ann. Chem.*, 1882, **215**, 316; (c) J. Simon and T. Urbański, *Roczniki. Chem.*, 1939, **19**, 487; (d) J. C. Drummond, *Chem. Ind (London)*, 1922, 338.
47. H. C. Lin and G. A. Olah, *Synthesis*, 1974, 444.
48. H. Adkins and L. B. Haines, *J. Am. Chem. Soc.*, 1925, **47**, 1419.
49. H. McCormack, *Ind. Eng. Chem.*, 1937, **29**, 1333; *Chem. Abstr.*, 1947, **41**, 153.
50. (a) L. Desvergnes, *Chimie et Industrie*, 1931, **25**, 3; *Chem. Abstr.*, 1931, **25**, 2699.

51. P. F. Frankland and F. H. Garner, *Chem. Ind (London)*, 1920, 259.
52. (a) N. Picton and J. J. Sudborough, *J. Chem. Soc.*, 1906, 589; (b) Nietzki and Schedler, *Chem. Ber.*, 1897, **30**, 1666; (c) *Beilstein, Handbuch der Organischen Chemie, 4th Edn, Vol. 13*, Spinger, Berlin, 60.
53. (a) T. M. Benziger, *International Annual Conference of ICT*, Karlsruhe, Germany, July 1–3, 1981, 491–493; (b) T. M. Benziger and R. K. Rohiver, 'Pilot Plant Production of TATB', *LASL-3632 (1973)*, Los Alamos National Laboratory, Los Alamos, New Mexico.
54. C. L. Jackson and P. S. Smith, *J. Am. Chem. Soc.*, 1910, **32**, 168.
55. H. G. Adolph, W. M. Koppes, G. W. Lawrence and M. E. Sitzmann, *J. Chem. Soc. Perkin Trans. 1*, 1981, 1815.
56. D. L. Seaton and G. C. Shaw, *J. Org. Chem.*, 1961, **26**, 5227.
57. H. H. Hodgson and E. R. Ward, *J. Chem. Soc.*, 1946, 533.
58. L. A. Day, C. D. Johnson and E. R. Ward, *J. Chem. Soc.*, 1959, 487.
59. C. D. Johnson and E. R. Ward, *J. Chem. Soc.*, 1961, 4314.
60. L. M. Casilio, E. J. Fendler and J. H. Fendler, *J. Org. Chem.*, 1971, **36**, 1750.
61. H. H. Hodgson, A. P. Mahadevan and E. R. Ward, in *Organic Syntheses, Coll. Vol. III*, Ed. E. C. Horning, John Wiley & Sons, Inc., New York, 341 (1955).
62. H. H. Hodgson, A. P. Mahadevan and E. R. Ward, *J. Chem. Soc.*, 1947, 1392.
63. H. H. Hodgson and E. Marsden, *J. Chem. Soc.*, 1944, 22.
64. H. H. Hodgson, F. Heyworth and E. R. Ward, *J. Chem. Soc.*, 1948, 1512.
65. J. G. Hawkins, G. D. Johnson and E. R. Ward, *J. Chem. Soc.*, 1960, 894.
66. J. Chatt and W. P. Wynne, *J. Chem. Soc.*, 1943, 33.
67. R. Q. Brewster, R. Phillips and B. Williams, in *Organic Syntheses, Coll. Vol. III*, Ed. E. C. Horning, John Wiley & Sons, Inc., New York, 337 (1955).
68. P. H. Griffiths, W. A. Walkey and H. B. Watson, *J. Chem. Soc.*, 1934, 631.
69. F. Arnall and T. Lewis, *Chem. Ind (London)*, 1958, 2987.
70. R. O. C. Norman and G. K. Radda, *J. Chem. Soc.*, 1961, 3030.
71. K. Halvarson and L. Melander, *Arkiv. Kemi.*, 1957, **11**, 77.
72. W. M. Weaver, in *The Chemistry of the Nitro and Nitroso Groups, Part 2, Organic Nitro Chemistry Series*, Ed. H. Feuer, Wiley-Interscience, New York, 'Introduction of the Nitro Group into Aromatic Systems', Chapter 1, 1–48 (1970).
73. J. J. Hiller. Jr and P. Kovacic, *J. Org. Chem.*, 1965, **30**, 2871.
74. (a) E. Bamberger and K. Landsteiner, *Chem. Ber.*, 1893, **26**, 485; (b) E. Bamberger, *Chem. Ber.*, 1894, **27**, 359, 384; 1895, **28**, 399; 1897, **30**, 1248.
75. K. Schofield, *Aromatic Nitration*, Cambridge University Press, Cambridge, 240–245 (1980).
76. H. M. Fitch, in *Organic Syntheses, Coll. Vol. III*, Ed. E. C. Horning, John Wiley & Sons, Inc., New York, 658 (1955).
77. R. J. Gillespie and D. G. Norton, *J. Chem. Soc.*, 1953, 971.
78. S. J. Kuhn and G. A. Olah, in *Friedel-Crafts and Related Reactions, Vol. 2*, Ed. G. A. Olah, Wiley-Interscience, New York (1964).
79. M. R. Draper and J. H. Ridd, *J. Chem. Soc. Perkin Trans. II*, 1981, 94.
80. H. Martinsen, *Z. Phys. Chem.*, 1905, **50**, 425; 1907, **59**, 605.
81. (a) R. B. Moodie, K. Schofield and P. N. Thomas, *J. Chem. Soc. Perkin Trans. II*, 1978, 318; (b) A. Hantzsch, *Chem. Ber.*, 1925, **58**, 941; *Z. Phys. Chem.*, 1930, **149**, 161.
82. R. J. Gillsepie and D. J. Millen, *Chem. Soc. Quart. Rev.*, 1948, **2**, 277.
83. (a) R. B. Moodie, K. Schofield and A. R. Wait, *J. Chem. Soc. Perkin Trans. II*, 1984, 921; (b) M. P. DiGiaimo, A. P. Paul and S. M. Tsang, *J. Org. Chem.*, 1964, **29**, 3387.
84. (a) H. C. Brown and R. A. Wirkkali, *J. Am. Chem. Soc.*, 1966, **88**, 1447; (b) R. B. Moodie, K. Schofield and G. D. Tobin, *J. Chem. Soc. Perkin Trans. II*, 1977, 1688.
85. H. C. Lin and G. A. Olah, *J. Am. Chem. Soc.*, 1974, **96**, 549.
86. W. G. Blucher, C. L. Coon and M. E. Hill, *J. Org. Chem.*, 1973, **38**, 4243.

87. J. W. Barnett, R. B. Moodie, K. Schofield, P. G. Taylor and J. B. Weston, *J. Chem. Soc. Perkin Trans. I*, 1979, 747.
88. R. Malhotra, S. C. Narang and G. A. Olah, *J. Org. Chem.*, 1978, **43**, 4628.
89. (a) R. Aniszfeld, G. A. Olah, G. K. Surya Prakash and R. Rasul, *J. Am. Chem. Soc.*, 1992, **114**, 5608; (b) A. Burrichter, G. A. Olah, G. K. Surya Prakash and G. Rasul, in *Nitration: Recent Laboratory and Industrial Developments, ACS Symposium Series 623*, Eds. L. F. Albright, R. V. C. Carr and R. J. Schmitt, American Chemical Society, Washington, DC, Chapter 2, 10–18 (1996).
90. (a) R. D. Chapman, R. D. Gilardi, C. B. Kreutzberger and M. F. Welker, *J. Org. Chem.*, 1999, **64**, 960; (b) T. Axenrod, R. D. Chapman, R. D. Gilardi, X. -P. Guan, L. Qi and J. Sun, *Tetrahedron Lett*, 2001, **42**, 2621.
91. (a) W. F. Anzilotti, G. F. Hennion and R. J. Thomas, *Ind. Eng. Chem.*, 1940, **32**, 408; (b) G. F. Hennion, *US Pat.* 2 314 212 (1943).
92. (a) S. J. Kuhn and G. A. Olah, *Chem. Ind (London)*, 1956, 98; (b) S. J. Kuhn, A. Mlinko and G. A. Olah, *J. Chem. Soc.*, 1956, 4257.
93. (a) O. Boeters and R. Wolffenstein, *Chem. Ber.*, 1913, **46**, 586; (b) O. Boeters and R. Wolffenstein, *Ger. Pat.* 194 883 (1906).
94. R. Schramm, E. Segel and F. H. Westheimer, *J. Am. Chem. Soc.*, 1947, **69**, 773.
95. L. M. Stock and T. L. Wright, *J. Org. Chem.*, 1977, **42**, 2875; 1979, **44**, 3467.
96. T. L. Davis, N. L. Drake, R. W. Helmkamp, D. E. Worall and A. M. Young, *J. Am. Chem. Soc.*, 1921, **43**, 594.
97. F. Blechta and K. Pátek, *Z. Ges. Schiess-Sprengstoffw*, 1927, **22**, 314.
98. (a) P. Laszlo and J. Vandormael, *Chem. Lett*, 1988, 1843; (b) G. A. DeBoos, A. Musson and K. Smith, *J. Org. Chem.*, 1998, **63**, 8448; (c) A. Fischer, J. Packer, J. Vaughan and G. J. Wright, *J. Chem. Soc.*, 1964, 3687; (d) Y. Mizuno and O. Simamura, *J. Chem. Soc.*, 1958, 3875; (e) J. R. Knowles, R. O. C. Norman and G. K. Radda, *J. Chem. Soc.*, 1960, 4885.
99. (a) J. H. Robson, *J. Am. Chem. Soc.*, 1955, **77**, 107; (b) J. Reinhart and J. H. Robson, *J. Am. Chem. Soc.*, 1955, **77**, 2453; (c) E. E. Gilbert, J. R. Leccacorvi and M. Warman, in *Industrial and Laboratory Nitrations, ACS Symposium Series 22*, Eds. L. F. Albright and C. Hanson, American Chemical Society, Washington, DC, Chapter 23, 327–340 (1976); (d) E. S. Jessop, A. R. Mitchell and P. F. Pagoria, *Propell. Explos. Pyrotech.*, 1996, **21**, 14.
100. (a) V. P. Iushin, M. S. Komelin and V. A. Tartakovskii, *Zh. Org. Khim*, 1999, **35**, 489; (b) D. W. Moore and R. L. Willer, *J. Org. Chem.*, 1985, **50**, 5123; (c) R. L. Willer, *US Pat.* 4 539 405; *Chem. Abstr.*, 1986, **104**, 91609k.
101. S. J. Kuhn and G. A. Olah, in *Friedel-Crafts and Related Reactions, Vol. 3, Part II*, Ed. G. A. Olah, Wiley-Interscience, New York, 1393–1491 (1964).
102. G. A. Olah, in *Industrial and Laboratory Nitration, ACS Symposium Series 22*, Eds. L. F. Albright and C. Hanson, American Chemical Society, Washington, DC, Chapter 1, 1–47 (1976).
103. B. V. Gidaspov, E. L. Golod, Yu. V. Guk and M. A. Ilyushin, *Russ. Chem. Rev.*, 1983, **52**, 284.
104. S. J. Kuhn, *Can. J. Chem.*, 1962, **40**, 1660.
105. I. I. Maletina, V. V. Orda and L. M. Yagupolskii, *Zh. Org. Khim.*, 1974, **10**, 2226; *Engl. Transl.*, 2240.
106. F. Effenberger and J. Geke, *Synthesis*, 1975, 40.
107. H. C. Lin, *Ph.D. Thesis*, Case Western Reserve University, Cleveland, Ohio (1972).
108. (a) A. N. Banyshnikova and A. I. Titov, *Zh. Obshch. Khim*, 1936, **6**, 1800; (b) L. A. Pinck, *J. Am. Chem. Soc.*, 1927, **49**, 2536.
109. A. I. Titov, *Zh. Obshch. Khim*, 1937, **7**, 667.
110. G. B. Bachman and C. M. Vogt, *J. Am. Chem. Soc.*, 1958, **80**, 2987.
111. A. Schaarschmidt, *Chem. Ber.*, 1924, **57**, 2065.
112. (a) K. Maeda, T. Mori and H. Suzuki, *Synthesis*, 1994, 841; (b) T. Mori, H. Suzuki and T. Takeuchi, *J. Org. Chem.*, 1996, **61**, 5944; (c) T. Ishibashi, T. Murashima, H. Suzuki and K. Tsukamoto,

Tetrahedron Lett, 1991, **32**, 6591; (d) K. Maeda, T. Mori, H. Suzuki and S. Yonezawa, *J. Chem. Soc. Perkin Trans. I*, 1994, 1367.
113. (a) T. Mori and H. Suzuki, *Synlett.*, 1995, 383; (b) I. Kozai, T. Murashima, H. Suzuki and A. Tatsumi, *Chem. Lett.*, 1993, 1421.
114. P. E. Eaton, K. Lukin and K. Tani, *J. Am. Chem. Soc.*, 1997, **119**, 1476.
115. (a) H. Grubert and K. L. Rinehart Jr, *Tetrahedron Lett.*, 1959, **1**, 16; (b) J. F. Helling and H. Shechter, *Chem. Ind (London)*, 1959, 1157.
116. N. A. Kudav and M. P. Majumdar, *Ind. J. Chem. Sec. B*, 1976, **14B**, 1012.
117. J. V. Crivello, *J. Org. Chem.*, 1981, **46**, 3056.
118. J. B. Menke, *Rec. Trav. Chim. Pays-Bas*, 1925, **44**, 141, 270; 1928, **48**, 618.
119. M. B. Anderson, L. C. Christie, T. Goetzen, M. C. Guzman, M. A. Hananel, W. D. Kornreich, H. Li, V. P. Pathak, A. K. Rabinovich, R. J. Rajapakse, S. A. Shackelford, L. K. Truesdale, S. M. Tsank and H. N. Vazir, *J. Org. Chem.*, 2003, **68**, 267.
120. (a) P. Laszlo, *Acc. Chem. Res.*, 1986, **19**, 121; (b) P. Laszlo, *Preparative Chemistry Using Supported Reagents*, Academic Press, San Diego (1987); (c) P. Laszlo and J. Vandormael, *Chem. Lett.*, 1988, 1843.
121. (a) W. J. Donaldson and H. R. Wright, *US Pat.* 2 416 974 (1947); *Chem. Abstr.*, 1947, **41**, 3485; (b) E. Plazak and S. Ropuszynski, *Roczniki. Chem.*, 1958, **32**, 681; *Chem. Abstr.*, 1959, **53**, 3111.
122. (a) A. I. Titov, *Zh. Obshch. Khim*, 1948, **18**, 2190; (b) V. P. Alanya, G. S. Shnayder and A. V. Topchiev, *Dokl. Akad. Nauk USSR*, 1954, **195**, 89; (c) H. C. Lin and G. A. Olah, *J. Am. Chem. Soc.*, 1974, **96**, 2892.
123. R. Malhotra, S. C. Narang and G. A. Olah, *J. Org. Chem.*, 1978, **43**, 4628.
124. (a) S. J. Kuhn and G. A. Olah, *J. Am. Chem. Soc.*, 1961, **83**, 4564; (b) C. C. Price and C. A. Sears, *J. Am. Chem. Soc.*, 1953, **75**, 3276; (c) G. A. Olah, P. Ramaiah, G. Sandford, A. Orlinkov and G. K. Surya Prakash, *Synthesis*, 1994, 468.
125. (a) T. C. Bruice, M. J. Gregory and S. L. Walters, *J. Am. Chem. Soc.*, 1968, **90**, 1612; (b) J. F. Riordan, M. Sokolovsky and B. L. Vallee, *J. Am. Chem. Soc.*, 1966, **88**, 4104.
126. (a) G. A. Olah, J. A. Olah and N. A. Overchuk, *J. Org. Chem.*, 1965, **30**, 3373; (b) C. A. Cupas and R. L. Pearson, *J. Am. Chem. Soc.*, 1968, **90**, 4742; (c) C. A. Cupas, S. C. Narang, G. A. Olah, J. A. Olah and R. L. Pearson, *J. Am. Chem. Soc.*, 1980, **102**, 3507; (d) A. P. Fung, S. C. Narang and G. A. Olah, *J. Org. Chem.*, 1981, **46**, 2706.
127. D. V. Nightingale, *Chem. Rev*, 1947, **40**, 117.
128. C. Hanson, T. Kaghazchi and M. W. T. Pratt, in *Industrial and Laboratory Nitration, ACS Symposium Series 22*, Eds. L. F. Albright and C. Hanson, American Chemical Society, Washington, DC, Chapter 8, 132–155 (1976).
129. N. A. Kirshen and D. S. Ross, in *Industrial and Laboratory Nitration, ACS Symposium Series 22*, Eds. L. F. Albright and C. Hanson, American Chemical Society, Washington, DC, Chapter 7, 114–131 (1976).
130. (a) H. T. Clarke and W. W. Hartman, in *Organic Syntheses, Coll. Vol. 1*, Ed. H. Gilman, John Wiley & Sons, Inc., New York, 541 and 543 (1941).
131. M. Giua, *Gazz. Chim. Ital.*, 1915, **45[I]**, 348, 557; 1915, **54[II]**, 306, 351; 1922, **52[I]**, 186.
132. T. L. Davis, *Chemistry of Powder and Explosives, Coll. Vol.*, Angriff Press, Hollywood, CA, 170 (reprinted 1992, first printed 1943).
133. Y. Takayama and Y. Tsubuku, *Bull. Chem. Soc. Japan*, 1942, **17**, 109; *Chem. Abstr.*, 1947, **41**, 4471b.
134. W. M. Weaver, in *The Chemistry of the Nitro and Nitroso Groups, Part 2, Organic Nitro Chemistry Series*, Ed. H. Feuer, Wiley-Interscience, New York, 11–12 (1970).
135. (a) D. Gaude, R. Le Goaller and J. L. Pierre, *Synth. Commun*, 1986, **16**, 63; (b) M. J. Thompson and P. J. Zeegers, *Tetrahedron Lett.*, 1988, **29**, 2471.
136. E. Bamberger and K. Landsteiner, *Chem. Ber.*, 1893, **26**, 485.
137. D. V. Banthorpe, E. D. Hughes and D. L. H. Williams, *J. Chem. Soc.*, 1964, 5349.

138. (a) J. Glazer, E. D. Hughes, C. K. Ingold, A. T. James, G. T. Jones and E. Roberts, *J. Chem. Soc.*, 1950, 2657; (b) E. D. Hughes and G. T. Jones, *J. Chem. Soc.*, 1950, 2678.
139. R. L. Atkins, C. L. Coon, A. T. Nielsen, W. P. Norris and M. E. Sitzmann, *J. Org. Chem.*, 1980, **45**, 2341.
140. B. Flürscheim and E. L. Holmes, *J. Chem. Soc.*, 1928, 3041.
141. R. M. Hainer, G. R. Handrick and W. C. Lothrop, *J. Am. Chem. Soc.*, 1951, **73**, 3581.
142. T. E. Browne, A. A. DeFusco and A. T. Nielsen, *J. Org. Chem.*, 1985, **50**, 4211.
143. R. L. Atkins, R. A. Hollins and W. S. Wilson, *J. Org. Chem.*, 1986, **51**, 3261.
144. R. L. Atkins and W. S. Wilson, *J. Org. Chem.*, 1986, **51**, 2572.
145. A. P. Chafin, S. L. Christian, A. T. Nielsen and W. S. Wilson, *J. Org. Chem.*, 1994, **59**, 1714.
146. E. B. Starkey, in *Organic Syntheses, Coll. Vol. II*, Ed. A. H. Blatt, John Wiley & Sons, Inc., New York, 225 (1943).
147. A. Hardy and E. R. Ward, *J. Chem. Soc.*, 1957, 2634.
148. A. Contardi and G. Körner, *Atti. Accad. Naz. Lincei, Roma [5]*, 1914, **23**, **I**, 633; 1914, **23**, **II**, 464.
149. A. Contardi and G. Körner, *Atti. Accad. Naz. Lincei, Roma [5]*, 1914, **23**, **II**, 466.
150. A. Contardi and G. Körner, *Atti. Accad. Naz. Lincei, Roma [5]*, 1915, **24**, **I**, 891.
151. A. Contardi and G. Körner, *Atti. Accad. Naz. Lincei, Roma [5]*, 1915, **24**, **I**, 345.
152. A. Holleman, *Rec. Trav. Chim. Pays-Bas*, 1930, **49**, 112; 1930, **49**, 501.
153. R. L. Atkins, A. T. Nielsen and W. P. Norris, *J. Org. Chem.*, 1979, **44**, 1181.
154. R. L. Atkins, A. T. Nielsen, W. P. Norris and W. R. Vuono, *J. Org. Chem.*, 1983, **48**, 1056.
155. A. T. Nielsen, *Nitrocarbons*, Wiley-VCH, Weinheim, 97 (1995).
156. W. D. Langley, in *Organic Syntheses, Col Vol. III*, Ed. E. C. Horning, John Wiley & Sons, Inc., New York, 334 (1955).
157. H. H. Hodgson and J. S. Wignall, *J. Chem. Soc.*, 1927, 2216.
158. W. Bohm and A. Kirpal, *Chem. Ber.*, 1932, **65**, 680.
159. E. V. Brown, *J. Am. Chem. Soc.*, 1957, **79**, 3565.
160. W. Bradley and E. Leete, *J. Chem. Soc.*, 1951, 2129.
161. W. Herz and D. R. K. Murty, *J. Org. Chem.*, 1961, **26**, 122.
162. J. L. Hartman and R. H. Wiley, *J. Am. Chem. Soc.*, 1951, **73**, 494.
163. E. E. Gilbert and J. R. Leccacorvi, *Propell. Explos. Pyrotech.*, 1976, **1**, 89.
164. W. D. Emmons, *J. Am. Chem. Soc.*, 1957, **79**, 5528.
165. E. E. Gilbert, *Synthesis*, 1977, 315.
166. W. D. Emmons and R. W. White, *Tetrahedron*, 1962, **17**, 31.
167. W. D. Emmons, *J. Am. Chem. Soc.*, 1954, **76**, 3470.
168. L. I. Khmelnitskii, S. S. Novikov and T. S. Novikova, *Izv. Akad. Nauk USSR, Ser. Khim*, 1962, 516.
169. R. P. Bayer and R. R. Holmes, *J. Am. Chem. Soc.*, 1960, **82**, 3454.
170. J. O. Edwards and K. M. Ibne-Rasa, *J. Am. Chem. Soc.*, 1962, **84**, 763.
171. F. P. Greenspan, *Ind. Eng. Chem*, 1947, **39**, 847.
172. J. D'Ans and A. Kneip, *Chem. Ber.*, 1915, **48**, 1144.
173. W. G. Blucher, C. L. Coon, W. H. Dennis Jr and D. H. Rosenblatt, *J. Chem. Eng. Data*, 1975, **20**, 202.
174. R. Jeyaraman and R. W. Murray, *J. Org. Chem.*, 1985, **50**, 2847.
175. L. Mohan, R. W. Murray and S. N. Rajadhyaksha, *J. Org. Chem.*, 1989, **54**, 5783.
176. R. Jeyaraman, L. Mohan and R. W. Murray, *Tetrahedron Lett*, 1986, **27**, 2335.
177. K. R. Beck Jr, A. E. Moormann and D. L. Zabrowski, *Tetrahedron Lett*, 1988, **29**, 4501.
178. G. B. Bachman and K. G. Strawn, *US Pat.* 3 377 387 (1968).
179. E. Keinan and Y. Mazur, *J. Org. Chem.*, 1977, **42**, 844.
180. W. T. Borden and K. E. Gilbert, *J. Org. Chem.*, 1973, **44**, 659.
181. O. Fischer and I. Frost, *Chem. Ber.*, 1893, **26**, 3083.
182. H. Hock and H. Kropf, *Chem. Ber.*, 1956, **89**, 2436.
183. E. Bamberger and T. Tschirner, *Chem. Ber.*, 1899, **32**, 342.

184. E. Bamberger and T. Tschirner, *Chem. Ber.*, 1898, **31**, 1522.
185. J. H. Boyer, in *Heterocyclic Compounds, Vol. 8*, Ed. R. C. Elderfield, John Wiley & Sons, Inc., New York, 463–522 (1961).
186. J. H. Boyer, in *The Chemistry of the Nitro and Nitroso Groups, Part 1, Organic Nitro Chemistry Series*, Ed. H. Feuer, Wiley-Interscience, New York, 'Methods of Formation of the Nitroso Group and Its Reactions', Chapter 5, 215–300 (1969).
187. E. Bamberger, *Liebigs Ann. Chem.*, 1901, **316**, 281
188. W. D. Langley in *Organic Syntheses, Coll. Vol. III*, Ed. E. C. Horning, John Wiley & Sons, Inc., New York, 334 (1955).
189. K. Bertels, *Chem. Ber.*, 1904, **37**, 2276.
190. A. Angeli and F. Angelico, *Gazz. Chim. Ital*, 1900, **30(II)**, 268.
191. R. Kuhn and W. Van Kalvaren, *Chem. Ber.*, 1938, **71**, 779.
192. C. J. Abshire, P. S. Bailey, R. E. Erickson and A. H. Riebel, *J. Am. Chem. Soc.*, 1960, **82**, 1801.
193. W. Borsche, *Chem. Ber.*, 1923, **56**, 1494.
194. W. Borsche, *Chem. Ber.*, 1923, **56**, 1939.
195. W. Borsche and E. Feske, *Chem. Ber.*, 1926, **59**, 815.
196. M. R. Crampton and M. El. Ghariani, *J. Chem. Soc. (B)*, 1970, 391.
197. W. Borsche and E. Feske, *Chem. Ber.*, 1926, **59**, 683.
198. (a) T. Urbański, *Chemistry and Technology of Explosives, Vol. 1*, Pergamon Press, Oxford, 259 (1964); (b) E. Yu. Orlova, *Chemistry and Technology of High Energy Explosive Substances*, Khimia, Leningrad, 140 (1973).
199. E. Haeffen and A. Hollander, *C. R. Acad. Sci., Ser. III*, 1920, 338.
200. T. Urbański, *Chemistry and Technology of Explosives, Vol. 1*, Pergamon Press, Oxford, 467 (1964).
201. J. Miller and K. W. Wong, *J. Chem. Soc.*, 1965, 5454.
202. J. F. Bunnett and W. D. Merrit. Jr, *J. Am. Chem. Soc.*, 1957, **79**, 5967.
203. A. L. Beckwith, G. D. Leahy and J. Miller, *J. Chem. Soc.*, 1952, 3552.
204. G. P. Briner, M. Liveris, P. G. Lutz and J. Miller, *J. Chem. Soc.*, 1954, 1265.
205. Th. J. de Boer and I. P. Dirkx, in *The Chemistry of the Nitro and Nitroso Groups, Part 1, Organic Nitro Chemistry Series*, Ed. H. Feuer, Wiley-Interscience, New York, 'Activating Effects of the Nitro Group in Aromatic Substitution', Chapter 8, 487–612 (1969).
206. (a) R. Boyer, E. Y. Spencer and G. F. Wright, *Can. J. Res*, 1946, **24B**, 200; (b) B. I. Buzykin, I. E. Moisak, V. V. Nurgatin and G. P. Sharnin, *Zh. Org. Khim.*, 1967, **3**, 82.
207. (a) J. Murto, *Act Chem. Scand*, 1966, **20**, 310; (b) Z. Talik, *Roczniki. Chem.*, 1960, **34**, 917; *Chem. Abstr.*, 1961, **55**, 10434; (c) T. Abe, *Bull. Chem. Soc. Japan.*, 1961, **34**, 21.
208. (a) C. W. L. Bevan and J. Hirst, *J. Chem. Soc.*, 1956, 254; (b) N. F. Levchenko and L. M. Litvinenko, *Zh. Obshch. Khim.*, 1959, **29**, 924; *Chem. Abstr.*, 1960, **54**, 1247; (c) N. F. Levchenko and L. M. Litvinenko, *Zh. Obshch. Khim.*, 1960, **30**, 1673; *Chem. Abstr.*, 1961, **55**, 1520; (d) R. S. Cheshko and L. M. Litvinenko, *Zh. Obshch. Khim.*, 1960, **30**, 3682; *Chem. Abstr.*, 1961, **55**, 19846.
209. (a) R. Foster and R. K. Mackie, *Tetrahedron*, 1962, **18**, 161; (b) Z. Talik, *Roczniki. Chem.*, 1960, **34**, 917; *Chem. Abstr.*, 1961, **55**, 10434.
210. (a) C. W. L. Bevan and J. Hirst, *J. Chem. Soc.*, 1956, 254; (b) W. K. Yan and J. Miller, *J. Chem. Soc.*, 1963, 3492.
211. P. Von Romburgh, *Recl. Trav. Chim. Pays-Bas*, 1883, **2**, 31, 103, 304; 1884, **3**, 392; 1889, **8**, 215.
212. (a) A. G. Carbonit and A. G. Sprengstoff, *Ger. Pat.* 269 826 (1912); 275 037 (1912); 286 543 (1912); (b) D. Twiss, *J. Chem. Soc.*, 1914, 1675; (c) B. M. Ginzburg, V. V. Nurgatin and G. P. Sharnin, *Tr. Kazan. Khim.*; *Chem. Abstr.*, 1975, **82**, 86157j; (d) T. A. Eneikina, V. V. Nurgatin, G. P. Sharnin and L. A. Trutneva, *Izv. Vyssh. Ucheb. Zaved, Khim. Khim, Tekhnol.*, 1974, **17**(12), 1817; *Chem. Abstr.*, 1975, **83**, 9358u.
213. D. M. O'Keefe, 'HNAB: Synthesis and Characterization', *SAND-74-0239 (1976)*, Sandia National Laboratory, Albuquerque, New Mexico.
214. (a) E. Grandmougin and H. Leeman, *Chem. Ber.*, 1906, **39**, 4385; 1908, **41**, 1297; (b) T. Urbański, *Chemistry and Technology of Explosives, Vol. 1*, Pergamon Press, Oxford, 574 (1964).

215. (a) L. Desvergnes, *Mém. Poudres*, 1918–22, **19**, 217, 269; 1931, **25**, 3, 291, 507, 1271; (b) L. Desvergnes, *Chimie et Industrie*, 1931, **25**, 291.
216. E. Wedekind, *Liebigs Ann. Chem.*, 1902, **323**, 246.
217. (a) K. Okoń, *Roczniki. Chem.*, 1958, **32**, 213, 713; (b) K. Okoń, *Bull. Acad. Polon. Sci., Sér. Chim.*, 1958, **6**, 319; (c) R. J. W. Le Fèvre, *J. Chem. Soc.*, 1931, 813.
218. J. M. Emeury and H. Girardon, *Ger. Pat.* 2 221 406 (1973); *Chem. Abstr.*, 1973, **78**, 97304y.
219. R. Dietschy and R. Nietzki, *Chem. Ber.*, 1901, **34**, 55.
220. L. J. Andrews, C. E. Castro and R. M. Keefer, *J. Am. Chem. Soc.*, 1958, **80**, 2322.
221. J. Bielecki and F. Ullmann, *Chem. Ber.*, 1901, **34**, 2180.
222. J. Meyer, *Ger. Pat.* 234 726 (1909).
223. (a) F. E. Condon and J. P. Trivedi, *J. Org. Chem.*, 1971, **36**, 1926; (b) I. F. Falyakhov and G. P. Sharnin, *Izv. Vyssh. Ucheb. Zaved, Khim. Khim, Tekhnol*, 1969, **12**(8), 1057; *Chem. Abstr.*, 1970, **72**, 31346h.
224. (a) Griesheim, *Ger. Pat.* 86 295 (1895); (b) J. Marshall, *Ind. Eng. Chem.*, 1920, **12**, 336; (c) T. Urbański, *Chemistry and Technology of Explosives, Vol. 1*, Pergamon Press, Oxford, 562–566 (1964); (d) T. L. Davis, *Chemistry of Powder and Explosives, Coll. Vol.*, Angriff Press, Hollywood, CA, 184–187 (reprinted 1992, first printed 1943).
225. T. Urbański, *Chemistry and Technology of Explosives, Vol. 1*, Pergamon Press, Oxford, 484–486 and 519–520 (1964).
226. T. Urbański, *Chemistry and Technology of Explosives, Vol. 1*, Pergamon Press, Oxford, 547–548 (1964).
227. (a) A. A. Ashtou, *J. Chem. Educ*, 1963, **40**, 545; (b) C. F. H. Allen and F. B. Wells, in *Organic Syntheses, Col Vol. II*, Ed. A. H. Blatt, John Wiley & Sons, Inc., New York, 221 (1943).
228. (a) T. Urbański, *Chemistry and Technology of Explosives, Vol. III*, Pergamon Press, Oxford, 61–62 (1967); (b) I. G. Farbenindustrie, 'Manufacture of Intermediates for Dyestuffs', *BIOS Final Report No. 986, Part II*; (c) G. Desseigne, *Mém. Poudres*, 1938, **28**, 156.
229. E. Vongerichten, *Chem. Ber.*, 1899, **32**, 2571.
230. (a) T. Zincke, *Liebigs Ann. Chem.*, 1904, **330**, 361; (b) W. Würker and T. Zincke, *Liebigs Ann. Chem.*, 1905, **341**, 365.
231. G. Weisspfennig and T. Zincke, *J. Prakt. Chem.*, 1912, **2**, 85, 211.
232. F. H. Kendall and J. Miller, *J. Chem. Soc. (B)*, 1967, 119.
233. J. Bielecki and F. Ullmann, *Chem. Ber.*, 1901, **34**, 2177.
234. (a) S. Iyer, *Propell. Explos. Pyrotech.*, 1982, **7**, 37; (b) S. Iyer, *J. Energ. Mater*, 1984, **2**, 151; (c) C. D. Hutchinson, R. W. Millar and V. K. Mohan, *Propell. Explos. Pyrotech.*, 1984, **9**, 161.
235. (a) T. R. Gibbs and A. Popolato, *LASL Explosive Property Data*, University California Press, Los Angeles, 38 and 157 (1980); (b) R. Meyer, *Explosives*, Verlag Chimie, Weinheim (1981); (c) S. Zeman, *Thermochim. Acta*, 1993, **216**, 157.
236. A. Bailey, J. M. Bellerby and S. A. Kinloch, *Phil. Trans. R. Soc. Lond.*, 1992, **A339**, 321.
237. V. I. Siele and M. Warman, *J. Org. Chem.*, 1961, **26**, 2997.
238. (a) S. B. Wright, *US Pat.* 3 173 817 (1965); *Chem. Abstr.*, 1965, **62**, 12968g; (b) S. B. Wright, *US Pat.* 3 296 041 (1967); *Chem. Abstr.*, 1967, **66**, 87227p.
239. J. P. Agrawal, Mehilal. R. B. Salunke and A. K. Sikder, *J. Hazard. Mater.*, 2001, **A84**, 117.
240. (a) J. P. Agrawal, Mehilal, U. S. Prasad and R. N. Surve, *Def. Sci. J.*, 1998, **48**, 323; (b) R. L. Atkins, C. L. Coon, R. A. Henry, A. H. Lepine, D. W. Moore, A. T. Nielsen, W. P. Norris, D. V. H. Son and R. J. Spanggord, *J. Org. Chem.*, 1979, **44**, 2499 and references therein.
241. C. X. Lu, *Kogyo. Kayaku.*, 1990, **51**, 275.
242. A. M. Deng, C. X. Lu and Z. S. Lu, *Kogyo. Kayaku.*, 1990, **51**, 281; *Chem. Abstr.*, 1991, **115**, 32128g.
243. M. D. Coburn, B. W. Harris, H. H. Hayden, K. -Y. Lee and M. M. Stinecipher, *Ind. Eng. Chem. Prod. Res. Dev.*, 1986, **25**, 68.
244. (a) M. D. Coburn and T. E. Jackson, *J. Heterocycl. Chem.*, 1968, **5**, 199; (b) M. D. Coburn, *US Pat.* 3 483 211 (1969); *Chem. Abstr.*, 1970, **72**, 55458x.

245. J. P. Agrawal, Mehilal, U. S. Prasad and R. N. Surve, *New. J. Chem.*, 2000, **24**, 583.
246. J. P. Agrawal, V. K. Bapat and R. N. Surve, 'Synthesis, Characterization and Evaluation of Explosive Properties of 5-Picrylamino-1,2,3,4-Tetrazole', *2nd High Energy Materials Conference and Exhibits, IIT*, Madras, 8–10 December, 1998, 403.
247. J. P. Agrawal, V. K. Bapat, R. R. Mahajan, Mehilal and P. S. Makashir, 'A Comparative Study of Thermal and Explosive Behaviour of 5- Picrylamino-1,2,3,4-Tetrazole (PAT) and 5,5′-Styphnylamino-1,2,3,4-tetrazole (SAT)', *International Work-Shop on Unsteady Combustion and Interior Ballistics*, St. Petersburg, Russia, Vol. 1, June 25–30, 2000, 199.
248. J. P. Agrawal, V. K. Bapat, Mehilal, B. G. Polke and A. K. Sikder, *J. Energ. Mater.*, 2000, **18**, 299.
249. I. Dunstan, *Chem. Br.*, 1971, **7**, 62.
250. T. Urbański and S. K. Vasudeva, *J. Sci. Ind. Res.*, 1978, **37**, 250.
251. J. P. Agrawal, *Prog. Energy Combust. Sci.*, 1998, **24**, 1.
252. T. Urbański, *Chemistry and Technology of Explosives, Vol. 1*, Pergamon Press, Oxford, 258 (1964).
253. T. Urbański, *Chemistry and Technology of Explosives, Vol. 1*, Pergamon Press, Oxford, 259 (1964).
254. E. Yu. Orlova, G. M. Shutov, V. L. Zbarskii and V. F. Zhilin, *J. Gen. Chem. USSR*, 1965, **35**, 1363; *Zh. Obshch. Khim.*, 1963, **33**, 3210.
255. T. Urbański, *Chemistry and Technology of Explosives, Vol. 1*, Pergamon Press, Oxford, 326–337 (1964).
256. (a) T. Urbański, *Chemistry and Technology of Explosives, Vol. 1*, Pergamon Press, Oxford, 'TNT Manufacture', Chapter 9, 345–394 (1964); (b) M. H. Muraour, *Bull. Soc. Chim. France*, 1924, **35**, 367.
257. P. T. Izzo, *J. Org. Chem.*, 1959, **24**, 2026.
258. (a) J. F. Bunnett, E. W. Garbisch. Jr and K. M. Pruitt, *J. Am. Chem. Soc.*, 1957, **79**, 385; (b) R. E. Parker and T. O. Read, *J. Chem. Soc.*, 1962, 3149.
259. T. Urbański, *Chemistry and Technology of Explosives, Vol. 1*, Pergamon Press, Oxford, 465 (1964).
260. J. Meisenheimer and E. Patzig, *Chem. Ber.*, 1906, **39**, 2533.
261. A. R. Mitchell, P. F. Pagoria and R. D. Schmidt, *US Pat.* 5 633 406 (1997).
262. (a) M. Makosza and J. Winiarski, *Acc. Chem. Res*, 1987, **20**, 282; (b) M. Makosza and K. Wojciechowski, *Liebigs Ann. Chem.*, 1997, 1805.
263. (a) M. Bialecki and M. Makosza, *J. Org. Chem.*, 1992, **57**, 4784; 1998, **63**, 4878; (b) N. Kawamura and S. Seko, *J. Org. Chem.*, 1996, **61**, 442; (c) N. Kawamura, K. Miyake and S. Seko, *J. Chem. Soc. Perkin Trans. 1*, 1999, 1437.
264. A. R. Mitchell, P. F. Pagoria and R. D. Schmidt, Presented at the 211th American Chemical Society National Meeting, New Orleans, LA, 24–28 March, 1996.
265. (a) S. Iyer, *J. Energ. Mater*, 1984, **2**, 151; (b) A. P. Marchand and G. M. Reddy, *Synthesis*, 1992, 261.
266. I. I. Chervin, A. R. Mitchell, V. V. Rozhkov, R. D. Schmidt and S. A. Shevelev, *J. Org. Chem.*, 2003, **68**, 2498.
267. (a) R. J. Pollitt and B. C. Saunders, *J. Chem. Soc.*, 1965, 4615; (b) R. Foster and C. A. Fyfe, *Tetrahedron*, 1965, **21**, 3363; (c) S. S. Gitis and I. G. L'vovich, *Zh. Obshch. Khim.*, 1964, **34**, 2250; *Chem. Abstr.*, 1964, **61**, 9420.
268. (a) R. Foster, *Nature*, 1955, **176**, 746; (b) T. Abe, *Bull. Chem. Soc. Japan*, 1959, **32**, 339; (c) M. R. Crampton and V. Gold, *J. Chem. Soc. Chem. Commun*, 1965, 549; (d) R. Foster, *Nature*, 1959, **183**, 1042; (e) E. Buncel and E. A. Symons, *Can. J. Chem.*, 1966, **44**, 771.
269. T. Urbański, *Chemistry and Technology of Explosives, Vol. 1*, Pergamon Press, Oxford, 545 (1964).
270. J. Meisenheimer, *Liebigs Ann. Chem.*, 1902, **323**, 205.
271. (a) R. J. W. Le Fevre, S. L. M. Saunders and E. E. turner, *J. Chem. Soc.*, 1927, 1168; (b) D. L. Fox and E. E. Turner, *J. Chem. Soc.*, 1930, 1853; (c) H. E. Ungnade, *Chem. Rev*, 1946, **38**, 405.
272. A. F. Holleman, *Rec. Trav. Chim. Pays-Bas*, 1930, **49**, 50.
273. (a) W. Borsche and E. Feske, *Chem. Ber.*, 1927, **60**, 157; (b) E. T. Borrows, J. C. Clayton, B. A. Hems and A. G. Long, *J. Chem. Soc.*, 1949, 190.

274. A. Bailway, J. M. Bellerby and S. A. Kiwloch, *Phil. Trans. R. Soc. Lond*, 1992, **A339**, 321.
275. B. M. Dodratz, 'Insensitive High Explosive Triaminotrinitrobenzene (TATB): Development and Characterization – 1888 to 1994', Los Alamos National Laboratory Report *LA-13014-H*, (1995); University of California, *Report UC-741*.
276. J. P. Agrawal, *Propell. Explos. Pyrotech.*, 2005, **30**, 316–328.
277. T. M. Benziger and D. G. Ott, *J. Energ. Mater*, 1987, **5**, 343; *US Pat.* 4 952 733 (1990).
278. (a) J. A. Hoffman and C. F. McDonough, *US Pat.* 3 278 604 (1966); (b) J. C. Deacons, M. J. Kamlet and D. C. Washington, *US Pat.* 3 394 183 (1968).
279. A. J. Bellamy, P. Golding, N. W. Mitchell and S. J. Ward, *J. Chem. Res. (S)*, 2002, 412.
280. T. Urbański, *Chemistry and Technology of Explosives, Vol. 1*, Pergamon Press, Oxford, 252 (1964).
281. R. M. Hainer, G. R. Handrick and W. C. Lothrop, *J. Am. Chem. Soc.*, 1951, **73**, 3581.
282. (a) A. C. Farthing and D. G. Parkes, *J. Chem. Soc.*, 1948, 1275; (b) J. B. Cohen and H. D. Dakin, *J. Chem. Soc.*, 1902, 26; (c) R. L. Atkins, A. T. Nielsen and W. P. Norris, *US Pat.* 116 351 (1980); *Chem. Abstr.*, 1981, **94**, 33126Q; (d) R. L. Atkins, C. L. Coon, R. A. Henry, A. H. Lepie, D. W. Moore, A. T. Nielsen, W. P. Norris, R. J. Spaggord and D. V. H. Son, *J. Org. Chem.*, 1979, **44**, 2499.
283. A. R Mitchell, P.F. Pagoria and R.D. Schmidt, 'A New Synthesis of TATB using Inexpensive Starting Materials and Mild Reaction Conditions', *27th International Annual Conference of ICT*, Karlsruhe, Germany, 25–28 June, 1996, 29/1–29/11.
284. L. C. Behr and R. H. Wiley, *J. Am. Chem. Soc.*, 1950, **72**, 1822.
285. Z. Bonecki and T. Urbański, *Bull. Acad. Polon. Ser Chim*, 1961, **9**, 463.
286. E. E. Gilbert, *J. Energ. Mater.*, 1984, **2**, 215.
287. T. Urbański, *Chemistry and Technology of Explosives, Vol. 1*, Pergamon Press, Oxford, 309 (1964).
288. L. A. Kaplan and K. G. Shipp, *J. Org. Chem.*, 1966, **31**, 857.
289. K. G. Shipp, *J. Org. Chem.*, 1964, **29**, 2620.
290. G. P. Sollott, *J. Org. Chem.*, 1982, **47**, 2471.
291. E. E. Gilbert, *Propell. Explos. Pyrotech.*, 1980, **5**, 15.
292. G. Benez, J. Deres, L. Hajos and T. Kompolthy, *Hungarian. Pat.* 9639 (1975); *Chem. Abstr.*, 1976, **84**, 58886u.
293. P. Golding and G. F. Hayes, *Propell. Explos. Pyrotech.*, 1979, **4**, 115.
294. L. J. Bement, Applications of Temperature Resistant Explosives to NASA Missions, in *Proceedings of the Symposium on Thermally Stable Explosives*, Whiteoak, MD, USA, NOL, 1970.
295. A. P. Chafin, S. L. Christian, R. A. Hollins, A. T. Nielsen and W. P. Norris, 'Energetic Materials Research at NWC', *Proc. ADPA Meetings on Compatibility of Plastics and Other Materials with Explosives, Propellants, Pyrotechnics and Processing of Explosives, Propellants and Ingredients*, Long Beach, CA, 1986, 122–125.
296. J. P. Agrawal, Mehilal, A. K. Sikder and N. Sikder, *Ind. J. Eng. Mater. Sci.*, 2004, **11**, 516.
297. A. J. Bell, E. Eadie, R. W. Read, B. W. Skelton and A. H. White, *Aust. J. Chem.*, 1987, **40**, 175.
298. J. C. Decans, L. A. Kaplan and R. E. Oesterling, *US Pat.* 3 404 184 (1968); *Chem. Abstr.*, 1969, **70**, 37444u.
299. T. Urbański, *Chemistry and Technology of Explosives, Vol. 4*, Pergamon Press, Oxford, 206 (1984).
300. J. C. Dacons, *US Pat.* 3 755 471 (1973); *Chem. Abstr.*, 1974, **80**, 49964h.
301. R. Hagel and K. H. Redecker, *Propell. Explos. Pyrotech.*, 1987, **12**, 196.
302. B. Berger, B. Hass, G. Reinhard, *Proc. 26th International Annual Conference of ICT*, Karlsruhe, Germany, 1995, Paper 2.
303. J. P. Agrawal, V. K. Bapat, R. Daniel and R. G. Sarawadekar, 'A Comparative Study of Properties of PNP and NC Based Pyrotechnic Formulations', *Proc. 34th International Annual Conference of ICT*, Karlsruhe, Germany, June 24–27, 2003, p 64/1–64/13.
304. R. S. Riggs, *US Pat.* 4 861 924 (1989).
305. J. C. Dacons, *US Pat.* 3 592 860 (1971); *Chem. Abstr.*, 1971, **75**, 89737g.

5

Synthetic Routes To *N*-Nitro Functionality

5.1 INTRODUCTION

Aromatic and aliphatic nitro compounds are of huge industrial importance and are invaluable intermediates in organic synthesis. This has in part fuelled the enormous amount of research into nitration. Compounds resulting from the nitration of nitrogen are of far less use for mainstream organic synthesis. However, the $N–NO_2$ (*N*-nitro) group is an important 'explosophore' and is present in many energetic materials. Consequently, research into *N*-nitration has been exclusively driven by the use of the products as energetic materials. Some of these compounds are in wide use today as high explosives and ingredients of propellants and are manufactured on an enormous scale.

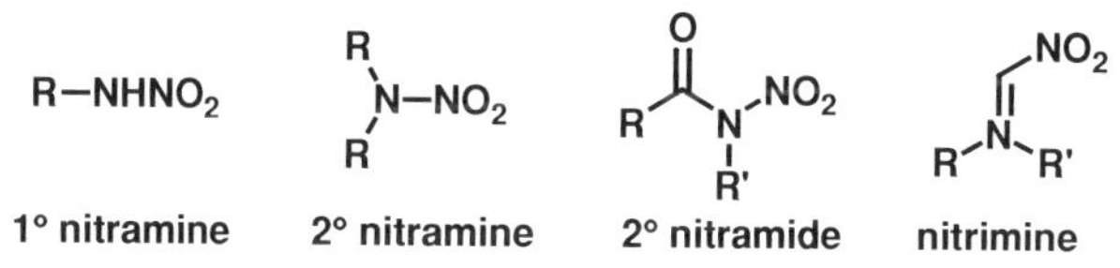

$R–NHNO_2$ — 1° nitramine; $R_2N–NO_2$ — 2° nitramine; 2° nitramide; nitrimine

Figure 5.1

There are four important groups of *N*-nitro compounds which are relevant to energetic materials synthesis.[1] These are primary nitramines, secondary nitramines, secondary nitramides (including *N*-nitroureas and *N,N′*-dinitroureas) and nitrimines. The synthesis and incorporation of these *N*-nitro functionalities into organic compounds is the focus of this chapter.

The replacement of amine and amide hydrogen with a nitro group via direct nitration is an important route to *N*-nitro functionality. However, the cleavage of other bonds is also important. In the case of C–N bond cleavage the process is known as 'nitrolysis' and is an invaluable route to many energetic materials (Section 5.6). The nitrolysis of hexamine and the syntheses of the important explosives HMX and RDX are discussed in Section 5.15. This area of chemistry could easily demand a separate chapter of its own and is the most complex and diverse in the field of nitramine chemistry.

Organic Chemistry of Explosives J. P. Agrawal and R. D. Hodgson

Nitrimines are a relatively minor group of *N*-nitro compound. The nitro derivatives of guanidine and related compounds are the most important examples of the nitrimine group.

Primary nitramines have acidic protons and are able to undergo condensation reactions to form functionalized nitramines. These reactions are discussed in Section 5.13 because the products have potential application as energetic polymer precursors or find use for the synthesis of other explosives.

5.2 NITRAMINES, NITRAMIDES AND NITRIMINES AS EXPLOSIVES

The energetic nature of the N–NO_2 group means that *N*-nitro-based explosives are some of the most powerful explosives available and these have largely superseded aromatic *C*-nitro compounds for military applications. Many nitramines exhibit high brisance and high chemical stability in combination with a favourable low sensitivity to impact and friction compared to nitrate ester explosives of similar power.

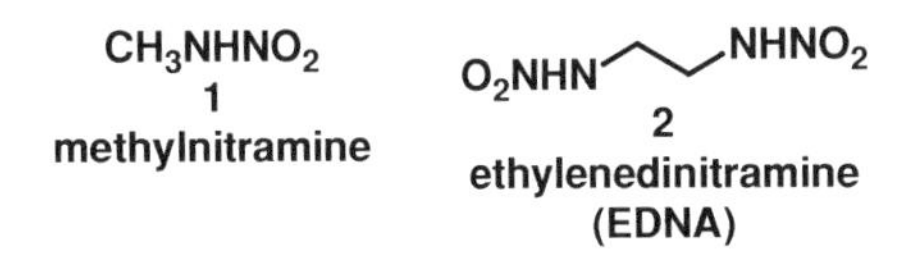

Figure 5.2

The chemical properties of primary and secondary nitramines are important in relation to their use as explosives. Primary nitramines contain acidic hydrogen in the form of $-N\mathit{H}NO_2$ and, consequently, in the presence of moisture, primary nitramines corrode metals and form metal salts, some of which are primary explosives. This is one reason why powerful explosives like methylnitramine (1) have not found practical use. Ethylenedinitramine (EDNA) (2) suffers from similar problems but its high brisance (VOD $\sim$ 8240 m/s, $d = 1.66$ g/cm^3) and low sensitivity to impact have seen it used for some applications.

In contrast, secondary nitramines have no acidic hydrogen and often exhibit a high chemical stability in combination with acceptable thermal and impact sensitivity. Consequently, secondary nitramines are often the explosives of choice for military use.

1,3,5-Trinitro-1,3,5-triazacyclohexane (RDX) (3) and 1,3,5,7-tetranitro-1,3,5,7-tetraazacyclooctane (HMX) (4) are the most important of the secondary nitramine explosives. RDX exhibits both high brisance (VOD $\sim$ 8440 m/s, $d = 1.70$ g/cm^3) and stability, finding extensive use as a military explosive in the form of compressed or cast mixtures with other explosives, or in the form of PBXs (plastic bonded explosives) where it is incorporated into a polymer matrix with added plasticizer. HMX (VOD $\sim$ 9110 m/s, $d = 1.90$ g/cm^3) exhibits higher performance than RDX due to its higher density, but this is offset by its higher cost of production compared to RDX. Consequently, HMX is restricted to military use, finding use in high performance propellant and explosive formulations. Both RDX and HMX are discussed further in Section 5.15.

Although RDX and HMX are adequate for military applications, they are by no means perfect. The risk of premature detonation increases when such explosives are used in shells for high calibre guns due to the higher 'set-back' force. Also of concern is the risk of catastrophic

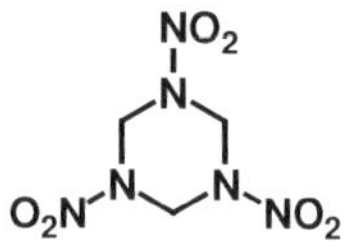

3
1,3,5-trinitro-1,3,5-triazacyclohexane (RDX)

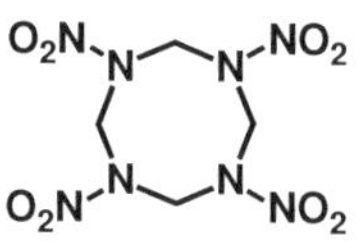

4
1,3,5,7-tetranitro-1,3,5,7-tetraazacyclooctane (HMX)

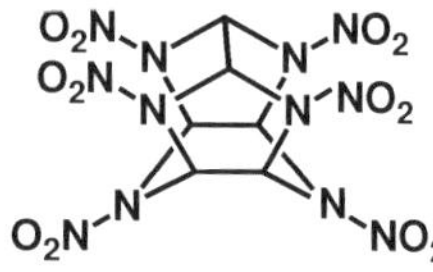

5
2,4,6,8,10,12-hexanitro-2,4,6,8,10,12-hexaazaisowurtzitane (CL-20)

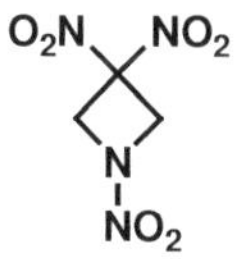

6
1,3,3-trinitroazetidine (TNAZ)

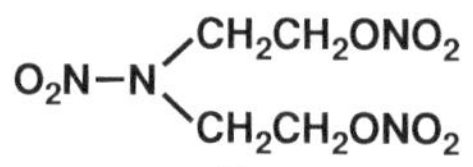

7
N-nitrodiethanolamine dinitrate (DINA)

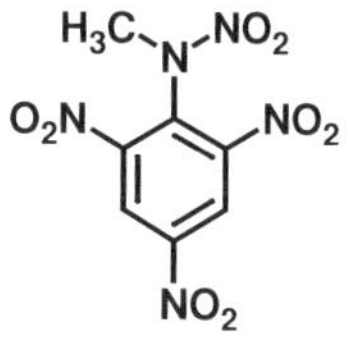

8
N,2,4,6-tetranitro-*N*-methylaniline (tetryl)

Figure 5.3

explosion in the magazine of ships. The use of explosives in a polymeric matrix greatly reduces the risk of premature detonation from impact and friction, but the performance suffers when the polymer and plasticizer are inert. The use of energetic polymers and plasticizers is one field of research. The main field of research is the synthesis of new high performance energetic materials with favourable physical properties. Much of the continuing research into nitramines is focused on finding and synthesizing materials for use in insensitive munitions. This has led to the synthesis of high performance explosives like CL-20 (VOD $\sim$ 9380 m/s, $d = 2.04$ g/cm^3) (5) and TNAZ (6). TNAZ is a potential melt-cast explosive (m.p. 101 °C). Both TNAZ and CL-20 are now synthesized on a pilot plant scale. These and other caged and strained ring nitramines are discussed more fully in Chapter 6.

A number of important explosives contain nitramino functionality in conjunction with nitrate ester or *C*-nitro functionality. *N*-Nitrodiethanolamine dinitrate (DINA) (7) is a powerful explosive which can be melt-cast into charges. *N*,2,4,6-Tetranitro-*N*-methylaniline (tetryl) (8) exhibits high brisance (VOD $\sim$ 7920 m/s, $d = 1.73$ g/cm^3) and has found application in both detonators and boosters, in addition to being a component of some composite high explosives.

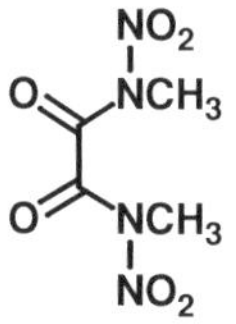

9
N,N'-dinitro-*N,N'*-dimethyloxamide

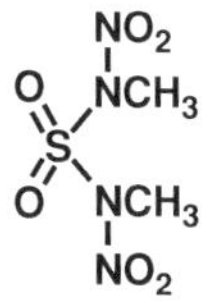

10
N,N'-dinitro-*N,N'*-dimethylsulfamide

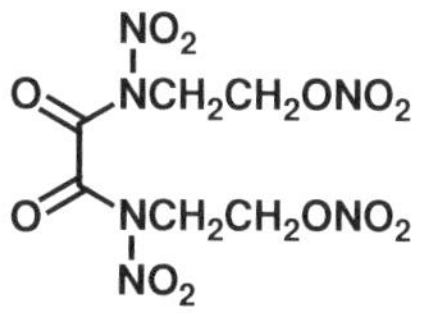

11
N,N'-dinitro-*N,N'*-bis(2-hydroxyethyl)-oxamide dinitrate (NENO)

Figure 5.4

Some compounds containing the secondary nitramide group have found limited use as explosives. *N,N'*-Dinitro-*N,N'*-dimethyloxamide (9), *N,N'*-dinitro-*N,N'*-dimethylsulfamide (10) and *N,N'*-dinitro-*N,N'*-bis(2-hydroxyethyl)oxamide dinitrate (NENO) (11) are powerful secondary high explosives. Both (10) (m.p. 90 °C) and (11) (m.p. 91 °C) have conveniently low melting points which allows the melt casting of charges. NENO is particularly useful in this respect because its slow crystallization from the molten state results in the formation of homogeneous charges free of cavities. The facile hydrolysis of secondary nitramides to primary nitramines in the presence of aqueous acid or base and, in some cases, in prolonged contact with hot water, has undoubtedly limited their use as practical explosives.

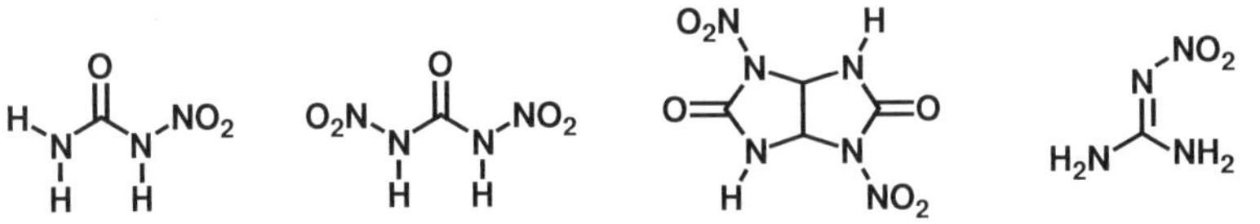

12 nitrourea

13 ***N,N'*-dinitrourea (DNU)**

14 1,4-dinitroglycouril (DINGU)

15 nitroguanidine

Figure 5.5

N-Nitroureas are an interesting group of compounds. The simplest member, 'nitrourea' (12), is a labile substance and readily decomposes in the presence of water. *N,N'*-Dinitrourea (DNU) (13), although a powerful explosive, shows similar properties. *N*-Substituted-*N*-nitroureas are more hydrolytically stable. Much interest has focused on the incorporation of the *N*-nitrourea functionality into cyclic and caged structures because of the increase in performance observed. This is due to increased crystal density, which is attributed to the rigidity of the urea functionality. The *N*-substituted-*N*-nitrourea functionality is also associated with a low sensitivity to impact, a property possibly due to intramolecular hydrogen bonding in the nitrourea framework. One such compound, 1,4-dinitroglycouril (DINGU) (14), is classified an insensitive high explosive (IHE) and exhibits good performance (VOD $\sim$ 7580 m/s, $d = 1.99$ g/cm^3). *N,N'*-Disubstituted-*N,N'*-dinitroureas are also associated with high performance. However, many cyclic *N,N'*-dinitroureas are hydrolytically unstable and decompose on contact with water. Such compounds will never find use as practical explosives.

Some compounds can be drawn as primary nitramines or as nitrimines. The two groups are tautomeric but have very different properties. Nitrimines do not contain acidic hydrogen and so their solutions are neutral. Nitroguanidine (15) exists in the nitrimine form under normal conditions, and although its structure can be drawn as a primary nitramine, its properties are not consistent with such a structure. Nitroguanidine (15) is a compound of some importance in the explosives industry as a component of triple-base propellants and also as a precursor to other explosives. The low combustion temperature of nitroguanidine-containing gun propellants makes them both 'flashless' and less erosive to gun barrels, a consequence of the high nitrogen content of nitroguanidine ($CH_4N_4O_2$ = 54 % N). Although nitroguanidine is an explosive, its fibrous nature imparts an extremely low density to the compound even on compression, and consequently, it exhibits low performance; this factor alone limits the use of nitroguanidine as an explosive. If nitroguanidine were to exist as a primary nitramine under normal conditions it is unlikely it would have found wide applications in explosive technologies.

5.3 DIRECT NITRATION OF AMINES

5.3.1 Nitration under acidic conditions

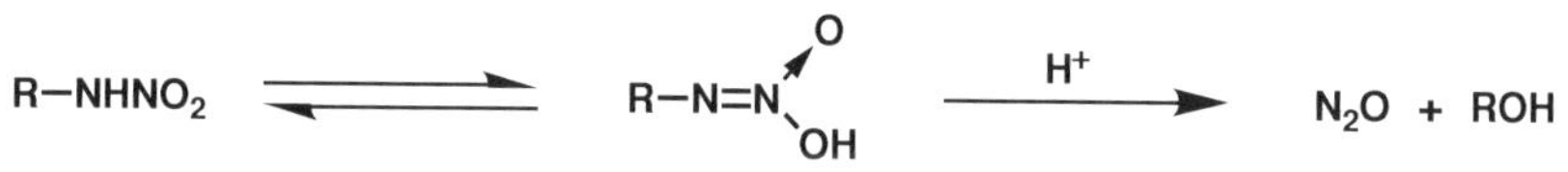

$$R{-}NHNO_2 \rightleftharpoons R{-}N{=}N(\rightarrow O){-}OH \xrightarrow{H^+} N_2O + ROH \quad \text{(Eq. 5.1)}$$

Figure 5.6

The direct nitration of a primary amine to a nitramine with nitric acid or mixtures containing nitric acid is not possible due to the instability of the tautomeric isonitramine in strongly acidic solution (Equation 5.1). Secondary amines are far more stable under strongly acidic conditions and some of these can undergo electrophilic nitration with nitric acid in a dehydrating medium like acetic anhydride.

Experimentally, it is found that only weakly basic amines can be nitrated under highly acidic conditions.[2,3] Strong amine bases are largely protonated under such conditions with the positive charge on nitrogen repelling electrophilic nitrating species, whereas weaker bases have a higher proportion of free amine base present under the same conditions. It follows that very weakly basic amines or those with a poor affinity for protons are nitrated more readily than stronger bases.[2,3] This is reflected in the fact that treatment of iminodiacetonitrile,[3] piperidine,[4,5] dimethylamine,[4,5] and diisopropylamine[3] with a mixture of nitric acid in acetic anhydride generates the corresponding secondary nitramines in 93 %, 22 %, 6 % and negligible yields, respectively. Electron-withdrawing groups on the carbon α to the amino group reduce proton affinity and help nitration. Accordingly, amines of the type RCH_2NHCH_2R, where $R = CN$, COOH or $CONH_2$, are readily nitrated with nitric acid–acetic anhydride mixtures.[6]

A number of other amines containing electron-withdrawing groups undergo facile *N*-nitration (Table 5.1). The direct *N*-nitration of aniline derivatives is limited to amines of

Table 5.1
***N*-Nitration of amines with nitric acid and its mixtures**

Entry	Substrate	Conditions	Product	Yield (%)
1	$(CF_3)_2NH$	70% HNO_3, TFAA	$(CF_3)_2NNO_2$	78[7]
2	$(CF_3CH_2)_2NH.HNO_3$	HNO_3, Ac_2O	$(CF_3CH_2)_2NNO_2$	84[8]
3	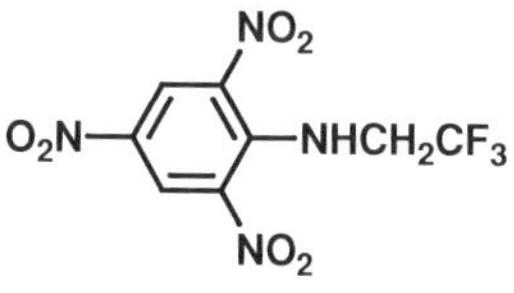 2,6-$(NO_2)_2$-4-O_2N-C_6H_2-$NHCH_2CF_3$	HNO_3, H_2SO_4	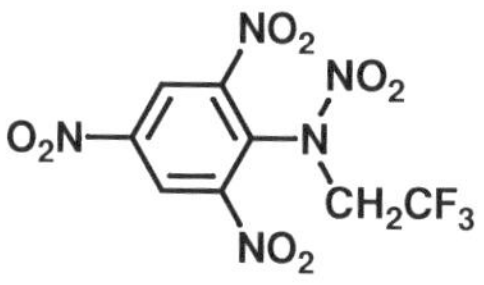 2,6-$(NO_2)_2$-4-O_2N-C_6H_2-$N(NO_2)CH_2CF_3$	98[9]
4	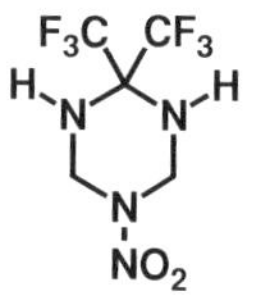 2,2-bis(CF_3)-1,3-H-5-NO_2 hexahydrotriazine	100% HNO_3, TFAA	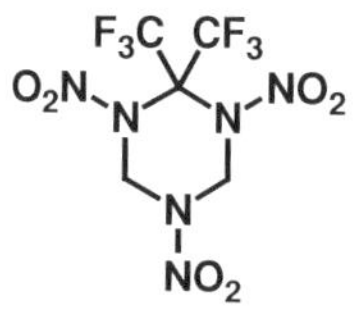 2,2-bis(CF_3)-1,3,5-tri-NO_2 hexahydrotriazine	70[10]

low basicity i.e. those containing one or more nitro groups on the aromatic ring (Table 5.1, Entry 3).[9,11]

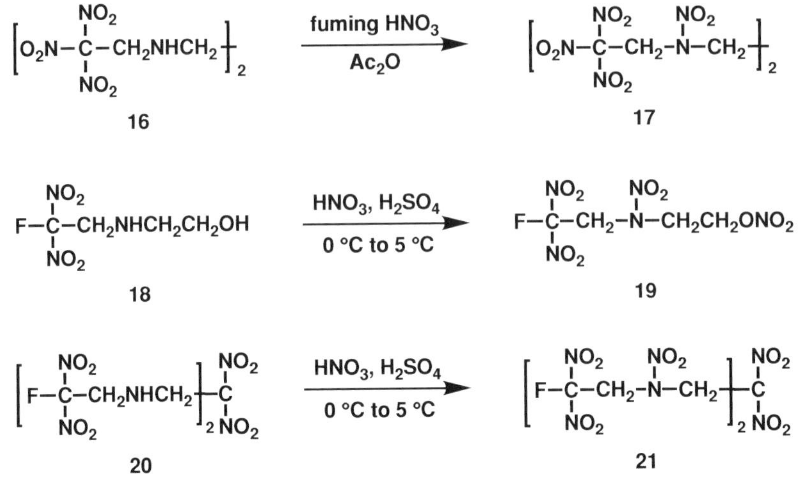

$[O_2N-C(NO_2)_2-CH_2NHCH_2]_2$ (16) $\xrightarrow[Ac_2O]{\text{fuming } HNO_3}$ $[O_2N-C(NO_2)_2-CH_2-N(NO_2)-CH_2]_2$ (17)

$F-C(NO_2)_2-CH_2NHCH_2CH_2OH$ (18) $\xrightarrow[0\,°C \text{ to } 5\,°C]{HNO_3, H_2SO_4}$ $F-C(NO_2)_2-CH_2-N(NO_2)-CH_2CH_2ONO_2$ (19)

$[F-C(NO_2)_2-CH_2NHCH_2]_2C(NO_2)_2$ (20) $\xrightarrow[0\,°C \text{ to } 5\,°C]{HNO_3, H_2SO_4}$ $[F-C(NO_2)_2-CH_2-N(NO_2)-CH_2]_2C(NO_2)_2$ (21)

Figure 5.7

A large number of nitramine-based explosives have been synthesized via Mannich-type condensation reactions (Section 5.13.2).[12–18] The amines generated from these reactions often have the powerful electron-withdrawing trinitromethyl or fluorodinitromethyl groups positioned on the carbon α to the amino group. This reduces amine basicity to an extent that *N*-nitration becomes facile. The energetic nitramines (17)[12], (19)[13] and (21)[14] have been synthesized from the condensation of ethylenediamine with 2,2,2-trinitroethanol, 2-fluoro-2,2-dinitroethanol with ethanolamine, and 2-fluoro-2,2-dinitroethylamine with 2,2-dinitro-1,3-propanediol respectively, followed by *N*-nitration of the resulting amine bases (16), (18) and (20), respectively.

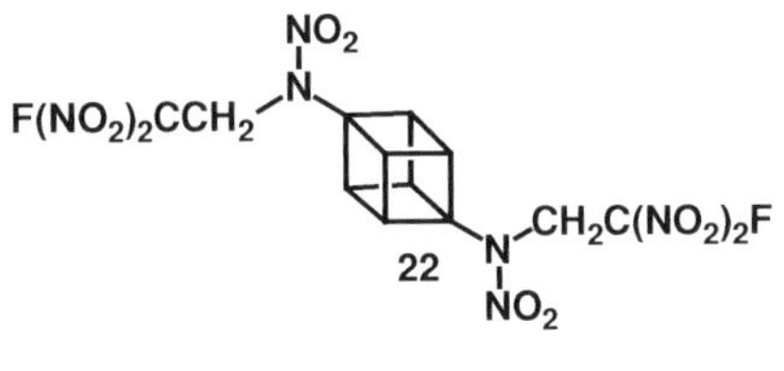

NO_2
$F(NO_2)_2CCH_2-N$
$N-CH_2C(NO_2)_2F$
NO_2
22

Figure 5.8

The energetic cubane-based nitramine (22) is also synthesized from the direct *N*-nitration of the parent amine using a mixture of trifluoroacetic anhydride and nitric acid.[15]

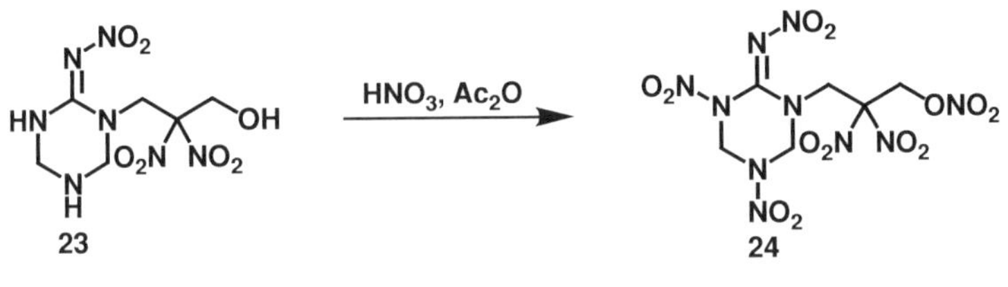

$N-NO_2$, HN, N, OH, O_2N, NO_2, N, H
23
$\xrightarrow{HNO_3, Ac_2O}$
$N-NO_2$, O_2N-N, N, ONO_2, O_2N, NO_2, N, NO_2
24

Figure 5.9

The direct N-nitration of the amino groups of the hexahydrotriazine (23) is only possible due to the inherent low basicity of the methylenediamine functionality.[19] The methylenediamine unit is present in many cyclic and bicyclic polyamines and these are potential precursors to energetic polynitramines. Unfortunately, this route to polynitramines is rarely possible because such polyamines are usually intrinsically unstable and will readily equilibrate to a lower energy, less strained system. For the same reason, polyamines containing the methylenediamine functionality are difficult to prepare and isolate, often rapidly decomposing in both aqueous and acidic solution. A far more common route involves the preparation of N-protected versions of the polyamine followed by 'nitrolysis' (Section 5.6). Even so, examples of heterocyclic methylenediamine N-nitration exist.

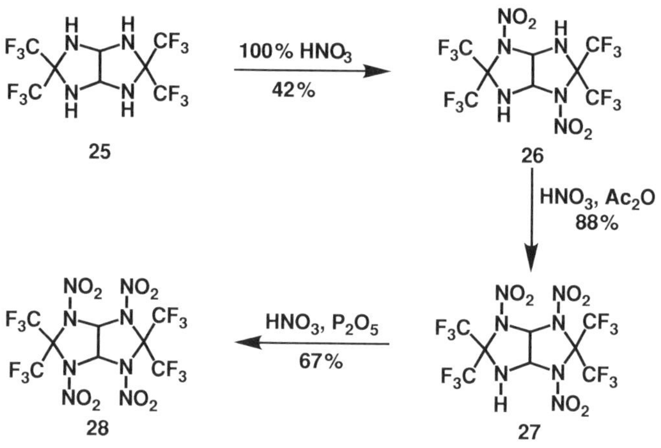

25 —(100% HNO_3, 42%)→ 26 —(HNO_3, Ac_2O, 88%)→ 27 —(HNO_3, P_2O_5, 67%)→ 28

Figure 5.10

3,3,7,7-Tetrakis(trifluoromethyl)-2,4,6,8-tetraazabicyclo[3.3.0]octane (25) is an interesting example.[20] This polyamine can be isolated as the free base and is fairly stable because of the '*gem*-dimethyl effect'. Stability is further enhanced by the replacement of the methyl protons with fluorine atoms. The introduction of the first nitro groups into (25) requires particular care and a low nitration temperature, and even then, the yield is low (42 %) because of acid-catalyzed ring-opening. These nitro groups protect (26) from further ring-opening and so more vigorous nitrating conditions are employed for the synthesis of the tetranitramine (28).

29 —(TFAA, HNO_3 or N_2O_5 in 100% HNO_3)→ 30

Figure 5.11

The N-nitration of the furazan-based heterocycle (29) has been reported.[21] The corresponding tetranitramine (30) is an unstable substance, but obtained on treating (29) with either trifluoroacetic anhydride (TFAA) in nitric acid or dinitrogen pentoxide in nitric acid. In this case the furazan rings stabilize the 1,4,5,8-tetraazadecalin structure and further reduce the basicity of the amidine amino groups. A number of other furazan and nitrogen-rich nitramines have been reported and these are discussed in Chapter 6 together with other energetic bicyclic and caged nitramines.

If nitration under acidic conditions could only be used for the nitration of the weakest of amine bases its use for the synthesis of secondary nitramines would be severely limited. An important discovery by Wright and co-workers[3,4,22] found that the nitrations of the more basic amines are strongly catalyzed by chloride ion. This is explained by the fact that chloride ion, in the form of anhydrous zinc chloride, the hydrochloride salt of the amine, or dissolved gaseous hydrogen chloride, is a source of electropositive chlorine under the oxidizing conditions of nitration and this can react with the free amine to form an intermediate chloramine. The corresponding chloramines are readily nitrated with the loss of electropositive chlorine and the formation of the secondary nitramine in a catalytic cycle (Equations 5.2, 5.3 and 5.4).[22] The mechanism of this reaction is proposed[22] to involve chlorine acetate as the source of electropositive chlorine but other species may play a role. The success of the reaction appears to be due to the chloramines being weaker bases than the parent amines.

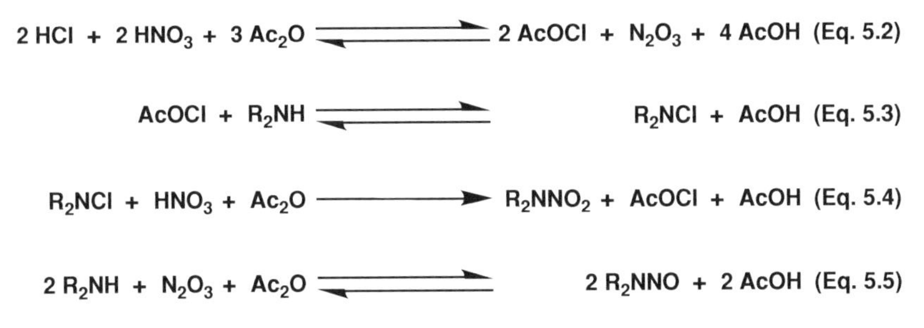

Figure 5.12 Proposed mechanism for chloride-catalyzed nitration[22,24]

Wright illustrated the effectiveness of chloride-catalyzed nitration for a number of amines of different basicity. Wright showed that weakly basic amines like iminodiacetonitrile and its dimethyl and tetramethyl derivatives are all nitrated in high yield with nitric acid–acetic anhydride mixtures in the absence of chloride ion.[3] In contrast, the slightly more basic 3,3′-iminodipropionitrile is not appreciably nitrated with acetic anhydride–nitric acid, but the inclusion of a catalytic amount of the hydrochloride salt of the amine base generates the corresponding nitramine in 71 % yield.[3]

Wright found it more suitable to use the nitrate salts of the more basic amines, like in the case of morpholine, which ignites at ambient temperature in the presence of nitric acid–acetic anhydride mixtures. Morpholine nitrate is not nitrated with acetic anhydride–nitric acid in the absence of chloride ion at room temperature, but the addition of 4 mole % of zinc chloride generates a 65 % yield of *N*-nitromorpholine, and this yield rises to 93 % if the hydrochloride salt of morpholine is directly nitrated.[3] While morpholine nitrate is unaffected by treatment with acetic anhydride–nitric acid at room temperature, the same reaction in the presence of ammonium nitrate at 65 °C is reported to yield *N*-nitromorpholine in 48 % yield.[23]

Wright found that as amine basicity increases a larger amount of chloride ion is needed to effectively catalyze the nitration. The nitrate salt of diethylamine needs the addition of 4 mole % of zinc chloride to generate a 60 % yield of the corresponding nitramine.[3] In the extreme case of diisopropylamine, where the isopropyl groups strongly donate electron density to nitrogen and make it a relatively strong base, a full equivalent of chloride ion is needed to attain an acceptable yield.[3] Some nitramines can be prepared via the sole addition of zinc chloride to a slurry of the amine nitrate in acetic anhydride, which is a method discussed in Section 5.11.[4]

The chloride-catalyzed nitration of secondary amines has its disadvantages. The most serious disadvantage being the formation of nitrosamines as by-products, a consequence of the redox reaction between chlorine anion and nitric acid (Equation 5.5). The toxicity of nitrosamines cannot be stressed enough, and in the case of diisopropylamine, the corresponding nitrosamine is exceedingly toxic and a very powerful carcinogen. The same is true of *N,N*-dimethylnitrosamine which is biologically reduced to the powerful carcinogen *N,N*-dimethylhydrazine. It is found that an increase in the amount of chloride ion present in these nitration reactions increases the amount of nitrosamine formed, a scenario which is grim for the nitration of diisopropylamine. Additionally, the nitration of strongly basic amines, even under chloride ion catalysis, is slow and allows competition with slow side-reactions such as nitrosation and acetylation of the amino functionality.[3]

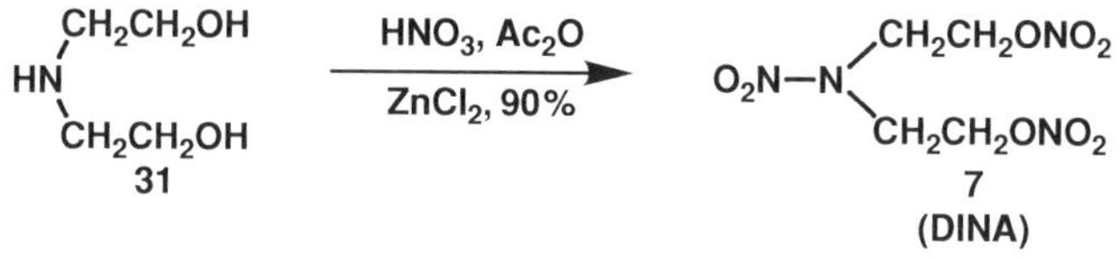

$$HN(CH_2CH_2OH)_2 \ (\mathbf{31}) \xrightarrow[ZnCl_2,\ 90\%]{HNO_3,\ Ac_2O} O_2N{-}N(CH_2CH_2ONO_2)_2 \ (\mathbf{7}, \text{DINA})$$

Figure 5.13

Chloride-catalyzed nitration is used for the industrial synthesis of *N*-nitrodiethanolamine dinitrate (DINA) (7), an explosive plasticizer with VOD ~ 7580 m/s. The synthesis of DINA (7) uses a mixture of nitric acid in acetic anhydride to nitrate diethanolamine (31) in the presence of catalytic zinc chloride or a small portion of the diethanolamine as the hydrochloride salt; the yield of DINA from this process is usually about 90 %.[4] Note that diethanolamine is only a moderately basic amine, but enough so to prevent its direct nitration with acetic anhydride–nitric acid mixtures in the absence of chloride anion. As expected, a small amount of nitrosamine impurity (<5 %) is formed during the synthesis of DINA. Fortunately, this nitrosamine is unstable and is readily decomposed by the action of boiling water or the injection of stream into the crude molten product. The important role of chloride anion in the synthesis of DINA is illustrated by the following observation – if the reaction mixture is vented, so that the volatile sources of electropositive chlorine can escape, the yields of DINA are very poor and the reaction can cease altogether.[4] DINA is a desirable explosive with a conveniently low melting point to permit melt casting. DINA is also an excellent gelatinizing agent for nitrocellulose.

The founder of the chloride-catalyzed nitration of amines, G. F. Wright, has extensively reviewed chloride-catalyzed nitration.[24] Even 50 years later, the importance and success of this method as a route to nitramines is illustrated by the number of examples which can be found in the literature.

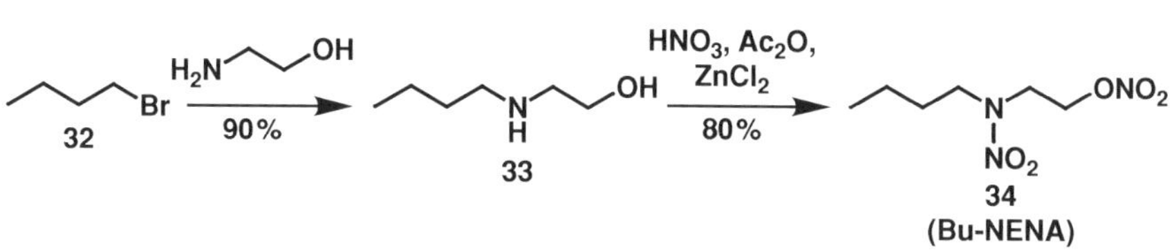

$$\mathbf{32}\ (n\text{-}C_4H_9Br) \xrightarrow[90\%]{H_2NCH_2CH_2OH} \mathbf{33}\ (n\text{-}C_4H_9NHCH_2CH_2OH) \xrightarrow[80\%]{HNO_3,\ Ac_2O,\ ZnCl_2} \mathbf{34}\ (n\text{-}C_4H_9N(NO_2)CH_2CH_2ONO_2,\ \text{Bu-NENA})$$

Figure 5.14

One simple example of chloride-catalyzed nitration is the synthesis of the energetic plasticizer Bu-NENA (34) from the nitration of *n*-butyl ethanolamine (33) with a mixture of acetic anhydride–nitric acid to which catalytic zinc chloride has been added.[25]

Many cyclic, bicyclic and caged polyamines are synthesized via condensation reactions and these are usually isolated as their salts. The free bases are usually difficult to isolate and readily decompose in aqueous or acid solution. Since hydrochloric acid is frequently employed in these condensations the hydrochloride salt of the amine is usually used directly for the *N*-nitration.

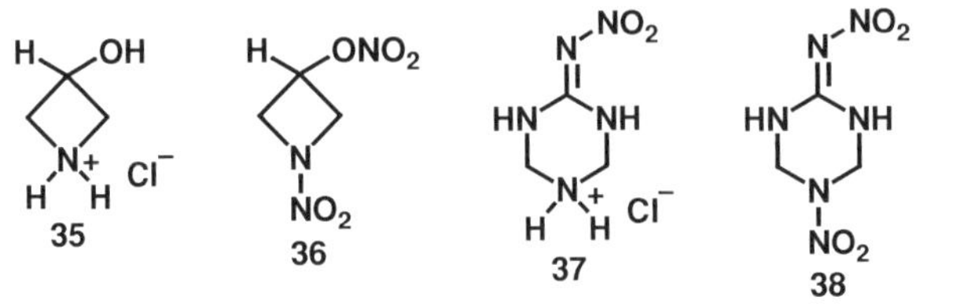

Figure 5.15

The strained *N*-nitroazetidine (36) has been synthesized from the nitration of the hydrochloride salt of the corresponding azetidine (35) with acetic anhydride–nitric acid.[26] The heterocyclic guanidine (38) is synthesized from (37) in a similar way.[27]

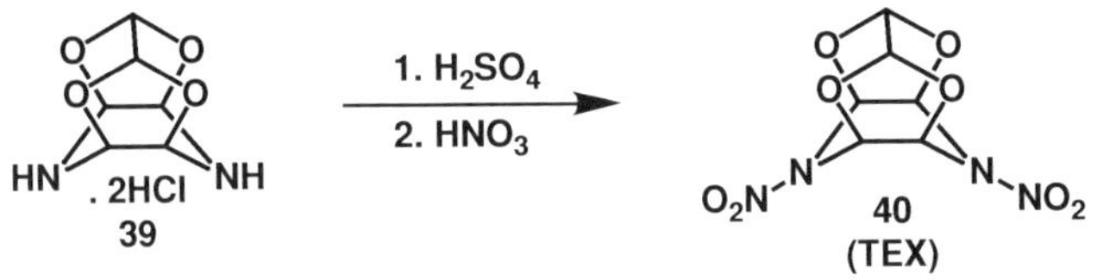

Figure 5.16

Acetic anhydride–nitric acid mixtures are extensively used for chloride-catalyzed nitrations. Other nitrating agents have been used and involve similar sources of electropositive chlorine for intermediate chloramine formation. 4,10-Dinitro-4,10-diaza-2,6,8,12-tetraoxaisowurtzitane (TEX) (40), an insensitive high performance explosive (VOD $\sim$ 8665 m/s, $d = 1.99$ g/cm^3), is synthesized by treating the dihydrochloride salt of the corresponding amine (39) with strong mixed acid.[28]

As previously discussed, heterocyclic polyamines containing methylenediamine functionality are usually unstable if unprotected. In contrast, the presence of a urea group stabilizes this functionality and allows the isolation of a number of heterocyclic amines. These are usually synthesized via a condensation reaction and isolated as the hydrochloride salt. The *N*-nitration of the 2,5,7,9-tetraazabicyclo[4.3.0]nonan-8-one[29] and 2,4,6,8-tetraazabicyclo[3.3.0]octan-3-one[30] ring systems has been investigated and serve as valuable examples.

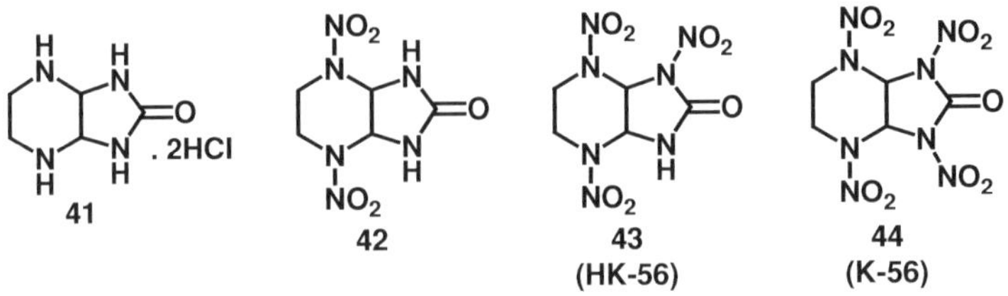

Figure 5.17

Table 5.2 *N*-Nitration of 2,5,7,9-tetraazabicyclo[4.3.0]nonan-8-one. dihydrochloride (41) with various nitrating agents (ref. 29)

Nitrating agent	Product	Yield (%)
NO_2BF_4/CH_3NO_2	42	86
HNO_3/Ac_2O (20 °C)	43/44 (7:93)	53
HNO_3/Ac_2O (60 °C)	44	78
100 % HNO_3 (60 °C)	43/44 (10:90)	40
HNO_3/N_2O_5 (20 °C)	44	82

Source: Reprinted with permission from H. R. Graindorge, P. A. Lescop, M. J. Pouet and F. Terrier, in *Nitration: Recent Laboratory and Industrial Developments, ACS Symposium Series 623*, Eds. L. F. Albright, R. V. C. Carr and R. J. Schmitt, American Chemical Society, Washington, DC, 46 (1996); Copyright 1996 American Chemical Society.

Three products have been observed from the *N*-nitration of the dihydrochloride salt of 2,5,7,9-tetraazabicyclo[4.3.0]nonan-8-one (41) with various nitrating agents, and two products observed for the *N*-nitration of 2,4,6,8-tetraazabicyclo[3.3.0]octan-3-one dihydrochloride (45). Table 5.2 and Table 5.3 give a comparative idea of the efficiency and nitrating power of different nitrating agents.

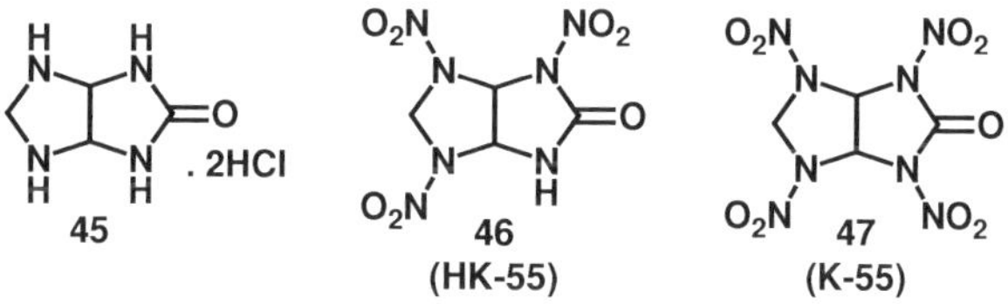

Figure 5.18

Table 5.3 *N*-Nitration of 2,4,6,8-tetraazabicyclo[3.3.0]octan-3-one. dihydrochloride (45) with various nitrating agents (ref. 30a)

Nitrating agent	Product	Yield (%)
90 % HNO_3/Ac_2O (0 °C to 40 °C)	46	72
100 % HNO_3/Ac_2O (< 15 °C)	46	53
100 % HNO_3/Ac_2O (20 °C to 50 °C)	47	49
100 % HNO_3/TFAA (0 °C to 40 °C)	47	51

Source: Reprinted with permission from C. L. Coon, E. S. Jessop, A. R. Mitchell, P. F. Pagoria and R. D. Schmidt, in *Nitration: Recent Laboratory and Industrial Developments, ACS Symposium Series 623*, Eds. L. F. Albright, R. V. C. Carr and R. J. Schmitt, American Chemical Society, Washington, DC, 153 (1996); Copyright 1996 American Chemical Society.

It is observed that the nitration of the amino nitrogens of both (41) and (45) is considerably more facile than the urea nitrogens, and this is generally observed with other compounds of similar structure;[31] the *N*-nitration of amides and ureas is discussed in Section 5.5. It is also apparent that nitronium tetrafluoroborate is milder and more selective under these conditions compared to other highly acidic nitrating agents. A mixture of acetic anhydride and fuming nitric acid is a common reagent used for *N*-nitration. This reagent is relatively mild and excellent results are frequently observed. A solution of trifluoroacetic anhydride and nitric acid is a more powerful nitrating agent and frequently used for difficult substrates. A solution of 20–30 % dinitrogen pentoxide in absolute nitric acid is a powerful *N*-nitrating agent and mainly finds use for nitrolysis reactions (Section 5.6). Using the most powerful nitrating agent is not always the best strategy for an *N*-nitration. This can give a lower yield than a less powerful nitrating agent due to side-reactions such as competing acid-catalyzed ring-opening. This is particularly noticeable during the nitrolysis of some *N*-protected heterocyclic amines (Section 5.6).

The synthesis and properties of nitramines derived from strained and bicyclic amines are discussed in more detail in Chapter 6. Such compounds often exhibit high performance resulting from high crystal densities and/or high heats of formation due to internal strain.

5.3.2 Nitration with nonacidic reagents

Unlike the direct nitration of amines under acidic conditions, nucleophilic nitration is an excellent route to both primary and secondary nitramines. In these reactions the amine or the conjugate base of the amine is used to attack a source of NO_2. This source may be a nitrogen oxide, nitronium salt, cyanohydrin nitrate, alkyl nitrate ester or any other similar source of nitronium ion.

5.3.2.1 Alkyl nitrates

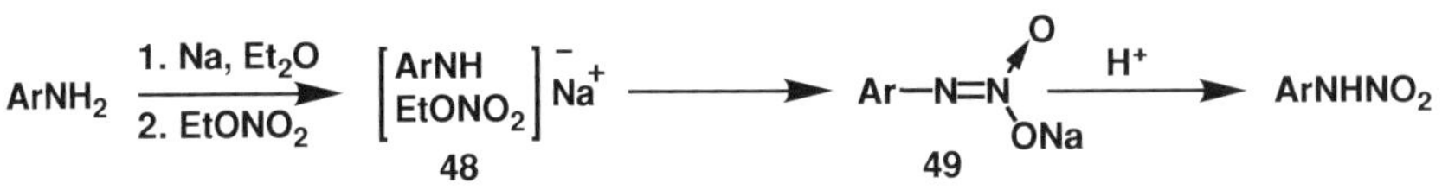

$$ArNH_2 \xrightarrow[2.\ EtONO_2]{1.\ Na,\ Et_2O} \underset{\mathbf{48}}{\left[\begin{matrix} ArNH \\ EtONO_2 \end{matrix}\right]^{-} Na^{+}} \longrightarrow \underset{\mathbf{49}}{Ar{-}N{=}N(\rightarrow O)ONa} \xrightarrow{H^+} ArNHNO_2 \quad \text{(Eq. 5.6)}$$

Figure 5.19

Angeli[32] pioneered the concept of nucleophilic amine nitration by *N*-nitrating aniline with ethyl nitrate in the presence of potassium or sodium metal. In this reaction a complex anion (48) of ethyl nitrate and the deprotonated amine decomposes to the metal salt of the nitramine (49) which can be freed and the nitramine isolated by acidification at 0 °C (Equation 5.6). It should be noted that the acidification of aromatic nitramines or their metal salts with aqueous acid can cause rearrangement to the ring-nitrated product. This problem can be prevented by first isolating the nitramine as its barium salt, followed by suspending the salt in water and aspirating with carbon dioxide to release the free nitramine.[33]

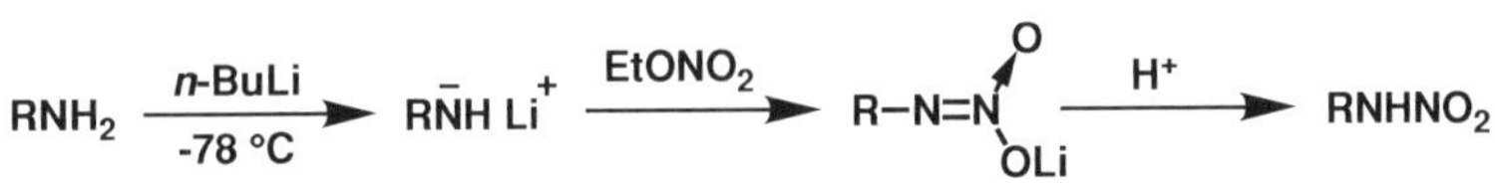

$$RNH_2 \xrightarrow[-78\ ^\circ C]{n\text{-}BuLi} R\overset{-}{N}H\ \overset{+}{Li} \xrightarrow{EtONO_2} R{-}N{=}N(\rightarrow O)OLi \xrightarrow{H^+} RNHNO_2 \quad \text{(Eq. 5.7)}$$

Figure 5.20

Bamberger[34] improved upon the method of Angeli[32] by using alkali metal alkoxides as bases rather than the alkali metal. However, this method is still a low yielding route to nitramines and is a consequence of nitrate esters functioning more like alkylating agents than nitrating agents. This problem is largely overcome by using lithium bases, which react irreversibly with the amine to form the conjugate base (Equation 5.7). Both butyl lithium and phenyl lithium are commonly used as bases in these reactions and, in some reactions, nitrate esters other than ethyl nitrate are used. The reaction of a series of primary amines with butyl lithium in the presence of ethyl nitrate at −78 °C in diethyl ether or hexane solvents has been used to prepare primary nitramines in moderate to good yield, including: methylnitramine (35 %), *iso*-propylnitramine (58 %), *n*-butylnitramine (49 %), *sec*-butylnitramine (45 %) and *tert*-butylnitramine (37 %).[35] This procedure also works well for the synthesis of nitramines from secondary amine substrates.

$$R^1R^2NH \xrightarrow{EtMgBr} R^1R^2N{-}MgBr \xrightarrow{n\text{-}BuONO_2} R^1R^2NNO_2 \quad \text{(Eq. 5.8)}$$

Figure 5.21

More recently, Polish chemists[36] have reported a synthesis of both aryl and aliphatic secondary nitramines by treating amine substrates with ethyl magnesium bromide followed by reaction with *n*-butyl nitrate (Equation 5.8). This method, which uses nonpolar solvents like hexane or benzene, has been used to synthesize aliphatic secondary nitramines, and *N*-nitro-*N*-methylanilines which otherwise undergo facile Bamberger rearrangement in the presence of acid. The direct nitration of *N*-unsubstituted arylamines usually requires the presence of an electron-withdrawing group. Reactions are retarded and yields are low for sterically hindered amines.

Despite the moderate to good yields obtained for a range of primary and secondary nitramines, the above methods have not found wide use. Their use in organic synthesis is severely limited by the incompatibility of many functional groups in the presence of strong bases. This is particularly relevant to the synthesis of explosive materials, where nitrate ester and *C*-nitro functionality are incompatible with strong bases.

5.3.2.2 Electron-deficient nitrate esters

Ordinarily, alkyl nitrate esters will not nitrate amines under neutral conditions. However, Schmitt, Bedford and Bottaro[37] have reported the use of some novel electron-deficient nitrate esters for the direct *N*-nitration of secondary amines. The most useful of these is 2-(trifluoromethyl)-2-propyl nitrate, which nitrates a range of aliphatic secondary amines to the corresponding nitramines in good to excellent yields. Nitrosamine formation is insignificant in these reactions. 2-(Trifluoromethyl)-2-propyl nitrate cannot be used for the nitration of primary amines, or secondary amines containing ethylenediamine functionality like that in piperazine. Its use is limited with highly hindered amines or amines of diminished nucleophilicity due to inductive or steric effects.

5.3.2.3 Cyanohydrin nitrates

Acetone cyanohydrin nitrate, a reagent prepared by treating acetone cyanohydrin with nitric acid–acetic anhydride, has been used for the *N*-nitration of aliphatic and alicyclic secondary

$$R^1R^2NH \xrightarrow[80\ ^\circ C]{NC-C(CH_3)_2-ONO_2} R^1R^2NNO_2$$

R^1 = H, R^2 = *n*-Pr (50%)
R^1 = H, R^2 = *i*-Bu (54%)
R^1 = Me, R^2 = Me (76%)
R^1 = Et, R^2 = Et (60%)
R^1 = *n*-Pr, R^2 = *n*-Pr (42%)

Figure 5.22

amines (55–80 % yield) and primary amines (50–60 % yield).[38] In these reactions an excess of the amine is heated with acetone cyanohydrin nitrate at 80 °C, neat in the case of liquid secondary amines, or in acetonitrile or THF in the case of primary amines or those that are solids at the reaction temperature.

Acetone cyanohydrin nitrate will not nitrate amines with branching on the carbon α to the nitrate group. For these substrates the use of ethyl nitrate and lithium bases is favoured. α-Aminonitriles are frequently observed as impurities under the reaction conditions because of the slow decomposition of acetone cyanohydrin nitrate to hydrogen cyanide and acetone. The need for an excess of amine during these reactions is wasteful and only practical if this component is cheap and widely available. Other cyanohydrin nitrates are less efficient *N*-nitrating agents.[38]

5.3.2.4 Dinitrogen pentoxide

The first use of dinitrogen pentoxide as an *N*-nitrating agent appears to have been for the conversion of aromatic amines to arylnitramines.[39] Difficulties in preparing pure dinitrogen pentoxide meant that reactions with aliphatic amines were not properly examined for another 60 years.

$$2\ R^1R^2NH + N_2O_5 \xrightarrow[-30\ ^\circ C]{CCl_4} R^1R^2NNO_2 + R^1R^2NH.\ HNO_3 \quad (Eq.\ 5.9)$$

Figure 5.23

Emmons and co-workers[40] prepared a series of aliphatic secondary nitramines by treating amines with a solution of dinitrogen pentoxide in carbon tetrachloride at −30 °C (Equation 5.9). The amine component needs to be in excess of two equivalents relative to the dinitrogen pentoxide if high yields of nitramine are to be attained. This is wasteful because at least half the amine remains unreacted. However, yields are high and there is no reason why the amine cannot be recovered as the nitrate salt. The method is particularly useful for the nitration of hindered secondary amines substrates such as those with branching on the α carbon.

The reaction of dinitrogen pentoxide with primary aliphatic nitramines and amines leads to deamination and the formation of a nitrate ester as the major product. Consequently, dinitrogen pentoxide cannot be used for the synthesis of primary nitramines. In contrast, both primary and secondary arylamines undergo efficient *N*-nitration with dinitrogen pentoxide in chlorinated solvents.[41]

$$R_3N + N_2O_5 \xrightarrow[-20\,°C\ \text{to}\ -30\,°C]{CH_2Cl_2} R_3\overset{+}{N}NO_2\ \overset{-}{NO_3} \quad \text{(Eq. 5.10)}$$

Figure 5.24

Tertiary alkylamines react with dinitrogen pentoxide in carbon tetrachloride at subambient temperature to give tertiary alkylnitramine nitrate salts (Equation 5.10).[40]

5.3.2.5 Dinitrogen tetroxide

Dinitrogen tetroxide can react as a nitrating or nitrosating agent depending on the conditions of the reaction.[42,43] Solutions of dinitrogen tetroxide in methylene chloride react with secondary aliphatic amines at −80 °C to give the nitramine and the nitrite salt of the amine without appreciable nitrosamine formation.[42] At higher temperatures nitrosamine formation becomes increasingly competitive and dominates at room temperature. The solvent appears to be important; the reaction of diethylamine with dinitrogen tetroxide in diethyl ether at −80 °C results in a 1:1 mixture of nitramine and nitrosamine, whereas the same reaction in methylene chloride exclusively yields the nitramine. The reaction of dinitrogen tetroxide with primary amines is not a feasible route to primary nitramines due to competing deamination of both the starting material and primary nitramine product, leading to alcohol and nitrate ester products.[42,44]

Solutions of dinitrogen tetroxide in carbon tetrachloride react with aliphatic cyclic and acyclic tertiary amines between 0 °C and 45 °C to give nitrosamines via nitrosolysis.[45]

5.3.2.6 Nitronium tetrafluoroborate

$$2\ R^1R^2NH + NO_2BF_4 \xrightarrow{CH_2Cl_2,\ 20\,°C} R^1R^2NNO_2 + R^1R^2NH.\ HBF_4 \quad \text{(Eq. 5.11)}$$

Figure 5.25

While nitramines are formed from the reaction of secondary amines with nitronium salts the success of the reaction depends on the basicity of the amine (Equation 5.11).[46–49] Thus, amines of low to moderate basicity are *N*-nitrated in good yields. The nitration of more basic amines is slow and the nitrosamine is often observed as a significant by-product, a consequence of the partial reduction of the nitronium salt to the nitrosonium salt during the reaction.[48] Increased reaction temperature is also found to increase the amount of nitrosamine formed.[48] The amine substrate is usually used in excess to compensate for the release of the strong mineral acid formed during the reactions. Both nitronium tetrafluoroborate and the more soluble hexafluorophosphate are commonly used for *N*-nitrations. Solvents like acetonitrile, methylene chloride, nitromethane, dioxane, sulfolane, ethyl acetate and esters of phosphoric acid are commonly used.

The reaction of nitronium salts with primary amines is not usually a feasible route to primary nitramines, except in the case of some electron-deficient arylamines[50,51]. Picramide

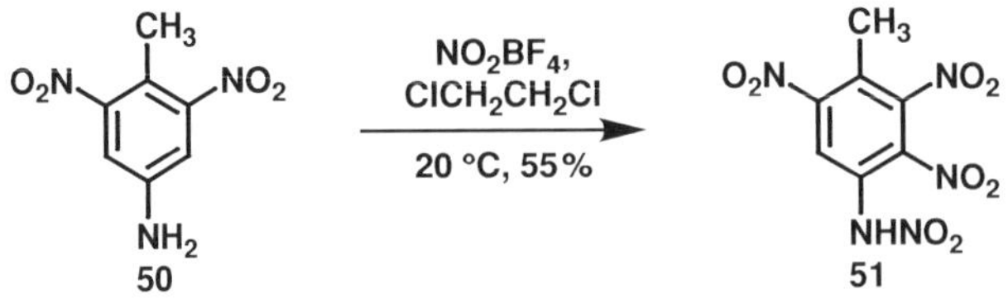

$$\text{50 (4-amino-2,6-dinitrotoluene)} \xrightarrow[20\,°C,\ 55\%]{NO_2BF_4,\ ClCH_2CH_2Cl} \text{51 (4-amino-}N\text{,2,3,6-tetranitrotoluene)}$$

Figure 5.26

and 4-amino-2,6-dinitrotoluene (50) react with nitronium tetrafluoroborate to give *N*,2,4,6-tetranitroaniline (85 %)[50] and 4-amino-*N*,2,3,6-tetranitrotoluene (51) (55 %)[51] respectively. Simple primary aliphatic amines give a low yield of the corresponding nitrate ester through deamination; *n*-butylamine gives a 20 % yield of *n*-butyl nitrate on treatment with nitronium tetrafluoroborate in acetonitrile.[46]

Ridd and Yoshida[52] explored the *N*-nitration of aromatic nitroanilines with nitronium hexafluorophosphate in nitromethane with the aim of studying the Bamberger rearrangement (Chapter 4, Section 4.5). In was found that amines of low basicity essentially undergo complete *N*-substitution if the amine is in excess; 2,4-dinitroaniline and 2,3-dinitroaniline form *N*,2,4-trinitroaniline and *N*,2,3-trinitroaniline, respectively, as the sole products. More basic nitroanilines give mixtures of *C*-substituted, *N*-substituted and poly-substituted materials.

The use of nitronium salts for *N*-nitration has been extensively reviewed.[53,54]

5.3.2.7 *Transfer nitration*

The efficient *N*-nitration of secondary amines has been achieved by transfer nitration with 4-chloro-5-methoxy-2-nitropyridazin-3-one, a reagent prepared from the nitration of the parent 4-chloro-5-methoxypyridazin-3-one with copper nitrate trihydrate in acetic anhydride.[55] Reactions have been conducted in methylene chloride, ethyl acetate, acetonitrile and diethyl ether where yields of secondary nitramine are generally high. Homopiperazine is selectively nitrated to *N*-nitrohomopiperazine or *N,N′*-dinitrohomopiperazine depending on the reaction stoichiometry. *N*-Nitration of primary amines or aromatic secondary amines is not achievable with this reagent.

5.3.2.8 *Other N-nitrating agents*

Other reagents have been used for the direct nitration of amines under nonacidic conditions. Mandel[56] reported the synthesis of methylnitramine, dimethylnitramine and diisopropylnitramine from the reaction of nitryl fluoride with the parent amine. The presence of an excess of amine appears to be of prime importance given that other reports give the nitrosamine as a major by-product.[37]

Nitryl chloride[57] and tetranitromethane[58] also *N*-nitrate amines. However, competing *N*-nitrosation is a particular problem and results from a redox reaction between the nitrating agent and the amine.

A solution of tetra-*n*-butylammonium nitrate and triflic anhydride in methylene chloride is reported to give a 43 % yield of *N*-nitropyrrolidine from pyrrolidine.[59] This reagent has been successfully used for cyclic amide *N*-nitration and for nitrolysis.[59] More work needs to be reported before the scope of this reagent for nitramine synthesis can be assessed. The same reagent has proven particularly successful for aromatic nitration.[60]

Despite the amount of research focused on the *N*-nitration of amines with nonacidic reagents, the use of conventional acidic reagents based on nitric acid and acid anhydrides has been far more extensive for the synthesis of energetic materials.

5.4 NITRATION OF CHLORAMINES

5.4.1 Nitration of dialkylchloramines

$$R^1R^2NH \xrightarrow{HOCl} R^1R^2NCl \xrightarrow{HNO_3,\ Ac_2O} R^1R^2NNO_2 \quad \text{(Eq. 5.12)}$$

Figure 5.27

The role of dialkylchloramines as intermediates in the chloride-catalyzed nitration of secondary amines is discussed in Section 5.3.1. Wright and co-workers[49] studied this reaction further and prepared a number of dialkylchloramines by treating secondary amines with aqueous hypochlorous acid (Equation 5.12). Treatment of these dialkylchloramines with nitric acid in acetic anhydride forms the corresponding secondary nitramine, a result consistent with the chloride-catalyzed nitration of amines.[49]

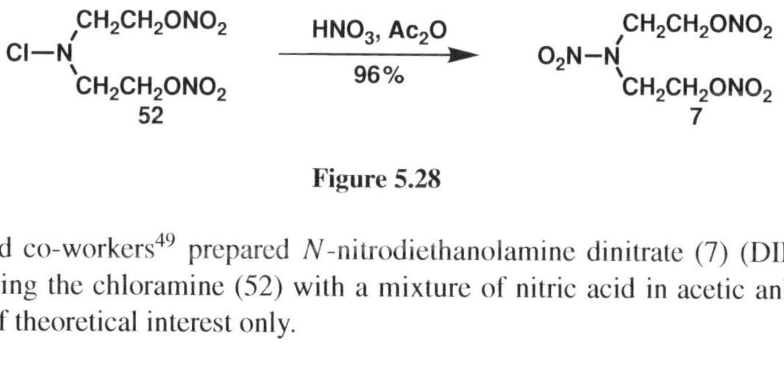

$$Cl{-}N(CH_2CH_2ONO_2)_2\ (\mathbf{52}) \xrightarrow[96\%]{HNO_3,\ Ac_2O} O_2N{-}N(CH_2CH_2ONO_2)_2\ (\mathbf{7})$$

Figure 5.28

Wright and co-workers[49] prepared *N*-nitrodiethanolamine dinitrate (7) (DINA) in 96 % yield by treating the chloramine (52) with a mixture of nitric acid in acetic anhydride. The synthesis is of theoretical interest only.

5.4.2 Nitration of alkyldichloramines

$$RNH_2 \xrightarrow{2\ HOCl} RNCl_2 \xrightarrow{HNO_3,\ Ac_2O} RN(NO_2)Cl \xrightarrow{NaHSO_3\ (aq)} RNHNO_2 \quad \text{(Eq. 5.13)}$$

Figure 5.29

Wright and co-workers[61] prepared a number of alkyldichloramines from the action of hypochlorous acid on primary amines and found these stable enough in acidic solution to undergo nitration with acetic anhydride–nitric acid mixtures to give the corresponding *N*-chloronitramines (Equation 5.13). *N*-Chloronitramines are isolatable intermediates and stable under acidic conditions, although some are sensitive and violent explosives. The presence of acetic anhydride during the nitration of alkyldichloramines is essential in keeping the reaction conditions anhydrous – *N*-chloronitramines are readily hydrolyzed in the presence of aqueous acid and it is unlikely that the resulting primary nitramine would survive the acidic conditions of the nitration.

N-Chloronitramines can be isolated or the diluted nitration liquors treated directly with an aqueous solution of sodium bisulfite to generate the corresponding nitramine in good yield. The presence of a reducing agent during hydrolysis is essential because the conversion of a *N*-chloronitramine to a nitramine and hypochlorous acid is a reversible process.

5.5 *N*-NITRATION OF AMIDES AND RELATED COMPOUNDS

The direct *N*-nitration of *N*-alkylamides is the most important of the synthetic routes to secondary nitramides. Hydrolysis of these secondary nitramides is an important but indirect route to primary nitramines, a reaction discussed in Section 5.10.

Numerous studies have shown that amide substrates containing the – $CONHCH_2$ – group are readily nitrated to the corresponding nitramide with a range of nitrating agents. The facile nitration of such amides is mainly due to the weakly basic nature of the nitrogen atom in these substrates. Substrates containing the – CONHCO – urea group are far less susceptible to *N*-nitration. Heterocyclic *N*-nitroureas and *N,N′*-dinitroureas are recognised as a class of energetic materials (Chapter 6). Consequently, the nitration of ureas has been well studied.

5.5.1 Nitration with acidic reagents

Secondary nitramides are relatively stable in highly acidic media and so their synthesis from the direct nitration of *N*-substituted amides with nitric acid and its mixtures is feasible. The synthesis of primary nitramides from the nitration of *N*-unsubstituted amides is usually not possible in acidic media, although this class of compounds have no practical value as explosives anyway.

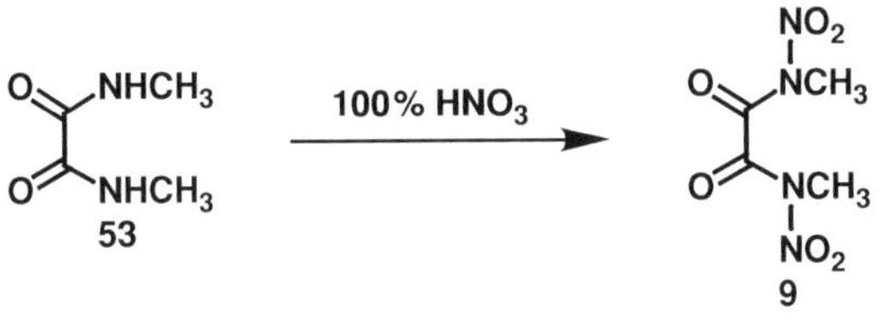

Figure 5.30

Pure nitric acid (98–100 %) has long been the classic reagent for the *N*-nitration of *N*-alkylamides, and in some cases, this reagent gives excellent yields of the corresponding secondary nitramide. Many *N*-alkyl carbamates and sulfonamides have been successfully nitrated in this way.[62–70] Treatment of the oxamide (53),[69] the sulfonamide (54)[69] and the carbamate (55)[70] with pure nitric acid at 0 °C yields the secondary high explosives, (9), (10) and (56) respectively. *N,N′*-Dinitro-*N,N′*-dimethyloxamide (9) (m.p. = 124 °C) forms a convenient eutectic with TNT and the melt solidifies to give charges free of gross cavities. A eutectic mixture of 70 % PETN and 30 % *N*, *N'*-dinitro-*N*, *N'*-dimethyloxamide melts at 100 °C and has a VOD ~ 8500 m/s. Replacing 10 % of the PETN with dimethyl oxalate lowers the melting point to 82 °C but the VOD drops to 7900 m/s.

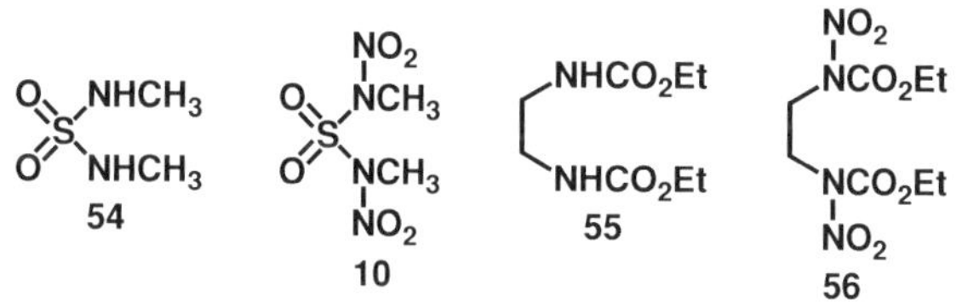

Figure 5.31

A solution of fuming or absolute nitric acid in acetic anhydride is another commonly used reagent for the *N*-nitration of *N*-alkylamides. Curry and Mason[71] used this reagent to synthesize a series of aliphatic and alicyclic *N*-nitro-*N*-alkylcarbamates from the corresponding *N*-alkylcarbamates. Aromatic ring nitration is observed if aryl groups are present in the carbamate substrate.

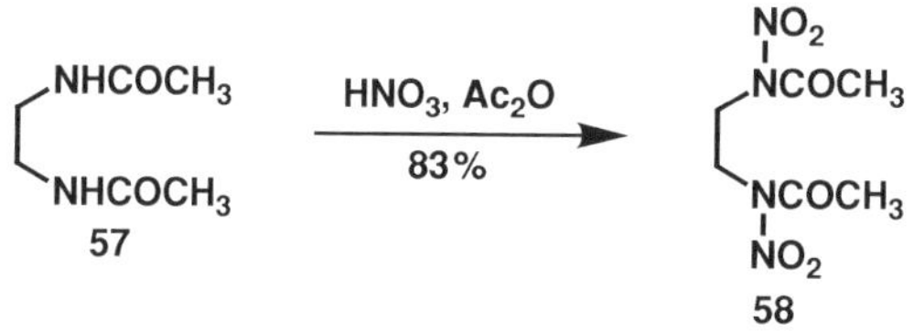

Figure 5.32

Nitric acid–acetic anhydride reagent has been used to synthesize *N*, *N'*-dinitro-*N*, *N'*-ethylenebisacetamide (58) from *N*, *N'*-ethylenebisacetamide (57); the former is a secondary high explosive and a precursor to the powerful explosive ethylenedinitramine (Section 5.10).[70] It is interesting to note that (58) is not formed when (57) is treated with nitric acid alone or with strong mixed acid.

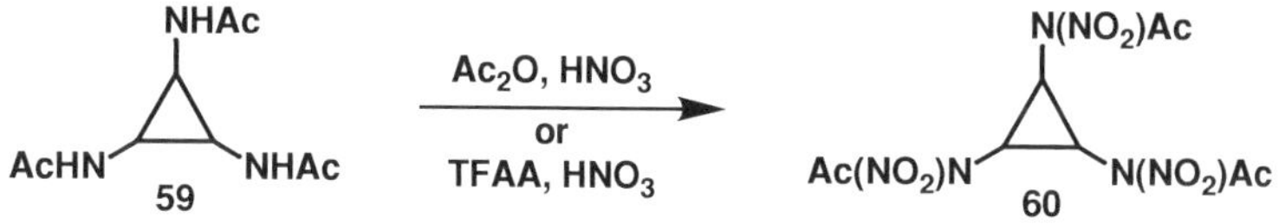

Figure 5.33

The energetic nitramide (60) has been prepared from the nitration of the tris-acetamide (59) with nitric acid and acetic anhydride or trifluoroacetic anhydride.[72] Nitric acid–trifluoroacetic anhydride mixtures are powerful nitrating agents and well suited for amide and urea *N*-nitration.

Nitric acid–acetic anhydride mixtures give poor yields for the nitration of amides with groups that hinder the amide nitrogen against electrophilic attack. The use of higher temperatures in these reactions leads to variable amounts of the *N*-nitroso compound as a by-product.[71]

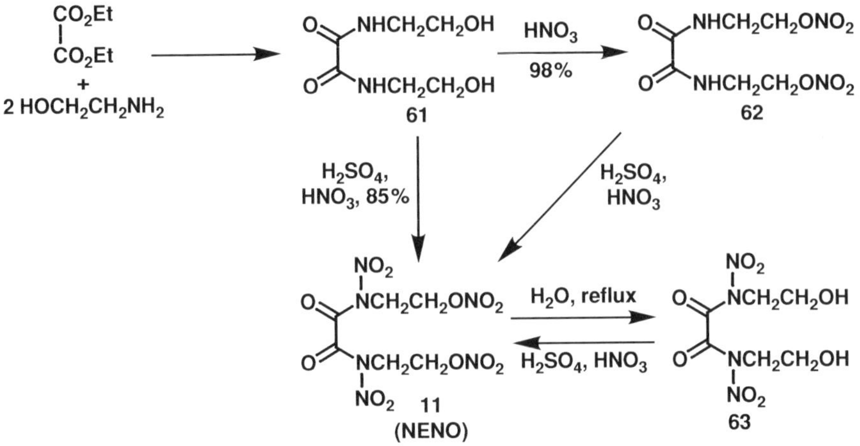

Figure 5.34

A mixture of concentrated sulfuric and nitric acids has been used for the *N*-nitration of amides and ureas. *N,N'*-Dinitro-*N,N'*-bis(2-hydroxyethyl)oxamide dinitrate (NENO) (11) is prepared from the action of mixed acid on *N,N'*-bis(2-hydroxyethyl)oxamide (61), itself prepared from the condensation of diethyloxalate with two equivalents of ethanolamine.[73] Nitrosylsulfuric acid is an inhibitor of *N*-nitration and so nitrous acid should be rigorously excluded. The reaction of (61) with absolute nitric acid results in *O*-nitration of the hydroxy groups but no *N*-nitration, and consequently, (62) is isolated as the sole product.

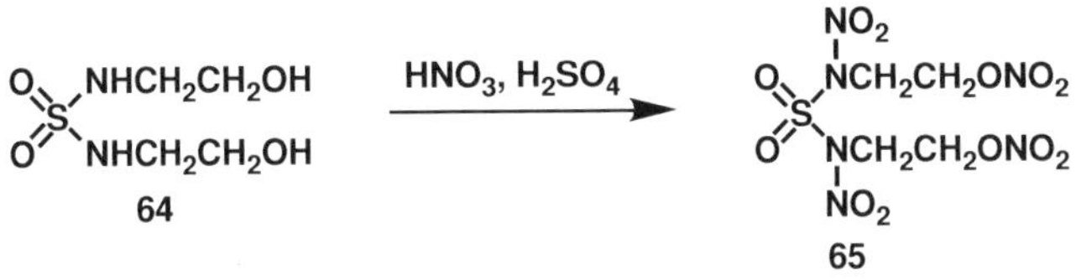

Figure 5.35

A number of other secondary nitramide explosives have been prepared from the action of mixed acid on the parent amide. Treatment of sulfuryl chloride with two equivalents of ethanolamine, followed by nitration of the resulting sulfamide (64) with mixed acid, yields the nitramide explosive *N,N'*-dinitro-*N,N'*-bis(2-hydroxyethyl)sulfamide dinitrate (65).[74]

Although a large number of secondary nitramides have been prepared they have not found wide use as explosives because of their facile hydrolysis to acidic primary nitramines in the presence of water. Research has focused on the synthesis of cyclic and bicyclic *N*-nitroureas and *N,N'*-dinitroureas because of their high performance.

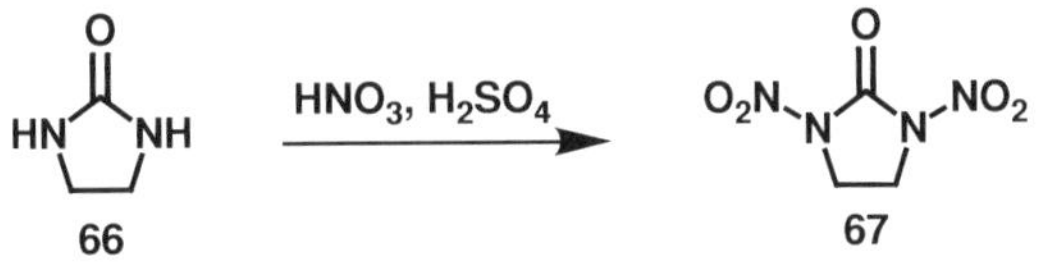

Figure 5.36

N,N'-Dinitro-2-imidazolidone (67), a precursor to ethylenedinitramine (EDNA), is synthesized from the nitration of 2-imidazolidinone (66) with mixed acid.[70]

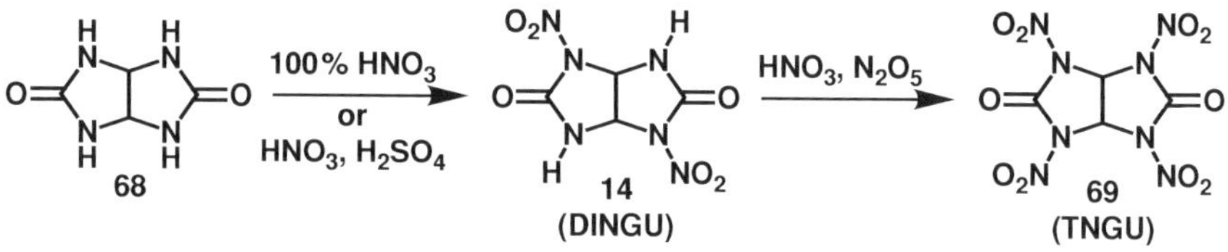

Figure 5.37

Glycouril (68) is converted to its dinitro derivative (14) (DINGU) on treatment with either absolute nitric acid or strong mixed acid. Further nitration to the tetranitro derivative (69) (TNGU) requires the use of a more powerful nitrating agent in the form of a 20 % solution of dinitrogen pentoxide in absolute nitric acid.[75] The latter reagent has only fairly recently found wide use for the synthesis of energetic materials, but the number of examples of its use for nitrolysis (Section 5.6) is already extensive. In the laboratory a solution of dinitrogen pentoxide in nitric acid is readily synthesized by the addition of phosphorous pentoxide[76] to absolute nitric acid or from the anodic oxidation of dinitrogen tetroxide[77] in absolute nitric acid (Chapter 9).

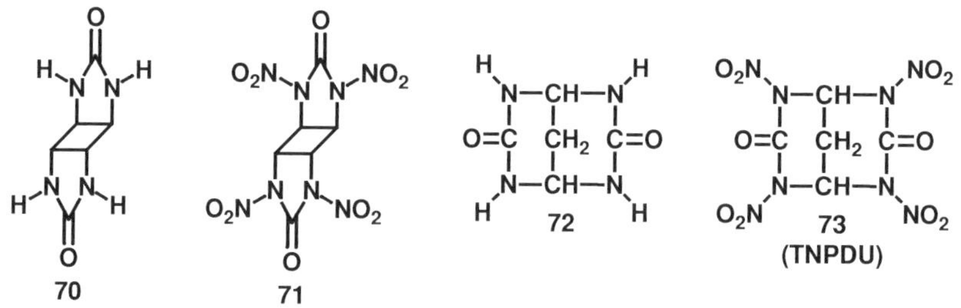

Figure 5.38

The tricycle (71) has been synthesized in 97 % yield via the nitration of the corresponding urea (70) with dinitrogen pentoxide in nitric acid.[78] The high performance bicyclic *N,N'*-dinitrourea known as TNPDU (73) is synthesized from the reaction of propanediurea (72) with acetic anhydride–nitric acid.[79]

N,N'-Dinitrourea (DNU) has been prepared by the nitration of urea with a mixture of 98 % nitric acid and 20 % oleum between −10 °C and −15 °C. *N,N'*-Dinitrourea is unstable at room temperature. However, the diammonium and dipotassium salts are more stable and decompose at 110 °C and 135 °C respectively.[80] *N,N'*-Dinitrourea may find future use for the synthesis of bicyclic and caged heterocyclic *N,N'*-dinitroureas.

The synthesis, properties and potential uses of high performance *N*-nitroureas and *N,N'*-dinitroureas are discussed in more detail in Chapter 6.

5.5.2 Nitration with nonacidic reagents

The *N*-nitration of amides and related compounds with nonacidic nitrating agents has received far less attention for the synthesis of energetic materials compared to the use of acidic reagents employing nitric acid and its mixtures. Even so, reagents like dinitrogen pentoxide in chlorinated solvents and nitronium salts, are efficient and mild nitrating agents for amides and ureas.

5.5.2.1 Dinitrogen pentoxide

Von Runge and Triebs[81] used a solution of dinitrogen pentoxide in chloroform for the *N*-nitration of both amides and imides. Solutions of dinitrogen pentoxide in chlorinated solvents are not neutral nitrating agents when amides and imides are nitrated – the presence of acidic N–H protons in these substrates leads to the formation of nitric acid. Sodium fluoride acts like a 'base' towards nitric acid and so its addition to these reactions can increase product yield.[82] Sodium acetate has been used for the same purpose during the nitration of *n*-butyl-*N,N'*-dimethylurea.[83] The effectiveness of dinitrogen pentoxide for the *N*-nitration of ureas is further illustrated by its use in the conversion of 2-imidazolidinone to *N,N'*-dinitro-2-imidazolidinone in 90 % yield.[82] In the presence of sodium fluoride the yield for this reaction exceeds 90 %.

Solutions of dinitrogen pentoxide in methylene chloride or chloroform have been used for the *N*-nitration of the sodium salts of some *N*-alkylsulfamides.[84] Sulfonamide substrates with both alkyl and aryl substituents are *N*-nitrated in excellent yields with this reagent. Aromatic ring nitration occurs when aryl substituents are present.

A practical application of dinitrogen pentoxide in methylene chloride reagent involves the nitration of either ammonium carbamate or nitrourethane, followed by ammonolysis to yield ammonium dinitramide, an energetic oxidizer with enormous potential for use in future high performance propellant compositions.[85] This important reaction is discussed in more detail in Chapter 9.

5.5.2.2 Nitronium salts

Olsen and co-workers[46] used a solution of nitronium tetrafluoroborate in acetonitrile for the *N*-nitration of acetamides and urethanes at −30 °C. The following nitramides were obtained by this method: *N*-nitroacetamide (13 %), *N*-nitro-2-chloroacetamide (55 %), *N*-nitro-*n*-butylacetamide (40 %), *N*-nitrobenzamide (53 %), ethyl *N*-nitro-*n*-butylcarbamate (91 %) and *N*-nitrosuccinimide (43 %). The low yield of *N*-nitroacetamide, a primary nitramide, is attributed to competing hydrolysis due to the release of tetrafluoroboric acid as the reaction progresses. The scope of the reaction is improved by moving to more basic solvents like ethyl acetate, 1,4-dioxane and trimethyl phosphate.[86]

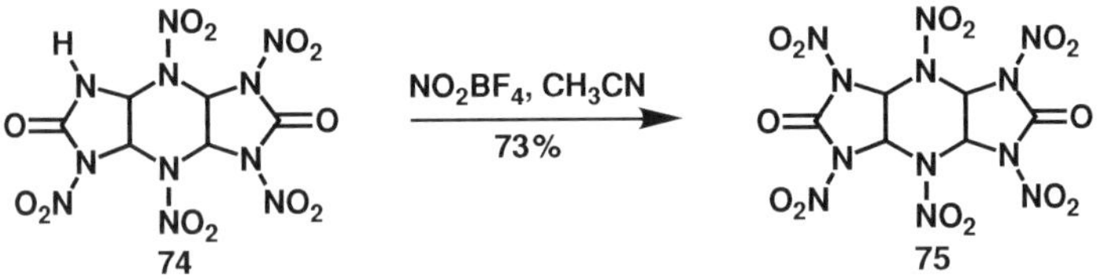

Figure 5.39

Boyer and co-workers[87] used a solution of nitronium tetrafluoroborate in acetonitrile for the introduction of the final nitro group into the energetic dinitrourea (75).

5.5.2.3 Other nitrating agents

A number of reagents derived from nitrate salts and acid anhydrides have been reported for the *N*-nitration of amides and related compounds. Crivello[88] first reported the use of metal nitrates in trifluoroacetic anhydride (TFAA) for the nitration of aromatic systems. Chapman and Suri[89] used the same system, which probably involves trifluoroacetyl nitrate as the active nitrating agent, for the *N*-nitration of some cyclic amides, imides and ureas (66). For this purpose, ammonium nitrate–TFAA in nitromethane was used. A comprehensive study of this nitrating system in relation to amides was subsequently conducted, including optimization of conditions and exploring other related nitrating agents.[90]

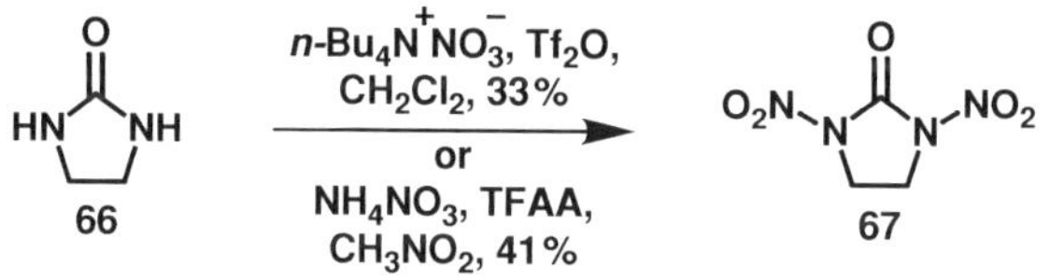

Figure 5.40

A reagent composed of tetra-*n*-butylammonium nitrate and TFAA in methylene chloride has been used to nitrate a series of *N*-alkyl and *N*-aryl amides (40–90 %).[91] The formation of significant amounts of *N*-nitrosamides was noted. Tetra-*n*-butylammonium nitrate and triflic anhydride in methylene chloride has been used to successfully nitrate a variety of heterocyclic amides, imides and ureas (66).[92]

5.6 NITROLYSIS

'Nitrolysis' is a term originally used for the rupture of a N–C bond leading to the formation of the $N\text{–}NO_2$ group. A prime example is the nitrolysis of the $N\text{–}CH_2$ bonds of hexamine to form the important military explosives RDX and HMX. Nitrolysis is the most important route available to polynitramine energetic materials.

The scope of nitrolysis is huge, with examples of nitramine formation from the cleavage of tertiary amines, methylenediamines, carbamates, ureas, formamides, acetamides and other amides. The definition of nitrolysis must be extended to the nitrative cleavage of other nitrogen bonds because sulfonamides and nitrosamines are also important substrates for these reactions. The nitrative cleavage of silylamines and silylamides is also a form of nitrolysis (Section 5.7).

The range of reagents used for nitrolysis is also vast and includes absolute nitric acid, acid anhydride–nitric acid, mixed acid, dinitrogen pentoxide–nitric acid, nitronium salts and many more.

5.6.1 Nitrolysis of amides and their derivatives

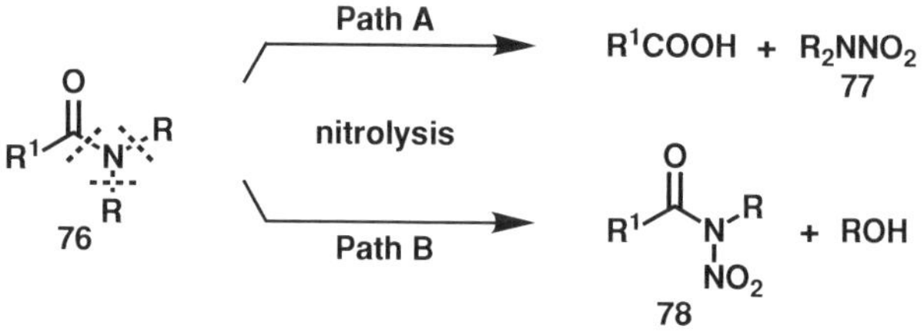

Figure 5.41

Substrates containing *N*,*N*-disubstituted amide functionality (76) can undergo nitrolysis by two pathways leading to different products – a secondary nitramine (77) can be formed from the breakage of the amide bond in path A, whereas rupture of either of the other two C–N bonds in path B leads to a secondary nitramide (78). Examples exist of both types of reaction, but generally speaking, the secondary nitramine is the observed product from the nitrolysis of *N,N*-disubstituted amides.

5.6.1.1 Nitrolysis with acidic reagents

Acidic reagents composed of nitric acid and its mixtures are by far the most important reagents used in nitrolysis reactions. Early work by Robson and Reinhart[93,94] showed that *N,N*-disubstituted amides in the form of formamides, acetamides and sulfonamides containing straight chain alkyl groups in association with the acyl group undergo efficient nitrolysis to the corresponding secondary nitramines when treated with a solution of fuming nitric acid in trifluoroacetic anhydride. The use of nitric acid in acetic anhydride is generally less efficient, although good yields of nitramine product can be attained for some substrates.[93,94] A number of important observations were made: (1) alkyl groups with branching on the α carbon lead to greatly reduced yields, (2) substrates containing electronegative substituents next to the acyl group give a greatly reduced yield of nitramine product due to a reduction in electron density at the acyl nitrogen and (3) *N,N*-dialkylureas and *N,N*-dialkylcarbamates are found to be poor substrates for nitrolysis with acid anhydride–nitric acid mixtures. While *N,N*-dialkylacetamides undergo facile nitrolysis with nitric acid–acid anhydride mixtures, the corresponding *N*-alkyl-*N,N*-diacylamines are inert to nitrolysis under these conditions, a consequence of the further reduction in electron density at the acyl nitrogen.[95]

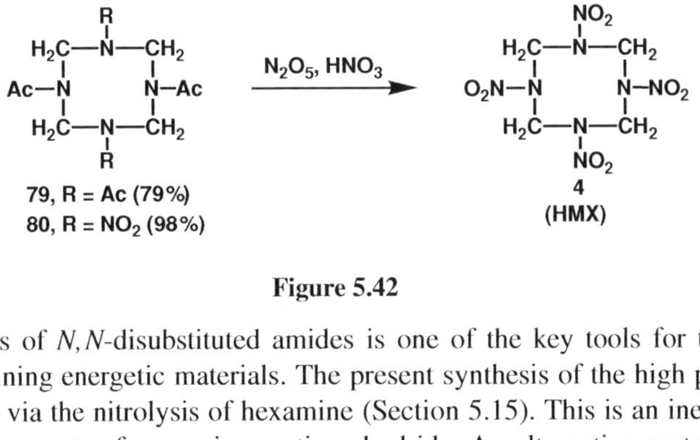

N_2O_5, HNO_3

79, R = Ac (79%)
80, R = NO_2 (98%)

4
(HMX)

Figure 5.42

The nitrolysis of *N,N*-disubstituted amides is one of the key tools for the synthesis of nitramine containing energetic materials. The present synthesis of the high performance explosive HMX is via the nitrolysis of hexamine (Section 5.15). This is an inefficient reaction requiring large amounts of expensive acetic anhydride. An alternative route to HMX (4) is via the nitrolysis of either 1,3,5,7-tetraacetyl-1,3,5,7-tetraazacyclooctane (79) (79 %) or 1,5-dinitro-3,7-diacetyl-1,3,5,7-tetraazacyclooctane (80) (98 %) with dinitrogen pentoxide in absolute nitric acid.[96] These reactions are discussed in more detail in Section 5.15.

Nitrolysis reactions employing 1,3,5-trisubstituted-1,3,5-triazacyclohexanes have been explored as alternative routes to RDX.[30,97] Some of the results are illustrated in Table 5.4 and show the difference in the efficiency of the three nitrolysis agents used, namely, absolute nitric acid, phosphorus pentoxide–nitric acid and trifluoroacetic anhydride–nitric acid. The acetamide derivative (81) (TRAT) undergoes incomplete nitrolysis on treatment with absolute nitric acid and trifluoroacetic anhydride–nitric acid to give a crude product containing some 1-acetyl-3,5-dinitro-1,3,5-triazacyclohexane (TAX) (82) (Table 5.4, Entry 1); the latter can be preferentially formed in 93 % by suitably modifying the reaction conditions.[97] Interestingly, the nitrolysis of

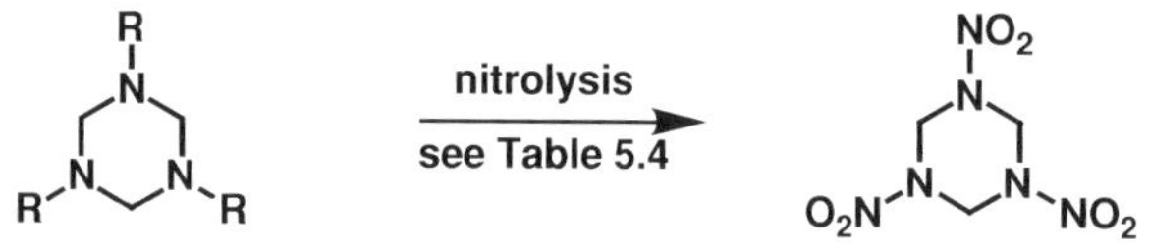

nitrolysis
see Table 5.4

Table 5.4
Nitrolysis of 1,3,5-trisubstituted-1,3,5-triazacyclohexanes

Entry	R (see above)	Yield (%) of RDX (3)			Ref.
		HNO_3–P_2O_5	100% HNO_3	HNO_3–TFAA	
1	CH_3CO	95	15[a]	80[b]	97
2	C_2H_5CO	80	55	98	97
3	*n*-C_3H_7CO	60	40	94	97
4	$(CH_3)_2CHCO$	50	30	94	97
5	$(CH_3)_3CCO$	0	0	15	97
6	CH_3OCO	0	---	0	97
7	*t*-BuOCO	---	---	5	30
8	*p*-$CH_3C_6H_4SO_2$	0	---	---	97

[a] Crude product contains 25% TAX (82). [b] Crude product contains 4% TAX (82).

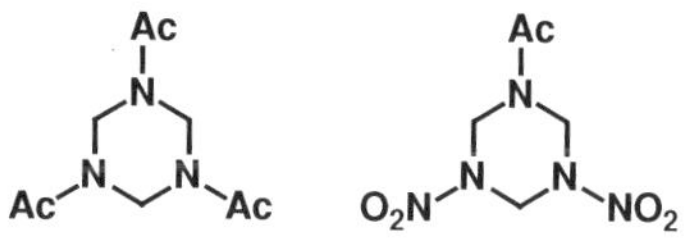

81 (TRAT) 82 (TAX)

Figure 5.43

similar substrates, where R = alkanoyl, yield RDX as the sole product (Table 5.4, Entries 2, 3 and 4).[97] Carbamate derivatives derived from simple straight chain aliphatic alcohols are found to be inert to nitrolysis, even with the powerful TFAA–nitric acid and phosphorus pentoxide–nitric acid mixtures (Table 5.4, Entry 6).[97] The nitrolysis of the *tert*-butoxycarbonyl (BOC) derivative gives a 5 % yield of RDX (Table 5.4, Entry 7).[30] Sulfonamide derivatives like the tosylate appear to be fairly inert to nitrolysis in this particular case (Table 5.4, Entry 8).

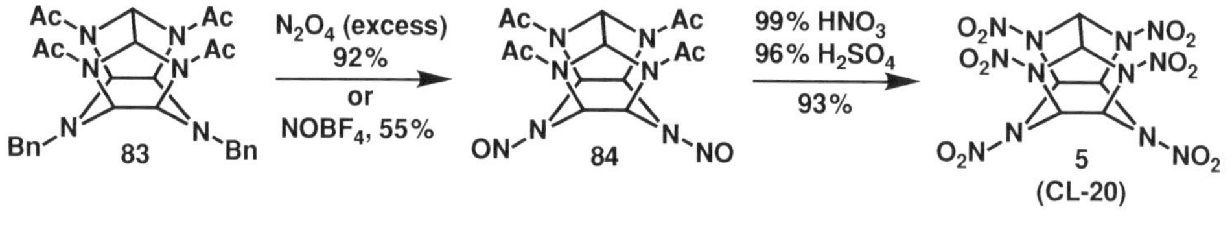

83 — N_2O_4 (excess) 92% or $NOBF_4$, 55% → 84 — 99% HNO_3 96% H_2SO_4 93% → 5 (CL-20)

Figure 5.44

The high performance nitramine explosive known as CL-20 (5) has been synthesized via a two-stage nitrolysis starting from the key intermediate (83).[98,99] The first stage uses dinitrogen tetroxide or nitrosonium tetrafluoroborate for nitrosolysis.[98] The second step, involving nitrolysis of the acetamide and nitrosamine bonds, is achieved with nitronium tetrafluoroborate (>90 %)[98] or mixed acid at 75 °C to 80 °C (93 %).[99] The synthesis of CL-20 is discussed in more detail in Chapter 6.

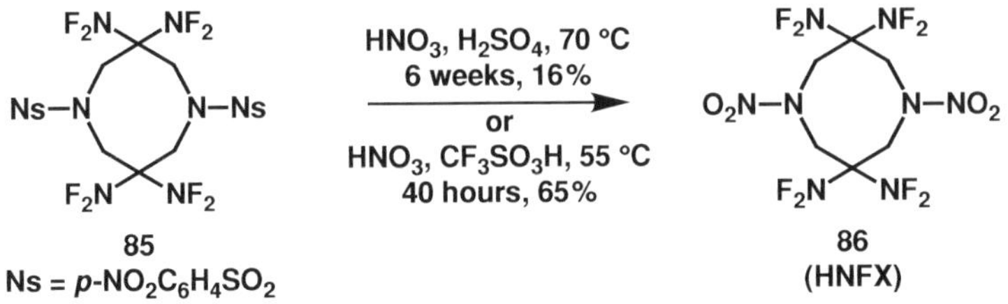

Figure 5.45

Inductive effects can have very pronounced effects on the reactivity of amides and similar substrates towards nitrolysis. Chemists at the Naval Air Warfare Center (NAWC) have reported an extreme case encountered during the synthesis of the energetic 1,5-diazocine known as HNFX (86).[100] A key step in this synthesis involves a very difficult nitrolysis of the electron deficient N-nosyl (4-nitrobenzenesulfonyl) bonds of (85). Nitrolysis with strong mixed acid requires a temperature of 70 °C for 6 weeks to achieve a yield of 16 %. The same reaction with nitric acid–triflic acid requires a temperature of 55 °C for 40 hours to achieve a 65 % yield of HNFX. The same chemists reported a similar case of N-nosyl bond nitrolysis which needed a nitrating agent composed of nitric acid–triflic acid–antimony pentafluoride.[101]

5.6.1.2 Nitrolysis with nonacidic reagents

$$R^1C(O)NR_2 + NO_2BF_4 \longrightarrow R_2NNO_2 + R^1\overset{+}{C}O\,\overset{-}{B}F_4 \quad \text{(Eq. 5.14)}$$

Figure 5.46

Secondary nitramines are conveniently prepared from the nitrolysis of *N*, *N*-dialkylamides with nitronium salts in acetonitrile or ethyl acetate at 20 °C where the acyl group is converted into an acylium tetrafluoroborate (Equation 5.14).[102] Problems can occur if commercial nitronium salts like the tetrafluoroborate are used without purification. The presence of nitrosonium salts can then lead to nitrosamines via nitrosolysis. Yields of secondary nitramine up to 90 % have been reported[102] with solutions of nitronium tetrafluoroborate in acetonitrile; di-*n*-butylnitramine is obtained in 82 % yield from the nitrolysis of corresponding acetamide.[46]

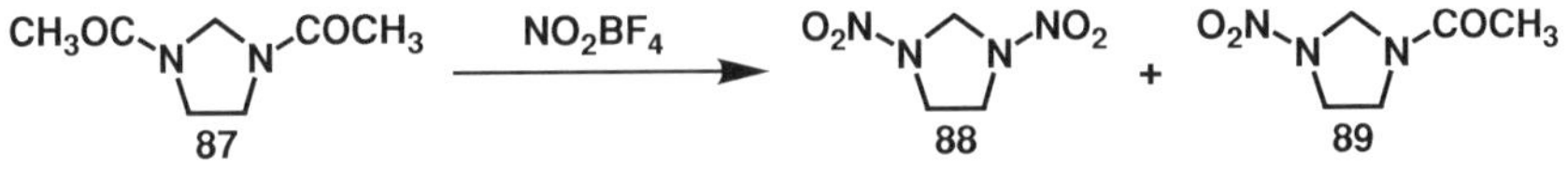

Figure 5.47

N,N′-Diacetylimidazolidine (87) undergoes nitrolysis on treatment with nitronium tetrafluoroborate to yield either *N,N′*-dinitroimidazolidine (88) or *N*-acetyl-*N′*-nitroimidazolidine (89) depending on the reaction stoichiometry and conditions used.[103] *N,N*-Diacylmethylamines are inert to reagents like trifluoroacetic anhydride–nitric acid but undergo nitrolysis with nitronium salts, and hence, provide a route to alkyl-*N,N*-dinitramines.[103] The nitrolysis of *N*-alkylamides with nitronium tetrafluoroborate is more of a problem due to the instability of the primary nitramine product in the presence of the strong acid liberated during the reactions.[104]

5.6.2 Nitrolysis of N-alkyl bonds

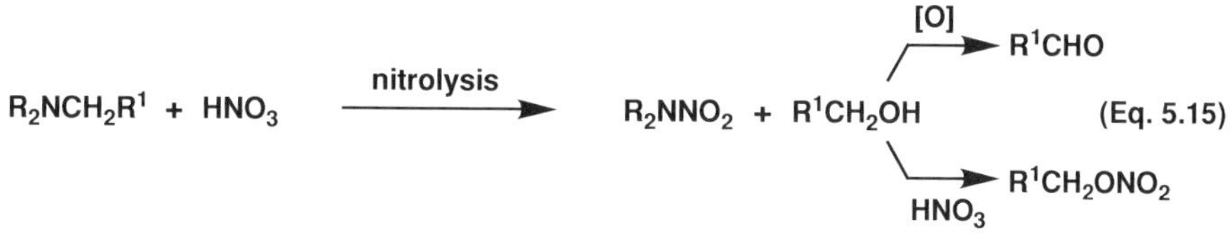

$$R_2NCH_2R^1 + HNO_3 \xrightarrow{\text{nitrolysis}} R_2NNO_2 + R^1CH_2OH \quad \begin{cases} \xrightarrow{[O]} R^1CHO \\ \xrightarrow{HNO_3} R^1CH_2ONO_2 \end{cases} \qquad \text{(Eq. 5.15)}$$

Figure 5.48

The nitrolysis of tertiary amines in the form of *tert*-butylamines and methylenediamines has been used to synthesize numerous polynitramine-based energetic materials. In these reactions one of the N–C bonds is cleaved to generate a secondary nitramine and an alcohol; the latter is usually *O*-nitrated or oxidized under the reaction conditions (Equation 5.15). The ease in which nitrolysis occurs is related to the stability of the expelled alkyl cation. Consequently, the *tert*-butyl group and the iminium cation from methylenediamines are excellent leaving groups.

5.6.2.1 Primary and secondary alkyl groups

Tertiary amine substrates containing only primary or secondary alkyl functionality are reluctant to undergo nitrolysis and often need pushing by increasing the reaction temperature. The reaction of nitric acid–acetic anhydride mixtures with tertiary alkylamines containing only primary and secondary alkyl groups usually generates the corresponding nitrosamine as the main product.[105] Although nitrosamines can be oxidized[106] to the corresponding nitramine their high toxicity makes their formation undesirable.

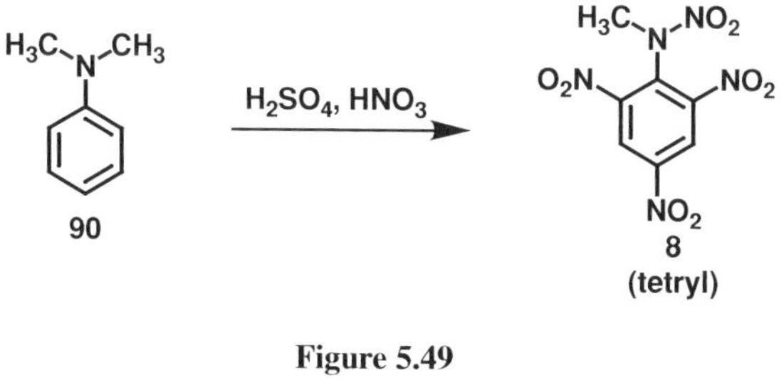

H_3C–N–CH_3 (phenyl) **90** $\xrightarrow{H_2SO_4,\ HNO_3}$ H_3C–N–NO_2, O_2N, NO_2, NO_2 **8** **(tetryl)**

Figure 5.49

A important exception is the synthesis of the practical high explosive known as tetryl (8), prepared by treating *N,N*-dimethylaniline (90) with nitric acid, mixed acid or acetic anhydride–nitric acid mixtures.[107] In these reactions, ring nitration to 2,4-dinitro-*N,N*-dimethylaniline is followed by cleavage of one of the N–CH_3 bonds to form the nitramino functionality; subsequent ring nitration yields tetryl (Section 5.14).

5.6.2.2 Tert-Butyl groups

The nitrolysis of tertiary amines containing a *tert*-butyl group is synthetically useful and often gives a high yield of the corresponding secondary nitramine. The *tert*-butyl group can be used as a blocking group for the Mannich condensation of polynitroalkanes with amines and formaldehyde. The need for a blocking group is essential to avoid multiple condensations which can lead to a complex mixture of products or even polymerization. Mannich condensations employing *tert*-butylamine have provided many polynitro derivatives of aliphatic amines in which a *tert*-butyl group blocks the amine functionality.

Table 5.5
Synthesis of secondary nitramines from the nitrolysis of *tert*-butylamines (ref. 108)

Entry	Substrate	Nitrolysis agent	Product	Yield (%)
1a	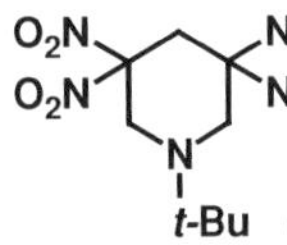 (91)	H_2SO_4, HNO_3	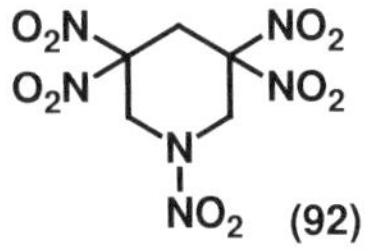 (92)	81
1b	(91)	100% HNO_3	(92)	96
2	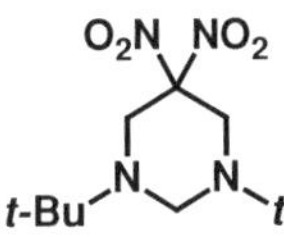 (93)	H_2SO_4, HNO_3	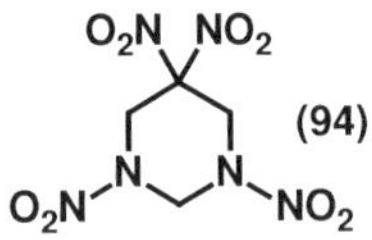 (94)	87
3	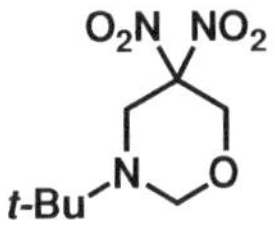 (95)	Ac_2O, HNO_3	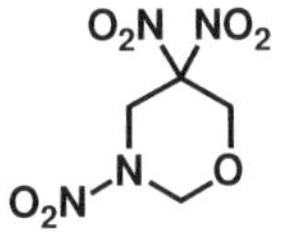 (96)	54
4	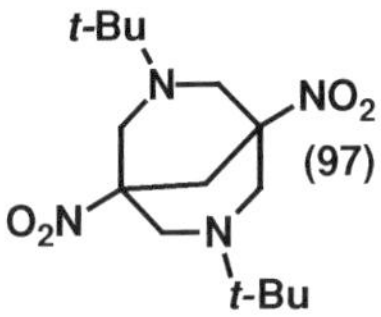 (97)	100% HNO_3	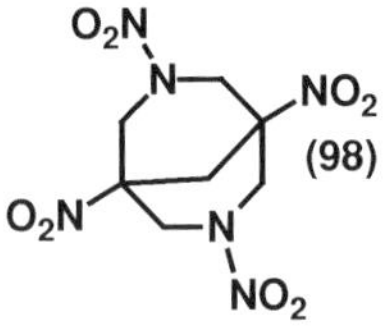 (98)	59
5	t-BuN$(CH_3)_2$. HCl (99)	Ac_2O, HNO_3	$(CH_3)_2NNO_2$ (100)	55

Adolph and Cichra[108] prepared a number of cyclic nitramines from the nitrolysis of *tert*-butyl protected Mannich products (Table 5.5). Nitrolysis of the *tert*-butyl groups was achieved with mixed acid, pure nitric acid or a mixture of nitric acid in acetic anhydride depending on the substrate. Pure nitric acid was found to affect the nitrolysis of both the *tert*-butyl groups of (97), (Table 5.5, Entry 4) whereas the use of mixed acid led to the isolation of the product where only one of the *tert*-butyl groups had undergone nitrolysis. Some of the cyclic nitramine products from this study were found to be powerful explosives. Adolph and Cichra[108] found that the hydrochloride salt of *tert*-butyldimethylamine (99) undergoes nitrolysis to dimethylnitramine (100) in good yield when treated with nitric acid in acetic anhydride (Table 5.5, Entry 5); the free amine base gives a much lower yield of nitramine product.

In some related work, Lear and co-workers[109] prepared the powerful cyclic nitramine explosive Keto-RDX (102) from the nitrolysis–nitration of 2-oxa-5-*tert*-butyltriazone (101). Pagoria and co-workers[30a,110] conducted a full study on the effect of different nitrating agents on the yield of Keto-RDX (Table 5.6).

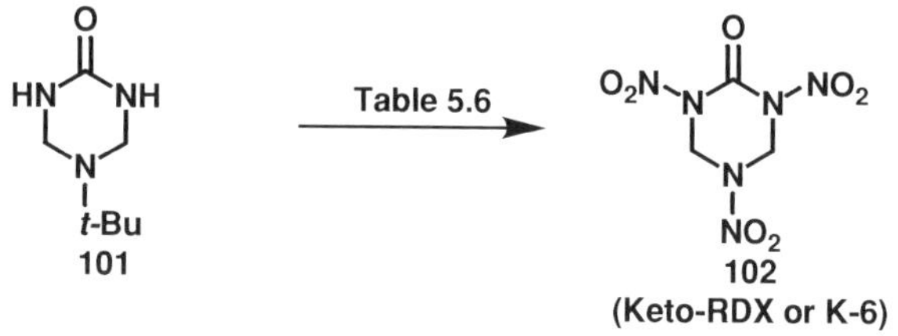

101 → (Table 5.6) → 102 (Keto-RDX or K-6)

Table 5.6
Effect of nitrolysis agent on the yield of Keto-RDX (ref. 30a)

Nitrating agent	Yield (%)
$TFAA/N_2O_5/HNO_3$	57
$TFAA/100\%\ HNO_3$	43
$Ac_2O/100\%\ HNO_3$	57
$Ac_2O/90\%\ HNO_3$	21
$Ac_2O/70\%\ HNO_3$	0
H_2SO_4/HNO_3	0
NO_2BF_4	40

Source: Reprinted with permission from C. L. Coon, E. S. Jessop, A. R. Mitchell, P. F. Pagoria and R. D. Schmidt, in *Nitration: Recent Laboratory and Industrial Developments, ACS Symposium Series* 623, Ed. L. F. Albright, R. V. C. Carr and R. J. Schmitt, American Chemical Society, Washington, DC, 155 (1996); Copyright 1996 American Chemical Society.

103 → (HNO_3, Ac_2O, 35% or NH_4NO_3, Ac_2O, 90%) → 6 (TNAZ)

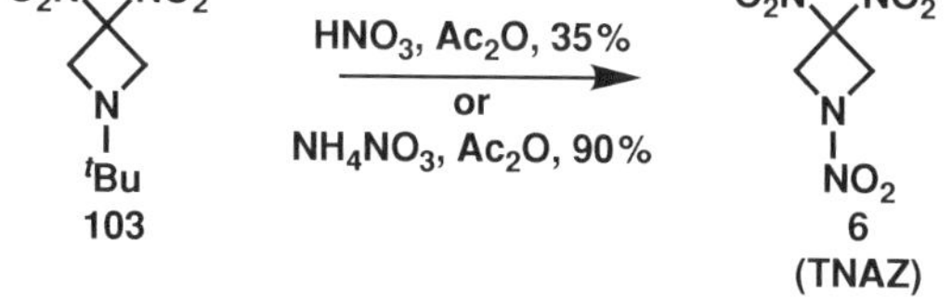

Figure 5.50

Nitrolysis of a *tert*-butyl group is also a key step in the synthesis of the high performance explosive known as TNAZ (6). The nitrolysis of the *N*-*tert*-butylazetidine (103) has been achieved with acetic anhydride–nitric acid[111] and acetic anhydride–ammonium nitrate.[112]

5.6.2.3 Methylenediamines

Methylenediamines are readily synthesized from the reaction of secondary amines with formaldehyde. Many aliphatic amines are too basic for direct nitration without a chloride catalyst, and even then, nitrosamine formation can be a problem. Their conversion into intermediate methylenediamines before nitration is therefore a useful route to secondary nitramines. The success of these nitrolysis reactions is attributed to the inherent low basicity of the methylenediamine nitrogens.

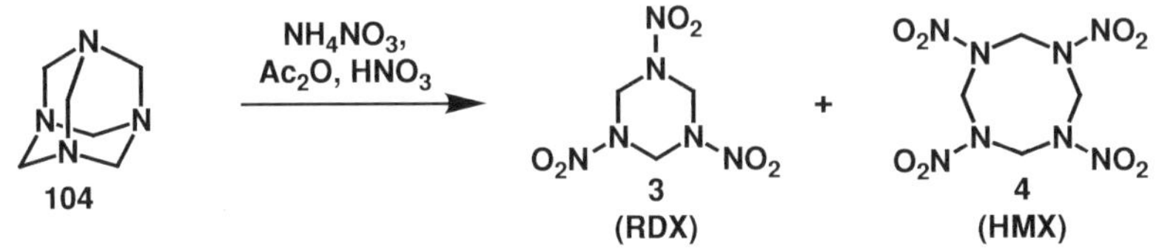

NH_4NO_3, Ac_2O, HNO_3

104 → 3 (RDX) + 4 (HMX)

Figure 5.51

The most important nitrolysis reaction to date is the formation of RDX (3) and HMX (4) from the caged methylenediamine known as hexamine (104). These important military explosives were first mass manufactured by this route towards the end of the Second World War and they are still prepared by this route today. The process uses a mixture of acetic anhydride, ammonium nitrate and nitric acid. The nitrolysis of hexamine is one of the most widely studied reactions in the history of explosives. Many other cyclic and linear polynitramines have been isolated from these reactions and this rich chemistry is discussed in more detail in Section 5.15.

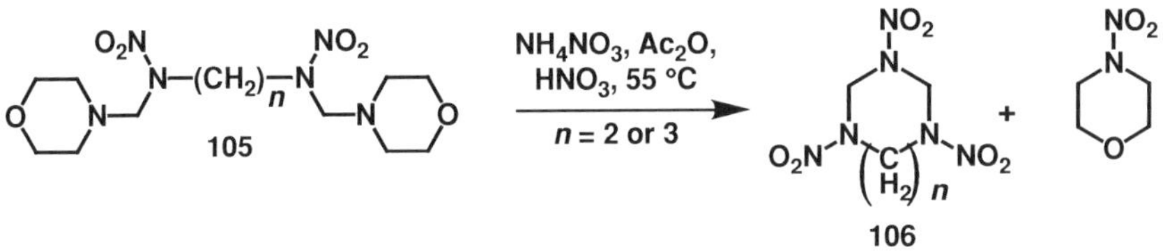

NH_4NO_3, Ac_2O, HNO_3, 55 °C; n = 2 or 3

105 → 106 + N-nitromorpholine

Figure 5.52

Chapman[113] studied the nitrolysis of symmetrical methylenediamines. The nitrolysis of N,N,N',N'-tetramethylmethylenediamine with nitric acid–acetic anhydride–ammonium nitrate mixtures gives both dimethylnitramine and RDX; the latter probably arises from the nitrolysis of hexamine formed from the reaction of ammonium nitrate and formaldehyde released from the hydrolysis of the methylenediamine. The same reaction with some morpholine-based methylenediamines (105) allows the synthesis of 1,3,5-trinitro-1,3,5-triazacycloalkanes (106).

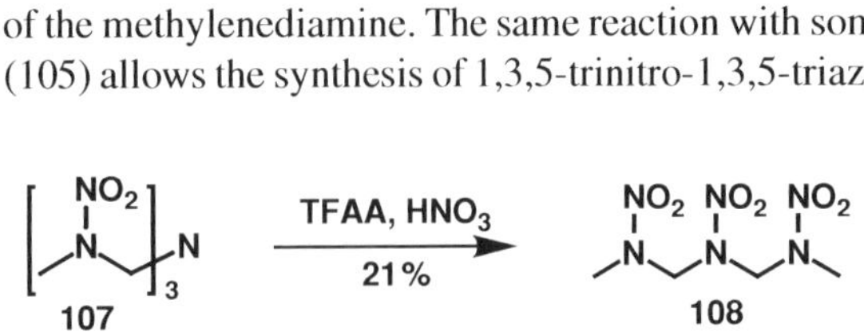

TFAA, HNO_3; 21%

107 → 108

Figure 5.53

The nitrolysis of (107) with nitric acid–trifluoroacetic anhydride yields 2,4,6-trinitro-2,4,6-triazaheptane (108).[114]

The nitrolysis of substituted methylenediamines with nitronium salts can lead to a number of products depending on the nature of the substituents within the substrate. Electron-withdrawing or resonance-stabilizing groups favour the expulsion of an immonium ion and the formation of a secondary nitramine in yields between 58 % and 78 %.[103]

A solution of dinitrogen pentoxide in methylene chloride–acetonitrile also yields secondary nitramines from symmetrical methylenediamines. When the substituent is aliphatic or heterocyclic the nitrolysis occurs specifically at the aminal methylene and yields of secondary nitramine between 25 % and 54 % are reported.[103a]

5.6.3 Nitrolysis of nitrosamines

The conversion of a nitrosamine to a nitramine can be affected by either nitrolysis or oxidation. While the results of these two reactions are identical, they are mechanistically very different. For this reason, the oxidation of nitrosamines is discussed separately in Section 5.9.

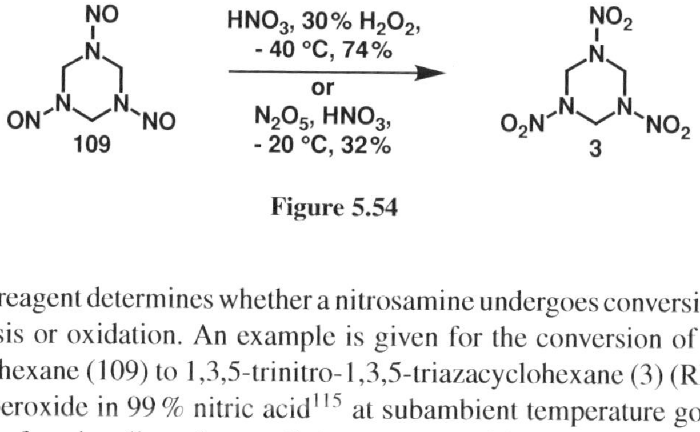
NO
ON
NO
109
HNO3, 30% H2O2,
- 40 °C, 74%
or
N2O5, HNO3,
- 20 °C, 32%
NO2
O2N
NO2
3

Figure 5.54

The choice of reagent determines whether a nitrosamine undergoes conversion to a nitramine by either nitrolysis or oxidation. An example is given for the conversion of 1,3,5-trinitroso-1,3,5-triazacyclohexane (109) to 1,3,5-trinitro-1,3,5-triazacyclohexane (3) (RDX) – the use of 30 % hydrogen peroxide in 99 % nitric acid[115] at subambient temperature goes via oxidation of the nitrosamine functionality, whereas dinitrogen pentoxide in pure nitric acid[116] makes use of a nitrolysis pathway via C–N bond cleavage.

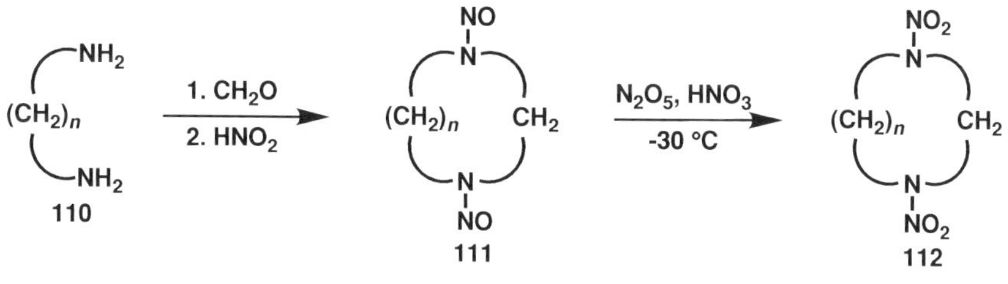
NH2
(CH2)n
NH2
110
1. CH2O
2. HNO2
NO
(CH2)n
CH2
NO
111
N2O5, HNO3
-30 °C
NO2
(CH2)n
CH2
NO2
112

Figure 5.55

The nitrolysis of nitrosamines has been relatively unexplored because of the high toxicity of such substrates. Atkins and Willer[117,118] have prepared a number of cyclic 1,3-dinitrosamines

Table 5.7
Synthesis of cyclic nitramines using the method of Atkins and Willer (ref. 117)

Entry	Substrate	Nitrosamine (% yield)	Nitramine (% yield)
1	$H_2N(CH_2)_2NH_2$	 113 (90)	 88 (85)
2	$H_2NCH_2CH(CH_3)NH_2$	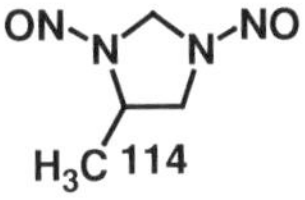 114 (93)	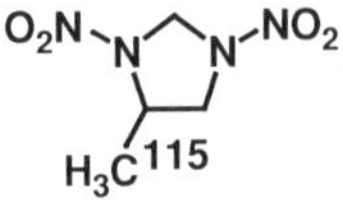 115 (75)
3	$H_2N(CH_2)_3NH_2$	 116 (95)	 117 (30)
4	$C(CH_2NH_2)_4$	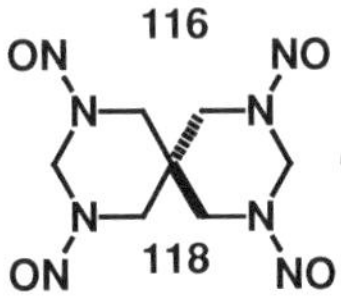 118 (>55)	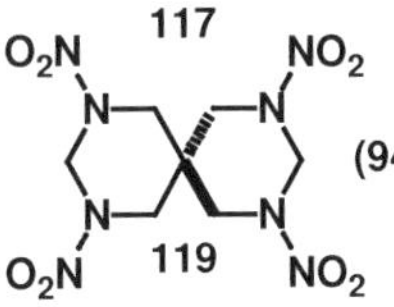 119 (94)

(111) from the reaction of linear aliphatic diamines (110) with formaldehyde followed by trapping the resulting 1,3-diazacycloalkanes with nitrous acid. High yields are reported for the synthesis of both 5- and 6- membered ring 1,3-dinitrosamines, but yields are much lower for 7-membered ring analogs (Table 5.7). The spirocycle (118) is prepared by treating a mixture of 2,2-bis(methylamino)-1,3-propanediamine and formaldehyde with nitrous acid.

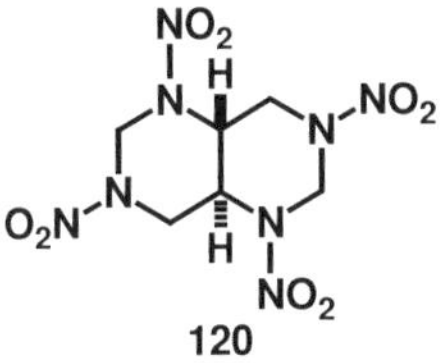

Figure 5.56

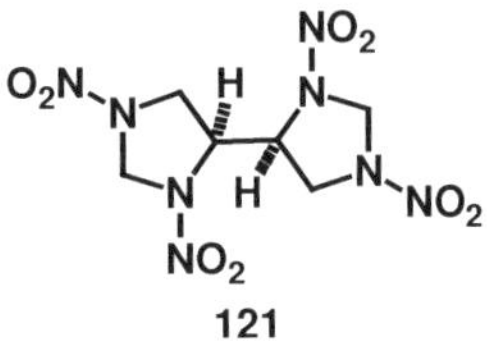

Atkins and Willer[117,118] were unable to oxidize their cyclic 1,3-dinitrosamines to the corresponding 1,3-dinitramines with peroxytrifluoroacetic acid but found that nitrolysis with dinitrogen pentoxide in nitric acid was more successful (Table 5.7). Willer[118] used the same methodology to prepare bicyclic 1,3-dinitramines like (120) and (121). It was found necessary to blow dry nitrogen gas through these reactions in order to remove any dinitrogen tetroxide formed during the nitrolysis, otherwise the 1,3-dinitramine products were found to contain 1-nitrosamine-3-nitramines as contaminants. 1-Nitrosamine-3-nitramines are found to be the main products when the cyclic 1,3-dinitrosamines are treated with 100 % nitric acid.

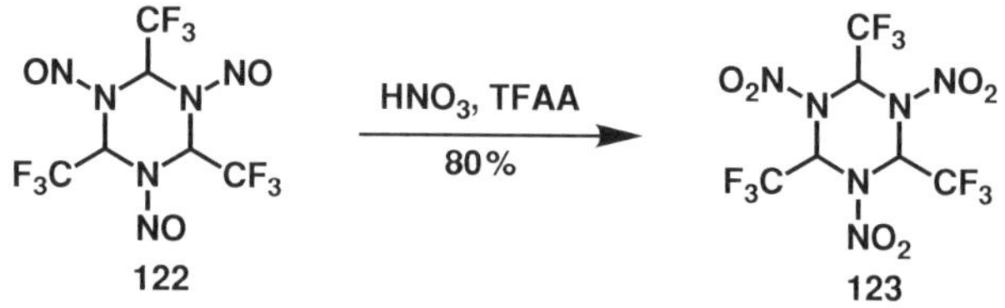

Figure 5.57

Krimmel and co-workers[10] have reported the synthesis of 2,4,6-tris(trifluoromethyl)-1,3,5-trinitro-1,3,5-triazacyclohexane (123) from the nitrolysis of the nitrosamine (122).

5.7 NITRATIVE CLEAVAGE OF OTHER NITROGEN BONDS

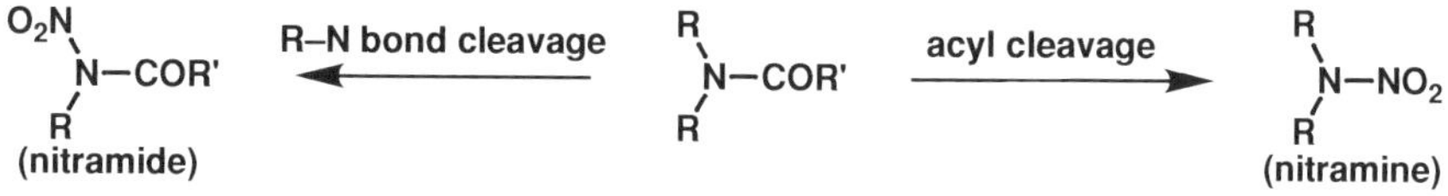

Figure 5.58

In the previous section we discussed the nitrolysis of *N,N*-dialkylamides as a common route to secondary nitramines, especially when the parent amine is too basic to allow for direct *N*-nitration. This route is usually very successful, but it has shortfalls, including the potential for nitrolysis of one of the C–N bonds other than the acyl linkage. This leads to a reduced yield and lower product purity. The by-product of amide nitrolysis is an acyl nitrate which also poses safety and disposal problems on a large scale. This problem is encountered during the synthesis of the high explosive HMX via the nitrolysis of DADN (Section 5.15.2).[119] Problems with amide nitrolysis are partly attributed to the reduction of electron density on nitrogen due to the electron-withdrawing acyl group. Consequently, some acyl derivatives of polycyclic polyamines are inert or extremely sluggish to nitrolysis.

The nitrolysis of *tert*-butyl substituted amines with dinitrogen pentoxide in nitric acid solves many of these problems. However, these substrates are not always accessible via the usual condensation routes. The acidic reagents frequently used for these nitrolysis reactions are not always suitable for substrates with acid-sensitive or easily oxidized functionality.

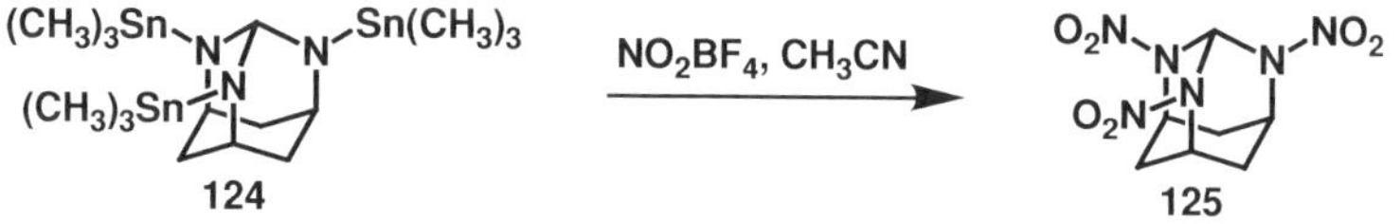

Figure 5.59

Problems associated with amide nitrolysis have been approached in a very logical way. One strategy is to make the nitrogen atom more susceptible to electrophilic attack and hence facilitate nitrolysis. An electron-donating group on nitrogen would achieve this. Accordingly, the nitrolysis of stannylamines like (124) have been reported.[98] Nitrative cleavage of a N–P bond has been reported in the form of HMPA, which forms dimethylnitramine on reaction with nitric acid. Some nitrosamine impurity is formed during the reaction.[120]

The use of silicon in the form of silylamines is more amenable for synthetic use. Olah and co-workers[121] have reported on the cleavage of *N*-silyl compounds with nitronium tetrafluoroborate. Dimethylnitramine, diethylnitramine, *N,N′*-dinitropiperazine and *N*-nitromorpholine have been prepared via this route, which is known as nitrodesilylation. Also reported are the syntheses of *N*-nitrotriazoles from the nitrodesilyation of the corresponding *N*-trimethylsilyl derivatives with nitronium salts. This is a useful route to such compounds, acidic reagents leading to the facile N → C migration of the nitro group.[121]

Millar and Philbin[122] have explored the nitrodesilylation of silylamines with dinitrogen pentoxide for the synthesis of nitramines and their derivatives. These reactions, which involve nitrative Si–N heteroatom cleavage, are conducted in methylene chloride at subambient temperature. Trimethylsilylamines give high yields of nitramine product and reactions are clean (Table 5.8).

$$RR'N{-}SiR''_3 \xrightarrow{N_2O_5,\ CH_2Cl_2} RR'N{-}NO_2 + R''_3SiONO_2$$

Table 5.8
Synthesis of nitramines, nitramides and nitroureas via the nitrodesilylation of *N*-trimethylsilyl compounds with dinitrogen pentoxide (ref. 122)

Entry	Substrate	Product	Yield (%)
1	$(H_3C)_2N{-}Si(CH_3)_3$ (126)	$(H_3C)_2N{-}NO_2$ (100)	78
2	$(i\text{-}Bu)_2N{-}Si(CH_3)_3$ (127)	$(i\text{-}Bu)_2N{-}NO_2$ (128)	87
3	$(CH_3)_3Si{-}N$(piperazine)$N{-}Si(CH_3)_3$ (129)	$O_2N{-}N$(piperazine)$N{-}NO_2$ (130)	91
4	1,3-bis$(Si(CH_3)_3)$ hexahydropyrimidine (131)	1,3-dinitro hexahydropyrimidine (117)	69
5	$H_3C{-}CO{-}N(CH_3){-}Si(CH_3)_3$ (132)	$H_3C{-}CO{-}N(CH_3){-}NO_2$ (133)	79
6	cyclic urea, N,N′-bis$(Si(CH_3)_3)$ (134)	cyclic urea, N,N′-bis(NO_2) (135)	82

Good yields are attainable for some sterically hindered nitramines like di-*iso*-butylnitramine (128) (Table 5.8, Entry 2). The synthesis of *N,N'*-dinitrohexahydropyrimidine (117) from the corresponding disilylamine (131) is of note (Table 5.8, Entry 4); this compound was previously synthesized[117] from the nitrative cleavage of *N,N'*-dinitrosohexahydropyrimidine with dinitrogen pentoxide–nitric acid in only 30 % yield (Table 5.7, Entry 3). Yields are lower for compounds with silyl groups containing secondary and tertiary alkyl groups like TIPS and TBDMS respectively. Secondary nitramides and nitroureas are also obtained in good yield from the corresponding TMS derivatives (Table 5.8, Entries 5 and 6).

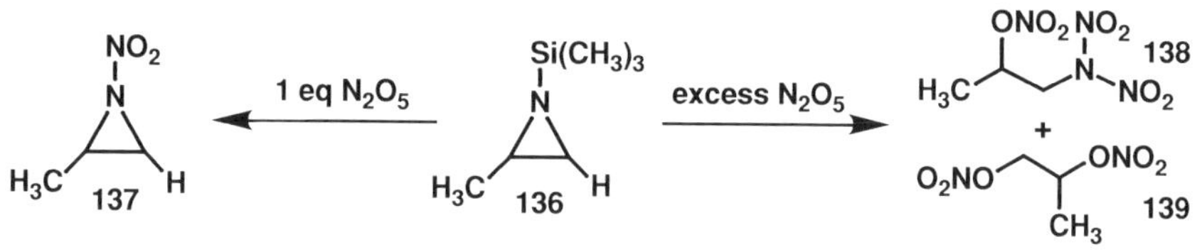

Figure 5.60

The reaction of the silylaziridine (136) with one equivalent of dinitrogen pentoxide in methylene chloride yields the *N*-nitroaziridine (137), whereas with excess reagent a mixture of the *N,N*-dinitramine-nitrate (138) and the dinitrate ester (139) is obtained; the former is a high-energy compound and of some difficulty to prepare via other routes.[122]

Nitrodesilylation with dinitrogen pentoxide is an important route to nitramine-based explosives and may find future industrial use given its low environmental impact. Providing that anhydrous conditions are maintained (dinitrogen pentoxide prepared from the ozonolysis of dinitrogen tetroxide[123]), reactions are suitable for the synthesis of products containing acid-sensitive functionality. The trimethylsilyl precursors are readily prepared from chlorotrimethylsilane in the presence of triethylamine for strong amine bases; weaker amine bases and amides/ureas require lithiation with an alkyl lithium reagent before treatment with the chlorosilane.

5.8 RING-OPENING NITRATION OF STRAINED NITROGEN HETEROCYCLES

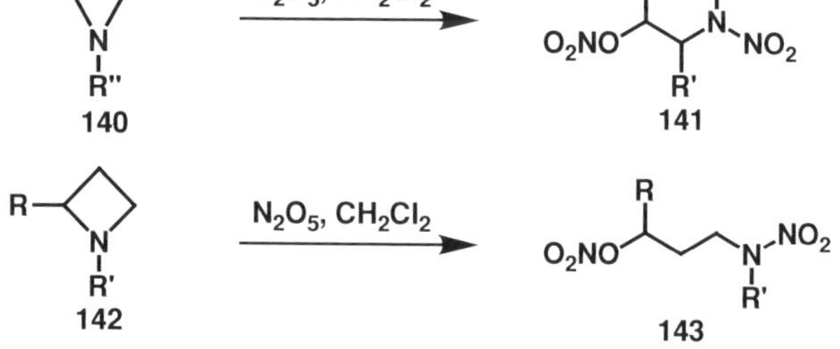

Figure 5.61

Millar and co-workers[124–126] studied the ring-opening nitration of strained nitrogen heterocycles with dinitrogen pentoxide as a route to compounds containing both nitrate ester and nitramino functionality. The reaction of aziridines (140) and azetidines (142) with dinitrogen pentoxide in inert solvents causes ring C–N bond cleavage and can be used to form 1,2- (141) and 1,3- (143) nitramine-nitrates respectively. This is analogous to work in which epoxides and oxetanes are reacted with dinitrogen pentoxide to form 1,2- and 1,3-dinitrate esters respectively (Section 3.4). Such reactions are chemically very efficient because the whole of the nitrogen pentoxide molecule is incorporated into the product. The 1,2-nitramine-nitrate system is present in high explosives such as DINA and pentryl.

5.8.1 Aziridines

Millar and co-workers[124–126] studied the reaction of dinitrogen pentoxide in chlorinated solvents with a number of different *N*-substituted aziridines and found that reactions are highly dependent on the nature of the *N*-substituent on the aziridine nitrogen (Table 5.9). *N*-Alkylaziridines give good yields of 1,2-nitramine-nitrate product (Table 5.9, Entry 1). These

Table 5.9
1,2-Nitramine-nitrates from the reaction of aziridines with dinitrogen pentoxide (ref. 124)

Entry	Substrate	Mole ratio N_2O_5 : aziridine	Product	Yield (%)
1	N, *n*-Bu (144)	1.0 : 1.0	$CH_2(CH_2)_2CH_3$, O_2NO, N, NO_2 (145)	69
2a	N, CO_2Et (146)	1.1 : 1.0	CO_2Et, O_2NO, N, NO_2 (147)	82[a]
2b		1.0 : 1.0		67[b]
3	N–C(=O)–N, H_3C, CH_3 (148)	2.2 : 1.0	CH_3 NO_2 NO_2 CH_3, O_2NO, N, N, O, ONO_2 (149)	50[c]
4	N, CH_2CH_2OH (150)	2.2 : 1.0	O_2N–N($CH_2CH_2ONO_2$)$_2$ (7)	< 10[d]
5	NO_2, O_2N, N, NO_2 (151)	1.13 : 1.0	NO_2, O_2N, N(NO_2)$CH_2CH_2ONO_2$, NO_2 (152)	76

[a] Product contains ~10% acyl nitrate impurity. [b] Product contains < 5% acyl nitrate impurity. [c] Product is readily hydrolyzed. [d] Bulk of the product mass is polymeric material.

compounds, known as NENAs, find use as energetic plasticizers; the *n*-butyl analogue (145) is known as Bu-NENA and finds use in some LOVA propellant formulations.[46,127] Carbamate derivatives of aziridine also give good yields of 1,2-nitramine-nitrate product, although the formation of acyl nitrate impurities is found to be a problem when using an excess of dinitrogen pentoxide (Table 5.9, Entries 2a and 2b). Such impurities are derived from the trans-esterification of the carbamate ester functionality and can be considerably reduced by limiting the amount of dinitrogen pentoxide used in the reaction. Carbamates like (147) are excellent plasticizers for use in energetic formulations.[46] High yields of 1,2-nitramine-nitrate product are also obtained from *N,N*-dialkylureas (Table 5.9, Entry 3). Amides are found to be poor substrates in these reactions.

Substrates with acidic/labile hydrogens are found to give low yields of 1,2-nitramine-nitrate product because of the liberation of nitric acid on reaction with dinitrogen pentoxide, which in turn, promotes acid-catalyzed aziridine ring opening and the formation of polymeric material. Consequently, *N*-alkylureas containing free N–H bonds give poor yields of nitramine product. Similar results are seen for other substrates containing labile hydrogen; 2-aziridineethanol (150) reacts with dinitrogen pentoxide to generate DINA (7) but the yield is less than 10 % (Table 5.9, Entry 4).

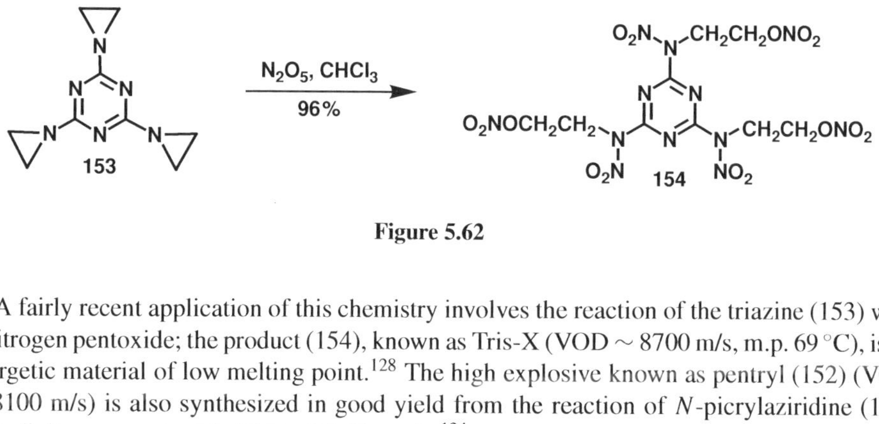

N_2O_5, $CHCl_3$, 96%; 153 → 154 (O_2N–N–$CH_2CH_2ONO_2$ substituents)

Figure 5.62

A fairly recent application of this chemistry involves the reaction of the triazine (153) with dinitrogen pentoxide; the product (154), known as Tris-X (VOD ~ 8700 m/s, m.p. 69 °C), is an energetic material of low melting point.[128] The high explosive known as pentryl (152) (VOD ~ 8100 m/s) is also synthesized in good yield from the reaction of *N*-picrylaziridine (151) with dinitrogen pentoxide (Table 5.9, Entry 5).[124]

Recent advances in the technology used to synthesize dinitrogen pentoxide means that such reactions may achieve industrial importance, although aziridines are not as readily available as the corresponding epoxides.

5.8.2 Azetidines

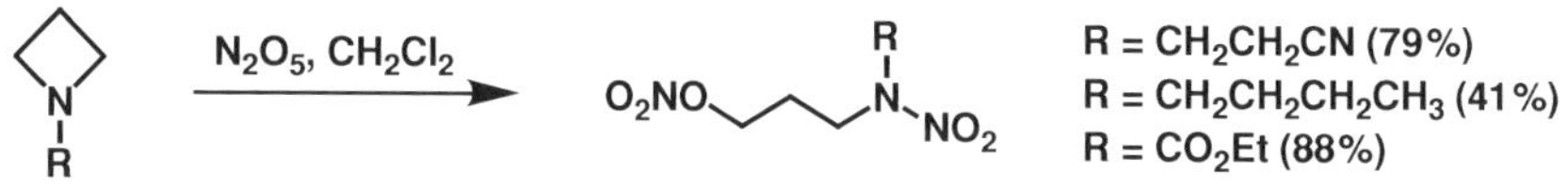

N_2O_5, CH_2Cl_2

R = CH_2CH_2CN (79%)
R = $CH_2CH_2CH_2CH_3$ (41%)
R = CO_2Et (88%)

Figure 5.63

The reaction of azetidines with dinitrogen pentoxide is found to reflect the reduced ring strain in this system compared to aziridines.[125,126] Accordingly, while the carbamate and *N*-alkyl derivatives of azetidine are found to yield the expected 1,3-nitramine-nitrates, the *N*-acyl derivatives undergo nitrolysis to yield *N*-nitroazetidine. Azetidine also undergoes *N*-nitration to give *N*-nitroazetidine on reaction with dinitrogen pentoxide.

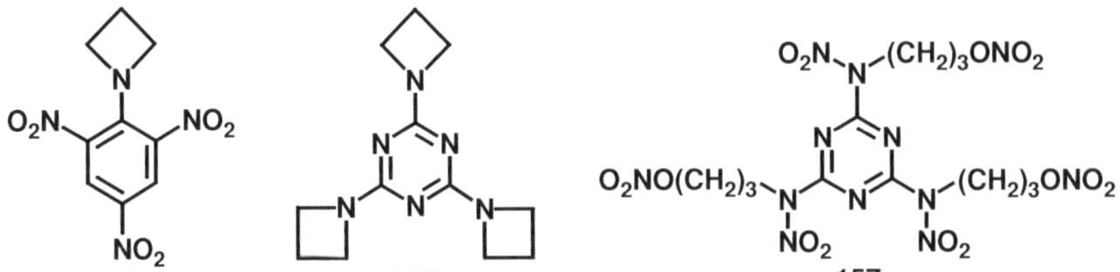

Figure 5.64

N-Picrylazetidine (155) is found to be inert to attack by dinitrogen pentoxide, a consequence of the electron-withdrawing picryl group. The triazine (156) gave a 60 % crude yield of the energetic 1,3-nitramine-nitrate (157) which is an analogue of Tris-X.

5.9 NITROSAMINE OXIDATION

In view of the highly carcinogenic nature of many nitrosamines any experiments involving their isolation must be discouraged, and for this reason, this section has only been written for completeness. This high toxicity is unfortunate because the preparation of nitrosamines from the parent amines is often facile and they provide a route to highly pure nitramines. Other equally useful methods for the synthesis of nitramines, such as the chloride-catalyzed nitration of secondary amines, also suffer from the formation of nitrosamines in appreciable amounts and must also be viewed with caution.

$$R^1R^2NH \xrightarrow{HNO_2} R^1R^2NNO \xrightarrow[CH_2Cl_2]{CF_3CO_3H} R^1R^2NNO_2 \quad \text{(Eq. 5.16)}$$

Figure 5.65

Secondary nitrosamines are readily synthesized in excellent yield by treating secondary amines with aqueous nitrous acid (Equation 5.16).[106] Emmons found that these secondary nitrosamines can be oxidized to the corresponding nitramines in good to excellent yield on treatment with peroxytrifluoroacetic acid (Equation 5.16); the latter is prepared from the addition of 90 % hydrogen peroxide to an excess of trifluoroacetic acid or trifluoroacetic anhydride.[106] The method is found to yield the corresponding nitramine in a state of high purity with highest yield attainable under anhydrous conditions using methylene chloride as solvent.[106,129,130] Other oxidants have been reported although their use is more limited.[4,105]

5.10 HYDROLYSIS OF NITRAMIDES AND NITROUREAS

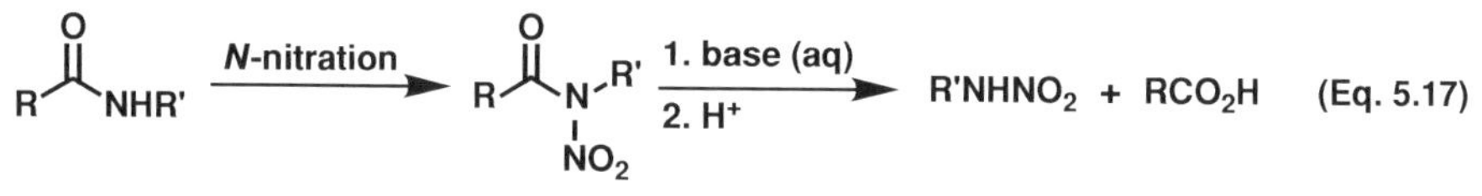

$$RC(O)NHR' \xrightarrow{N\text{-nitration}} RC(O)N(NO_2)R' \xrightarrow[2.\ H^+]{1.\ \text{base (aq)}} R'NHNO_2 + RCO_2H \quad \text{(Eq. 5.17)}$$

Figure 5.66

The instability of primary nitramines in acidic solution means that the nitration of the parent amine with nitric acid or its mixtures is not a feasible route to these compounds. The hydrolysis of secondary nitramides is probably the single most important route to primary nitramines. Accordingly, primary nitramines are often prepared by an indirect four step route: (1) acylation of a primary amine to an amide, (2) *N*-nitration to a secondary nitramide, (3) hydrolysis or ammonolysis with aqueous base and (4) subsequent acidification to release the free nitramine (Equation 5.17). Substrates used in these reactions include sulfonamides, carbamates (urethanes), ureas and carboxylic acid amides like acetamides and formamides etc. The nitration of amides and related compounds has been discussed in Section 5.5.

A particularly useful synthesis of primary nitramines involves the nitration of the appropriate carbamate ester followed by ammonolysis with gaseous ammonia in diethyl ether. The ammonium salt of the nitramine precipitates in pure form and is carefully acidified to give the free nitramine. The corresponding carbamate esters are readily synthesized from the action of chlorocarboxylic acid esters on alkylamines in the presence of alkali hydroxides.

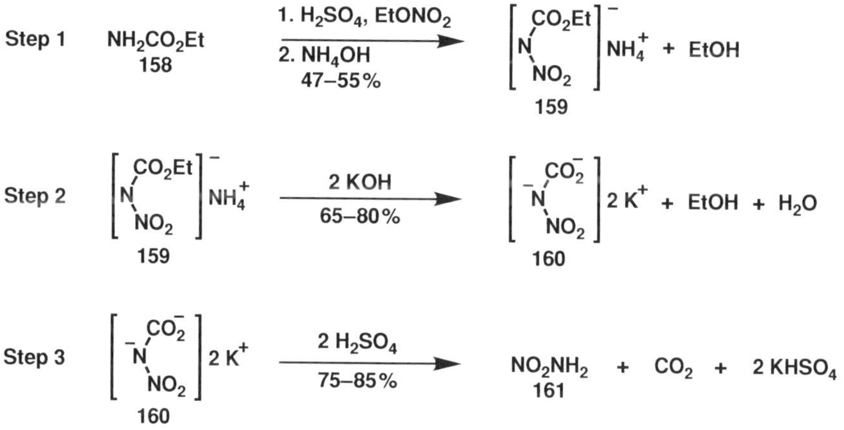

Step 1 $$\underset{\mathbf{158}}{NH_2CO_2Et} \xrightarrow[\substack{2.\ NH_4OH \\ 47\text{–}55\%}]{1.\ H_2SO_4,\ EtONO_2} \underset{\mathbf{159}}{[N(CO_2Et)(NO_2)]^- NH_4^+} + EtOH$$

Step 2 $$\underset{\mathbf{159}}{[N(CO_2Et)(NO_2)]^- NH_4^+} \xrightarrow[65\text{–}80\%]{2\ KOH} \underset{\mathbf{160}}{[{}^-N(CO_2^-)(NO_2)]\ 2K^+} + EtOH + H_2O$$

Step 3 $$\underset{\mathbf{160}}{[{}^-N(CO_2^-)(NO_2)]\ 2K^+} \xrightarrow[75\text{–}85\%]{2\ H_2SO_4} \underset{\mathbf{161}}{NO_2NH_2} + CO_2 + 2\ KHSO_4$$

Figure 5.67 Synthesis of nitramine

Nitramine (161), the simplest member of the nitramines, can be prepared from ethyl carbamate (158) in three synthetic steps; nitration of the latter with mixed acid or a solution of ethyl nitrate in concentrated sulfuric acid, followed by isolation of the resulting ethyl *N*-nitrocarbamate as the ammonium salt (159), hydrolysis with aqueous potassium hydroxide and subsequent acidification, yields nitramine (161).[131]

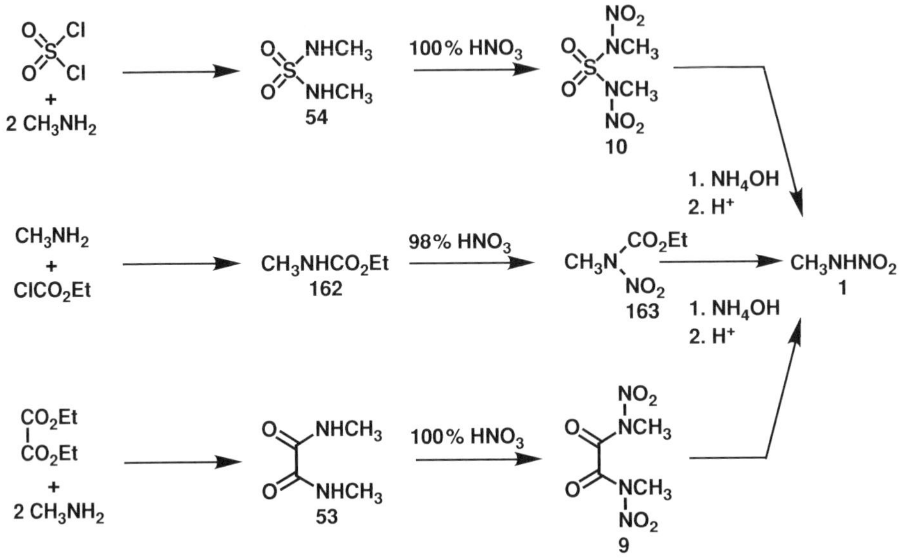

Figure 5.68

Methylnitramine (1) can be prepared from the hydrolysis of an appropriate secondary nitramide. One route involves the nitration of *N,N'*-dimethylsulfamide (54) to *N,N'*-dinitro-*N,N'*-dimethylsulfamide (10) with absolute nitric acid, followed by ammonolysis and subsequent acidification.[132]

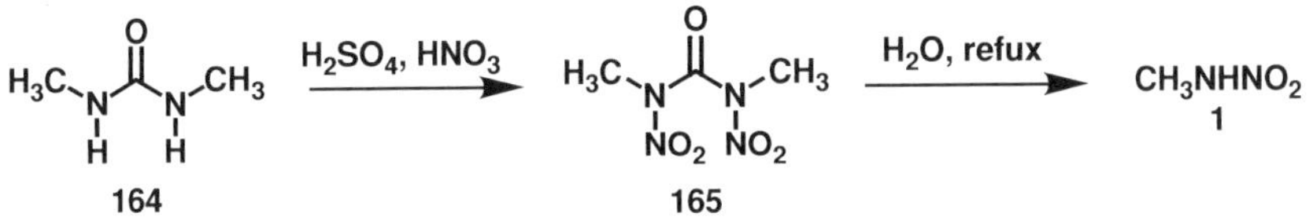

Figure 5.69

Methylnitramine (1) has also been synthesized from the hydrolysis of ethyl *N*-methyl-*N*-nitrocarbamate (163),[133] *N,N'*-dinitro-*N,N'*-dimethyloxamide (9)[69] and *N,N'*-dinitro-*N,N'*-dimethylurea (165);[134] the latter, synthesized from the mixed acid nitration of *N,N'*-dimethylurea (164), has been suggested as a possible industrial route to methylnitramine.[134]

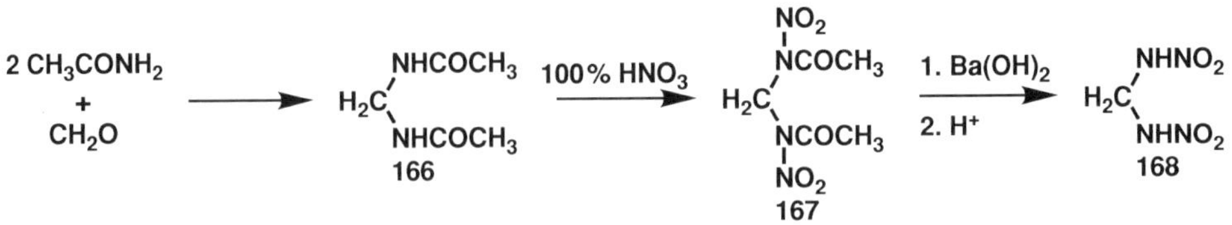

Figure 5.70

Methylenedinitramine (1,3-dinitro-1,3-diazapropane) (168) can be prepared from the nitration of the bis-acetamide (166), followed by hydrolysis of the product (167) with barium hydroxide and subsequent acidification.[135] Methylenedinitramine has no practical value as an explosive due to its facile decomposition in solution.

Bachman and co-workers[70] studied the synthesis of the high explosive ethylenedinitramine (2) (EDNA) from the hydrolysis of secondary nitramides. One of the oldest routes to EDNA (2) involves the nitration of 2-imidazolidone (66) with mixed acid, followed by hydrolysis of the resulting *N,N′*-dinitro-2-imidazolidone (67) with boiling water.[136] 2-Imidazolidone (66) is readily synthesized from the reaction of urea with ethylene glycol,[137] or by treating either urea[137] or ethyl carbonate[70] with ethylenediamine.

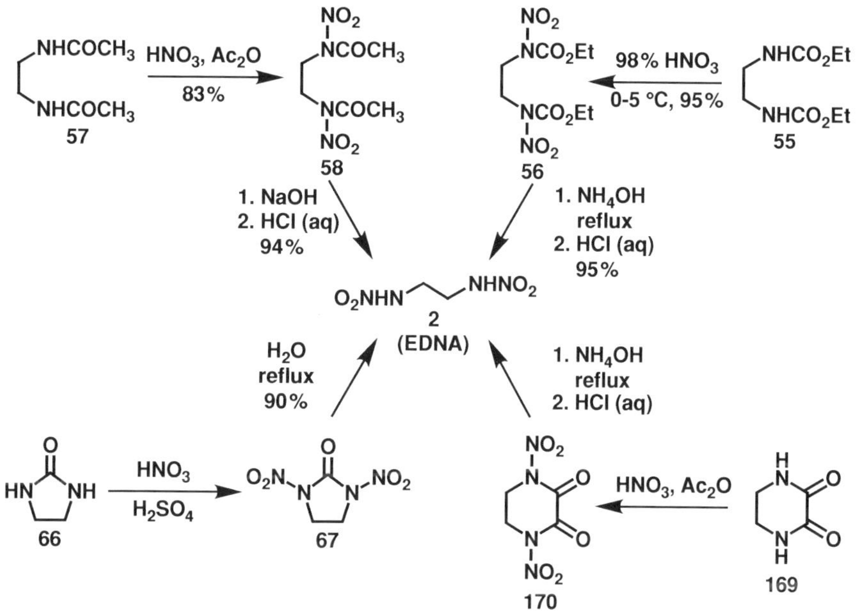

Figure 5.71 Synthetic routes to ethylenedinitramine (EDNA) via nitramide hydrolysis

The nitration of the carbamate (55) with pure nitric acid, followed by hydrolysis of the product (56) and subsequent acidification, also yields EDNA (2).[70,136] Acetylation of ethylenediamine, followed by nitration of the resulting product, *N,N′*-ethylenebisacetamide (57), yields *N,N′*-dinitro-*N,N′*-ethylenebisacetamide (58), an explosive which also yields EDNA (2)[70] on treatment with aqueous base followed by acidification. EDNA (2) has also been synthesized from ethyleneoxamide (169); the latter is prepared from the condensation of ethylenediamine with diethyloxalate.[70]

The nitration and hydrolysis of ethyl carbamates like (55) has been used to synthesize other nitramines, including: 1,3-dinitraminopropane, 1,4-dinitraminobutane, 1,5-dinitraminopentane and 1,6-dinitraminohexane.[138,139]

The facile hydrolysis of *N,N′*-dinitro-2-imidazolidone to ethylenedinitramine (EDNA) shows that *N,N′*-dinitroureas can be useful precursors to linear dinitramines. Cyclic *N,N′*-dinitroureas can be prepared indirectly from nitroguanidine. Nitroguanidine is found to

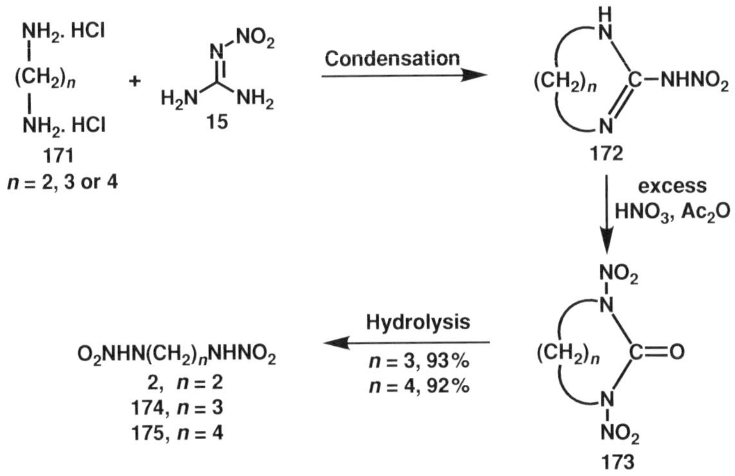

$NH_2 \cdot HCl$–$(CH_2)_n$–$NH_2 \cdot HCl$ (171, n = 2, 3 or 4) + H_2N–C(=N–NO_2)–NH_2 (15) —Condensation→ 172 (cyclic $(CH_2)_n$, NH, C–$NHNO_2$, N)

172 —excess HNO_3, Ac_2O→ 173 (cyclic $(CH_2)_n$, N–NO_2, C=O, N–NO_2)

173 —Hydrolysis (n = 3, 93%; n = 4, 92%)→ $O_2NHN(CH_2)_nNHNO_2$

2, n = 2
174, n = 3
175, n = 4

Figure 5.72

decompose in the presence of primary alkylamines with the loss of ammonia and the formation of 3-alkyl-substituted nitroguanidines.[140] The same reaction in the presence of a linear diamine (171) forms a 2-nitramino-1,3-diazacycloalk-2-ene (172), which is readily nitrated to the corresponding cyclic *N,N'*-dinitrourea (173) with an excess of nitric acid in acetic anhydride.

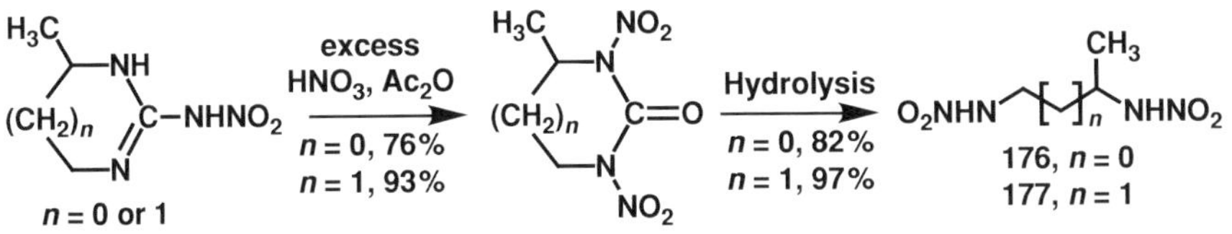

H_3C-substituted cyclic $(CH_2)_n$, NH, C–$NHNO_2$, N (n = 0 or 1) —excess HNO_3, Ac_2O (n = 0, 76%; n = 1, 93%)→ H_3C-substituted cyclic $(CH_2)_n$, N–NO_2, C=O, N–NO_2 —Hydrolysis (n = 0, 82%; n = 1, 97%)→ O_2NHN–[CH_2]$_n$–$CH(CH_3)$–$NHNO_2$

176, n = 0
177, n = 1

Figure 5.73

McKay and co-workers[141,142] prepared a number of cyclic *N,N'*-dinitroureas via the nitroguanidine route; hydrolysis to the parent linear dinitramines was effected with boiling water or aqueous sodium hydroxide. This route was used to synthesize: ethylenedinitramine (2), 1,2-dinitraminopropane (176), 1,3-dinitraminopropane (174), 1,3-dinitraminobutane (177) and 1,4-dinitraminobutane (175).

5.11 DEHYDRATION OF NITRATE SALTS

The dehydration of the nitrate salts of some primary and secondary amines can yield the corresponding nitramine. Dimethylnitramine has been prepared in 65 % yield from the dehydration of dimethylamine nitrate in acetic anhydride to which 4 mole % of anhydrous zinc chloride has been added.[4] The same reaction in the absence of chloride ion only generates a 5 % yield of dimethylnitramine.[4] Some arylnitramines derived from weakly basic amines have been prepared via the addition of the amine nitrate salts to acetic anhydride.[143]

Davis studied the dehydration of urea nitrates as a route to *N*-nitroureas.[144] The nitrate salt of *N*-methylurea undergoes dehydration–rearrangement on treatment with concentrated sulfuric acid to give *N*-nitro-*N*-methylurea in 42 % yield. In this compound the nitro and methyl groups are attached to the same nitrogen and so its hydrolysis can provide a route to methylnitramine. In contrast, the nitrate salts of ethyl, *n*-propyl, *n*-butyl and *n*-amyl ureas, give *N*-nitro-*N*′-ethylurea (49 %), *N*-nitro-*N*′-propylurea (60 %), *N*-nitro-*N*′-butylurea (67 %) and *N*-nitro-*N*′-amylurea (67 %), respectively, on treatment with concentrated sulfuric acid.

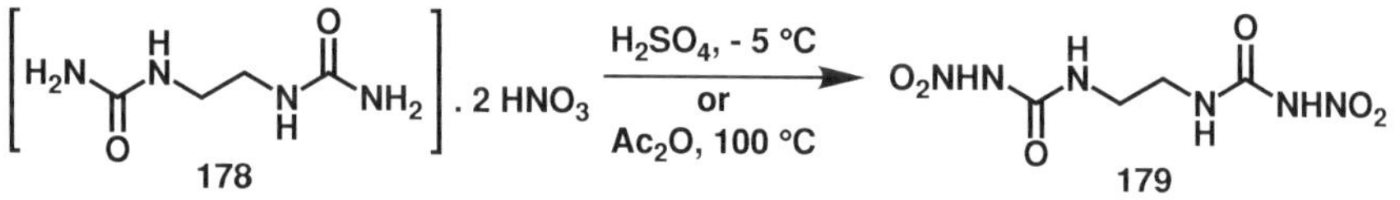

Figure 5.74

Bachman and co-workers[70] reported the synthesis of 3,3′-dinitro-1,1′-ethylenebisurea (179) by treating the dinitrate salt of ethylenebisurea (178) with cold concentrated sulfuric acid or by heating with acetic anhydride.

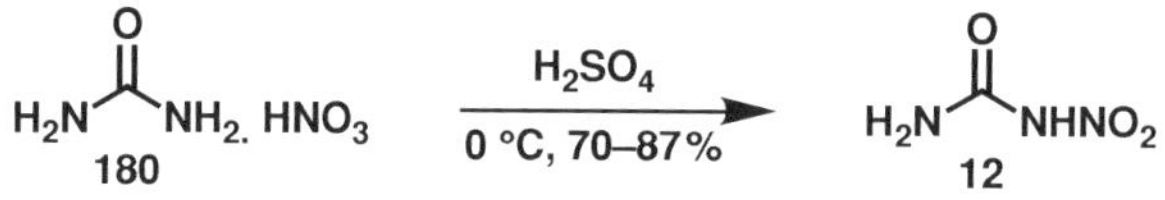

Figure 5.75

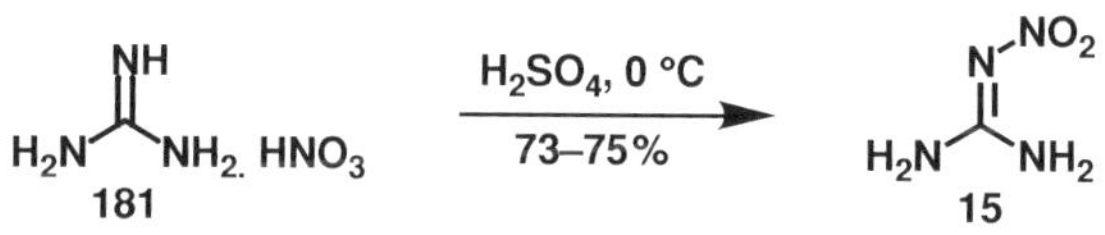

Figure 5.76

A standard method for the synthesis of nitrourea (12) involves treating urea nitrate (180) with 90 % sulfuric acid.[145] On a laboratory and industrial scale, nitroguanidine (15) is conveniently prepared from the dehydration–rearrangement of guanidine nitrate (181), a process accomplished by stirring the latter as a suspension in concentrated sulfuric acid at subambient temperature; note that nitroguanidine is actually a nitrimine and not a monobasic acid like nitrourea.[146] The guanidine nitrate used for this synthesis is conveniently prepared on a laboratory scale by fusing a mixture of dicyandiamide with two mole equivalents of ammonium nitrate at 160 °C for 2 hours.[147] However, this preparation of guanidine nitrate has quite rightly aroused concern due to the potentially explosive nature of this mixture.

5.12 OTHER METHODS

- The oxidation of aryl diazoates with oxidants like hypochlorite, permanganate and ferricyanide anion has seen some limited use for the synthesis of nitramines.[148] This method finds use for the synthesis of arylnitramines where aromatic ring nitration is not required and so excludes the use of standard nitrating agents.

- Secondary nitramines have been prepared from the metathesis of dialkylcarbamyl chlorides with silver nitrate in acetonitrile followed by the spontaneous decomposition of the resulting dialkylcarbamyl nitrates.[149] Yields of nitramine are low and accompanied by nitrosamine impurities.
- Primary nitramines have been formed from the reaction of aliphatic isocyanates with nitronium tetrafluoroborate in ethyl acetate or acetonitrile, followed by hydrolysis.[150]
- Metathesis reactions between *N*-chloramines and silver nitrite in alkaline solution are reported to give the silver salt of the corresponding primary nitramine. The method is of little synthetic value.[151]
- Nitramines have been prepared from the reaction of *N*-chloramines with dinitrogen tetroxide.[152]
- The reaction of substituted diazomethanes with ethyl *N*-nitrocarbamate, followed by hydrolysis of the ethyl *N*-alkyl-*N*-nitrocarbamate, has been used to prepare some primary arylnitramines where aromatic ring nitration is not required, and so limits the use of conventional *N*-nitrating agents.[153] The method has not been fully investigated.
- The rearrangement of imidoyl nitrates yields nitramides via an O→N migration.[154] This reaction is instantaneous when imidoyl chlorides are treated with silver nitrate in acetonitrile.

5.13 PRIMARY NITRAMINES AS NUCLEOPHILES

Primary nitramines contain an acidic proton which enables them to behave as nucleophiles and undergo addition and condensation reactions. These reactions are extremely useful in two respects. Firstly, these reactions convert primary nitramino functionality into secondary nitramino functionality, which is no longer acidic and much more chemically stable. Secondly, these addition and condensation reactions can be used to prepare functionalized derivatives of polynitramines which can be used to synthesize energetic polymers and other explosive compounds.

5.13.1 1,4-Michael addition reactions

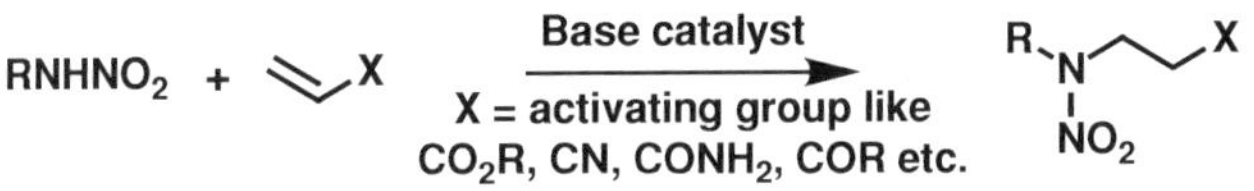

RNHNO$_2$ + CH$_2$=CH–X —(Base catalyst; X = activating group like CO$_2$R, CN, CONH$_2$, COR etc.)→ R–N(NO$_2$)–CH$_2$CH$_2$–X (Eq. 5.18)

Figure 5.77

The anions of primary nitramines, like other nucleophiles, can undergo Michael 1,4-addition reactions with a range of α,β-unsaturated substrates to form secondary nitramines of varying molecular complexity (Equation 5.18). Kissinger and Schwartz[155] prepared a number of secondary nitramines from the condensation of primary nitramines with α,β-unsaturated ketones, esters, amides and cyanides. In a standard experiment a solution of the primary nitramine and an excess of the Michael acceptor is warmed in the presence of a catalytic amount of Triton-B base.

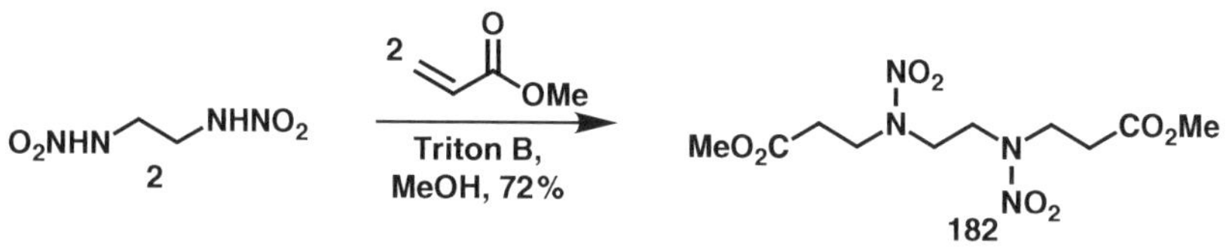

Figure 5.78

1,4-Michael addition reactions are particularly useful when linear aliphatic bis-nitramines are used because the products contain two terminal functional groups like in the diester (182).[155] The terminal functionality of such products can be used, or modified by simple functional group conversion, to provide oligomers for the synthesis of energetic polymers; such oligomers often use terminal alcohol, isocyanate or carboxy functionality for this purpose.

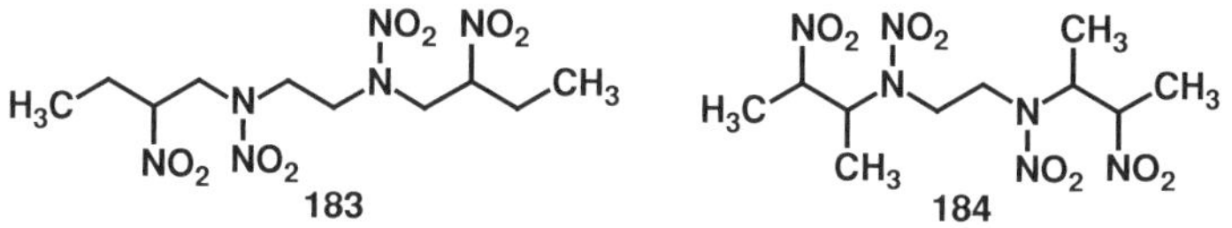

Figure 5.79

Feuer and Miller[156] synthesized 3,5,8,10-tetranitro-5,8-diazadodecane (183) and 3,8-dimethyl-2,4,7,9-tetranitro-4,7-diazadecane (184) from the reactions of ethylenedinitramine with 2-nitrobutyl acetate and 3-nitro-2-butyl acetate respectively; the latter reagents readily undergo elimination in the presence of sodium acetate base to give the corresponding α-nitroalkenes.

5.13.2 Mannich condensation reactions

Primary nitramines react with amines in the presence of an aldehyde to form 1,3-amino-nitramines in a reaction analogous to the Mannich condensation. In these reactions the amine and aldehyde component combine to form an intermediate imine which is then attacked by the nitramine nucleophile.

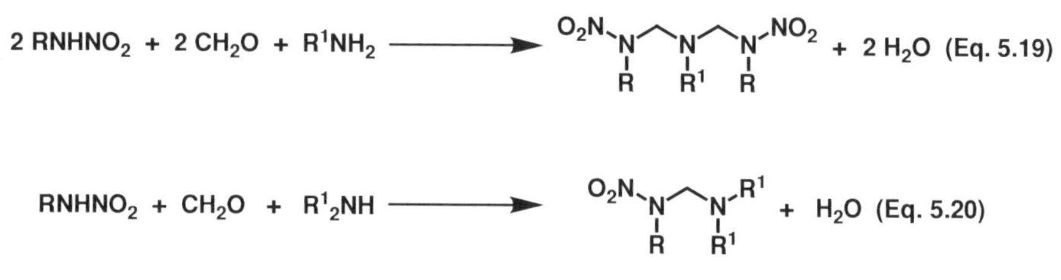

$$2\ RNHNO_2 + 2\ CH_2O + R^1NH_2 \longrightarrow O_2N\text{-}N(R)\text{-}CH_2\text{-}N(R^1)\text{-}CH_2\text{-}N(R)\text{-}NO_2 + 2\ H_2O \quad \text{(Eq. 5.19)}$$

$$RNHNO_2 + CH_2O + R^1_2NH \longrightarrow O_2N\text{-}N(R)\text{-}CH_2\text{-}NR^1_2 + H_2O \quad \text{(Eq. 5.20)}$$

Figure 5.80

$$2\,CH_3NHNO_2 + 2\,CH_2O + CH_3NH_2 \longrightarrow O_2N\text{-}N(CH_3)CH_2N(CH_3)CH_2N(CH_3)\text{-}NO_2 + 2\,H_2O$$

185

Figure 5.81

The products obtained from these condensations are predictable for simple mono-functional substrates; primary amines usually form compounds where two equivalents of primary nitramine are incorporated into the product (Equation 5.19), whereas simple secondary amines can only combine with one equivalent of nitramine (Equation 5.20).[157,158] The bis-nitramine (185) is formed from the reaction of methylamine with two mole equivalents each of methylnitramine and formaldehyde.

$H_2C(NHNO_2)_2$ (168) → 1. EtOAc, CH_2O; 2. EtOH, $MeNH_2$ → 186

Figure 5.82

Mannich condensations involving polyamines or polynitramines are more complex, and in the case of linear dinitramines, leads to very interesting and diverse chemistry, enabling the synthesis of many cyclic nitramine products. The reaction of methylenedinitramine (168) with various primary amines in the presence of formaldehyde leads to 1,5-dinitro-1,3,5,7-tetraazacyclooctanes; the 3,7-dimethyl analogue (186) is isolated when methylamine is used.[158]

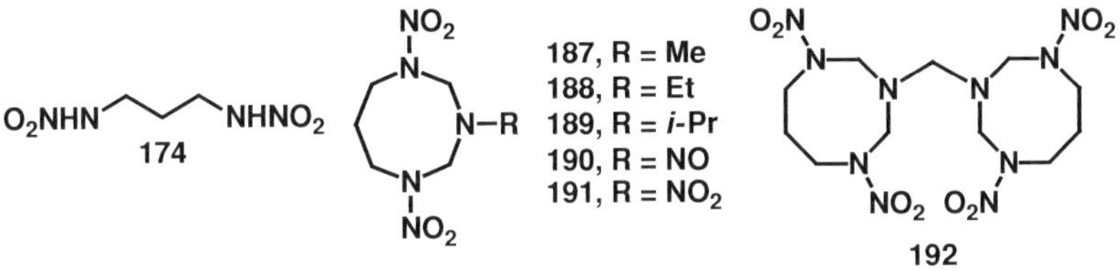

$O_2NHN(CH_2)_3NHNO_2$ (174)

187, R = Me
188, R = Et
189, R = *i*-Pr
190, R = NO
191, R = NO_2

192

Figure 5.83

1,3-Dinitraminopropane (174) reacts with a range of primary amines in the presence of formaldehyde to yield 1,5-dinitro-1,3,5-triazacyclooctanes like (187) (R = methyl), (188) (R = ethyl), and (189) (R = *iso*-propyl).[159] When ammonia is used as the amine component in these reactions the bicycle (192) is formed.[159] Reaction of the bicycle (192) with nitrous acid and absolute nitric acid leads to C–N bond cleavage with the formation of (190) and (191), respectively.[159]

Ethylenedinitramine (EDNA) (2) condenses with ammonia in the presence of formaldehyde to form the triazacycloheptane bicycle (193).[160] The chemistry of the bicycle (193) is very diverse and is only partly summarized in this discussion.[160]

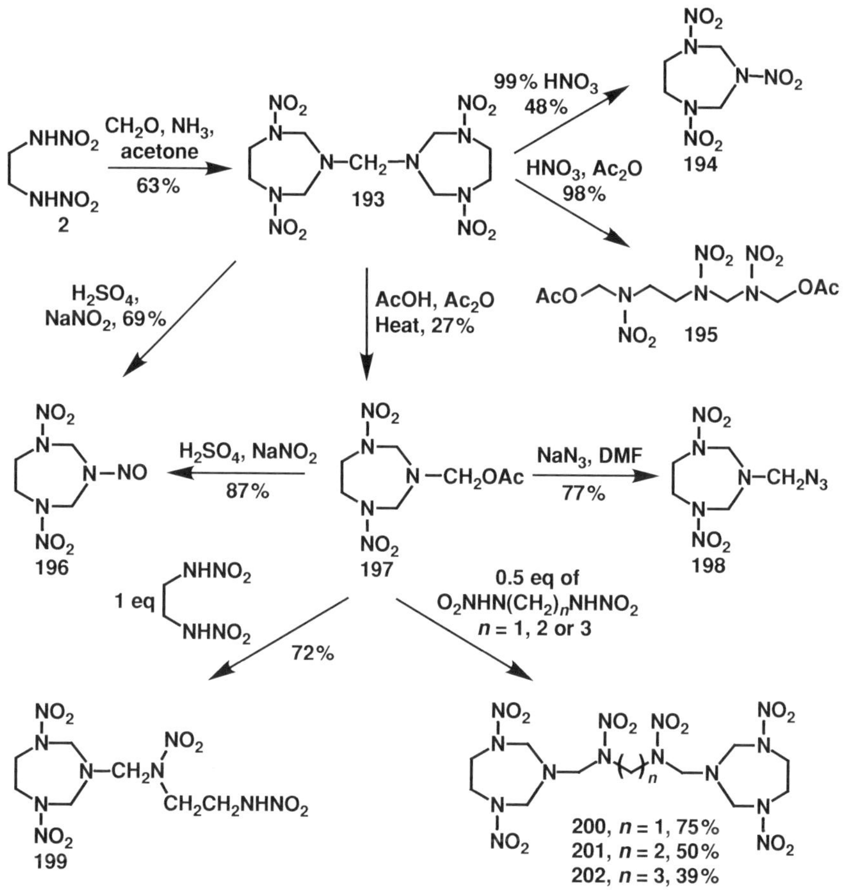

Figure 5.84

Nitrolysis of the bicycle (193) with absolute nitric acid, followed by quenching with water, yields 1,3,5-trinitro-1,3,5-triazacycloheptane (194), whereas the same reaction with fuming nitric acid in acetic anhydride leads to the formation of the linear diacetate (195), a feature which is consistent with the nitrolysis of hexamine (Section 5.15). Reaction of (193) with a solution of nitrous acid leads to C–N bond cleavage and yields the nitrosamine (196). Cleavage of (193) is also seen on treatment with hot acetic anhydride in acetic acid; the product of this reaction, the acetate (197), readily undergoes displacement of the acetate group with a range of nucleophiles, of which products the azide (198) is probably the most interesting as a potential explosive. The acetate (197) reacts with one equivalent of ethylenedinitramine in DMF to form the tetranitramine (199). The reaction of two equivalents of the acetate (197) with one equivalent of methylenedinitramine, ethylenedinitramine or 1,3-dinitraminopropane leads to the formation of the bicycles (200), (201), or (202) respectively.

Some interesting nitramine products are derived from the reaction of ethylenedinitramine (2) with formaldehyde in the presence of various linear aliphatic diamines; the bicycles (203) and

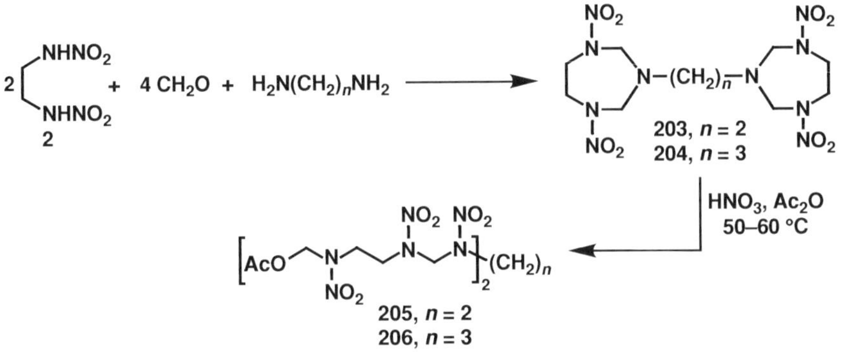

Figure 5.85

(204) are formed when the amine components are ethylenediamine and 1,3-diaminopropane respectively.[161] Nitrolysis of the bicycles (203) and (204) with nitric acid in acetic anhydride yields the linear diacetates (205) and (206), respectively.[161] Note that only bicycle (193), where the triazacycloheptane units are separated by a single methylene group, leads to 1,3,5-trinitro-1,3,5-triazacycloheptane (194) on nitrolysis. 1,3,5-Trinitro-1,3,5-triazacycloheptane (194) has never been isolated from the nitrolysis of extended bicycles like (203) and (204).

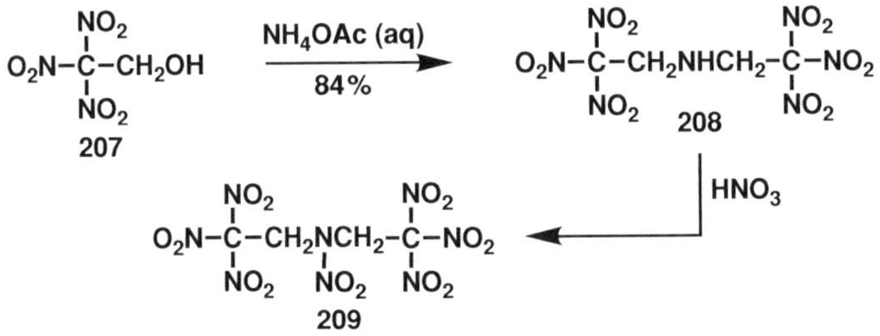

Figure 5.86

The use of primary nitramines in Mannich reactions is an important route to numerous secondary nitramines. However, a far more common route to such nitramines involves the Mannich condensation of a terminal *gem*-dinitroalkane, formaldehyde, and an amine, followed by *N*-nitration of the resulting polynitroalkylamine.[162] The preformed methylol derivative of the *gem*-dinitroalkane is often used in these reactions and so formaldehyde can be omitted. This route has been used to synthesize explosives like (92)[163] and (209).[164]

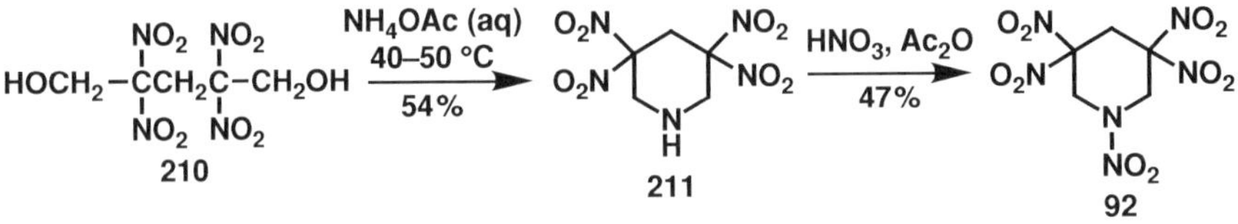

Figure 5.87

5.13.3 Condensations with formaldehyde

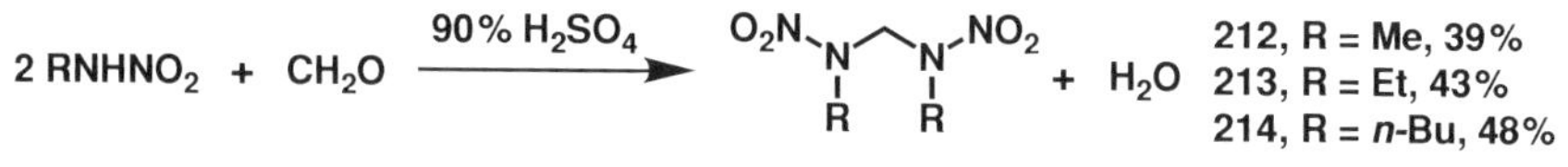

2 $RNHNO_2$ + CH_2O →(90% H_2SO_4) O_2N–N(R)–CH_2–N(R)–NO_2 + H_2O
212, R = Me, 39%
213, R = Et, 43%
214, R = *n*-Bu, 48%

Figure 5.88

The condensation of a primary nitramine with formaldehyde in the presence of concentrated sulfuric acid is a useful route to 1,3-dinitramines. A number of linear dinitramines have been prepared via this route including 2,4-dinitro-2,4-diazapentane (212), 3,5-dinitro-3,5-diazaheptane (213), and 5,7-dinitro-5,7-diazaundecane (214).[165]

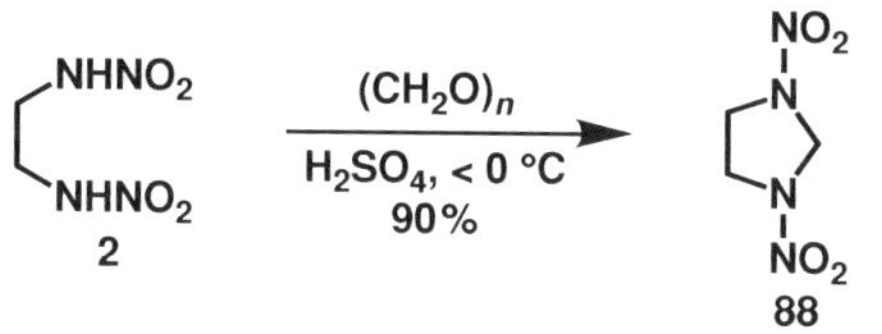

$NHNO_2$ / $NHNO_2$ (2) →($(CH_2O)_n$; H_2SO_4, < 0 °C; 90%) NO_2 / N / N / NO_2 (88)

Figure 5.89

The condensation of ethylenedinitramine (2) with paraformaldehyde in the presence of sulfuric acid at subambient temperature yields *N,N′*-dinitroimidazolidine (88).[165]

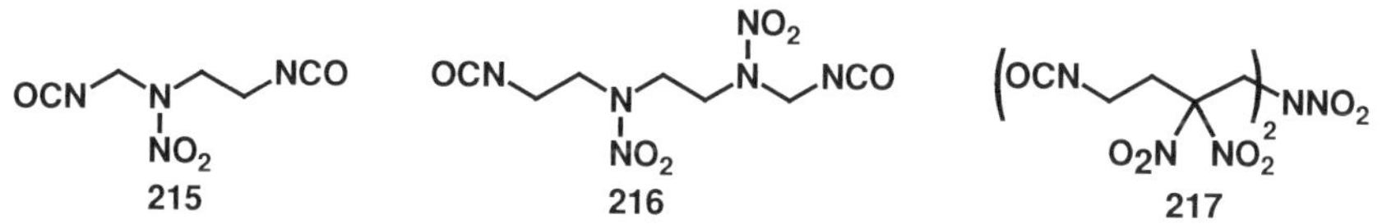

OCN–N(NO_2)–NCO (215); OCN–N(NO_2)–N(NO_2)–NCO (216); (OCN–C(NO_2)$_2$)$_2$NNO$_2$ (217)

Figure 5.90

Under aqueous conditions formaldehyde reacts with primary nitramines to form the corresponding methylol derivatives. The versatility of the terminal hydroxy group of these methylol derivatives is illustrated by their facile conversion to more reactive functional groups, like isocyanates, which can then be reacted with compounds containing hydroxy or carboxy functionality. Diisocyanates like (215), (216) and (217) have been reacted with various polynitroaliphatic diols for the synthesis of energetic polymers.[166]

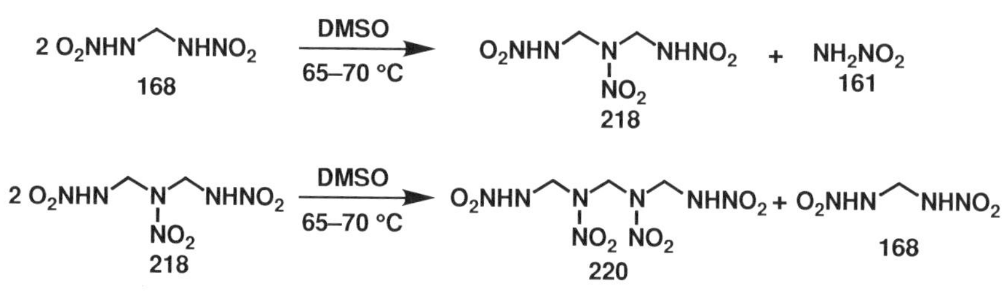

2 O_2NHN–$NHNO_2$ (168) →(DMSO, 65–70 °C) O_2NHN–N(NO_2)–$NHNO_2$ (218) + NH_2NO_2 (161)

2 O_2NHN–N(NO_2)–$NHNO_2$ (218) →(DMSO, 65–70 °C) O_2NHN–N(NO_2)–N(NO_2)–$NHNO_2$ (220) + O_2NHN–$NHNO_2$ (168)

Figure 5.91

An interesting reaction has been reported for the synthesis of polymethylenenitramines and also illustrates the instability of these compounds. Such compounds can undergo nitromethylene transfer reactions when heated in polar solvents like DMSO.[167] Methylenedinitramine (168) reacts under these conditions to generate both nitramine (161) and 1,3,5-trinitro-1,3,5-triazapentane (218); reaction of the latter under similar conditions allows the synthesis of 1,3,5,7-tetranitro-1,3,5,7-tetraazaheptane (220) and so on. The primary nitramino groups of these polymethylenenitramines can undergo condensation reactions and may be useful for the synthesis of energetic polymers.

5.13.4 Nucleophilic displacement reactions

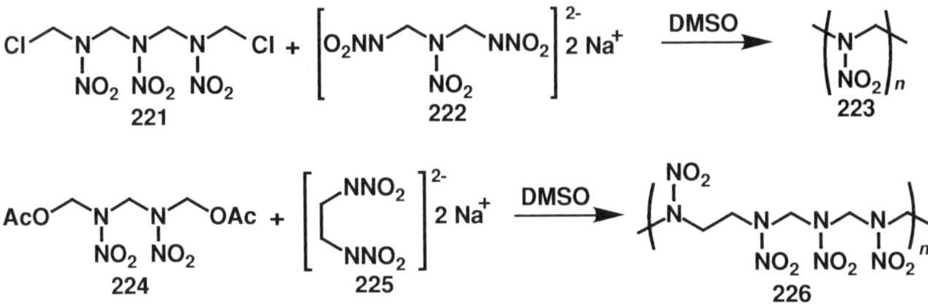

Figure 5.92

The readiness with which primary nitramines form terminal methylol derivatives on reaction with formaldehyde has been utilized for the synthesis of some energetic polymers by converting the terminal hydroxy functionality into a better leaving group such as an acetate ester or a halogen. The alkali metal salts of some linear polynitramines have been used as nucleophiles to displace such leaving groups and form energetic polymers like (223)[168] and (226).[169]

The alkylation of primary nitramines with alkyl halides is of little preparative value for the synthesis of secondary nitramines. Such reactions often result in a mixture of *N*- and *O*-alkylated products. The product distribution appears to be very dependent on the nature of the cation of the nitramine used, with silver salts[170] favouring *O*-alkylation and alkali metal salts[171] usually giving *N*-alkylation as the predominant product. However, this is not always the case.

5.14 AROMATIC NITRAMINES

Numerous aromatic nitramines have been synthesized but only *N*,2,4,6-tetranitro-*N*-methylaniline (tetryl) and 1-(2-nitroxyethylnitramino)-2,4,6-trinitrobenzene (pentryl) have found practical use as explosives. Both tetryl and pentryl are more powerful than TNT. Tetryl is widely used in boosters and as a component of explosive formulations like tetrytol (tetryl/TNT), PTX-1 (tetryl/RDX/TNT) and Composition C-3 (tetryl/RDX/TNT/DNT/MNT/NC).

Tetryl (8) can be prepared from the nitration–oxidation of *N*,*N*-dimethylaniline (90) with a variety of nitrating agents and conditions, including the use of a large excess of 70 % nitric

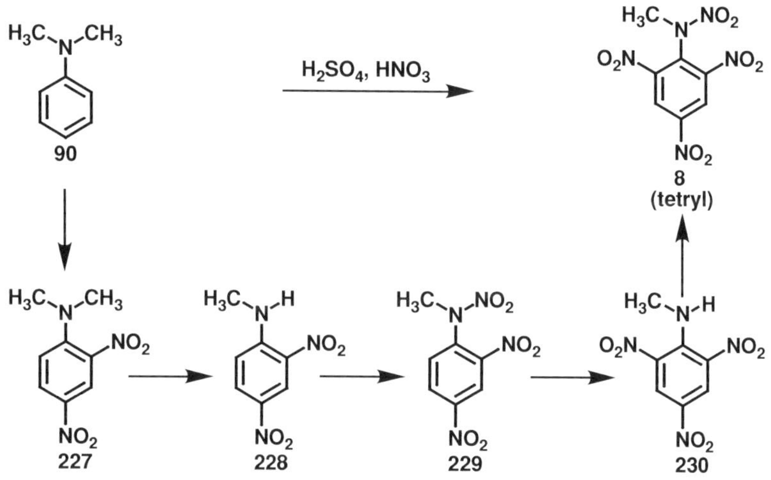

Figure 5.93

acid at 80 °C,[172] more concentrated nitric acid at lower temperature,[173] mixed acid[174] and a mixture of nitric acid in acetic anhydride.[175] Studies have shown that the nitration of *N,N*-dimethylaniline with mixed acid proceeds in five steps – *N,N*-dimethylaniline (90) first undergoes aromatic ring nitration to (227), followed by oxidation of one of the methylamino groups to a carbamic acid group which subsequently loses carbon dioxide to form (228), further nitration now generates nitramine (229) which undergoes N → C nitro group rearrangement to give (230), and further nitration yields tetryl (8).[176] The *N*-nitration of (228) to (229), and (230) to (8) is promoted by the presence of the aromatic ring nitro groups which reduce the basicity of the amino group.

The nitration of *N,N*-dimethylaniline inevitably produces copious amount of nitrogen oxides from the methyl group oxidation step and other side-reactions. Oxidation of the methyl group is favoured by high reaction temperature but so are other oxidative side-reactions. Consequently, the highest yields for the nitration of *N,N*-dimethylaniline to tetryl are achieved with mild nitrating agents like acetic anhydride in nitric acid; nitrating systems with more powerful oxidizing potential give lower yields.

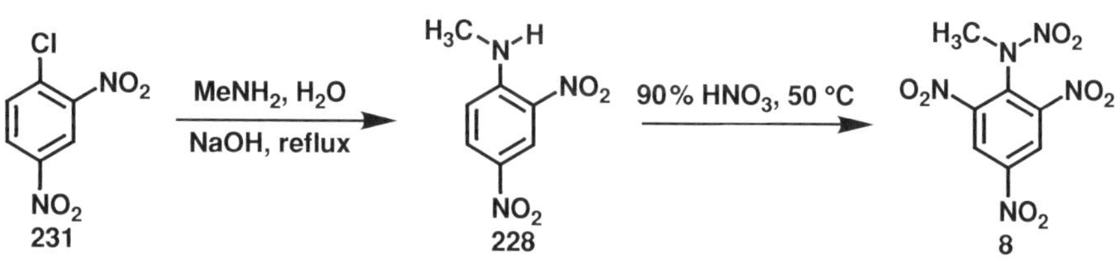

Figure 5.94

Tetryl (8) can be synthesized in two steps from 2,4-dinitrochlorobenzene (231). Thus, reaction of the latter with methylamine under aqueous conditions yields 2,4-dinitro-*N*-methylaniline (228), which readily undergoes nitration to tetryl (8) on treatment with mixed acid.[177] This route is favoured on both industrial and laboratory scales – the two nitro groups of 2,4-dinitro-*N*-methylaniline make this substrate less susceptible to oxidative side-reactions and so yields of tetryl are considerably higher than for the direct nitration of *N,N*-dimethylaniline.

Tetryl has been synthesized by treating picryl chloride with the potassium salt of methylnitramine but the reaction is of theoretical interest only.[178]

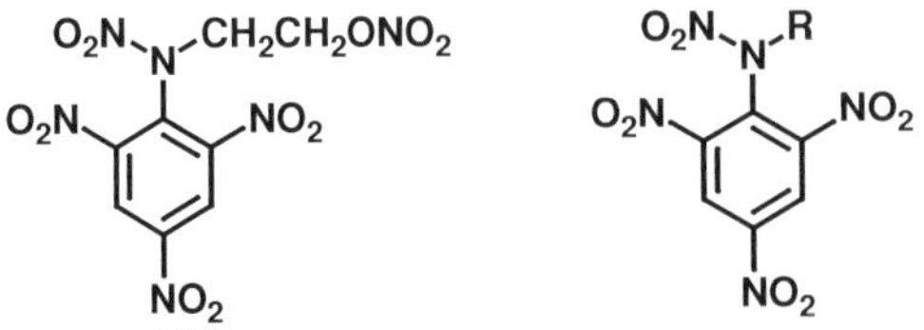

152 (pentryl)

232, R = Et (ethyltetryl)
233, R = *n*-Bu (butyltetryl)

Figure 5.95

Pentryl (152) is obtained from the action of fuming nitric acid or mixed acid on *N*-(2,4-dinitrophenyl)ethanolamine, itself obtained from the reaction of 2,4-dinitrochlorobenzene with ethanolamine.[179] Another route to pentryl (152) involves the nitration of *N*-phenylethanolamine, which is obtained from the reaction of aniline with ethylene oxide.[74]

Other aromatic nitramines have not found use as practical explosives. Ethyltetryl (232) is prepared from the nitration of 2,4-dinitro-*N*-ethylaniline, *N,N*-diethylaniline or *N*-ethylaniline.[178] Butyltetryl (233) can be synthesized from the nitration of 2,4-dinitro-*N*-butylaniline, which is attainable from the reaction of *n*-butylamine with 2,4-dinitrochlorobenzene.[178]

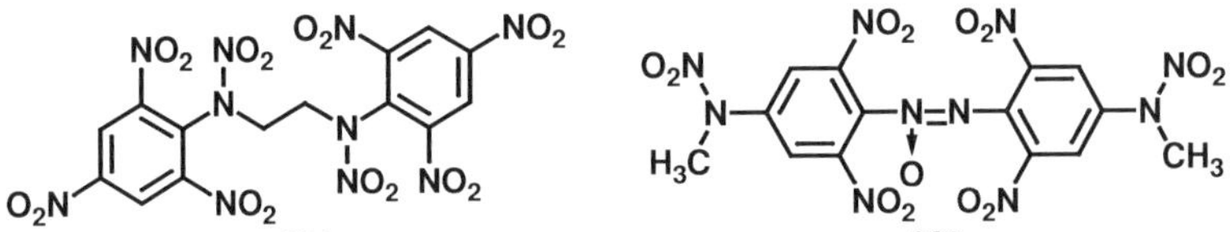

234

235

Figure 5.96

The aromatic nitramine (234) can be prepared by nitrating *N,N'*-diphenylethylenediamine[180] or 2,2',4,4'-tetranitro-*N,N'*-diphenylethylenediamine with mixed acid,[181] the latter synthesized from 2,4-dinitrochlorobenzene. The azoxy-nitramine (235) is prepared by nitrating 4,4'-bis(dimethylamino)azoxybenzene with mixed acid.[182,183]

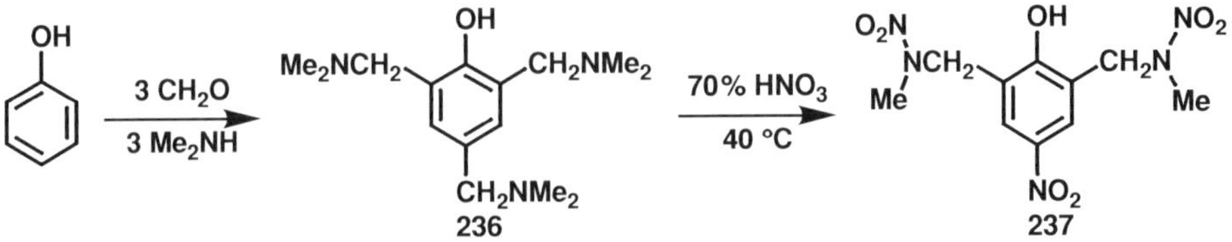

3 CH_2O, 3 Me_2NH → 236; 70% HNO_3, 40 °C → 237

Figure 5.97

The phenol (237) can be synthesized via a Mannich condensation between phenol, formaldehyde and dimethylamine,[184] followed by nitration–nitrolysis of the product (236) with concentrated nitric acid.[185] The phenolic group of (237) is acidic, and so enables the formation of metal salts which are very impact sensitive explosives.

5.15 THE NITROLYSIS OF HEXAMINE

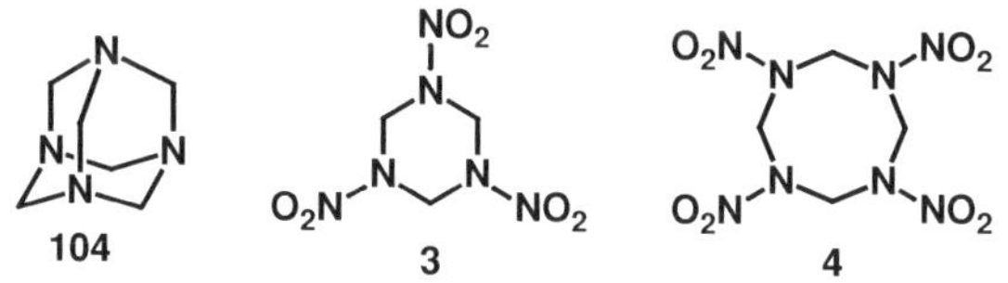

104

3
(RDX)

4
(HMX)

Figure 5.98

The nitrolysis of hexamine (104) is one of the most complex and widely studied processes in the history of energetic materials synthesis. There are twelve CH_2–N bonds in hexamine (104) which can undergo scission, and with nitrolysis occurring to varying extents, it is unsurprising that many different compounds can be produced during these reactions. Many of the possible compounds are transient intermediates or are too unstable for isolation, as in the case of many linear methylol nitramines whose existence is known from their reactions in solution or from the isolation of their acetyl or nitrate ester derivatives. With so many reaction routes available during the nitrolysis of hexamine it may seem strange that the cyclic nitramines RDX (3) and HMX (4) can be isolated in such high yields. The reason for this observation is that both RDX and HMX are very chemically stable and by far the most stable compounds present under the harsh conditions of the nitrolysis. If fact, studies have shown that various fragments from the nitrolysis of hexamine can recombine to form either RDX or HMX and so provide a sink for active methylene and nitrogen containing fragments. Linear methylol nitramines and their derivatives are unstable under the nitrolysis conditions and at reaction temperatures of 80 °C they can fragment into smaller molecules which are also capable of forming RDX or HMX.

The importance of cyclic nitramines as military explosives has meant that an enormous amount of research has been conducted in this area. Only some of the rich array of products and by-products obtainable from hexamine nitrolysis are discussed in this section. For mechanistic studies and detailed analysis of these reactions the primary research papers should be consulted.[24,186–197]

5.15.1 The synthesis of RDX

1,3,5-Trinitro-1,3,5-triazacyclohexane (3), also known as Cyclonite, Hexogen, T4 or more commonly as RDX (research department explosive), was first discovered by Henning[198] and is the most important military explosive in modern day use. The high brisance exhibited by RDX, its stability on storage and low sensitivity to impact and friction in comparison to many nitrate ester explosives, makes it a desirable secondary high explosive. The use of RDX as a component of composite explosives like Composition B, Torpex, Cyclotols, DBX, HBX, Hex-24, PTX-1 etc. and plastic explosives like Composition C-4, Semtex-H, PVA-4 etc. is extensive.

RDX is usually prepared from the nitrolysis of hexamine and this is the most widely studied reaction for any explosive. As previously discussed, the nitrolysis of the CH_2–N bonds of hexamine can produce products other than RDX. The higher homolog, HMX is a common impurity in crude RDX prepared via this route. However, the presence of HMX is not undesirable if the RDX is to be used as an explosive. Other impurities are less desirable, such as complex linear nitramine-nitrates, which lower the melting point of RDX in addition to increasing impact sensitivity and lowering thermal stability. However, these nitramine-nitrate impurities are less stable than RDX and can be hydrolyzed on prolonged treatment with boiling water, a process known as 'degassing', which releases toxic volatiles such as nitrogen oxides and formaldehyde. Studies have shown that these linear nitramine-nitrate by-products can be re-subjected to the conditions of nitrolysis[192] and form additional RDX; this process is practised on an industrial scale.

RDX has been synthesized by the different methods discussed below. Only methods 5.15.1.2 and 5.15.1.3 have received industrial importance. Method 5.15.1.7 is a convenient laboratory route to analytically pure RDX.

5.15.1.1 Treatment of hexamine with nitric acid

In this method, first established by Herz[199] and later studied by Hale,[200] hexamine is introduced into fuming nitric acid which has been freed from nitrous acid. The reaction is conducted at 20–30 °C and on completion the reaction mixture is drowned in cold water and the RDX precipitates. The process is, however, very inefficient with some of the methylene and nitrogen groups of the hexamine not used in the formation of RDX. The process of nitrolysis is complex with formaldehyde and some other fragments formed during the reaction undergoing oxidation in the presence of nitric acid. These side-reactions mean that up to eight times the theoretical amount of nitric acid is needed for optimum yields to be attained.

$$(CH_2)_6N_4 + 4\ HNO_3 \longrightarrow RDX + 3\ CH_2O + NH_4NO_3 \qquad \text{(Eq. 5.21)}$$

$$(CH_2)_6N_4 + 6\ HNO_3 \longrightarrow RDX + 6\ H_2O + 3\ CO_2 + 2\ N_2 \qquad \text{(Eq. 5.22)}$$

Figure 5.99

The stoichiometry of the Hale nitrolysis reaction is very dependent on reaction conditions. Even so, this reaction has been postulated to conform to the stoichiometry in Equation (5.21)[200] and Equation (5.22).[201] Based on the assumption that one mole of hexamine produces one mole of RDX the Hale nitrolysis reaction commonly yields 75–80 % of RDX.

5.15.1.2 Nitrolysis of hexamine dinitrate with nitric acid – ammonium nitrate – acetic anhydride

Both Köffler[201] in Germany (1943) and Bachmann[195,197] in the US (1941) discovered this method independently. In Germany the reaction was known as the KA-process. The process is much more efficient than method 5.15.1.1 and works on the principle that one mole of hexamine can produce two moles of RDX. However, hexamine is deficient in nitrogen for this to occur, so in this process, ammonium nitrate is used as a reagent to supplement nitrogen and allow the reaction to follow Equation (5.23).

$$(CH_2)_6N_4.2HNO_3 + 2\ HNO_3 + 2\ NH_4NO_3 + 6\ Ac_2O \longrightarrow 2\ RDX + 12\ AcOH \quad \text{(Eq. 5.23)}$$

Figure 5.100

An unusual feature of the KA-process is that the reaction is conducted at 60–80 °C. Solutions of nitric acid in acetic anhydride are known to be prone to dangerous 'fume off' at temperatures above ambient. However, a saturated solution of ammonium nitrate in fuming nitric acid can be added to warmed acetic anhydride without such danger. In fact, these reactions are commonly conducted at 60–80 °C as a matter of safety by preventing a build-up of unreacted starting material. The hexamine used in these reactions is in the form of the dinitrate salt, which is formed as a crystalline salt on addition of a saturated aqueous solution of hexamine to concentrated nitric acid below 15 °C. The use of hexamine dinitrate in this process reduces the amount of nitric acid needed for the nitrolysis.

The order and timing of the addition of reagents in the KA-process is varied but in a typical procedure three reagents, namely, acetic anhydride, a solution of ammonium nitrate in nitric acid, and solid hexamine dinitrate, are added slowly, in small portions and in parallel, into the reaction vessel which is preheated to 60–80 °C. On completion the reaction mixture is often cooled to 50–60 °C and the RDX filtered and sometimes washed with acetic acid. This process produces a product which melts over a 2 °C range but the RDX still contains up to 10 % HMX as a by-product. Dilution of the reaction mixture with water before removing the RDX produces a very impure product containing numerous unstable linear nitramine-nitrates. Based on the assumption that one mole of hexamine dinitrate produces two mole of RDX the KA-process commonly yields 75–80 % of RDX.

5.15.1.3 Nitrolysis of hexamine with ammonium nitrate – nitric acid

$$(CH_2)_6N_4 + 2\ NH_4NO_3 + 4\ HNO_3 \longrightarrow 2\ RDX + 6\ H_2O \quad \text{(Eq. 5.24)}$$

Figure 5.101

This method is known as the K-process after its discoverer Köffler.[201] Like method 5.15.1.2 it uses ammonium nitrate to compensate for the nitrogen deficiency in hexamine and works to Equation (5.24) where two moles of RDX are produced per mole of hexamine. As observed with method 5.15.1.2, the addition of ammonium nitrate to nitric acid appears to prevent dangerous oxidation reactions from occurring. In fact, this nitrolysis reaction only occurs at elevated temperature and so a constant temperature of 80 °C is usually maintained throughout the reaction. Yields of approximately 90 % are attainable based on one mole of hexamine producing two moles of RDX.

5.15.1.4 Reaction of paraformaldehyde with ammonium nitrate in the presence of acetic anhydride

$$3\,CH_2O + 3\,NH_4NO_3 + 6\,Ac_2O \longrightarrow RDX + 12\,AcOH \quad \text{(Eq. 5.25)}$$

Figure 5.102

Paraformaldehyde and ammonium nitrate undergo dehydration in the presence of acetic anhydride in a reaction known as the E-method after its discoverer M. Ebele.[201] Ross and Schiessler[202] discovered the same reaction independently in Canada in 1940. The reaction proceeds according to the stoichiometry in Equation (5.25). Studies conducted by Wright and co-workers[192] and later by Winkler and co-workers[203] suggest the reaction occurs in the two discreet steps shown in Equations (5.26) and (5.27).

$$6\,CH_2O + 4\,NH_4NO_3 + 6\,Ac_2O \longrightarrow (CH_2)_6N_4 + 4\,HNO_3 + 12\,AcOH \quad \text{(Eq. 5.26)}$$

$$(CH_2)_6N_4 + 4\,HNO_3 \longrightarrow RDX + 3\,CH_2O + NH_4NO_3 \quad \text{(Eq. 5.27)}$$

Figure 5.103

The E-method is more expensive than other methods of RDX production because of the relatively large amount of acetic anhydride used. However, the E-method uses no nitric acid for the formation of RDX, and so must be considered the safest of the methods. The process can, however, be dangerous if the reaction is not initiated before the reagent addition starts, due to a potential build up of unreacted material. On safety grounds the paraformaldehyde and ammonium nitrate are added to previously warmed acetic anhydride, or alternately, a catalytic amount of boron trifluoride catalyst (~0.4 %) is added at the start of the reaction, the latter initiating the reaction at room temperature and lowering the proportion of undesirable *N*-acetyl linear nitramines formed during the reaction. The low acidity of the reaction mixtures in the E-process favours the formation of HMX as a by-product (~6 % HMX in the RDX product). The reaction based on Equation (5.25) gives yields of RDX up to 80 %.

5.15.1.5 Reaction of sulfamic acid, formaldehyde and nitric acid

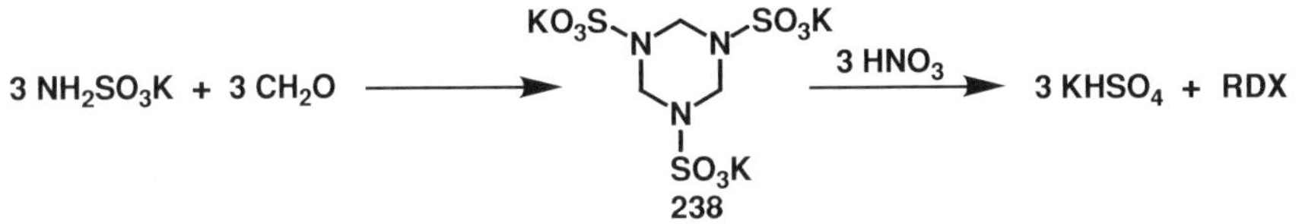

$$3\,NH_2SO_3K + 3\,CH_2O \longrightarrow \textbf{238} \xrightarrow{3\,HNO_3} 3\,KHSO_4 + RDX$$

(238: 1,3,5-tris(SO_3K)-hexahydro-1,3,5-triazine ring; labels: KO_3S, SO_3K, SO_3K, 238)

Figure 5.104

This unusual process known as the W-method was discovered in Germany by Wolfram[201] and involves the condensation of the potassium salt of sulfamic acid with formaldehyde to form the heterocycle (238) followed by treatment with nitric acid. The extreme sensitivity of (238) to hydrolysis means that nitrolysis has to be conducted under anhydrous conditions using sulfur trioxide[201] or phosphorous pentoxide[204] dissolved in fuming nitric acid. The yield of RDX from the W-method is 80–90 %.

5.15.1.6 Reaction of hexamine with dinitrogen pentoxide

The nitrolysis of hexamine with 40 equivalents of a 25 % solution of dinitrogen pentoxide in absolute nitric acid in carbon tetrachloride at −20 °C is reported to give a 57 % yield of RDX. The product is free from HMX as determined by NMR (≥95 % pure).[116]

5.15.1.7 Nitrosation–oxidation/nitrolysis of hexamine

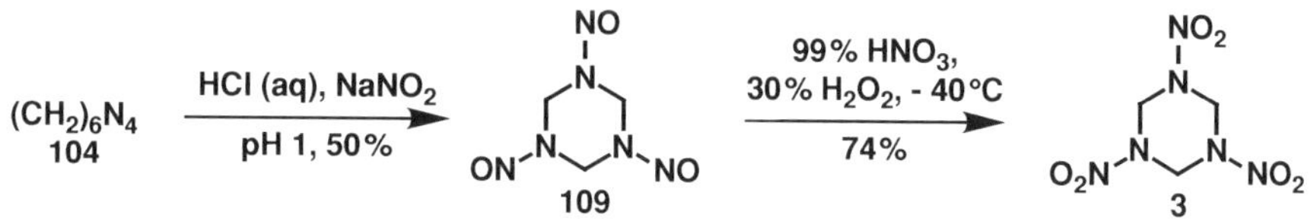

Figure 5.105

The presence of HMX as an impurity in RDX is not a problem when the product is used as an explosive. However, the need for an analytical sample of RDX makes other more indirect methods feasible. One such method involves the oxidation of 1,3,5-trinitroso-1,3,5-triazacyclohexane (109) ('R-salt') with a mixture of hydrogen peroxide in nitric acid at subambient temperature and yields analytical pure RDX (74 %) free from HMX.[115] The same conversion has been reported in 32 % yield with three equivalents of a 25 % solution of dinitrogen pentoxide in absolute nitric acid.[116] 1,3,5-Trinitroso-1,3,5-triazacyclohexane (109) is conveniently prepared from the reaction of hexamine with nitrous acid at high acidity.[205] 1,3,5-Trinitroso-1,3,5-triazacyclohexane is itself a powerful explosive but has a low chemical stability.

5.15.1.8 Nitrolysis of 1,3,5-triacyl-1,3,5-triazacyclohexanes

The acetolysis of hexamine has proven a diverse route to cyclic polyamides. 1,3,5-Triacetyl-1,3,5-triazacyclohexane (TRAT) has been synthesized in 63 % yield by heating hexamine with acetic anhydride at 98 °C.[206] Analogous acyl derivatives are prepared by heating with other acid anhydrides.[206]

While the present industrial synthesis of RDX via the Bachmann process has many faults, it is high yielding and would be difficult to match by an alternative synthesis. Even so, Gilbert and co-workers[97] reported on a study investigating the nitrolysis of a series of 1,3,5-trisubstituted-1,3,5-triazacyclohexanes for the preparation of RDX. In the series, which includes sulfonamides, amides and carbamates, only in the cases where R = alkanoyl (acylamines) was efficient nitrolysis to RDX observed. These reactions are discussed in Section 5.6.1.1 and illustrated in Table 5.4.

5.15.2 The synthesis of HMX

1,3,5,7-Tetranitro-1,3,5,7-tetraazacyclooctane (4), commonly known as Octogen or HMX (high melting explosive), is the most powerful military explosive in current use. However, the higher performance of HMX compared to RDX is offset by its higher cost of production. Consequently, HMX has been restricted to military use where it has found application as a component of some high performance propellants and powerful explosive compositions. Octol is a cast explosive containing HMX (75 %) and TNT (25 %), whereas Octal consists of wax-desensitized HMX (70 %) and aluminium powder (30 %) and is pressed into charges. HMX has also been incorporated into some high performance plastic bonded explosives (PBXs). One such explosive, PBX-9404, contains 94 % HMX, 3 % nitrocellulose and 3 % tris(chloroethyl)phosphate.

HMX can be synthesized from hexamine by any of the routes discussed below. Both methods 5.15.2.1 and 5.15.2.2 have been used for the industrial synthesis of HMX. However, the recent commercial availability of dinitrogen pentoxide means that method 5.15.2.3 is achieving industrial importance.

5.15.2.1 Synthesis of HMX from the nitrolysis of hexamine

Variations in the conditions used for the nitrolysis of hexamine have a profound effect on the nature and distribution of isolated products, including the ratio of RDX to HMX. It has been shown that lower reaction acidity and a reduction in the amount of ammonium nitrate used in the Bachmann process increases the amount of HMX formed at the expense of RDX.[196,207] Bachmann and co-workers[208] were able to tailor the conditions of hexamine nitrolysis to obtain an 82 % yield of a mixture containing 73 % HMX and 23 % RDX. Continued efforts to provide a method for the industrial synthesis of HMX led Castorina and co-workers[187] to describe a procedure which produces a 90 % yield of a product containing 85 % HMX and 15 % RDX. This procedure conducts nitrolysis at a constant reaction temperature of 44 °C and treats hexamine, in the presence of a trace amount of paraformaldehyde, with a mixture of acetic acid, acetic anhydride, ammonium nitrate and nitric acid. Bratia and co-workers[209] used a three stage 'aging process' and a boron trifluoride catalyst to obtain a similar result. A procedure reported by Picard[210] uses formaldehyde as a catalyst and produces a 95 % yield of a product containing 90 % HMX and 10 % RDX.

The different solubilities of HMX and RDX in organic solvents means that pure HMX is easily isolated from RDX–HMX mixtures; the higher solubility of RDX in acetone means that recrystallization of such mixtures from hot acetone yields pure HMX.[187,207]

5.15.2.2 Synthesis of HMX from the nitrolysis of DPT

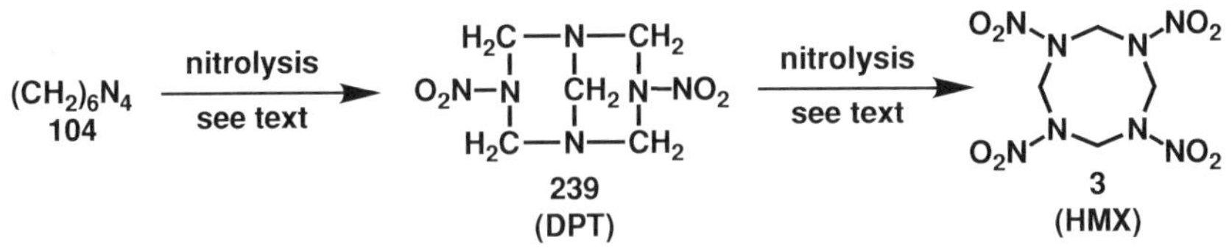

Figure 5.106

HMX (3) can be synthesized from the nitrolysis of 1,5-dinitroendomethylene-1,3,5,7-tetraazacyclooctane (DPT) (239). Wright and co-workers[189] reported that the reaction of

DPT (239) with a mixture of acetic anhydride, ammonium nitrate and nitric acid at 65–70 °C furnishes HMX in 65 % yield after purification. Bachmann and co-workers[196] repeated the method of Wright and co-workers but obtained a slightly lower yield of HMX (59 %) after purification of the crude product from both RDX (16 % total of the crude) and linear *N*-acetyl nitramine by-products. Another procedure treated DPT (239) with 1.6 mole equivalents of ammonium nitrate and 3.2 equivalents of fuming nitric acid at 60–65 °C and is reported to furnish HMX in 75 % yield after purification.[211]

1,5-Dinitroendomethylene-1,3,5,7-tetraazacyclooctane (DPT) (239) was first isolated from the partial nitrolysis of hexamine; the reaction of hexamine with acetic acid, acetic anhydride and nitric acid at 15–30 °C, followed by quenching with water and neutralization of the liquors with aqueous ammonia to pH 5.5–6.5, leads to the precipitation of DPT (239).[187,188,196] However, the yield of DPT from such reactions is often poor (15–25 %). DPT has also been synthesized from the reaction of hexamine dinitrate with acetic anhydride or cold 90 % aqueous sulphuric acid. Both methods under optimum conditions give yields of DPT of approximately 31 %.[188] The reaction of nitramine (NH_2NO_2) with aqueous formaldehyde, followed by neutralization of the reaction mixture with ammonia to pH 5.5–6.5, gives DPT in 73 % yield based on the nitramine starting material.[188] This last reaction presumably involves the formation of dimethylolnitramine as an intermediate (Section 5.15.4.2).

5.15.2.3 Other synthetic routes to HMX

Gilbert and co-workers[97] showed that the nitrolysis of 1,3,5-triacyl-1,3,5-triazacyclohexanes offered little benefit over the conventional synthesis of RDX via the nitrolysis of hexamine. This is not the case for HMX where its synthesis via the Bachmann process is far from perfect. This process and its modifications are expensive, requiring large amounts of acetic anhydride. The rate of production is slow and the maximum attainable yield is 75 %. In fact, HMX is five times as expensive as RDX to produce by this process and this prevents the widespread use of this high performance explosive. Many efforts have focused on finding more economical routes to HMX.

The nitrolysis of cyclic polyamides offers a possible alternative industrial synthesis of HMX. The nitrolysis of 1,3,5,7-tetraacetyl-1,3,5,7-tetraazacyclooctane (TAT) (79) and 1,5-diacetyl-3,7-dinitro-1,3,5,7-tetraazacyclooctane (DADN) (80) with a solution of dinitrogen pentoxide in anhydrous nitric acid gives HMX in 79 % and 98 % yields, respectively.[96] Interestingly, the same reactions with nitric acid–acetic anhydride fail at room temperature.

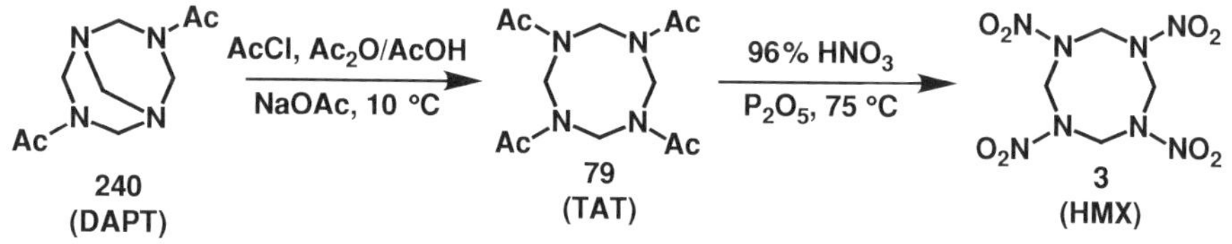

240 (DAPT) → AcCl, Ac_2O/AcOH, NaOAc, 10 °C → 79 (TAT) → 96% HNO_3, P_2O_5, 75 °C → 3 (HMX)

Figure 5.107

The acetolysis of hexamine has been extensively studied[96,206,212–216] and reviewed[217] and the synthesis of two key intermediates optimized. 3,7-Diacetyl-1,3,5,7-tetraazabicyclo

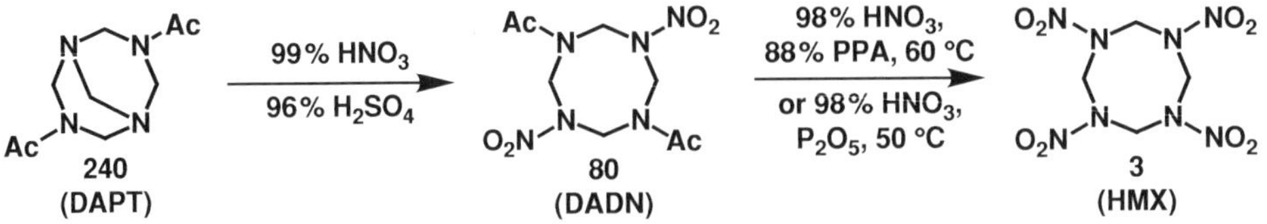

Figure 5.108

[3.3.1]nonane (DAPT) (240) can be prepared in yields above theoretical (119 % based on hexamine) by slowly adding acetic anhydride to a slurry of hexamine, water and ammonium acetate.[213] Further studies show that acetic anhydride can be replaced by ketene.[213] Gilbert and co-workers[214] isolated 1,3,5,7-tetraacetyl-1,3,5,7-tetraazacyclooctane (TAT) (79) in 70 % yield by heating DAPT (240) with acetic anhydride for 3 hours at 110 °C. Further improvement using a mixture of acetyl chloride, acetic anhydride, acetic acid and sodium acetate gave yields of TAT (79) between 75 and 90 %.[215] The direct preparation of TAT (79) from hexamine has also been described.[216]

The acetolysis of hexamine closely resembles that of hexamine nitrolysis. Accordingly, acidity is of key importance, with high concentrations of acetic acid favouring the formation of the 6-membered 1,3,5-triacetyl-1,3,5-triazacyclohexane (TRAT) and weakly acidic conditions favouring the 8-membered 1,3,5,7-tetraacetyl-1,3,5,7-tetraazacyclooctane (TAT).[217]

Gilbert and co-workers[214] conducted extensive studies into finding better routes to HMX. The direct nitrolysis of TAT (79) with phosphorous pentoxide in nitric acid is reported to give a 79 % yield of HMX. The same reaction with DAPT (240) is much lower yielding (maximum 11 %). However, a more satisfactory route is via the nitrolysis of the half-way intermediate, 1,5-diacetyl-3,7-dinitro-1,3,5,7-tetraazacyclooctane (DADN) (80). DADN (80) can be prepared from the nitrolysis of DAPT (240) or directly from hexamine. In the latter process, hexamine is treated with aqueous ammonium acetate and acetic anhydride, and the resulting solution of DAPT (240) added to a mixed acid composed of 99 % nitric acid and 96 % sulfuric acid, a process giving DADN (80) in 95 % yield. Extensive studies were conducted into the best conditions and reagents needed for DADN (80) nitrolysis, nitric acid–PPA (99 % yield, 100 % purity) and nitric acid–phosphorous pentoxide (99 % yield, 100 % purity) proving the most efficient. Other reagents gave poorer yields and include nitric acid–TFAA (82 %), nitrogen pentoxide–nitric acid (82 %), nitric acid–sulfur trioxide (60 %) and neat nitric acid (44 %).

The synthesis of HMX via the nitrolysis of DADN (80) with dinitrogen pentoxide in nitric acid is being utilized in the UK on pilot plant scale and is under development for large-scale production in the US.[214] The synthesis is a three-stage process: (1) hexamine is reacted with acetic anhydride and ammonium acetate to give DAPT (240), (2) mild nitration with mixed acid and (3) more vigorous nitration to HMX with nitrogen pentoxide in nitric acid. The latter reagent can be prepared *in situ* by using a mixture of phosphorous pentoxide in nitric acid[218] or via the electrochemical oxidation[219] of nitric acid–dinitrogen tetroxide mixtures.

5.15.3 Effect of reaction conditions on the nitrolysis of hexamine

Reaction conditions such as temperature, concentration, reaction acidity, stoichiometry and reactants used, together with their order of addition, all have a profound effect on the outcome of hexamine nitrolysis and the distribution of products obtained. Variations in such conditions have been extensively studied and have allowed the synthesis of a wealth of cyclic and linear nitramines.

5.15.3.1 *Low temperature nitrolysis of hexamine*

The nitrolysis of hexamine at low temperature has led to the synthesis of a number of cyclic nitramines. The reaction of hexamine dinitrate (241) with 88 % nitric acid at −40 °C, followed by quenching the reaction mixture onto crushed ice, leads to the precipitation of 3,5-dinitro-3,5-diazapiperidinium nitrate (242) (PCX) in good yield;[193] PCX is an explosive equal in power to RDX but is slightly more sensitive to impact. The reaction of PCX (242) with sodium acetate in acetic anhydride yields 1-acetyl-3,5-dinitro-1,3,5-triazacyclohexane (82) (TAX), which on further treatment with dilute alkali in ethanol yields the bicycle (243).[220]

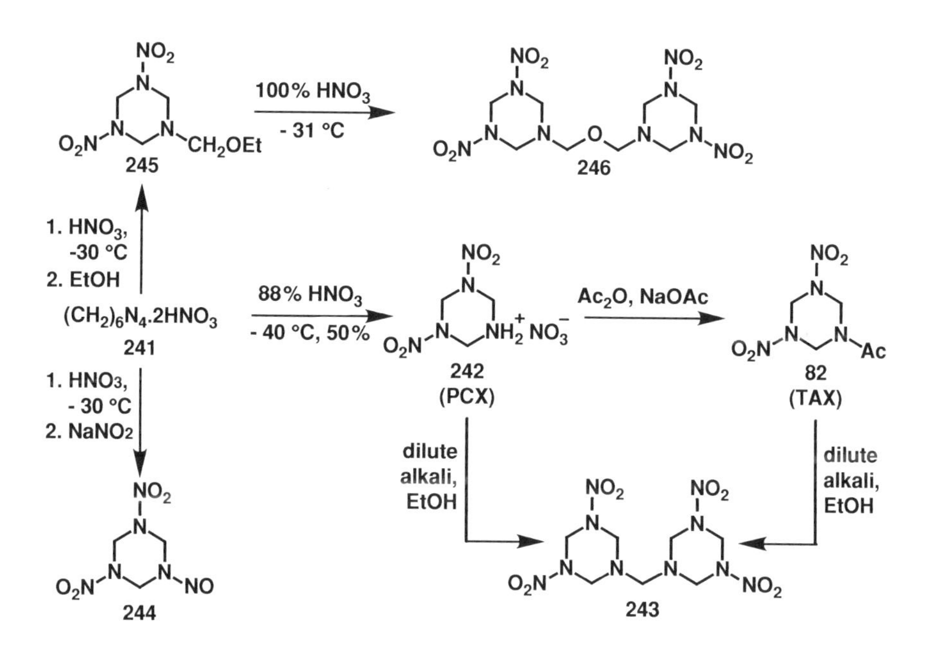

Figure 5.109 Low temperature nitrolysis of hexamine

The reaction of hexamine dinitrate (241) with 98 % nitric acid at −30 °C, followed by quenching with aqueous sodium nitrate, yields the nitrosamine (244).[220] When the same reaction is cautiously quenched with ethanol the ethoxyether (245) is obtained.[220] Treatment of the ethoxyether (245) with cold absolute nitric acid yields the bicyclic ether (246).[220] Treatment of any of the cyclic nitramines (242)–(246) with nitric acid and ammonium nitrate in acetic anhydride yields RDX.[220] Hexamine dinitrate is often used in low temperature nitrolysis experiments in order to avoid the initial exotherm observed on addition of hexamine to nitric acid.

5.15.3.2 Effect of acidity and the presence of ammonium nitrate on the nitrolysis of hexamine

Bachmann and co-workers[196] noted that hexamine can undergo two major types of cleavage, leading to the formation of compounds containing either three- or four-amino nitrogen atoms. Bachmann and co-workers[196] also noted that the products obtained from the nitrolysis of hexamine under the KA-process are dependent on the acidity and/or the activity of the nitrating agent. Under conditions of high acidity it was noted that RDX and the linear nitramine (247), with its three amino nitrogens, are the major products of the nitrolysis. In comparison, under conditions of low acidity, HMX and its linear nitramine analogue (248), with its four amino nitrogens, are the main products of nitrolysis. Bachmann and co-workers[196] also observed that the nitrolysis of hexamine with acetic anhydride and nitric acid in the presence of ammonium nitrate greatly favours the formation of the cyclic nitramines RDX and HMX, whereas in the absence of ammonium nitrate the linear nitramines (247) and (248) are favoured.

NO_2 NO_2 NO_2
AcO N N N OAc
247

NO_2 NO_2 NO_2 NO_2
AcO N N N N OAc
248

Figure 5.110

The above observations allow the selective formation of RDX, HMX or the two linear nitramines (247) and (248) by choosing the right reaction conditions. For the synthesis of the linear nitramine (247), with its three amino nitrogens, we would need high reaction acidity, but in the absence of ammonium nitrate. These conditions are achieved by adding a solution of hexamine in acetic acid to a solution of nitric acid in acetic anhydride and this leads to the isolation of (247) in 51 % yield. Bachmann and co-workers[196] also noted that (247) was formed if the hexamine nitrolysis reaction was conducted at 0 °C even in the presence of ammonium nitrate. This result is because ammonium nitrate is essentially insoluble in the nitrolysis mixture at this temperature and, hence, the reaction is essentially between the hexamine and nitric acid–acetic anhydride. If we desire to form linear nitramine (248) the absence of ammonium nitrate should be coupled with low acidity. These conditions are satisfied by the simultaneous addition of a solution of hexamine in acetic acid and a solution of nitric acid in acetic anhydride, into a reactor vessel containing acetic acid.

5.15.4 Other nitramine products from the nitrolysis of hexamine

5.15.4.1 The chemistry of DPT (239)

The chemistry of 1,5-dinitroendomethylene-1,3,5,7-tetraazacyclooctane (239) (DPT) is interesting in the context of the nitramine products which can be obtained from its nitrolysis under different reaction conditions. The nitrolysis of DPT (239) with acetic anhydride–nitric acid mixtures in the presence of ammonium nitrate is an important route to HMX (4) and this has been discussed in Section 5.15.2. The nitrolysis of DPT (239) in the absence of ammonium nitrate leads to the formation of 1,9-diacetoxy-2,4,6,8-tetranitro-2,4,6,8-tetraazanonane (248);[189] the latter has found use in the synthesis of energetic polymers.

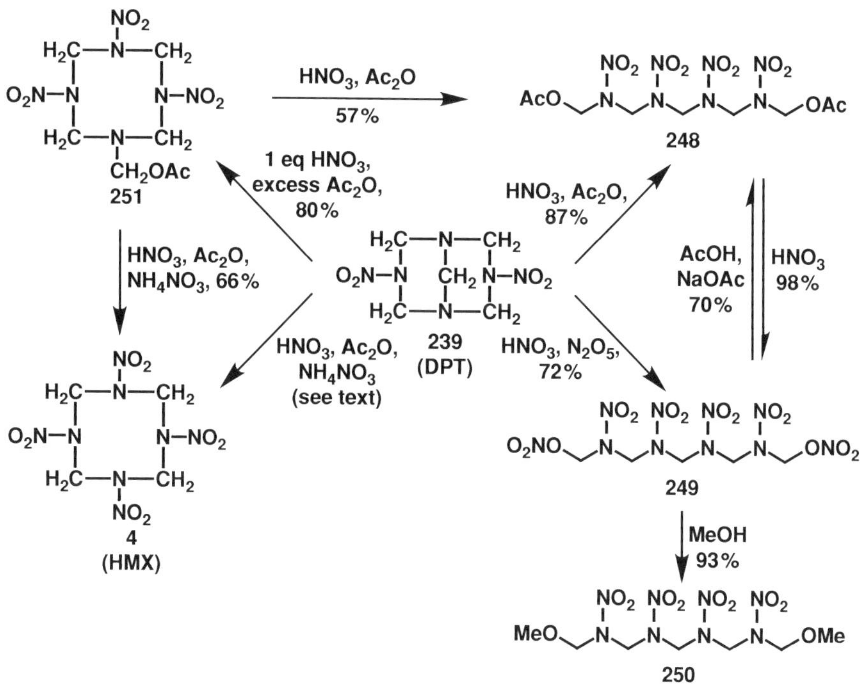

Figure 5.111 The chemistry of DPT (239)

Treatment of DPT (239) with dinitrogen pentoxide in pure nitric acid leads to the isolation of the nitrate ester (249), an unstable explosive which is highly sensitive to impact and readily undergoes hydrolysis.[189] A low nitration temperature favours the formation of (249) and its presence during the nitrolysis of hexamine is clearly undesirable. The nitrolysis of DPT (239) with one equivalent of pure nitric acid in an excess of acetic anhydride yields 1-acetomethyl-3,5,7-trinitro-1,3,5,7-tetraazacyclooctane (251),[189] a useful starting material for the synthesis of other explosives.[189,221]

5.15.4.2 The chemistry of dimethylolnitramine

Dimethylolnitramine (252) is known to be present under the conditions of the Hale nitrolysis. If the Hale nitrolysis reaction is quenched, the RDX removed by filtration and the aqueous liquors neutralized to remove DPT, the remaining filtrate can be extracted into ether and that solution evaporated over water to give an aqueous solution of dimethylolnitramine (252).[188]

Dimethylolnitramine (252) readily participates in Mannich condensation reactions; treatment of a aqueous solution of (252) with methylamine, ethylenediamine and Knudsen's base (254) (generated from fresh solutions of ammonia and formaldehyde) yields (253), (255) and (239) (DPT) respectively.[188] The cyclic ether (258) is formed from the careful dehydration of dimethylolnitramine (252) under vacuum.[188]

Dimethylolnitramine (252) is inevitably present as its dinitrate ester (256) under the conditions of hexamine nitrolysis. This compound is extremely sensitive to hydrolysis but can be

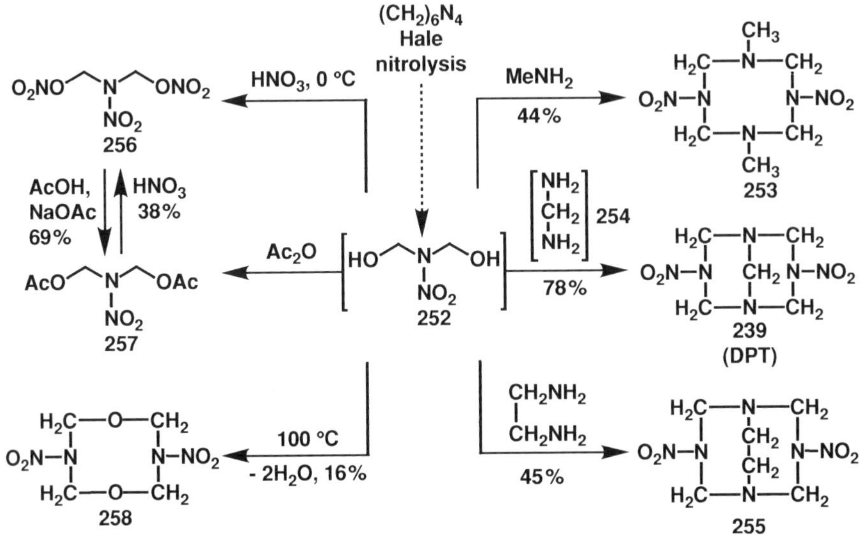

Figure 5.112 The chemistry of dimethylolnitramine (252)

converted to the more stable diacetate ester (257) on reaction with sodium acetate in acetic anhydride.[188] 1,3-Dinitroxydimethylnitramine (256) is present in the aqueous filtrate from both the KA-process and E-process (Section 5.15.1).

5.15.4.3 The chemistry of linear nitramines

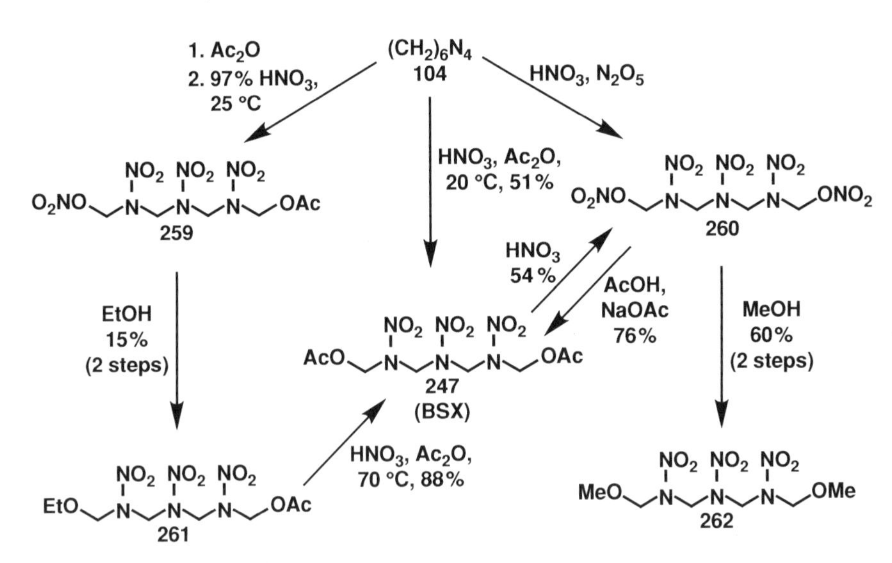

Figure 5.113 Linear nitramines from the nitrolysis of hexamine

The nitrolysis of hexamine can be used to obtain the linear nitramines (247), (259) and (260) depending on the conditions and reagents used. Thus, the nitrolysis of hexamine with a mixture of fuming nitric acid in acetic anhydride leads to the isolation of (247) (BSX), whereas the addition of 97 % nitric acid to a solution of hexamine in acetic anhydride forms the mixed nitrate–acetate ester (259).[191] The reaction of hexamine with dinitrogen pentoxide in absolute nitric acid leads to the formation of the dinitrate ester (260).[191] The nitrate ester groups of (259) and (260) are readily displaced and on reaction with sodium acetate in acetic acid form the corresponding acetate esters; the same reaction with low molecular weight alcohols forms the corresponding alkoxy ethers.[191]

REFERENCES

1. (a) T. Urbański, *Chemistry and Technology of Explosives, Vol. 3*, Pergamon Press, Oxford (1967); (b) T. Urbański, *Chemistry and Technology of Explosives, Vol. 4*, Pergamon Press, Oxford (1984); (c) *Kirk-Othmer Encyclopedia of Chemical Technology, 3rd Edn, Vol. 9*, Ed. M. Grayson, Wiley-Interscience, New York, 581–587 (1980).
2. G. E. Dunn, J. C. MacKenzie and G. F. Wright, *Can. J. Res.*, 1948, **26B**, 104.
3. W. J. Chute, G. E. Dunn, J. C. MacKenzie, G. S. Myers, G. N. R. Smart, J. W. Suggitt and G. F. Wright, *Can. J. Res.*, 1948, **26B**, 114.
4. W. J. Chute, K. G. Herring, L. E. Toombs and G. F. Wright, *Can. J. Res.*, 1948, **26B**, 89.
5. E. Bamberger and A. Kirpal, *Chem. Ber.*, 1895, **28**, 535.
6. J. V. Dubsky and A. P. N. Franchimont, *Recl. Trav. Chim. Pays-Bas*, 1916, **36**, 80.
7. A. H. Dinwoodie and R. N. Haszeldine, *J. Chem. Soc.*, 1965, 1675.
8. R. H. Meen and G. F. Wright, *J. Org. Chem.*, 1954, **19**, 391.
9. D. N. Gray and J. J. Schmidt-Collerus, *US Pat.* 3 562 333 (1971).
10. J. A. Krimmel, J. J. Schmidt-Collerus and J. A. Young, *J. Org. Chem.*, 1971, **36**, 347.
11. E. Macciotta, *Gazz. Chim. Ital.*, 1941, **71**, 81; *Chem. Abstr.*, 1942, **36**, 1593.
12. K. Shimo, *Tokyo Kogyo Shikensho Hokoku*, 1970, **65**, 46; *Chem. Abstr.*, 1971, **74**, 140812u.
13. W. H. Gilligan, *J. Org. Chem.*, 1971, **36**, 2138.
14. M. B. Frankel and E. F. Witucki, *US Pat.* 4 701 557 (1987); *Chem. Abstr.*, 1988, **108**, 97345a.
15. J. C. Bottaro, P. E. Penwell and R. J. Schmitt, '*Synthesis of Cubane Based Energetic Materials, Final Report, December 1989*', SRI International, Menlo Park, CA [AD-A217 147/8/XAB].
16. M. B. Frankel and E. F. Witucki, *J. Chem. Eng. Data.*, 1982, **27**, 94.
17. M. B. Frankel, *US Pat. Appl.* 361 643 (1982); *Chem. Abstr.*, 1983, **98**, 74896s.
18. P. Ai, Z. Tian, J. Yang and T. Zhang, 'Synthesis of some Polynitrodiols and their Acetal Compounds', *Proc. 17th International Pyrotechnics Seminar (Combined with 2nd Beijing International Symposium on Pyrotechnics and Explosives).*, Beijing Institute Technical Press, Beijing, China, Vol. 1, 261–268 (1991).
19. D. Wan, Q. Xu, G. Yao and Y. Yu, *Kogyu Kayaku*, 1982, **43**, 2; *Chem. Abstr.*, 1982, **97**, 23746q.
20. H. G. Adolph, M. Chaykovsky, C. George, R. Gilardi and W. M. Koppes, *J. Org. Chem.*, 1987, **52**, 1113.
21. D. W. Moore and R. L. Willer, *J. Org. Chem.*, 1985, **50**, 5123.
22. T. Connor, G. N. R. Smart and G. F. Wright, *Can. J. Res.*, 1948, **26B**, 294.
23. F. Chapman, *J. Chem. Soc.*, 1949, 1631.
24. G. F. Wright, in *The Chemistry of the Nitro and Nitroso Groups, Part 1, Organic Nitro Chemistry Series*, Ed. H. Feuer, Wiley-Interscience, New York, 'Methods of Formation of the Nitramino Group, its Properties and Reactions', Chapter 10, 614–684 (1969).
25. M. M. Bhalerao, B. R. Gandhe, M. A. Kulkarni, K. P. C. Rao and A. K. Sikder, *Propell. Explos. Pyrotech.*, 2004, **29**, 93.
26. T. G. Archibald, K. Baum and L. C. Garver, *Synth. Commun.*, 1990, **20**, 407.
27. D. Huang and R. R. Rindone, 'NNHT : A New Low Cost Insensitive Cyclic Nitramine', *Proc. Joint International Symposium on Compatibility of Plastics and Other Materials with Explosives,*

Propellants, Pyrotechnics and Processing of Explosives, Propellants and Ingredients, San Diego, CA, 62–68 (1991).
28. J. H. Boyer, V. T. Ramakrishnan and M. Vedachalam, *Heterocycles*, 1990, **31**, 479.
29. H. R. Graindorge, P. A. Lescop, M. J. Pouet and F. Terrier, in *Nitration: Recent Laboratory and Industrial Developments, ACS Symposium Series 623*, Eds. L. F. Albright, R. V. C. Carr and R. J. Schmitt, American Chemical Society, Washington, DC, Chapter 5, 43–50 (1996).
30. (a) C. L. Coon, E. S. Jessop, A. R. Mitchell, P. F. Pagoria and R. D. Schmidt, in *Nitration: Recent Laboratory and Industrial Developments, ACS Symposium Series 623*, Eds. L. F. Albright, R. V. C. Carr and R. J. Schmitt, American Chemical Society, Washington, DC, Chapter 14, 151–164 (1996). (b) E. S. Jessop, A. R. Mitchell and P. F. Pagoria, *Propell. Explos. Pyrotech.*, 1996, **21**, 14.
31. H. G. Adolph, J. H. Boyer, I. J. Dagley, J. L. Flippen-Anderson, C. George, R. Gilardi, K. A. Nelson, V. T. Ramakrishnan and M. Vedachalam, *J. Org. Chem.*, 1991, **56**, 3413.
32. A. Angeli and M. V. Maragliano, *Atti. Acad. Linceri.*, 1905, **14 (II)**, 127.
33. C. Hathaway, J. Kindig, J. R. Klink, D. Lazdins, W. N. White and E. F. Wolfarth, *J. Org. Chem.*, 1961, **26**, 4124.
34. E. Bamberger, *Chem. Ber.*, 1920, **53**, 2321.
35. S. C. Desai, D. B. Learn and L. J. Winters, *J. Org. Chem.*, 1965, **30**, 2471.
36. A. Domański, Z. Daszkiewicz and J. B. Kyziol, *Org. Prep. Proc. Int.*, 1994, **26(3)**, 337.
37. C. D. Bedford, J. C. Bottaro and R. J. Schmitt, *J. Org. Chem.*, 1987, **52**, 2292.
38. W. D. Emmons and J. P. Freeman, *J. Am. Chem. Soc.*, 1955, **77**, 4387.
39. (a) E. Bamberger, *Chem. Ber.*, 1984, **27**, 584; (b) E. Hoff. *Liebigs Ann. Chem.*, 1900, **311**, 91.
40. W. D. Emmons, A. S. Pagano and T. E. Stevens, *J. Org. Chem.*, 1958, **23**, 311.
41. A. Yu. Cheburkov, I. L. Knunyants and N. Mukhamadaliev, *Izv. Akad. Nauk USSR, Ser. Khim.*, 1966, 2119; Engl. Transl., 2053.
42. W. R. Feldman and E. H. White, *J. Am. Chem. Soc.*, 1957, **79**, 5832.
43. (a) B. C. Challis and S. A. Kyrtopoulos, *J. Chem. Soc. Perkin Trans. II*, 1978, 1296; (b) B. C. Challis and S. A. Kyrtopoulos, *J. Chem. Soc. Perkin Trans. I*, 1979, 299.
44. T. B. K. Lee and F. Wudl, *J. Am. Chem. Soc.*, 1971, **93**, 271.
45. (a) J. H. Boyer and T. P. Pillai, *J. Chem. Soc. Perkin Trans. I*, 1985, 1661; (b) J. H. Boyer, G. Kumar and T. P. Pillai, *J. Chem. Soc. Perkin Trans. I*, 1986, 1751.
46. D. W. Fish, E. E. Hamel and R. E. Olsen, in *Advanced Propellant Chemistry, Advances in Chemistry Series. No 54*, Ed. R. E. Gould, American Chemical Society, Washington, DC, Chapter 6, 48–54 (1966).
47. S. J. Kuhn and G. A. Olah, in *Friedel–Craft and Related Reactions, Vol. 3*, Ed. G. A. Olah, Wiley-Interscience, New York, 43 (1964).
48. B. V. Gidaspov, E. L. Golod and M. A. Ilyushin, *Zh. Org. Khim.*, 1977, **13**, 11.
49. G. S. Myers and G. F. Wright, *Can. J. Res.*, 1948, **26B**, 257.
50. A. H. Blatt, *OSRD-2014* (1944).
51. R. L. Atkins, R. A. Hollins and W. S. Wilson, *J. Org. Chem.*, 1986, **51**, 3261.
52. J. H. Ridd and T. Yoshida, in *Industrial and Laboratory Nitrations, ACS Symposium Series 22*, Eds. L. F. Albright and C. Hanson, American Chemical Society, Washington, DC, Chapter 3, 103–113 (1976).
53. G. A. Olah, R. Malhotra and S. C. Narang, *Nitration: Methods and Mechanisms*, Wiley-VCH, Weinheim, 278–286 (1989).
54. B. V. Gidaspov, E. L. Golod, Yu. V. Guk and M. A. Ilyushin, *Russ. Chem. Rev.*, 1983, **52(3)**, 284.
55. S.-D. Cho, H.-K. Kim, J.-J. Kim, S.-K. Kim, Y.-D. Park, M. Shiro and Y.-J. Yoon, *J. Org. Chem.*, 2003, **68**, 9113.
56. H. Mandel, *US Pat.* 3 071 438 (1963); *Chem. Abstr.*, 1963, **59**, 447.
57. R. S. Shineman, *Ph.D. Thesis*, Ohio State University (1957); University Microfilm, Ann Arbour, *MI Card No.* M. C. 58-2104.
58. (a) T. C. Bruice, M. J. Gregory and S. L. Walters, *J. Am. Chem. Soc.*, 1968, **90**, 1612; (b) T. C. Bruice and S. L. Walters, *J. Am. Chem. Soc.*, 1971, **93**, 2269; (c) J. F. Riordan and B. L. Vallee,

Methods Enzymol., 1972, **25B**, 515; (d) A. Castonguay and H. Van Vunakis, *Toxicol. Lett.*, 1979, **4**, 475.
59. C. M. Adams, C. M. Sharts and S. A. Shackelford, *Tetrahedron Lett.*, 1993, **34**, 6669.
60. M. B. Anderson, L. C. Christie, T. Goetzen, M. C. Guzman, M. A. Hananel, W. D. Kornreich, H. Li, V. P. Pathak, A. K. Rabinovich, R. J. Rajapakse, S. A. Shackelford, L. K. Truesdale, S. M. Tsank and H. N. Vazir, *J. Org. Chem.*, 2003, **68**, 267.
61. G. N. R. Smart and G. F. Wright, *Can. J. Res.*, 1948, **26B**, 284.
62. A. P. N. Franchimont and E. A. Klobbie, *Recl. Trav. Chim. Pays-Bas*, 1888, **7**, 258 and 343.
63. A. P. N. Franchimont and E. A. Klobbie, *Recl. Trav. Chim. Pays-Bas*, 1889, **8**, 297.
64. A. Lachman and J. Thiele, *Chem. Ber.*, 1894, **27**, 1520.
65. A. Lachman and J. Thiele, *Liebigs Ann. Chem.*, 1895, **288**, 287.
66. J. C. A. S. Thomas, *Recl. Trav. Chim. Pays-Bas*, 1890, **9**, 969.
67. H. Van Erp, *Recl. Trav. Chim. Pays-Bas*, 1895, **14**, 21.
68. H. J. Baker, *Recl. Trav. Chim. Pays-Bas*, 1912, **31**, 12.
69. A. P. N. Franchimont, *Recl. Trav. Chim. Pays-Bas*, 1883, **2**, 96; 1885, **4**, 196; 1894, **13**, 308.
70. W. E. Bachmann, W. J. Horton, E. L. Jenner, N. W. MacNaughton and C. E. Maxwell, *J. Am. Chem. Soc.*, 1950, **72**, 3132.
71. H. M. Curry and J. P. Mason, *J. Am. Chem. Soc.*, 1951, **73**, 5043.
72. V. P. Iushin, M. S. Komelin and V. A. Tartakovsky, *Zh. Org. Khim.*, 1999, **35**, 489.
73. R. S. Stuart and G. F. Wright, *Can. J. Res.*, 1948, **26B**, 401.
74. E. Hertz, *Brit. Pat.* 367 713 (1932).
75. J. Boileau, J. M. L. Emeury and J. P. Kehren, *US Pat.* 4 487 938 (1974).
76. G. V. Caesar, M. L. Cushing, M. Goldfrank and N. S. Gruenhut, *Inorg. Synth.*, 1950, **3**, 78.
77. (a) J. E. Harrar and R. K. Pearson, *J. Electrochem. Soc.*, 1983, **130**, 108; (b) N. Logan, *Pure Appl. Chem.*, 1986, **58**, 1147.
78. R. D. Chapman, J. W. Fischer, R. A. Hollins, C. K. Lowe-Ma and R. A. Nissan, *J. Org. Chem.*, 1996, **61**, 9340.
79. (a) J. P. Agrawal, G. M. Bhokare, D. B. Sarwade and A. K. Sikder, *Propell. Explos. Pyrotech.*, 2001, **26**, 63; (b) J. Hong and C. Zhu, *Proc. 17th International Pyrotechnics Seminar (Combined with 2nd Beijing International Symposium on Pyrotechnics and Explosives)*, Beijing Institute Technical Press, Beijing, China, 193 (1991).
80. H. Boniuk, I. Cieślowska-Glińska and M. Syczewski, *Propell. Explos. Pyrotech.*, 1998, **23**, 155.
81. J. von. Runge and W. Triebs, *J. Prakt. Chem.*, 1962, **15**, 233.
82. G. V. Caesar and M. Goldfrank, *US Pat.*, 2 400 288 (1946).
83. R. H. Erickson, B. S. Hahn, T. J. Ryan and E. H. White, *J. Org. Chem.*, 1984, **49**, 4860.
84. V. P. Gorelik, O. A. Luk'yanov and V. A. Tartakovsky, *Ivz. Akad. Nauk USSR, Ser. Khim.*, 1971, 1804.
85. J. C. Bottaro, P. E. Penwell and R. J. Schmitt, *J. Am. Chem. Soc.*, 1997, **119**, 9405.
86. S. A. Andreev, B. A. Lebedev and I. V. Tselinskii, *Zh. Org. Khim.*, 1978, **14**, 2513.
87. H. G. Adolph, J. H. Boyer, I. J. Dagley, J. L. Flippen-Anderson, C. George, R. Gilardi, K. A. Nelson, V. T. Ramakrishnan and M. Vedachalam, *J. Org. Chem.*, 1991, **56**, 3413.
88. J. V. Crivello, *J. Org. Chem.*, 1981, **46**, 3056.
89. R. D. Chapman and S. C. Suri, *Synthesis*, 1988, 743.
90. M. Aragonès, G. Garcia, P. Romea and J. Vilarrasa, *J. Org. Chem.*, 1991, **56**, 7038.
91. E. Carvalho, J. Iley, F. Norberto and E. Rosa, *J. Chem. Res. (S)*, 1989, 260.
92. C. M. Adams, S. A. Shackelford and C. M. Sharts, *Tetrahedron Lett.*, 1993, **34**, 6669.
93. J. H. Robson, *J. Am. Chem. Soc.*, 1955, **77**, 107.
94. J. Reinhart and J. H. Robson, *J. Am. Chem. Soc.*, 1955, **77**, 2453.
95. M. B. Frankel, M. H. Gold, C. H. Tieman and C. R. Vanneman, *J. Org. Chem.*, 1960, **25**, 744.
96. E. E. Gilbert and V. I. Siele, *US Pat.* 3 939 148 (1976).
97. E. E. Gilbert, J. R. Leccacorvi and M. Warman, in *Industrial and Laboratory Nitrations, ACS Symposium Series 22*, Eds. L. F. Albright and C. Hanson., American Chemical Society, Washington, DC, Chapter 23, 327–340 (1976).

98. (a) A. T. Nielsen, *US Pat.* 253 106 (1988); *Chem. Abstr.*, 1998, **128**, 36971t; (b) A. T. Nielsen, 'Synthesis of Caged Nitramine Explosives', presented at Joint Army, Navy, NASA, Air Force (JANNAF) Propulsion Meeting, San Diego, CA, 17 December, 1987; (c) A. P. Chafin, S. L. Christian, J. L. Flippen-Anderson, C. F. George, R. D. Gilardi, D. W. Moore, M. P. Nadler, A. T. Nielsen, R. A. Nissan and D. J. Vanderah, *Tetrahedron*, 1998, **54**, 11793; (d) C. L. Coon, J. L. Flippen-Anderson, C. F. George, R. D. Gilardi, A. T. Nielsen, R. A. Nissan and D. J. Vanderah, *J. Org. Chem.*, 1990, **55**, 1459.
99. A. J. Bellamy, P. Goede, N. V. Latypov and U. Wellmar, *Org. Proc. Res. Dev.*, 2000, **4**, 156.
100. R. D. Chapman, R. D. Gilardi, C. B. Kreutzberger and M. F. Welker, *J. Org. Chem.*, 1999, **64**, 960.
101. T. Axenrod, R. D. Chapman, R. D. Gilardi, X.-P. Guan, L. Qi and J. Sun, *Tetrahedron Lett.*, 2001, **42**, 2621.
102. (a) S. A. Andreev, B. V. Gidaspov, B. A. Lebedev, L. A. Novik and I. V. Tselinskii, *Zh. Org. Khim.*, 1978, **14**, 240; (b) S. A. Andreev, I. N. Shohbor and I. V. Tselinskii, *Zh. Org. Khim.*, 1980, **16**, 1353; (c) S. A. Andreev, B. A. Lebedev and I. V. Tselinskii, *Zh. Org. Khim*, 1980, **16**, 1365.
103. (a) O. A. Luk'yanov, N. M. Seregina and V. A. Tartakovsky, *Ivz. Akad. Nauk USSR, Ser. Khim.*, 1976, 225; (b) O. A. Luk'yanov, T. G. Mel'nikova, N. M. Seregina and V. A. Tartakovsky, *'Tezisy VI Vsesoyuznogo Soveshchaniya Po Nitrosoedinenii'* (Abstracts of Reports at the 6th All-Union Conference on the Chemistry of Nitro-Compounds), Moscow, 33 (1977).
104. S. A. Andreev, B. V. Gidaspov, B. A. Lebedev, G. I. Koldobskii and I. V. Tselinskii, *Zh. Org. Khim.*, 1978, **14**, 907.
105. J. H. Boyer, T. P. Pillai and V. T. Ramakrishnan, *Synthesis*, 1985, 677.
106. W. D. Emmons, *J. Am. Chem. Soc.*, 1954, **76**, 3468.
107. T. Urbański, *Chemistry and Technology of Explosives, Vol. 3*, Pergamon Press, Oxford, 40–56 (1967).
108. H. G. Adolph and D. A. Cichra, *J. Org. Chem.*, 1982, **47**, 2474.
109. R .B. Crawford, L. de Vore, K. Gleason, K. Hendry, D. P. Kirvel, R. D. Lear, R. R. McGuire and R. D. Stanford, *Energy and Technology Review, Jan–Feb 1988, UCRL-52000-88-1/2*, Lawrence Livermore National Laboratory, Livermore, CA.
110. R. D. Breithaupt, C. L. Coon, E. S. Jessop, A. R. Mitchell, G. L. Moody, P. F. Pagoria, J. F. Poco and C. M. Tarver., *Propell. Explos. Pyrotech.*, 1994, **19**, 232; 'Synthesis, Scale-up, and Characterisation of K-6', *Report No UCRL-LR-109404* (1992), Lawrence Livermore National Laboratory, Livermore, CA.
111. T. G. Archibald, K. Baum, C. George and R. D. Gilardi, *J. Org. Chem.*, 1990, **55**, 2920.
112. T. G. Archibald, M. D. Coburn and M. A. Hiskey, *Waste Management*, 1997, **17**, 143.
113. F. Chapman, *J. Chem. Soc.*, 1949, 1631.
114. W. P. Norris, *J. Org. Chem.*, 1960, **25**, 1244.
115. F. J. Brockman, D. C. Downing and G. F. Wright, *Can. J. Res.*, 1949, **27B**, 469.
116. R. L. Atkins and J. W. Fischer, *Org. Prep. Proc. Int.*, 1986, **18**, 281.
117. R. L. Atkins and R. L. Willer, *J. Org. Chem.*, 1984, **49**, 5147.
118. R. L. Willer, *J. Org. Chem.*, 1984, **49**, 5150.
119. T. Benzinger, M. D. Coburn, R. K. Davey, E. E. Gilbert, R. W. Hutchinson, J. Leccacorvi, R. Motto, R. K. Rohwer, V. I. Siele and M. Warman, *Propell. Explos. Pyrotech.*, 1981, **6**, 67.
120. C. D. Bedford, J. C. Bottaro, D. F. McMillen and R. J. Schmitt, *J. Org. Chem.*, 1988, **53**, 4140.
121. Work conducted by G. A. Olah, G. K. Surya Prakash and N. Trivedi, as reported by G. A. Olah, in *Chemistry of Energetic Materials*, Ed. G. A. Olah and D. R. Squire, Academic Press, 197 (1991).
122. R. W. Millar and S. P. Philbin, *Tetrahedron*, 1997, **53**, 4371.
123. (a) A. D. Harris, H. B. Jonassen and J. C. Trebellas, *Inorg. Synth.*, 1967, **9**, 83; (b) R. W. Millar, N. C. Paul and D. H. Richards, *UK Pat. Appl.* 2 181 124 (1987) and 2 181 139 (1987); (c) T. E. Devendorf and J. R. Stacy, in *Nitration: Recent Laboratory and Industrial Developments, ACS Symposium Series 623*, Eds. L. F. Albright, R. V. C. Carr and R. J. Schmitt, American Chemical Society, Washington, DC, Chapter 8, 68–77 (1996).

124. P. Golding, R. W. Millar, N. C. Paul and D. H. Richards, *Tetrahedron*, 1993, **49**, 7063.
125. P. Golding, R. W. Millar, N. C. Paul and D. H. Richards, *Tetrahedron Lett.*, 1988, **29**, 2735.
126. P. Golding, R. W. Millar, N. C. Paul and D. H. Richards, *Tetrahedron*, 1995, **51**, 5073.
127. (a) L. A. Fang, S. Q. Hua, V. G. Ling and L. Xin, Preliminary Study on Bu-NENA Gun Propellants, *27th International Annual Conference of ICT*, Karlsruhe, Germany, 25–28 June, 1996, 51; (b) N. F. Stanley and P. A. Silver, 'Bu-NENA Gun Propellants', *JANNAF Propulsion Meetings*, Vol. 2, 10 September 1990, 515; (c) R. A. Johnson and J. J. Mulley, 'Stability and Performance Characteristics of NENA Materials and Formulations', *Joint International Symposium on Energetic Materials Technology*, New Orleans, Louisiana, 5–7 October, 1992, 116.
128. P. Bunyan, P. Golding, R. W. Millar, N. C. Paul, D. H. Richards and J. A. Rowley, *Propell. Explos. Pyrotech.*, 1993, **18**, 55.
129. R. D. Dresdner, J. A. Young and S. N. Tsoukalas, *J. Am. Chem. Soc.*, 1960, **82**, 396.
130. E. R. Bissel and M. Finger, *J. Org. Chem.*, 1959, **24**, 1256.
131. A. Lachman and J. Thiele, *Liebigs Ann. Chem.*, 1895, **288**, 275.
132. A. P. N. Franchimont, *Recl. Trav. Chim. Pays-Bas*, 1883, **3**, 275.
133. A. P. N. Franchimont and E. A. Klobbie, *Recl. Trav. Chim. Pays-Bas*, 1887, **7**, 354.
134. (a) M. S. Chang and R. S. Orndoff, *US Pat. Appl.* 394 218 (1983); *Chem. Abstr.*, 1983, **98**, 182088p; (b) G. W. Naufleet, *US Pat. Appl.* 394 084 (1983); *Chem. Abstr.*, 1983, **98**, 182087n.
135. R. C. Brian and A. H. Lamberton, *J. Chem. Soc.*, 1949, 1633.
136. A. P. N. Franchimont and E. A. Klobbie, *Recl. Trav. Chim. Pays-Bas*, 1886, **5**, 280; 1888, **7**, 17; 1888, **7**, 239.
137. C. E. Schweitzer, *J. Org. Chem.*, 1950, **15**, 471.
138. (a) P. Dekkers, *Recl. Trav. Chim. Pays-Bas*, 1890, **9**, 92; (b) R. J. Hardy, W. S. Lindsay and G. C. Mees, *Brit. Pat.* 997 826 (1965); *Chem. Abstr.*, 1965, **63**, 13075e.
139. H. M. Curry and J. P. Mason, *J. Am. Chem. Soc.*, 1951, **73**, 5043.
140. A. F. McKay, *Chem. Rev.*, 1952, **51**, 301.
141. A. F. McKay and G. F. Wright, *J. Am. Chem. Soc.*, 1948, **70**, 3990.
142. A. F. McKay and D. F. Manchester, *J. Am. Chem. Soc.*, 1949, **71**, 1970.
143. E. Hoff, *Liebigs Ann. Chem.*, 1900, **311**, 99.
144. T. L. Davis and N. D. Constan, *J. Am. Chem. Soc.*, 1936, **58**, 1800.
145. A. W. Ingersoll and B. F. Armendt, in *Organic Syntheses, Coll. Vol. 1*, Ed. A. H. Blatt, John Wiley & Sons, Inc., New York, 417 (1941).
146. T. L. Davis, in *Organic Syntheses, Coll. Vol. 1*, Ed. A. H. Blatt, John Wiley & Sons, Inc., New York, 399 (1941).
147. T. L. Davis, in *Organic Syntheses, Coll. Vol. 1*, Ed. A. H. Blatt, John Wiley & Sons, Inc., New York, 302 (1941).
148. (a) E. Bamberger, *Chem. Ztg.*, 1892, **16**, 185; (b) O. Hinsberg, *Chem. Ber.*, 1892, **25**, 1092; (c) J. Thiele, *Liebigs Ann. Chem.*, 1910, **376**, 239; (d) A. Kuchenbecker and T. Zinke, *Liebigs Ann. Chem.*, 1904, **330**, 1.
149. W. P. Norris, *J. Am. Chem. Soc.*, 1959, **81**, 3346.
150. B. V. Gidaspov, B. A. Lebedev and L. V. Cherednichenko, *Zh. Org. Khim.*, 1978, **14**, 735.
151. A. Berg, *Ann. Chim. Phys.*, 1894, **3**, 357.
152. D. D. DesMarteau and J. Foropoulos Jr, *Inorg. Chem.*, 1984, **23**, 3720.
153. M. I. Gillibrand and A. H. Lamberton, *J. Chem. Soc.*, 1949, 1883.
154. (a) E. De Carvalho, J. Iley, F. Norberto and E. Rosa, *J. Chem. Soc. Perkin Trans. II*, 1992, 281; (b) E. De Carvalho, J. Iley, F. Norberto, P. Patel and E. Rosa, *J. Chem. Res. (S)*, 1985, 132.
155. L. W. Kissinger and M. Schwartz, *J. Org. Chem.*, 1958, **23**, 1342.
156. H. Feuer and R. Miller, *J. Org. Chem.*, 1961, **26**, 1348.
157. F. Chapman, P. G. Owston and P. Woodcock, *J. Chem. Soc.*, 1949, 1647.
158. F. Chapman, P. G. Owston and P. Woodcock, *J. Chem. Soc.*, 1949, 1638.
159. J. A. Bell and I. Dunstan, *J. Chem. Soc. (C)*, 1966, 870.

160. J. A. Bell and I. Dunstan, *J. Chem. Soc. (C)*, 1966, 862.
161. J. A. Bell and I. Dunstan, *J. Chem. Soc. (C)*, 1967, 562.
162. (a) K. Klager, *J. Org. Chem.*, 1958, **23**, 1519; (b) M. B. Frankel, *J. Chem. Eng. Data.*, 1962, **7**, 410; (c) M. B. Frankel and K. Klager, *J. Chem. Eng. Data.*, 1962, **7**, 412; (d) L. W. Kissinger and H. E. Ungnade, *J. Org. Chem.*, 1965, **30**, 354; (e) M. B. Frankel and E. F. Witucki, *US Pat.* 4 701 557 (1987); *Chem. Abstr.*, 1988, **108**, 97345a; (f) W. H. Gilligan and M. E. Sitzman, *J. Energ. Mater.*, 1985, **3**, 293; (g) L. T. Eremenko, R. G. Gafurov, F. Ya. Natsibullin and S. I. Sviridor, *Izv. Akad. Nauk USSR, Ser. Khim* (Engl. Transl.), 1970, 329; (h) L. T. Eremenko, R. G. Gafurov and E. M. Sogomonyan, *Izv. Akad. Nauk USSR, Ser. Khim* (Engl. Transl.), 1971, 2480.
163. M. B. Frankel, *J. Org. Chem.*, 1961, **26**, 4709.
164. F. R. Schenck and G. A. Wetterholm, *US Pat.* 2 731 460 (1956); *Chem. Abstr.*, 1956, **50**, 7125g.
165. L. Goodman, *J. Am. Chem. Soc.*, 1953, **75**, 3019.
166. H. L. Herman, in *Encyclopedia of Explosives and Related Items. Vol. 8*, Ed. S. M. Kaye, ARRADCOM, Dover, New Jersey, 138 (1978).
167. V. V. Arakcheeva, L. T. Eremenko, B. S. Fedorov and G. V. Lagodzinskaya, *Izv. Akad. Nauk USSR, Ser. Khim.*, 1984, 2407.
168. G. D. Sammons, *US Pat.* 3 151 165 (1964); *Chem. Abstr.*, 1965, **62**, 2662.
169. H. Feuer and R. Millar, *US Pat.* 3 040 099 (1962); *Chem. Abstr.*, 1962, **57**, 12778.
170. H. Backer, *Recl. Trav. Chim. Pays-Bas*, 1912, **31**, 142.
171. (a) A. P. N. Franchimont, *Recl. Trav. Chim. Pays-Bas*, 1910, **29**, 296; (b) A. P. N. Franchimont and H. van Erp, *Recl. Trav. Chim. Pays-Bas*, 1895, **14**, 235; (c) A. P. N. Franchimont and H. Umbgrowe, *Recl. Trav. Chim. Pays-Bas*, 1896, **15**, 195; (d) P. Bruck and A. Lamberton, *J. Chem. Soc.*, 1955, 3997.
172. A. Semeńczuk and T. Urbański, *Bull. Acad. Polon. Sci., Cl. III*, 1957, **5**, 649.
173. H. H. Hodgson and G. Turner, *J. Chem. Soc.*, 1942, 584.
174. K. H. Mertens, *Chem. Ber.*, 1877, **10**, 995.
175. A. Semeńczuk and T. Urbański, *Bull. Acad. Polon. Sci., Cl. III*, 1958, **6**, 309.
176. C. E. Clarkson, I. G. Holden and T. Malkin, *J. Chem. Soc.*, 1950, 1556.
177. T. Urbański, *Chemistry and Technology of Explosives, Vol. 3*, Pergamon Press, Oxford, 47–48 (1967).
178. P. Van Romburgh, *Recl. Trav. Chim. Pays-Bas*, 1883, **2**, 31, 108 and 103; 1884, **3**, 392; 1889, **8**, 215.
179. R. C. Moran, *US Pat.* 1 560 427 (1925).
180. G. M. Bennett, *J. Chem. Soc.*, 1919, 576.
181. R. F. B. Cox, *US Pat.* 2 125 221 (1938).
182. J. Urbański and T. Urbański, *Bull. Acad. Polon. Sci., Cl. III*, 1958, **6**, 307.
183. J. Urbański and T. Urbański, *Roczniki. Chem.*, 1959, **33**, 693.
184. H. A. Bruson and C. W. MacMullen, *J. Am. Chem. Soc.*, 1941, **63**, 270.
185. A. Semeńczuk, *Biul. WAT 5, XXII.*, 1956, 58.
186. J. R. Autera and T. C. Castorina, *I and EC Product Research and Development*, 1965, **4**, 170.
187. T. C. Castorina, R. J. Graybush, S. Helf, F. S. Holahan and J. V. R. Kaufman, *J. Am. Chem. Soc.*, 1960, **82**, 1617.
188. W. J. Chute, D. C. Downing, A. F. McKay, G. S. Myers and G. F. Wright, *Can. J. Res.*, 1949, **27B**, 218.
189. A. F. McKay, H. H. Richmond and G. F. Wright, *Can. J. Res.*, 1949, **27B**, 462.
190. G. S. Myers and G. F. Wright, *Can. J. Res.*, 1949, **27B**, 489.
191. W. J. Chute, A. F. McKay, R. H. Meen, G. S. Myers and G. F. Wright, *Can. J. Res.*, 1949, **27B**, 503.
192. E. Aristoff, J. A. Graham, R. H. Meen, G. S. Myers and G. F. Wright, *Can. J. Res.*, 1949, **27B**, 520.
193. A. H. Vroom and C. A. Winkler, *Can. J. Res.*, 1950, **28B**, 701.
194. M. Kirsch and C. A. Winkler, *Can. J. Res.*, 1950, **28B**, 715.
195. W. E. Bachmann and J. C. Sheehan, *J. Am. Chem. Soc.*, 1949, **71**, 1842.
196. W. E. Bachmann, W. J. Horton, E. L. Jenner, N. W. MacNaughton and L. B. Scott, *J. Am. Chem. Soc.*, 1951, **73**, 2769.
197. W. E. Bachmann and E. L. Jenner, *J. Am. Chem. Soc.*, 1951, **73**, 2773.

198. G. F. Henning, *Ger. Pat.* 104 280 (1899).
199. E. V. Herz, *Brit. Pat.* 145 791 (1920); *US Pat.* 1 402 693 (1922).
200. G. C. Hale, *J. Am. Chem. Soc.*, 1925, **47**, 2754.
201. (a) 'General Summary of Explosive Plants', *Technical Report P.B. 925*, US Department of Commerce, Washington, DC, 1945; (b) 'RDX Manufacture in Germany', *Technical Report P.B. 262*, US Department of Commerce, Washington, DC, 1945; (c) W. de. C. Crater, *Ind. Eng. Chem.*, 1949, **40**, 1627.
202. J. H. Ross and R. W. Schiessler, *Brit. Pat.* 595 354 (1947); *US Pat.* 2 434 230 (1948).
203. A. Gillies, H. L. Williams and C. A. Winkler, *Can. J. Chem.*, 1951, **29**, 377.
204. W. P. Binnie, H. L. Cohen and G. F. Wright, *J. Am. Chem. Soc.*, 1950, **72**, 4457.
205. W. E. Bachmann and N. C. Deno, *J. Am. Chem. Soc.*, 1951, **73**, 2777.
206. E. E. Gilbert, V. I. Siele and M. Warman, *J. Heterocycl. Chem.*, 1973, **10**, 97.
207. S. Epstein and C. A. Winkler, *Can. J. Chem.*, 1952, **30**, 734.
208. W. E. Bachmann, *OSRD Report No. 1981*.
209. P. S. Bratia, B. S. Singh and H. Singh, *Def. Sci. J.*, 1982, **32**, 297.
210. J. P. Picard, *US Pat.* 2 983 725 (1961); *Chem. Abstr.*, 1961, **55**, 20436.
211. E. Ju. Orlova, N. A. Orlova, G. M. Shuter, V. L. Zbarskii, V. F. Zhilin and L. I. Vitkovskaya, *Nedra.*, Moscow (1975).
212. (a) J. Chen and S.-F. Wang, *Propell. Explos. Pyrotech.*, 1984, **9**, 58; (b) J. Chen and S. Wang, *Kexue Tongbao*, 1984, **29**, 595; (c) J. Chen, F. Li and S. -F. Wang, *Proc. International Symposium on Pyrotechnics and Explosives*, China Academic Publishers, Beijing, China, 197–202 (1987); (d) A. P. Cooney, M. R. Crampton, and P. Golding, *J. Chem. Soc. Perkin Trans. II*, 1986, 835; (e) A. P. Cooney, M. R. Crampton, M. Jones and P. Golding, *J. Heterocycl. Chem.*, 1987, **24**, 1163.
213. E. E. Gilbert, V. I. Siele and M. Warman, *J. Heterocycl. Chem.*, 1974, **11**, 237.
214. (a) T. M. Benzinger, M. D. Coburn, R. K. Davey, E. E. Gilbert, R. W. Hutchinson, J. Leccacorvi, R. Motto, R. K. Rohwer, V. I. Siele and M. Warman, *Propell. Explos. Pyrotech.*, 1981, **6**, 67; (b) E. E. Gilbert, R. W. Hutchinson, J. Leccacorvi, R. Motto, V. I. Siele and M. Warman, 'Alternative Processes for HMX Manufacture', *Technical Report ARLCD-TR-78008*, US Army Research and Development Command, Dover, New Jersey, October 1979; *AD-A083 793*, available from Defence Documentation Center, Cameron Station, Alexandria, VA 22314.
215. V. I. Siele, *US Pat.* 3 979 379 (1976).
216. L. Fuping and W. Shaofang, 'One-Step Process for TAT Preparation from Hexamethylenetetramine', *International Annual Conference of ICT*, Karlsruhe, Germany, 1–3 July, 1981, 589–601.
217. J. Chen, F. Li and S. Wang, *Propell. Explos. Pyrotech.*, 1990, **15**, 54.
218. S. Baryla, L. Gerlotka and Z. Kurnatowki, *Polish. Pat.* 152 897 (1991); *Chem. Abstr.*, 1991, **115**, 283096q.
219. C. L. Coon, J. E. Harrar, R. R. McGuire and R. K. Pearson, *US Pat. Appl.* 399 948; *Chem. Abstr.*, 1984, **100**, 123574f.
220. K. W. Dunning and W. J. Dunning, *J. Chem. Soc.*, 1950, 2920, 2925 and 2928.
221. M. B. Frankel and D. O. Woolery, *J. Org. Chem.*, 1983, **48**, 611.

6

Energetic Compounds 2: Nitramines and Their Derivatives

In Chapter 5 we discussed the methods used to incorporate *N*-nitro functionality into compounds in addition to the synthesis of the heterocyclic nitramine explosives RDX and HMX. The high performance of such heterocyclic nitramines has directed considerable resources towards the synthesis of compounds containing strained or caged skeletons in conjunction with *N*-nitro functionality. These compounds derive their energy release on detonation from both the release of molecular strain and the combustion of the carbon skeleton. Some nitramine compounds contain heterocyclic structures with little to no molecular strain. Even so, such skeletons often lead to an increase in crystal density relative to the open chain compounds and this usually results in higher explosive performance. A common feature of explosives containing *N*-nitro functionality is their higher performance compared to standard *C*-nitro explosives like TNT. Compounds containing strained or caged skeletons in conjunction with *N*-nitro functionality are some of the most powerful explosives available.

6.1 CYCLOPROPANES

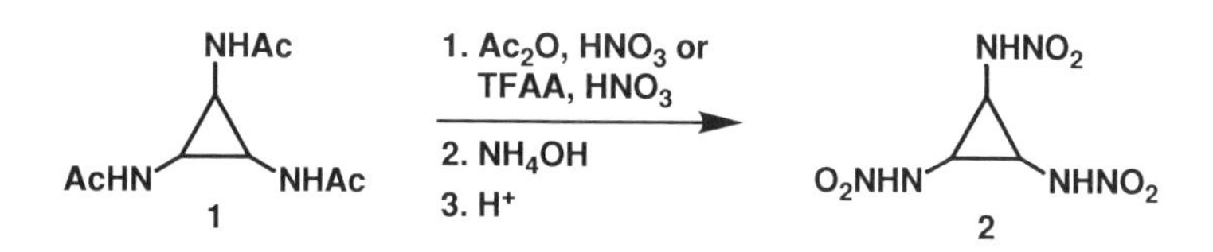

Figure 6.1

1,2,3-Tris(nitramino)cyclopropane (2) has been synthesized via the nitration of 1,2,3-tris(acetamido)cyclopropane (1) with acetic anhydride–nitric acid, followed by ammonolysis of the resulting secondary nitramide and subsequent acidification of the ammonium salt.[1] This strategy is a common route to primary nitramines (see Section 5.10). 1,2,3-Tris(nitramino)cyclopropane has a favourable oxygen balance and is predicted to exhibit high performance.[1]

Organic Chemistry of Explosives J. P. Agrawal and R. D. Hodgson

6.2 CYCLOBUTANES

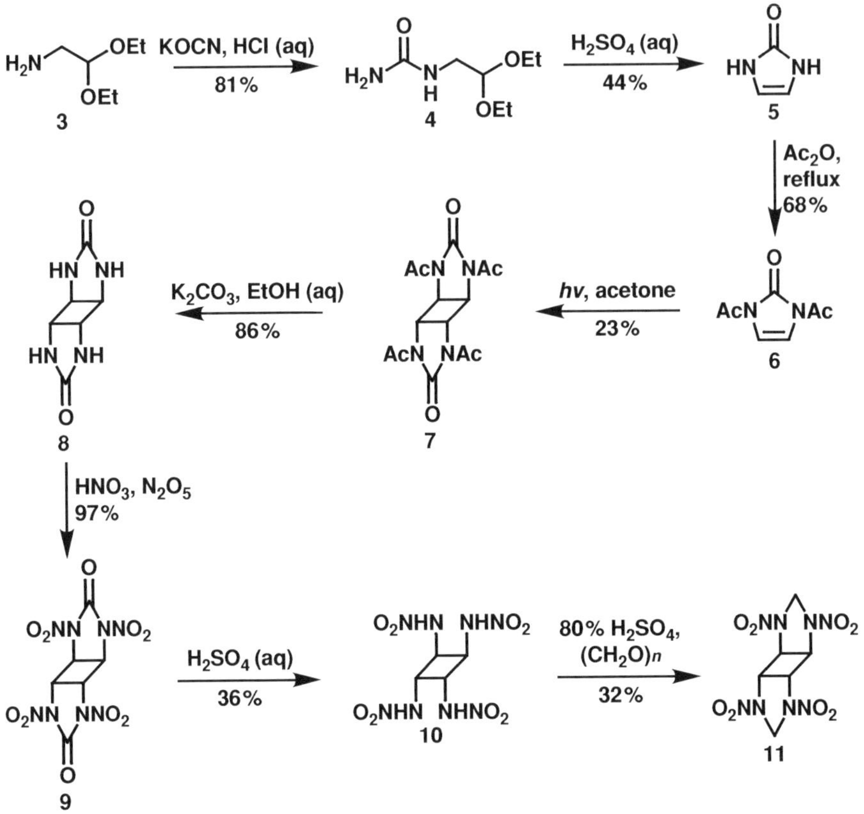

Figure 6.2

Chapman and co-workers[2] have synthesized nitramino derivatives of cyclobutane. Their synthesis starts from the reaction of aminoacetaldehyde diethylacetal (3) with potassium cyanate in aqueous hydrochloric acid to give ureidoacetaldehyde diethylacetal (4) which undergoes ring closure to the imidazolinone (5) on treatment with aqueous sulfuric acid. Acetylation of the imidazolinone (5) with acetic anhydride, followed by a photo-induced [2 + 2] cycloaddition, yields the cyclobutane derivative (7). Deacetylation of (7) with ethanolic potassium carbonate, followed by treatment of the resulting bis-urea (8) with absolute nitric acid or dinitrogen pentoxide in fuming nitric acid, yields octahydro-1,3,4,6-tetranitro-3a,3b,6a,6b-cyclobuta[1,2-*d*:3,4-*d'*]diimidazole-2,5-dione (9), a powerful explosive with a detonation velocity of 8400 m/s and a high crystal density of 1.99 g/cm^3, both properties typical of the energetic and structurally rigid nature of cyclic *N*,*N'*-dinitroureas.

The *N*,*N'*-dinitrourea (9) is a precursor to the nitramine explosives (10) and (11).[2] Thus, refluxing (9) in aqueous sulfuric acid yields *N*,*N'*,*N''*,*N'''*-tetranitro-1,2,3,4-cyclobutanetetramine (10), an explosive which is isomeric with HMX. Treatment of (10) with paraformaldehyde in 80 % aqueous sulfuric acid yields octahydro-1,3,4,6-tetranitro-3a,3b,6a,6b-cyclobuta[1,2-*d*:3,4-*d*′]diimidazole (11).

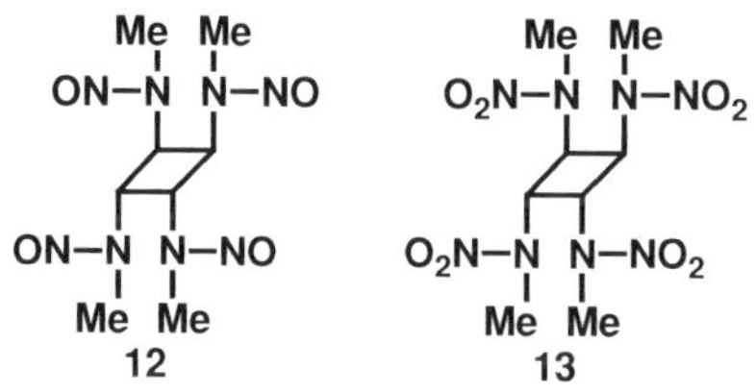

Figure 6.3

The tetranitrosamine (12) and the tetranitramine (13) are also synthesized from the bis-urea (8), although these are less energetic and have less favourable oxygen balances than (9), (10) and (11).[2]

6.3 AZETIDINES – 1,3,3-TRINITROAZETIDINE (TNAZ)

1,3,3-Trinitroazetidine (TNAZ) (18) is the product of a search for high performance explosives which also exhibit desirable properties, such as high thermal stability and low sensitivity to shock and impact. TNAZ is a powerful explosive which exhibits higher performance than RDX and HMX in the low vulnerability ammunition XM-39 gun-propellant formulations, while also showing low sensitivity to impact and good thermal stability.[3] TNAZ has a convenient low melting point (101 °C) which allows for the melt casting of charges. TNAZ is also fully miscible in molten TNT. These favourable properties have meant that TNAZ has been synthesized by numerous routes[4–9] and is now manufactured on a pilot plant scale.

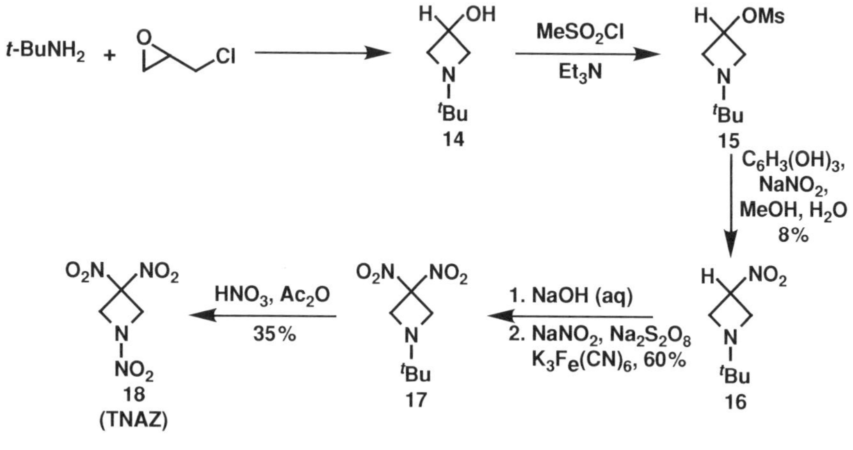

Figure 6.4 Archibald and co-workers route to TNAZ[4]

Archibald and co-workers[4] reported the first synthesis of TNAZ (18) in 1989. This route uses the reaction between *tert*-butylamine and epichlorohydrin to form the required azetidine ring. The *N-tert*-butyl-3-hydroxyazetidine (14) formed from this reaction is treated with methanesulfonyl chloride and the resulting mesylate (15) reacted with sodium nitrite in the presence of phloroglucinol to yield *N-tert*-butyl-3-nitroazetidine (16), the phloroglucinol used in this reaction preventing the formation of nitrite ester by-product. Oxidative nitration of *N-tert*-butyl-3-nitroazetidine (16) to *N-tert*-butyl-3,3-dinitroazetidine (17) is achieved in 39 % yield with a mixture of sodium nitrite and silver nitrate, and in 60 % yield with sodium nitrite and sodium persulfate in the presence of potassium ferricyanide. The synthesis of TNAZ (18) is completed by nitrolysis of the *tert*-butyl group of (17) with nitric acid in acetic anhydride. Unfortunately, this synthesis provides TNAZ in less than 20 % overall yield, a consequence of the low yields observed for both the initial azetidine ring-forming reaction and the reaction of (15) with nitrite ion.

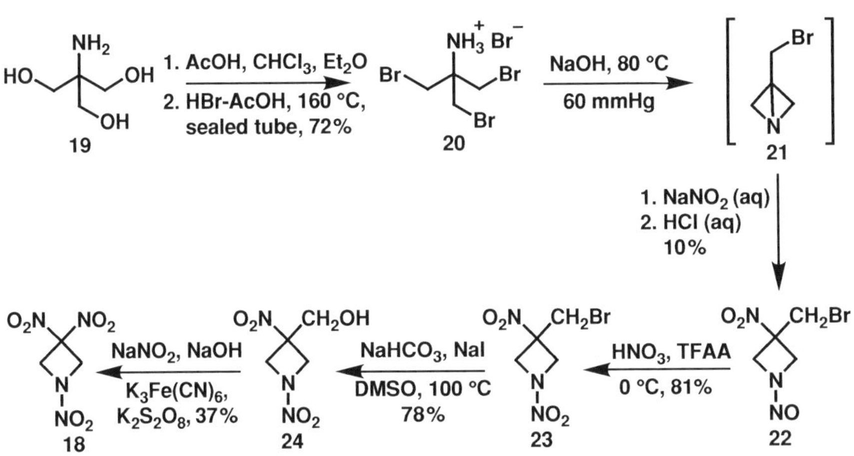

Figure 6.5 Marchand and co-workers route to TNAZ[5]

Marchand and co-workers[5] reported a synthetic route to TNAZ (18) involving a novel electrophilic addition of $NO^+ NO_2^-$ across the highly strained C(3)–N bond of 3-(bromomethyl)-1-azabicyclo[1.1.0]butane (21), the latter prepared as a nonisolatable intermediate from the reaction of the bromide salt of tris(bromomethyl)methylamine (20) with aqueous sodium hydroxide under reduced pressure. The product of this reaction, *N*-nitroso-3-bromomethyl-3-nitroazetidine (22), is formed in 10 % yield but is also accompanied by *N*-nitroso-3-bromomethyl-3-hydroxyazetidine as a by-product. Isolation of (22) from this mixture, followed by treatment with a solution of nitric acid in trifluoroacetic anhydride, leads to nitrolysis of the *tert*-butyl group and yields (23). Treatment of (23) with sodium bicarbonate and sodium iodide in DMSO leads to hydrolysis of the bromomethyl group and the formation of (24). The synthesis of TNAZ (18) is completed by deformylation of (24), followed by oxidative nitration, both processes achieved in 'one pot' with an alkaline solution of sodium nitrite, potassium ferricyanide and sodium persulfate. This route to TNAZ gives a low overall yield and is not suitable for large scale manufacture.

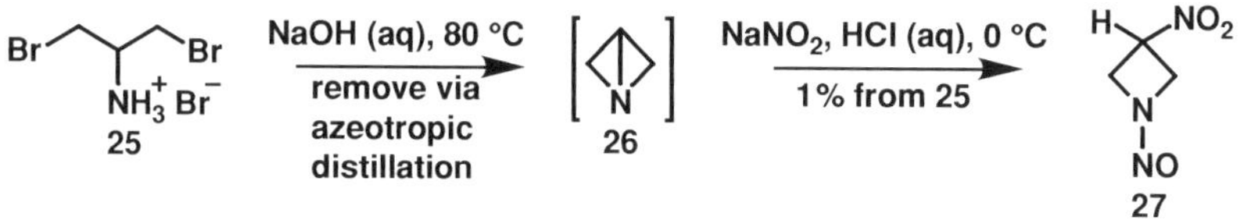

Figure 6.6

The synthesis of TNAZ (18) via the electrophilic addition of $NO^+NO_2^-$ across the C(3)–N bond of 1-azabicyclo[1.1.0]butane (26) was found to be very low yielding (∼1 %) and impractical.[5] Nagao and workers[6] reported a similar synthesis of TNAZ via this route but the overall yield was low.

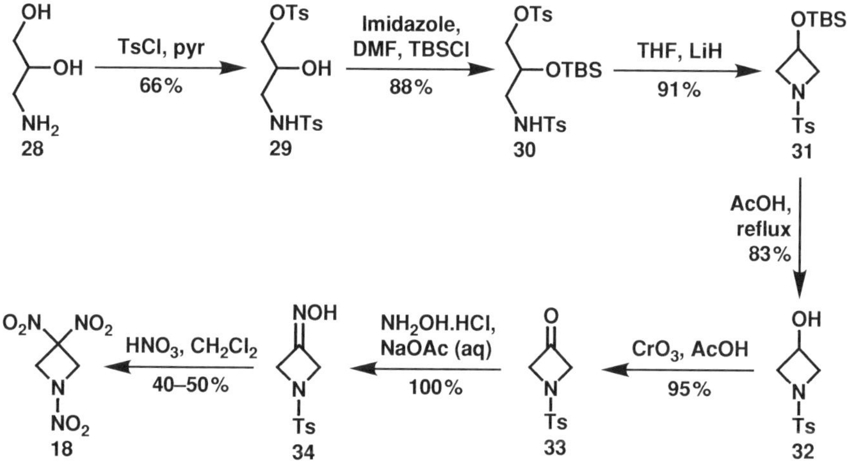

Figure 6.7 Axenrod and co-workers route to TNAZ[7,8]

Axenrod and co-workers[7,8] reported a synthesis of TNAZ (18) starting from 3-amino-1,2-propanediol (28). Treatment of (28) with two equivalents of *p*-toluenesulfonyl chloride in the presence of pyridine yields the ditosylate (29), which on further protection as a TBS derivative, followed by treatment with lithium hydride in THF, induces ring closure to the azetidine (31) in excellent yield. Removal of the TBS protecting group from (31) with acetic acid at elevated temperature is followed by oxidation of the alcohol (32) to the ketone (33). Treatment of the ketone (33) with hydroxylamine hydrochloride in aqueous sodium acetate yields the oxime (34). The synthesis of TNAZ (18) is completed on treatment of the oxime (34) with pure nitric acid in methylene chloride, a reaction leading to oxidation–nitration of the oxime group to *gem*-dinitro functionality and nitrolysis of the *N*-tosyl bond. This synthesis provides TNAZ in yields of 17–21 % over the seven steps.

Archibald, Coburn, and Hiskey[9] at Los Alamos National Laboratory (LANL) have reported a synthesis of TNAZ (18) that gives an overall yield of 57 % and is suitable for large scale manufacture. Morton Thiokol in the US now manufactures TNAZ on a pilot plant scale via this route. This synthesis starts from readily available formaldehyde and nitromethane, which under base catalysis form tris(hydroxymethyl)nitromethane (35), and without isolation from

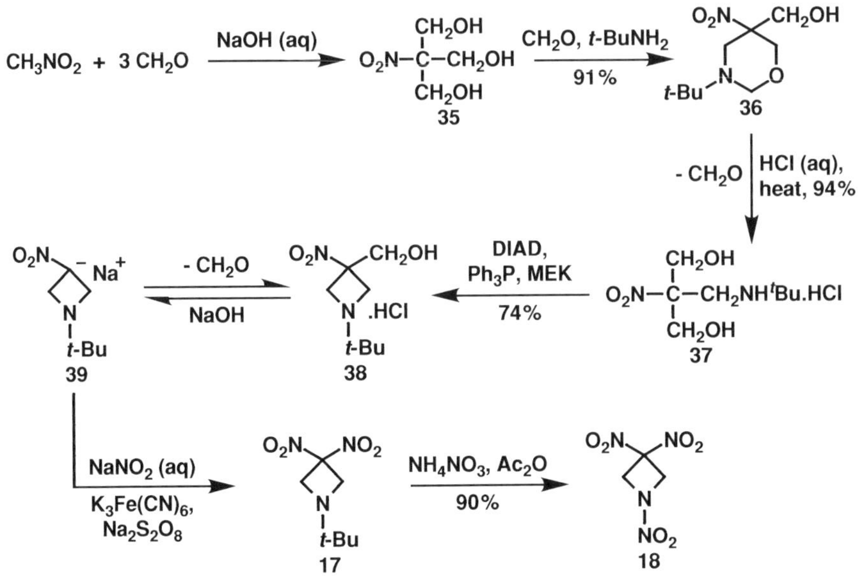

Figure 6.8 Archibald, Coburn and Hiskey's route to TNAZ[9]

solution, the latter is treated with formaldehyde and *tert*-butylamine to form the 1,3-oxazine (36). Reaction of the oxazine (36) with one equivalent of hydrochloric acid, followed by heating under reflux leads to ring cleavage, elimination of formaldehyde, and the formation of the aminodiol (37), which on reaction with DIAD and triphenylphosphine under Mitsonubu conditions forms the hydrochloride salt of azetidine (38) in good yield. Reaction of the azetidine (38) with an alkaline solution of sodium persulfate and sodium nitrite in the presence of catalytic potassium ferricyanide leads to tandem deformylation–oxidative nitration to yield 1-*tert*-butyl-3,3-dinitroazetidine (17). The nitrolysis of (17) with a solution of ammonium nitrate in acetic anhydride completes the synthesis of TNAZ (18).

6.4 CUBANE–BASED NITRAMINES

The incorporation of the nitramino group into the core of cubane has not yet been achieved. However, a number of cubane-based energetic nitramines and nitramides have been synthesized.

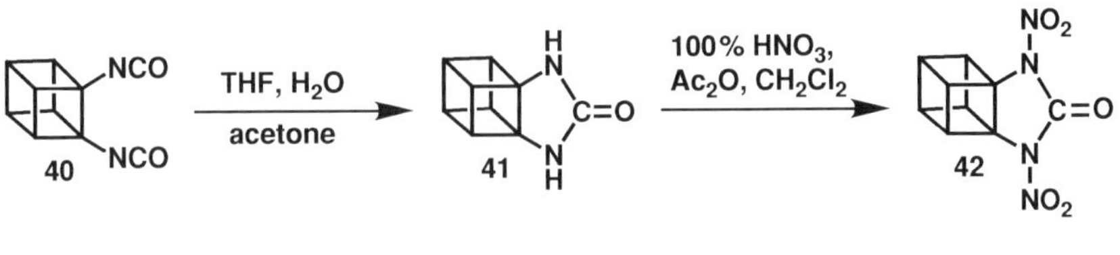

Figure 6.9

Eaton and co-workers[10] synthesized the cubane-based dinitrourea (42) via *N*-nitration of the cyclic urea (41) with nitric acid–acetic anhydride. Cubane-based nitramide (43) is prepared from the *N*-nitration of the corresponding bis-amide with acetic anhydride–nitric acid.[11] Bis-nitramine (44) is prepared from the *N*-nitration of the corresponding diamine with TFAA–nitric acid.[12]

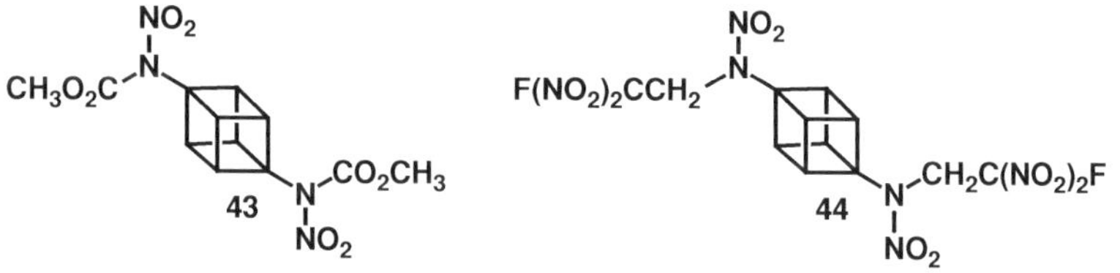

Figure 6.10

6.5 DIAZOCINES

Diazocines are eight-membered heterocycles containing two nitrogen atoms. The *N*-nitro and *N*-nitroso derivatives of 1,5-diazocines are energetic materials with potential for use in high-energy propellants.

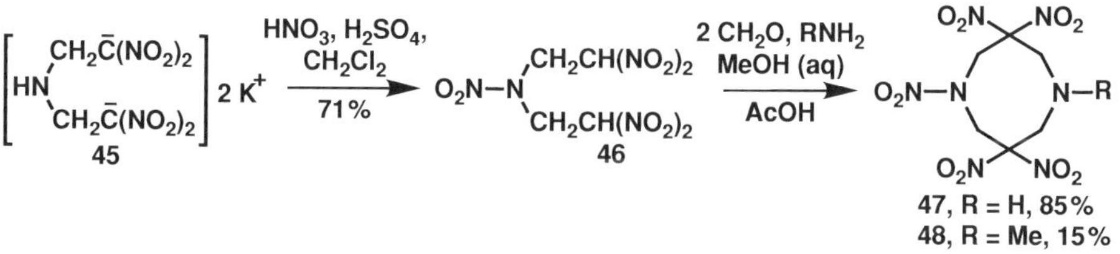

Figure 6.11

Adolph and Cichra[13] synthesized a number of polynitroperhydro-1,5-diazocines and compared their properties with the powerful military explosive HMX. A type of Mannich condensation was used to form the 1,5-diazocine rings; the condensation of ammonia and methylamine with formaldehyde and bis(2,2-dinitroethyl)nitramine (46)[14] forming diazocines (47) and (48) respectively. 1,3,3,7,7-Pentanitrooctahydro-1,5-diazocine (47) is *N*-nitrated to 1,3,3,5,7,7-hexanitrooctahydro-1,5-diazocine (52) in near quantitative yield using mixed acid.

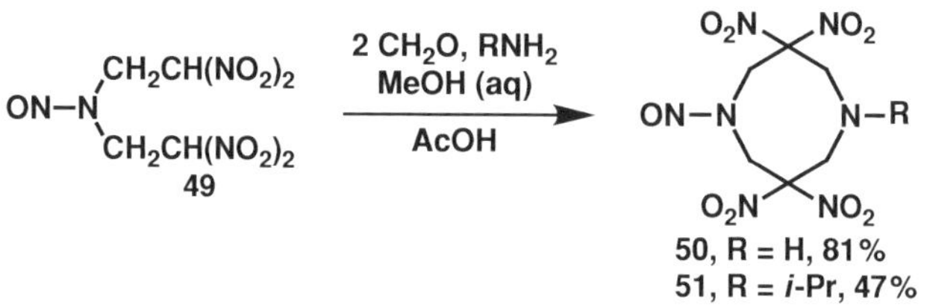

Figure 6.12

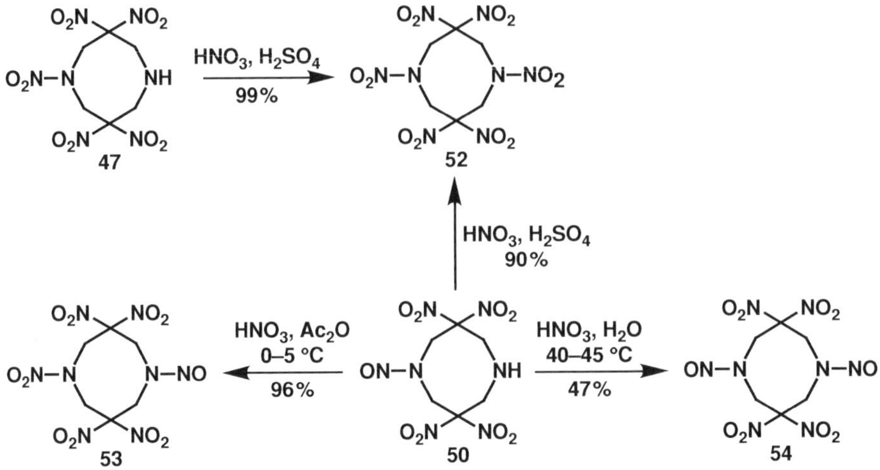

Figure 6.13

Adolph and Cichra[13] prepared some *N*-nitroso-1,5-diazocines from the condensation of bis(2,2-dinitroethyl)nitrosoamine (49) with formaldehyde and various amines. 3,3,7,7-Tetranitro-1-nitrosooctahydro-1,5-diazocine (50), the product obtained from the Mannich condensation of (49), formaldehyde and ammonia, was used to prepare nitro- and nitroso- 1,5-diazocines (52), (53), and (54).

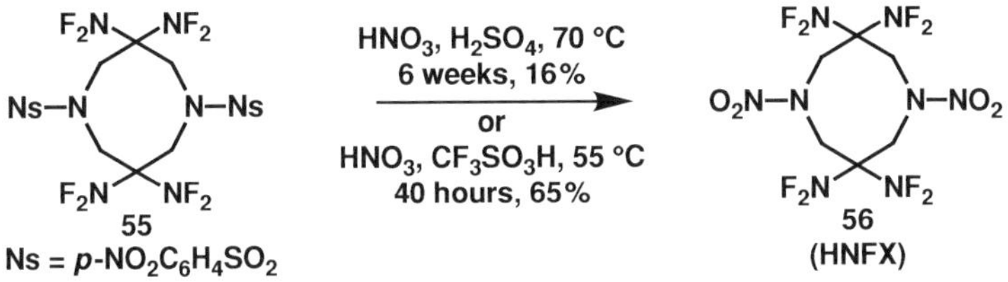

Figure 6.14

The search for new high-energy compounds has led to the incorporation of difluoramino (NF_2) functionality into 1,5-diazocines. Chapman and co-workers[15] synthesized the energetic heterocycle 3,3,7,7-tetrakis(difluoroamino)octahydro-1,5-dinitro-1,5-diazocine (56) (HNFX) from the nitrolysis of the *N*-nosyl derivative (55). This nitrolysis is very difficult because the amide bonds of (55) are highly deactivated, and the problem is made worst by the steric hindrance at both amide bonds. Treatment of (55) with standard mixed acid requires both elevated temperature and up to 6 weeks reaction time for complete amide nitrolysis and formation of HNFX (56). Chapman and co-workers found that a solution of nitric acid in triflic acid led to complete amide nitrolysis within 40 hours at 55 °C. Solutions of nitric acid in superacids like triflic acid are powerful nitrating agents with the protonitronium cation[16] (NO_2H^{2+}) as the probable active nitrating agent.

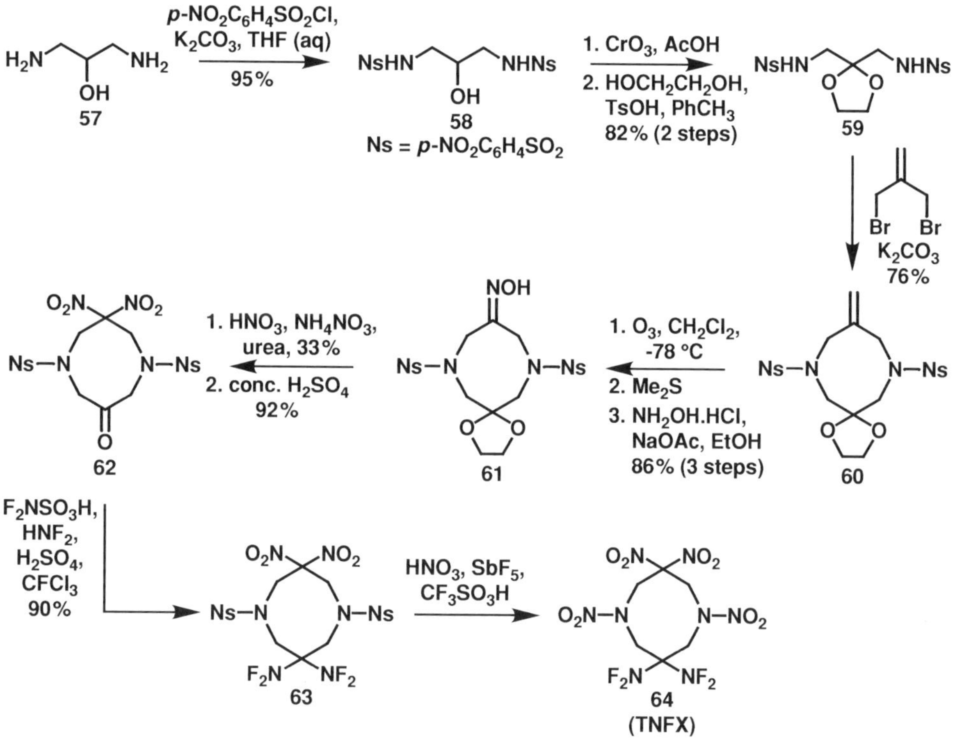

Figure 6.15

Chapman and co-workers[17] also reported the synthesis of 3,3-bis(difluoroamino)octahydro-1,5,7,7-tetranitro-1,5-diazocine (64) (TNFX). The synthesis of TNFX (64) starts from commercially available 1,3-diamino-2-propanol (57), which is elaborated in seven steps using standard organic reactions to give the oxime (61). Oxidation–nitration of the oxime (61) with ammonium nitrate in absolute nitric acid, followed by hydrolysis of the 1,3-dioxalane functionality with concentrated sulfuric acid, yields the required 1,5-diazocin-3-(2*H*)-one (62). Introduction of difluoroamino functionality into the 1,5-diazocine ring is achieved by treating the ketone (62) with a mixture of difluoramine and difluorosulfamic acid in sulfuric acid. Nitrolysis of the *N*-nosyl amide bonds of (63) was found to be challenging – treatment of (63) with a solution of nitric acid in triflic acid is not sufficient to effect the nitrolysis of both *N*-nosyl amide bonds. However, the addition of the Lewis acid, antimony pentafluoride, to this nitrating mixture was found to affect nitrolysis within a reasonable reaction time, possibly by increasing the concentration of protonitronium ion presence in solution.

6.6 BICYCLES

2,4,6,8-Tetranitro-2,4,6,8-tetraazabicyclo[3.3.0]octane (bicyclo-HMX) (69) has seen considerable research efforts focused into its preparation.[18–21] Interest in bicyclo-HMX arises from its increased rigidity compared to HMX, a property which should result in higher density and

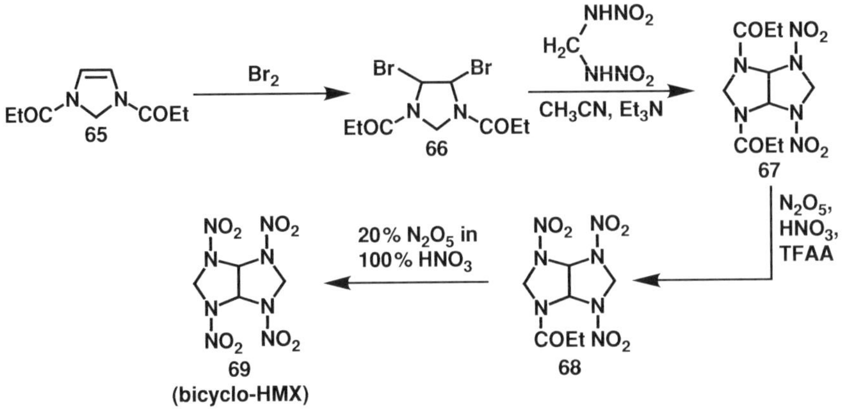

Figure 6.16

performance. Many of the problems with the synthesis of bicyclo-HMX arise from the ease with which the bis-imidazolidine ring opens during nitration. The only reported successful synthesis of bicyclo-HMX is from chemists at the Lawrence Livermore National Laboratory (LLNL).[20,21] This synthesis starts with the bromination of *N,N′*-dipropanoyl-1,2-dihydroimidazole (65). The product of this reaction, the dibromide (66), is treated with methylenedinitramine to effect a displacement of the halogen atoms and form the bicycle (67). Nitrolysis of the bicycle (67) is effected with an unusual but powerful nitrating agent composed of dinitrogen pentoxide, absolute nitric acid and TFAA. This reaction gives the trinitramine (68) in 90 % yield; further reaction with 20 % dinitrogen pentoxide in absolute nitric acid yields bicyclo-HMX (69).

The above synthesis has a few noteworthy points. The nitrolysis of bicyclic amides like (67) are frequently problematic in terms of inertness towards nitrolysis and the ease with which ring decomposition occurs. This synthesis is an interesting balancing act. Ring decomposition results when the bicycle (67) is treated with absolute nitric acid, mixed acid or nitronium salts. When the diacetyl equivalent of the bicycle (67) is treated with dinitrogen pentoxide–absolute nitric acid–TFAA reagent, the yield drops to 10 %.

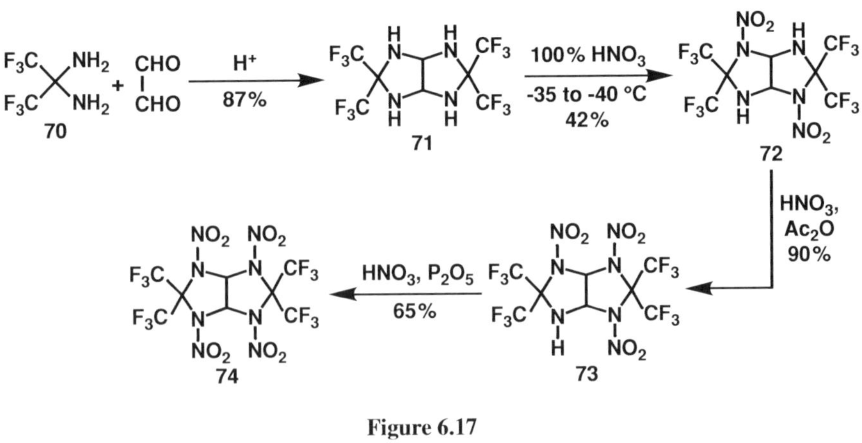

Figure 6.17

The energetic tetranitramine (74) is prepared from the sequential *N*-nitration of the bicycle (71); the latter prepared from the acid-catalyzed condensation of 2,2-diaminohexafluoropropane (70) with glyoxal.[18] The crystal density of (74) (2.18 g/cm^3) is one of the highest reported for an explosive containing an organic skeleton. Accordingly, its performance is expected to be high.

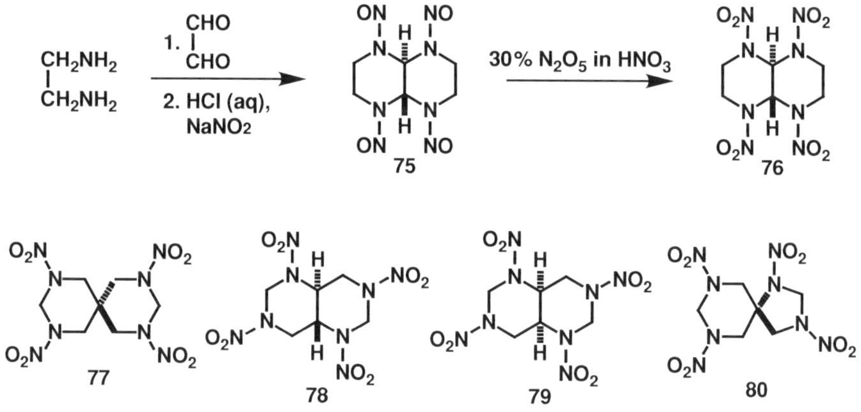

Figure 6.18

Trans-1,4,5,8-Tetranitro-1,4,5,8-tetrazadecalin (76) (TNAD) has been synthesized from the condensation of ethylenediamine with glyoxal, followed by *in situ* nitrosation of the resulting *trans*-1,4,5,8-tetraazadecalin and treatment with a 30 % solution of dinitrogen pentoxide in absolute nitric acid.[22,23] TNAD has been classified an insensitive high explosive (IHE) and exhibits similar performance to RDX. Willer and Atkins[23,24] used the same strategy to synthesize the cyclic nitramine explosives (77), (78), (79), and (80).

6.7 CAGED HETEROCYCLES – ISOWURTZITANES

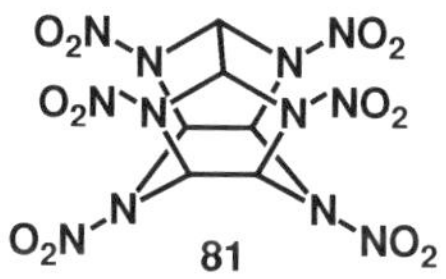

Figure 6.19

2,4,6,8,10,12-Hexanitro-2,4,6,8,10,12-hexaazaisowurtzitane (HNIW) (81), known as CL-20, was first synthesized by Nielsen and co-workers[25] at the Naval Air Warfare Center (NAWC) and is currently the most powerful nonnuclear explosive (VOD $\sim$ 9380 m/s, $\Delta H_f = +410$ kJ/mol) being synthesised on a pilot plant scale.[26] The compact caged structure of the isowurtzitane skeleton is reflected in the high crystal density (2.04 g/cm^3) of CL-20. CL-20 is now finding application in high performance propellants and its use is expected to result in major technological advances in future weapon systems.

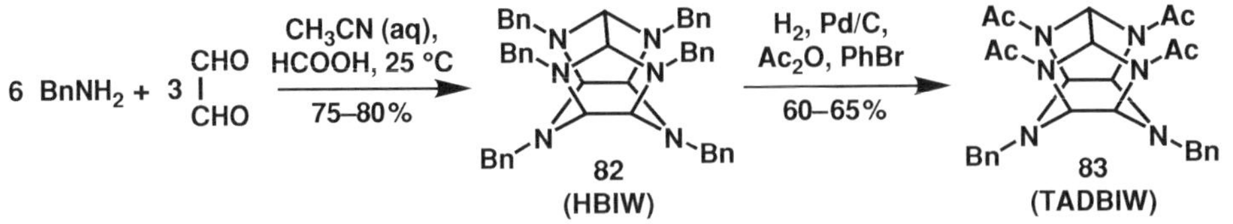

Figure 6.20

The synthesis of energetic materials containing strained or caged structures frequently requires many synthetic steps which can offset the gain in explosive performance. Nielsen and co-workers[27] have shown that this is not always the case in finding that 2,4,6,8,10,12-hexabenzyl-2,4,6,8,10,12-hexaazaisowurtzitane (HBIW) (82) can be synthesized in high yield, and in one step, from the reaction of benzylamine and glyoxal in aqueous acetonitrile in the presence of catalytic amounts of formic acid. Reductive acetylation of HBIW (82) with palladium hydroxide on carbon in acetic anhydride in the presence of catalytic bromobenzene yields the tetraacetate (83) (TADBIW). Treatment of TADBIW (83) with 3 mole equivalents of nitrosonium tetrafluoroborate in sulfolane, followed by 12 mole equivalents of nitronium tetrafluoroborate in the same pot, gives CL-20 (81) in 90 % yield.

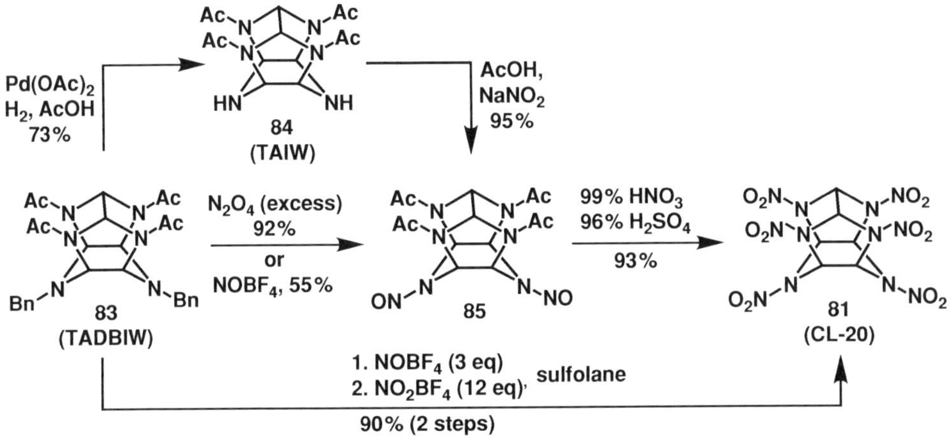

Figure 6.21

While the route described above is highly convenient for the synthesis of CL-20 on a laboratory scale, the availability of nitronium tetrafluoroborate makes further research and development essential.[28] Further studies have shown that the dinitrosamine (85) can be obtained in high yield from the reaction of TADBIW (83) with excess dinitrogen tetroxide,[25,28a] or from its reductive debenzylation with hydrogen and palladium acetate in acetic acid followed by nitrosation with sodium nitrite in acetic acid.[28b] The dinitrosamine (85) is readily converted to CL-20 (81) in high yield on reaction with mixed acid at 75–80 °C.[28a] Several other studies and modifications to the original route have been reported including: (1) the synthesis of HBIW (82) from benzylamine and glyoxal in the presence of mineral acid,[28c] (2) reductive debenzylation of HBIW (82) under a variety of conditions,[28d] (3) hydrogenation of TADBIW (83) in acetic anhydride-acetic acid[28d] and formic acid[28e] with a palladium catalyst to yield

4,10-diethyl- and 4,10-diformyl- 2,6,8,12-tetraacetyl-2,4,6,8,10,12-hexaazaisowurtzitanes respectively, (4) synthesis of TADBIW (83) (75 %) via the reductive debenzylation of HBIW (82) with a mixture of palladium on carbon, acetic anhydride and *N*-acetoxysuccinimide in ethylbenzene,[28b28f] (5) nitrolysis of the dinitrosamine (85) with nitronium tetrafluoroborate[25] (59 %) or absolute nitric acid[28b] (95 %) to yield 4,10-dinitro-2,6,8,12-tetraacetyl-2,4,6,8,10,12-hexaazaisowurtzitane, followed by its nitrolysis to CL-20 (81) on treatment with mixed acid,[28b] (6) nitration of 2,6,8,12-tetraacetyl-2,4,6,8,10,12-hexaazaisowurtzitane (84) (TAIW) to CL-20 (81) with mixed acid at 60 °C,[28g] (7) debenzylation of TADBIW (83) with ceric ammonium nitrate (CAN) followed by nitration of the dinitrate salt of TAIW (84) with mixed acid,[28h] (8) acetylation of TAIW (84) with acetic anhydride[28b28h] followed by nitrolysis of the resulting 2,4,6,8,10,12-tetraacetyl-2,4,6,8,10,12-hexaazaisowurtzitane with mixed acid,[28h] (9) oxidative debenzylation–acetylation of HBIW (82) with potassium permanganate and acetic anhydride followed by nitrosolysis and nitrolysis of the resulting TADBIW (83) to give CL-20 (81) in fair yield.[28i] Many of these nitrolysis reactions may be achieved with dinitrogen pentoxide in absolute nitric acid (Section 5.6). Agrawal and co-workers[29] synthesized CL-20 via the original route specified by Nielsen and co-workers[25] and conducted a comprehensive study into its characterization, thermal properties and impact sensitivity.

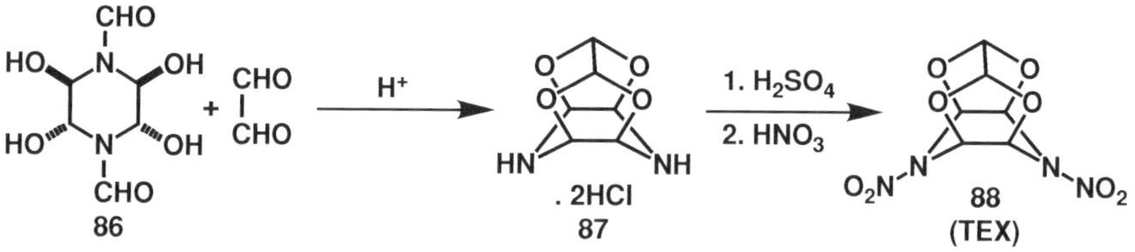

Figure 6.22

4,10-Dinitro-4,10-diaza-2,6,8,12-tetraoxaisowurtzitane (TEX) (88) was synthesized by Boyer and co-workers[30] from the condensation of 1,4-diformyl-2,3,5,6-tetrahydroxypiperazine (86) with glyoxal trimer, followed by *in situ* nitration of the resulting isowurtzitane dihydrochloride (87) by slow sequential addition of sulfuric acid followed by nitric acid. TEX (88) is less energetic (VOD ~ 8665 m/s) than β-HMX but has a high crystal density (1.99 g/cm^3) and has been suggested as an energetic additive in high performance propellants. At the time of discovery of TEX, the US military was considering its use in insensitive munitions.

Strategies used for the synthesis of polyazapolycyclic-caged nitramines and nitrosamines are the subject of an excellent review by A. T. Nielsen.[31] Nielsen identified three routes to such compounds:[25c]

(1) ‘Proceeding from a preformed polyazapolycyclic caged structure which precisely incorporates the desired final heterocyclic ring.’ The syntheses of CL-20 (81) and TEX (88) are examples.

(2) ‘Proceeding from a precursor polyaza-caged structure, which may be different from the desired product, but includes the final structure within the cage.’ Although not a caged compound the synthesis of RDX from the nitrolysis of hexamine would fit this category.

(3) ‘Cyclisation of a precursor polynitramine to produce the desired final cage structure.’

6.8 HETEROCYCLIC NITRAMINES DERIVED FROM MANNICH REACTIONS

Hybrid compounds containing heterocyclic nitramine and *gem*-dinitro functionality represent a class of high performance energetic materials. Such compounds frequently exhibit higher heats of formation, crystal density, detonation velocity and pressure, and better oxygen balance compared to analogous aromatic compounds.

The Mannich reaction has been used to synthesize numerous heterocyclic nitramine explosives. Adolph and Cichra[32] prepared a number of *N*-heterocycles containing *tert*-butyl *N*-blocking groups. The nitrolysis of these *t*-butyl groups provides the corresponding *N*-nitro derivatives in excellent yields (Section 5.6.2.2). Some of the nitramine products from these reactions are powerful, energetic explosives with attractive properties.

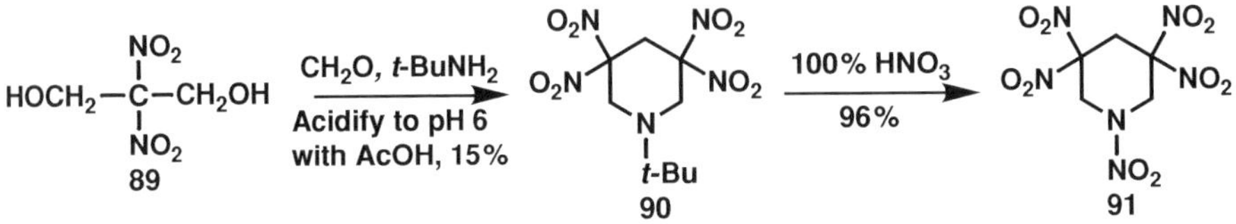

$HOCH_2-C(NO_2)_2-CH_2OH$ (89) → [CH_2O, *t*-$BuNH_2$; Acidify to pH 6 with AcOH, 15%] → 90 (*t*-Bu) → [100% HNO_3, 96%] → 91 (NO_2)

Figure 6.23

1,3,3,5,5-Pentanitropiperidine (91) is prepared from the condensation of 2,2-dinitro-1,3-propanediol (89) with formaldehyde and *t*-butylamine under slightly acidic conditions, followed by nitrolysis of the *t*-butyl group of the resulting piperidine (90) with mixed acid or absolute nitric acid.[32]

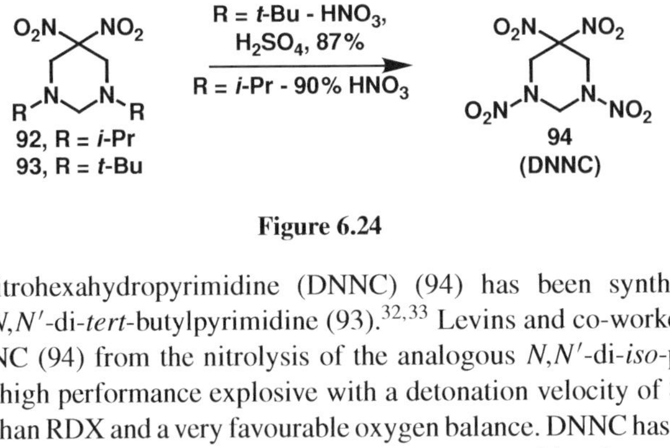

92, R = *i*-Pr; 93, R = *t*-Bu → [R = *t*-Bu - HNO_3, H_2SO_4, 87%; R = *i*-Pr - 90% HNO_3] → 94 (DNNC)

Figure 6.24

1,3,5,5-Tetranitrohexahydropyrimidine (DNNC) (94) has been synthesized from the nitrolysis of the *N,N'*-di-*tert*-butylpyrimidine (93).[32,33] Levins and co-workers[34] reported the synthesis of DNNC (94) from the nitrolysis of the analogous *N,N'*-di-*iso*-propylpyrimidine (92). DNNC is a high performance explosive with a detonation velocity of 8730 m/s, impact sensitivity lower than RDX and a very favourable oxygen balance. DNNC has been suggested[34] for use as an oxidizer in propellant compositions. This is also considered as an excellent oxidant for pyrotechnic compositions.[33]

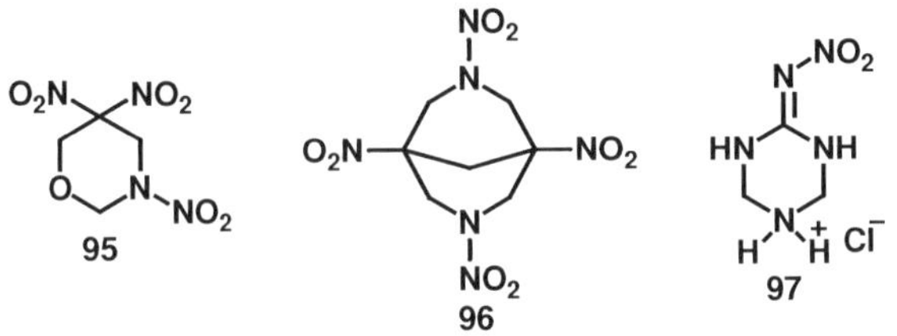

95 96 97

Figure 6.25

Adolph and Cichra[32] used a similar strategy of *tert*-butyl nitrolysis to synthesize 1,5,5-trinitro-1,3-oxazine (95) and the bicycle (96).

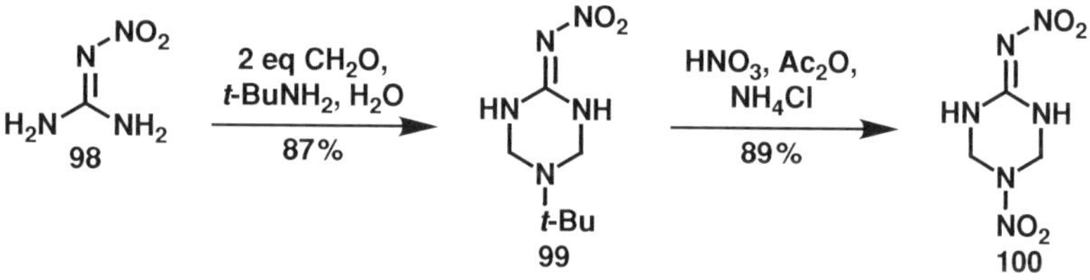

Figure 6.26

Dagley and co-workers[35] reported the synthesis of 2-nitrimino-5-nitrohexahydro-1,3,5-triazine (100) from the Mannich condensation of nitroguanidine (98), formaldehyde and *t*-butylamine, followed by nitrolysis of the *t*-butyl group of the resulting product, 2-nitrimino-5-*tert*-butylhexahydro-1,3,5-triazine (99). The triazine (100) has also been synthesized from the reaction of nitroguanidine and hexamine in aqueous hydrochloric acid, followed by nitration of the resulting product (97) with a solution of nitric acid in acetic anhydride.[36]

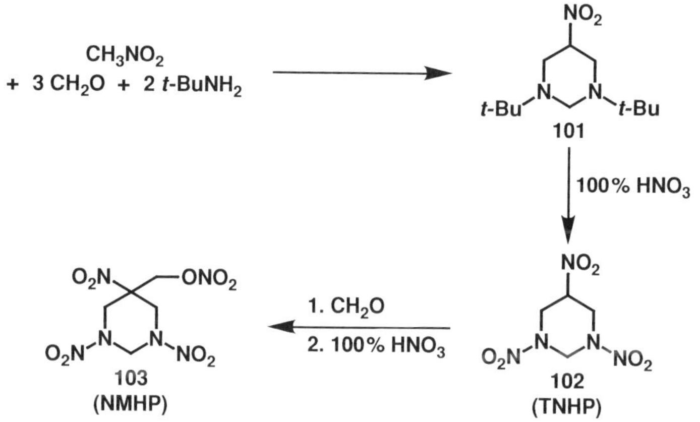

Figure 6.27

The Mannich condensation between nitromethane, formaldehyde and *t*-butylamine, followed by nitrolysis of the resulting product (101), has been used to synthesize 1,3,5-trinitrohexahydropyrimidine (102) (TNHP); treatment of the latter with formaldehyde in a Henry type methylolation, followed by *O*-nitration with nitric acid, yields the nitrate ester (103).[37]

6.9 NITROUREAS

As early as 1974 French chemists[38] reported the synthesis of the nitrourea explosives 1,4-dinitroglycouril (DINGU) (105) and 1,3,4,6-tetranitroglycouril (TNGU or Sorguyl) (106). Their synthesis is both short and efficient: the reaction of urea with glyoxal forming glycouril (104), which is then treated with absolute nitric acid or mixed acid to produce DINGU (105); reaction of the latter with dinitrogen pentoxide in nitric acid yields TNGU (106).

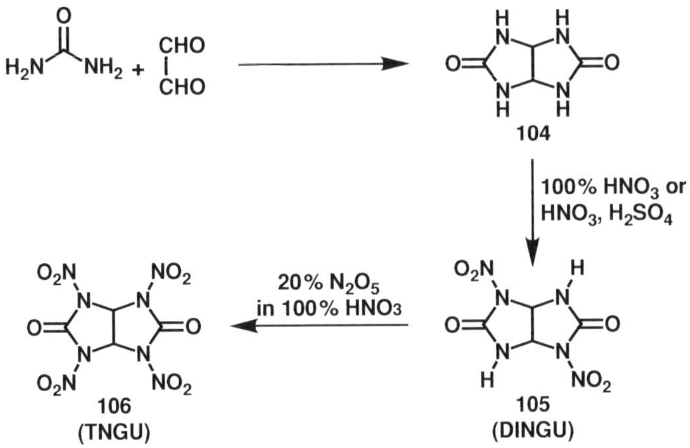

Figure 6.28

TNGU (106) is a powerful explosive with a detonation velocity of 9150 m/s and one of the highest crystal densities (2.04 g/cm^3) reported for known C,H,N,O-based energetic materials.[38,39] However, like all *N,N′*-dinitroureas, TNGU is readily hydrolyzed by cold water and of limited use as a practical explosive. DINGU (105), being an *N*-nitrourea, is more hydrolytically stable than TNGU and decomposes only slowly on treatment with boiling water. DINGU has been classified as an insensitive high explosive[40] (IHE) but is less energetic than TNGU, having a detonation velocity of 7580 m/s and a density of 1.99 g/cm^3. This insensitivity to impact is attributable to intramolecular hydrogen bonding in the nitrourea framework. The simplicity with which DINGU is synthesized from cheap and readily available starting materials has prompted research into its use in PBXs and LOVA munitions.[41]

Chinese chemists[42] reported the base hydrolysis of TNGU. The product, 1,1,2,2-tetranitraminoethane, has been used to prepare a series of heterocyclic nitramines via condensation reactions and may find future use for the synthesis of heterocyclic caged nitramines.

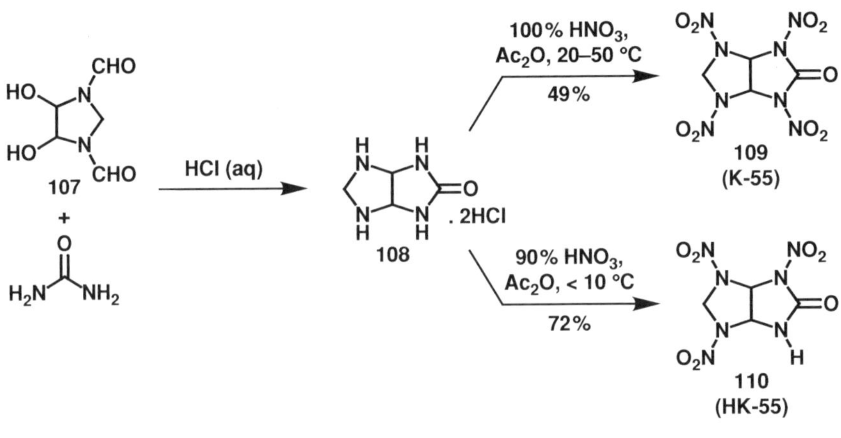

Figure 6.29

Li and co-workers[43] recognised the potential of cyclic *N*-nitroureas as energetic materials and reported the synthesis of 2,4,6,8-tetranitro-2,4,6,8-tetraazabicyclo[3.3.0]octane-3-one (109) (K-55) from the nitration of 2,4,6,8-tetraazabicyclo[3.3.0]octane-3-one dihydrochloride (108) with absolute nitric acid in acetic anhydride at room temperature; the latter obtained from the condensation of *N,N′*-diformyl-4,5-dihydroxyimidazolidine (107) with urea in aqueous hydrochloric acid. Pagoria and co-workers[21,44] reported the synthesis of 2,4,6-trinitro-2,4,6,8-tetraazabicyclo[3.3.0]octane-3-one (110) (HK-55) in 72 % yield from the nitration of (108) with 90 % nitric acid in acetic anhydride at subambient temperature (Table 5.3). HK-55 has a relatively high density (1.905 g/cm^3) coupled with a low sensitivity to shock.

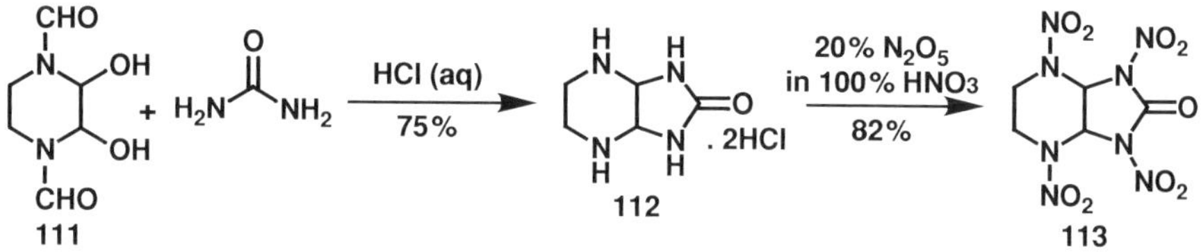

Figure 6.30

Graindorge and co-workers[45] reported the synthesis of 2,5,7,9-tetranitro-2,5,7,9-tetraazabicyclo[4.3.0]nonane-8-one (113) (K-56, TNABN) from the nitration of 2,5,7,9-tetraazabicyclo[4.3.0]nonane-8-one dihydrochloride (112) with dinitrogen pentoxide in absolute nitric acid, the latter obtained from the condensation of urea with 1,4-diformyl-2,3-dihydroxypiperazine (111) in hydrochloric acid.[31] Treatment of (112) with nitronium tetrafluoroborate in nitromethane results in the nitration of the piperazine ring nitrogens only and the isolation of (114) in 86 % yield (Table 5.2).

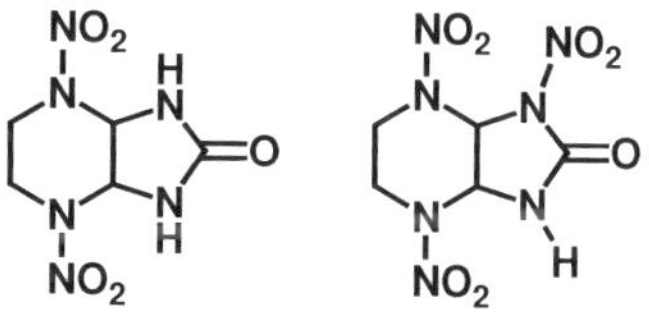

Figure 6.31

Agrawal and co-workers[46] also conducted extensive studies into the synthesis, characterization and thermal and explosive behaviour of (113) (K-56, TNABN). 2,5,7,9-Tetraazabicyclo[4.3.0]nonane-8-one (112) was synthesized from the direct reaction of ethylenediamine with glyoxal, followed by reaction of the resulting cyclic imine with urea in concentrated hydrochloric acid; nitration of (112) was achieved in 51 % yield with a mixture of nitric acid–acetic anhydride. Agrawal showed that K-56/TNABN is significantly more resistant to hydrolytic destruction than TNGU.

Pagoria and co-workers[21,44] also reported the synthesis of (113) (K-56, TNABN) and the trinitrated derivative, 2,5,7-trinitro-2,5,7,9-tetraazabicyclo[4.3.0]nonane-8-one (115) (HK-56). Their route to the bicycle (112) was via bromination of 1,3-diacetyl-2-imidazolone, followed by reaction with ethylenedinitramine and nitrolysis of the acetyl groups.

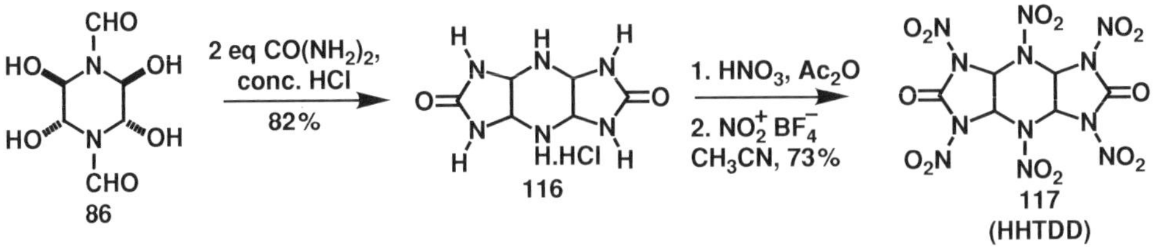

Figure 6.32

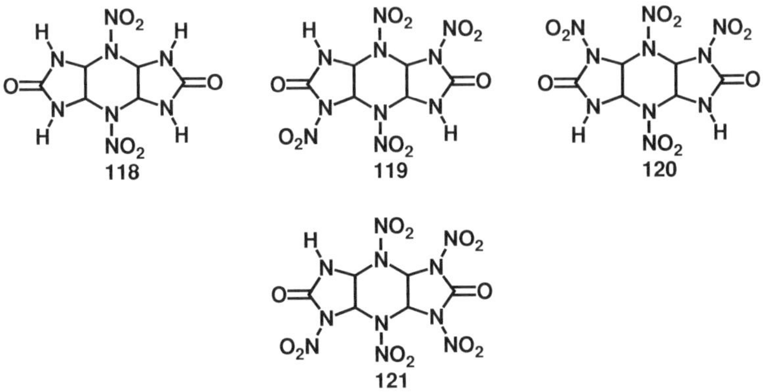

Figure 6.33

Boyer and co-workers[47] reported the synthesis of 2,6-dioxo-1,3,4,5,7,8-hexanitrodecahydro-1*H*,5*H*-diimidazo[4,5-*b*:4′,5′-*e*]pyrazine (117) (HHTDD). The hydrochloride salt of the tricycle (116) was synthesized from the reaction of 1,4-diformyl-2,3,5,6-tetrahydroxypiperazine (86) with a solution of urea in concentrated hydrochloric acid, followed by recrystallization of the product from methanol. The nitration of the tricycle (116) was studied in some detail. The low temperature nitration of (116) with pure nitric acid leads to the nitration of the piperazine nitrogens only and the isolation of the 4,8-dinitro derivative (118) in 28 % yield. Nitration of the urea nitrogens proves more difficult with (116) yielding a mixture of tetranitro derivatives, (119) and (120), on nitration with nitric acid in acetic anhydride. Further treatment of this mixture with excess nitric acid in acetic or trifluoroacetic anhydrides for a prolonged period yields the pentanitro derivative (121). Treatment of (119), (120) or (121) with nitronium tetrafluoroborate in acetonitrile produces HHTDD (117). The direct nitration of (116) with a solution of 20 % dinitrogen pentoxide in nitric acid gives HHTDD (117) in 74 % crude yield. HHTDD (117) has an excellent oxygen balance and exhibits high performance (calculated VOD ~ 9700 m/s, 2.07 g/cm^3). However, the hydrolytic stability of HHTDD is poor and so limits its value as a practical explosive.

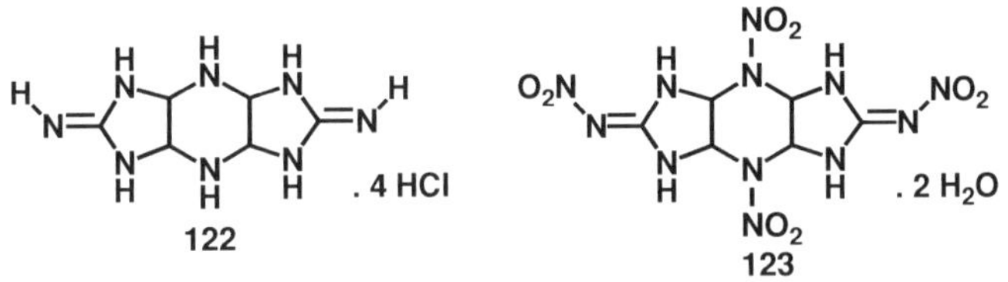

Figure 6.34

Boyer and co-workers[48] also reported the synthesis of the guanidine tricycle (122), prepared as the tetrahydrochloride salt from the condensation of two equivalents of guanidine with 1,4-diformyl-2,3,5,6-tetrahydroxypiperazine in concentrated hydrochloric acid. Treatment of the tricycle (122) with absolute nitric acid yields the bis-nitrimine (123), whereas the same reaction with nitric acid–acetic anhydride yields HHTDD (117).

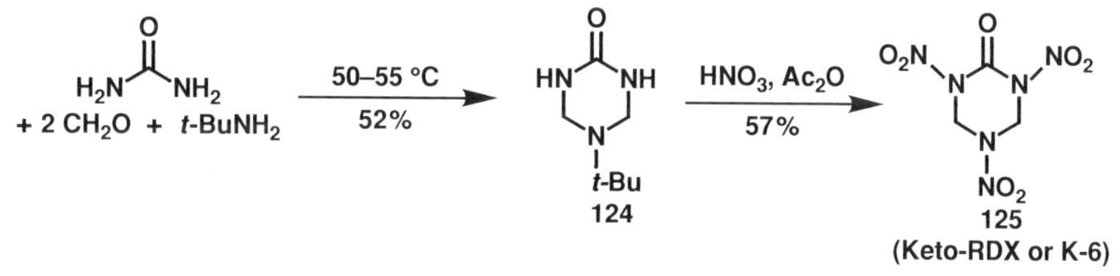

Figure 6.35

Chemists at Lawrence Livermore National Laboratory (LLNL) synthesized the RDX analogue 1,3,5-trinitro-2-oxo-1,3,5-triazacyclohexane (125) (Keto-RDX or K-6) from a Mannich reaction between urea, formaldehyde and *t*-butylamine, followed by nitrolysis of the resulting 2-oxo-5-*tert*-butyltriazone (124) with nitric acid in acetic anhydride or dinitrogen pentoxide in absolute nitric acid.[20,21] Nitrolysis with other nitrating agents has also been reported, including nitronium tetrafluoroborate (40 %), TFAA–nitric acid (43 %) and mixed acid (0 %) – see Table 5.6.[21,49] Keto-RDX is not as hydrolytically labile as other *N*,*N*′-dinitroureas and its ease of preparation and relatively high performance (4 % > HMX) makes its future application attractive.

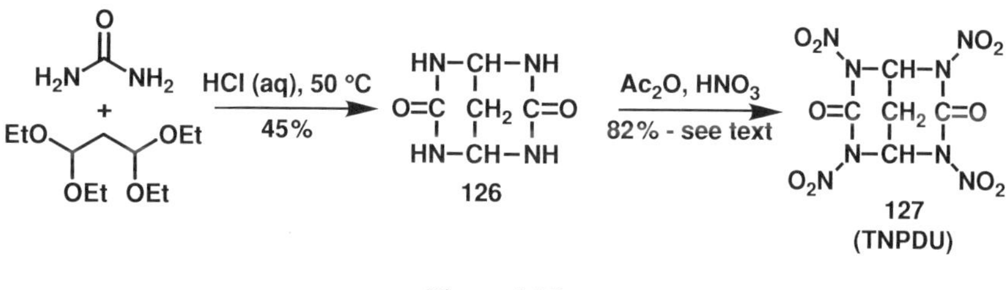

Figure 6.36

Tetranitropropanediurea (127) (TNPDU) is a high performance *N*,*N*′-dinitrourea explosive (VOD ~ 9030 m/s) synthesized from the nitration of propanediurea (126) with nitric acid in acetic anhydride,[50] the latter readily synthesized from the condensation of urea with 1,1,3,3-tetraethoxypropane. Agrawal and co-workers[46] conducted extensive studies into the synthesis, characterization and thermal behaviour of TNPDU. The nitration step was significantly improved by using a 'slow nitration procedure' which involves the slow addition of propanediurea to 98 % nitric acid followed by slow addition of acetic anhydride. This gave a higher yield of TNPDU than previously reported, and excellent product purity which avoids the need for a lengthy purification step. Agrawal noted that the hydrolytic stability of TNPDU is better than similar compounds and, in particular, TNGU. The impact and friction sensitivity of TNPDU and its formulations were also explored.

6.10 OTHER ENERGETIC NITRAMINES

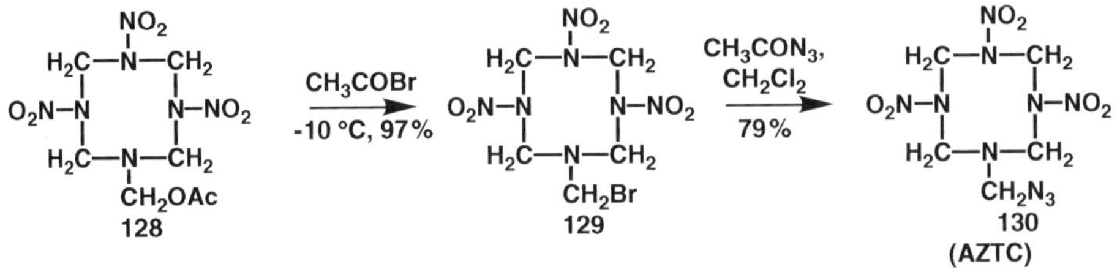

Figure 6.37

Some energetic compounds are engineered to contain two or more different energetic functionalities. The azido group has a high heat of formation and so its presence in energetic materials is favorable on thermodynamic grounds. However, compounds containing only the azido 'explosophore' rarely find use as practical explosives. More common is the incorporation of other functionality into such compounds. In the case of 1-(azidomethyl)-3,5,7-trinitro-1,3,5,7-tetraazacyclooctane (130) (AZTC), an azido derivative of HMX, the azidomethyl group triggers initial thermal decomposition and makes AZTC much more sensitive to initiation than HMX. AZTC (130) is prepared from the reaction of the acetate ester (128) with acetyl bromide, followed by treating the resulting bromide (129) with a solution of acetyl azide.[51] Direct treatment of the acetate ester (128) with azide nucleophile leads to decomposition of the eight-membered ring. The azido groups of the energetic azido-nitramine (131), known as DATH, are a similar trigger for its decomposition.[52]

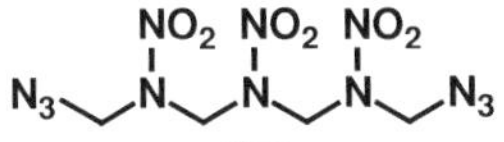

Figure 6.38

Some energetic materials contain both nitramine and nitrate ester functionality. Tris-X (132), a high performance explosive (VOD ~ 8700 m/s) with a low melting point (69 °C), is synthesized from the reaction of 2,4,6-tris(aziridino)-1,3,5-triazine with dinitrogen pentoxide in chloroform at subambient temperature (Section 5.8.1).[53] A homologue of Tris-X, known as Methyl Tris-X, has been synthesized using the same methodology.[53] However, the thermal stability of Tris-X is only marginally acceptable suggesting that this family of explosives is unlikely to be used for munitions.

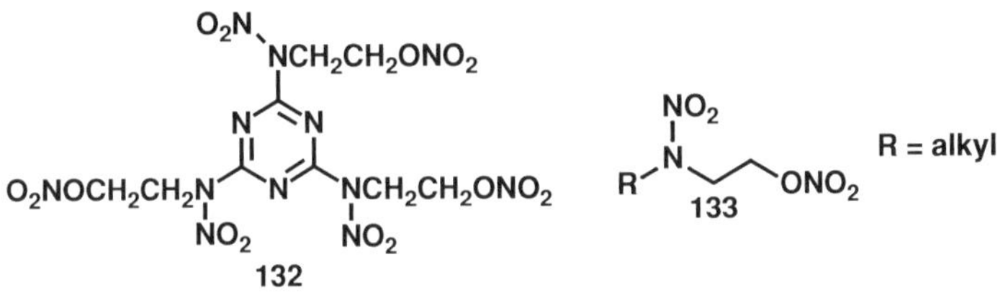

Figure 6.39

Nitramine-nitrates of general structure (133) are known as NENAs and are conveniently prepared from the nitrative cleavage of *N*-alkylaziridines[53,54] with dinitrogen pentoxide or from the direct nitration of the corresponding aminoalcohols.[55] These compounds find use as energetic plastisizers in explosive and propellant formulations; Bu-NENA (R = *n*-Bu) is a component of some LOVA (low vulnerability ammunition) propellants.[56]

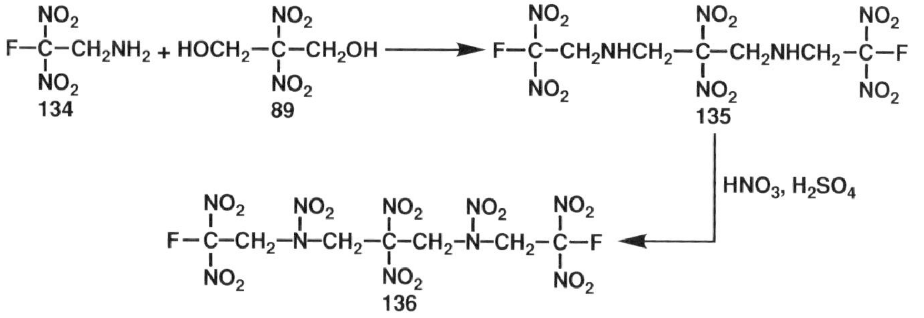

Figure 6.40

A large number of energetic materials containing nitramino functionality in conjunction with aliphatic *C*-nitro groups have been reported. Many of these contain dinitromethyl, trinitromethyl or fluorodinitromethyl functionality. The bis-nitramine (136) has been synthesized from the mixed acid nitration of the diamine (135), the latter being the condensation product of 2-fluoro-2,2-dinitroethylamine (134) with 2,2-dinitro-1,3-propanediol (89). Bis-nitramine (136) has been suggested as a high-energy oxidizer in propellants.[57]

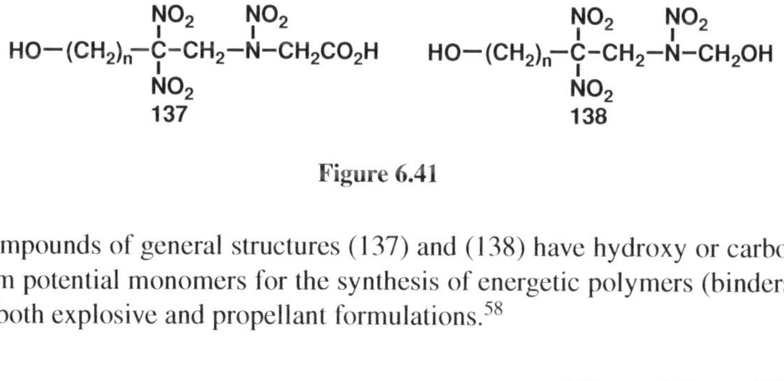

Figure 6.41

Some compounds of general structures (137) and (138) have hydroxy or carboxy termini, making them potential monomers for the synthesis of energetic polymers (binders) and plasticizers for both explosive and propellant formulations.[58]

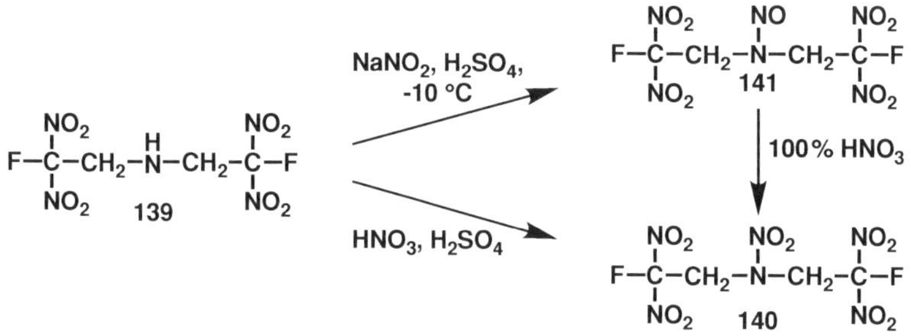

Figure 6.42

N-Nitration of the amine (139) with mixed acid yields the energetic nitramine (140).[59] The same reaction with sodium nitrite in sulfuric acid, or with nitrosyl fluoride in methylene chloride, yields the nitrosamine (141), which is also an energetic high explosive.[60]

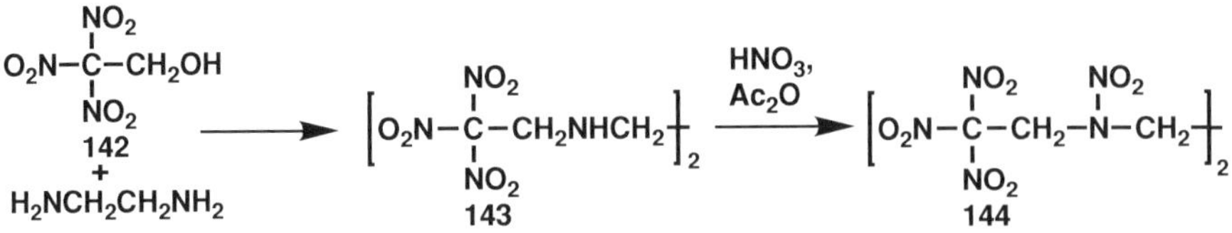

$$O_2N-C(NO_2)_2-CH_2OH\ (142) + H_2NCH_2CH_2NH_2 \longrightarrow [O_2N-C(NO_2)_2-CH_2NHCH_2]_2\ (143) \xrightarrow{HNO_3,\ Ac_2O} [O_2N-C(NO_2)_2-CH_2-N(NO_2)-CH_2]_2\ (144)$$

Figure 6.43

The trinitromethyl group is often incorporated into explosive molecules to increase oxygen balance. In fact, the six oxygen atoms present in the trinitromethyl group often give rise to a positive oxygen balance. The energetic nitramine (144) is an example of an explosive with an excellent oxygen balance.[61] *N*-Nitro-*N*′-(2,2,2-trinitroethyl)guanidine (TNENG) (145) has been prepared[62] from the reaction of nitroguanidine, formaldehyde and nitroform. TNENG has attracted interest as a burn rate accelerator in energetic propellants, the trigger for its decomposition being the trinitromethyl group.

$O_2NHN-C(=NH)-NHCH_2C(NO_2)_3$
145
(TNENG)

$O_2N-N[CH_2C(NO_2)_2NF_2]_2$
146
(DFAP)

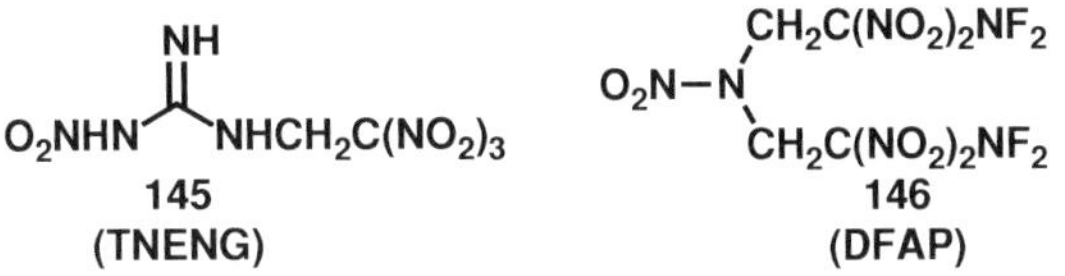

Figure 6.44

DFAP (146) is a high-energy material with potential as an oxidizer in energetic propellants. DFAP has been prepared from the reaction of bis(2,2-dinitroethyl)nitramine with NF_2OSO_2F.[63]

A number of energetic heterocycles containing both furazan and nitramine functionality have been reported – these are discussed in Section 7.3.4. There are many other examples of compounds containing nitramino functionality in conjunction with other explosophores. These are too numerous to discuss fully in this text. Many of these compounds are discussed in three major reviews.[64–66]

6.11 ENERGETIC GROUPS

6.11.1 Dinitramide anion

The dinitramide anion (147) was first synthesized[67–73] at the Zelinsky Institute in Russia in 1971 and is one of the most significant discoveries in the field of energetic materials. Ammonium dinitramide (ADN) has attracted particular interest as a chlorine free, and hence, environmentally friendly alternative to ammonium perchlorate in composite propellants. The absence of carbon and chlorine in its structure reduces the radar signature in the exhaust plume of ADN-based propellants in rockets/missiles. The amount of 'free oxygen' in ammonium dinitramide is also high, allowing for formulations with powerful reducing agents like aluminium and boron.

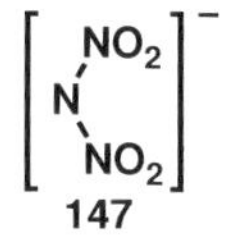

Figure 6.45

Many studies into the dinitramide anion (147) have looked at the effect the counterion has on physical properties. The ammonium, alkali metal, guanidinium, biguanidinium, aminoguanidinium, hydroxylammonium, 1,2-ethanediammonium and tetraammonium-1,2,4,7-cubane salts of dinitramide have been prepared. Various metal salts of dinitramide are conveniently prepared by ion exchange of the cesium or ammonium salts on polymer resins. The *N*-guanylurea salt of dinitramide, known as FOX-12, has been prepared from the addition of an aqueous solution of ammonium dinitramide to the sulfate salt of guanylurea; the low solubility of FOX-12 in cold water leading to its precipitation in 81 % yield.[74] FOX-12 is a very insensitive explosive with potential for use as an ingredient in energetic propellants, or for use in insensitive explosive munitions. The synthesis of materials like FOX-12 reflects the increased need for insensitive explosives and propellants for modern applications. Although nitrocellulose–nitroglycerine double-base propellants are still widely used for military applications, most exhibit a high sensitivity to shock or impact which can sometimes lead to premature explosion.

The dinitramide ion is stable in both acidic and basic solutions between pH 1–15 at room temperature but is slowly decomposed in the presence of strong concentrated acid. In contrast to alkyl *N,N*-dinitramines (Section 6.11.2) where the central nitrogen atom is highly electron deficient, the dinitramide anion has its negative charge delocalized over both nitrogen and oxygen atoms with the consequence that the N–N bonds are less susceptible to rupture. However, the dinitramide anion is not as stable as the nitrate anion; ammonium dinitramide melts at 92 °C and decomposition starts at 130 °C.

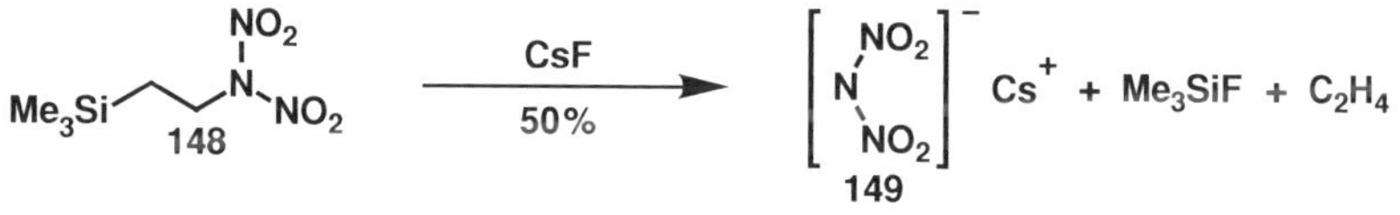

Figure 6.46

Numerous synthetic routes to the dinitramide anion have been reported.[75] Cesium dinitramide (149) has been synthesized via the fluoride-catalyzed β-elimination of 1-(*N,N*-dinitramino)-2-trimethylsilylethane (148) with cesium fluoride; the latter prepared by treating 2-(trimethylsilyl)ethyl isocyanate with a solution of nitronium tetrafluoroborate and pure nitric acid in acetonitrile.[75]

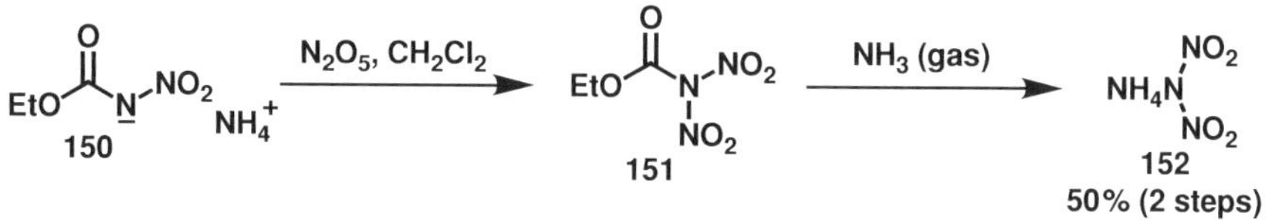

Figure 6.47

Ammonium dinitramide (152) is synthesized by treating a solution of ammonium nitrourethane (150) with nitronium tetrafluoroborate or dinitrogen pentoxide in methylene chloride at $-30\,^{\circ}C$, followed by ammonolysis of the resulting ethyl *N,N*-dinitrourethane (151).[75] Ammonium dinitramide can be prepared from the nitration of ethyl carbamate and ammonium carbamate with the same reagents. This is currently the most efficient route to ammonium dinitramide and is used for its manufacture (Section 9.11).

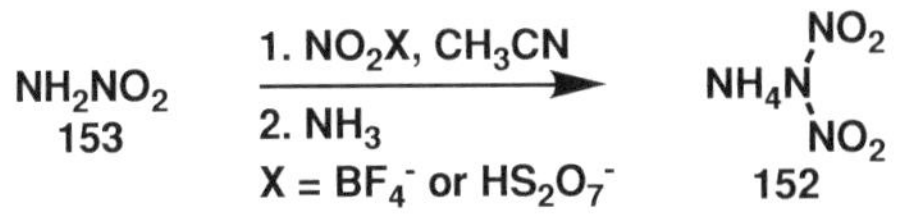

$$NH_2NO_2\ (\mathbf{153}) \xrightarrow[\text{2. } NH_3;\ X = BF_4^- \text{ or } HS_2O_7^-]{\text{1. } NO_2X,\ CH_3CN} NH_4N(NO_2)_2\ (\mathbf{152})$$

Figure 6.48

The nitration of nitramine (153) with nitronium tetrafluoroborate, followed by neutralization of the resulting dinitraminic acid with ammonia, also generates ammonium dinitramide (152).[75] Neutralization of this reaction with alkylamines, instead of ammonia, yields the corresponding alkylammonium salts of dinitramide. The nitration of ammonia with dinitrogen pentoxide (15 %) or nitronium salts like the tetrafluoroborate (25 %) yield ammonium dinitramide (152) through the initial formation of nitramine.

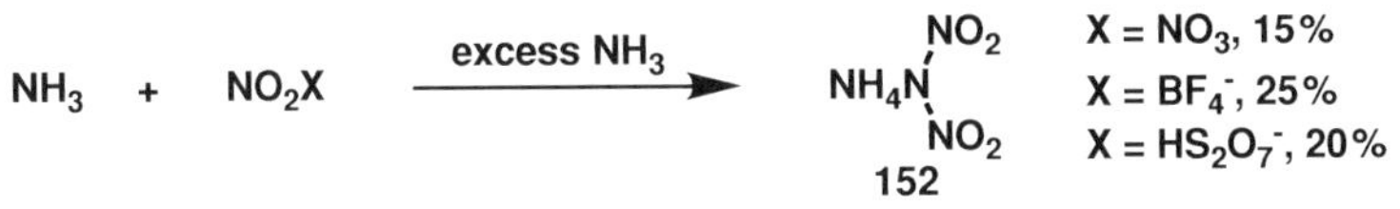

$$NH_3 + NO_2X \xrightarrow{\text{excess } NH_3} NH_4N(NO_2)_2\ (\mathbf{152})$$

$X = NO_3$, 15%
$X = BF_4^-$, 25%
$X = HS_2O_7^-$, 20%

Figure 6.49

Ammonium dinitramide has been synthesized from the nitration of ammonium sulfamate with strong mixed acid at -35 to $-45\,^{\circ}C$ followed by neutralization of the resulting dinitraminic acid with ammonia.[76] The yield is ~ 45 % when the mole ratios of sulfuric acid to nitric acid is 2:1 and ammonium sulfamate to total acid is 1:6. The nitration of other sulfonamide derivatives, followed by hydrolysis with metal hydroxides, also yields dinitramide salts.[77]

6.11.2 Alkyl *N,N*-dinitramines

Alkyl *N,N*-dinitramines belong to a class of highly energetic materials. However, their use is limited by poor thermal stability and a high sensitivity to shock and impact. These undesirable properties result from the high electron deficiency on the central nitrogen atom of the *N,N*-dinitramino group which makes the N–N bonds highly susceptible to cleavage.

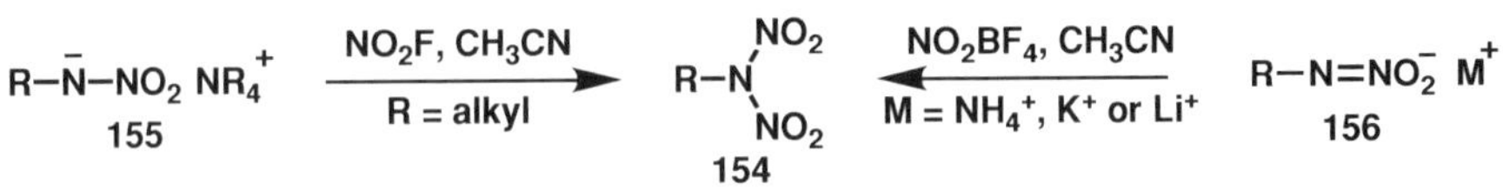

$$R{-}\bar{N}{-}NO_2\ NR_4^+\ (\mathbf{155}) \xrightarrow[R = \text{alkyl}]{NO_2F,\ CH_3CN} R{-}N(NO_2)_2\ (\mathbf{154}) \xleftarrow[M = NH_4^+,\ K^+ \text{ or } Li^+]{NO_2BF_4,\ CH_3CN} R{-}N{=}NO_2^-\ M^+\ (\mathbf{156})$$

Figure 6.50

Alkyl *N*,*N*-dinitramines (154) have been prepared from the reaction of the tetraalkylammonium salts (155) of primary nitramines with nitryl fluoride in acetonitrile at subambient temperature.[78] The same reaction with the primary nitramine or its alkali metal salts yields the corresponding nitrate ester.[79] Treatment of the ammonium, potassium, or lithium salts of primary nitramines (156) with a solution of nitronium tetrafluoroborate in acetonitrile at subambient temperature yield alkyl *N*,*N*-dinitramines.[80,81] The same reactions in ether or ester solvents enables the free nitramine to be used.[82] The nitrolysis of *N*-alkylnitramides (157)[83] and *N*,*N*-diacylamines[84] with nitronium tetrafluoroborate in acetonitrile, and the nitration of aliphatic isocyanates[85] with nitronium tetrafluoroborate and nitric acid in acetonitrile, also yield alkyl *N*,*N*-dinitramines (154).

157 + $NO_2^+ BF_4^-$ → 154 + $R'CO^+ BF_4^-$

R, R' = alkyl

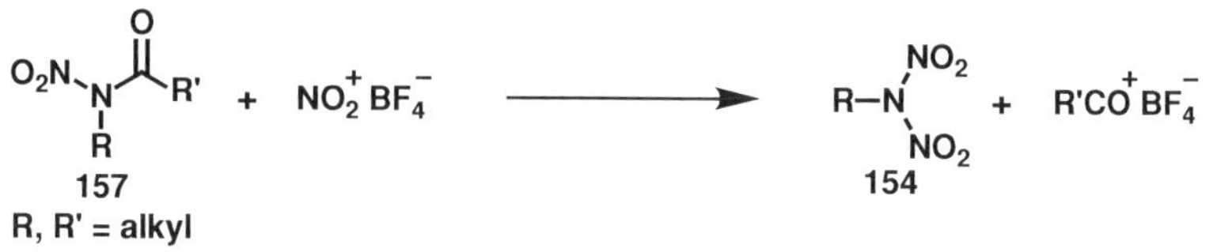

Figure 6.51

6.11.3 *N*-Nitroimides

The *N*-nitroimide functionality is a stable but highly energetic group which has been incorporated into some heterocycles in the search for new energetic materials. Katritzky and co-workers[86,87] synthesized *N*-nitroimides by treating alkylhydrazinium nitrates[88] with nitronium tetrafluoroborate in acetonitrile or with solutions of acyl nitrates prepared from the addition of nitric acid to mixtures of TFA–TFAA or acetic acid–acetic anhydride. Olah and co-workers[89] synthesized the *N*-nitroimides (160) and (161) by treating the corresponding tertiary amines, DABCO (158) and *N*,*N*,*N*′,*N*′-tetramethyl-1,3-propanediamine, respectively, with an aqueous solution of barium oxide, barium nitrate and hydroxylamine-*O*-sulfonic acid, followed by *N*-nitration of the resulting hydrazinium nitrates with TFA–TFAA.

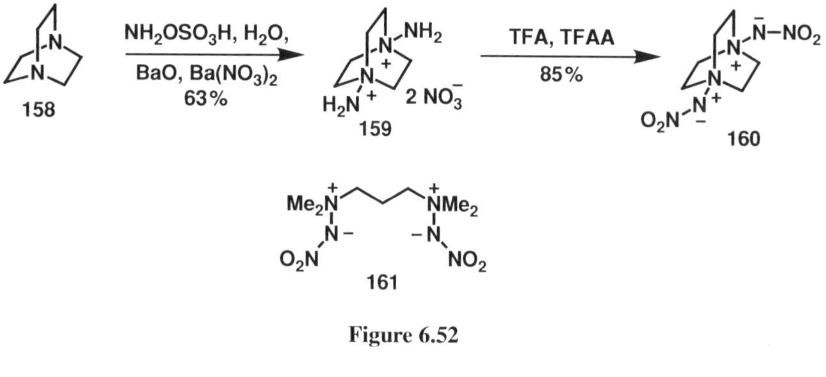

158 → (NH_2OSO_3H, H_2O, BaO, $Ba(NO_3)_2$, 63%) → 159 ($2\,NO_3^-$) → (TFA, TFAA, 85%) → 160

161

Figure 6.52

N-Nitroimides derived from tertiary amines contain a quaternary nitrogen atom which has a zwitterionic structure with the negative charge on one nitrogen atom stabilized by the electron-withdrawing effect of the adjacent nitrogen atom. *N*-Nitroimides derived from secondary amines have no quaternary nitrogen and have the negative charge counterbalanced with a positively charged species.

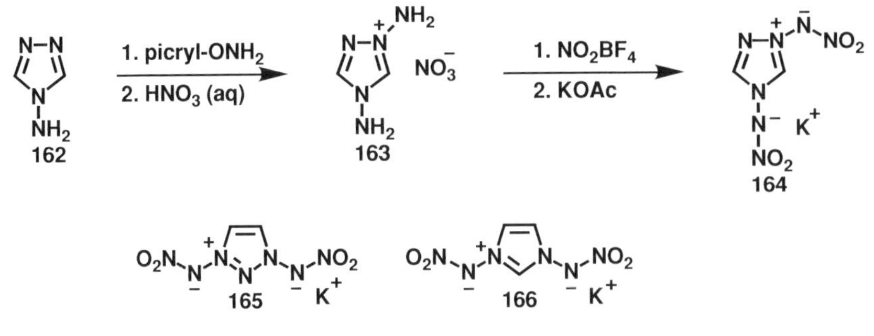

Figure 6.53

The nitrogen atoms of heterocycles like imidazoles and triazoles have been converted into N-nitroimide groups. The N-nitroimide (164) is synthesized from 1-amino-1,3,4-triazole (162) by N-amination of the tertiary nitrogen with O-picrylhydroxylamine, addition of nitric acid to give the nitrate salt (163), followed by N-nitration with nitronium tetrafluoroborate in acetonitrile.[90] The 1,2,3-triazole (165)[91] and the imidazole (166)[90] are synthesized in a similar way. The synthesis of N-nitroimides has been the subject of an excellent review.[92]

REFERENCES

1. V. P. Iushin, M. S. Komelin and V. A. Tartakovsky, *Zh. Org. Khim.*, 1999, **35**, 489.
2. R. D. Chapman, J. W. Fischer, R. A. Hollins, C. K. Lowe-Ma and R. A. Nissan, *J. Org. Chem.*, 1996, **61**, 9340.
3. J. O. Doali, R. A. Fifer, D. I. Kruezynski and B. J. Nelson, *Technical Report No. BRL-MR-378/5*, US Ballistic Research Laboratory, MD (1989).
4. T. G. Archibald, K. Baum, C. George and R. Gilardi, *J. Org. Chem.*, 1990, **55**, 2920.
5. T. G. Archibald, S. G. Bott, A. P. Marchand and D. Rajagopal, *J. Org. Chem.*, 1995, **60**, 4943.
6. K. Hayashi, T. Kumagai and Y. Nagao, *Heterocycles*, 2000, **53**, 447.
7. T. Axenrod, P. R. Dave, C. Watnick and H. Yazdekhasti, *J. Org. Chem.*, 1995, **60**, 1959.
8. T. Axenrod, P. R. Dave, C. Watnick and H. Yazdekhasti, *Tetrahedron Lett.*, 1993, **34**, 6677.
9. T. G. Archibald, M. D. Coburn and M. A. Hiskey, *Waste Management*, 1997, **17**, 143.
10. P. E. Eaton, K. Pramod and R. Gilardi, *J. Org. Chem.*, 1990, **55**, 5746.
11. G. T. Cunkle and R. L. Willer, 'Cubanes as Solid Propellants Ingredients', *SPIE Proceedings*, 1988, **872**, 24.
12. J. C. Bottaro, P. E. Penwell and R. J. Schmitt, *'Synthesis of Cubane Based Energetic Materials, Final Report, December 1989'*, SRI International, Menlo Park, CA [AD-A217 147/8/XAB].
13. H. G. Adolph and D. A. Cichra, *Synthesis*, 1983, 830.
14. K. Klager, *J. Org. Chem.*, 1958, **23**, 1519.
15. R. D. Chapman, R. D. Gilardi, C. B. Kreutzberger and M. F. Welker, *J. Org. Chem.*, 1999, **64**, 960.
16. R. Aniszfeld, G. A. Olah, G. Rasul and G. K. Surya Prakash, *J. Am. Chem. Soc.*, 1992, **114**, 5608.
17. T. Axenrod, R. D. Chapman, R. D. Gilardi, X. -P. Guan, L. Qi and J. Sun, *Tetrahedron Lett.*, 2001, **42**, 2621.
18. H. G. Adolph, M. Chaykovsky, C. George, R. Gilardi and W. M. Koppes, *J. Org. Chem.*, 1987, **52**, 1113.

19. C. L. Coon, 'Research on the Synthesis of Heterocyclic Explosives', in *Proc. International Symposium on Pyrotechnics and Explosives*, China Academic Publishers, Beijing, China., 183–186 (1987).
20. R .B. Crawford, L. de Vore, K. Gleason, K. Hendry, D. P. Kirvel, R. D. Lear, R. R. McGuire and R. D. Stanford, 'Energy and Technology Review, Jan–Feb 1988', *UCRL-52000-88-1/2*, Lawrence Livermore National Laboratory, Livermore, CA.
21. C. L. Coon, E. S. Jessop, A. R. Mitchell, P. F. Pagoria and R. D. Schmidt, in *Nitration: Recent Laboratory and Industrial Developments, ACS Symposium Series 623*, Eds. L. F. Albright, R. V. C. Carr and R. J. Schmitt, American Chemical Society, Washington, DC, Chapter 14, 151–164 (1996).
22. R. L. Willer, *US Pat.* 4 443 602 (1984); *Chem. Abstr.*, 1984, **101**, 72759f.
23. R. L. Willer, *J. Org. Chem.*, 1984, **49**, 5150.
24. R. L. Atkins and R. L. Willer, *J. Org. Chem.*, 1984, **49**, 5147.
25. (a) A. T. Nielsen, *US Pat.* 253 106 (1988); *Chem. Abstr.*, 1998, **128**, 36971t; (b) A. T. Nielsen, 'Synthesis of Caged Nitramine Explosives', presented at Joint Army, Navy, NASA, Air Force (JANNAF) Propulsion Meeting, San Diego, CA, 17 December, 1987; (c) A. P. Chafin, S. L. Christian, J. L. Flippen-Anderson, C. F. George, R. D. Gilardi, D. W. Moore, M. P. Nadler, A. T. Nielsen, R. A. Nissan and D. J. Vanderah, *Tetrahedron*, 1998, **54**, 11793.
26. Thiokol Corporation in the US, as reported by P. Braithwaite, S. Collignon, J. C. Hinshaw, G. Johnstone, R. Jones, V. A. Lyon, K. Poush and R. B. Wardle, in *Proc. International Symposium on Energetic Materials Technology*, American Defence Preparedness Association (1994); *Chem. Abstr.*, 1996, **125**, 172464v.
27. C. L. Coon, J. L. Flippen-Anderson, C. F. George, R. D. Gilardi, A. T. Nielsen, R. A. Nissan and D. J. Vanderah, *J. Org. Chem.*, 1990, **55**, 1459.
28. (a) A. J. Bellamy, P. Goede, N. V. Latypov and U. Wellmar, *Org. Process Res. Dev.*, 2000, **4**, 156; (b) M. Ikeda, T. Kodama and M. Tojo, *PCT Int. Appl.* WO 96/23792 (1996); *Chem. Abstr.*, 1996, **125**, 275920v; (c) L. F. Cannizzo, W. W. Edwards, T. K. Highsmith and R. B. Wardle, *PCT Int. Appl.* WO 97/00873 (1997), *US Pat. Appl.* 493 627 (1995); *Chem. Abstr.*, 1997, **126**, 145956w; (d) A. J. Bellamy, *Tetrahedron*, 1995, **51**, 4711; (e) W. W. Edwards and R. B. Wardle, *PCT Int. Appl.* WO 97/20785 (1997), *US Pat. Appl.* 568 451 (1995); *Chem. Abstr.*, 1997, **127**, 110983w; (f) M. Ikeda, T. Kodama and M. Tojo, *Jpn. Kokai Tokkyo Koho* JP 08/208655 [96/208655] (1996); *Chem. Abstr.*, 1996, **125**, 301030b; (g) T. Kodama, S. Kawabe, H. Mira and M. Miyake, *PCT Int. Appl.* WO 98/05666 (1998), *Jpn. Pat. Appl.* 96/223239 (1996); *Chem. Abstr.*, 1998, **128**, 167451w; (h) B. R. Gandhe, G. M. Gore, R. Sivabalan and S. Venugopalan, in *Proc. 5th International High Energy Materials Conference and Exhibit*, 23–25 Nov, 2005, DRDL, Hyderabad; (i) S. P. Pang, Y. Z. Yu and X. Q. Zhao, *Propell. Explos. Pyrotech.*, 2005, **30**, 442.
29. J. P. Agrawal, B. R. Gandhe, H. Singh, A. K. Sikder and N. Sikder, *Def. Sci. J.*, 2002, **52(2)**, 135.
30. J. H. Boyer, V. T. Ramakrishnan and M. Vedachalam, *Heterocycles*, 1990, **31**, 479.
31. A. T. Nielsen, in *Chemistry of Energetic Materials*, Eds. G. A. Olah and D. R. Squire, Academic Press, San Diego, CA, Chapter 5, 95–124 (1991).
32. H. G. Adolph and D. A. Cichra, *J. Org. Chem.*, 1982, **47**, 2474.
33. J. Boileau, G. Jacob and M. Piteau, *Propell. Explos. Pyrotech.*, 1990, **15**, 38.
34. C. D. Bedford, C. L. Coon, S. Jose and D. A. Levins, *US Pat.* 4 346 222 (1982); *Chem. Abstr.*, 1983, **98**, 18971a.
35. M. D. Cliff, I. J. Dagley, R. P. Parker and G. Walker, *Propell. Explos. Pyrotech.*, 1998, **23**, 179.
36. D. Huang and R. R. Rindone, 'NNHT: A New Low Cost Insensitive Cyclic Nitramine', in *Proc. Joint International Symposium on Compatibility of Plastics and Other Materials with Explosives, Propellants, Pyrotechnics and Processing of Explosives, Propellants and Ingredients*, San Diego, CA, 62–68 (1991).
37. H. H. Licht and H. Ritter, *Propell. Explos. Pyrotech.*, 1985, **10**, 147.
38. J. Boileau, J. M. L. Emeury and J. P. Kehren, *US Pat.*, 4 487 938 (1974).
39. (a) J. Boileau, J. M. L. Emeury and J. P. Kehren, *Ger. Pat.* 2 435 651 (1975); (b) *Encyclopaedia of Explosives and Related Items*, Eds. H. A. Aaronson, G. D. Clift, B. T. Fedoroff, E. F. Reese and O. E. Sheffield., Picatinny Arsenal, Dover, New Jersey, Vol. 1, A65 (1960).

40. M. D. Coburn, B. W. Harris, H. H. Hayden, K. Y. Lee and M. M. Stinecipher, *Ind. Eng. Chem. Prod. Res. Dev.*, 1986, **25**, 68.
41. J. Li, in *Proc. 17th International Pyrotechnics Seminar (Combined with 2nd Beijing International Symposium on Pyrotechnics and Explosives)*, Beijing Institute Technical Press, Beijing, China, 322–332 (1991).
42. M. Zhang, Y. Zheng, D. Zhou and J. Zhou, *Binggong Zuebao*, 1988, 59; *Chem. Abstr.*, 1988, **109**, 189782q.
43. M. Chen, G. Hua and W. Li, 'Synthesis and properties of 2,4,6,8-Tetranitro-2,4,6,8-tetraazabicyclo[3.3.0]octan-3-one', in *Proc. International Symposium on Pyrotechnics and Explosives*, China Academic Publishers, Beijing, China, 187–189 (1987).
44. E. S. Jessop, A. R. Mitchell and P. F. Pagoria., *Propell. Explos. Pyrotech.*, 1996, **21**, 14.
45. H. R. Graindorge, P. A. Lescop, M. J. Pouet and F. Terrier, in *Nitration: Recent Laboratory and Industrial Developments, ACS Symposium Series 623*, Eds. L. F. Albright, R. V. C. Carr and R. J. Schmitt, American Chemical Society, Washington, DC, Chapter 5, 43–50 (1996).
46. J. P. Agrawal, G. M. Bhokare, D. B. Sarwade and A. K. Sikder, *Propell. Explos. Pyrotech.*, 2001, **26**, 63.
47. H. G. Adolph, J. H. Boyer, I. J. Dagley, J. L. Flippen-Anderson, C. George, R. Gilardi, K. A. Nielsen, V. T. Ramakrishnan and M. Vedachalam, *J. Org. Chem.*, 1991, **56**, 3413.
48. J. H. Boyer, V. T. Ramakrishnan and M. Vedachalam, *Heteroatom. Chem.*, 1991, **2**, 313.
49. (a) R. D. Breithaupt, C. L. Coon, E. S. Jessop, A. R. Mitchell, G. L. Moody, P. F. Pagoria, J. F. Poco and C. M. Tarver, *Propell. Explos. Pyrotech.*, 1994, **19**, 232; (b) 'Synthesis, Scale-up, and characterization of K-6', *Report No UCRL-LR-109404* (1992), Lawrence Livermore National Laboratory, Livermore, CA.
50. J. Hong and C. Zhu, in *Proc. 17th International Pyrotechnics Seminar (Combined with 2nd Beijing International Symposium on Pyrotechnics and Explosives)*, Beijing Institute Technical Press, Beijing, China, 193 (1991).
51. M. B. Frankel and D. O. Woolery, *J. Org. Chem.*, 1983, **48**, 611.
52. T. B. Brill, Y. Oyumi and A. L. Rheingold, *J. Phys. Chem.*, 1987, **91**, 920.
53. P. Bunyan, P. Golding, R. W. Millar, N. C. Paul, D. H. Richards and J. A. Rowley, *Propell. Explos. Pyrotech.*, 1993, **18**, 55.
54. (a) P. Golding, R. W. Millar, N. C. Paul and D. H. Richards, *Tetrahedron*, 1993, **49**, 7063; (b) P. Golding, R. W. Millar, N. C. Paul and D. H. Richards, *Tetrahedron. Lett*, 1988, **29**, 2735.
55. M. M. Bhalerao, B. R. Gandhe, M. A. Kulkarni, K. P. C. Rao and A. K. Sikder, *Propell. Explos. Pyrotech.*, 2004, **29(2)**, 93.
56. (a) D. W. Fish, E. E. Hamel and R. E. Olsen, in *Advanced Propellant Chemistry, Advances in Chemistry Series. No 54*, Ed. R. E. Gould., American Chemical Society, Washington DC, Chapter 6, 48–54 (1966); (b) L. A. Fang, S. Q. Hua, V. G. Ling and L. Xin, Preliminary Study on Bu-NENA Gun Propellants, *27th International Annual Conference of ICT*, Karlsruhe, Germany, June 25–28, 1996, 51; (c) N. F. Stanley and P. A. Silver, Bu-NENA Gun Propellants, *JANNAF Propulsion Meetings*, 10 September 1990, Vol. 2, 515; (d) R. A. Johnson and J. J. Mulley, Stability and Performance Characteristics of NENA Materials and Formulations, *Joint International Symposium on Energetic Materials Technology*, New Orleans, Louisiana, 5–7 October, 1992, 116.
57. M. B. Frankel and E. F. Witucki, *US Pat.* 4 701 557 (1987); *Chem. Abstr.*, 1988, **108**, 97345a.
58. W. H. Gilligan and M. E. Sitzman, *J. Energ. Mater.*, 1985, **3**, 293.
59. L. T. Eremenko, R. G. Gafurov, F. Ya. Natsibullin and S. I. Sviridor, *Izv. Akad. Nauk USSR, Ser. Khim.*, 1970, **19**, 329.
60. L. T. Eremenko, R. G. Gafurov and E. M. Sogomonyan, *Izv. Akad. Nauk USSR, Ser. Khim.*, 1971, **20**, 2480.
61. K. Shimo, *Toyko Kogyo Shikensho Hokoku*, 1970, **65**, 46; *Chem. Abstr.*, 1971, **74**, 140812u.
62. T. B. Brill and Y. Oyumi, *J. Phys. Chem.*, 1987, **91**, 3657.
63. A. A. Fainzilberg, B. V. Litvinov, B. G. Loboiko, G. M. Nazin, V. I. Pepekin, S. A. Shevelev and S. P. Smirnov, *Dokl. Akad. Nauk USSR*, 1994, **336**, 86.

64. R. J. Spear and W. S. Wilson, *J. Energ. Mater.*, 1984, **2**, 61.
65. I. J. Dagley and R. J. Spear, in *Organic Energetic Compounds*, Ed. P. L. Marinkas., Nova Science Publishers, Inc., New York, Chapter 2, 47–163 (1996).
66. H. G. Adolph and W. M. Koppes, in *Nitro Compounds: Recent Advances in Synthesis and Chemistry., Organic Nitro Chemistry Series.*, Eds. H. Feuer and A. T. Neilsen., VCH Publishers., Chapter 4, 367–605 (1990).
67. O. V. Anikin, V. P. Gorelik, O. A. Luk'yanov and V. A. Tartakovsky, *Izv. Akad. Nauk USSR, Ser. Khim.*, 1994, **43**, 1457.
68. N. O. Cherskaya, V. P. Gorelik, O. A. Luk'yanov, V. A. Shlyapochnikov and V. A. Tartakovsky, *Izv. Akad. Nauk USSR, Ser. Khim.*, 1994, **43**, 1522.
69. O. A. Luk'yanov, N. I. Shlykova and V. A. Tartakovsky, *Izv. Akad. Nauk USSR, Ser. Khim.*, 1994, **43**, 1680.
70. V. P. Gorelik, O. A. Luk'yanov and V. A. Tartakovsky, *Izv. Akad. Nauk USSR, Ser. Khim.*, 1994, **43**, 89.
71. T. A. Klimova, Y. V. Konnova, O. A. Luk'yanov and V. A. Tartakovsky, *Izv. Akad. Nauk USSR, Ser. Khim.*, 1994, **43**, 1200.
72. A. R. Agevnin, A. A. Leichenko, O. A. Luk'yanov, N. M. Seregina and V. A. Tartakovsky, *Izv. Akad. Nauk USSR, Ser. Khim.*, 1995, **44**, 108.
73. O. V. Anikin, N. O. Cherskaya, V. P. Gorelik, O. A. Luk'yanov, G. I. Oleneva, V. A. Shlyapochnikov and V. A. Tartakovsky, *Izv. Akad. Nauk USSR, Ser. Khim.*, 1995, **44**, 1449.
74. U. Bemm, H. Bergman, A. Langlet and H. Östmark, *Thermochim. Acta*, 2002, **384**, 253.
75. J. C. Bottaro, P. E. Penwell and R. J. Schmitt, *J. Am. Chem. Soc.*, 1997, **119**, 9405.
76. (a) M. Kanakeval, K. N. Ninan, G. Santhosh and S. Venkatachalam, *Ind. J. Chem. Tech.*, 2002, **9**, 223; (b) A. Langlet, H. Östmark and H. Wingborg, *US Pat.* 5 976 483 (1999).
77. A. Langlet, H. Östmark and H. Wingborg, *PCT Int. Appl.* WO 97/06099 (1996).
78. L. T. Eremenko, B. S. Fedorov and R. G. Gafurov, *Izv. Akad. Nauk USSR, Ser. Khim.*, 1979, **28**, 2111.
79. (a) L. T. Eremenko, B. S. Fedorov and R. G. Gafurov, *Izv. Akad. Nauk USSR, Ser. Khim.*, 1977, **26**, 345; (b) L. T. Eremenko, B. S. Fedorov and R. G. Gafurov, *Izv. Akad. Nauk USSR, Ser. Khim.*, 1971, **20**, 1501.
80. E. E. Hamel, C. Heights and R. E. Olsen, *Brit. Pat.* 1 126 5591 (1968); *Chem. Abstr.*, 1969, **70**, 67584.
81. J. D. Malley, *Brit. Pat.* 1 126 591 (1968); *Chem. Abstr.*, 1969, **70**, 67584g.
82. S. A. Andrew, B. V. Gidaspov, M. A. Ilyusin and B. A. Lebedev, *Zh. Org. Khim.*, 1978, **14**, 2055.
83. S. A. Andrew and B. A. Lebedev, *Zh. Org. Khim.*, 1978, **14**, 907.
84. O. A. Luk'yanov, T. G. Melnikova, N. M. Seregina and V. A. Tartakovsky, *Abstracts of Reports on the Chemistry of Nitro-Compounds*, Moscow, 33 (1977).
85. J. C. Bottaro, P. E. Penwell and R. J. Schmitt, *Synth. Commun.*, 1991, **21**, 945.
86. J. Epsztajn and A. R. Katritzsky, *Tetrahedron Lett.*, 1969, **10**, 4739.
87. J. Epsztajn, A. R. Katritzsky, E. Lunt, J. W. Mitchell and G. J. Roche, *J. Chem. Soc. Perkin Trans. 1*, 1973, 2622.
88. (a) G. L. Omietanski and H. H. Sisler, *J. Am. Chem. Soc.*, 1956, **57**, 1585; (b) R. Gösl and A. Meusen, *Angew. Chem.*, 1957, **69**, 754; (c) R. Gösl and A. Meusen, *Chem. Ber.*, 1959, **92**, 2521.
89. J. L. Flippen-Anderson, C. George, R. Gilardi, G. A. Olah, G. K. Surya Prakash, C. B. Roa, M. B. Sassaman and M. Zuanic, *J. Org. Chem.*, 1992, **57**, 1585.
90. V. A. Myasnikov, O. P. Shitov, V. A. Tartakovsky, V. A. Vyazkov and I. L. Yudin, *Izv. Akad. Nauk USSR, Ser. Khim.*, 1991, **40**, 1239.
91. O. P. Shitov, V. A. Tartakovsky and V. A. Vyazkov, *Izv. Akad. Nauk USSR, Ser. Khim.*, 1989, **38**, 2654.
92. E. T. Apazov, S. L. Ioffe, A. V. Kalnin, Y. N. Strelenso and V. A. Tartakovsky, *Mendeleev Commun.*, 1991, 95.

7

Energetic Compounds 3: *N*-Heterocycles

7.1 INTRODUCTION

Until the end of the First World War the main filling for mass ordnance was TNT and its mixtures with ammonium nitrate known as Amatols. During this period improved methods for the manufacture of RDX enabled its inclusion in munitions, usually in formulation with TNT in the form of the Cyclotols. RDX and its mixtures are still the most widely used explosives for military use.

RDX by any measure is a high performance explosive. However, rapid advances in warfare technology demand even higher performance materials coupled with low sensitivities to impact, shock and friction. Most secondary high explosives in wide use today are vulnerable to premature detonation when used in high demand applications such as the warheads of high-speed guided missiles and high-calibre guns. The incorporation of such explosives into a polymeric matrix (PBX) has been a common strategy to reduce sensitivity and this is generally successful. However, such explosives are still susceptible to detonation from the shock of another explosive. The risk of catastrophic explosion in the magazine of a ship or similar munitions storage areas cannot be ignored. Many countries have an ongoing research program to find new energetic materials with a low vulnerability to accidental initiation. The intention is to gradually phase out current explosives for insensitive high explosives (IHEs). Another area of research involves finding and synthesizing thermally stable explosives. Such materials have commercial value for applications involving high temperatures like the drilling of deep oil wells and for the space programmes.

Many of the aforementioned properties are present in nitrogen heterocycles and these are the discussion point of this chapter. Many of the *N*-heterocycles described in this chapter have high percentages of nitrogen in their skeletal structure, and consequently, have exceptionally high heats of formation and are highly endothermic in nature. Such compounds are classically energetic and release large amounts of energy on combustion and often exhibit high performance. The high nitrogen content of these compounds often leads to a high crystal density which is itself associated with increased performance. Research into this class of energetic materials is still strong and many *N*-heterocycles have found specialized applications. Unlike caged polynitropolycycloalkanes and polynitramines, many *N*-heterocycles are fairly

Organic Chemistry of Explosives J. P. Agrawal and R. D. Hodgson

easy to synthesize and, coupled with their high performance, it is probable that some of these compounds may eventually replace common high explosives like RDX.

Vast research efforts have been pooled into finding new energetic *N*-heterocycles over the past 30 years and, consequently, the number of reported compounds is huge. It is quite impossible to discuss all the materials reported in this area in the space available. We personally believe that *N*-heterocycles should be the subject of its own book and this may well be the case in the future. We draw the reader to a number of excellent reviews,[1] which together cover most of the past and present literature on *N*-heterocycles.

7.2 5-MEMBERED RINGS – 1N – PYRROLES

Nitro derivatives of pyrrole are not considered practical explosives for two reasons. Firstly, the heat of formation of the pyrrole ring offers no benefits over standard arylene hydrocarbons. Secondly, during nitration, pyrroles, like thiophenes and furans, are much more prone to oxidation and acid-catalyzed ring-opening than arylene hydrocarbons. A common strategy for the synthesis of highly nitrated pyrroles is to conduct the nitration in stages, the initial mono-nitration using a mild nonacidic nitrating agent. As more nitro groups are introduced the pyrrole ring becomes more electron deficient and less prone to oxidation and so allows for the use of harsher and more acidic nitrating agents for further nitration.

Pagoria and co-workers[2] reported the nitration of *N*-*tert*-butylpyrrole to *N*-*tert*-butyl-2,3,4-trinitropyrrole in 40 % yield over three steps. Stegel and co-workers[3] reported the same synthesis but conducted the nitration in two steps using mixed acid. Hinshaw and co-workers[4] used *N*-*tert*-butyl-2,3,4-trinitropyrrole for the synthesis of 2,3,4,5-tetranitropyrrole in a reaction involving initial deprotection followed by nitration with mixed acid at elevated temperature. 2,3,4,5-Tetranitropyrrole has a perfect oxygen balance but slowly decomposes on storage at room temperature. Stegel and co-workers[3] also reported the synthesis of *N*-methyl-2,3,4,5-tetranitropyrrole from the nitration of *N*-methyl-2,3,4-trinitropyrrole with mixed acid.

Russian chemists have reported the synthesis of *N*-alkyl-3,4-dinitropyrroles from the cyclization of primary amines, formaldehyde and the potassium salt of 2,3,3-trinitropropanol.[5]

7.3 5-MEMBERED RINGS – 2N

7.3.1 Pyrazoles

Heat of formation and density calculations correlate so well with performance parameter like detonation velocity that chemists have a good idea of the performance of an energetic material before its synthesis and testing. The pyrazolo[4,3-*c*]pyrazoles DNPP (9) and LLM-119 (10) were predicted[2] to exhibit performances equal to 85 % and 104 % relative to that of HMX.

Shevelev and co-workers[6] first synthesized DNPP (9) from 3,5-dimethylpyrazole. Subsequently, Pagoria and co-workers[7] improved the synthesis, obtaining DNPP (9) in 21 % overall yield from 2,4-pentanedione (1). An interesting feature of this synthesis is the tandem decarboxylation–nitration step which occurs on treating (8) with absolute nitric acid at elevated temperature. As predicted from theoretical calculations DNPP (9) is less energetic than HMX but exhibits higher thermal stability and lower sensitivity to impact. Amination of DNPP

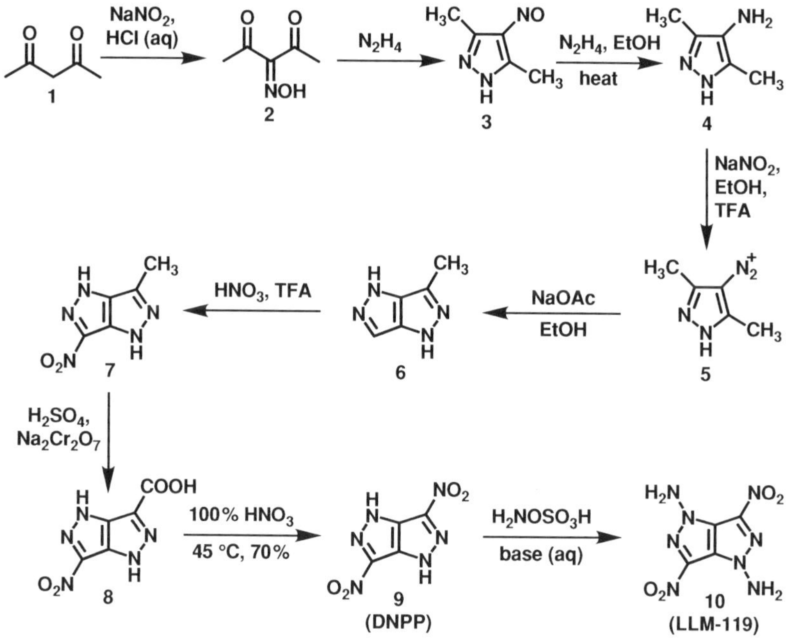

Figure 7.1

(9) with hydroxylamine-*O*-sulfonic acid[8] in aqueous base yields LLM-119 (10);[7] the latter exhibits higher performance than DNPP and a lower sensitivity to impact.

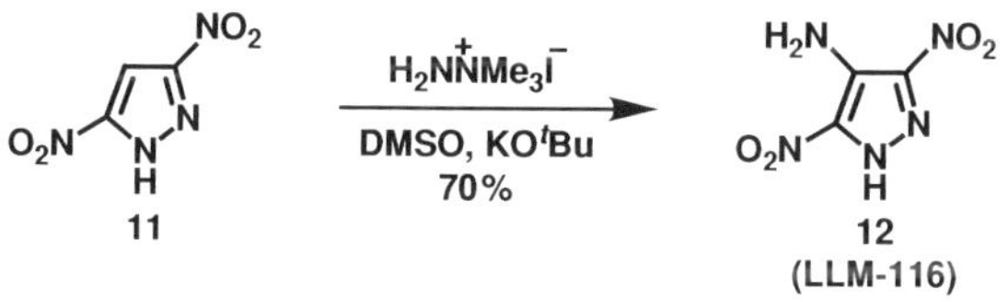

Figure 7.2

The increase in thermal stability and reduction in impact sensitivity observed on introducing amino groups adjacent to nitro groups in aromatic systems is known to result from intramolecular hydrogen bonding interactions (Section 4.8.1.4). This effect is also illustrated in 4-amino-3,5-dinitropyrazole (LLM-116) (12), an energetic material showing a lower sensitivity to impact than 3,5-dinitropyrazole (11). LLM-116 (12) is synthesized from the *C*-amination of 3,5-dinitropyrazole (11) with 1,1,1-trimethylhydrazinium iodide (TMHI) in the presence of potassium *tert*-butoxide base.[9]

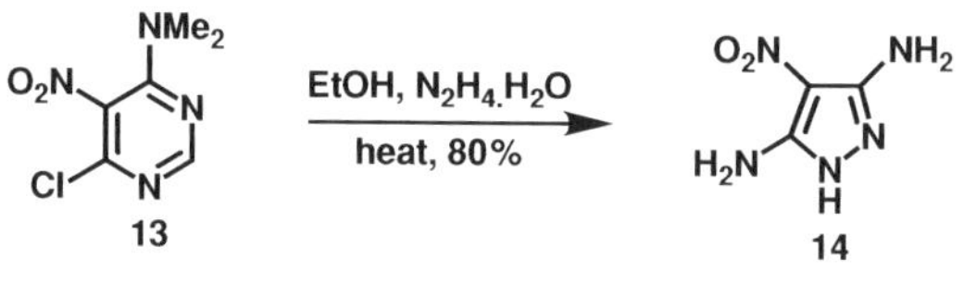

Figure 7.3

Poullain and co-workers[10] synthesized *N*-substituted-3,5-diamino-4-nitropyrazoles by treating substituted pyrimidines with *N*-alkylhydrazines. 3,5-Diamino-4-nitropyrazole (14) has been synthesized by treating the pyrimidine (13) with hydrazine hydrate. This methodology is limited in scope but the products are useful intermediates for the synthesis of insensitive high explosives.

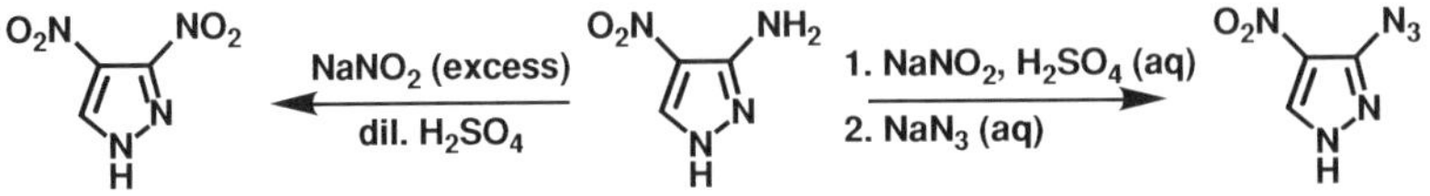

NaNO$_2$ (excess), dil. H$_2$SO$_4$; 1. NaNO$_2$, H$_2$SO$_4$ (aq), 2. NaN$_3$ (aq)

Figure 7.4

A common route to nitropyrazoles involves the diazotization of the corresponding amino derivatives in the presence of excess sodium nitrite.[11] Diazotization in the presence of sodium azide allows the introduction of the azido 'explosophore'.[11]

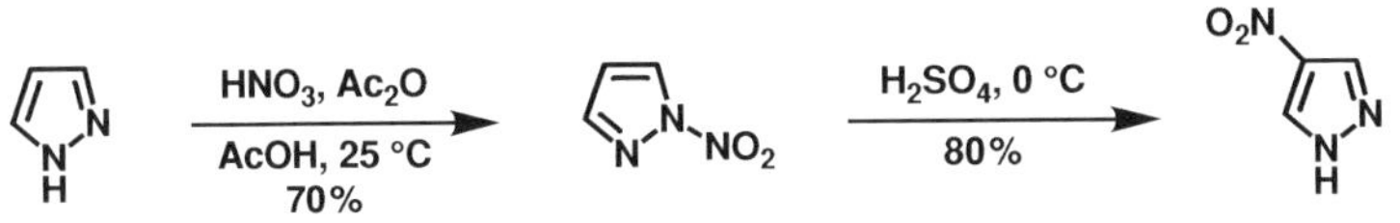

HNO$_3$, Ac$_2$O, AcOH, 25 °C, 70% ; H$_2$SO$_4$, 0 °C, 80%

Figure 7.5

Initial nitration of pyrazole derivatives with nitric acid in acetic or trifluoroacetic anhydrides leads to *N*-nitropyrazoles, which rearrange to the *C*-nitrated product on stirring in concentrated sulfuric acid at subambient temperature. This N $\rightarrow$ C nitro group rearrangement often occurs *in situ* when pyrazoles are nitrated with mixed acid.

7.3.2 Imidazoles

The direct nitration of imidazole with acidic reagents is difficult due to facile nitrogen protonation (p$K_{aH} \sim 7$). Nitration of imidazoles proceeds in the 4- and 5-positions with the amidine 2-position being quite inert. Imidazole can be directly nitrated to 4,5-dinitroimidazole but no further.[12] 2,4,5-Trinitroimidazole (TNI) can be prepared from the successive nitration of 2-nitroimidazole; the latter synthesized from the diazotization of 2-aminoimidazole in the presence of excess sodium nitrite and a copper salt.[12] The nitrative cleavage of polyiodoimidazoles also provides a route to polynitroimidazoles.[12,13]

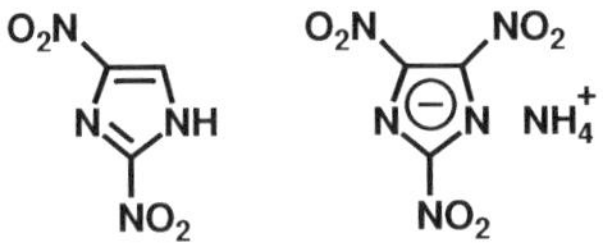

15 (2,4-DNI) ; 16 (ANTI)

Figure 7.6

2,4-Dinitroimidazole (2,4-DNI) (15) is readily prepared from the nitration of 2-nitroimidazole.[12,14] 2,4-DNI exhibits moderate performance and is regarded as a shock insensitive explosive. The relatively low cost and facile synthesis makes 2,4-DNI a realistic alternative to TNT for mass use in ordnance. At present, chemists at Lawrence Livermore National Laboratory (LLNL) are working with ARDEC and LANL on optimizing synthesis and lowering production costs.[14] The ammonium salt of 2,4,5-trinitroimidazole (ATNI) (16) has received some interest but its inability to form a useful eutectic with ammonium nitrate means that it is unlikely to find application.[13]

7.3.3 1,3,4-Oxadiazoles

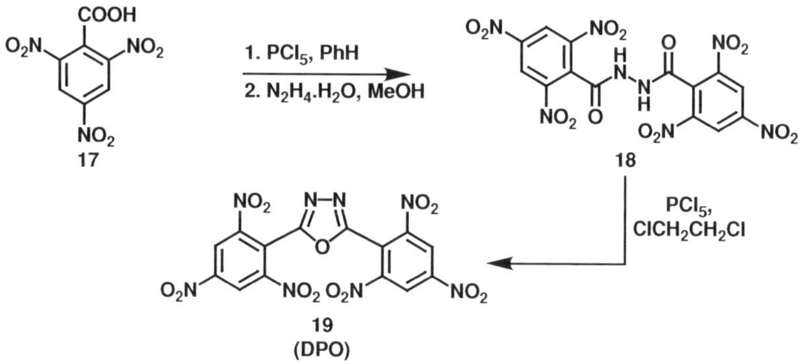

1. PCl_5, PhH
2. $N_2H_4.H_2O$, MeOH

PCl_5, $ClCH_2CH_2Cl$

17 18 19 (DPO)

Figure 7.7

The dehydration of *N,N′*-diacylhydrazines is a standard method for the formation of the 1,3,4-oxadiazole ring. 2,5-Dipicryl-1,3,4-oxadiazole (DPO) (19) is synthesized by treating 2,4,6-trinitrobenzoic acid (17) with phosphorous pentachloride, followed by treatment with hydrazine to give the *N,N′*-diacylhydrazine (18) which undergoes dehydration on further reaction with phosphorous pentachloride in 1,2-dichloroethane.[15] DPO exhibits high thermal stability but is very sensitive to impact and shock, making it useful in detonation transfer compositions.

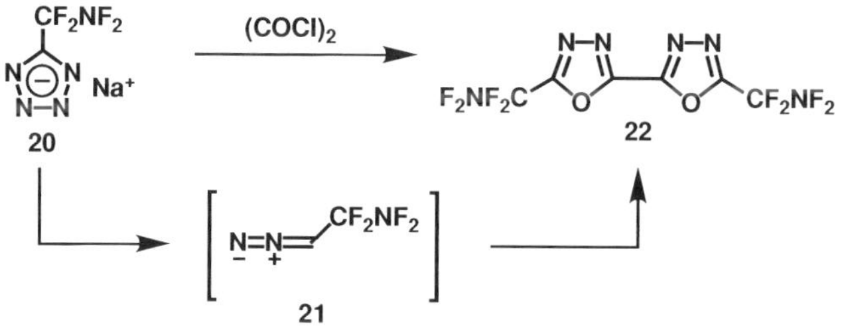

$(COCl)_2$

20 21 22

Figure 7.8

The energetic 1,3,4-oxadiazole (22) is synthesized from the reaction of the tetrazole (20) with oxalyl chloride.[16] In this reaction the tetrazole (20) undergoes a reverse cycloaddition with the expulsion of nitrogen and the formation of the 1,3-dipolar diazoalkane (21) which reacts with the carbonyl groups of oxalyl chloride to form the 1,3,4-oxadiazole rings.

7.3.4 1,2,5-Oxadiazoles (furazans)

Nitro and amino derivatives of the furazan ring (1,2,5-oxadiazole) are nitrogen-rich energetic materials with potential use in both propellant and explosive formulations. Some nitro-substituted furazans have excellent oxygen balance and exhibit detonation velocities close to very powerful military explosives.

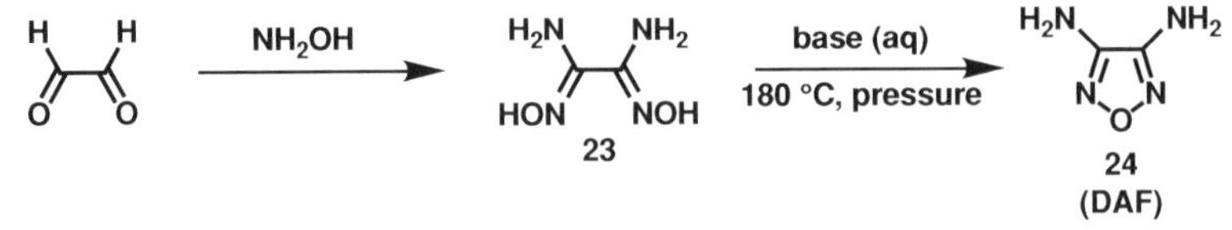

Figure 7.9

3,4-Diaminofurazan (DAF) (24) is a starting material for the synthesis of many nitro-substituted furazans and is readily prepared from the cyclization of 1,2-diaminoglyoxime (23) in the presence of aqueous base under pressure at 180 °C;[17] the latter prepared from the reaction of glyoxal,[18] glyoxime,[19] cyanogen[20] or dithiooxamide[21] with hydroxylamine.

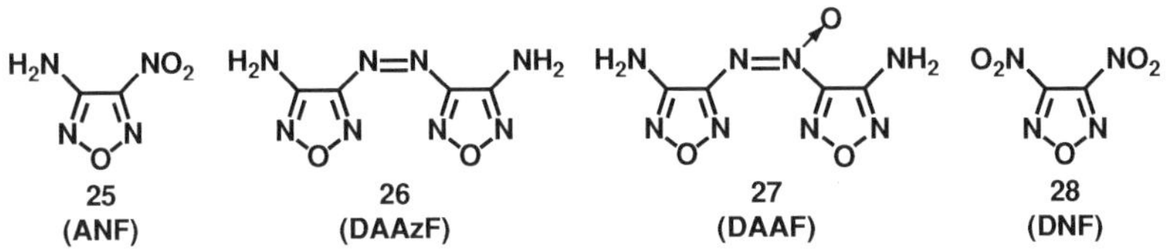

Figure 7.10

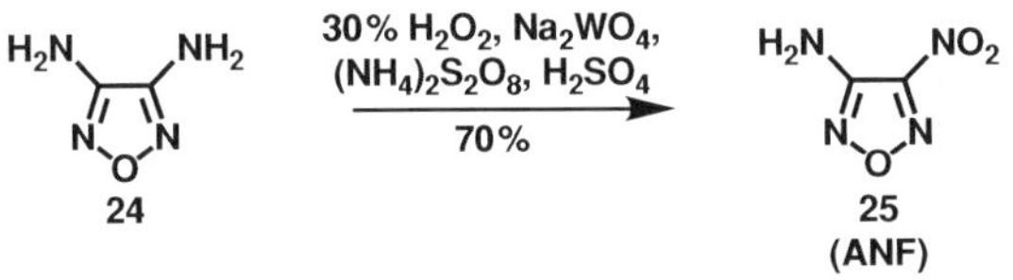

Figure 7.11

The oxidation of DAF (24) with hydrogen peroxide can yield 3-amino-4-nitrofurazan (ANF) (25), 4,4′-diamino-3,3′-azofurazan (DAAzF) (26), or 4,4′-diamino-3,3′-azoxyfurazan (DAAF) (27) depending on the conditions employed.[22] The most convenient route to ANF (25) involves treating DAF (24) with a mixture of 30 % aqueous hydrogen peroxide, sodium tungstate and ammonium persulfate in concentrated sulfuric acid.[23,24] Both of the amino groups of DAF (24) are oxidized to give 3,4-dinitrofurazan (DNF) (28) if 30 % hydrogen peroxide is replaced by 90 % hydrogen peroxide.[24] DNF (28) is a very powerful explosive with a positive oxygen balance but it is too reactive and shock sensitive to be considered for use as a practical explosive.

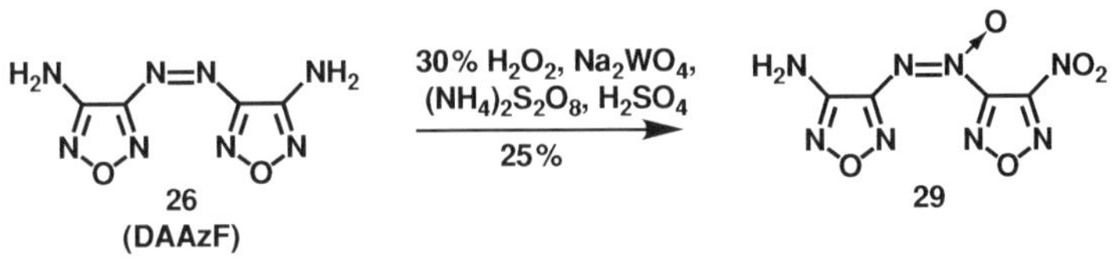

Figure 7.12

DAAzF (26) can be oxidized with a mixture of 30 % aqueous hydrogen peroxide, sodium tungstate and ammonium persulfate in concentrated sulfuric acid to yield

4-amino-4′-nitro-3,3′-azofurazan (29).[25] The use of stronger hydrogen peroxide solutions can oxidize both amino groups and yield either DNAzBF (30)[24] or DNABF (31).[24]

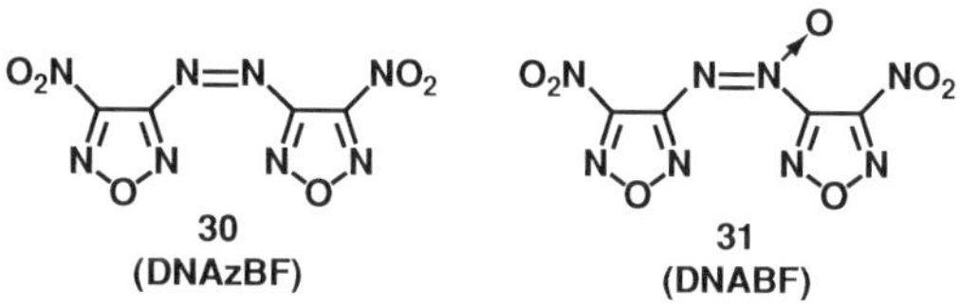

30 (DNAzBF)

31 (DNABF)

Figure 7.13

Oxidation of 4,4′-diamino-3,3′-bifurazan (DABF) (32) with 90% hydrogen peroxide in trifluoroacetic acid yields 4,4′-dinitro-3,3′-bifurazan (DNBF) (33).[17] DNBF exhibits high performance (VOD ∼ 8800 m/s, $d = 1.92\ g/cm^3$) coupled with a conveniently low melting point (85 °C) permitting the casting of charges. However, it is very sensitive to impact, demanding stringent safety measures during synthesis and handling.

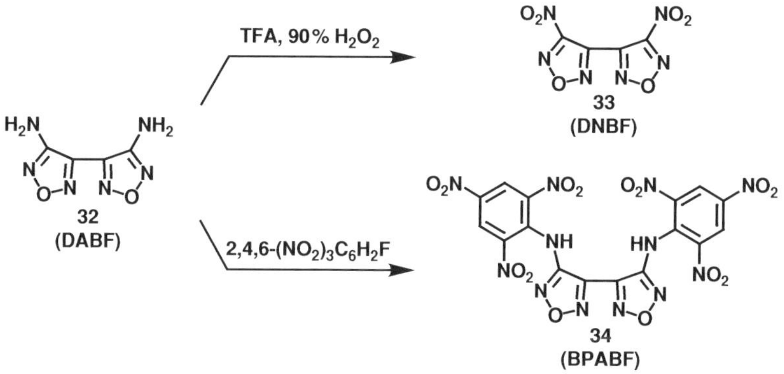

TFA, 90% H_2O_2

33 (DNBF)

32 (DABF)

2,4,6-$(NO_2)_3C_6H_2F$

34 (BPABF)

Figure 7.14

Some energetic compounds have picryl groups (2,4,6-trinitrophenyl-) introduced as substituents in the 3- and 4-positions of the furazan ring. Coburn[17] synthesized a series of picrylamino-substituted furazans, including 4,4′-bis(picrylamino)-3,3′-bifurazan (BPABF) (34) from the reaction of 4,4′-diamino-3,3′-bifurazan (DABF) (32) with two equivalents of picryl fluoride.

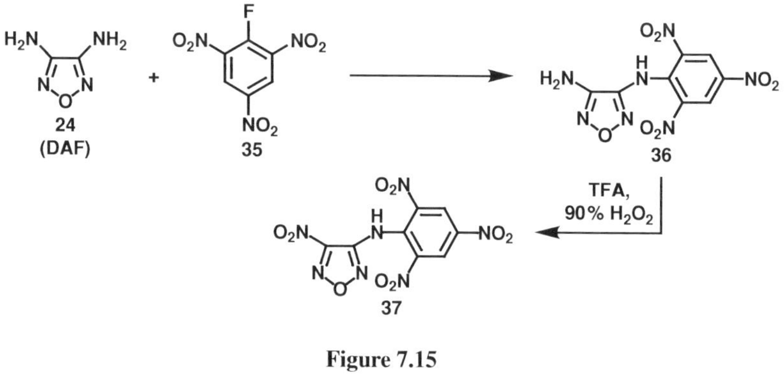

24 (DAF) + 35 → 36

TFA, 90% H_2O_2

37

Figure 7.15

Coburn[17] synthesized 3-nitro-4-(picrylamino)furazan (37) from the reaction of DAF (24) with one equivalent of picryl fluoride (35) followed by oxidation of the remaining amino group with hydrogen peroxide in trifluoroacetic acid.

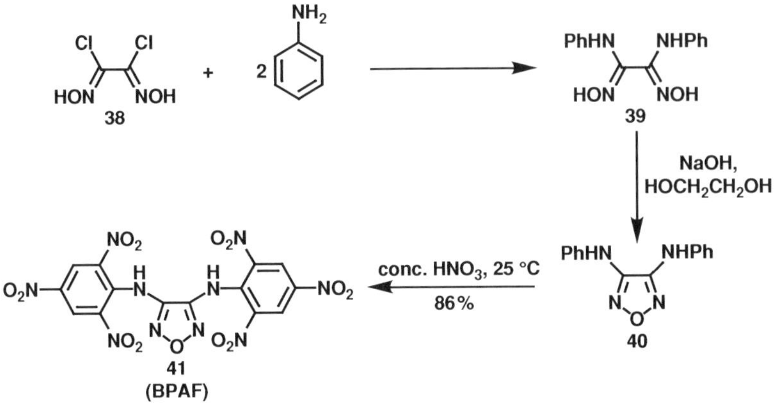

Figure 7.16

Coburn[17] also reported the synthesis of BPAF (41), the 3,4-bis-picrylamino derivative of furazan. Thus, reaction of two equivalents of aniline with 1,2-dichloroglyoxime (38) yields the bis-aniline (39), which on treatment with sodium hydroxide in ethylene glycol undergoes cyclization to the furazan (40), and nitration of the latter with concentrated nitric acid at room temperature yields BPAF (41).

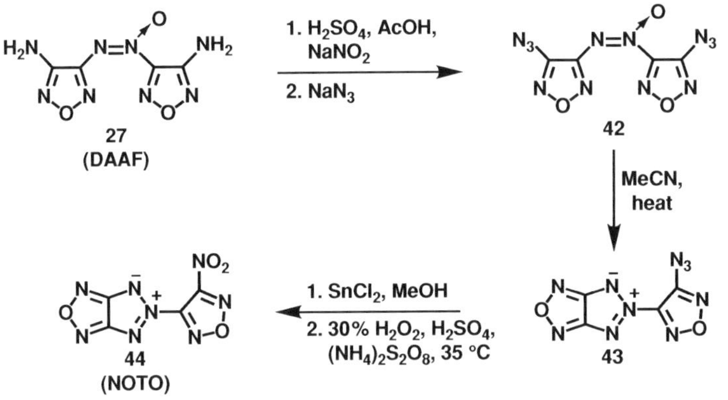

Figure 7.17

Boyer and Gunasekaren[26] reported the synthesis of the furazan-based heterocycle NOTO (44), which contains 50 % by mass of nitrogen and is a liquid at room temperature. The five-step synthesis of NOTO (44) starts from the diazotization of 4,4′-diamino-3,3′-azoxyfurazan (DAAF) (27), followed by reaction with sodium azide to form the diazide (42). Heating the diazide (42) as a solution in acetonitrile induces cyclization to the triazole (43) and this is followed by reduction and oxidation of the remaining azide group to complete the synthesis of NOTO (44).

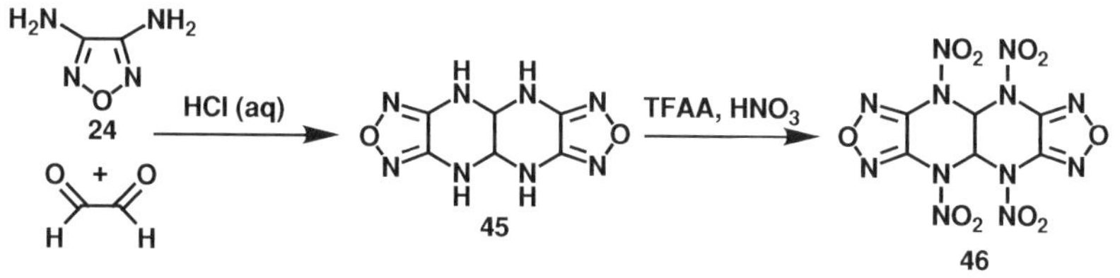

Figure 7.18

Moore and Willer[27–29] reported the synthesis of some nitramine explosives containing a furazan ring fused to a piperazine ring. The tetranitramine (46) is synthesized from the condensation of 3,4-diaminofurazan (DAF) (24) with glyoxal under acidic conditions followed by *N*-nitration of the resulting heterocycle (45). The calculated performance for the tetranitramine (46) is very high but the compound proves to be unstable at room temperature. Instability is a common feature of heterocyclic nitramines derived from the nitration of aminal nitrogens.

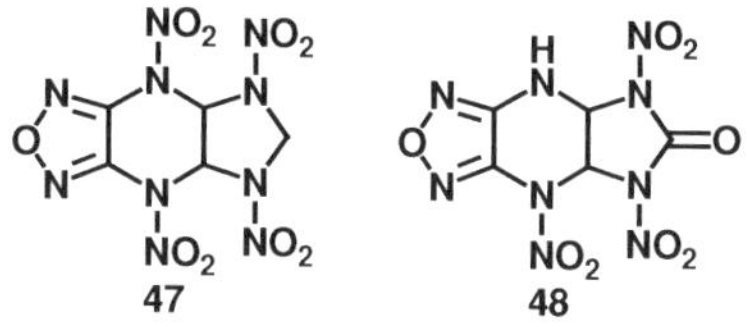

Figure 7.19

Sun and co-workers[30] synthesized the furazans (47) and (48) from the nitration of the products derived from the reaction of 3,4-diaminofurazan (DAF) (24) with *N*,*N*′-diformyl-4,5-dihydroxyimidazole and 4,5-dihydroxyimidazolid-3-one, respectively.

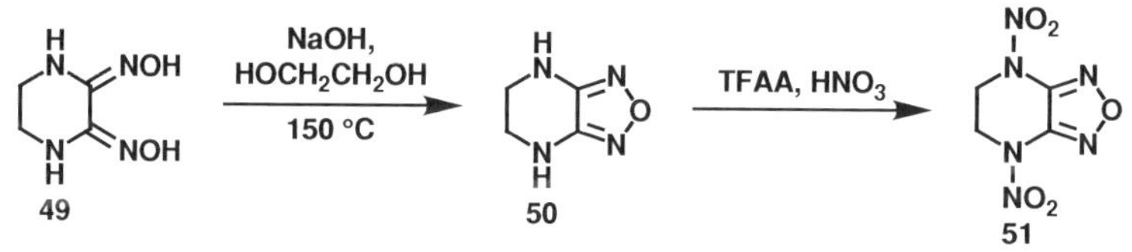

Figure 7.20

Willer[31] synthesized the bis-nitramine (51) via the cyclodehydration of the dioxime (49) with sodium hydroxide in ethylene glycol followed by subsequent nitration of the resulting heterocycle (50).

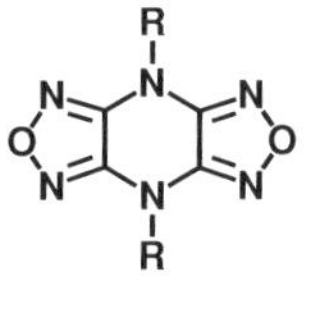

Figure 7.21

Tselinskii and co-workers[32] reported the synthesis of the bis(furazano)piperazine (52) and its nitration to the energetic bis-nitramine (53) (calculated VOD $\sim$ 9700 m/s) with nitrogen dioxide in acetonitrile. The bis-nitramine (53) is too reactive for practical use, although the bis-picryl derivative (54), formed from the dianion of (52) and picryl chloride, is more stable.

7.3.5 Benzofurazans

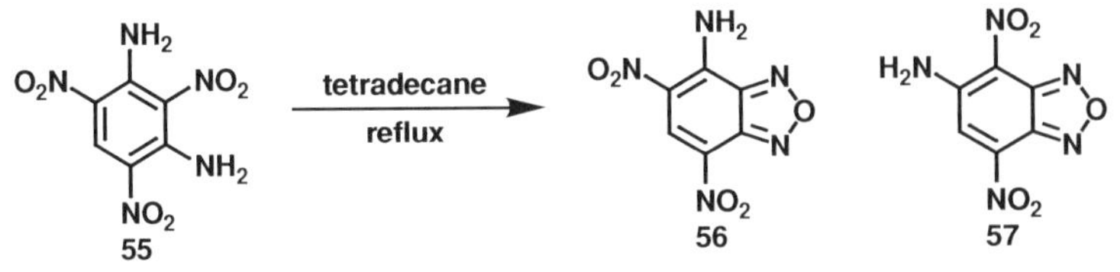

Figure 7.22

Some nitro derivatives of benzofurazan have been investigated for their explosive properties. 4-Amino-5,7-dinitrobenzofurazan (56) has been prepared[33] by a number of routes including: (1) the thermally induced cyclodehydration of 1,3-diamino-2,4,6-trinitrobenzene (55), (2) the nitration of 4-amino-7-nitrobenzofurazan and (3) the reduction of 4-amino-5,7-dinitrobenzofuroxan with triphenylphoshine. The isomeric 5-amino-4,7-dinitrobenzofurazan (57) has been prepared along similar routes.[33]

7.3.6 Furoxans

The furoxan ring is a highly energetic heterocycle whose introduction into organic compounds is a known strategy for increasing crystal density and improving explosive performance.

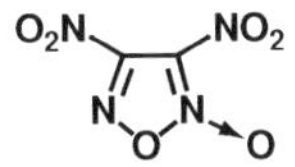

Figure 7.23

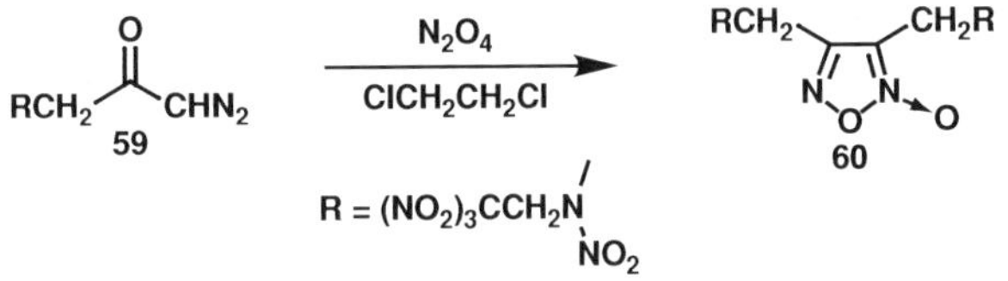

Figure 7.24

Simple nitro derivatives of furoxan have not attracted much interest for use as practical energetic materials, a consequence of their poor thermal stability and the reactivity of the nitro groups to nucleophilic displacement. 3,4-Dinitrofuroxan (DNFX) (58) has been prepared from the nitration of glyoxime followed by cyclization of the resulting dinitroglyoxime.[34] DNFX is unstable at room temperature and highly sensitive to impact. 3-Nitro-4-methylfuroxan is formed in low yield from the reaction of dinitrogen tetroxide with propylene at subambient temperature.[35] The reaction of diazoketones with dinitrogen tetroxide has been used to synthesize energetic 3,4-disubstituted furoxans like (60).[36]

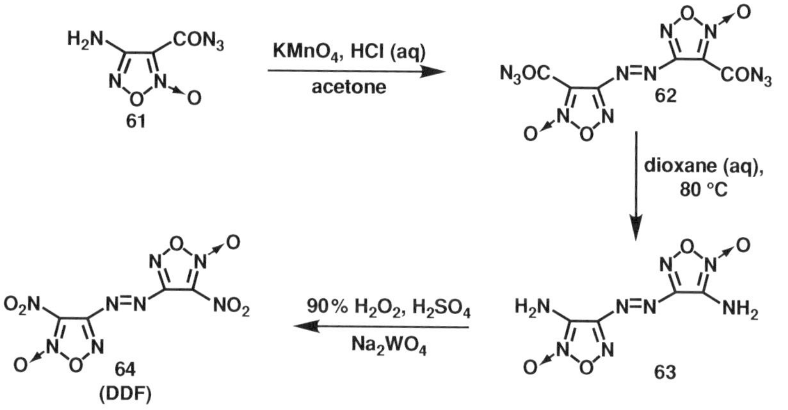

Figure 7.25

A high-energy material has recently been reported in the form of 4,4'-dinitro-3,3'-diazenofuroxan (DDF) (64).[37] This material was synthesized from the oxidative coupling of 4-amino-3-(azidocarbonyl)furoxan (61), followed by Curtius rearrangement and oxidation of the resulting amino groups to nitro groups. The experimental detonation velocity of DDF (64) reaches 10000 m/s at a crystal density of 2.02 g/cm^3. The high density of DDF is due to very efficient crystal packing.

7.3.7 Benzofuroxans

Benzofuroxans are far more stable than simple furoxans and are more favourable for practical applications. There are two standard methods for the synthesis of the benzofuroxan skeleton: (1) treating an *ortho*-nitroarylamine with a mild oxidant like sodium hypochlorite and (2) either heating or irradiating an *ortho*-nitroarylazide with UV light. Benzofuroxan itself has also been prepared by treating 1,2-benzoquinone dioxime with alkaline hypochlorite or alkaline potassium ferricyanide solution.[38]

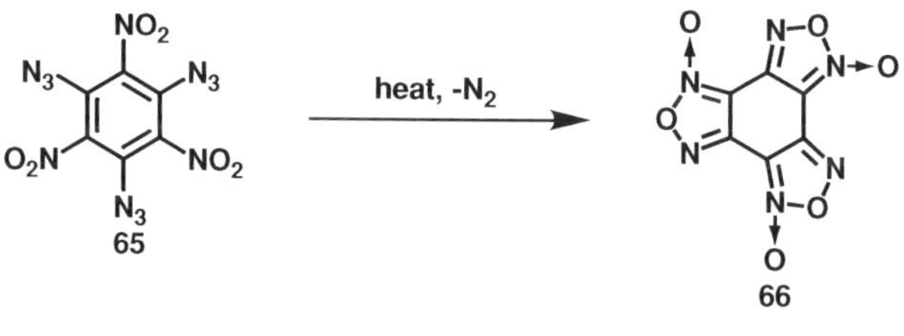

Figure 7.26

Benzenetrifuroxan (66) is a powerful explosive which was first synthesised by Turek[39] in 1931 by heating 1,3,5-triazido-2,4,6-trinitrobenzene (65) to its melting point (131 °C); the latter prepared from the reaction of 1,3,5-trichloro-2,4,6-trinitrobenzene with sodium azide in aqueous ethanol.

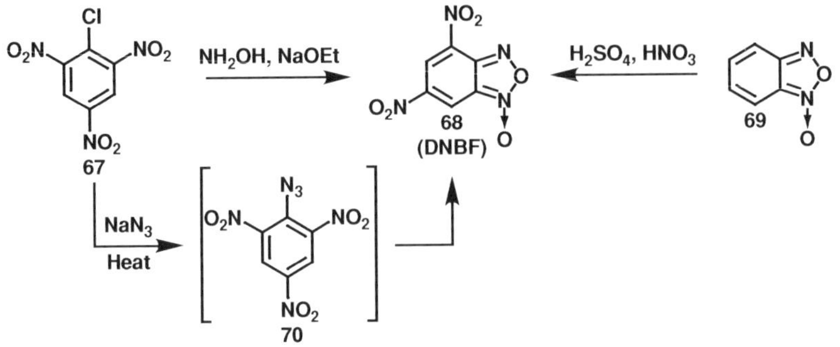

Figure 7.27

4,6-Dinitrobenzofuroxan (DNBF) (68) has been prepared from the nitration of benzofuroxan (69) with mixed acid,[40] and by treating picryl chloride (67) with sodium azide and heating the resulting picryl azide (70) in an inert solvent.[41,42]

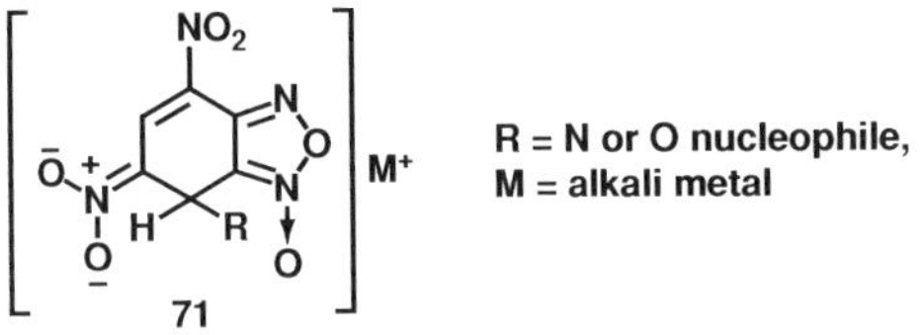

Figure 7.28

DNBF (68) readily forms stable Meisenheimer complexes (71) with numerous oxygen and nitrogen nucleophiles and some of these are primary explosives with useful initiating properties.[42–46]

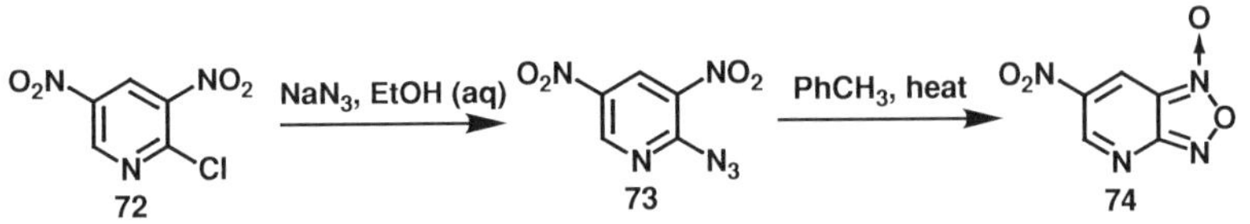

Figure 7.29

The fused pyridine-furoxan (74) has been synthesized from 3,5-dinitro-2-chloropyridine (72) via the azide (73).[47]

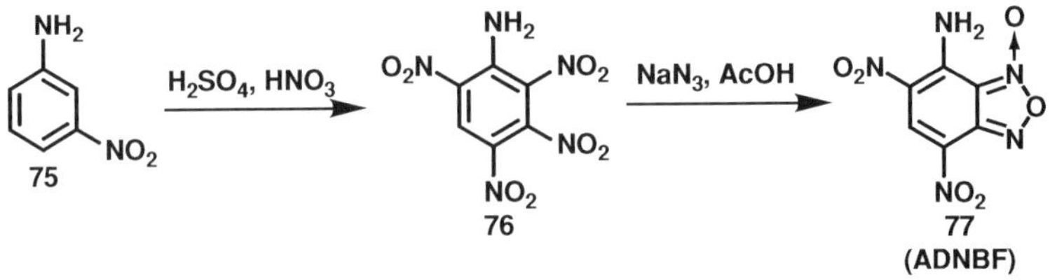

Figure 7.30

The introduction of amino functionality adjacent to a nitro group in a benzofuroxan is known to reduce impact sensitivity and increase thermal stability. Accordingly,

7-amino-4,6-dinitrobenzofuroxan (77) (ADNBF) is more impact insensitive than 4,6-dinitrobenzofuroxan. ADNBF (77) has been prepared[48] from the nitration of *m*-nitroaniline (75), followed by reaction of the product, 2,3,4,6-tetranitroaniline (76), with sodium azide in acetic acid; the latter reagent resulting in displacement of the labile *m*-nitro group of (76) and spontaneous *in situ* cyclization of the resulting *o*-nitroarylazide. ADNBF has also been synthesized from the nitration of 6-chlorobenzofuroxan, followed by reaction with ammonia in methylene chloride.[48] A more unusual route to ADNBF involves the reaction of 4,6-dinitrobenzofuroxan (DNBF) with a source of *N*-formyl anion and oxidation of the resulting Meisenheimer complex with nitric acid.[49] ADNBF has excellent overall properties (calculated VOD ~ 7900 m/s, $d = 1.90$ g/cm^3 and m.p. 270 °C) and is under advanced development in the US.

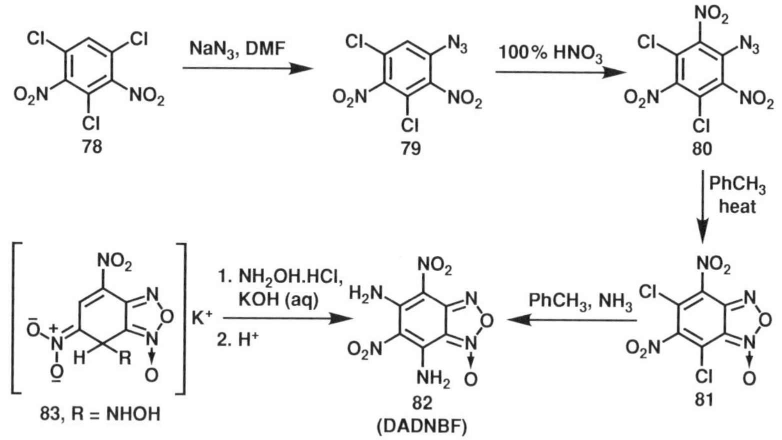

Figure 7.31

5,7-Diamino-4,6-dinitrobenzofuroxan (DADNBF) (82), an impact insensitive high performance explosive (VOD ~ 8050 m/s, $d = 1.91$ g/cm^3), has been prepared in four steps from 1,3,5-trichloro-2,4-dinitrobenzene (78),[50] and also by treating the Meisenheimer complex (83)[51] with excess hydroxylamine hydrochloride in aqueous base. DADNBF has also been synthesized in five steps starting from 2-nitroaniline.[51]

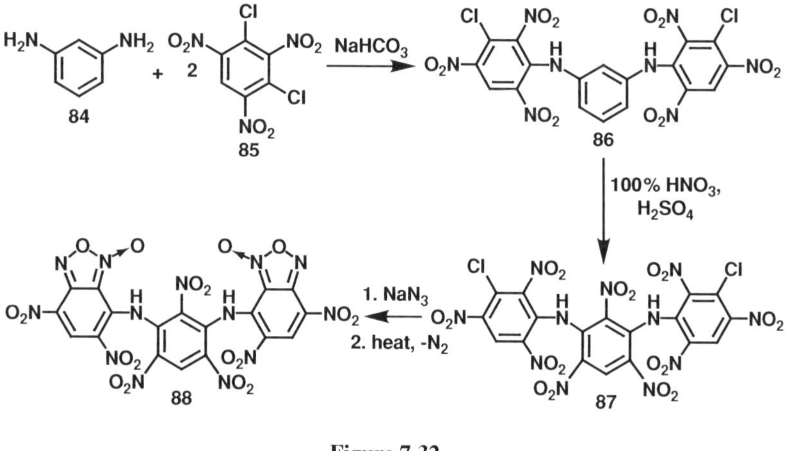

Figure 7.32

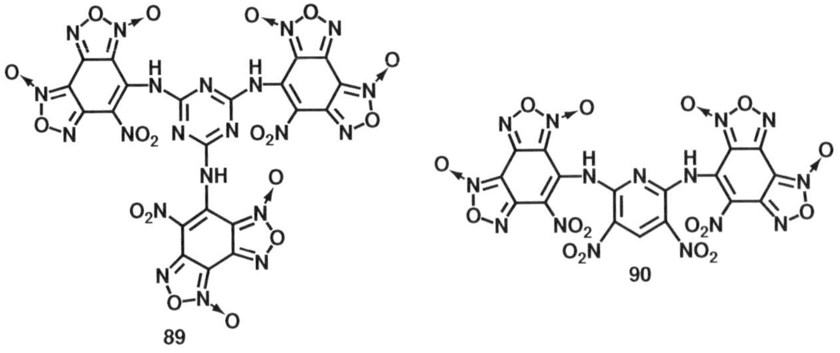

Figure 7.33

The incorporation of amino groups into high molecular weight benzofuroxans has been explored in an attempt to improve thermal stability and lower sensitivity to impact. Some of these compounds have been specifically designed to exhibit these desirable properties, and although of high molecular weight, many of these compounds can be synthesized in relatively few synthetic steps. The synthesis of (88) (calculated VOD ~ 8570 m/s, $d = 1.92$ g/cm^3) starts from the condensation of *p*-phenylenediamine (84) with two equivalents of styphnyl chloride (85), tri-nitration of the phenylenediamine rings, displacement of the chloro groups of (87) with azide nucleophile, followed by thermolysis of the resulting azide.[52,53] Other complex benzofuroxans prepared by similar routes include the triazine (89)[54] (calculated VOD ~ 8630 m/s, $d = 1.90$ g/cm^3, m.p. 231 °C) and the pyridine (90)[55] (calculated VOD ~ 8695 m/s, $d = 1.90$ g/cm^3, m.p. 316 °C).

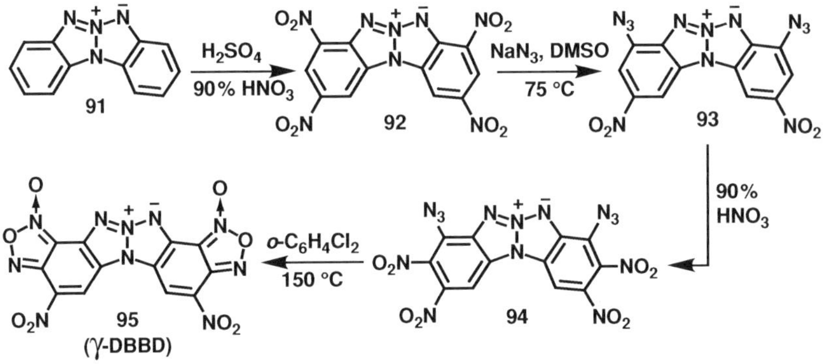

Figure 7.34

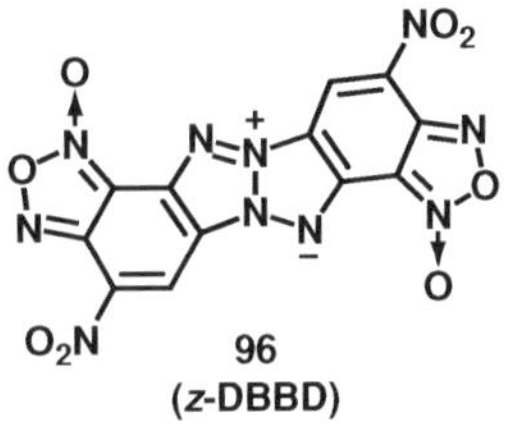

Figure 7.35

The tetraazapentalene ring system forms the core of the thermally insensitive explosive TACOT (Section 7.10) and so its fusion with the furoxan ring would be expected to enhance thermal stability and lead to energetic compounds with a high density. γ-DBBD (95) is prepared from the nitration of tetraazapentalene (91), nucleophilic displacement of the *o*-nitro groups with azide anion, further nitration to (94), followed by furoxan formation on heating in *o*-dichlorobenzene at reflux.[56] The isomeric explosive z-DBBD (96) has been prepared via a similar route.[57]

7.4 5-MEMBERED RINGS – 3N

7.4.1 Triazoles

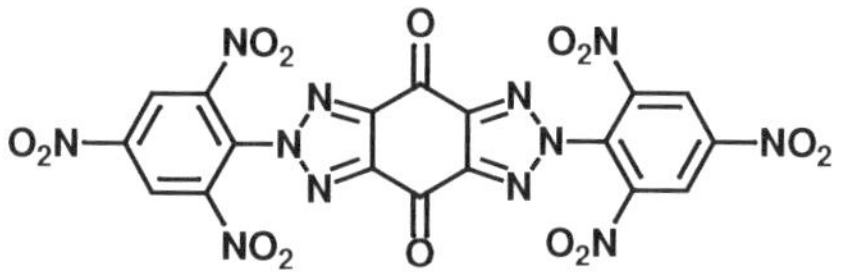

Figure 7.36

Incorporation of a triazole ring into a compound is a known strategy for increasing thermal stability. Many triazole compounds show high thermal stability coupled with a low sensitivity to shock and impact. 2,6-Dipicrylbenzo[1,2-*d*][4,5-*d*′]bistriazole-4,8-dione (97) (m.p. 430 °C) is one such example.[58]

Analyses of the structures and properties of a large number of energetic materials reveal that a combination of amino and nitro groups in a molecule often leads to better thermal stability, lower sensitivity to shock and impact, and increased explosive performance because of an increase in crystal density. Such observations are attributed to both intermolecular and intramolecular hydrogen bonding interactions between adjacent amino and nitro groups. Some modern triazole-based explosives have been designed and synthesized with this in mind.

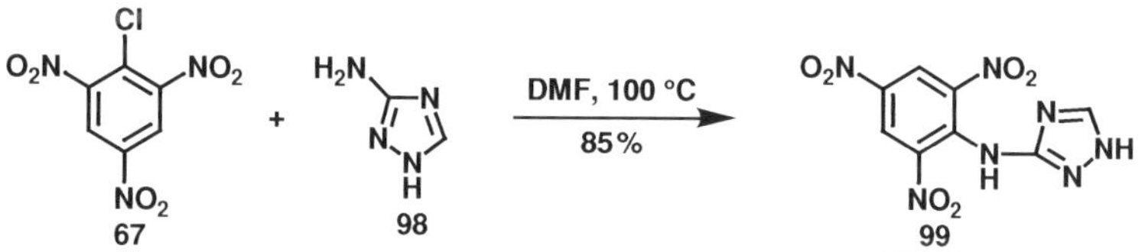

Figure 7.37

3-Amino-1,2,4-triazole is a useful starting material for the synthesis of many 1,2,4-triazole-based explosives. Jackson and Coburn[59] synthesized a number of picryl- and picrylamino-substituted 1,2,4-triazoles. PATO (99) is synthesized from the reaction of 3-amino-1,2,4-triazole (98) with picryl chloride (67).[59,60] PATO has also been synthesized from the reaction of 3-amino-1,2,4-triazole with *N*,2,4,6-tetranitromethylaniline (tetryl).[61] PATO has a low sensitivity to impact and is thermally stable up to 310 °C. PATO (VOD ~ 7469 m/s) exhibits lower performance to TATB (VOD ~ 8000 m/s) which is the common benchmark standard for thermal stability and insensitivity in explosives.

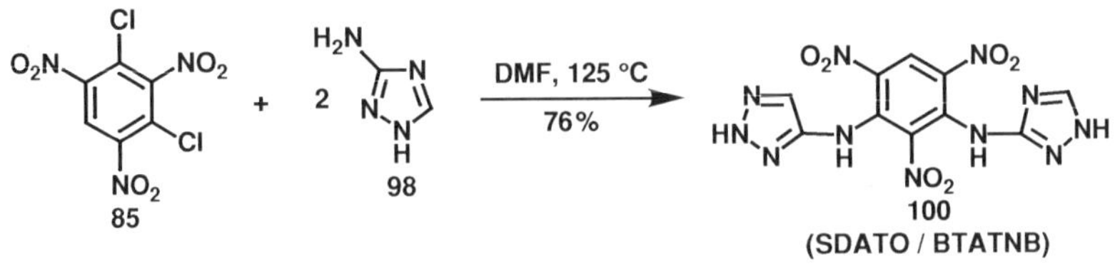

Figure 7.38

Agrawal and co-workers[62] synthesized 1,3-bis(1,2,4-triazol-3-amino)-2,4,6-trinitrobenzene (SDATO or BTATNB) (100) from the reaction of two equivalents of 3-amino-1,2,4-triazole (98) with styphnyl chloride (85). The performance of SDATO (calculated VOD ~7609 m/s) is slightly higher than PATO while showing more insensitivity to impact.

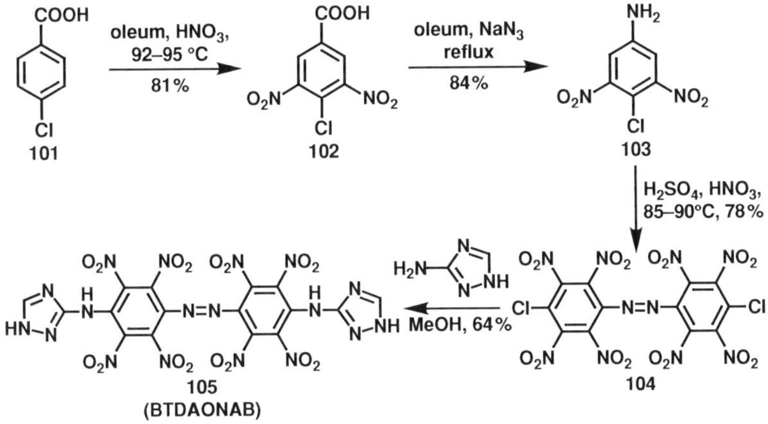

Figure 7.39

N,*N*′-Bis(1,2,4-triazol-3-yl)-4,4′-diamino-2,2′,3,3′,5,5′,6,6′-octanitroazobenzene (BTDAONAB) (105) has recently been synthesized by Agrawal and co-workers[63] by tandem nitration–oxidative coupling of 4-chloro-3,5-dinitroaniline (103) followed by displacement of the chloro groups with 3-amino-1,2,4-triazole. This is a thermally stable explosive with some impressive properties, exceeding TATB in both thermal stability and explosive performance (VOD ~ 8321 m/s, $d = 1.97\ g/cm^3$). This compound doesn't melt and the DTA exotherm is not seen until 550 °C.

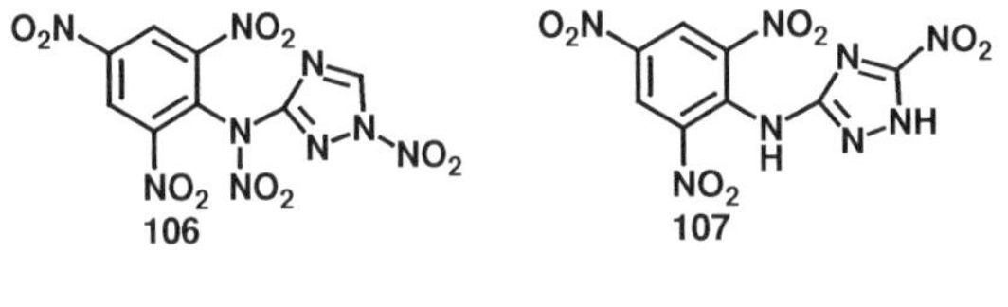

Figure 7.40

C-Nitration of 1,2,3-triazole and 1,2,4-triazole rings can be achieved with either mixed acid or solutions of nitric acid in acetic anhydride. *N*-Nitration is usually achieved with nitric acid in acetic anhydride at ambient to subambient temperatures. Thermal rearrangement of the *N*-nitro product to the more stable *C*-nitro product often occurs at higher nitration temperature.

This is seen during the nitration of PATO (99), which on treatment with a mixture of nitric acid in acetic anhydride at 40 °C for 30 minutes yields the *N*-nitro product (106), whereas the same reaction at 60 °C for 1.5 hours yields the *C*-nitro product (107).[61] An explosive known as PANT (109) has been prepared from the reaction of 4-amino-1,2,3-triazole (108) with picryl chloride followed by *C*-nitration of the 1,2,3-triazole ring with mixed acid at room temperature.[64]

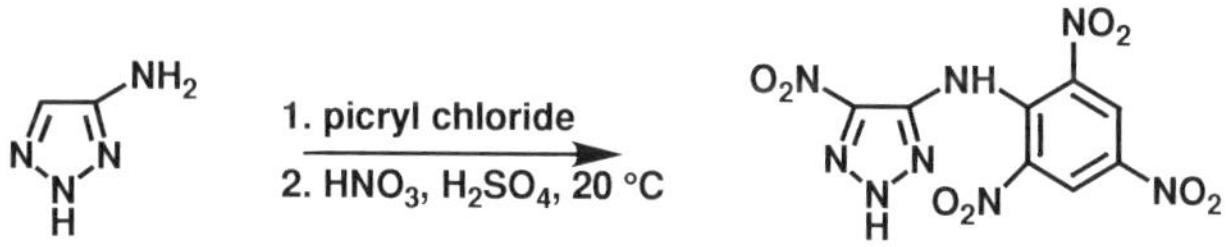

1. picryl chloride
2. HNO_3, H_2SO_4, 20 °C

108

109 (PANT)

Figure 7.41

Amino derivatives of 1,2,3- and 1,2,4-triazoles are useful precursors to the corresponding nitro-substituted triazoles. 3-Amino-1,2,4-triazole (98) undergoes diazotization on reaction with nitrous acid; the resulting diazonium salt (110) can react with a range of nucleophiles, including an aqueous solution of sodium nitrite which yields 3-nitro-1,2,4-triazole (111).[65,66] Diazotization of 3,5-diamino-1,2,4-triazole (112), followed by heating with an aqueous solution of sodium nitrite, yields 3,5-dinitro-1,2,4-triazole (113).[67,68]

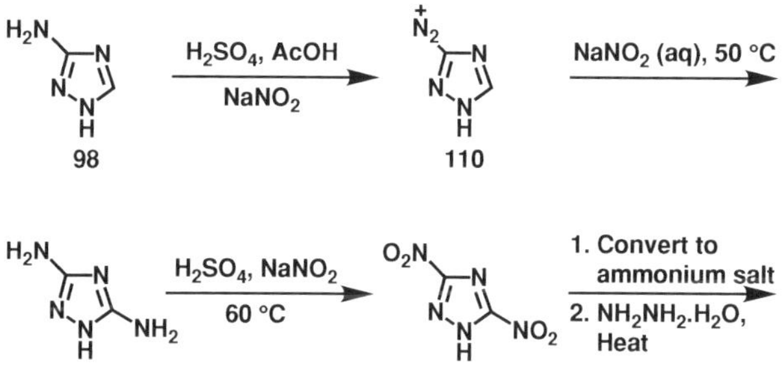

H_2SO_4, AcOH / $NaNO_2$

$NaNO_2$ (aq), 50 °C

98 110 111 57% (2 steps)

H_2SO_4, $NaNO_2$ / 60 °C

1. Convert to ammonium salt
2. $NH_2NH_2.H_2O$, Heat

112 113 114 (ANTA) 50% (3 steps)

Figure 7.42

Treatment of the ammonium salt of 3,5-dinitro-1,2,4-triazole (113) with hydrazine hydrate leads to selective reduction of one of the nitro groups to yield 3-amino-5-nitro-1,2,4-triazole (ANTA) (114), a high performance explosive (calculated VOD ~ 8460 m/s) possessing thermal stability (m.p. 238 °C) and an extremely low sensitivity to impact.[68] ANTA (114) is also synthesized[69] from the nitration of 3-acetyl-1,2,4-triazole with anhydrous nitric acid in acetic anhydride at subambient temperature followed by hydrolysis of the acetyl functionality. The ammonium salt of 3,5-dinitro-1,2,4-triazole (113) is itself a useful explosive which forms a eutectic with ammonium nitrate.[70]

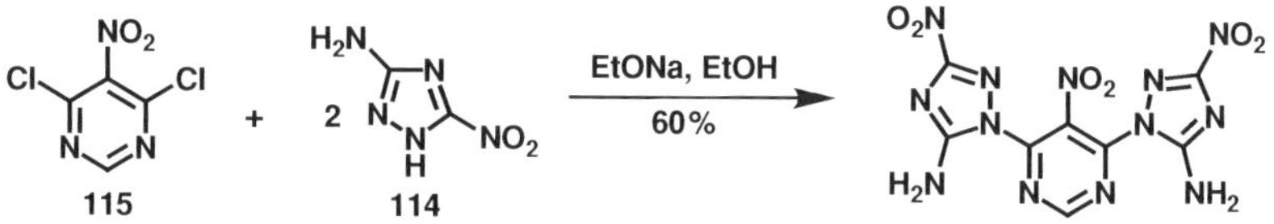

115 + 2 114

EtONa, EtOH / 60%

116 (DANTNP)

Figure 7.43

ANTA (114) readily forms a stable anion on reaction with bases like sodium ethoxide and this anion has been used as a nucleophile for the synthesis of many ANTA derivatives. Laval and co-workers[71] synthesized DANTNP (116) (calculated VOD ~ 8120 m/s, $d = 1.84\ g/cm^3$, m.p. > 330 °C) from the reaction of 4,6-dichloro-5-nitropyrimidine (115) with two equivalents of ANTA (114) in the presence of sodium ethoxide. Agrawal and co-workers[72] studied the thermal and explosive properties of both ANTA and DANTNP and suggested their use for applications in propellant/explosive formulations where insensitivity coupled with thermal stability is of prime importance. The activation energies of ANTA and DANTNP indicate that DANTNP is more thermally stable than ANTA.

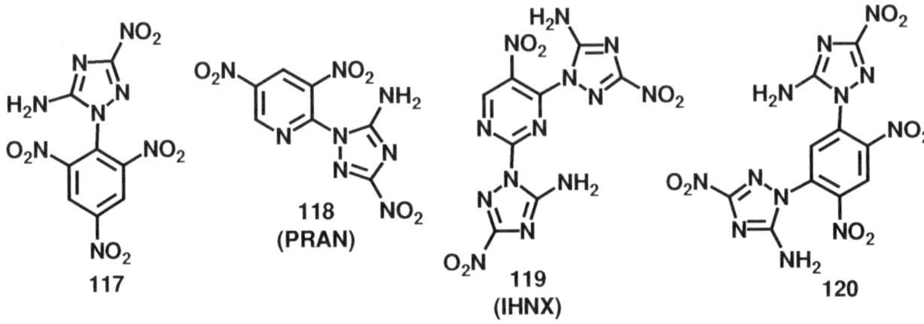

117 118 (PRAN) 119 (IHNX) 120

Figure 7.44

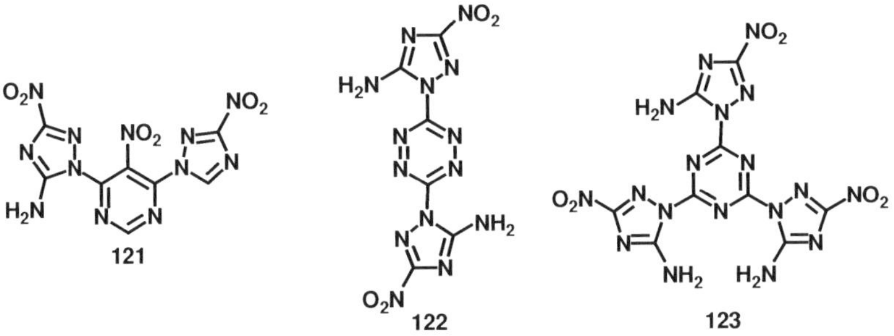

121 122 123

Figure 7.45

Pagoria and co-workers[2] synthesized a number of thermally stable explosives from the reaction of the sodium salt of ANTA with chloro-substituted arylenes and *N*-heterocycles. These include the synthesis of (117) from picryl chloride, PRAN (118) from 2-chloro-3,5-dinitropyridine, IHNX (119) from 2,4-dichloro-5-nitropyrimidine, (120) from 1,5-dichloro-2,4-dinitrobenzene, and (121) from 4-chloro-6-(3-nitro-1,2,4-triazolyl)-5-nitropyrimidine. Coburn and co-workers[68] reported the synthesis of the tetrazine (122) and the triazine (123) from the reaction of the sodium salt of ANTA with 3,6-dichlorotetrazine and cyanuric chloride respectively.

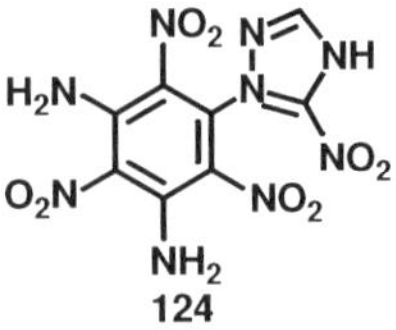

124

Figure 7.46

Laval and Vignane[73] reported the synthesis of the nitrotriazole (124) from the reaction of 3-nitro-1,2,4-triazole with 3,5-diamino-1-chloro-2,4,6-trinitrobenzene. The nitrotriazole (124) is a useful secondary high explosive, exhibiting high performance and a low sensitivity to impact.

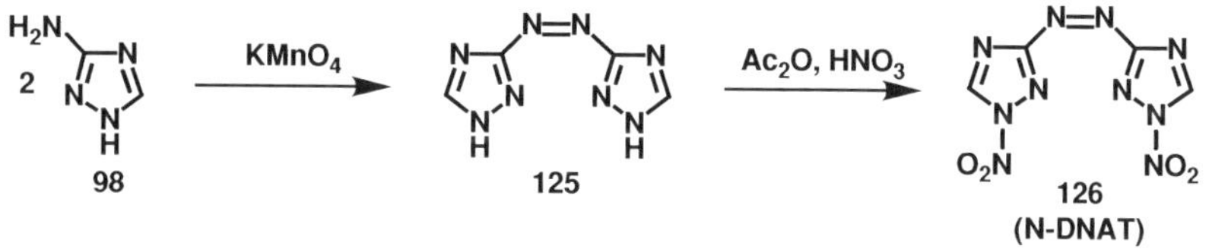

Figure 7.47

Chemists at Los Alamos National Laboratory (LANL) have shown interest in the 1,2,4-triazole ring as a component of new energetic materials. During their study 1,1'-dinitro-3,3'-azo-1,2,4-triazole (N-DNAT) (126), a potential additive for high-energy propellant formulations, was synthesized from the oxidative coupling of 3-amino-1,2,4-triazole (98), followed by *N*-nitration of the resulting product (125) with nitric acid in acetic anhydride.[74] *N*-Nitrotriazoles are less thermally stable than their *C*-nitro isomers and so the isomeric C-DNAT (127) is an important target. The thermal rearrangement of N-DNAT (126) to C-DNAT (127) was not successful under the conditions employed.[75] C-DNAT (127) was later synthesized from the oxidative coupling of the potassium salt of ANTA with potassium permanganate. C-DNAT (127) burns without smoke or residue and has high potential for use in advanced propellant formulations.[68]

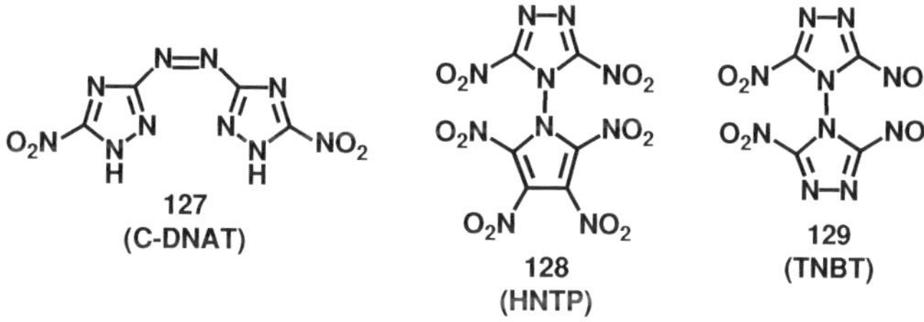

Figure 7.48

Thermodynamic calculations conducted at Los Alamos National Laboratory also identified HNTP (128) and TNBT (129) as target molecules likely to exhibit high performance.[75]

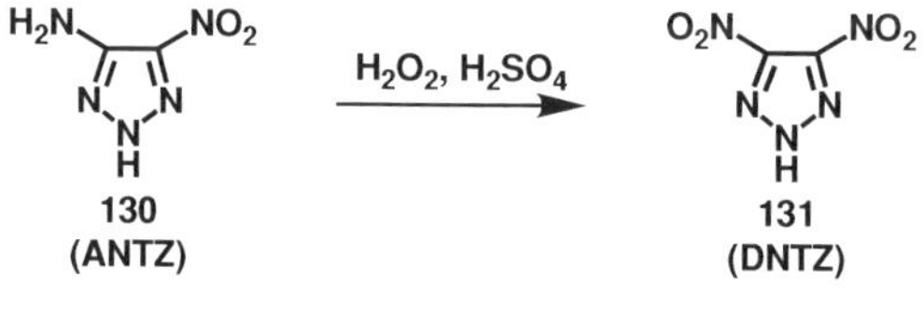

Figure 7.49

Baryshnikov and co-workers[76] synthesized some nitro-substituted 1,2,3-triazoles using an ingenious cycloaddition reaction between sodium azide and 1,1-dinitroethene; the latter prepared *in situ* from a number of precursors including 2,2-dinitroethyl acetate (Section 1.10.2.3).

4-Amino-5-nitro-1,2,3-triazole (ANTZ) (130), an explosive showing high thermal stability, has been synthesized via this route; the reaction of sodium azide, acetaldehyde and 2,2-dinitroethyl acetate forming 4-methyl-5-nitro-1,2,3-triazole, which on conversion of the methyl group to an amino group yields ANTZ (130). Treatment of ANTZ (130) with hydrogen peroxide in sulfuric acid yields 4,5-dinitro-1,2,3-triazole (DNTZ) (131).

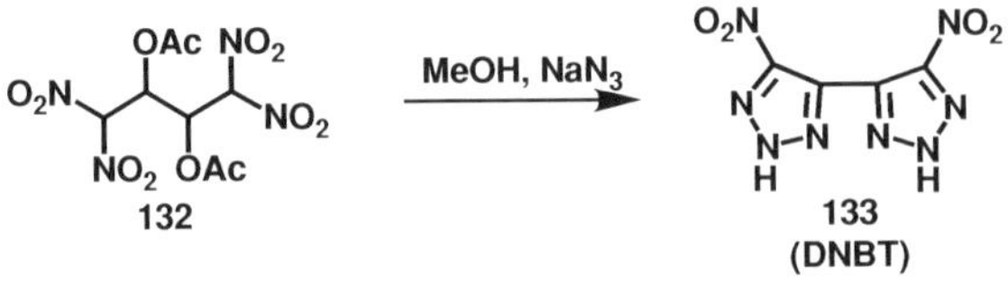

OAc NO2 O2N NO2 NO2 OAc 132 —MeOH, NaN3→ O2N NO2 N N N-N N-N H H 133 (DNBT)

Figure 7.50

Baryshnikov and co-workers[77] used the same methodology for the synthesis of 5,5'-dinitro-4,4'-bis(1,2,3-triazole) (133) (DNBT) from 1,1,4,4-tetranitro-2,3-butanediol diacetate (132) in the presence of sodium azide.

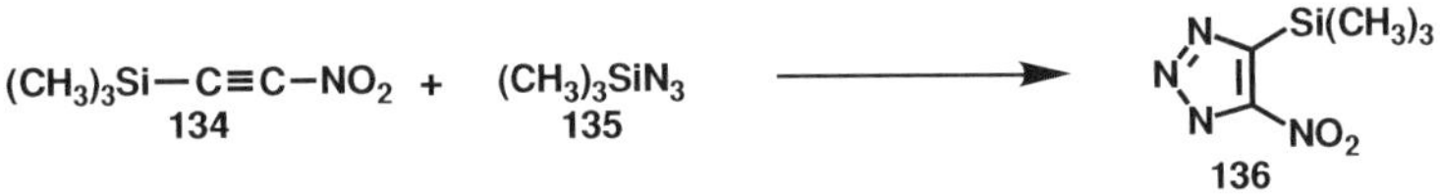

$(CH_3)_3Si-C{\equiv}C-NO_2$ + $(CH_3)_3SiN_3$ → 136
134 135

Figure 7.51

Gilardi and co-workers[78] reported a synthesis of 4-(trimethylsilyl)-5-nitro-1,2,3-triazole (136) via a cycloaddition between 1-nitro-2-(trimethylsilyl)acetylene (134) and trimethylsilyl azide (135). This may provide a route to 4,5-dinitro-1,2,3-triazole via nitrodesilylation or lead to the synthesis of 4-amino-5-nitro-1,2,3-triazole, an isomer of ANTA.

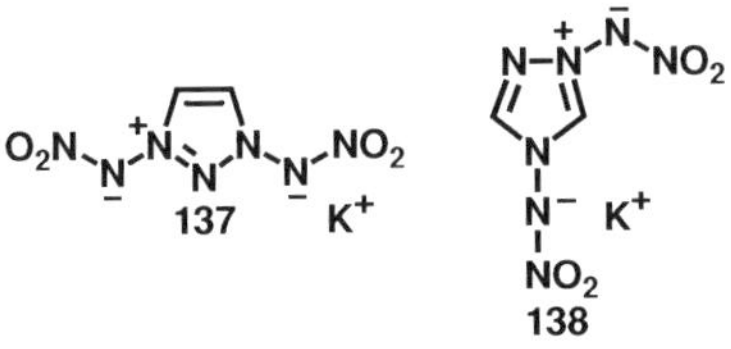

137 K^+ 138 K^+

Figure 7.52

The two energetic *N*-nitroimide explosives (137) and (138) have been prepared from 1,2,3-triazole and 1,2,4-triazole respectively.[79,80] These and other *N*-nitroimide-based energetic compounds are discussed in more detail in Section 6.11.3.

7.4.2 Triazolones

3-Nitro-1,2,4-triazol-5-one (NTO) (140) is prepared from the nitration of 1,2,4-triazol-5-one (139) under a variety of conditions; the latter prepared from the condensation of semicarbazide

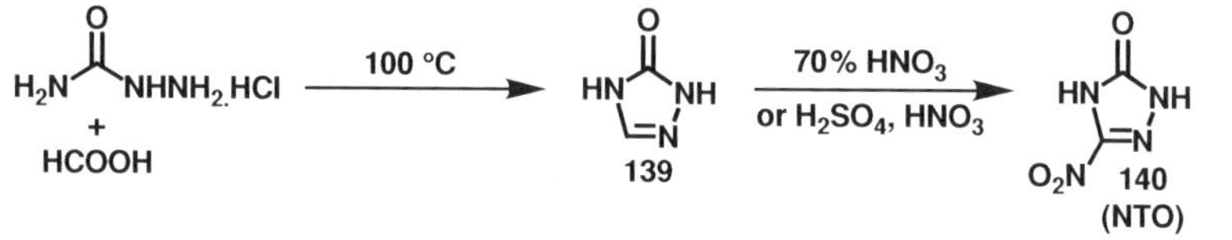

Figure 7.53

hydrochloride with formic acid.[81–83] The facile synthesis of NTO from readily available starting materials, coupled with its high performance (VOD $\sim$ 8510 m/s, d = 1.91 g/cm^3), and properties which classify it as an insensitive high explosive (IHE), makes NTO a very attractive explosive for use in insensitive munitions, either cast, pressed or in a plastic bonded matrix. NTO is much less sensitive to impact than both the widely used military explosives RDX and HMX.[84] The French first used NTO in combination with HMX in plastic bonded explosives.[85] NTO is now widely used alone or in combination with HMX or RDX for use as a filling for insensitive munitions.[81,86] AFX644 is an explosive composition based on NTO (40 %), TNT (30 %), Wax (10 %) and Al powder (20 %) which has recently been used by the USAF as a low vulnerability general purpose filler for bombs and is classified as an extremely insensitive detonating substance (EIDS). Agrawal[87] has recently reviewed the advances made in the use of NTO in munitions and explosive formulations. Spear and co-workers[83] have reviewed the history of NTO and its structural, chemical, explosive and thermal properties, together with its synthesis.

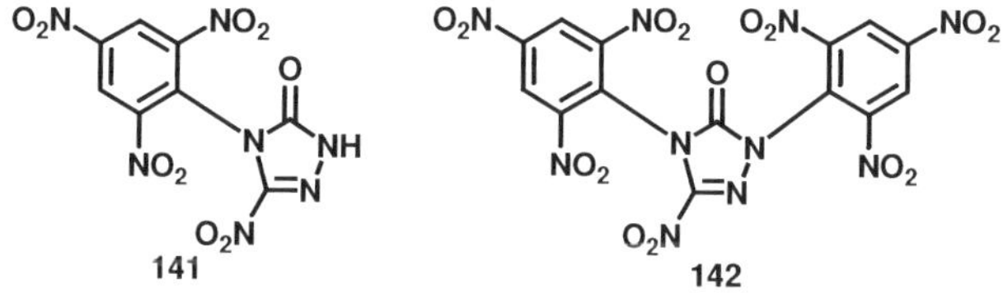

Figure 7.54

The acidity of 1-H in NTO (140) has been exploited by a number of researchers in this field. Many amine salts of NTO have been reported including the ammonium, hydrazinium, guanidinium, aminoguanidinium, diaminoguanidinium, triaminoguanidinium and 1,2-ethylenediammonium salts.[88] Metal salts of NTO have been prepared by the addition of metal hydroxides or carbonates to aqueous solutions of NTO.[89] The physical properties of these salts have been thoroughly investigated[89]. Coburn and Lee[90] used NTO as a nucleophile, treatment of NTO (140) with a mixture of picryl fluoride and 1-methyl-2-pyrrolidinone yielding either (141) or (142) depending on the reaction stoichiometry.

7.4.3 Benzotriazoles

1-Picryl-5,7-dinitro-2H-benzotriazole (BTX) (145) was first synthesized by Coburn and co-workers[91] from the reaction of benzotriazole with picryl chloride in DMF, followed by nitration with mixed acid. Gilardi and co-workers[92] reported an alternative synthesis of BTX involving treatment of 2-amino-4,6-dinitrodiphenylamine (143) with a solution of nitrous acid, which leads to an intramolecular cyclisation via the diazonium salt to give (144), nitration of the

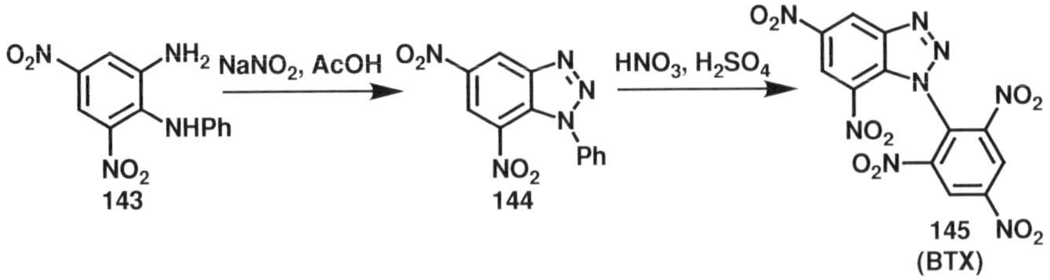

Figure 7.55

latter with mixed acid yields BTX (145). The reaction of (144) with weaker nitrating agents has been used to synthesize derivatives where the phenyl ring is either mono- or di-nitrated.[92] BTX has been suggested for use in thermally stable detonators, although its high sensitivity to impact is a distinct disadvantage.

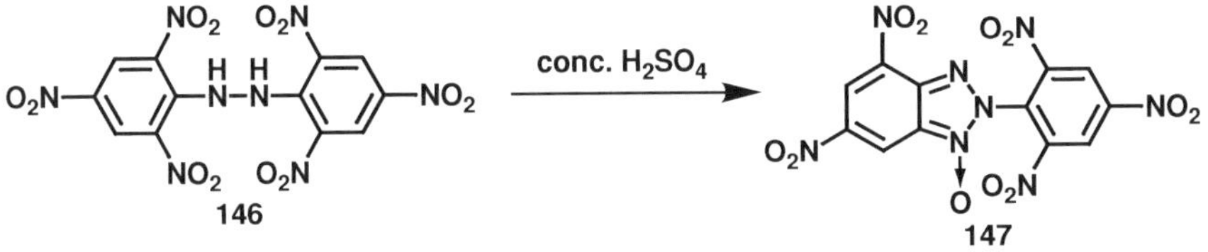

Figure 7.56

The benzotriazene-1-oxide (147) has been synthesized from the intramolecular acid-induced cyclisation of *N,N′*-dipicrylhydrazine (146).[93]

7.5 5-MEMBERED RINGS – 4N

The high nitrogen content and the endothermic nature of the tetrazole ring lends itself to the synthesis of energetic materials. Compounds such as 1-*H* tetrazole and 5-aminotetrazole can be used as nucleophiles to incorporate the tetrazole ring into other molecules. 5-Aminotetrazole is synthesized from the reaction of dicyandiamide with sodium azide in hydrochloric acid.

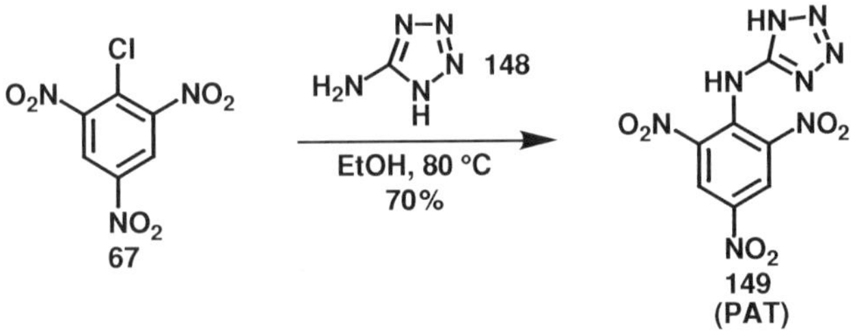

Figure 7.57

Agrawal and co-workers reported the synthesis of two tetrazole-based explosives, namely, 5-picrylamino-1,2,3,4-tetrazole[94] (PAT) (149) and 5,5'-styphnylamino-1,2,3,4-tetrazole[95] (SAT) (150) from the reaction of 5-amino-1,2,3,4-tetrazole (148) with picryl chloride and styphnyl chloride respectively. These explosives have been studied for their thermal and explosive properties. The thermal stability of SAT (exotherm peak at 123 °C) is lower than PAT (exotherm peak at 185 °C), which is possibly attributed to the decreased electron-withdrawing power of the picryl group by being attached to two tetrazole units. PAT and SAT have calculated VODs of 8126 m/s and 8602 m/s respectively.[95]

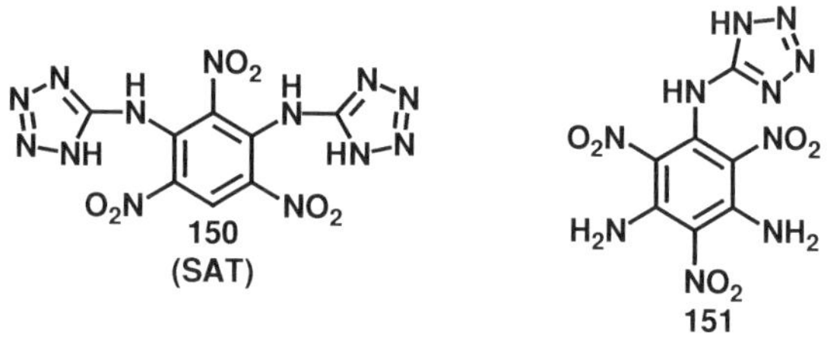

150 (SAT)

151

Figure 7.58

The reaction of 5-aminotetrazole with 3,5-diamino-2,4,6-trinitrofluorobenzene generates the energetic tetrazole (151).[96]

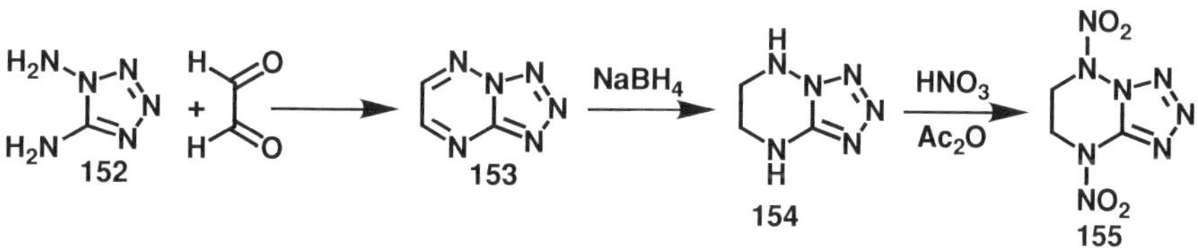

152 153 154 155

NaBH$_4$ HNO$_3$ Ac$_2$O

Figure 7.59

The bis-nitramine (155) has been prepared from the reaction of 1,5-diaminotetrazole (152) with glyoxal, followed by reduction of the resulting fused heterocycle (153) with sodium borohydride and subsequent *N*-nitration of the piperazine nitrogens of (154).[97]

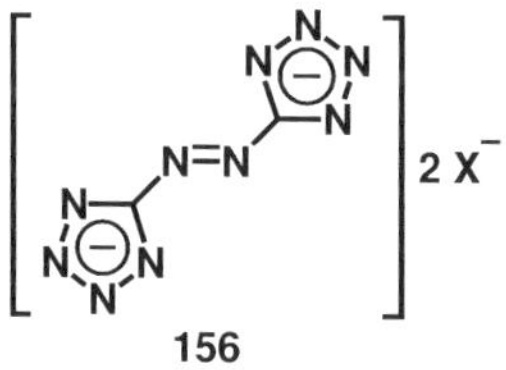

156

Figure 7.60

Salts of azotetrazole (156) are energetic compounds, the guanidinium and methylammonium salts finding use as gas generators when mixed with inorganic oxidizers.[98,99] Thiele[100,101] prepared alkali, alkaline earth and heavy metal salts of azotetrazole, the lead salt exhibiting initiating properties. The dihydrazinium salt of azotetrazole is prepared[102] from the reaction of the barium salt of azotetrazole with an aqueous solution of hydrazinium sulfate. This salt is highly energetic with a very high heat of formation and one of the highest nitrogen contents (85 %) reported for any organic compound.

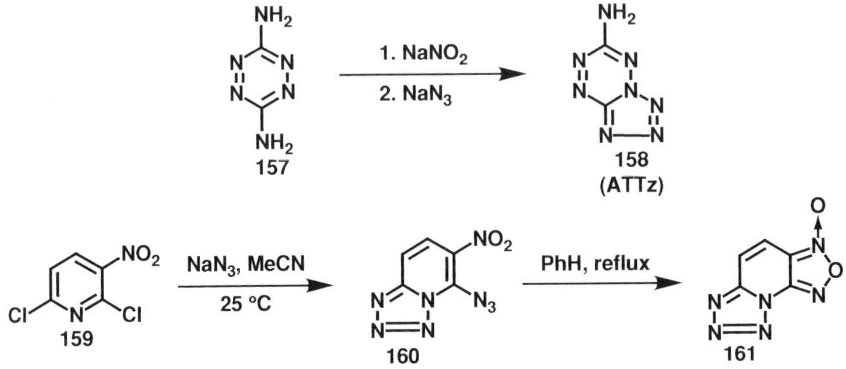

Figure 7.61

The tetrazole ring can be synthesized from the 1,3-dipolar cycloaddition of an azide with a cyanide group or a similar nitrogen dipolarophile. Tri-*n*-butyltin azide and trimethylsilyl azide have been used as organic alternatives to the azide anion. Ammonium azide is frequently used and is generated *in situ* from sodium azide and ammonium chloride in DMF. The fused heterocycle (158) (ATTz) has been synthesized from the reaction of 3,6-diaminotetrazine (157) with nitrous acid and then sodium azide.[103] The heterocycle (161), incorporating both a tetrazole and furoxan group, has been prepared from the reaction of 3-nitro-2,6-dichloropyridine (159) with two equivalents of azide anion followed by heating the product (160) under reflux in benzene.[47]

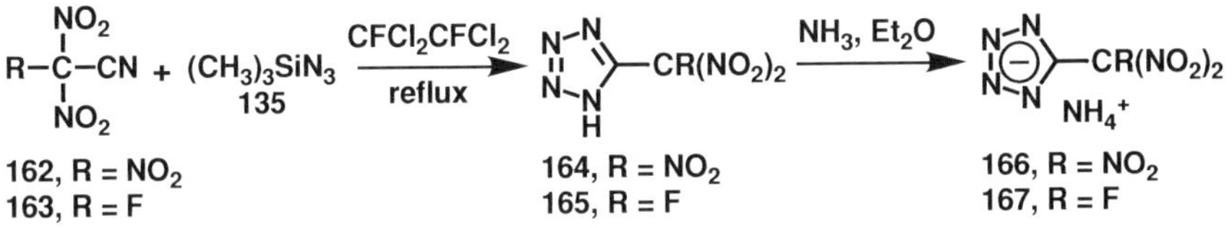

Figure 7.62

5-(Trinitromethyl)tetrazole (166) and 5-(dinitrofluoromethyl)tetrazole (167) have been synthesized and isolated as their ammonium salts from the reactions of trimethylsilyl azide with the corresponding nitriles followed by reaction with ammonia in diethyl ether.[104]

5-Nitrotetrazole is readily prepared from the diazotization of 5-aminotetrazole in the presence of excess sodium nitrite and is best isolated as the copper salt complex with ethylenediamine.[105] The salts of 5-nitrotetrazole have attracted interest for their initiating properties. The mercury salt is a detonating primary explosive.[105] The amine salts of 5-nitrotetrazole are reported to form useful eutectics with ammonium nitrate.[106]

7.6 6-MEMBERED RINGS – 1N – PYRIDINES

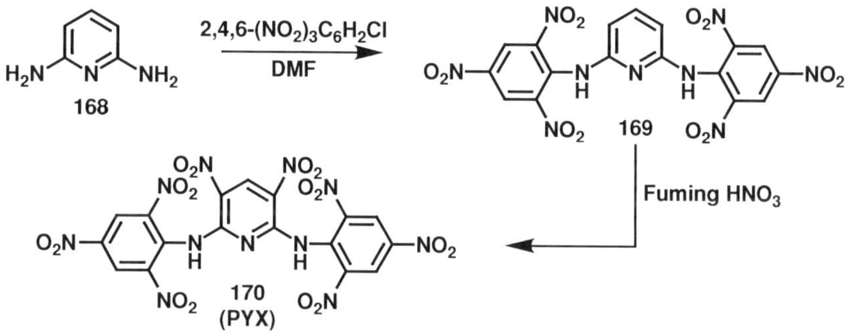

Figure 7.63

The synthesis of polynitro derivatives of pyridine and other six-membered nitrogen heterocycles by electrophilic aromatic substitution is often not feasible due to electron deficiency in these rings; pyridine is nitrated to 3-nitropyridine only under the most vigorous nitrating conditions. A solution to this problem is to incorporate electron-releasing groups into the pyridine ring. Such an approach is seen during the synthesis of the thermally stable explosive known as PYX (170); 2,6-bis(picrylamino)pyridine (169) is di-nitrated at the 3- and 5-positions under relatively mild nitrating conditions.[107,108] PYX (VOD ~ 7450 m/s, $d = 1.75$ g/cm^3, m.p. 460 °C) is now commercialized in the US and is manufactured by Chemtronics under licence from Los Alamos National Laboratory for use in thermally stable perforators for oil and gas wells.[109]

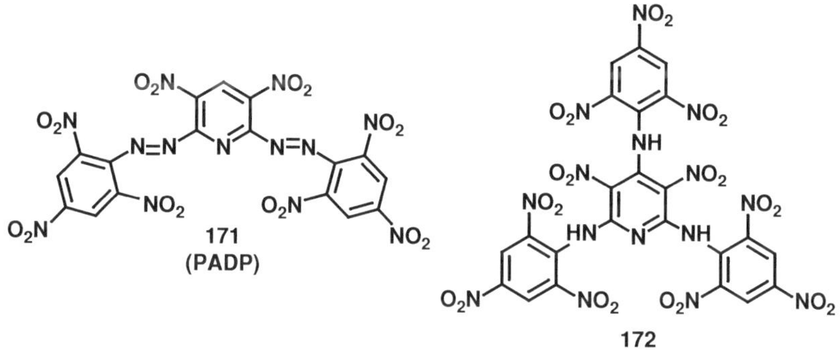

Figure 7.64

PADP (171), an explosive synthesized from the reaction of 3,5-dinitro-2,6-bis-(hydrazino)pyridine with picryl chloride in DMF followed by oxidation with nitric acid, also exhibits high thermal stability.[110] 2,4,6-Tris(picrylamino)-3,5-dinitropyridine (172) exhibits much lower thermal stability than both PYX (170) and PADP (171), a consequence of increased steric crowding around the pyridine ring.[108]

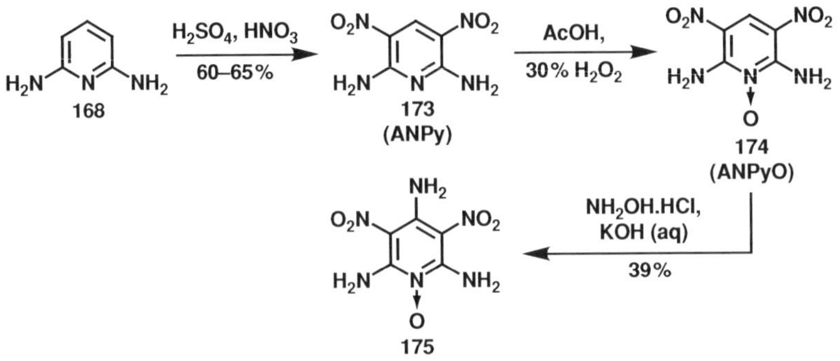

Figure 7.65

The direct nitration of 2,6-diaminopyridine (168) with mixed acid yields 2,6-diamino-3,5-dinitropyridine (ANPy) (173).[111] Oxidation of ANPy (173) with peroxyacetic acid yields ANPyO (174) (calculated VOD $\sim$ 7840 m/s, $d = 1.88$ g/cm^3).[112] *C*-Amination of ANPyO (174) with hydroxylamine hydrochloride in aqueous base yields the triamine (175), an impact insensitive explosive of high thermal stability.[113]

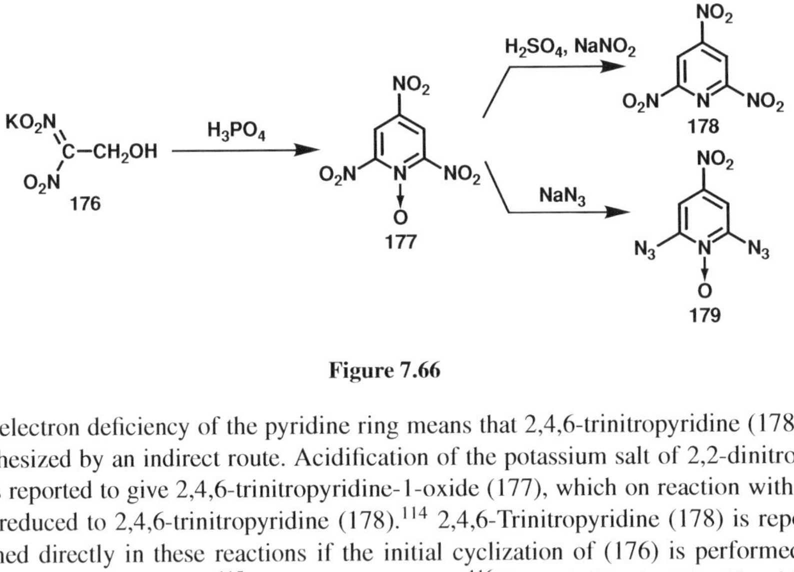

Figure 7.66

The electron deficiency of the pyridine ring means that 2,4,6-trinitropyridine (178) has to be synthesized by an indirect route. Acidification of the potassium salt of 2,2-dinitroethanol (176) is reported to give 2,4,6-trinitropyridine-1-oxide (177), which on reaction with nitrous acid is reduced to 2,4,6-trinitropyridine (178).[114] 2,4,6-Trinitropyridine (178) is reported to be formed directly in these reactions if the initial cyclization of (176) is performed in the presence of dilute nitric acid[115] or 2,2-dinitroethanol[116] is used directly. The *N*-oxide (177) is susceptible to nucleophilic substitution at the 2- and 6-positions, treatment of (177) with sodium azide yielding the energetic diazide (179).[114]

7.7 6-MEMBERED RINGS – 2N

Pyrazine and pyrimidine heterocycles, like pyridine, are electron deficient and need the presence of an activating/electron-releasing group to allow efficient electrophilic nitration to occur. An example of this strategy is seen during the synthesis of 2,6-diamino-3,5-dinitropyrazine (ANPz) (183) where one of the chloro groups of 2,6-dichloropyrazine (180) is substituted for a methoxy group which then allows nitration in the 3- and 5-positions under moderate conditions to yield (182), nucleophilic displacement of both methoxy and chloro groups with ammonium hydroxide in acetonitrile yielding 2,6-diamino-3,5-dinitropyrazine (ANPz) (183).[7] Treatment of ANPz (183) with aqueous peroxytrifluoroacetic acid yields the *N*-oxide (LLM-105) (184), an explosive showing high thermal stability (decomposition point of 354 °C).[7] The conversion of ANPz (183) ($d = 1.84\ g/cm^3$) to the *N*-oxide (184) ($d = 1.918\ g/cm^3$) illustrates a useful strategy for increasing the crystal density of nitrogen heterocycles.

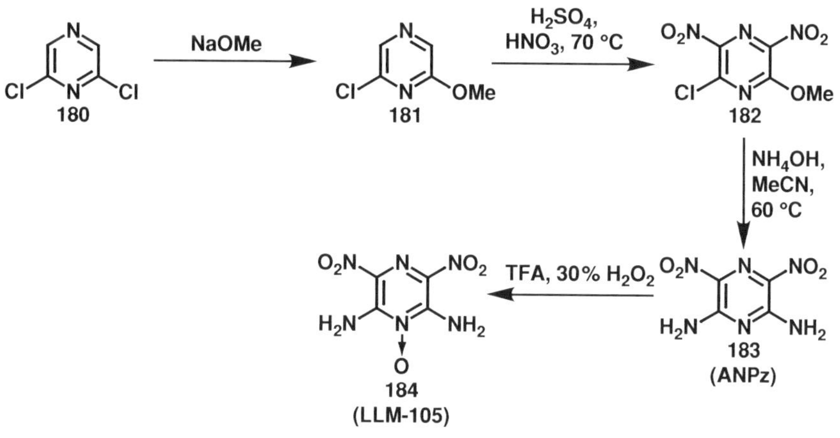

Figure 7.67

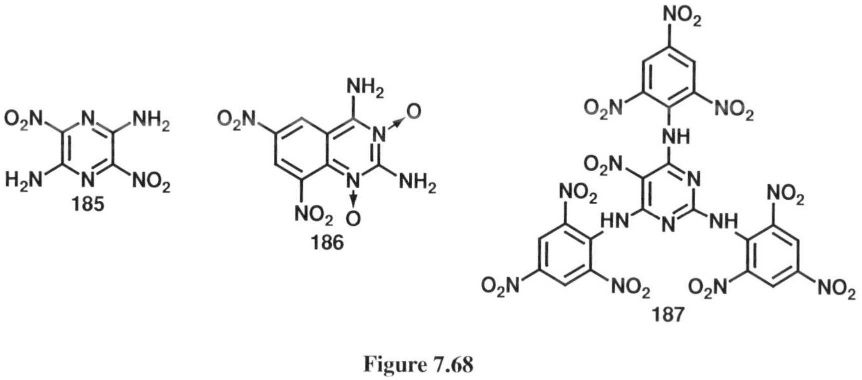

Figure 7.68

Millar and co-workers reported the synthesis of a number of energetic pyrimidines, pyrazines and their bicyclic analogues, including 2,5-diamino-3,6-dinitropyrazine (185) and the quinazoline (186).[117,118]

Chemists at Los Alamos National Laboratory synthesized a series of picrylamino-substituted pyrimidines as part of a research effort to find new thermally stable explosives. The pyrimidine-based explosive (187) is synthesized via the reaction of 2,4,6-triaminopyrimidine with picryl fluoride followed by subsequent nitration with nitric acid.[119]

7.8 6-MEMBERED RINGS – 3N

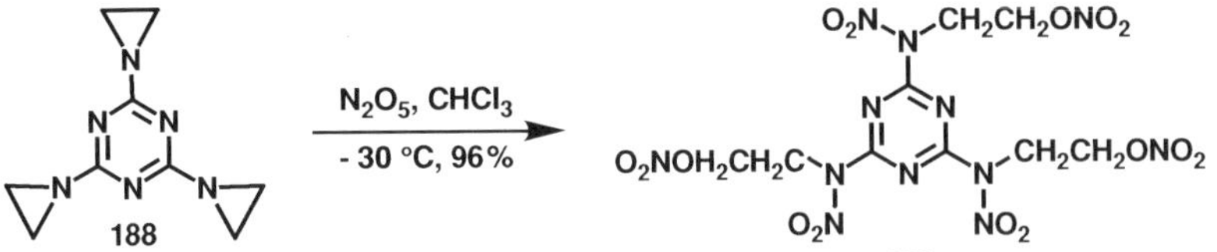

189
(Tris-X)

Figure 7.69

The 1,2-nitramine-nitrate Tris-X (189) (VOD ~ 8700 m/s, m.p. 69 °C) has been synthesized from the reaction of 2,4,6-tris(aziridino)-1,3,5-triazine (188) with dinitrogen pentoxide in chloroform or methylene chloride at subambient temperature (Section 5.8.1).[120]

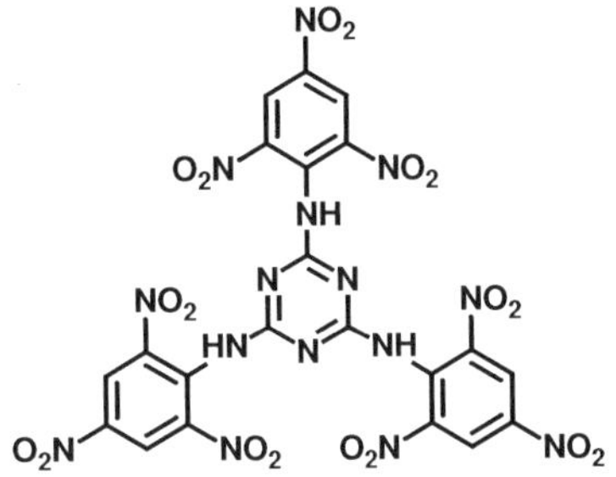

190
(TPM)

Figure 7.70

Coburn[121] synthesized 2,4,6-tris(picrylamino)-1,3,5-triazine (TPM) (190) from the reaction of aniline with cyanuric chloride followed by nitration of the product with mixed acid.[121] Treatment of TPM (190) with acetic anhydride–nitric acid leads to *N*-nitration and the isolation of the corresponding tris-nitramine. The high thermal stability of TPM (m.p. 316 °C) coupled with its facile synthesis and low sensitivity to impact has led to its large scale manufacture in the US by Hercules Inc. China has reported a low-cost synthetic route to TPM but this has a limited production capacity.[1d]

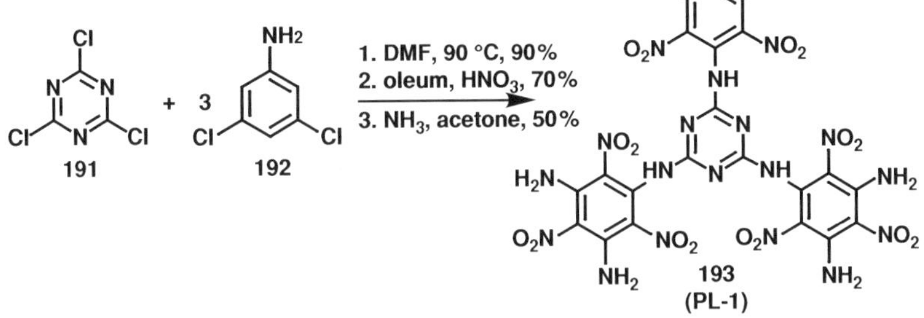

193
(PL-1)

Figure 7.71

Agrawal and co-workers[122] synthesized the insensitive triazine-based explosive PL-1 (193) from the reaction of cyanuric chloride (191) with 3,5-dichloroaniline (192), followed by nitration and displacement of the chloro groups with ammonia in acetone. PL-1 (193) (VOD ~ 7861 m/s, $d = 2.02$ g/cm^3, DTA exotherm at 335 °C) has overall comprehensive properties close to TATB.

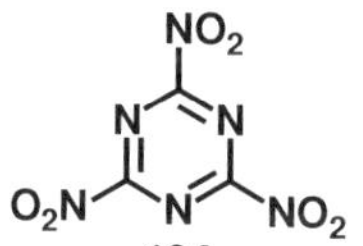

Figure 7.72

2,4,6-Trinitro-1,3,5-triazine (194) has seen numerous attempts[123] at its synthesis, but it remains elusive and is probably too unstable for isolation under normal conditions.

7.9 6-MEMBERED RINGS – 4N

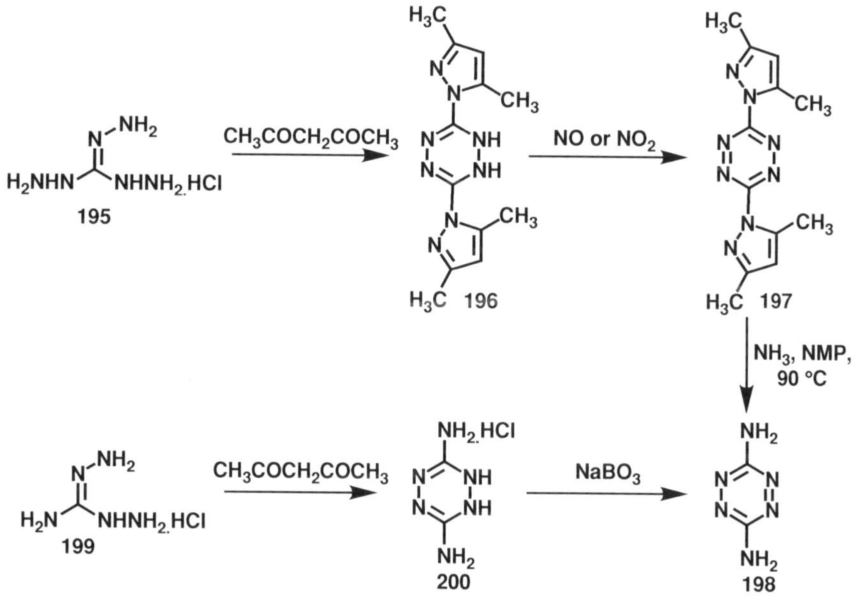

Figure 7.73

Tetrazine-based explosives are often highly energetic. The high nitrogen content of such compounds often results in high crystal density and explosive performance. 3,6-Diamino-1,2,4,5-tetrazine (198) is a starting material for the synthesis of other 1,2,4,5-tetrazines and is itself synthesized from the condensation of triaminoguanidine hydrochloride (195) with 2,4-pentanedione, followed by oxidation of the resulting dihydrotetrazine (196) to the tetrazine (197) and treatment of the latter with ammonia under pressure.[124] 3,6-Diamino-1,2,4,5-tetrazine (198) is also synthesized from the condensation of 1,3-diaminoguanidine hydrochloride (199) with 2,4-pentanedione, followed by oxidation of the resulting dihydrotetrazine (200) with sodium perborate.[125]

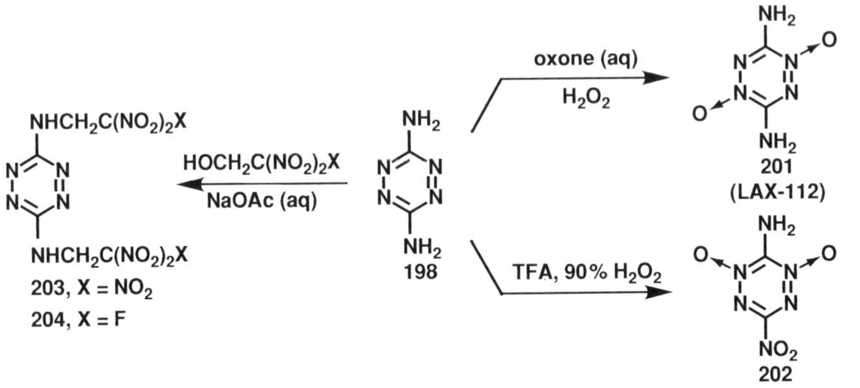

Figure 7.74

Oxidation of 3,6-diamino-1,2,4,5-tetrazine (198) with oxone in the presence of hydrogen peroxide yields 3,6-diamino-1,2,4,5-tetrazine-2,4-dioxide (201) (LAX-112).[126] The same reaction with 90 % hydrogen peroxide in trifluoroacetic acid yields 3-amino-6-nitro-1,2,4,5-tetrazine-2,4-dioxide (202).[126] Treatment of 3,6-diamino-1,2,4,5-tetrazine (198) with 2,2,2-trinitroethanol and 2,2-dinitro-2-fluoroethanol generates the Mannich condensation products (203) and (204) respectively.[127,128]

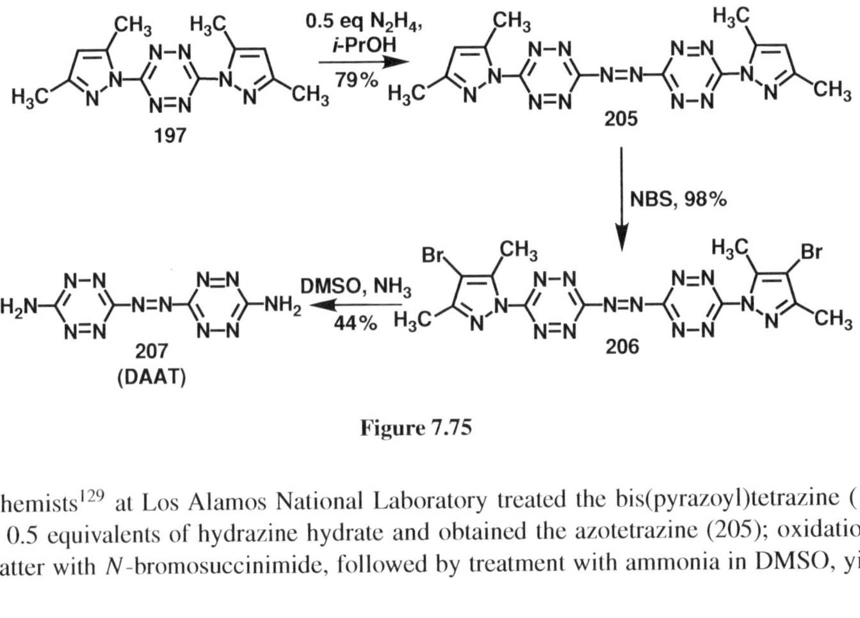

Figure 7.75

Chemists[129] at Los Alamos National Laboratory treated the bis(pyrazoyl)tetrazine (197) with 0.5 equivalents of hydrazine hydrate and obtained the azotetrazine (205); oxidation of the latter with *N*-bromosuccinimide, followed by treatment with ammonia in DMSO, yields

3,3'-azo-bis(6-amino-1,2,4,5-tetrazine) (207) (DAAT). DAAT has a very large positive heat of formation, initially measured[129] as 862 kJ/mol, although recent work[130] reports this could be as high as 1035 kJ/mol.

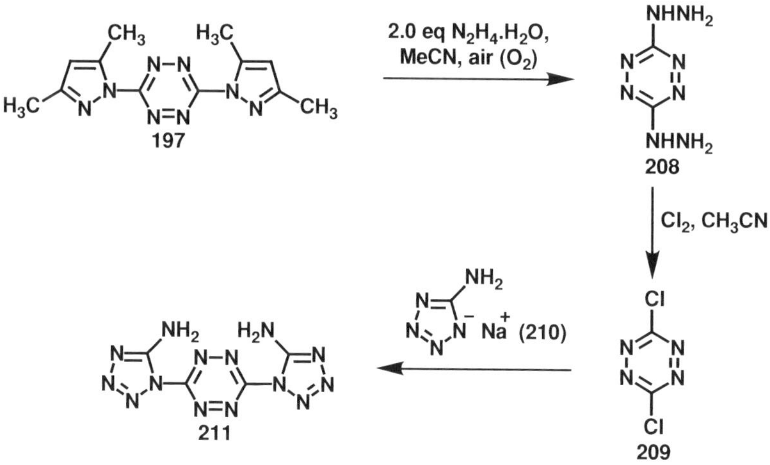

Figure 7.76

Treatment of the bis(pyrazoyl)tetrazine (197) with an excess of hydrazine hydrate generates 3,6-bis(hydrazino)-1,2,4,5-tetrazine (208), a compound which might find use as an energetic additive in high performance propellants.[131] Several salts of (208) have been reported, including the dinitrate and diperchlorate. 3,6-Dichloro-1,2,4,5-tetrazine (209), the product from treating (208) with chlorine in acetonitrile, reacts with the sodium salt of 5-aminotetrazole (210) to yield (211) ($C_4H_4N_{14}$ – 79 % N).[131]

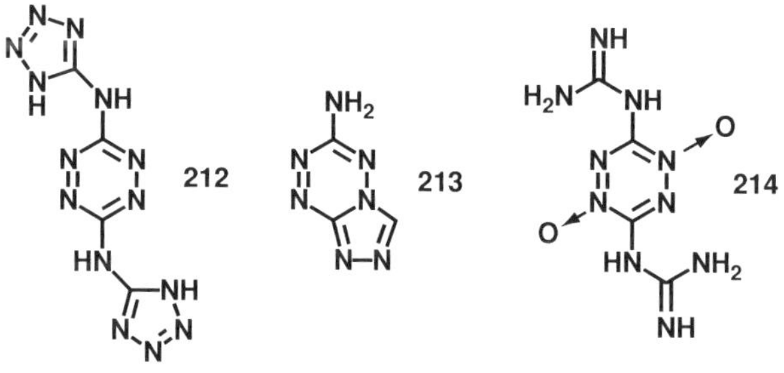

Figure 7.77

Hiskey and co-workers at Los Alamos National Laboratory conducted extensive studies into the synthesis of tetrazine-based energetic materials. Compounds (212), (213) and (214) are some of the energetic tetrazines synthesized so far.[129,131,132]

Tartakovsky and co-workers[133] reported the synthesis of the highly energetic 1,2,3,4-tetrazino[5,6-*f*]benzo-1,2,3,4-tetrazine 1,3,7,9-tetra-*N*-oxide (TBTDO) (219). The synthesis

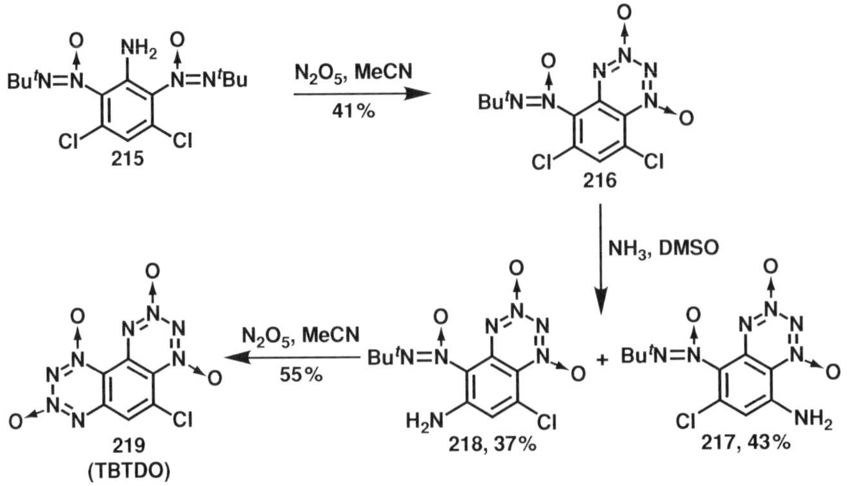

Figure 7.78

of the 1,2,3,4-tetrazine-1,3-dioxide ring system starts from (215) and utilizes the reaction of the *tert*-butyl-*NNO*-azoxy group with an adjacent amino group in the presence of dinitrogen pentoxide. The *tert*-butyl-*NNO*-azoxy groups of (215) are introduced by treating the corresponding nitroso derivative with *N,N*-dibromo-*tert*-butylamine. Tartakovsky and Churakov[134] recently reviewed the chemistry of 1,2,3,4-tetrazines.

7.10 DIBENZOTETRAAZAPENTALENES

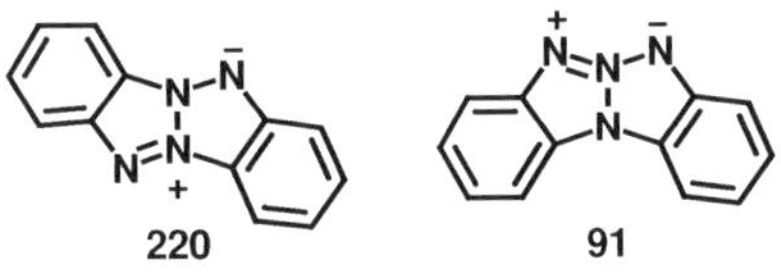

Figure 7.79

The dibenzotetraazapentalene ring system was first discovered by chemists at DuPont[135–138] and is a planar system with six electrons delocalized over four nitrogen atoms. There are two isomeric arrangements of these four nitrogens which lead to the 1,3a,4,6a- (220) and 1,3a,6,6a- (91) ring systems. Nitro derivatives of both isomeric dibenzotetraazapentalenes have been explored as thermally stable explosives.

Isomeric 1,3a,4,6a- (220) and 1,3a,6,6a- (91) dibenzotetraazapentalenes can be prepared from the thermal decomposition of 2-(*o*-azidophenyl)-2*H*-benzotriazole (224) and 1-(*o*-azidophenyl)-2*H*-benzotriazole (230), respectively, in high boiling solvents such as *o*-dichlorobenzene and decalin.[135–138] This synthesis was improved upon when it was found that (220) and (91) can be prepared from the reactions of 2-(*o*-nitrophenyl)-2*H*-benzotriazole (226) and 1-(*o*-nitrophenyl)-2*H*-benzotriazole (229), respectively, with triethyl phosphite in refluxing xylene.[138]

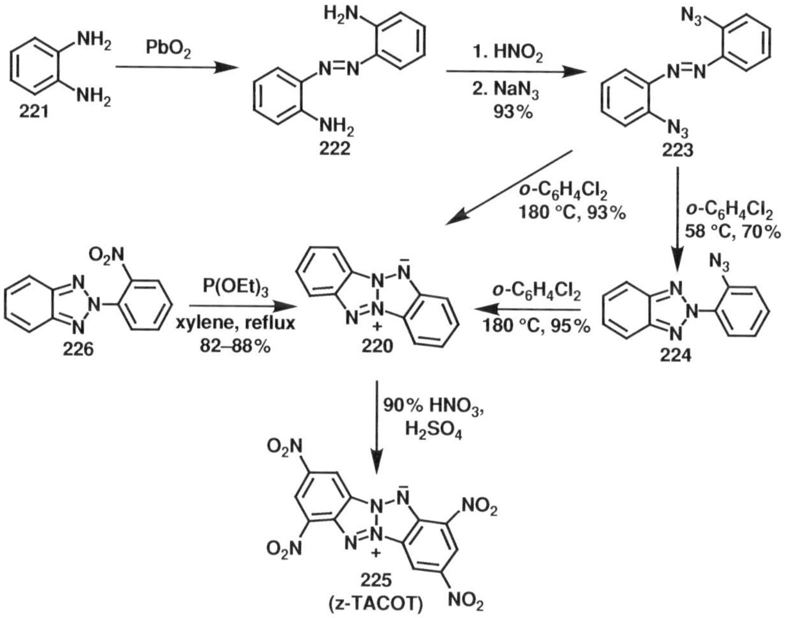

Figure 7.80

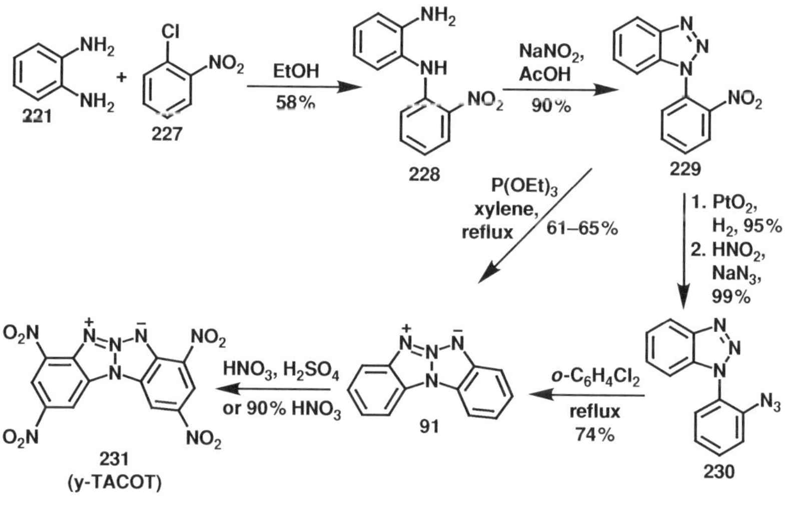

Figure 7.81

Treatment of the isomeric 1,3a,4,6a- (220) and 1,3a,6,6a- (91) dibenzotetraazapentalenes with mixed acid or fuming nitric acid forms z-TACOT (225) and y-TACOT (231), respectively.[137,138] z-TACOT (225) and y-TACOT (231) are highly insensitive to impact and electrostatic charge and exhibit extremely high thermal stability (ignition temperature ~ 494 °C, DTA curve 354 °C).

z-TACOT (225) and y-TACOT (231) are single isomers but their syntheses are rather long for cost effective commercial manufacture. Industrially produced TACOT is a mixture of these two isomers and the synthesis is much shorter and more convenient than the production of the single isomers. In this synthesis benzotriazole is treated with *o*-nitrochlorobenzene to produce an isomeric mixture of 1-(*o*-nitrophenyl)-2*H*-benzotriazole and 2-(*o*-nitrophenyl)-2*H*-benzotriazole, which on treatment with triethyl phosphite in refluxing xylene yields a mixture of isomeric 1,3a,4,6a- and 1,3a,6,6a-dibenzotetraazapentalenes and nitration of this mixture produces commercial TACOT. This mixture (VOD $\sim$ 7250 m/s, $d = 1.64$ g/cm^3) finds extensive use for applications where high temperature resistance is needed. TACOT is available commercially in the form of flexible linear shaped charges (FLSC), plastic bonded explosive (PBX), high-density charges and in mild detonating fuze. TACOT is also reported to be used in the manufacture of high-temperature resistant detonators.

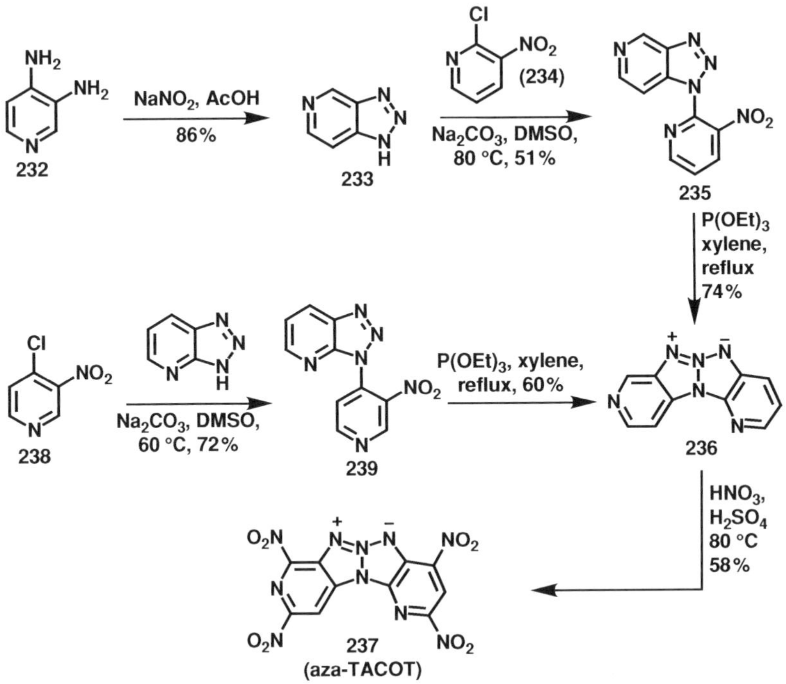

Figure 7.82

The dipyridotetraazapentalene ring (236) also exhibits high thermal stability. The tetranitro-derivative of this ring system, aza-TACOT (237), is synthesized by a similar route to that of y-TACOT, but is more energetic than the latter.[139,140]

REFERENCES

1. (a) J. P. Agrawal, *Propell. Explos. Pyrotech.*, 2005, **30**, 316; (b) A. K. Sikder and N. Sikder, *J. Haz. Mater.*, 2004, **A112**, 1; (c) G. S. Lee, A. R. Mitchell, P. F. Pagoria and R. D. Schmidt, *Thermochim. Acta.*, 2002, **384**, 187; (d) J. P. Agrawal, *Prog. Energy Combust. Sci.*, 1998, **24**, 1; (e) I. J. Dagley

and R. J. Spear, 'Synthesis of Organic Energetic Compounds', in *Organic Energetic Compounds*, Ed. P. L. Marinkas, Nova Science Publishers, Inc., New York, Chapter 2, 47–163 (1996); (f) R. J. Spear and W. S. Wilson, *J. Energ. Mater.*, 1984, **2**, 61.
2. G. S. Lee, A. R. Mitchell, P. F. Pagoria and R. D. Schmidt, *Thermochim. Acta.*, 2002, **384**, 187.
3. G. Doddi, P. Mencarelli, A. Razzini and F. Stegel, *J. Org. Chem.*, 1979, **44**, 2321.
4. W. W. Edwards, C. George, R. Gilardi and J. C. Hinshaw, *J. Heterocycl. Chem.*, 1992, **29**, 1721.
5. V. W. Belikov, Y. P. Egorov, S. S. Novikov, E. N. Safonova and L. V. Semenov, *Russ. Chem. Bull.*, 1959, **8**, 1386.
6. I. L. Dalinger, V. I. Gulevskaya, M. I. Kanishchev, S. A. Shevelev, T. K. Shkineva and B. I. Ugrak, *Russ. Chem. Bull.*, 1993, **42**, 1063.
7. J. Cutting, J. Forbes, F. Garcia, R. Lee, D. M. Hoffman, A. R. Mitchell, P. F. Pagoria, R. D. Schmidt, R. L. Simpson and R. L. Swansiger, Presented at the *Insensitive Munitions and Energetic Materials Technology Symposium*, San Diego, CA, 1998.
8. I. L. Dalinger, S. A. Shevelev and V. M. Vinogradov, *Mendeleev Commun.*, 1993, 111.
9. A. R. Mitchell, P. F. Pagoria and R. D. Schmidt, Presented at the *211th American Chemical Society National Meeting*, New Orleans, LA, 24–28 March, 1996.
10. P. Badol, F. Goujon, J. Guillard and D. Poullain, *Tetrahedron Lett.*, 2003, **44**, 5943.
11. P. A. Ivanov, N. V. Latypov, M. S. Pevzner and V. A. Silevich, *Chem. Heterocycl. Compd.*, 1976, **12**, 1355.
12. L. V. Epishina, O. V. Lebedev, L. I. Khmelnitskii, S. S. Novikov and V. V. Sevastyanova, *Chem. Heterocycl. Compd.*, 1970, **6**, 465 and 614.
13. (a) H. H. Cady, M. D. Coburn, B. W. Harris and R. N. Rogers, 'Synthesis and Thermochemistry of Ammonium Trinitroimidazole', *LA-6802-MS* (1977), Los Alamos National Laboratory, New Mexico; (b) M. D. Coburn, *US Pat.* 4 028 154 (1977); *Chem. Abstr.*, 1978, **88**, 9198v.
14. C. L. Coon, M. F. Folts, P. F. Pagoria and R. L. Simpson, 'A New Insensitive Explosive that is Low Cost: 2,4-Dinitroimidazole', *LLNL Report 1–7*, Lawrence Livermore National Laboratory, CA.
15. M. E. Sitzman, *J. Energ. Mater.*, 1988, **6**, 129.
16. E. O. John, R. L. Kirchmeier and J. M. Shreeve, *J. Fluorine Chem.*, 1989, **47**, 333.
17. (a) M. D. Coburn, *J. Heterocycl. Chem.*, 1968, **5**, 83; (b) *US Pat.* 3 414 570 (1968); *Chem. Abstr.*, 1969, **70**, 47505d.
18. M. L. Trudell and A. K. Zelinin, *J. Heterocycl. Chem.*, 1997, **34**, 1057.
19. J. H. Boyer, A. Gunasekaren, T. Jayachandran and M. L. Trudell, *J. Heterocycl. Chem.*, 1995, **32**, 1405.
20. D. C. Barham, L. W. Kissinger, A. Narath and H. E. Ungnade, *J. Org. Chem.*, 1963, **28**, 134.
21. G. A. Pearse Jr and R. T. Pflaum, *J. Am. Chem. Soc.*, 1959, **81**, 6505.
22. M. D. Boldyrev, B. V. Gidaspov, V. D. Nikolaev and G. D. Soludyuk, *J. Org. Chem. (USSR)*, 1981, **17**, 756.
23. G. S. Lee, A. R. Mitchell, P. F. Pagoria and R. D. Schmidt, *J. Heterocycl. Chem.*, 2001, **38**, 1227.
24. N. S. Aleksandrova, O. V. Kharitonova, L. I. Khmelnitskii, V. O. Kulagina, T. M. Melnikova, S. S. Novikov, T. S. Novikova, T. S. Pivina and A. B. Sheremetev, *Mendeleev Commun.*, 1994, 230.
25. R. D. Gilardi, M. L. Trudell and A. K. Zelinin, *J. Heterocycl. Chem.*, 1998, **35**, 151.
26. J. H. Boyer and A. Gunasekaren, *Heteroatom. Chem.*, 1993, **4**, 521.
27. R. L. Willer, 'Synthesis of a New Energetic Material, 1,4,5,8-Tetranitro-1,4,5,8-tetraazadifurano[3,4-c][3,4-h]decalin (CL-15)', *NWCTP 6397* (1982), Naval Weapons Center, China Lake, CA.
28. R. L. Willer, *US Pat.* 4 503 229 (1985); *Chem. Abstr.*, 1985, **103**, 54099c.
29. D. W. Moore and R. L. Willer, *J. Org. Chem.*, 1985, **50**, 5123.
30. Y. Du, M. Jiang, Q. Sun and X. Fu, 'Detonation Properties and Thermal Stabilities of Furazano-Fused Cyclic Nitramines', in *Proc. International Symposium on Pyrotechnics and Explosives*, China Academic Publishers, Beijing, China, 412 (1987).
31. R. L. Willer, *US Pat.* 4 539 405; *Chem. Abstr.*, 1986, **104**, 91609k.

32. I. S. Abdrakhmanov, G. K. Khisamutdinov, I. Z. Kondyukov, V. L. Korolev, S. F. Melnikov, T. A. Mratkhuzina, S. V. Pirogov, T. V. Romanova, S. P. Smirnov and I. V. Tselinskii, *Zh. Org. Khim.*, 1997, **33**, 1739.
33. T. P. Hobin, *Tetrahedron*, 1968, **24**, 6145.
34. (a) T. I. Godovikova, S. P. Golova, L. I. Khmelnitskii, O. A. Rakitin and S. A. Vozchikova, *Mendeleev Commun.*, 1993, 209; (b) M. Y. Antipin, T. I. Godovikova, S. P. Golova, L. I. Khmelnitskii, Y. A. Strelenko and Y. T. Struchkov, *Mendeleev Commun.*, 1994, 7.
35. S. Fumasoni, G. Giacobe, R. Martinelli and G. Schippa, *Chim. Ind. (Milan)*, 1966, **47**, 1064.
36. A. L. Fridman, G. S. Ismagilova and A. D. Nikolaeva, *Chem. Heterocycl. Compd*, 1971, **7**, 804.
37. A. N. Binnikov, A. S. Kulikov, N. N. Makhov, I. V. Orchinnikov and T. S. Pivina, '4-Amino-3-azidocarbonyl Furoxan as an Universal Synthon for the Synthesis of Energetic Compounds of the Furoxan Series',' *30th International Annual Conference of ICT*, Karlsruhe, Germany, 1999, 58/1–58/10.
38. J. Boyer, *Heterocyclic Compounds, Vol. 7*, Ed. R. C. Elderfield, John Wiley & Sons, Inc., New York, 463–522 (1961).
39. O. Turek, *Chimie et Industrie*, 1931, **26**, 781.
40. A. J. Green and F. M. Rowe, *J. Chem. Soc.*, 1913, 2023.
41. (a) E. Schrander, *Chem. Ber.*, 1917, **50**, 777; (b) R. Dietschy and R. Nietzki, *Chem. Ber.*, 1901, **34**, 55.
42. J. P. Agrawal, Mehilal, R. B. Salunke and P. D. Shinde, *Propell. Explos. Pyrotech.*, 2003, **28**, 77.
43. W. P. Norris and R. J. Spear, *Propell. Explos. Pyrotech.*, 1983, **8**, 85.
44. W. P. Norris, R. W. Read and R. J. Spear, *Aust. J. Chem.*, 1984, **37**, 985.
45. W. P. Norris, R. W. Read and R. J. Spear, *Aust. J. Chem.*, 1983, **36**, 297.
46. W. P. Norris and R. J. Spear, 'Potassium 4-Hydroxyamino-5,7-dinitro-4,5-dihydrobenzofurazanide-3-oxide, the First in a Series of New Primary Explosives', *MRL-TR-870* (1983), Materials Research Laboratory, Victoria, Australia.
47. C. K. Lowe-Ma, R. A. Nissan and W. S. Wilson, 'Tetrazolo[1,5-a]pyridines and Furazano[4,5-b]pyridine-1-oxides as Energetic Materials,' *NWC TP 6985* (1989), Naval Weapons Center, China Lake, CA.
48. W. P. Norris, '7-amino-4,6-dinitrobenzofuroxan, an Insensitive High Explosive', *NWC TP 6522* (1984), Naval Weapons Center, China Lake, CA.
49. B. Gohrmann and H. -J. Niclas, *J. Prakt. Chem.*, 1989, **331**, 819.
50. C. Boren and L. Zhiyuan, 'Synthesis of 5,7-Diamino-4,6-dinitro-benzofuroxan', *21st International Annual Conference of ICT*, Karlsruhe, Germany, 1990, Paper 58.
51. M. P. Kramer, W. P. Norris and D. J. Vanderah, 'CL-14, a High Performance Insensitive Explosive', *NWC TP 6555* (1989), Naval Weapons Center, China Lake, CA.
52. C. Boren, W. Naixing and O. Yuxiang, 'Synthesis of *N,N'*-Bis-(2,4-dinitrobenzofuroxan)-1,3,5-trinitro-2,6-diaminobenzene', in *Proc. 17th International Pyrotechnics Seminar (Combined with 2nd Beijing International Symposium on Pyrotechnics and Explosives)*, Beijing Institute Technical Press, Beijing, China, 235 (1991).
53. B. Chen, Y. Ou and N. Wang, *Propell. Explos. Pyrotech.*, 1993, **18**, 111.
54. B. Chen, Y. Ou and N. Wang, *Propell. Explos. Pyrotech.*, 1994, **19**, 145.
55. B. Chen, Y. Ou and N. Wang, *J. Energ. Mater.*, 1993, **11**, 47.
56. J. H. Boyer, G. Eck, E. D. Stevens, G. Subramanian and M. L. Trudell, *J. Org. Chem.*, 1996, **61**, 5801.
57. J. H. Boyer, D. Buzatu, E. D. Stevens, G. Subramanian and M. L. Trudell, *J. Org. Chem.*, 1995, **60**, 6110.
58. J. K. Berlin and M. D. Coburn, *J. Heterocycl. Chem.*, 1975, **12**, 235.
59. M. D. Coburn and T. E. Jackson, *J. Heterocycl. Chem.*, 1968, **5**, 199.
60. M. D. Coburn, *US Pat.* 3 483 211 (1969).

61. B. Chen, J. Li and Y. Ou, 'Modified Preparation and Nitration of 3-Picrylamino-1,2,4-triazole', in *Proc. 17th International Pyrotechnics Seminar (Combined with 2nd Beijing International Symposium on Pyrotechnics and Explosives)*, Beijing Institute Technical Press, Beijing, China, 196 (1991).
62. J. P. Agrawal, Mehilal, U. S. Prasad and R. N. Surve, *New. J. Chem.*, 2000, **24**, 583.
63. J. P. Agrawal, Mehilal, A. K. Sikder and N. Sikder, *Indian. J. Eng. Mater. Sci.*, 2004, **11**, 516.
64. P. N. Neuman, *J. Heterocycl. Chem.*, 1970, **7**, 1159.
65. R. V. Kendall and R. A. Olofson, *J. Org. Chem.*, 1970, **35**, 2246.
66. E. J. Browne, *Aust. J. Chem.*, 1969, **22**, 2251.
67. M. M. Stinecipher, *Proc. 7th International Symposium on Detonation*, US Naval Academy, Annapolis, Maryland, 1981, Vol. 1, 733.
68. M. D. Coburn, M. A. Hiskey, K.-Y. Lee and C. B. Storm, *J. Energ. Mater.*, 1991, **9**, 415.
69. L. V. Kilina, T. N. Kulibabina, M. S. Pevzner and N. A. Povarova, *Chem. Heterocycl. Compd.*, 1979, **15**, 929.
70. M. M. Stinecipher, 'Eutectic Explosives Containing Ammonium Nitrate', *Final Report, Oct 1979–Sept 1981, LA-9973-MS*, Los Alamos National Laboratory, New Mexico.
71. P. Charrue, F. Laval and Ch. Wartenberg, *Propell. Explos. Pyrotech.*, 1995, **20**, 23.
72. J. P. Agrawal, M. Geetha, D. B. Sarwade and A. K. Sikder, *J. Haz. Mater.*, 2001, **A82**, 1.
73. F. Laval and P. Vignane, *European Pat.* 320 368 A1 (1989).
74. K. -Y. Lee, Los Alamos National Laboratory, *Report LA-10346-MS* (1985).
75. M. D. Coburn, B. W. Harris, H. H. Hayden, K.-Y. Lee and M. M. Stinecipher, *Ind. Eng. Chem. Prod. Res. Dev.*, 1986, **25**, 68.
76. A. T. Baryshnikov, V. I. Erashko, A. A. Fainzilberg, A. L. Laikhter, L. G. Melnikova, V. V. Semenov, S. A. Shevelev, B. I. Ugrak and N. I. Zubanova, *Russ. Chem. Bull.*, 1992, **41**, 751.
77. A. T. Baryshnikov, V. I. Erashko, A. A. Fainzilberg, V. V. Semenov, S. A. Shevelev, B. I. Ugrak and N. I. Zubanova, *Russ. Chem. Bull.*, 1992, **41**, 1657.
78. C. D. Bedford, J. C. Bottaro, C. George, R. Gilardi and R. J. Schmitt, *J. Org. Chem.*, 1990, **55**, 1916.
79. V. A. Myasnikov, O. P. Shitov, V. A. Tartakovsky, V. A. Vyazkov and I. L. Yudin, *Izv. Acad. Nauk USSR, Ser. Khim.*, 1991, **40**, 1239.
80. O. P. Shitov, V. A. Tartakovsky and V. A. Vyazkov, *Izv. Acad. Nauk USSR, Ser. Khim.*, 1989, **38**, 2654.
81. M. D. Coburn, L. B. Chapman and K.-Y. Lee, *J. Energ. Mater.*, 1987, **5**, 27.
82. (a) M. T. Adeline, L. Le. Campion and J. Ouazzani, *Propell. Explos. Pyrotech.*, 1997, **22**, 233; (b) A. Langlet, '3-Nitro-1,2,4-triazole-5-one (NTO), a New Explosive with High Performance and Low Sensitivity', *Report FAO-29787-23* (1990).
83. C. N. Lovey, R. J. Spear and M. J. Wolfson, 'A Preliminary Assessment of NTO as an Insensitive High Explosive', *MRL-TR-89-18*, Materials Research Laboratory, Victoria, Australia.
84. J. P. Agrawal, J. E. Field and S. M. Walley, *Combust. Flame.*, 1998, **112**, 62.
85. A. Becuwe, 'The use of 5-Oxa-3-nitro-1,2,4-triazole as an Explosive and Explosives Containing this Substance', *French. Pat.* FR 2 584 066 (1987).
86. M. D. Coburn and K.-Y. Lee, '3-Nitro-1,2,4-triazol-5-one, a Less Sensitive Explosive', *LA-10302-MS*, Los Alamos National Laboratory, New Mexico.
87. J. P. Agrawal, *Propell. Explos. Pyrotech.*, 2005, **30**, 316.
88. C.-W. Chang, T.-C. Chang, C. Chen and Y. -M. Wong, *Propell. Explos. Pyrotech.*, 1997, **22**, 240.
89. (a) G. Singh, I. P. S. Kapoor, S. Prem Felix and J. P. Agrawal, *Propell. Explos. Pyrotech.*, 2002, **27**, 16; (b) L. D. Redman and R. J. Spear, 'An Evaluation of Metal Salts of NTO as Potential Primary Explosives', *MRL-TR-563* (1989), Materials Research Laboratory, Victoria, Australia.
90. M. D. Coburn and K. -Y. Lee, *J. Heterocycl. Chem.*, 1990, **27**, 575.

91. M. D. Coburn, *J. Heterocycl. Chem.*, 1973, **10**, 743.
92. J. L. Flippen-Anderson, R. D. Gilardi, A. M. Pitt and W. S. Wilson, *Aust. J. Chem.*, 1992, **45**, 513.
93. E. Buncel, S. Cohen, R. A. Renfrow and M. J. Strauss, *Aust. J. Chem.*, 1983, **36**, 1843.
94. J. P. Agrawal, V. K. Bapat and R. N. Surve, 'Synthesis, Characterization and Evaluation of Explosive Properties of 5-Picrylamino-1,2,3,4-Tetrazole', *2nd High Energy Materials Conference and Exhibits, IIT*, Madras, 8–10 December, 1998, 403.
95. J. P. Agrawal, V. K. Bapat, R. R. Mahajan, Mehilal and P. S. Makashir, 'A Comparative Study of Thermal and Explosive Behavior of 5-Picrylamino-1,2,3,4-Tetrazole (PAT) and 5,5'-Styphnylamino-1,2,3,4-tetrazole (SAT)', *International Work-Shop on Unsteady Combustion and Interior Ballistics*, St. Petersburg, Russia, June 25–30, 2000, Vol. 1, 199.
96. H. G. Adolph and M. Chaykovsky, *J. Energ. Mater.*, 1990, **8**, 392.
97. R. A. Henry and R. L. Willer, *J. Org. Chem.*, 1988, **53**, 5371.
98. M. Ara, M. Itoh, T. Matsuzawa and M. Tamura, *Proc. 22nd International Pyrotechnics Seminar*, 317 (1996).
99. M. Temblay, *Can. J. Chem.*, 1965, **43**, 1154.
100. J. T. Marais and J. Thiele, *Liebigs Ann. Chem.*, 1893, **273**, 144.
101. J. Thiele, *Liebigs Ann. Chem.*, 1898, **303**, 57.
102. A. Hammerl, G. Holl, M. Kaiser, T. M. Klapötke, H. Nöth, U. Ticmanis and M. Warchhold, *Inorg. Chem.*, 2001, **40**, 3570.
103. H. H. Licht and H. Ritter, *J. Energ. Mater.*, 1994, **12**, 223.
104. A. H. Albert and V. Grakauskas, 'Polynitroalkyltetrazoles', *UCRL-15187* (1980), Lawrence Livermore National Laboratory, CA.
105. M. J. Kamlet and W. H. Gilligan, 'Synthesis of Mercuric 5-Nitrotetrazole', *NSWC/WOL/TR-76-146, AD-A036086* (1976), Naval Surface Weapons Centre, White Oak, Maryland.
106. M. D. Coburn and K. -Y. Lee, *J. Energ. Mater.*, 1983, **1**, 109.
107. M. D. Coburn, *US Pat.* 3 678 061 (1972); *Chem. Abstr.*, 1972, **77**, 139812z.
108. M. D. Coburn and J. L. Singleton, *J. Heterocycl. Chem.*, 1972, **9**, 1039.
109. (a) M. D. Coburn, B. W. Harris, H. H. Hayden, K. -Y. Lee and M. M. Stinecipher, *Ind. Eng. Chem. Prod. Res. Dev.*, 1986, **25**, 68; (b) J. H. Chen, C. H. Lin, S. Y. Liu, G. S. Shaw and M. D. Wu, *Huoyao Jishi*, 1991, **7**, 53.
110. M. D. Coburn, *J. Heterocycl. Chem.*, 1974, **11**, 1099.
111. H. H. Licht and H. Ritter, *J. Heterocycl. Chem.*, 1995, **32**, 585.
112. (a) R. Gilardi, R. A. Hollins, L. H. Merwin, R. A. Nissan and W. S. Wilson, *J. Heterocycl. Chem.*, 1996, **33**, 895; (b) R. A. Nissan and W. S. Wilson, '2,6-Diamino-3,5-dinitropyridine-1-oxide and New Explosives', Naval Air Warfare Centre, Weapons Division TR CA 93555-6001 (1995).
113. R. Gilardi, R. A. Hollins, L. H. Merwin, R. A. Nissan and W. S. Wilson, *Material Research Society Symposium Proceedings*, Pittsburg, PA, Vol. 418 (1996).
114. H. H. Licht and H. Ritter, *Propell. Explos. Pyrotech.*, 1988, **13**, 25.
115. L. Minxiu and L. Hengyuan, 'A Study of Synthesis and Thermal Stabilities of 2,4,6-Trinitropyridine and its Derivatives', in *Proc. International Symposium on Pyrotechnics and Explosives*, China Academic Publishers, Beijing, China, 214–218 (1987).
116. W. Daozheng, *Kexue Tongbao (Scientia)*, 1986, **31**, 1034.
117. R. G. Coombes, R. W. Millar and S. P. Philbin, *Propell. Explos. Pyrotech.*, 2000, **25**, 302.
118. R. P. Claridge, J. Hamid, R. W. Millar and S. P. Philbin, *Propell. Explos. Pyrotech.*, 2004, **29**, 81.
119. M. D. Coburn, B. W. Harris and J. L. Singleton, *J. Heterocycl. Chem.*, 1973, **10**, 167.
120. P. Bunyan, P. Golding, R. W. Millar, N. C. Paul, D. H. Richards and J. A. Rowley, *Propell. Explos. Pyrotech.*, 1993, **18**, 55.
121. M. D. Coburn, *J. Heterocycl. Chem.*, 1966, **11**, 365.
122. J. P. Agrawal, V. K. Bapat, Mehilal, B. G. Polke and A. K. Sikder, *J. Energ. Mater.*, 2000, **18**, 299.
123. (a) M. D. Coburn, C. L. Coon, H. H. Hayden and A. R. Mitchell, *Synthesis*, 1986, 490; (b) G. D. Hartman, R. D. Hartman and J. E. Schwering, *Tetrahedron Lett*, 1983, **24**, 1011.

124. G. A. Buntain, M. D. Coburn, B. W. Harris, M. A. Hiskey, K. -Y. Lee and D. G. Ott, *J. Heterocycl. Chem.*, 1991, **28**, 2049.
125. M. D. Coburn and D. G. Ott, *J. Heterocycl. Chem.*, 1990, **27**, 1941.
126. M. D. Coburn, M. A. Hiskey, K. -Y. Lee, D. G. Ott and M. M. Stinecipher, *J. Heterocycl. Chem.*, 1993, **30**, 1593.
127. Z. Xizeng and T. Ye, 'Synthesis and Properties of Tetrazine Explosives', in *Proc. International Symposium on Pyrotechnics and Explosives*, China Academic Publishers, Beijing, China, 241 (1987).
128. J. C. Bottara and R. J. Schmitt, *New Nitration Concepts. Final Report 1987*, SRI International, Menlo Park, CA [AD-A187518].
129. D. E. Chavez, R. D. Gilardi and M. A. Hiskey, *Angew. Chem. Int. Ed.*, 2000, **39**, 1791.
130. S. Löbbecke and J. Kerth, *Propell. Explos. Pyrotech.*, 2002, **27**, 111.
131. D. E. Chavez and M. A. Hiskey, *J. Heterocycl. Chem.*, 1998, **35**, 1329.
132. D. E. Chavez, M. A. Hiskey and D. L. Naud, *Propell. Explos. Pyrotech.*, 2004, **29**, 209.
133. A. M. Churakov, A. E. Frumkin, V. V. Kachala, Y. A. Strelenko and V. A. Tartakovsky, *Org. Lett.*, 1999, **1**, 5.
134. A. M. Churakov and V. A. Tartakovsky, *Chem. Rev.*, 2004, **104**, 2601.
135. R. A. Carboni and J. E. Castle, *J. Am. Chem. Soc.*, 1962, **84**, 2453.
136. R. A. Carboni, J. E. Castle, J. C. Kauer and H. E. Simmons, *J. Am. Chem. Soc.*, 1967, **89**, 2618.
137. R. A. Carboni, J. C. Kauer, R. J. Harder and W. R. Hatchard, *J. Am. Chem. Soc.*, 1967, **89**, 2626.
138. R. A. Carboni and J. C. Kauer, *J. Am. Chem. Soc.*, 1967, **89**, 2633.
139. D. Balachari and M. L. Trudell, *Tetrahedron Lett.*, 1997, **38**, 8607.
140. D. Balachari, D. Beardall, E. D. Stevens, M. L. Trudell and C. A. Wright, *Propell. Explos. Pyrotech.*, 2000, **25**, 75.

8

Miscellaneous Explosive Compounds

The vast majority of organic compounds finding wide use as commercial or military explosives contain nitro functionality in the form of polynitroarylenes, nitramines and nitrate esters. These groups include important explosives like TNT, TATB, RDX, HMX and NG etc. Energetic compounds containing other 'explosophoric' groups tend to find only limited use because of poor thermal or chemical stability, high sensitivity to impact or friction, unsuitable physical properties or difficultly in synthesis. Some of these compounds include: azides, peroxides, diazophenols and numerous nitrogen-rich compounds derived from guanidine and its derivatives.

8.1 ORGANIC AZIDES

8.1.1 Alkyl azides

Dagley and Spear[1] noted that the introduction of an azido group into an organic compound increases its energy by $\sim$355 kJ/mol and so its presence in energetic compounds is clearly favorable on thermodynamic grounds. However, many organic compounds containing azido groups have not found wide application as practical energetic materials because of their poor thermal stability and relatively high sensitivity to mechanical stimuli. The explosive properties of inorganic azides are well known, lead azide finding wide use as the initiating charge in detonators.

Alkyl azides are conveniently prepared from the reaction of alkali metal azides with an alkyl halide, tosylate, mesylate, nitrate ester or any other alkyl derivative containing a good leaving group. Reactions usually work well for primary and secondary alkyl substrates and are best conducted in polar aprotic solvents like DMF and DMSO. The synthesis and chemistry of azido compounds is the subject of a functional group series.[2]

Frankel and co-workers[3] prepared a series of alkyl diazides from the reaction of dihaloalkanes with sodium azide in DMF at 95 °C, including 1,3-diazidopropane, 1,4-diazidobutane and 1,3-diazido-2,2-dimethylpropane. Tris(azidomethyl)amine, an energetic fuel with potential for use in bipropellant propulsion systems, is synthesized from the reaction of tris(chloroethyl)amine with sodium azide.[4,5]

Organic Chemistry of Explosives J. P. Agrawal and R. D. Hodgson

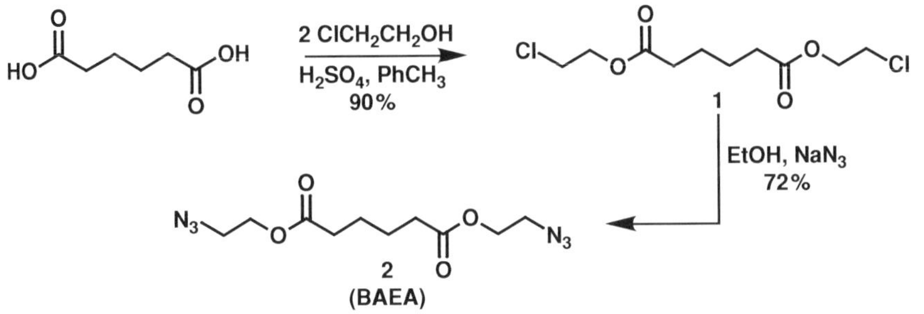

Figure 8.1

Agrawal and co-workers[6] synthesized the energetic plasticizer bis(2-azidoethyl)adipate (2) (BAEA) from the reaction of bis(2-chloroethyl)adipate (1) with sodium azide in ethanol (72 %).

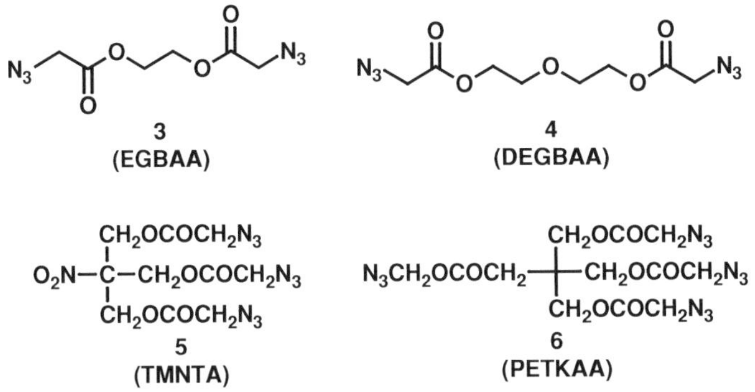

Figure 8.2

Drees and co-workers[7] synthesized a number of energetic azido plasticizers whose structures are based on those of known nitrate ester plasticizers. EGBAA (3), DEGBAA (4), TMNTA (5) and PETKAA (6) are synthesized from the corresponding chloroacetate esters with sodium azide in DMSO.

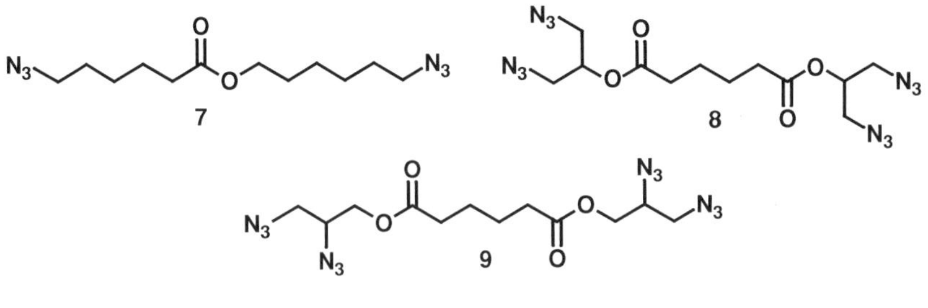

Figure 8.3

Several other azido esters has been reported, including (7),[5] (8)[8] and (9),[9] which are synthesized along similar routes of ester formation followed by substitution of halogen with azide anion or in the reverse order.

Many energetic compounds have been reported where the azido group is in conjunction with another 'explosophore'. This has been a popular approach to new energetic materials. 2-Azidoethyl nitrate, an explosive resembling nitroglycerine (NG) in its properties, was synthesized some time ago from the reaction of 2-chloroethanol with sodium azide followed by *O*-nitration of the product, 2-azidoethanol, with nitric acid.[10]

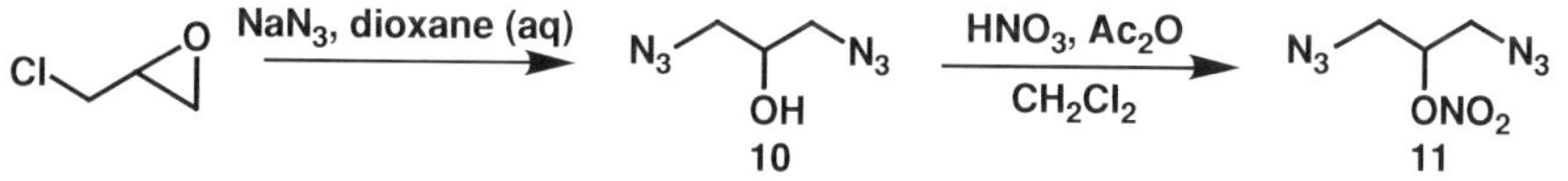

Figure 8.4

1,3-Diazido-2-propanol (10) has been synthesized from the reaction of epichlorohydrin with sodium azide in aqueous dioxane.[11] 1,3-Diazido-2-propanol (10) reacts with acetic anhydride–nitric acid to yield the nitrate ester (11),[12] whereas its reactions with polynitro derivatives of acid chlorides and alkyl chlorides has been used to synthesize a range of energetic materials.[8,12,13]

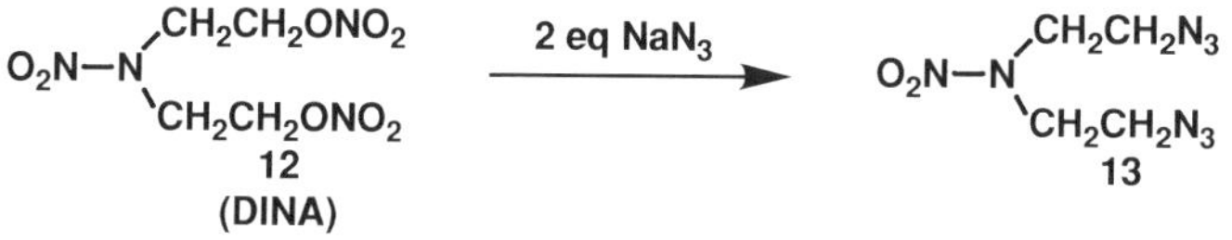

Figure 8.5

Displacements with azido anion are tolerant of many pre-existing 'explosophoric' groups but the nitrate ester group readily undergoes displacement as seen for the synthesis of bis(2-azidoethyl)nitramine (13) from *N*-nitrodiethanolamine dinitrate (12) (DINA).[14]

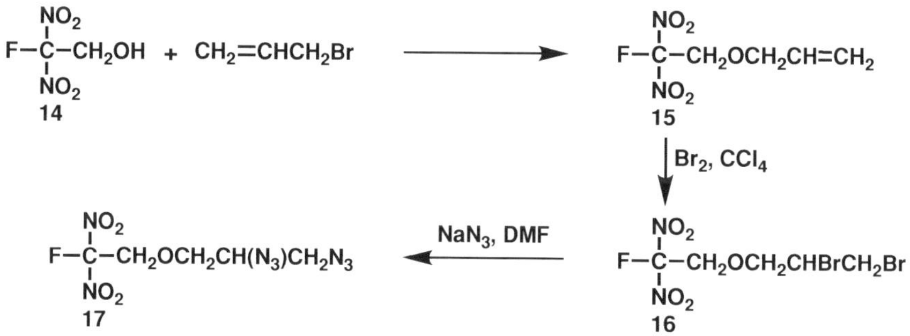

Figure 8.6

The diazide (17) is obtained from the reaction of the dibromide (16) with sodium azide in DMF.[13,15] The fluorodinitrobutyrate ester (18) is synthesized from the corresponding allyl fluorodinitrobutyrate via a similar route of bromination and displacement with azide.[16] The bis(fluorodinitroethyl)amine (19) is also obtained from the corresponding bromide.[16]

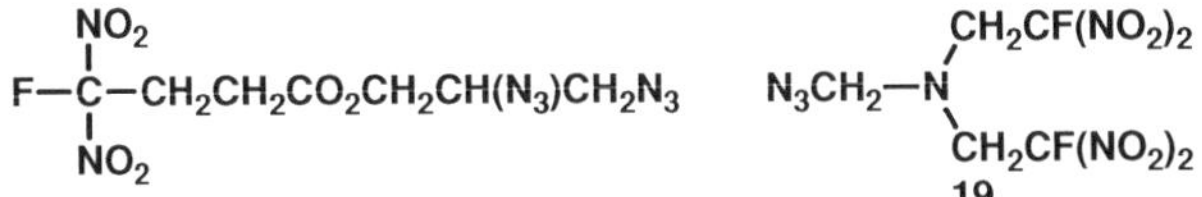

18

19

Figure 8.7

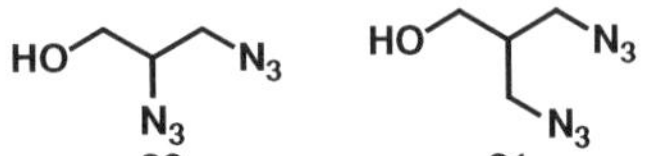

20

21

Figure 8.8

The azidoalcohols (20) and (21) have been reacted with acid chlorides and alkyl chloride to give esters containing both azido and fluorodinitromethyl functionality.[16,17]

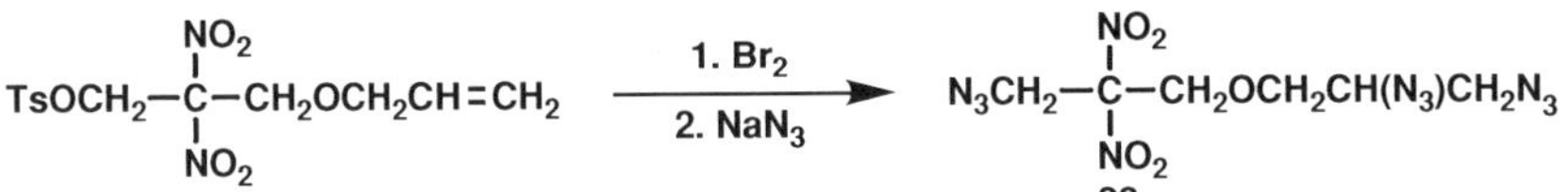

22

23

Ts = p-$CH_3C_6H_4SO_2$

Figure 8.9

The triazide (23) has been synthesized from the bromination of the tosylate (22) followed by displacement of both bromide and tosylate functionality with sodium azide.[16]

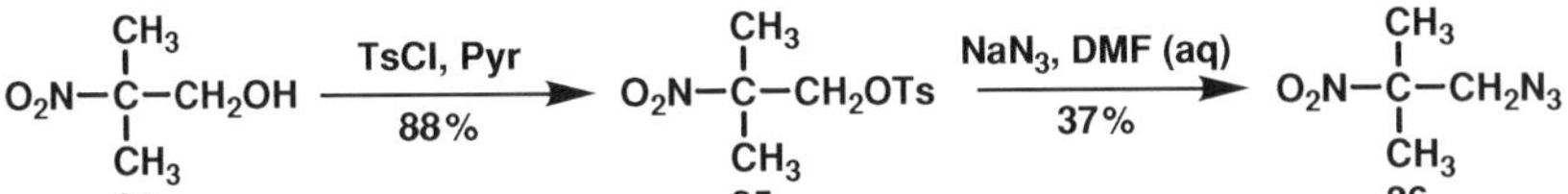

24

25

26

Figure 8.10

2-Methyl-2-nitro-1-azidopropane (26), synthesized from the tosylate (25), has been suggested for use as an energetic plasticizer.[18]

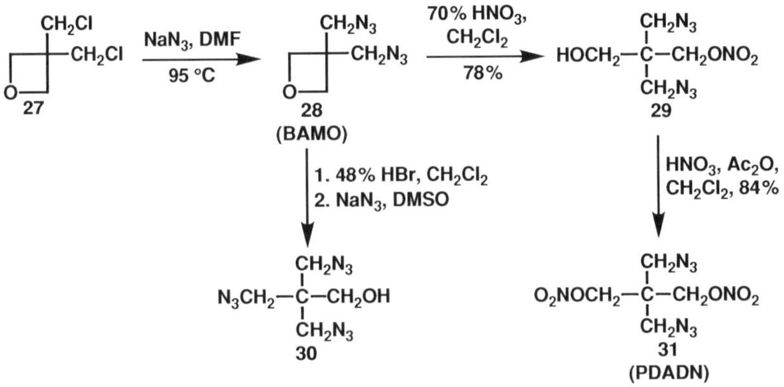

27

28 (BAMO)

29

30

31 (PDADN)

Figure 8.11

3,3-Bis(azidomethyl)oxetane (BAMO) (28), the product from treating 3,3-bis (chloromethyl)oxetane (27) with sodium azide in DMF,[19] undergoes acid-catalyzed ring opening on reaction with 70 % nitric acid to give the nitrate ester (29).[20] Treatment of (29) with nitric acid in acetic anhydride yields 2,2-bis(azidomethyl)-1,3-propanediol dinitrate (PDADN) (31).[20] Reaction of BAMO (28) with aqueous hydrobromic acid in methylene chloride, followed by treatment of the resulting bromide with sodium azide in DMSO, yields the triazide (30).[20] The hydroxy groups of (29) and (30) have been reacted with acid chlorides like 4,4,4-trinitrobutyryl chloride for the synthesis of energetic plasticizers.[21]

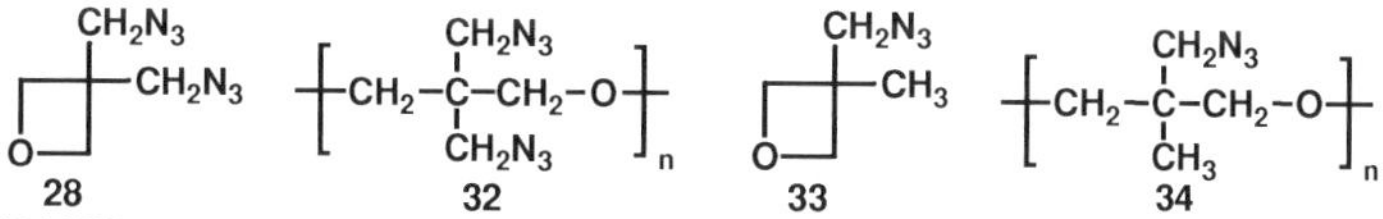

28 (BAMO) 32 Poly[BAMO] 33 (AMMO) 34 Poly[AMMO]

Figure 8.12

The energetic nature of the azido group makes its incorporation into energetic polymers and binders very desirable. 3,3-Bis(azidomethyl)oxetane (BAMO) (28) and 3-azidomethyl-3-methyloxetane (AMMO) (33) are energetic monomers which on polymerization result in the energetic polymers poly[BAMO] (32) and Poly[AMMO] (34), respectively, both of which are under evaluation as potential energetic alternatives to HTPB in composite propellant formulations.[22]

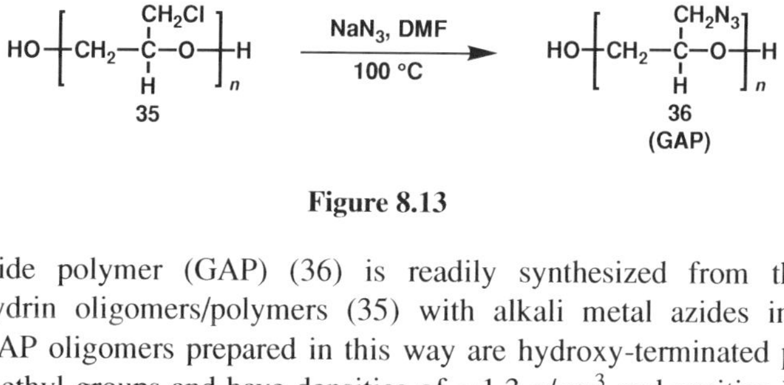

35 $\xrightarrow[100\ °C]{NaN_3,\ DMF}$ 36 (GAP)

Figure 8.13

Glycidyl azide polymer (GAP) (36) is readily synthesized from the reaction of polyepichlorohydrin oligomers/polymers (35) with alkali metal azides in polar aprotic solvents.[23,24] GAP oligomers prepared in this way are hydroxy-terminated polyethers with pendant azidomethyl groups and have densities of $\sim$1.3 g/cm^3 and positive heats of formation. GAP oligomers with number-average molar masses of 400–500 are used as plasticizers, whereas those between 3000 and 3500 are used as binders for propellants and undergo curing with isocyanates to give mixed polyether–carbamate energetic polymers. GAP is extremely insensitive to impact, even in the presence of additives like Pyrex powder and polycarbonate disc,[25] and has been used in the US in propellant formulations for a number of years. SNPE in France manufactures GAP for sale to European countries under licence from Rocketdyne in the US. Research at Fiat Avio in Italy and SNPE in France has focused on developing GAP-based propellants containing energetic nitrate ester plasticizers like MTN and BTTN, nitramines like RDX and CL-20, and energetic oxidizers like ammonium dinitramide (ADN) and hydrazinium nitroform (HNF).[22] The use of GAP in propellants has been reviewed.[24]

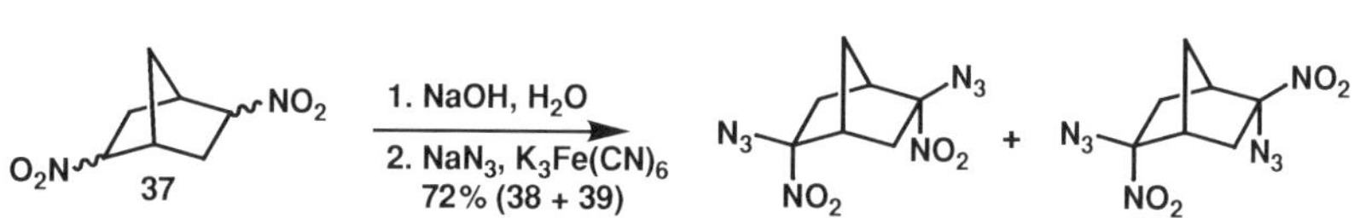

37 → 1. NaOH, H_2O; 2. NaN_3, $K_3Fe(CN)_6$; 72% (38 + 39) → 38 + 39

Figure 8.14

Bowman and co-workers[26] synthesized 2-azido-2-nitropropane by treating the sodium salt of 2-nitropropane with a mixture of sodium azide and potassium ferricyanide. Olah and co-workers[27] used the same methodology for the synthesis of alicyclic *gem*-azidonitroalkanes from secondary nitroalkanes. Isomeric azidonitronorbornanes (38) and (39) were synthesized from 2,5-dinitronorbornane (37). Some of the *gem*-azidonitroalkanes synthesized during this work have poor chemical and thermal stability.

1-Azido-1,1-dinitroalkanes have been synthesized from the electrolysis of 1,1-dinitroalkanes in alkaline solution containing sodium azide.[28] The reaction of trinitromethyl compounds[29] with lithium azide in DMF and DMSO, and the electrolysis of 1,1-dinitromethyl compounds[30] in the presence of azide anion, also generate 1-azido-1,1-dinitroalkanes.

8.1.2 Aromatic Azides

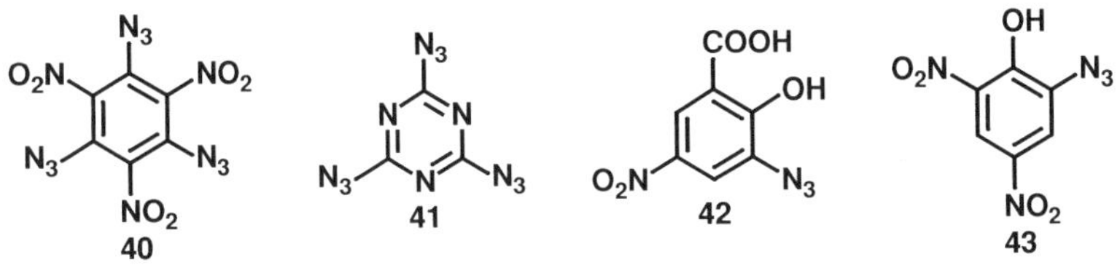

Figure 8.15

Azido groups are conveniently incorporated into aromatic rings via nucleophilic aromatic substitution of aryl halides containing nitro or other deactivating groups *o*/*p*- to the leaving group. 1,3,5-Triazido-2,4,6-trinitrobenzene (40), the product from the reaction of 1,3,5-trichloro-2,4,6-trinitrobenzene with excess sodium azide, is an explosive with VOD ~7500 m/s (at $d = 1.54$ g/cm^3) and has some prospects of practical use as a primary explosive.[31] Cyanuric triazide (41), prepared from cyanuric chloride and sodium azide, is a powerful initiator but must be considered a highly dangerous substance due to its high sensitivity to mechanical stimuli; large crystals of cyanuric triazide may detonate even on touch.[32] Lead salts of the arylazides (42) and (43) have been explored for use in detonators but prove to be poor initiators.[33]

Figure 8.16

Gilbert and Voreck[34] synthesized hexakis(azidomethyl)benzene (HAB) (45) from the reaction of hexakis(bromomethyl)benzene (44) with sodium azide in DMF. This azide has been comprehensively characterized for physical, thermochemical and explosive properties and stability. HAB is a thermally and hydrolytically stable solid and not highly sensitive to shock, friction or electrostatic charge but is sensitive to some types of impact. It shows preliminary promise for possible use as a substitute for lead styphnate in less sensitive bridgewire detonators and also for tetrazene in percussion detonators. HAB contains 62 % nitrogen and belongs to the class of 'planar radial' compounds, which have compact, symmetrical, disc-like structures, resulting in high melting point, good stability and low solubility in solvents.

8.2 PEROXIDES

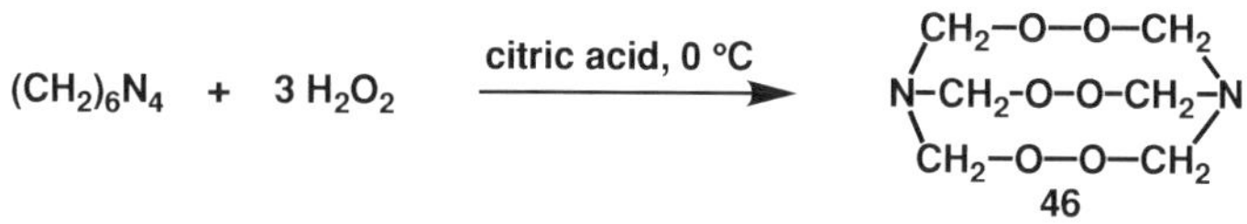

$(CH_2)_6N_4$ + 3 H_2O_2 —citric acid, 0 °C→ 46 (HMTD)

Figure 8.17

No peroxide has found practical use as an explosive, a consequence of the weak oxygen–oxygen bond leading to poor thermal and chemical stability and a high sensitivity to impact. Hexamethylenetriperoxidediamine (HMTD) (46) is synthesized from the reaction of hexamine with 30 % hydrogen peroxide in the presence of citric acid.[35] HMTD is a more powerful initiating explosive than mercury fulminate but its poor thermal and chemical stability prevents its use in detonators.

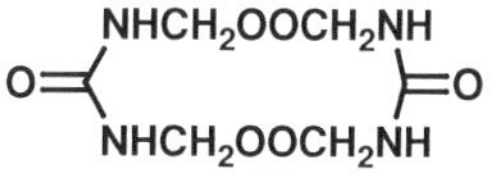

47

Figure 8.18

Another interesting dialkylperoxide explosive, which probably has the structure of (47), is synthesized by the addition of hydrogen peroxide and nitric acid to a solution of urea and formaldehyde.[36]

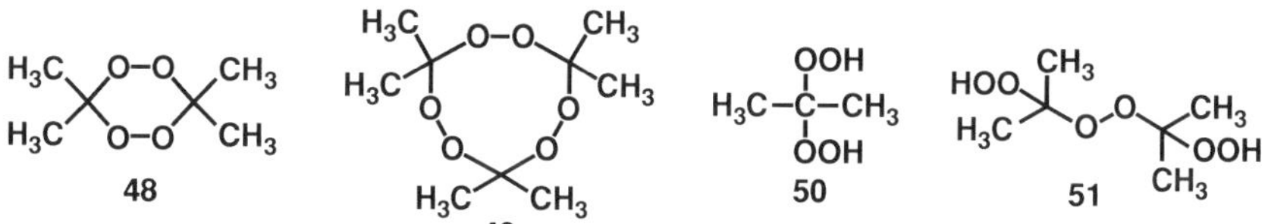

48 49 (TATP) 50 51

Figure 8.19

Some ketone-derived peroxides have explosive properties, of which the most interesting are obtained from acetone. Four acetone-derived peroxides have been synthesized. Acetone peroxide dimer (48) is obtained in 94 % yield by treating acetone with a slight excess of 86 % hydrogen peroxide in acetonitrile in the presence of concentrated sulfuric acid at subambient temperature.[37] The reaction of acetone with potassium persulfate in dilute sulfuric acid also yields acetone peroxide dimer (48).[38] Acetone peroxide trimer (49), also known as triacetone triperoxide (TATP), has been obtained as a by-product of these reactions or by the addition of acetone to a cooled solution of 1.0 equivalent of 50 % hydrogen peroxide and 0.25 equivalents of concentrated sulfuric acid. The latter experiment yields a mixture containing 90 % TATP (49) which can be purified by low temperature recrystallization from pentane.[39] In the absence of mineral acid the hydroperoxides (50) and (51) can be obtained from these reactions.[39] Thus, a mixture of acetone and 50 % hydrogen peroxide stirred at 0 °C for 3 hours produces 2,2-bis(hydroperoxy)propane (50) as the sole product. Longer reaction times lead to the formation of (51), which is formed to the extent of approximately 20 % after ten days and is isolated by low temperature recrystallization from pentane.[39]

Molecules such as TATP (49) possess explosive strength similar to TNT. Furthermore, TATP is extremely sensitive to heat and vibrational shock and can be ignited with an open flame or small electrical discharge i.e. does not need a primer unlike conventional explosives.

8.3 DIAZOPHENOLS

Diazophenols, also known as diazo oxides or diazonium phenolates, are thought to be zwitterions[40] with negative and positive charges localized on the oxo and diazo groups respectively, although tautomeric quinonoid[41] structures have not been ruled out. Most diazophenols are sensitive to impact and exhibit properties which are characteristic of primary explosives, the nitro derivatives exploding violently on the application of heat or mechanical stimuli. Studies of these compounds have focused on the synthesis of various nitro-substituted diazophenols in an attempt to fine tune properties such as chemical stability, impact and friction sensitivity, and initiating potential.

8.3.1 Diazophenols from the diazotization of aminophenols

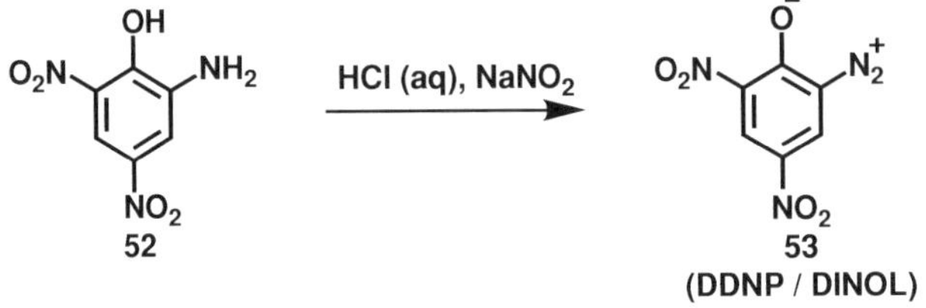

52 → 53 (DDNP / DINOL), HCl (aq), $NaNO_2$

Figure 8.20

2-Diazo-4,6-dinitrophenol (DDNP or DINOL) (53) can be prepared from the diazotization of 2-amino-4,6-dinitrophenol (52) (picramic acid) with nitrous acid;[42] the latter is obtained from the selective reduction of picric acid with ammonium sulfide.[43] 2-Diazo-4,6-dinitrophenol (53) is widely used as an initiating charge in detonators and caps.

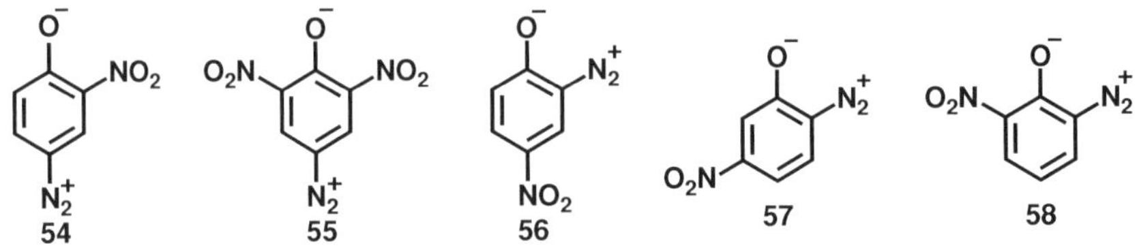

54 55 56 57 58

Figure 8.21

Vaughan and Phillips[44] prepared a number of nitro-substituted 4-diazophenol (54 and 55) and 2-diazophenol (56, 57 and 58) derivatives from the diazotization of the corresponding 4-amino- and 2-amino-phenols respectively. This work showed that nitro derivatives of 4-diazophenol are more stable than the corresponding 2-diazophenols and that the presence of a nitro group *o*/*p*- to the oxo group leads to higher stabilization compared to when the nitro group is positioned *m*- to the oxo group. Diazophenols containing two *o*/*p*- nitro groups to the oxo group show even higher stability; 2-diazo-4,6-dinitrophenol (53) and 4-diazo-2,6-dinitrophenol (55) falling into this category.

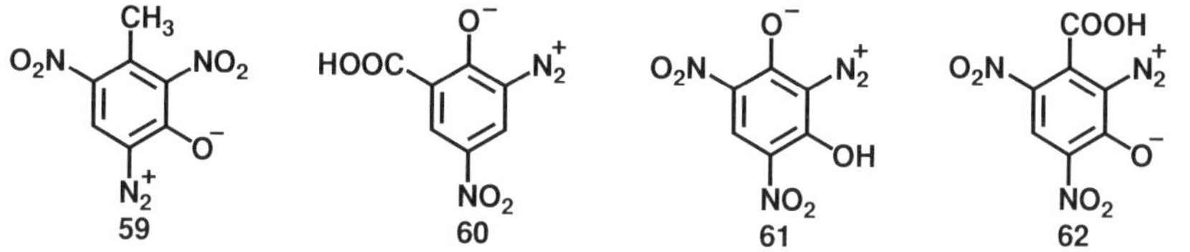

Figure 8.22

Glowiak[41] studied the stability of the four diazophenols (59–62), which he prepared from the diazotization of the corresponding aminophenols. It was noted that (61) and (62) show higher chemical stability than (59) and (60); the latter compounds were postulated to have a quinonoid structure rather than a zwitterionic diazophenol structure.

8.3.2 Diazophenols from the rearrangement of *o*-nitroarylnitramines

The nitration of anilines frequently involves the formation of a nitramine which can sometimes be isolated or may undergo a spontaneous $N \rightarrow C$ nitro group rearrangement depending on the conditions of the nitration. This is the basis of the Bamberger rearrangement (Section 4.5). However, when an aromatic nitramine contains a nitro substituent in an *ortho* position, an electrocyclic rearrangement can occur with the formal displacement of HNO_3 from the molecule and the formation of a diazophenol; in effect, the *ortho* nitro group undergoes an intramolecular displacement by the adjacent nitramino functionality.[45] Studies have shown that rearrangement is favoured when the intermediate nitramine contains a nitro group *ortho* to the nitro group being displaced i.e. *meta* to the nitramino functionality, this arrangement activating the nitro group towards displacement.[45] More than often in these cases the formation of the diazophenol is so favourable (an intramolecular process) that the intermediate nitramine is not isolated and the diazophenol is the sole product.

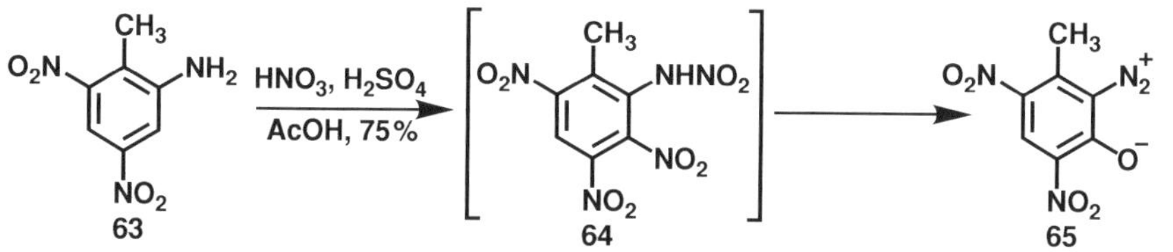

Figure 8.23

The nitration of 2-amino-4,6-dinitrotoluene (63) with a mixture of nitric acid in sulfuric and acetic acids yields 2-diazo-3-methyl-4,6-dinitrophenol (65) in 75 % yield without isolation of the intermediate nitramine (64).[46]

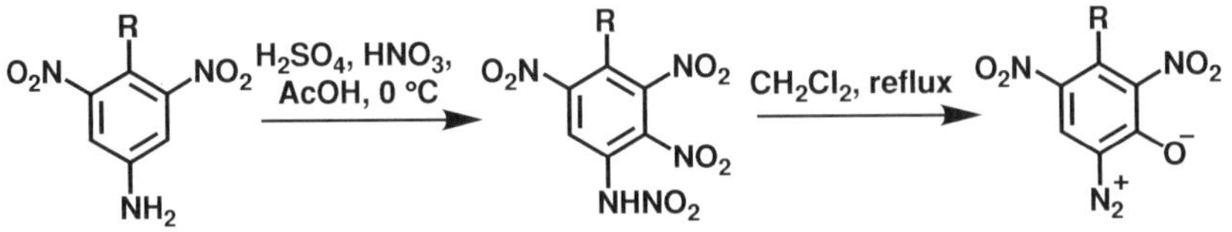

66, R = Me
67, R = Cl

68, R = Me
69, R = Cl

70, R = Me, 63% (2 steps)
71, R = Cl, 51% (2 steps)

Figure 8.24

The nitration of both 4-methyl- and 4-chloro-2,6-dinitrotoluenes (66 and 67) with mixed acid in acetic acid at subambient temperature allows the isolation of the nitramines, (68) and (69), respectively. Thermolysis of (68) and (69) in refluxing methylene chloride yields the corresponding diazophenols, (70) and (71), respectively.[46] Scilly and co-workers[47] isolated 2-diazo-4,6-dinitrophenol (DINOL) (53) from the thermolysis of *N*,2,3,5-tetranitroaniline (73) in ethyl acetate at 60 °C.

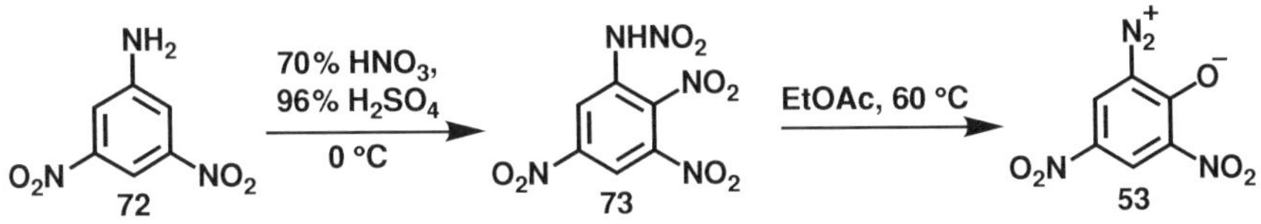

Figure 8.25

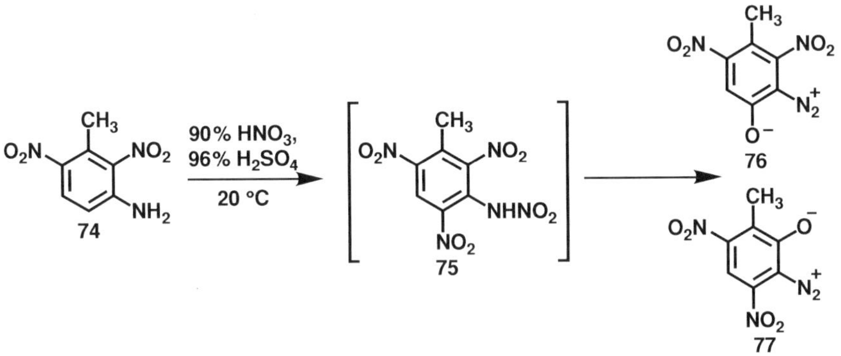

Figure 8.26

Unsymmetrical arylnitramines with two nitro groups positioned *ortho* to the nitramino functionality can yield two isomeric diazophenol products. The diazophenols (76) and (77) were isolated in a 4:1 ratio from the nitration of 3-amino-2,6-dinitrotoluene (74) with mixed acid, the reaction proceeding via the intermediate nitramine (75).[45]

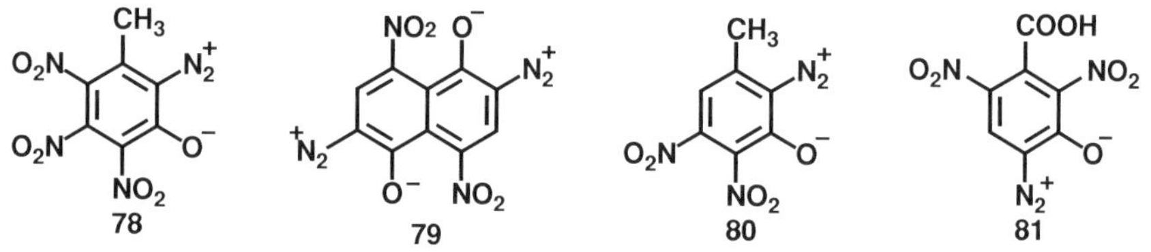

Figure 8.27

Chemists at the Naval Air Warfare Center (NAWC), China Lake, have conducted much research into the nitration of various substituted anilines as an indirect route to highly nitrated arylene hydrocarbons (Section 4.5). On numerous occasions these chemists found that diazophenols are formed as by-products and sometimes as the main or only product of a reaction. During these studies the diazophenols (65)[46] and (78–81)[45,48–50] were isolated and characterized. These diazophenols were screened for use as explosive components of both percussion and stab-sensitive primary explosive compositions.[45]

8.4 NITROGEN-RICH COMPOUNDS FROM GUANIDINE AND ITS DERIVATIVES

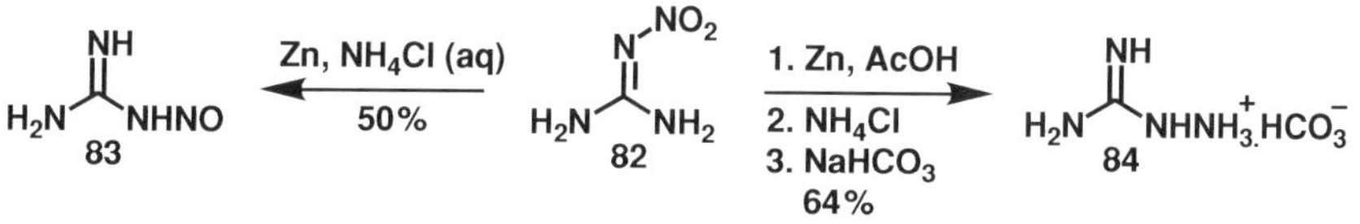

Figure 8.28

Nitroguanidine (82) is a starting material for the synthesis of a number of nitrogen-rich compounds of which many have explosive properties. Nitrosoguanidine (83) is prepared from the reduction of nitroguanidine (82) with zinc dust in the presence of aqueous ammonium chloride.[51] Nitrosoguanidine is a primary explosive but its slow decomposition on contact with water limits its use. The reduction of nitroguanidine (82) with zinc dust in aqueous acetic acid yields aminoguanidine which is usually isolated as the sparingly soluble bicarbonate salt (84).[52]

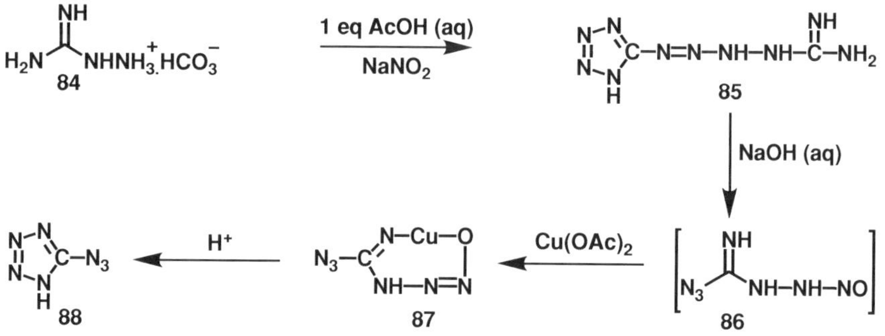

Figure 8.29

The reaction of aminoguanidine with sodium nitrite under neutral conditions yields tetrazolylguanyltetrazene hydrate (85), a primary explosive commonly known as tetrazene.[53] Tetrazene (85) is only formed in the absence of free mineral acid and so a common method for its preparation treats the bicarbonate salt of aminoguanidine (84) with one equivalent of acetic acid followed by addition of aqueous sodium nitrite.[54] Tetrazene (85) is decomposed by aqueous alkali to form triazonitrosoaminoguanidine (86) which is isolated as the cuprate salt (87) on addition of copper acetate to the reaction mixture.[55,56] Acidification of the copper salt (87) with mineral acid leads to the formation of 5-azidotetrazole (88) (CHN_7 = 88 % N).[55,56]

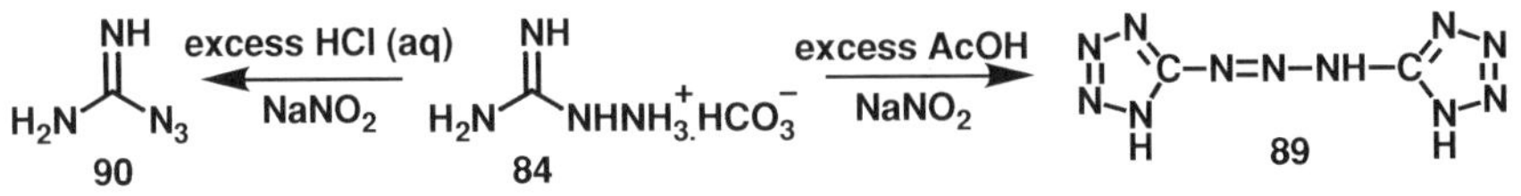

Figure 8.30

The reaction of aminoguanidine bicarbonate (84) with sodium nitrite in the presence of excess acetic acid produces 1,3-ditetrazolyltriazine (89), another nitrogen-rich heterocycle ($C_2H_3N_{11}$ = 85 % N) which readily forms explosive metal salts.[55,56] The reaction of aminoguanidine bicarbonate (84) with sodium nitrite in the presence of mineral acid yields guanyl azide (90), of which, the perchlorate and picrate salts are primary explosives.[55,56] Guanyl azide (90) reacts with sodium hydroxide to form sodium azide, whereas reaction with weak base or acid forms 5-aminotetrazole.[55,56]

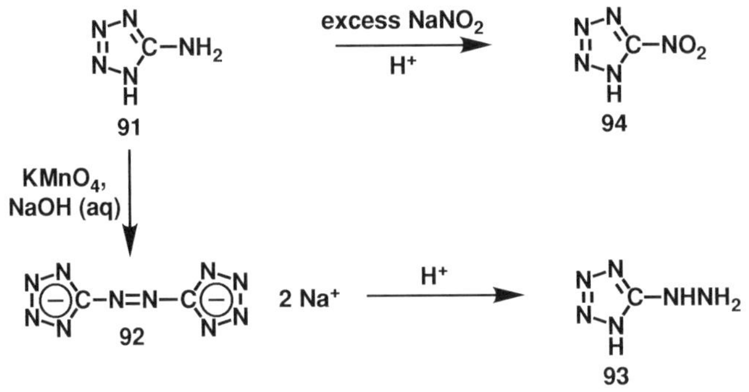

Figure 8.31

5-Aminotetrazole (91) reacts with potassium permanganate in excess aqueous sodium hydroxide to yield the disodium salt of 5-azotetrazole (92).[57] 5-Azotetrazole is unstable and attempts to isolate it by acidification yields 5-hydrazinotetrazole (93).[58] Diazotization of 5-aminotetrazole (91) in the presence of excess sodium nitrite yields 5-nitrotetrazole (94), a powerful explosive whose mercury and silver salts are primary explosives.[59]

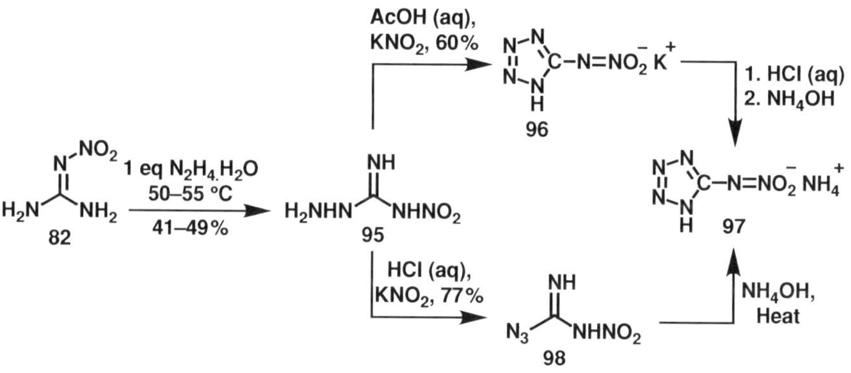

Figure 8.32

Nitroguanidine (82) undergoes hydrazinolysis on treatment with one equivalent of hydrazine hydrate to yield nitraminoguanidine (95), a compound which possesses explosive properties.[60,61] Nitraminoguanidine (95) reacts with potassium nitrite in the presence of acetic acid to yield the potassium salt of 5-nitraminotetrazole (96), whereas the same reaction in the presence of mineral acid yields the azide (98), the latter yielding the ammonium salt of 5-nitraminotetrazole (97) on heating with aqueous ammonia.[62] Reduction of nitraminoguanidine (95) with zinc dust in acetic acid yields diaminoguanidine (99).[60]

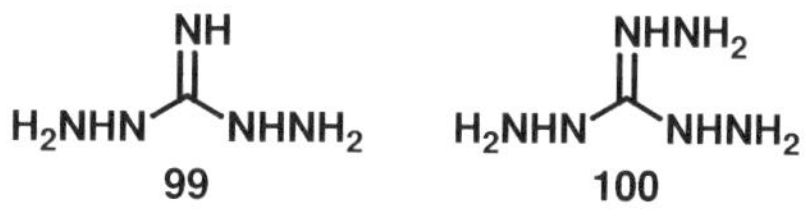

Figure 8.33

The salts formed between triaminoguanidine (100) and some oxidizing acids have attracted interest as potential components of energetic propellants. Triaminoguanidine (100) has been synthesized by treating dicyandiamide,[63] guanidine,[64] nitroguanidine[64] and diaminoguanidine[64] with an excess of hydrazine hydrate at reflux. The reaction between hydrazine hydrate and carbon tetrachloride at reflux is also reported to form triaminoguanidine (100).[65]

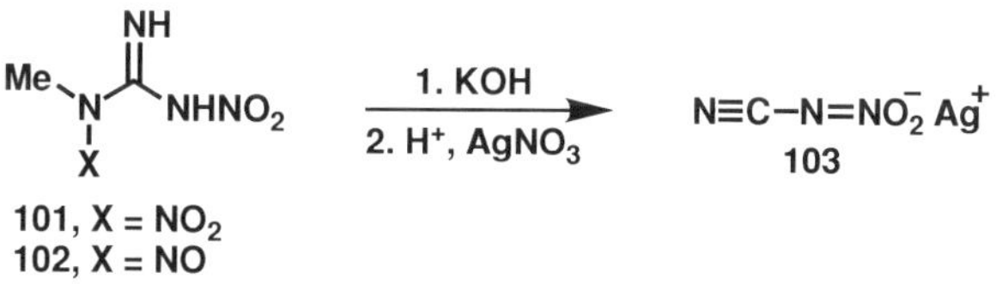

Figure 8.34

Hydrolysis of *N*′-nitro-*N*′-methylnitroguanidine (101) and *N*′-nitroso-*N*′-methylnitroguanidine (102) with aqueous potassium hydroxide results in the formation of potassium nitrocyanamide.[66] Addition of acidic silver nitrate solution to these reaction mixtures leads to the precipitation of the dangerously explosive silver salt (103) which has been used to prepare a variety of nitrocyanamide metal salts for testing as primary explosives.[67]

REFERENCES

1. I. J. Dagley and R. J. Spear, in *Organic Energetic Compounds*, Ed. P. L. Marinkas, Nova Science Publishers Inc., New York, Chapter 2, 135 (1996).
2. *The Chemistry of the Azido Group*, Ed. S. Patai, Interscience, London-New York (1971).
3. M. B. Frankel, E. R. Wilson and D. O. Woolery, 'Energetic Azido Compounds', *Ann. Rep. AD-A083770* (1980).
4. M. B. Frankel and E. R. Wilson, *US Pat.* 4 499 723 (1985), *Chem. Abstr.*, 1985, **103**, 8576h.
5. J. E. Flanagan, M. B. Frankel, E. R. Wilson and E. F. Witucki, *J. Chem. Eng. Data*, 1983, **28**, 285.
6. J. P. Agrawal, R. K. Bhongle, F. M. David and J. K. Nair, *J. Energ. Mater.*, 1993, **11**, 67.
7. D. Drees, D. Löffel, A. Messmer and K. Schmid, *Propell. Explos. Pyrotech.*, 1999, **24**, 159.
8. J. E. Flanagan, M. B. Frankel and E. R. Wilson, *US Pat. Appl.* 283 708 (1982); *Chem. Abstr.*, 1982, **96**, 165045p.
9. M. B. Frankel and E. R. Wilson, *J. Chem. Eng. Data*, 1982, **27**, 472.
10. A. Rusiecki and T. Urbański, *Wiad. Techn. Uzbr.*, 1934, **26**, 442.
11. R. V. Heisler, W. E. McEwen and C. A. Vander Werf, *J. Am. Chem. Soc.*, 1954, **76**, 1231.
12. M. B. Frankel and E. F. Witucki, *J. Chem. Eng. Data.*, 1979, **24**, 247.
13. (a) J. E. Flanagan and M. B. Frankel, *US Pat.* 4 085 123 (1978); *Chem. Abstr.*, 1978, **89**, 108104p; (b) J. E. Flanagan, M. B. Frankel and E. F. Witucki, *US Pat.* 4 141 910 (1979); *Chem. Abstr.*, 1979, **90**, 168043y.
14. E. F. Witucki, *US Pat. Appl.* 366 745 (1982); *Chem. Abstr.*, 1983, **98**, 74899v.
15. M. B. Frankel and E. F. Witucki, *US Pat.* 4 341 712 (1982); *Chem. Abstr.*, 1982, **97**, 165541t.
16. M. B. Frankel and E. F. Witucki, *J. Chem. Eng. Data*, 1982, **27**, 94.
17. M. B. Frankel, *US Pat. Appl.* 350 494 (1982); *Chem. Abstr.*, 1982, **97**, 200326a.
18. C. Boren, L. Jisheng, W. Naixing and O. Yuxiang, *Propell. Explos. Pyrotech.*, 1998, **23**, 46.
19. M. B. Frankel and E. R. Wilson, *J. Chem. Eng. Data.*, 1981, **26**, 219.
20. M. B. Frankel and E. R. Wilson, *J. Org. Chem.*, 1985, **50**, 3211.
21. (a) M. B. Frankel and E. R. Wilson, *US Pat.* 4 683 086 (1987); *Chem. Abstr.*, 1987, **107**, 179583v; (b) M. B. Frankel and E. R. Wilson, *US Pat.* 4 683 085 (1987); *Chem. Abstr.*, 1987, **107**, 217094w.
22. B. D'Andrea, A. Faure, F. Lillo and C. Perut, *Acta Astronautica*, 2000, **47**, 103.
23. (a) J. E. Flanagan and M. B. Frankel, *US Pat.* 4 268 450 (1980); *Chem. Abstr.*, 1980, **93**, 47806c; (b) J. E. Flanagan and M. B. Frankel, *Belgian. Pat.* 89 858 (1979); (c) E. Ahad, *US Pat.* 4 882 395 (1989); *Chem. Abstr.*, 1990, **112**, 159265t; (d) E. Ahad, *US Pat.* 4 891 438 (1990); *Chem. Abstr.*, 1988, **109**, 171140u; (e) J. E. Flanagan, L. R. Grant, R. I. Wagner and E. R. Wilson, *US Pat.* 4 937 361 (1981).
24. (a) J. E. Flanagan, M. B. Frankel and L. R. Grant, *J. Propulsion. Power*, 1992, **8**, 560; (b) S. N. Asthana, A. N. Nazare and H. Singh, *J. Energ. Mater.*, 1992, **10**, 43.
25. J. P. Agrawal, J. E. Field and S. M. Walley, *J. Propulsion Power*, 1997, **13**, 463.
26. S. I. Al-Khalil, W. R. Bowman and M. C. R. Symons, *J. Chem. Soc. Perkin Trans. I*, 1986, 555.
27. R. Bau, G. A. Olah, G. K. Surya Prakash, A. Schreiber, J. J. Struckhoff Jr and K. Weber, *J. Org. Chem.*, 1997, **62**, 1872.
28. C. M. Wright, *US Pat.* 3 883 377 (1975); *Chem. Abstr.*, 1975, **83**, 149779g.
29. M. B. Frankel and E. R. Wilson, *UK Pat. Appl.* GB 2 123 829 (1984); *Chem. Abstr.*, 1984, **100**, 194573r.

30. (a) M. B. Frankel and J. F. Weber, *Propell. Explos. Pyrotech.*, 1990, **15**, 26; (b) M. B. Frankel and J. F. Weber, *US Pat.* 4 795 593 (1989); *Chem. Abstr.*, 1989, **111**, 133640r; (c) M. B. Frankel and J. F. Weber, *US Pat.* 4 900 851 (1990); *Chem. Abstr.*, 1990, **113**, 23117j.
31. (a) O. Turek, *Chimie et Industrie*, 1931, **26**, 781; (b) D. Adam, K. Karaghiosoff and T. M. Klapotke, *Propell. Explos. Pyrotech.*, 2002, **27**, 7.
32. E. Ohse and E. Ott, *Chem. Ber.*, 1921, **54**, 179; *Chem. Zentr.*, 1922, 1118 and 1194; *Ger. Pat.* 350 564 (1919); *Ger. Pat.* 355 926 (1920); *US Pat.* 1 390 387 (1921); *Brit. Pat.* 170 359 (1921); *Chem. Abstr.*, 1923, **17**, 1242.
33. B. Glowiak, *Bull. Acad. Polon. Sci., Ser. Khim.*, 1960, **8**, 5.
34. E. E. Gilbert and W. E. Voreck, *Propell. Explos. Pyrotech.*, 1989, **14**, 19.
35. (a) C. Girsewald, *Chem. Ber.*, 1912, **45**, 2571; *Ger. Pat.* 274 522 (1912); (b) T. L. Davis *Chemistry of Powder and Explosives, Coll. Vol.*, Ed. T. L. Davis, Angriff Press, Hollywood, CA, 451 (reprinted 1992, first printed 1943); (c) T. Urbański, *Chemistry and Technology of Explosives, Vol. 4*, Pergamon Press, Oxford, 225–227 (1984).
36. C. Girsewald and S. Siegen, *Chem. Ber.*, 1914, **47**, 2464.
37. K. J. McCullough, A. R. Morgan, D. C. Nonhebel, P. L. Pauson and G. J. White, *J. Chem. Research (S)*, 1980, 34.
38. A. Baeyer and V. Villiger, *Chem. Ber.*, 1900, **33**, 2479.
39. A. Golubovic and N. A. Milas, *J. Am. Chem. Soc.*, 1956, **81**, 6461.
40. (a) L. A. Kazitsyna and N. D. Klyueva, *Izv. Akad. Nauk USSR, Ser. Khim.*, 1970, 192; (b) L. M. Buchneva, L. A. Kazitsyna, E. E. Milliaresi, V. E. Ruchkin and N. G. Ruchkina, *Vestn. Mosk Univ., Ser. Khim. (2)*, 1979, **20**, 465.
41. B. Glowiak, *Bull. Acad. Polon. Sci, Ser. Khim.*, 1960, **8**, 9.
42. (a) P. Griess, *Liebigs Ann. Chem.*, 1858, **106**, 123; (b) T. L. Davis, *Chemistry of Powder and Explosives, Coll. Vol.*, Ed. T. L. Davis, Angriff Press, Hollywood, CA, 443 (reprinted 1992, first printed 1943).
43. L. Molard and J. Voganay, *Mém. Poudres*, 1957, **39**, 123.
44. J. Vaughan and L. Phillips, *J. Chem. Soc.*, 1947, 1560.
45. R. L. Atkins and W. S. Wilson, *J. Org. Chem.*, 1986, **51**, 2572.
46. R. L. Atkins, R. A. Hollins and W. S. Wilson, *J. Org. Chem.*, 1986, **51**, 3261.
47. P. R. Mudge, D. A. Salter and N. R. Scilly, *J. Chem. Soc. Chem. Commun.*, 1975, 569.
48. A. P. Chafin, S. L. Christian, A. T. Nielsen and W. S. Wilson, *J. Org. Chem.*, 1994, **59**, 1714.
49. T. E. Browne, A. A. DeFusco and A. T. Nielsen, *J. Org. Chem.*, 1985, **50**, 4211.
50. R. L. Atkins, C. L. Coon, R. A. Henry, A. H. Lepie, D. W. Moore, A. T. Nielsen, W. P. Norris, D. V. H. Son and R. J. Spanggord, *J. Org. Chem.*, 1979, **44**, 2499.
51. T. L. Davis and E. N. Rosenquist, *J. Am. Chem. Soc.*, 1937, **59**, 2112.
52. R. L. Shriner and F. W. Neuman, in *Organic Syntheses, Coll. Vol. III*, Ed. E. C. Horning, John Wiley & Sons, Inc., New York, 73, (1955).
53. K. A. Hoffmann and R. Roth, *Chem. Ber.*, 1910, **43**, 682.
54. T. L. Davis, *Chemistry of Powder and Explosives, Coll. Vol.*, Ed. T. L. Davis, Angriff Press, Hollywood, CA, 449 (reprinted 1992, first printed 1943).
55. (a) H. Hock, K. A. Hoffmann and R. Roth, *Chem. Ber.*, 1910, **43**, 1087; (b) H. Hock and K. A. Hoffmann, *Chem. Ber.*, 1910, **43**, 1866; (c) H. Hock and K. A. Hoffmann, *Chem. Ber.*, 1911, **44**, 2946; (d) H. Hock, K. A. Hoffmann and K. Kirmreutler, *Liebigs Ann. Chem.*, 1911, **380**, 131.
56. G. B. L. Smith, *Chem. Rev.*, 1939, **25**, 213.
57. J. Thiele, *Liebigs Ann. Chem.*, 1892, **270**, 5, 46; 1898, **303**, 40, 57; 1899, **305**, 64.
58. (a) J. T. Marias and J. Thiele, *Liebigs Ann. Chem.*, 1893, **273**, 144; (b) H. Ingle and J. Thiele, *Liebigs Ann. Chem.*, 1895, **287**, 233; (c) J. Thiele, *Liebigs Ann. Chem.*, 1898, **303**, 66.
59. E. von Herz, *Ger. Pat.* 562 511 (1931); *Chem. Abstr.*, 1933, **27**, 1013.
60. R. Phillips and J. F. Williams, *J. Am. Chem. Soc.*, 1928, **50**, 2465.

61. R. A. Henry, R. C. Makosky and G. B. L. Smith, *J. Am. Chem. Soc.*, 1951, **73**, 474.
62. J. Cohen, R. A. Henry, F. Lieber and E. Sherman, *J. Am. Chem. Soc.*, 1951, **73**, 2327.
63. K. Krauch and R. Stollé, *J. Prakt. Chem.*, 1913, **88**, 306.
64. A. Gaiter and G. Pellizzari, *Gazz. Chim. Ital.*, 1914, **44**, 72.
65. R. Stollé, *Chem. Ber.*, 1904, **37**, 3548.
66. A. F. McKay, *Can. J. Res.*, 1950, **28B**, 683.
67. S. R. Harris, *J. Am. Chem. Soc.*, 1958, **80**, 2302.

9

Dinitrogen Pentoxide – An Eco-Friendly Nitrating Agent

9.1 INTRODUCTION

'The majority of energetic materials (explosives, propellants etc.) are organic compounds which derive their energy from the nitro (–NO_2) group. Thus the –NO_2 functionality is one of the excellent characteristics of a molecule to be considered as an explosive and therefore, nitrations or nitration reactions[1] have special place in the field of high energy materials.'[2,a]

'The nitro substituents are generally classified according to the nature of the atom to which they are attached: carbon, nitrogen or oxygen.'[2,a] Accordingly, in the previous chapters, the introduction of *C*-nitro, *N*-nitro (nitramine) and *O*-nitro (nitrate ester) functionality into organic compounds has been discussed.

'It has been usual practice to introduce such explosophoric groups in the structure of a molecule with the use of conventional nitrating agents such as: absolute nitric acid, sulfuric acid–nitric acid, acetic anhydride–nitric acid, ammonium nitrate–acetic anhydride-nitric acid etc.'[2,a]

'A large number of explosive molecules have been reported in the literature using these conventional nitrating agents and important examples are 2,4,6-trinitrotoluene (TNT), 1,3,5-trinitro-1,3,5-triazacyclohexane (RDX), 1,3,5,7-tetranitro-1,3,5,7-tetraazacyclooctane (HMX) and 2-nitro-1,2,4-triazol-5-one (NTO). Substantial progress has also been made in the synthesis of thermally stable explosives such as 1,3-diamino-2,4,6-trinitrobenzene (DATB), 1,3,5-triamino-2,4,6-trinitrobenzene (TATB), 3-picrylamino-1,2,4-triazole (PATO) and 1,3-bis(3-amino-1,2,4-triazole)-2,4,6-trinitrobenzene (BATNB) etc. with these conventional nitrating agents.'[2,a]

'However, problems associated with the use of these nitrating agents are manifold and important among them are: (1) handling of nitrating agents on large scale, (2) low yields coupled with formation of by-products, (3) maintenance/control of temperature because of the exothermic nature of nitration reactions, (4) disposal of spent acids, and lastly (5) limited use for the nitration of highly deactivated nitroaromatics.'[2,a] On an industrial scale, nitrations with conventional nitric acid-based nitrating agents require strict control, including remote handling,

[a] Reproduced by permission of Infomedia India Ltd., publisher of *Chemical World* magazine.

elaborate reactors and blast-proof buildings. Product separation is a frequent problem with the mixed acid nitration of polyols. The mixed acid residue and the aqueous washings often contain considerable amounts of dissolved nitrate ester and so, presenting both a safety and waste disposal problem. Similar problems are encountered during the nitrolysis of hexamine via the Bachmann process, with the waste liquors containing many unstable species including acetyl nitrate. Such liquors need additional plant to remove or destroy organic residues before recycling or discharging as effluents.

European and American countries are slowly moving towards the use of a new nitration methodology based on dinitrogen pentoxide (N_2O_5).[3–5] Agrawal has recently described the potential of dinitrogen pentoxide methodology for high energy materials synthesis.[2] It is now widely believed that this new nitration technology is the future for energetic materials synthesis. This chapter discusses the recent advances made in energetic materials synthesis with dinitrogen pentoxide as the nitrating agent.

9.2 NITRATIONS WITH DINITROGEN PENTOXIDE

Dinitrogen pentoxide was first prepared[6] over 150 years ago but received little attention as a nitrating agent. This is probably due to technical difficulties in preparation and its low thermal stability, requiring temperatures of $-60\,^\circ C$ for its long term storage.[4] Its real potential for the synthesis of polynitroarylenes, nitramines and nitrate esters, has only recently been recognized.[3,5] This is in large due to sustained research at the Defence Evaluation and Research Agency (DERA) in the UK. Dinitrogen pentoxide is now seen as a real alternative to nitration with conventional nitrating agents. Environmental restrictions on hazardous waste effluents are also pushing this technology forward, and in many cases, dinitrogen pentoxide constitutes an 'environmentally friendly' alternative.

There are several advantages of using dinitrogen pentoxide over mixed acid; (1) reactions are considerably faster, less exothermic and easier to control, and cleaner, often due to an absence of oxidation by-products, (2) yields are higher, (3) product isolation is often easier, (4) there is no acid waste which needs treatment and disposal, and lastly, (5) some high energy materials such as nitrated hydroxy-terminated polybutadiene (NHTPB), poly[NIMMO], poly[GLYN] and ammonium dinitramide (ADN), which cannot be synthesized with the use of conventional nitrating agents, are readily synthesized by this versatile nitrating agent. In fact, dinitrogen pentoxide can achieve all the reactions of conventional nitrating agents but usually more efficiently.

Nitrations with dinitrogen pentoxide are conducted with two types of reagent:

(1) Solutions of dinitrogen pentoxide in anhydrous nitric acid. This is a highly acidic and powerful nitrating agent. It can be used to synthesize all types of explosives i.e. polynitroarylenes, nitramines and nitrate esters. It is particularly valuable for the nitration of highly deactivated nitroaromatics, and the synthesis of energetic nitramines by nitrolysis where mixed acid fails in many cases. However, like nitrations with many conventional nitrating agents, it is unselective.

(2) Solutions of dinitrogen pentoxide in chlorinated solvents. These are less active nitrating agents but much milder and more selective in their nitrations. These reactions are more controlled and usually very clean, nonoxidizing and, consequently, high yielding. With

nonprotic substrates it constitutes a nonacidic nitrating agent of high efficiency. These reagents are well suited for the ring cleavage of strained oxygen and nitrogen heterocycles and for selective *O*-nitration in the presence of other potentially reactive functionality. These reagents can be regarded as environmentally friendly nitrating agents, largely avoiding the need for hazardous acidic waste treatment and disposal.

The nitrating properties of the above two reagent types are related to the degree of dissociation in solution. In nitric acid, dinitrogen pentoxide is highly dissociated to give a high concentration of nitronium cation.[7] In contrast, dinitrogen pentoxide is unprotonated and largely undissociated in chlorinated solvents and the concentration of nitronium cation is low.[8] These two reagents are very much complementary in nature.

9.3 THE CHEMISTRY OF DINITROGEN PENTOXIDE

Dinitrogen pentoxide is a colorless crystalline solid which sublimes without melting at 32.5 °C at atmospheric pressure.[4] Dinitrogen pentoxide is inherently unstable and readily decomposes to oxygen and dinitrogen tetroxide at room temperature as shown in Equation (9.1). The rate of decomposition is temperature dependent with a half-life of 10 days at 0 °C and 10 hours at 20 °C.[4,9] It is stable for 2 weeks at −20 °C and up to 1 year at temperatures below −60 °C.

$$2\,N_2O_5 \longrightarrow 2\,N_2O_4 + O_2 \quad \text{(Eq. 9.1)}$$

Figure 9.1

Dinitrogen pentoxide is readily soluble in absolute nitric acid and chlorinated solvents. The polarity of the solvent has a significant effect on the rate of decomposition in solution. The rate is fastest in nonpolar solvents like chloroform and slower in polar solvents like nitromethane.[10] The decomposition rate for solutions of dinitrogen pentoxide in nitric acid is very slow and these solutions are moderately stable at subambient temperatures.[10]

Being the ‘anhydride of nitric acid’, dinitrogen pentoxide readily reacts with moisture to form nitric acid.

In the solid state, dinitrogen pentoxide is ionic, existing as $NO_2^+NO_3^-$ and sometimes called nitronium nitrate.[11] The same is true of dinitrogen pentoxide in polar solvents like nitric acid where complete ionization to nitronium and nitrate ions is observed.[7] In the vapour phase, and in nonpolar solvents, a covalent structure is observed.[11] This dichotomy of behavior[12] in both physical state and in solution means that no single nitrating agent is as diverse and versatile as nitrogen pentoxide.

9.4 PREPARATION OF DINITROGEN PENTOXIDE

Deville[6] first synthesized dinitrogen pentoxide in 1849 by reacting silver nitrate with chlorine gas. This reaction probably involves the initial formation of nitryl chloride and, accordingly, dinitrogen pentoxide can also be formed from the reaction of nitryl chloride or nitryl fluoride with a metal nitrate.[13] These reactions are more of theoretical interest than of any practical value.

$$6\ HNO_3 + P_2O_5 \longrightarrow 3\ N_2O_5 + 2\ H_3PO_4 \quad \text{(Eq. 9.2)}$$

Figure 9.2

On a laboratory scale the dehydration of nitric acid with phosphorous pentoxide is a convenient route to dinitrogen pentoxide (Equation 9.2). Isolation is achieved by sublimation and collection in a cold trap at −78 °C,[14] but the quality and yield of dinitrogen pentoxide is poor if the system is not continually flushed with a stream of ozone, a consequence of facile decomposition to dinitrogen tetroxide.

Phosphorous pentoxide in absolute nitric acid is essentially a solution of dinitrogen pentoxide in nitric acid and has been used for some nitrolysis reactions.[15] Nitrogen pentoxide concentrations up to 30 % can be prepared according to the amount of phosphorous pentoxide used.

$$N_2O_4 + O_3 \longrightarrow N_2O_5 + O_2 \quad \text{(Eq. 9.3)}$$

Figure 9.3

Two routes are feasible for the industrial synthesis of dinitrogen pentoxide and both are economically dependent on the cost of electricity. The reaction of dinitrogen tetroxide with ozone (Equation 9.3) was initially reported by Guye[16] and latter by Harris and co-workers[17]. This was later exploited and developed into an industrially useful process by chemists at the Defence and Evaluation Research Agency (DERA) in the UK.[18,19] It is mainly due to the continued research effects of these chemists that dinitrogen pentoxide is now recognized as the future for energetic materials synthesis. In this process a commercial ozonizer generates a gas stream containing a 5–10 % mixture of ozone in oxygen and this is mixed in a flow reactor with dinitrogen tetroxide. This is an instantaneous reaction and the dinitrogen pentoxide is isolated by trapping as a solid in condenser tubes cooled by dry ice–acetone or a similar coolant. Dinitrogen pentoxide from this process is essentially acid-free and usually stored at a temperature of −60 °C or below until required for reaction. Chemists at DERA have also reported ozonolysis in the liquid phase, so enabling the direct production of 10 % (wt./wt.) solutions of dinitrogen pentoxide in methylene chloride.[18] The reagents prepared in this way have been used in batch and flow reactors for the pilot plant scale production of energetic materials. Devendorf and Stacy[20] working at the Naval Surface Warfare Centre (NSWC) have reported a very similar continuous process which produces 360 g of dinitrogen pentoxide per hour. The introduction of high capacity ozonizers means that there is no intrinsic limitation to scale-up.

$$N_2O_4 + 2\ NO_3^- \rightleftharpoons 2\ N_2O_5 + 2e^- \quad \text{(Eq. 9.4)}$$

$$2e^- + 4\ HNO_3 \rightleftharpoons N_2O_4 + 2\ NO_3^- + 2\ H_2O \quad \text{(Eq. 9.5)}$$

$$4\ HNO_3 \rightleftharpoons 2\ N_2O_5 + 2\ H_2O \quad \text{(Eq. 9.6)}$$

Figure 9.4

The second route, involving the electrolysis of dinitrogen tetroxide solutions in 100 % nitric acid (Equations 9.4, 9.5 and 9.6), was pioneered at the Lawrence Livermore National Laboratory (LLNL)[21] and enables the preparation of nitric acid solutions containing up to 30 % by weight strength[22] of dinitrogen pentoxide in specially designed cells. This technology was further developed into a commercial process by chemists at DERA[23] and is now used for energetic materials synthesis on an industrial scale. Rodgers and Swinton[24] working for ICI Explosives Division in collaboration with DERA have reported a procedure where the dinitrogen pentoxide generated in this system can be phase separated and isolated as the pure crystalline solid for use in nitrations in nonpolar solvents. Bloom, Fleischmann and Mellor[25] reported a modification of the electrolytic process and used tetrabutylammonium nitrate and tetrafluoroborate as electrolytes for the electrochemical oxidation of solutions of dinitrogen tetroxide in methylene chloride. Chapman and Smith[26] reported the use of sodium fluoride as a base to remove nitric acid from electrochemically generated solutions of dinitrogen pentoxide in nitric acid, so enabling the preparation of acid-free solutions of dinitrogen pentoxide in inert solvents.

Equipment needed for the above procedures is not always available in the standard laboratory. A useful and widely used method for preparing solutions of dinitrogen pentoxide in nitric acid involves the distillation of mixtures of oleum and potassium nitrate in absolute nitric acid.[27] Another method uses a solution of sulfur trioxide and ammonium nitrate in nitric acid.[27] Although the original report[27] states that solutions of 28–42 % dinitrogen pentoxide in nitric acid can be prepared via this method, a later report[22] suggests that concentrations higher than 30 % are not attainable.

Dinitrogen pentoxide has also been synthesized from the reaction of nitric oxide (NO) and dinitrogen trioxide (N_2O_3) with ozone.[28] The reaction of nitric acid containing a trace of picric acid also yields dinitrogen pentoxide.[29]

9.5 *C*-NITRATION

Dinitrogen pentoxide has been used as a solution in organic solvents and in absolute nitric acid for aromatic nitrations. The former are mild and nonoxidizing nitrating agents and well suited for the nitration of acid-sensitive or activated substrates prone to oxidation i.e. phenols. Solutions of dinitrogen pentoxide in absolute nitric acid are similar to powerful nitrating agents like anhydrous mixed acid and valuable for the synthesis of polynitroarylenes. In this way the systems are complimentary in nature and quite unique amongst conventional nitrating agents.

The nitration of aromatic substrates with dinitrogen pentoxide in various organic solvents has been part of an extensive study by Ingold and co-workers.[30] The kinetics and mechanism of these reactions are complex and beyond the scope of this discussion but some important observations have been made. In polar solvents, nitrations with dinitrogen pentoxide are believed to involve the nitronium ion[31] and the activity is directly dependent on the concentration of nitronium ion in that medium. In nonpolar solvents the nitration probably proceeds via unionized dinitrogen pentoxide. The addition of salts like tetraethylammonium nitrate has a large accelerating effect on reaction rates in organic solvents, a point explained by the greater ionization of dinitrogen pentoxide in this medium which leads to an increase in nitronium ion concentration. This effect is seen with dinitrogen pentoxide in nitric acid where the dinitrogen pentoxide becomes fully ionized and a powerful nitrating agent.[7,10,32] For an in depth mechanistic discussion the readers are directed to an excellent review by Fischer.[3]

Nitrations with dinitrogen pentoxide in organic solvents are limited to activated and moderately deactivated aromatic substrates. Solutions of dinitrogen pentoxide in acetonitrile convert resorcinol to 2,4,6-trinitroresorcinol (styphnic acid) in quantitative yield;[24] the heavy metal salts of the latter finding use in ignition and pyrotechnic delay compositions. Phenol is quantitatively converted to 2,4,6-trinitrophenol (picric acid) with a solution of dinitrogen pentoxide in carbon tetrachloride.[3] The same reagent can be used to convert benzene to nitrobenzene; further reaction generates 1,3-dinitrobenzene but no 1,3,5-trinitrobenzene, even at temperatures up to 76 °C.[33] The formation of by-products is not uncommon during slow nitrations, or those conducted above room temperature, and may result from the decomposition of dinitrogen pentoxide.[34]

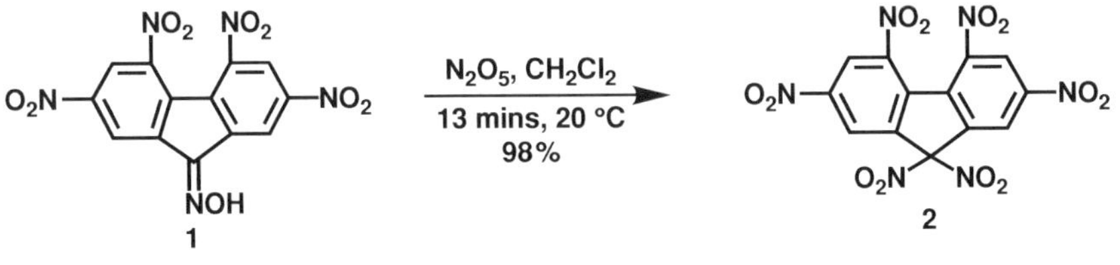

N_2O_5, CH_2Cl_2, 13 mins, 20 °C, 98%

1 → 2

Figure 9.5

The use of dinitrogen pentoxide in the Ponzio reaction for the oxidation–nitration of oximes to *gem*-dinitro groups has been reported by Russian chemists.[35] Millar and co-workers[36] extensively investigated these reactions and reported the synthesis of 2,4,5,7,9,9-hexanitrofluorene (2), a thermally stable explosive with an oxygen balance better than TNT. Other energetic materials containing *gem*-dinitro functionality were synthesized from the oximes of acetophenone, 4-nitroacetophenone, α-nitroacetophenone and 2-hydroxyacetophenone.

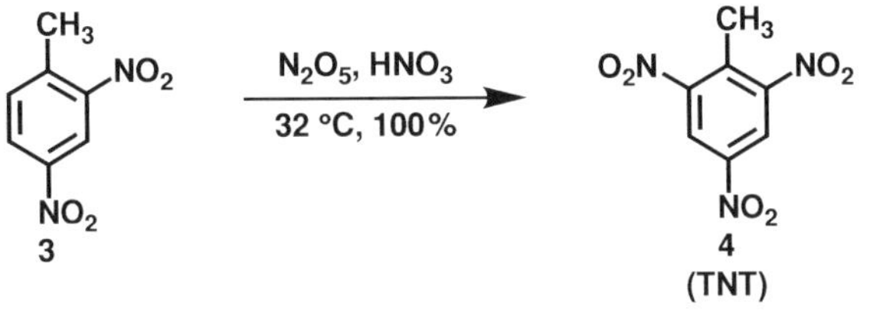

N_2O_5, HNO_3, 32 °C, 100%

3 → 4 (TNT)

Figure 9.6

Dinitrogen pentoxide in nitric acid is suitable for the synthesis of deactivated polynitroarylenes. This reagent has been used to convert 2,4-dinitrotoluene (3) to 2,4,6-trinitrotoluene (TNT) (4) in quantitative yield at a temperature of only 32 °C.[37] The fast rate of nitrations compared to those using conventional nitrating agents is notable. Moodie and co-workers[38] noted that at high concentrations of dinitrogen pentoxide in nitric acid i.e. 2.5–5 mol/L, the rate of nitration can increase some 15–30 times faster than nitrations at low concentrations of dinitrogen pentoxide and those using mixed acid of similar concentration.

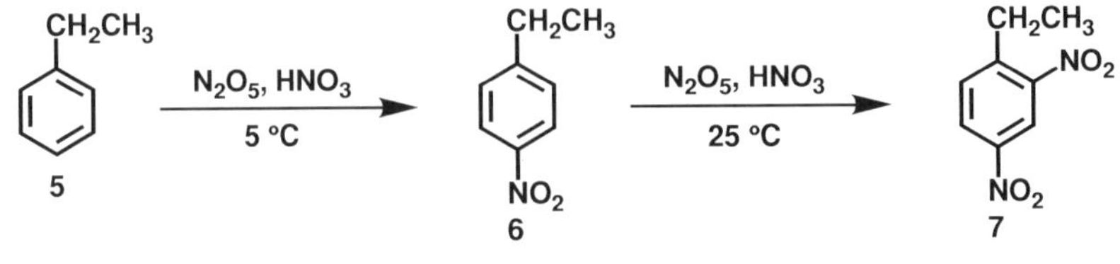

5 → (N_2O_5, HNO_3, 5 °C) → 6 → (N_2O_5, HNO_3, 25 °C) → 7

Figure 9.7

The nitration of ethylbenzene (5) to 4-nitroethylbenzene (6) is complete in 5 minutes at 5 °C and further nitration to 2,4-dinitroethylbenzene (7) requires only 10 minutes at 25 °C.[39,40] 2,4-Dinitroethylbenzene is a component of the energetic plasticizer known as K-10 (2:1 mixture of 2,4-dinitroethylbenzene and 2,4,6-trinitroethylbenzene).

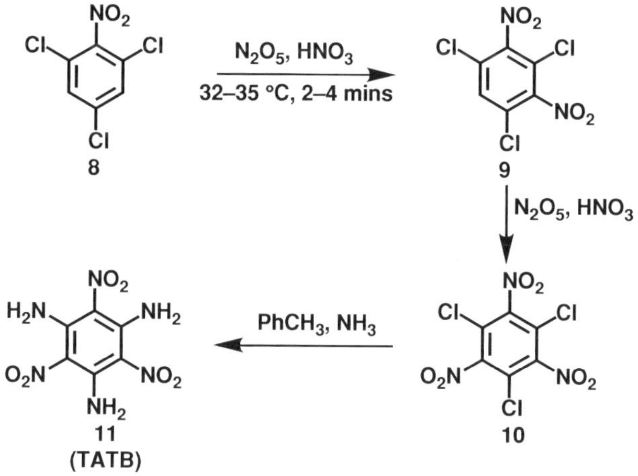

Figure 9.8

The nitration of 1,3,5-trichloro-2-nitrobenzene (8) to 1,3,5-trichloro-2,4-dinitrobenzene (9) with dinitrogen pentoxide in absolute nitric acid goes to completion in only 2–4 minutes at 32–35 °C.[38] Further nitration of (9) would yield 1,3,5-trichloro-2,4,6-trinitrobenzene (10) which undergoes ammonolysis on treatment with ammonia in toluene to give the thermally stable explosive 1,3,5-triamino-2,4,6-trinitrobenzene (TATB) (11). The same sequence of reactions with 1,3-dichloro-2-nitrobenzene provides a route to 1,3-diamino-2,4,6-trinitrobenzene (DATB). Such reactions are clean and occur in essentially quantitative yield.

Dinitrogen pentoxide has been used for aromatic nitration in other media and some are notable. While a solution of dinitrogen pentoxide in carbon tetrachloride will not convert 1,3-dinitrobenzene to 1,3,5-trinitrobenzene, a solution in concentrated sulfuric acid at 160 °C will effect this conversion.[33] In this medium dinitrogen pentoxide is fully ionized to nitronium and nitrate ions.[41] Solutions of dinitrogen pentoxide in liquid sulfur trioxide have been used for the nitration of some deactivated pyridines.[42]

The complex formed between dinitrogen pentoxide and boron trifluoride is a powerful nitrating agent and has been used in organic solvents like carbon tetrachloride, nitromethane and sulfolane for the nitration of deactivated aromatic substrates, including the conversion of benzoic acid to 1,3-dinitrobenzoic acid in 70 % yield.[43] Olah and co-workers[44] studied aromatic nitrations with dinitrogen pentoxide in the presence of other Lewis acids.

Nitrations with dinitrogen pentoxide in the presence of phosphorous pentoxide enable the complete utilization of available nitrogen.[45]

9.6 *N*-NITRATION

The direct *N*-nitration of secondary amines with acidic reagents is only possible in the case of substrates of low basicity. However, the presence of catalytic amounts of chloride ion in the form of zinc chloride or the hydrochloride salt of the amine does allow the *N*-nitration of more basic amines under acid conditions.[46] Such a route has been used to synthesize many energetic polynitramine explosives. The powerful nitrating agent composed of 20–30% dinitrogen pentoxide in nitric acid has found extensive use in this respect.

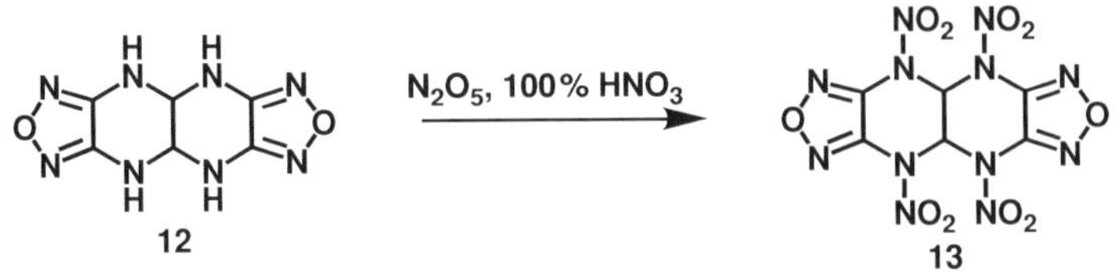

N_2O_5, 100% HNO_3

12 13

Figure 9.9

The furazan-based heterocycle (12) is *N*-nitrated to the corresponding nitramine (13) with nitrogen pentoxide in nitric acid without chloride catalyst because of the inherent low basicity of the methylenediamine functionality.[47]

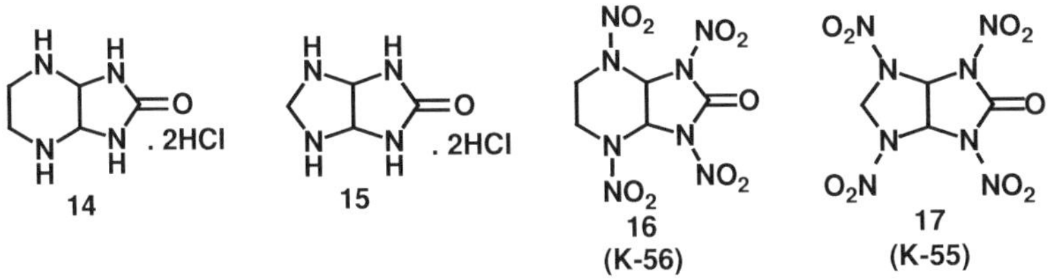

14 15 16 (K-56) 17 (K-55)

Figure 9.10

The *N*-nitration of urea and imide functionality is usually quite difficult compared to the analogous amides. Solutions of dinitrogen pentoxide in nitric acid are frequently employed to synthesize energetic *N*,*N*′-dinitroureas. Yields are often very high and go to completion unlike with many conventional nitrating agents which frequently yield partially nitrated products. Energetic bicycles K-56 (16)[48] and K-55 (17)[49] have been prepared from the dihydrochloride salts (14) and (15) respectively.

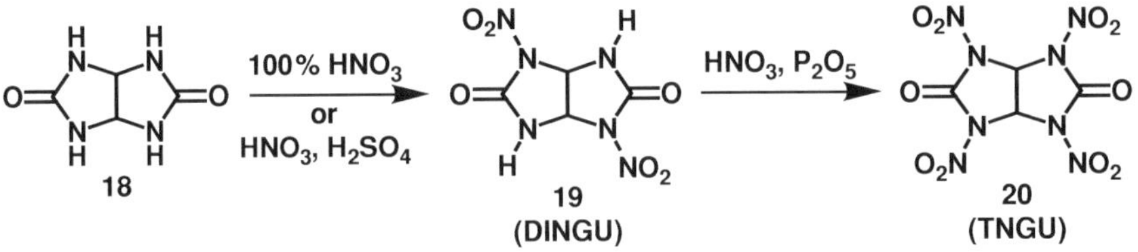

100% HNO_3 or HNO_3, H_2SO_4 ; HNO_3, P_2O_5

18 19 (DINGU) 20 (TNGU)

Figure 9.11

The synthesis of 1,3,4,6-tetranitroglycouril (20) (TNGU or Sorguyl) uses a mixture of phosphorous pentoxide in absolute nitric acid to effect the nitration of 1,4-dinitroglycouril (DINGU) (19).[50] This reagent is a convenient but crude source of dinitrogen pentoxide in nitric acid when the pure reagent is not readily available.

Solutions of dinitrogen pentoxide in aprotic solvents like methylene chloride, chloroform and carbon tetrachloride are efficient reagents for the *N*-nitration of secondary amines.[51] These reactions, known as 'nucleophilic nitrations', need excess amine present to react with the nitric acid formed during the reaction. Such nitrations are useful for the synthesis of secondary nitramines.[51]

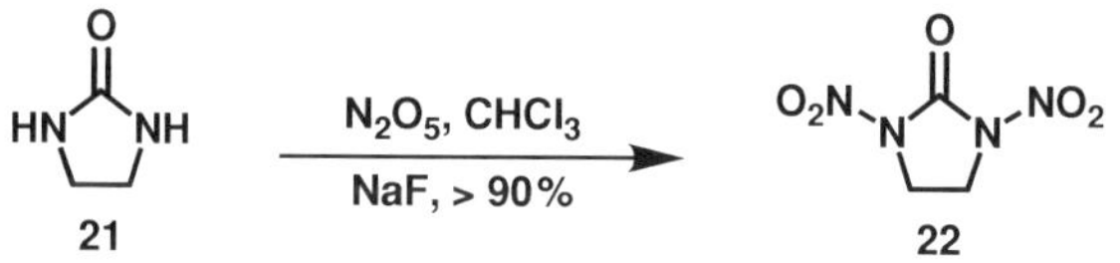

N_2O_5, $CHCl_3$; NaF, > 90%; 21 → 22

Figure 9.12

Dinitrogen pentoxide in chloroform has been used for the *N*-nitration of amides, imides and ureas, where yields are generally excellent.[52] 2-Imidazolidinone (21) is converted to *N*,*N*′-dinitro-2-imidazolidinone (22) in greater than 90% yield with this reagent in the presence of sodium fluoride.[53]

9.7 NITROLYSIS

Some of the major advances in the synthesis of energetic materials have arisen from the use of dinitrogen pentoxide–nitric acid solutions for nitrolysis reactions. These reactions are important routes to alicyclic, cyclic and caged energetic polynitramines.

The nitrolysis of hexamine is a direct route to the military high explosives 1,3,5-trinitro-1,3,5-triazacyclohexane (RDX) and 1,3,5,7-tetranitro-13,5,7-tetraazacyclooctane (HMX).[1,46] The direct nitrolysis of hexamine with dinitrogen pentoxide in absolute nitric acid provides RDX in 57 % yield.[54] RDX prepared by this process is exceptionally pure, but other reagents, like ammonium nitrate–nitric acid–acetic anhydride, give much higher yields, partly because they use ammonium nitrate to supplement for ammonium nitrogen deficiency in the reaction.

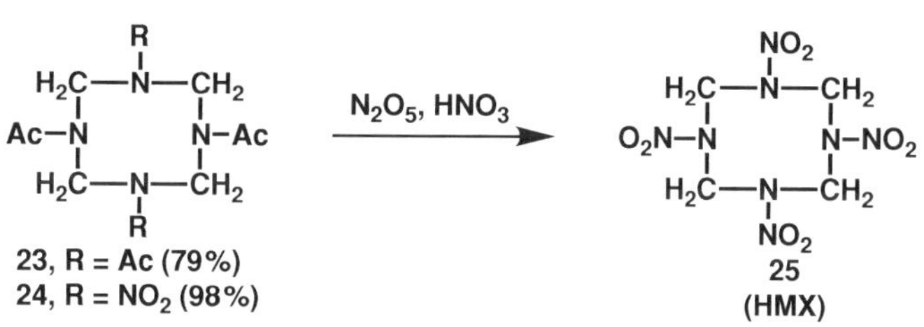

23, R = Ac (79%)
24, R = NO_2 (98%)
N_2O_5, HNO_3
25 (HMX)

Figure 9.13

The synthesis of HMX from the nitrolysis of hexamine with conventional reagents is far more problematic. However, HMX (25) is synthesized in high yield from the nitrolysis of 1,3,5,7-tetraacetyl-1,3,5,7-tetraazacyclooctane (TAT) (23) and 1,5-diacetyl-3,7-dinitro-1,3,5,7-tetraazacyclooctane (DADN) (24) with dinitrogen pentoxide in nitric acid.[55] DADN (24) is readily synthesized from the acetolysis of hexamine followed by mild nitration with mixed acid. The synthesis of HMX (25) via the nitrolysis of DADN (24) is now a pilot plant process in the U.S with consideration for scale-up to industrial production.[55] The industrial availability of the dinitrogen pentoxide–nitric acid reagent makes this a real possibility given that the overall yield for HMX in very high, and although this reagent is still expensive to manufacture, so is acetic anhydride which is used in large excess in the conventional Bachmann nitrolysis.[56] The nitrolysis of TAT (23) and DADN (24) fails with some conventional nitrating agents.

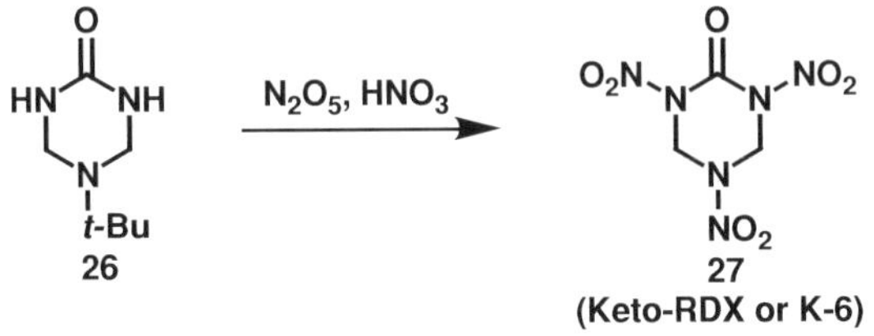

N_2O_5, HNO_3

26

27

(Keto-RDX or K-6)

Figure 9.14

Keto-RDX (27) has been synthesized from the nitrolysis–nitration of 2-oxa-5-*tert*-butyltriazone (26) with dinitrogen pentoxide in nitric acid.[57] Pagoria and co-workers[49] also reported the use of a very powerful nitrolysis agent for this purpose, composed of dinitrogen pentoxide–trifluoroacetic anhydride–nitric acid.

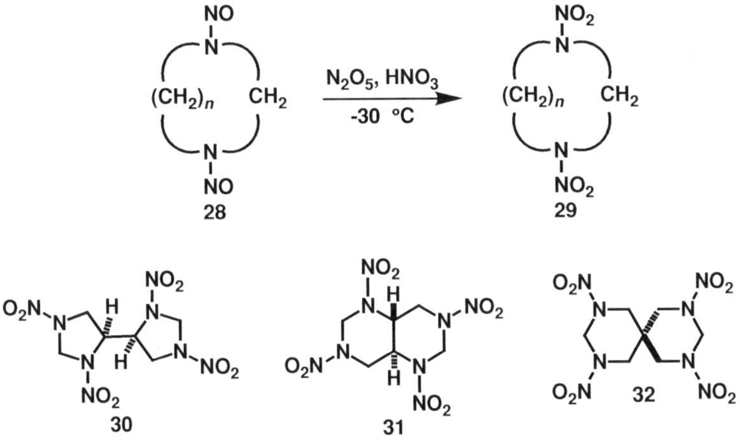

N_2O_5, HNO_3

-30 °C

28

29

30

31

32

Figure 9.15

Willer and Atkins[58] used solutions of 30 % nitrogen pentoxide in nitric acid for the nitrolysis of 1,3-dinitrosoamines (28); the latter synthesized from the reaction of 1,3-diamines with formaldehyde followed by *in situ* nitrosation of the resulting 1,3-diazacycles. Cyclic, bicyclic and spirocyclic polynitramines like (30), (31) and (32) have been synthesized via this method. Incomplete nitrolysis is observed when absolute nitric acid alone is used in these reactions.

Dinitrogen pentoxide in aprotic solvents is not usually used for nitrolysis reactions. The nitrogen atoms of amides and related compounds are usually quite electron deficient and these reagents are too mild to affect cleavage. However, the nitrolysis of symmetrical methylenediamines has been reported[59] with dinitrogen pentoxide in methylene chloride–acetonitrile, although yields are generally low.

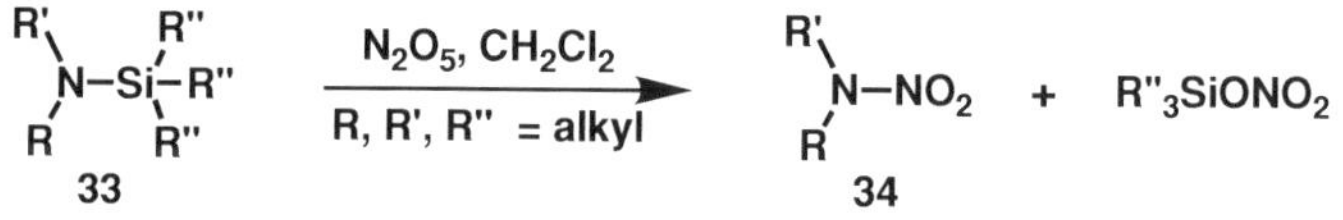

$$R'R N{-}SiR''_3 \xrightarrow[R,\ R',\ R''\ =\ alkyl]{N_2O_5,\ CH_2Cl_2} R'R N{-}NO_2 + R''_3SiONO_2$$

33 → **34**

Figure 9.16

Millar and Philbin[60] reported using dinitrogen pentoxide in methylene chloride at subambient temperatures for the nitrodesilylation of silylamines (33) and silylamides. The methodology, which results in nitramines (34), nitramides and nitroureas in excellent yields, is well suited for the synthesis of energetic materials and we believe it will find wide use in the future.

9.8 *O*-NITRATION

Solutions of dinitrogen pentoxide in anhydrous nitric acid are unlikely to find wide use for the industrial synthesis of nitrate esters – synthesis from the parent alcohol is relatively straightforward and so this reagent, which is more expensive than mixed acid, holds few advantages.

In contrast, the *O*-nitration of alcohols and polyols with a solution of dinitrogen pentoxide in chlorinated solvent has many advantages over conventional nitrating agents.[3,61] It is particularly useful for easily oxidized substrates and cases where the nitrate ester shows some solubility in mixed acid or water e.g. ethylene glycol dinitrate (0.5 g/100 ml of water). Aqueous liquors from such processes are hazardous and need treatment before disposal or acid recovery. *O*-Nitration with dinitrogen pentoxide enables the isolation of the nitrate ester as a solution in an organic solvent and, hence, greatly reduces the risk of accidental explosion. *O*-Nitrations with dinitrogen pentoxide in chlorinated solvents are much less exothermic and more controllable that those with mixed acid. Reactions are also very rapid and many are complete in seconds. Polyols which are only sparingly soluble in the reaction solvent are readily converted to the corresponding nitrate ester; mannitol (35) is converted to mannitol hexanitrate (36) in near quantitative yield.[62]

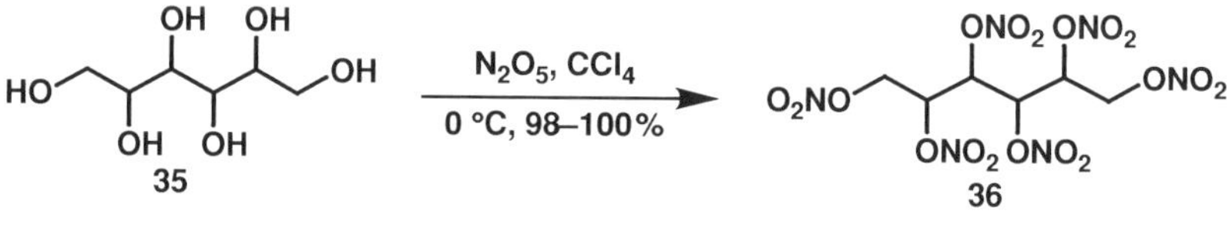

$$HOCH_2(CHOH)_4CH_2OH \xrightarrow[0\ °C,\ 98\text{–}100\%]{N_2O_5,\ CCl_4} O_2NOCH_2(CHONO_2)_4CH_2ONO_2$$

35 → **36**

Figure 9.17

Nitric acid is formed during these reactions and in some cases the addition of sodium fluoride as a base can be advantageous.[63] This is the case with higher carbohydrates like cellulose and starch.[63,64]

9.9 RING CLEAVAGE NITRATION

Chemists at DERA in the UK have reported some of the most important routes to nitramines and nitrate esters to appear in the past decade. This work involves the use of dinitrogen pentoxide in chlorinated solvents for the ring cleavage of strained ring nitrogen and oxygen heterocycles. Reactions proceed under nonacidic conditions and are incredibly efficient, incorporating all the dinitrogen pentoxide into the product.

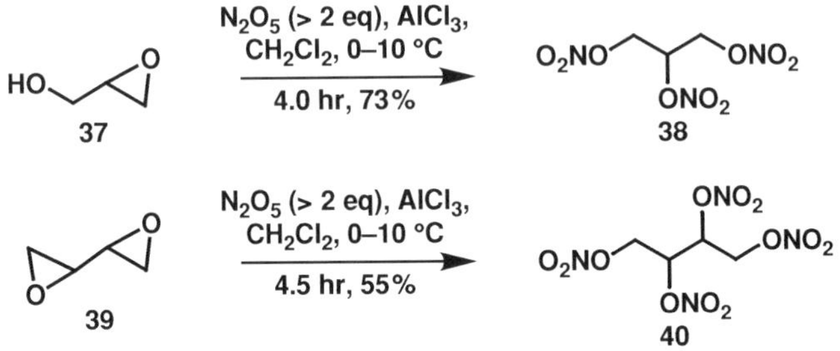

N_2O_5 (> 2 eq), $AlCl_3$, CH_2Cl_2, 0–10 °C, 4.0 hr, 73%: 37 → 38

N_2O_5 (> 2 eq), $AlCl_3$, CH_2Cl_2, 0–10 °C, 4.5 hr, 55%: 39 → 40

Figure 9.18

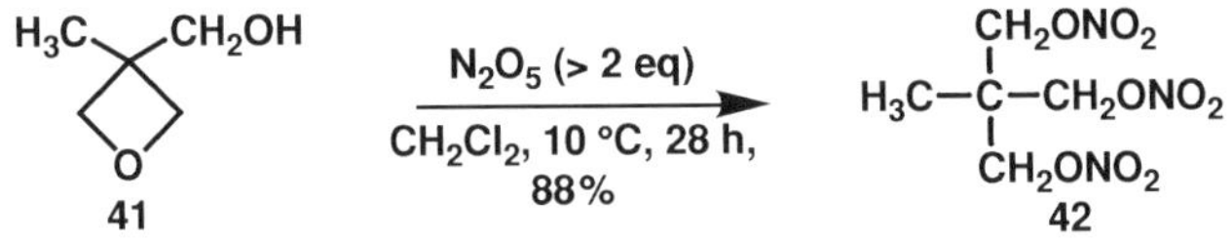

N_2O_5 (> 2 eq), CH_2Cl_2, 10 °C, 28 h, 88%: 41 → 42

Figure 9.19

Dinitrogen pentoxide in chlorinated solvents reacts with epoxides[65,66] and oxetanes[66,67] to yield 1,2-dinitrate and 1,3-dinitrate esters respectively. Reactions are notably clean and frequently high yielding for many substrates. Some epoxides containing Lewis basic oxygen functionality next to the epoxide ring react slowly and so aluminium chloride is added. The route is general and has been used to synthesize energetic nitrate ester plasticizers like nitroglycerine (NG) (38) and metriol trinitrate (MTN) (42) and high explosives like erythritol tetranitrate (40). The latter is not a practical explosive because the starting material, erythritol, is derived from scarce natural resources i.e. some types of seaweed. The scarcity of other polyols makes this route all the more attractive for the synthesis of their nitrate esters.

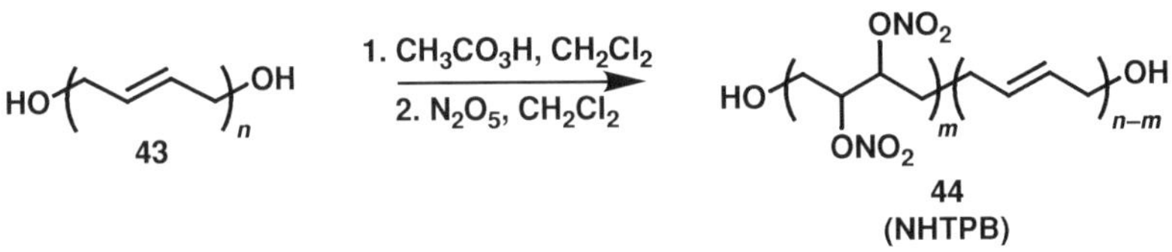

1. CH_3CO_3H, CH_2Cl_2; 2. N_2O_5, CH_2Cl_2: 43 → 44 (NHTPB)

Figure 9.20

The synthesis of nitrated hydroxy-terminated polybutadiene (NHTPB) (44) from the reaction of partially epoxidized hydroxy-terminated polybutadiene with dinitrogen pentoxide in methylene chloride is an interesting application of this chemistry.[68] NHTPB with ~10% of double bonds converted to dinitrate ester groups is an attractive energetic polymer with potential use as a binder in propellant/explosive formulations. The synthesis of NHTPB via this route has undergone scale-up and further evaluation.

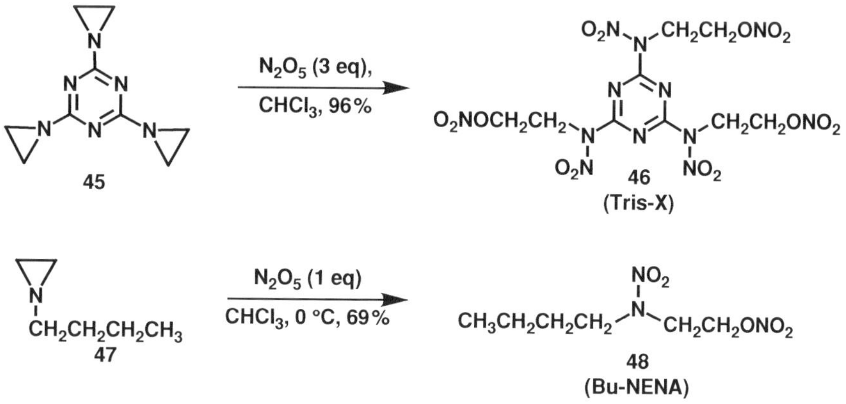

Figure 9.21

The reaction of aziridines[69,70] and azetidine[70,71] heterocycles with dinitrogen pentoxide in chlorinated solvents yields 1,2-nitramine nitrates and 1,3-nitramine nitrates respectively. In most cases yields are good to excellent, but, reactions are not as general as with the oxygen heterocycles; the outcome of reactions is heavily dependent on the nature of the substituent on the exocyclic nitrogen. Some of the products from these reactions find use as melt-castable explosives i.e. Tris-X (46)[72] and energetic plasticizers e.g. Bu-NENA (48);[73] the latter is a component of some LOVA (low vulnerability ammunition) propellant formulations.

9.10 SELECTIVE *O*-NITRATION

Chemists at DERA found that dinitrogen pentoxide in aprotic solvents, specifically chlorinated hydrocarbons, can be used for the selective *O*-nitration of substrates containing other potentially reactive functionality. This is only possible because such reagents are mild and nonacidic nitrating agents. The *O*-nitration of alcohols under these conditions is exceptionally fast and is complete, in many cases, within a matter of seconds. Other reactions like the ring-opening nitration of epoxides and oxetane heterocycles are much slower in comparison. This enables the synthesis of a class of interesting compounds containing the energetic nitrate ester functionality with a reactive strained ring heterocycle. Such compounds are precursors to energetic polymers and binders.

The success of these selective nitrations relies on the application on recent nitration technology – the use of flow reactors. The following is a discussion of some important work conducted at DERA and illustrates the research and development transition from laboratory to pilot plant scale synthesis of energetic materials.[19]

9.10.1 Glycidyl nitrate and NIMMO – batch reactor verses flow reactor[19]

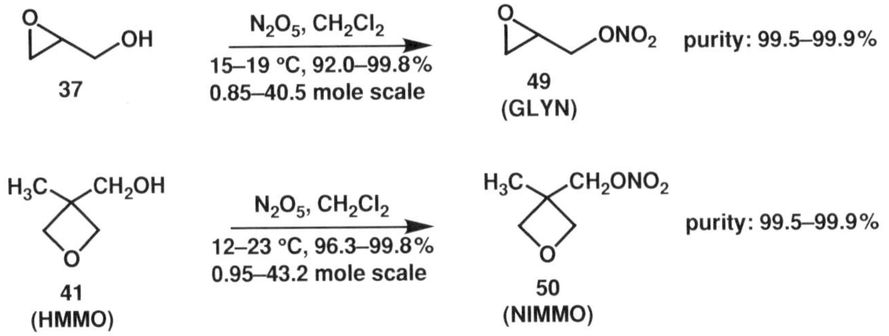

37 → N_2O_5, CH_2Cl_2, 15–19 °C, 92.0–99.8%, 0.85–40.5 mole scale → 49 (GLYN) purity: 99.5–99.9%

41 (HMMO) → N_2O_5, CH_2Cl_2, 12–23 °C, 96.3–99.8%, 0.95–43.2 mole scale → 50 (NIMMO) purity: 99.5–99.9%

Figure 9.22

The potential of dinitrogen pentoxide for selective nitration is illustrated during the semi-industrial synthesis of glycidyl nitrate (GLYN) (49) from glycidol (37) and 3-nitratomethyl-3-methyloxetane (NIMMO) (50) from 3-hydroxymethyl-3-methyloxetane (HMMO) (41). Glycidyl nitrate (49) and NIMMO (50) are precursors to the energetic plasticizers and binders known as poly[GLYN] and poly[NIMMO].

The nitrations of glycidol (37) and HMMO (41) have been conducted in flow reactors. The flow reactor consists of two reservoirs filled with the reactant and a solution of dinitrogen pentoxide in methylene chloride, respectively. These can be pumped at specific rates into the flow reactor – a packed column of glass beads surrounded by a cooling jacket and fitted with a temperature probe. The reactants are continuously fed through the packed column where they mix and react. The product stream is mixed with a continuous stream of aqueous alkali solution and run into a stirred vessel to neutralize excess dinitrogen pentoxide and nitric acid formed in the reaction. The product remains in the methylene chloride layer and is separated from the aqueous phase in a continuous separator. This is a continuous process but at any one time only a relatively small amount of reacting species and product are present in the reactor. This is clearly a very attractive manufacturing route for energetic materials because of the reduced hazards compared to batch reactors where all the materials are present in one pot until the completion of the reaction. This has other advantages relating to safety – exothermic reactions are easier to control because of the low reaction volume and reactions can be stopped immediately by ceasing to pump reactants from the reservoir.

Flow reactors are ideal for the synthesis of large amounts of material where the primary reaction is very fast and the secondary competing reaction is relatively slow. The *O*-nitration of glycidol (37) and HMMO (41) are therefore ideal reactions. Both of these compounds contain two potential reaction centres – the hydroxy groups and the strained heterocyclic rings. Initial *O*-nitration of these substrates is known to be extremely fast. The competing ring cleavage is suppressed in the flow reactor by immediately quenching the reaction stream as it leaves the reactor.

Reaction optimization was achieved by varying flow rate, concentration and the reaction exotherm temperature. In this way glycidol nitration reactions were scaled-up from 0.85 moles up to 40.5 moles. In a single run 4.64 kg of glycidyl nitrate (99.8 % yield) of 99.9 % purity was produced. Similar optimization for HMMO nitration produced 5.5 kg of NIMMO (99.1 % yield) of 99.6 % purity in a single run.

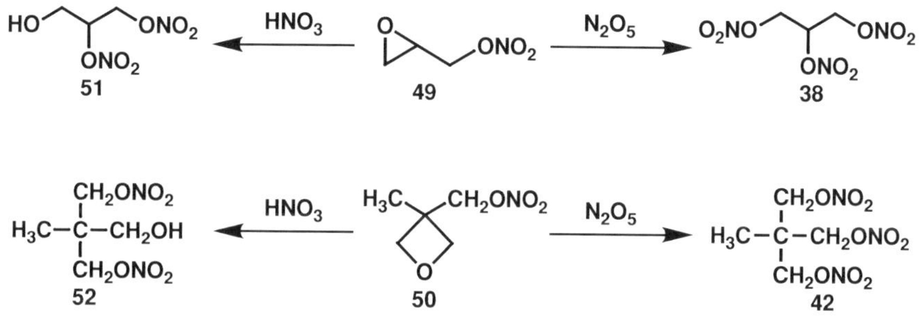

Figure 9.23

It is interesting to note that the nitration of HMMO (41) to NIMMO (50) in a batch reactor cannot be scaled-up above 1 mole because of competing ring cleavage leading to the production of metriol dinitrate (52) and trinitrate (42) as impurities. The nitration of glycidol (37) to glycidyl nitrate (49) under batch reactor conditions is unworkable, producing significant amounts of glycerol-1,2-dinitrate (51) and nitroglycerine (38).

9.11 SYNTHESIS OF THE HIGH PERFORMANCE AND ECO-FRIENDLY OXIDIZER – AMMONIUM DINITRAMIDE

Most composite propellants in wide use today use ammonium perchlorate as the oxidizer component. While such propellants benefit from high specific impulse and a lower vulnerability than double-base propellants based on nitroglycerine, they have a significant environmental impact, the exhaust fumes being rich in corrosive hydrogen chloride leading to acid rain after the launch of missiles/rockets.

The dinitramide anion is one of the most significant recent discoveries in the field of energetic materials. Ammonium dinitramide (ADN) (55) is of particular interest as a chlorine-free alternative to ammonium perchlorate. This is a high-energy oxidizer with a large amount of available oxygen for combustion. Propellants based on ammonium dinitramide exhibit higher specific impulse[74] than conventional propellants and the exhaust gases are rich in nitrogen, and so tactically reducing the exhausts radar signature.[75]

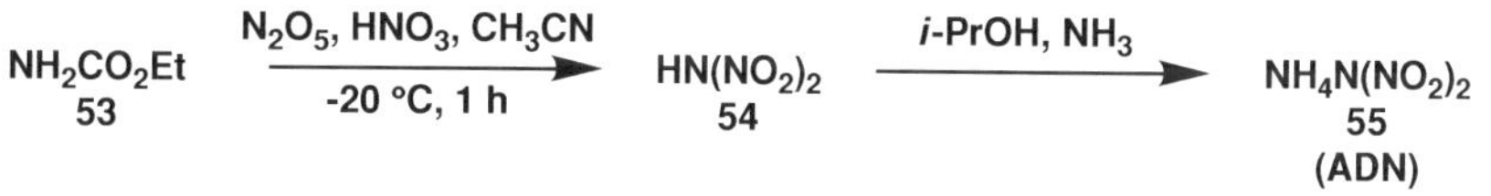

$$\underset{\mathbf{53}}{NH_2CO_2Et} \xrightarrow[\text{-20 °C, 1 h}]{N_2O_5,\ HNO_3,\ CH_3CN} \underset{\mathbf{54}}{HN(NO_2)_2} \xrightarrow{i\text{-PrOH},\ NH_3} \underset{\mathbf{55}\ (\mathbf{ADN})}{NH_4N(NO_2)_2}$$

Figure 9.24

Both Russian[76] and American[77] chemists have reported numerous routes to the ammonium dinitramide but many are generally not feasible for large scale production. One route involves treating ethyl or ammonium carbamate with nitronium tetrafluoroborate, followed by treatment with ammonia if ethyl carbamate (53) is used.[77] However, nitronium salts are not widely available because of the harsh conditions and reagents needed for their preparation. The same reaction is achieved with high efficiency with dinitrogen pentoxide–nitric acid reagent in acetonitrile at subambient temperature.[77] The industrial availability of this reagent has fuelled increased interest in the ammonium dinitramide as a component of futuristic propellants. The production cost of ammonium dinitramide continues to drop rapidly as dinitrogen pentoxide technology develops.

REFERENCES

1. G. A. Olah, R. Malhotra and S. C. Narang, *Nitration: Methods and Mechanisms*, Wiley-VCH, Weinheim (1989).
2. J. P. Agrawal, *Chemical World*, 2006, **5**, 40.
3. J. W. Fischer in *Nitro Compounds: Recent Advances in Synthesis and Chemistry*, Ed. H. Feuer and A. T. Nielsen, Wiley-VCH, Weinheim, Chapter 3, 267–359 (1990).
4. C. C. Addison and N. Logan, in *The Chemistry of Dinitrogen Pentoxide: Developments in Inorganic Nitrogen Chemistry*, Ed. C. B. Coburn, Elsevier, Amsterdam, Chapter 2 (1973).
5. *Nitration: Recent Laboratory and Industrial Developments, ACS Symposium Series 623*, Eds. L. F. Albright, R. V. C. Carr and R. J. Schmitt, American Chemical Society, Washington, DC (1996).
6. M. H. Deville, *Compt. Rend. Acad. Sci. Paris*, 1849, **28**, 257.
7. C. K. Ingold and D. J. Millen, *J. Chem. Soc.*, 1950, 2612.
8. J. Chedin, *Compt. Rend. Acad. Sci. Paris.*, 1935, **201**, 552.
9. F. Daniels and E. H. Johnston, *J. Am. Chem. Soc.*, 1921, **43**, 53.
10. H. Eyring and F. Daniels, *J. Am. Chem. Soc.*, 1930, **52**, 1486.
11. D. J. Millen, *J. Chem. Soc.*, 1950, 2606.
12. D. R. Goddard, E. D. Hughes and C. K. Ingold, *J. Chem. Soc.*, 1950, 2559.
13. K. O. Christe and W. W. Wilson, *Inorg. Chem.*, 1987, **26**, 1631.
14. (a) E. Pokorny and F. Russ, *Monatsh. Chem.*, 1934, **34**, 1051; (b) G. V. Caesar and M. Goldfrank, *J. Am. Chem. Soc.*, 1946, **68**, 372; (c) G. V. Caesar, M. L. Cushing, M. Goldfrank and N. S. Gruenhut, *Inorg. Synth.*, 1950, **3**, 78.
15. E. E. Gilbert and V. I. Siele, *US Pat.* 3 939 148 (1976); (b) T. M. Benzinger, M. D. Coburn, R. K. Davey, E. E. Gilbert, R. W. Hutchinson, J. Leccacorvi, R. Motto, R. K. Rohwer, V. I. Siele and M. Warman, *Propell. Explos. Pyrotech.*, 1981, **6**, 67.
16. P. A. Guye, *US Pat.* 1 348 873 (1920).
17. A. D. Harris, H. B. Jonassen and J. C. Trebellas, *Inorg. Synth.*, 1967, **9**, 83.
18. M. E. Colclough, H. Desai, P. Golding, P. J. Honey, R. W. Millar, N. C. Paul, A. J. Sanderson and M. J. Stewart, in *Nitration: Recent Laboratory and Industrial Developments, ACS Symposium Series 623*, Eds. L. F. Albright, R. V. C. Carr and R. J. Schmitt, American Chemical Society, Washington, DC, Chapter 11, 104–121 (1996).
19. N. C. Paul, in *Nitration: Recent Laboratory and Industrial Developments, ACS Symposium Series 623*, Eds. L. F. Albright, R. V. C. Carr and R. J. Schmitt, American Chemical Society, Washington, DC, Chapter 15, 165–173 (1996).
20. T. E. Devendorf and J. R. Stacy, in *Nitration: Recent Laboratory and Industrial Developments, ACS Symposium Series 623*, Eds. L. F. Albright, R. V. C. Carr and R. J. Schmitt, American Chemical Society, Washington, DC, Chapter 8, 68–77 (1996).
21. J. E. Harrar and R. K. Pearson, *J. Electrochem. Soc.*, 1983, **130**, 108.
22. N. Logan, *Pure Appl. Chem.*, 1986, **58**, 1147.
23. G. Bagg, *US Pat.* 5 181 996 (1990); *UK Pat.* 2 229 449 (1991) and 2 245 003 (1992).
24. M. J. Rodgers and P. F. Swinton, in *Nitration: Recent Laboratory and Industrial Developments, ACS Symposium Series 623*, Eds. L. F. Albright, R. V. C. Carr and R. J. Schmitt, American Chemical Society, Washington, DC, Chapter 7, 58–67 (1996).
25. A. J. Bloom, M. Fleischmann and J. M. Mellor, *Electrochim. Acta.*, 1987, **32**, 785.
26. R. D. Chapman and G. D. Smith, in *Nitration: Recent Laboratory and Industrial Developments, ACS Symposium Series 623*, Eds. L. F. Albright, R. V. C. Carr and R. J. Schmitt, American Chemical Society, Washington, DC, Chapter 9, 78–96 (1996).

27. C. Fréjaques, *French Pat.* 1 060 425 (1954).
28. S. Fraiberg, N. I. Kobozev and M. Temkin, *J. Gen. Chem. (USSR)*, 1933, **3**, 534.
29. K. Okón, *Biul. Wojskowej. Akad. Tech.*, 1958, **7**, 3; *Chem. Abstr.*, 1959, **53**, 2910a.
30. E. D. Hughes, V. Gold, C. K. Ingold and G. H. Williams, *J. Chem. Soc.*, 1950, 2452.
31. C. K. Ingold, *Structure and Mechanism in Organic Chemistry, 2nd Edn.*, Bell and Sons, London, Chapter 6 (1969).
32. K. Schofield, *Aromatic Nitration*, Cambridge University Press, New York, 23–43 (1980).
33. H. Adkins and L. B. Haines, *J. Am. Chem. Soc.*, 1925, **47**, 1419.
34. A. N. Baryshnikova and A. I. Titov, *Dokl. Akad. Nauk USSR*, 1957, **114**, 777.
35. (a) O. A. Luk'yanov and G. V. Pokhvisneva, *Izv. Akad. Nauk USSR, Ser. Khim.*, 1991, 2148; (b) O. A. Luk'yanov and T. L. Zhiguleva, *Izv. Akad. Nauk USSR, Ser. Khim.*, 1982, 1423; (c) L. I. Bagal, G. I. Kolesetskaya and I. V. Tselinskii, *Zh. Org. Khim.*, 1970, **6**, 334.
36. R. G. Coombes, P. J. Honey and R. W. Millar, in *Nitration: Recent Laboratory and Industrial Developments, ACS Symposium Series 623*, Eds. L. F. Albright, R. V. C. Carr and R. J. Schmitt, American Chemical Society, Washington, DC, Chapter 13, 134–150 (1996).
37. L. M. Efremova, E. Ya. Orlova, G. M. Shutov and O. G. Ulko, *Deposited Doc.*, 1982, VINITI 2349; *Chem. Abstr.*, 1983, **98**, 197337t.
38. (a) R. B. Moodie and R. J. Stephens, *J. Chem. Soc. Perkin Trans. II*, 1987, 1059; (b) R. B. Moodie, A. J. Sanderson and R. Willmer, *J. Chem. Soc. Perkin Trans. II*, 1990, 833; (c) R. B. Moodie, A. J. Sanderson and R. Willmer, *J. Chem. Soc. Perkin Trans. II*, 1991, 645.
39. P. J. Honey, *M. Phil. Thesis*, Hatfield Polytechnic, Hatfield, UK (1991).
40. M. Freemantle, *Chem. Eng. News.*, 1996, 7.
41. (a) D. C. Miller, *J. Chem. Soc.*, 1950, 2600; (b) A. Klemenc and K. Schöller, *Z. Anorg. Allg. Chem.*, 1924, **141**, 231.
42. J. M. Bakke and I. Hegbom, *Acta Chem. Scand.*, 1994, **48**, 181.
43. G. B. Bachman and J. L. Dever, *J. Am. Chem. Soc.*, 1958, **80**, 5871.
44. G. A. Olah, in *Industrial and Laboratory Nitrations, ACS Symposium Series 22*, Eds. L. F. Albright and C. Hanson, American Chemical Society, Washington, DC, Chapter 1, 1–47 (1976).
45. G. V. Caesar, *US Pat.* 2 400 287 (1946).
46. G. F. Wright, in *The Chemistry of the Nitro and Nitroso Groups, Part 1, Organic Nitro Chemistry Series*, Ed. H. Feuer, Wiley-Interscience, New York, Chapter 10, 614–684 (1969).
47. (a) R. L. Willer, 'Synthesis of a New Energetic Material: 1,4,5,8-Tetranitro-1,4,5,8-tetraazadifurano [3,4-c][3,4-h]decalin (CL-15)', *NWCTP 6397*, Naval Weapons Center, China Lake, CA (1982); (b) R. L. Willer, *US Pat.*, 4 503 229 (1985); *Chem. Abstr.*, 1985, **103**, 54099c; (c) D. W. Moore and R. L. Willer, *J. Org. Chem.*, 1985, **50**, 5123.
48. (a) H. R. Graindorge, P. A. Lescop, M. J. Pouet and F. Terrier, *Proceedings of the 21st American Chemical Society National Meeting*, Washington, DC (1996); (b) H. R. Graindorge, P. A. Lescop, M. J. Pouet and F. Terrier, in *Nitration: Recent Laboratory and Industrial Developments, ACS Symposium Series 623*, Eds. L. F. Albright, R. V. C. Carr and R. J. Schmitt, American Chemical Society, Washington, DC, Chapter 5, 43–50 (1996).
49. (a) C. L. Coon, E. S. Jessop, A. R. Mitchell, P. F. Pagoria and R. D. Schmidt, in *Nitration: Recent Laboratory and Industrial Developments, ACS Symposium Series 623*, Eds. L. F. Albright, R. V. C. Carr and R. J. Schmitt, American Chemical Society, Washington, DC, Chapter 14, 151–164 (1996); (b) E. S. Jessop, A. R. Mitchell and P. F. Pagoria., *Propell. Explos. Pyrotech.*, 1996, **21**, 14.
50. J. Boileau, J. M. L. Emeury and J. P. Kehren, *US Pat.* 4 487 938 (1974).
51. W. D. Emmons, A. S. Pagano and T. E. Stevens, *J. Org. Chem.*, 1958, **23**, 311.
52. S. Von Runge and W. Triebs, *J. Prakt. Chem.*, 1962, **15**, 223.
53. G. V. Caesar and M. Goldfrank, *US Pat.* 2 400 288 (1946).
54. R. L. Atkins and J. W. Fischer, *Org. Prep. Proc. Int.*, 1986, **18**, 281.
55. (a) E. E. Gilbert and V. I. Siele, *US Pat.* 3 939 148 (1976); (b) T. M. Benzinger, M. D. Coburn, R. K. Davey, E. E. Gilbert, R. W. Hutchinson, J. Leccacorvi, R. Motto, R. K. Rohwer, V. I. Siele and M. Warman, *Propell. Explos. Pyrotech.*, 1981, **6**, 67; (c) E. E. Gilbert, R. W. Hutchinson, J. Leccacorvi,

R. Motto, V. I. Siele and M. Warman, 'Alternative Processes for HMX Manufacture', *Technical Report ARLCD-TR-78008*, US Army Research and Development Command, Dover, NJ, October 1979; AD-A083 793 available from Defence Documentation Center, Cameron Station, Alexandria, VA 22314.

56. W. E. Bachmann, W. J. Horton, E. L. Jenner, N. W. MacNaughton and L. B. Scott, *J. Am. Chem. Soc.*, 1951, **73**, 2769.
57. R .B. Crawford, L. de Vore, K. Gleason, K. Hendry, D. P. Kirvel, R. D. Lear, R. R. McGuire and R. D. Stanford, 'Energy and Technology Review, Jan–Feb 1988', *UCRL-52000-88-1/2*, Lawrence Livermore National Laboratory, Livermore, CA.
58. (a) R. L. Atkins and R. L. Willer, *J. Org. Chem.*, 1984, **49**, 5147; (b) R. L. Willer, *J. Org. Chem.*, 1984, **49**, 5150.
59. O. A. Luk'yanov, N. M. Seregina and V. A. Tartakovsky, *Izv. Akad. Nauk USSR, Ser. Khim.*, 1976, 225.
60. R. W. Millar and S. P. Philbin, *Tetrahedron*, 1997, **53**, 4371.
61. G. B. Bachman and N. W. Connon, *J. Org. Chem.*, 1969, **34**, 4121.
62. W. E. Elias and L. D. Hayward., *Tappi J.*, 1958, **41**, 246.
63. G. V. Caesar and M. Goldfrank, *J. Am. Chem. Soc.*, 1946, **68**, 372.
64. (a) B. Vollmert, *Makromol. Chem. Phys.*, 1951, **6**, 78; (b) L. Brissard, J. Chedin and R. Dalmon, *Compt. Rend. Acad. Sci. Paris*, 1935, **201**, 664.
65. P. Golding, R. W. Millar, N. C. Paul and D. H. Richards, *Tetrahedron*, 1993, **49**, 7037.
66. P. Golding, R. W. Millar, N. C. Paul and D. H. Richards, *Tetrahedron Lett.*, 1988, **29**, 2731.
67. P. Golding, R. W. Millar, N. C. Paul and D. H. Richards, *Tetrahedron*, 1993, **49**, 7051.
68. M. E. Colclough and N. C. Paul, in *Nitration: Recent Laboratory and Industrial Developments, ACS Symposium Series 623*, Eds. L. F. Albright, R. V. C. Carr and R. J. Schmitt, American Chemical Society, Washington, DC, Chapter 10, 97–103 (1996).
69. P. Golding, R. W. Millar, N. C. Paul and D. H. Richards, *Tetrahedron*, 1993, **49**, 7063.
70. P. Golding, R. W. Millar, N. C. Paul and D. H. Richards, *Tetrahedron Lett.*, 1988, **29**, 2735.
71. P. Golding, R. W. Millar, N. C. Paul and D. H. Richards, *Tetrahedron*, 1995, **51**, 5073.
72. P. Bunyan, P. Golding, R. W. Millar, N. C. Paul, D. H. Richards and J. A. Rowley, *Propell. Explos. Pyrotech.*, 1993, **18**, 55.
73. (a) L. A. Fang, S. Q. Hua, V. G. Ling and L. Xin, 'Preliminary Study on Bu-NENA Gun Propellants', *Proc. 27th International Annual Conference of ICT*, Karlsruhe, Germany, June 25–28, 1996, 51; (b) N. F. Stanley and P. A. Silver, 'Bu-NENA Gun Propellants', *JANNAF Propulsion Meetings, Vol. 2*, 10 September 1990, 515; (c) R. A. Johnson and J. J. Mulley, Stability and Performance Characteristics of NENA Materials and Formulations, *Joint International Symposium on Energetic Materials Technology*, New Orleans, Louisiana, 5–7 October, 1992, 116.
74. J. C. Bottaro, *Chem. Ind. (London)*, 1996, 249.
75. (a) J. C. Bottaro, D. F. McMillen and M. J. Rossi, *J. Chem. Kinetics.*, 1993, **25**, 549; (b) T. B. Brill, P. J. Brush and D. G. Patil, *Combust. Flame.*, 1993, **92**, 178; (c) M. C. Lin, A. M. Medel and K. Morokuma, *J. Phys. Chem.*, 1995, **99**, 6842.
76. (a) O. V. Anikin, V. P. Gorelik, O. A. Luk'yanov and V. A. Tartakovsky, *Russ. Chem. Bull.*, 1994, **43**, 1457; (b) N. O. Cherskaya, V. P. Gorelik, O. A. Luk'yanov, V. A. Shlyapochnikov and V. A. Tartakovsky, *Russ. Chem. Bull.*, 1994, **43**, 1522; (c) O. A. Luk'yanov, N. I. Shlykova and V. A. Tartakovsky, *Russ. Chem. Bull.*, 1994, **43**, 1680; (d) V. P. Gorelik, O. A. Luk'yanov and V. A. Tartakovsky, *Russ. Chem. Bull.*, 1994, **43**, 89; (e) T. A. Klimova, Y. V. Konnova, O. A. Luk'yanov and V. A. Tartakovsky, *Russ. Chem. Bull.*, 1994, **43**, 1200; (f) A. R. Agevnin, A. A. Leichenko, O. A. Luk'yanov, N. M. Seregina and V. A. Tartakovsky, *Russ. Chem. Bull.*, 1995, **44**, 108; (g) O. V. Anikin, N. O. Cherskaya, V. P. Gorelik, O. A. Luk'yanov, G. I. Oleneva, V. A. Shlyapochnikov and V. A. Tartakovsky, *Russ. Chem. Bull.*, 1995, **44**, 1449.
77. J. C. Bottaro, P. E. Penwell and R. J. Schmitt, *J. Am. Chem. Soc.*, 1997, **119**, 9405.

Index

Note: Figures and Tables are indicated by *italic page numbers*

With kind thanks to Paul Nash for compilation of this index.

Printed in the USA/Agawam, MA
December 5, 2012
570946.066